Study and Solutions Guide

Precalculus with Limits:
A Graphing Approach

Fifth Edition

and

Precalculus Functions and Graphs:
A Graphing Approach

Fifth Edition

Larson/Hostetler/Edwards

Bruce H. Edwards

University of Florida
Gainesville, Florida

Houghton Mifflin Company   Boston   New York

Publisher: Richard Stratton
Sponsoring Editor: Cathy Cantin
Marketing Manager: Jennifer Jones
Editorial Associate: Jeannine Lawless
Project Editor/Editorial Assistant: Jill Clark
Marketing Associate: Mary Legere
New Title Project Manager: Susan Peltier

Printed in the United States of America.

ISBN 10: 0-618-85187-9

ISBN 13: 978-0-618-85187-4

3456789-CRS-11 10 09 08

Preface

This *Study and Solutions Guide* is a supplement to *Precalculus with Limits: A Graphing Approach*, Fifth Edition and *Precalculus Functions and Graphs: A Graphing Approach*, by Ron Larson, Robert Hostetler, and Bruce H. Edwards.

Solutions to the exercises in the text are given in two parts. Part I contains solutions to odd-numbered Section and Review Exercises; summaries of the chapters; and Practice Tests with solutions. Part II contains solutions to the Chapter and Cumulative Tests from the textbook.

This *Study and Solutions Guide* is the result of the efforts of Larson Texts, Inc. If you have any corrections or suggestions for improving this guide, we would appreciate hearing from you.

Bruce H. Edwards
358 Little Hall
University of Florida
Gainesville, FL 32611
be@math.ufl.edu

Contents

Part I **Solutions to Odd-Numbered Exercises and Practice Tests**

Chapter 1 Functions and Their Graphs 1
Chapter 2 Polynomial and Rational Functions 48
Chapter 3 Exponential and Logarithmic Functions 106
Chapter 4 Trigonometric Functions 149
Chapter 5 Analytic Trigonometry 205
Chapter 6 Additional Topics in Trigonometry 252
Chapter 7 Linear Systems and Matrices 297
Chapter 8 Sequences, Series, and Probability 373
Chapter 9 Topics in Analytic Geometry 420
Chapter 10 Analytic Geometry in Three Dimensions 485
Chapter 11 Limits and an Introduction to Calculus 509
Appendices 543

Solutions to Chapter Practice Tests 592

Part II **Solutions to Chapter and Cumulative Tests** 618

PART I

CHAPTER 1
Functions and Their Graphs

Section 1.1   Lines in the Plane 2

Section 1.2   Functions 8

Section 1.3   Graphs of Functions 13

Section 1.4   Shifting, Reflecting, and Stretching Graphs 19

Section 1.5   Combinations of Functions 23

Section 1.6   Inverse Functions 29

Section 1.7   Linear Models and Scatter Plots 36

Review Exercises 39

Practice Test 47

CHAPTER 1
Functions and Their Graphs

Section 1.1   Lines in the Plane

You should know the following important facts about lines.

■ The graph of $y = mx + b$ is a straight line. It is called a linear equation.

■ The slope of the line through (x_1, y_1) and (x_2, y_2) is

$$m = \frac{y_2 - y_1}{x_2 - x_1}.$$

■ (a) If $m > 0$ the line rises from left to right.   (b) If $m = 0$, the line is horizontal.

(c) If $m < 0$, the line falls from left to right.   (d) If m is undefined, the line is vertical.

■ Equations of Lines

(a) Slope-Intercept: $y = mx + b$ (b) Point-Slope: $y - y_1 = m(x - x_1)$

(c) Two-Point: $y - y_1 = \dfrac{y_2 - y_1}{x_2 - x_1}(x - x_1)$ (d) General: $Ax + By + c = 0$

(e) Vertical: $x = a$ (f) Horizontal: $y = b$

■ Given two distinct nonvertical lines

$$L_1: y = m_1 x + b_1 \quad \text{and} \quad L_2: y = m_2 x + b_2$$

(a) L_1 is parallel to L_2 if and only if $m_1 = m_2$ and $b_1 \neq b_2$.

(b) L_1 is perpendicular to L_2 if and only if $m_1 = -1/m_2$.

Vocabulary Check

1. (a) iii   (b) i   (c) v   (d) ii   (e) iv   **2.** slope

3. parallel   **4.** perpendicular   **5.** linear extrapolation

---

1. (a) $m = \frac{2}{3}$. Since the slope is positive, the line rises. Matches L_2.

(b) m is undefined. The line is vertical. Matches L_3.

(c) $m = -2$. The line falls. Matches L_1.

3.

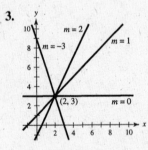

5. Slope $= \dfrac{\text{rise}}{\text{run}} = \dfrac{3}{2}$

7. Slope $= \dfrac{0 - (-10)}{-4 - 0} = \dfrac{10}{-4} = -\dfrac{5}{2}$

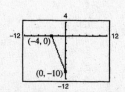

9.

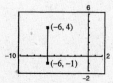

Slope is undefined.

11. Since $m = 0$, y does not change. Three points are $(0, 1)$, $(3, 1)$, and $(-1, 1)$.

13. Since m is undefined, x does not change and the line is vertical. Three points are $(1, 1)$, $(1, 2)$, and $(1, 3)$.

15. Since $m = -2$, y decreases 2 for every unit increase in x. Three points are $(1, -11)$, $(2, -13)$, and $(3, -15)$.

17. Since $m = \frac{1}{2}$, y increases 1 for every increase of 2 in x. Three points are $(9, -1)$, $(11, 0)$, and $(13, 1)$.

19. $5x - y + 3 = 0$
$$y = 5x + 3$$

(a) Slope: $m = 5$

y-intercept: $(0, 3)$

(b)

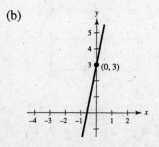

21. $5x - 2 = 0$
$$x = \tfrac{2}{5}$$

(a) Slope: undefined

No y-intercept

(b)

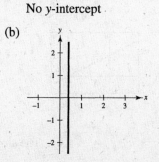

23. $3y + 5 = 0$
$$y = -\tfrac{5}{3}$$

(a) Slope: $m = 0$

y-intercept: $\left(0, -\tfrac{5}{3}\right)$

(b)

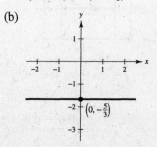

25. $y + 2 = 3(x - 0)$
$$y = 3x - 2 \implies 3x - y - 2 = 0$$

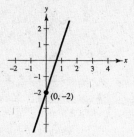

27. $y - (-3) = -\tfrac{1}{2}(x - 2)$
$$y + 3 = -\tfrac{1}{2}x + 1$$
$$2y + 4 = -x$$
$$x + 2y + 4 = 0$$

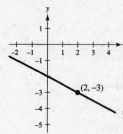

29. $x = 6$

$x - 6 = 0$

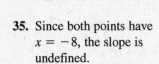

31. $y - \dfrac{3}{2} = 0\left(x + \dfrac{1}{2}\right)$

$y - \dfrac{3}{2} = 0$ horizontal line

33. $y + 1 = \dfrac{5 + 1}{-5 - 5}(x - 5)$

$y = -\dfrac{3}{5}(x - 5) - 1$

$y = -\dfrac{3}{5}x + 2$

35. Since both points have $x = -8$, the slope is undefined.

$x = -8$

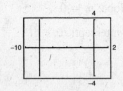

37. $y - \dfrac{1}{2} = \dfrac{\frac{5}{4} - \frac{1}{2}}{\frac{1}{2} - 2}(x - 2)$

$y = -\dfrac{1}{2}(x - 2) + \dfrac{1}{2}$

$y = -\dfrac{1}{2}x + \dfrac{3}{2}$

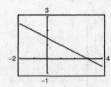

39. $y + \dfrac{3}{5} = \dfrac{-\frac{9}{5} + \frac{3}{5}}{\frac{9}{10} + \frac{1}{10}}\left(x + \dfrac{1}{10}\right)$

$y + \dfrac{3}{5} = -\dfrac{6}{5}\left(x + \dfrac{1}{10}\right)$

$y = -\dfrac{6}{5}x - \dfrac{18}{25}$

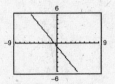

41. $y - 0.6 = \dfrac{-0.6 - 0.6}{-2 - 1}(x - 1)$

$y = 0.4(x - 1) + 0.6$

$y = 0.4x + 0.2$

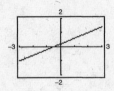

43. The slope is $\dfrac{-3 - (-7)}{1 - (-1)} = \dfrac{4}{2} = 2.$

$y - (-3) = 2(x - 1)$

$y + 3 = 2x - 2$

$y = 2x - 5$

45. Using the points (2004, 28,500) and (2006, 32,900), you have

$$m = \frac{32{,}900 - 28{,}500}{2006 - 2004} = \frac{4400}{2} = 2200$$

$$S - 28{,}500 = 2200(t - 2004)$$

$$S = 2200t - 4{,}380{,}300.$$

When $t = 2008$,

$$S = 2200(2008) - 4{,}380{,}300 = \$37{,}300.$$

47. $x - 2y = 4$

$-2y = -x + 4$

$y = \dfrac{1}{2}x - 2$

Slope: $\dfrac{1}{2}$

y-intercept: $(0, -2)$

The graph passes through $(0, -2)$ and rises 1 unit for each horizontal increase of 2.

49. $x = -6$

slope is undefined.

no y-intercept

The line is vertical and passes through $(-6, 0)$.

51. $y = 0.5x - 3$

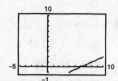 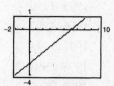

The second setting shows the x- and y-intercepts more clearly.

53. $m_{L_1} = \dfrac{9 + 1}{5 - 0} = 2$

$m_{L_2} = \dfrac{1 - 3}{4 - 0} = -\dfrac{1}{2} = -\dfrac{1}{m_{L_1}}$

L_1 and L_2 are perpendicular.

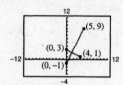

55. $m_{L_1} = \dfrac{0 - 6}{-6 - 3} = \dfrac{2}{3}$

$m_{L_2} = \dfrac{\frac{7}{3} + 1}{5 - 0} = \dfrac{2}{3} = m_{L_1}$

L_1 and L_2 are parallel.

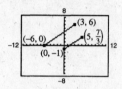

57. $4x - 2y = 3$

$y = 2x - \dfrac{3}{2}$

Slope: $m = 2$

(a) $y - 1 = 2(x - 2)$

$y = 2x - 3$

(b) $y - 1 = -\dfrac{1}{2}(x - 2)$

$y = -\dfrac{1}{2}x + 2$

59. $3x + 4y = 7$

$y = -\dfrac{3}{4}x + \dfrac{7}{4}$

Slope: $m = -\dfrac{3}{4}$

(a) $y - \dfrac{7}{8} = -\dfrac{3}{4}\left(x + \dfrac{2}{3}\right)$

$y = -\dfrac{3}{4}x + \dfrac{3}{8}$

(b) $y - \dfrac{7}{8} = \dfrac{4}{3}\left(x + \dfrac{2}{3}\right)$

$y = \dfrac{4}{3}x + \dfrac{127}{72}$

61. $x - 4 = 0$ vertical line

slope not defined

(a) $x - 3 = 0$ passes through $(3, -2)$

(b) $y = -2$ passes through $(3, -2)$ and is horizontal.

63. The slope is 2 and $(-1, -1)$ lies on the line. Hence,

$y - (-1) = 2(x - (-1))$

$y + 1 = 2(x + 1)$

$y = 2x + 1.$

65. The slope of the given line is 2. Then l has slope $-\dfrac{1}{2}$. Hence,

$y - 2 = -\dfrac{1}{2}(x - (-2))$

$y - 2 = -\dfrac{1}{2}(x + 2)$

$y = -\dfrac{1}{2}x + 1.$

67. (a) $y = 2x$ (b) $y = -2x$ (c) $y = \dfrac{1}{2}x$

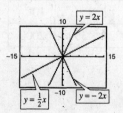

(b) and (c) are perpendicular.

69. (a) $y = -\dfrac{1}{2}x$ (b) $y = -\dfrac{1}{2}x + 3$

(c) $y = 2x - 4$

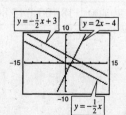

(a) and (b) are parallel.

(c) is perpendicular to (a) and (b).

71. (a)

Years	Slope
1995–1996	$0.69 - 0.91 = -0.22$
1996–1997	$0.57 - 0.69 = -0.12$
1997–1998	$0.74 - 0.57 = 0.17$
1998–1999	$1.60 - 0.74 = 0.86$
1999–2000	$0.82 - 1.60 = -0.78$
2000–2001	$0.92 - 0.82 = 0.10$
2001–2002	$0.20 - 0.92 = -0.72$
2002–2003	$0.00 - 0.20 = -0.20$
2003–2004	$0.31 - 0.00 = 0.31$

Greatest increase: 1998–1999 (0.86)

Greatest decrease: 1999–2000 (−0.78)

(b) $(5, 0.91)$, $(14, 0.31)$:

$$y - 0.91 = \frac{0.31 - 0.91}{14 - 5}(x - 5)$$

$$y = -\frac{1}{15}(x - 5) + \frac{91}{100} = -\frac{1}{15}x + \frac{373}{300}$$

$$y \approx -0.07x + 1.24$$

(c) Between 1995 and 2004, the earnings per share decreased at the rate of 0.07 per year.

(d) For 2010, $x = 20$ and
$y = -0.07(20) + 1.24 = -0.16$, which is reasonable.

73.
$$\frac{\text{rise}}{\text{run}} = \frac{3}{4} = \frac{x}{\frac{1}{2}(32)}$$

$$\frac{3}{4} = \frac{x}{16}$$

$$4x = 48$$

$$x = 12$$

The maximum height in the attic is 12 feet.

75. $(6, 2540)$, $m = 125$

$$V - 2540 = 125(t - 6)$$

$$V = 125t + 1790$$

77. $(6, 20,400)$, $m = -2000$

$$V - 20,400 = -2000(t - 6)$$

$$V = -2000t + 32,400$$

79. The slope is $m = -10$. This represents the decrease in the amount of the loan each week. Matches graph (b).

81. The slope is $m = 0.35$. This represents the increase in travel cost for each mile driven. Matches graph (a).

83. (a) $(0, 25,000)$, $(10, 2000)$

$$V - 25,000 = \frac{2000 - 25,000}{10 - 0}(t - 0)$$

$$V - 25,000 = -2300t$$

$$V = -2300t + 25,000$$

(c) $t = 0$: $V = -2300(0) + 25,000 = 25,000$

$t = 1$: $V = -2300(1) + 25,000 = 22,700$

etc.

(b)

t	0	1	2	3	4	5	6	7	8	9	10
V	25,000	22,700	20,400	18,100	15,800	13,500	11,200	8900	6600	4300	2000

85. (a) $C = 36,500 + 5.25t + 11.50t$

$= 16.75t + 36,500$

(b) $R = 27t$

(c) $P = R - C$

$= 27t - (16.75t + 36,500)$

$= 10.25t - 36,500$

(d) $0 = 10.25t - 36,500$

$36,500 = 10.25t$

$t \approx 3561$ hours

87. (a) $\dfrac{80,124 - 75,349}{2005 - 1991} = \dfrac{4775}{14} \approx 341$ students per year

(b) 1984: $75,349 - 341(7) \approx 72,962$ students

1997: $75,349 + 341(6) \approx 77,395$ students

2000: $75,349 + 341(9) \approx 78,418$ students

(Answers could vary.)

(c) Let $t = 0$ represent 1990.

$(1, 75,349), (15, 80,124)$

$y - 75,349 = \dfrac{80,124 - 75,349}{15 - 1}(t - 1)$

$y = \dfrac{4775}{14}(t - 1) + 75,349$

$y \approx 341t + 75,008$

The slope 341 represents the annual increase in students. It is positive, indicating that Penn State University increased its students from 1991 to 2005.

89. False. The slopes are different:

$\dfrac{4 - 2}{-1 + 8} = \dfrac{2}{7}$

$\dfrac{7 + 4}{-7 - 0} = -\dfrac{11}{7}$

91. $\dfrac{x}{5} + \dfrac{y}{-3} = 1$

$-3x + 5y + 15 = 0$

$a = 5$ and $b = -3$ are the x- and y-intercepts.

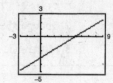

93. $\dfrac{x}{4} + \dfrac{y}{-\frac{2}{3}} = 1$

$-\dfrac{2}{3}x + 4y = \dfrac{-8}{3}$

$-2x + 12y = -8$

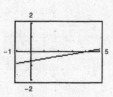

intercepts: $(4, 0), \left(0, -\dfrac{2}{3}\right)$

95. $\dfrac{x}{2} + \dfrac{y}{3} = 1$

$3x + 2y - 6 = 0$

97. $\dfrac{x}{-1/6} + \dfrac{y}{-2/3} = 1$

$-6x - \dfrac{3}{2}y = 1$

$12x + 3y + 2 = 0$

99. The slope is positive and the y-intercept is positive. Matches (a).

101. Both lines have positive slope, but their y-intercepts differ in sign. Matches (c).

103. No. The line $y = 2$ does not have an x-intercept.

105. Yes. Answers will vary.

107. Yes. $x + 20$

109. No. The term $x^{-1} = \dfrac{1}{x}$ causes the expression to not be a polynomial.

111. No. This expression is not defined for $x = \pm 3$.

113. $x^2 - 6x - 27 = (x - 9)(x + 3)$

115. $2x^2 + 11x - 40 = (2x - 5)(x + 8)$

117. Answers will vary.

Section 1.2   Functions

- ■   Given a set or an equation, you should be able to determine if it represents a function.
- ■   Given a function, you should be able to do the following.
  - (a) Find the domain.
  - (b) Evaluate it at specific values.

Vocabulary Check

1. domain, range, function

2. independent, dependent

3. piecewise-defined

4. implied domain

5. difference quotient

1. Yes, it does represent a function. Each domain value is matched with only one range value.

3. No, it does not represent a function. The domain values are each matched with three range values.

5. Yes, the relation represents y as a function of x. Each domain value is matched with only one range value.

7. No, it does not represent a function. The input values of 10 and 7 are each matched with two output values.

9. (a) Each element of A is matched with exactly one element of B, so it does represent a function.

   (b) The element 1 in A is matched with two elements, -2 and 1 of B, so it does not represent a function.

   (c) Each element of A is matched with exactly one element of B, so it does represent a function.

   (d) The element 2 of A is not matched to any element of B, so it does not represent a function.

11. Each are functions. For each year there corresponds one and only one circulation.

13. $x^2 + y^2 = 4 \implies y = \pm\sqrt{4 - x^2}$

Thus, y *is not* a function of x. For instance, the values $y = 2$ and -2 both correspond to $x = 0$.

15. $y = \sqrt{x^2 - 1}$

This *is* a function of x.

17. $2x + 3y = 4 \implies y = \frac{1}{3}(4 - 2x)$

Thus, y *is* a function of x.

19. $y^2 = x^2 - 1 \implies y = \pm\sqrt{x^2 - 1}$

Thus, y *is not* a function of x. For instance, the values $y = \sqrt{3}$ and $-\sqrt{3}$ both correspond to $x = 2$.

21. $y = |4 - x|$

This *is* a function of x.

23. $x = -7$ does not represent y as a function of x. All values of y correspond to $x = -7$.

25. $f(x) = \dfrac{1}{x + 1}$

(a) $f(4) = \dfrac{1}{(4) + 1} = \dfrac{1}{5}$

(b) $f(0) = \dfrac{1}{(0) + 1} = 1$

(c) $f(4t) = \dfrac{1}{(4t) + 1} = \dfrac{1}{4t + 1}$

(d) $f(x + c) = \dfrac{1}{(x + c) + 1} = \dfrac{1}{x + c + 1}$

27. $f(t) = 3t + 1$

(a) $f(2) = 3(2) + 1 = 7$

(b) $f(-4) = 3(-4) + 1 = -11$

(c) $f(t + 2) = 3(t + 2) + 1 = 3t + 7$

29. $h(t) = t^2 - 2t$

(a) $h(2) = 2^2 - 2(2) = 0$

(b) $h(1.5) = (1.5)^2 - 2(1.5) = -0.75$

(c) $h(x + 2) = (x + 2)^2 - 2(x + 2) = x^2 + 2x$

31. $f(y) = 3 - \sqrt{y}$

(a) $f(4) = 3 - \sqrt{4} = 1$

(b) $f(0.25) = 3 - \sqrt{0.25} = 2.5$

(c) $f(4x^2) = 3 - \sqrt{4x^2} = 3 - 2|x|$

33. $q(x) = \dfrac{1}{x^2 - 9}$

(a) $q(0) = \dfrac{1}{0^2 - 9} = -\dfrac{1}{9}$

(b) $q(3) = \dfrac{1}{3^2 - 9}$ is undefined.

(c) $q(y + 3) = \dfrac{1}{(y + 3)^2 - 9} = \dfrac{1}{y^2 + 6y}$

35. $f(x) = \dfrac{|x|}{x}$

(a) $f(3) = \dfrac{|3|}{3} = 1$

(b) $f(-3) = \dfrac{|-3|}{-3} = -1$

(c) $f(t) = \dfrac{|t|}{t} = \begin{cases} 1 & \text{if } t > 0 \\ -1 & \text{if } t < 0 \end{cases}$

$f(0)$ is undefined.

37. $f(x) = \begin{cases} 2x + 1, & x < 0 \\ 2x + 2, & x \geq 0 \end{cases}$

(a) $f(-1) = 2(-1) + 1 = -1$

(b) $f(0) = 2(0) + 2 = 2$

(c) $f(2) = 2(2) + 2 = 6$

39. $f(x) = \begin{cases} x^2 + 2, & x \le 1 \\ 2x^2 + 2, & x > 1 \end{cases}$

    (a) $f(-2) = (-2)^2 + 2 = 6$

    (b) $f(1) = (1)^2 + 2 = 3$

    (c) $f(2) = 2(2)^2 + 2 = 10$

41. $f(x) = \begin{cases} x + 2, & x < 0 \\ 4, & 0 \le x < 2 \\ x^2 + 1, & x \ge 2 \end{cases}$

    (a) $f(-2) = (-2) + 2 = 0$

    (b) $f(1) = 4$

    (c) $f(4) = 4^2 + 1 = 17$

43. $h(t) = \frac{1}{2}|t + 3|$

t	-5	-4	-3	-2	-1
$h(t)$	1	$\frac{1}{2}$	0	$\frac{1}{2}$	1

45. $f(x) = \begin{cases} -\frac{1}{2}x + 4, & x \le 0 \\ (x - 2)^2, & x > 0 \end{cases}$

x	-2	-1	0	1	2
$f(x)$	5	$\frac{9}{2}$	4	1	0

47. $f(x) = 15 - 3x = 0$

$$3x = 15$$
$$x = 5$$

49. $f(x) = \dfrac{3x - 4}{5} = 0$

$$3x - 4 = 0$$
$$3x = 4$$
$$x = \frac{4}{3}$$

51.

$$f(x) = g(x)$$
$$x^2 = x + 2$$
$$x^2 - x - 2 = 0$$
$$(x + 1)(x - 2) = 0$$
$$x = -1 \ \text{ or } \ x = 2$$

53. $f(x) = 5x^2 + 2x - 1$

Since $f(x)$ is a polynomial, the domain is all real numbers x.

55. $h(t) = \dfrac{4}{t}$

Domain: All real numbers except $t = 0$

57. $f(x) = \sqrt[3]{x - 4}$

Domain: all real numbers

59. $g(x) = \dfrac{1}{x} - \dfrac{3}{x + 2}$

Domain: All real numbers except

$x = 0, \ x = -2$

61. $g(y) = \dfrac{y + 2}{\sqrt{y - 10}}$

$$y - 10 > 0$$
$$y > 10$$

Domain: all $y > 10$.

63. $f(x) = \sqrt{4 - x^2}$

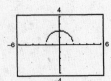

Domain: $[-2, 2]$

Range: $[0, 2]$

65. $g(x) = |2x + 3|$

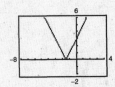

Domain: $(-\infty, \infty)$

Range: $[0, \infty)$

67. $f(x) = x^2$

    $\{(-2, 4), (-1, 1), (0, 0), (1, 1), (2, 4)\}$

69. $f(x) = |x| + 2$

    $\{(-2, 4), (-1, 3), (0, 2), (1, 3), (2, 4)\}$

71. $A = \pi r^2$, $C = 2\pi r$

$$r = \frac{C}{2\pi}$$

$$A = \pi\left(\frac{C}{2\pi}\right)^2 = \frac{C^2}{4\pi}$$

73. (a) According to the table, the maximum profit is 3375 for $x = 150$.

(b) 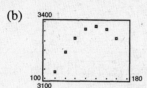   Yes, P is a function of x.

(c) Profit = Revenue − Cost

$\quad\quad\quad\quad$ = (price per unit)(number of units) − (cost)(number of units)

$\quad\quad\quad\quad$ = $[90 - (x - 100)(0.15)]x - 60x$

$\quad\quad\quad\quad$ = $(105 - 0.15x)x - 60x$

$\quad\quad\quad\quad$ = $45x - 0.15x^2$, $x > 100$

$$P = \begin{cases} 30x, & x \le 100 \\ 45x - 0.15x^2, & x > 100 \end{cases}$$

75. $A = \frac{1}{2}(\text{base})(\text{height}) = \frac{1}{2}xy.$

Since $(0, y)$, $(2, 1)$ and $(x, 0)$ all lie on the same line, the slopes between any pair of points are equal.

$$\frac{1 - y}{2 - 0} = \frac{1 - 0}{2 - x}$$

$$1 - y = \frac{2}{2 - x}$$

$$y = 1 - \frac{2}{2 - x} = \frac{x}{x - 2}$$

Therefore, $A = \frac{1}{2}xy = \frac{1}{2}x\left(\frac{x}{x - 2}\right) = \frac{x^2}{2x - 4}.$

The domain is $x > 2$, since $A > 0$.

77. (a) $V = (\text{length})(\text{width})(\text{height}) = yx^2$

$\quad\quad$ But, $y + 4x = 108$, or $y = 108 - 4x$.

$\quad\quad$ Thus, $V = (108 - 4x)x^2$.

$\quad\quad$ Since $y = 108 - 4x > 0$

$$4x < 108$$

$$x < 27.$$

$\quad\quad$ Domain: $0 < x < 27$

(b)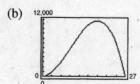

(c) The highest point on the graph occurs at $x = 18$. The dimensions that maximize the volume are $18 \times 18 \times 36$ inches.

79. The domain of $-1.97x + 26.3$ is $7 \leq x \leq 12$.

The domain of $0.505x^2 - 1.47x + 6.3$ is $1 \leq x \leq 6$.

You can tell by comparing the models to the given data. The models fit the data well on the domains above.

81. $f(11) = -1.97(11) + 26.3 = 4.63$

$4,630 in monthly revenue for November.

83. $n(t) = \begin{cases} -6.13t^2 + 75.8t + 577, & 0 \leq t \leq 6 \\ 24.9t + 672, & 6 < t \leq 13 \end{cases}$

$t = 0$ corresponds to 1990.

t	0	1	2	3	4	5	6	7	8	9	10	11	12	13
Model	577	647	704	749	782	803	811	846	871	896	921	946	971	996

85. (a) $F(y) = 149.76\sqrt{10}\,y^{5/2}$

y	5	10	20	30	40
$F(y)$	2.65×10^4	1.50×10^5	8.47×10^5	2.33×10^6	4.79×10^6

(Answers will vary.)

F increases very rapidly as y increases.

(b)

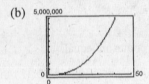

(c) From the table, $y \approx 22$ ft (slightly above 20). You could obtain a better approximation by completing the table for values of y between 20 and 30.

(d) By graphing $F(y)$ together with the horizontal line $y_2 = 1,000,000$, you obtain $y \approx 21.37$ feet.

87.
$$f(x) = 2x$$
$$\frac{f(x + c) - f(x)}{c} = \frac{2(x + c) - 2x}{c}$$
$$= \frac{2c}{c} = 2, \quad c \neq 0$$

89. $f(x) = x^2 - x + 1, \quad f(2) = 3$
$$\frac{f(2 + h) - f(2)}{h} = \frac{(2 + h)^2 - (2 + h) + 1 - 3}{h}$$
$$= \frac{4 + 4h + h^2 - 2 - h + 1 - 3}{h}$$
$$= \frac{h^2 + 3h}{h} = h + 3, \quad h \neq 0$$

91. $f(t) = \dfrac{1}{t}, \quad f(1) = 1$

$$\frac{f(t) - f(1)}{t - 1} = \frac{\dfrac{1}{t} - 1}{t - 1} = \frac{1 - t}{t(t - 1)} = \frac{-1}{t}, \quad t \neq 1$$

93. False. The range of $f(x)$ is $[-1, \infty)$.

95. $f(x) = \begin{cases} x + 4, & x \leq 0 \\ 4 - x^2, & x > 0 \end{cases}$

97. $f(x) = \begin{cases} 2 - x, & x \leq -2 \\ 4, & -2 < x < 3 \\ x + 1, & x \geq 3 \end{cases}$

99. The domain is the set of inputs of the function and the range is the set of corresponding outputs.

101. $12 - \dfrac{4}{x + 2} = \dfrac{12(x + 2) - 4}{x + 2} = \dfrac{12x + 20}{x + 2}$

103. $\dfrac{2x^3 + 11x^2 - 6x}{5x} \cdot \dfrac{x + 10}{2x^2 + 5x - 3} = \dfrac{x(2x^2 + 11x - 6)(x + 10)}{5x(2x - 1)(x + 3)}$

$\qquad\qquad\qquad\qquad = \dfrac{(2x - 1)(x + 6)(x + 10)}{5(2x - 1)(x + 3)}$

$\qquad\qquad\qquad\qquad = \dfrac{(x + 6)(x + 10)}{5(x + 3)}, x \neq 0, \dfrac{1}{2}$

Section 1.3   Graphs of Functions

- ■ You should be able to determine the domain and range of a function from its graph.
- ■ You should be able to use the vertical line test for functions.
- ■ You should be able to determine when a function is constant, increasing, or decreasing.
- ■ You should be able to find relative maximum and minimum values of a function.
- ■ You should know that f is
  (a) Odd if $f(-x) = -f(x)$.
  (b) Even if $f(-x) = f(x)$.

Vocabulary Check

1. ordered pairs

2. Vertical Line Test

3. decreasing

4. minimum

5. greatest integer

6. even

1. Domain: All real numbers

Range: $(-\infty, 1]$

$f(0) = 1$

3. Domain: $[-4, 4]$

Range: $[0, 4]$

$f(0) = 4$

5. $f(x) = 2x^2 + 3$

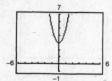

Domain: All real numbers

Range: $[3, \infty)$

7. $f(x) = \sqrt{x - 1}$

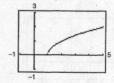

Domain: $x - 1 \geq 0 \implies x \geq 1$
or $[1, \infty)$

Range: $[0, \infty)$

9. $f(x) = |x + 3|$

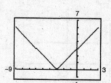

Domain: All real numbers

Range: $[0, \infty)$

11. $f(x) = x^2 - x - 6$

  (a) Domain: all real numbers

  (b) $f(x) = x^2 - x - 6 = (x - 3)(x + 2) = 0 \implies x = 3, -2$

  (c) These are the x-intercepts of f.

  (d) $f(0) = -6$

  (e) This is the y-intercept of f.

  (f) $f(1) = 1^2 - 1 - 6 = -6$. The coordinates are $(1, -6)$

  (g) $f(-1) = (-1)^2 - (-1) - 6 = -4$. The coordinates are $(-1, -4)$.

  (h) $f(-3) = (-3)^2 - (-3) - 6 = 6$. $(-3, f(-3)) = (-3, 6)$.

13. $f(x) = |x - 1| - 2$

  (a) Domain: all x

  (b) $|x - 1| - 2 = 0 \implies |x - 1| = 2 \implies x = -1, 3$

  (c) x-intercepts

  (d) $f(0) = |0 - 1| - 2 = -1$

  (e) y-intercept

  (f) $f(1) = |1 - 1| - 2 = -2, \quad (1, -2)$

  (g) $f(-1) = |-1 - 1| - 2 = 0, \quad (-1, 0)$

  (h) $f(-3) = |-3 - 1| - 2 = 2, \quad (-3, 2)$

15. $y = \frac{1}{2}x^2$

A vertical line intersects the graph just once, so y is a function of x. Graph $y_1 = \frac{1}{2}x^2$.

17. $x^2 + y^2 = 25$

A vertical line intersects the graph more than once, so y is not a function of x. Graph the circle as

$$y_1 = \sqrt{25 - x^2}$$
$$y_2 = -\sqrt{25 - x^2}.$$

19. $f(x) = \frac{3}{2}x$

f is increasing on $(-\infty, \infty)$.

21. $f(x) = x^3 - 3x^2 + 2$

f is increasing on $(-\infty, 0)$ and $(2, \infty)$.

f is decreasing on $(0, 2)$.

23. $f(x) = 3$

  (a)

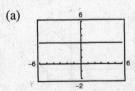

  (b) f is constant on $(-\infty, \infty)$.

25. $f(x) = x^{2/3}$

  (a)

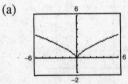

  (b) Increasing on $(0, \infty)$

   Decreasing on $(-\infty, 0)$

27. $f(x) = x\sqrt{x + 3}$

  (a)

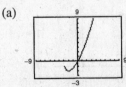

  (b) Increasing on $(-2, \infty)$

   Decreasing on $(-3, -2)$

29. $f(x) = |x + 1| + |x - 1|$

(a)

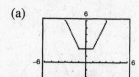

(b) Increasing on $(1, \infty)$,
constant on $(-1, 1)$,
decreasing on $(-\infty, -1)$

31. $f(x) = x^2 - 6x$

Relative minimum: $(3, -9)$

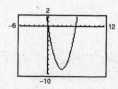

33. $y = 2x^3 + 3x^2 - 12x$

Relative minimum: $(1, -7)$

Relative maximum: $(-2, 20)$

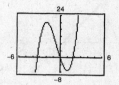

35. $h(x) = (x - 1)\sqrt{x}$

Relative minimum: $(0.33, -0.38)$

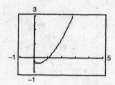

$(0, 0)$ is not a relative maximum because it occurs at the endpoint of the domain $[0, \infty)$.

37. $f(x) = x^2 - 4x - 5$

(a)

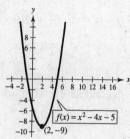

Minimum: $(2, -9)$

(b)

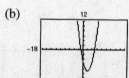

Minimum: $(2, -9)$

(c) Answers are the same.

39. $f(x) = x^3 - 3x$

(a)

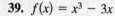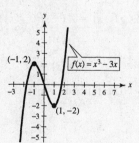

Relative maximum: $(-1, 2)$

Relative minimum: $(1, -2)$

(b) Relative maximum: $(-1, 2)$

Relative minimum: $(1, -2)$

(c) Answers are the same.

41. $f(x) = 3x^2 - 6x + 1$

(a)

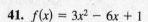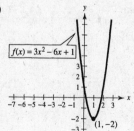

Relative minimum: $(1, -2)$

(b) Relative minimum: $(1, -2)$

(c) Answers are the same.

43. $f(x) = \begin{cases} 2x + 3, & x < 0 \\ 3 - x, & x \ge 0 \end{cases}$

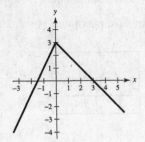

45. $f(x) = \begin{cases} \sqrt{x + 4}, & x < 0 \\ \sqrt{4 - x}, & x \ge 0 \end{cases}$

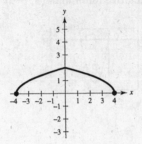

47. $f(x) = \begin{cases} x + 3, & x \le 0 \\ 3, & 0 < x \le 2 \\ 2x - 1, & x > 2 \end{cases}$

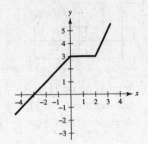

49. $f(x) = \begin{cases} 2x + 1, & x \le -1 \\ x^2 - 2, & x > -1 \end{cases}$

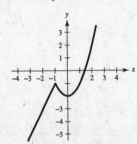

51. $f(x) = [\![x]\!] + 2$

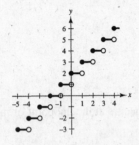

53. $f(x) = [\![x - 1]\!] + 2$

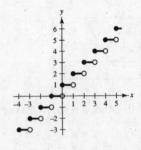

55. $f(x) = [\![2x]\!]$

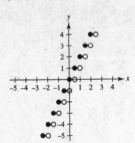

57. $s(x) = 2\left(\frac{1}{4}x - [\![\frac{1}{4}x]\!]\right)$

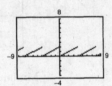

Domain: $(-\infty, \infty)$

Range: $[0, 2)$

Sawtooth pattern

59. $f(-t) = (-t)^2 + 2(-t) - 3$

$= t^2 - 2t - 3$

$\ne f(t) \ne -f(t)$

f is neither even nor odd.

61. $g(-x) = (-x)^3 - 5(-x)$

$= -x^3 + 5x$

$= -g(x)$

g is odd.

63. $f(-x) = (-x)\sqrt{1 - (-x)^2}$

$= -x\sqrt{1 - x^2}$

$= -f(x)$

The function is odd.

65. $g(-s) = 4(-s)^{2/3}$

$= 4s^{2/3}$

$= g(s)$

The function is even.

67. $\left(-\frac{3}{2}, 4\right)$

(a) If f is even, another point is $\left(\frac{3}{2}, 4\right)$.

(b) If f is odd, another point is $\left(\frac{3}{2}, -4\right)$.

69. $(4, 9)$

(a) If f is even, another point is $(-4, 9)$.

(b) If f is odd, another point is $(-4, -9)$.

71. $(x, -y)$

(a) If f is even, another point is $(-x, -y)$.

(b) If f is odd, another point is $(-x, y)$.

73. $f(x) = 5$, even

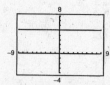

75. $f(x) = 3x - 2$ is neither even nor odd.

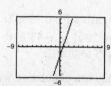

77. $h(x) = x^2 - 4$, even

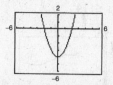

79. $f(x) = \sqrt{1 - x}$ is neither even nor odd.

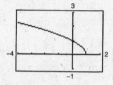

81. $f(x) = |x + 2|$ is neither even nor odd.

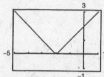

83. $f(x) = 4 - x \geq 0$

$$4 \geq x$$

$$(-\infty, 4]$$

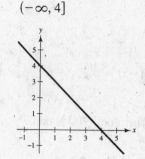

85. $f(x) = x^2 - 9 \geq 0$

$$x^2 \geq 9$$

$$x \geq 3 \quad \text{or} \quad x \leq -3$$

$$[3, \infty) \quad \text{or} \quad (-\infty, -3]$$

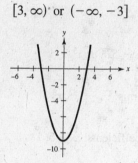

87. (a) The second model is correct. For instance,

$$C_2\left(\tfrac{1}{2}\right) = 1.05 - 0.38\left[\left[-\left(\tfrac{1}{2} - 1\right)\right]\right]$$

$$= 1.05 - 0.38\left[\left[\tfrac{1}{2}\right]\right] = 1.05.$$

(b)

The cost of an 18-minute 45-second call is

$$C_2\left(18\tfrac{45}{60}\right) = C_2(18.75) = 1.05 - 0.38[[-(18.75 - 1)]]$$

$$= 1.05 - 0.38[[-17.75]] = 1.05 - 0.38(-18)$$

$$= 1.05 + 0.38(18) = \$7.89.$$

89. $h = \text{top} - \text{bottom}$

$$= (-x^2 + 4x - 1) - 2$$

$$= -x^2 + 4x - 3, \quad 1 \leq x \leq 3$$

91. $P(t) = 0.0108t^4 - 0.211t^3 + 0.40t^2 + 7.9t + 1791$

$0 \le t \le 14$

(a)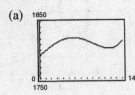

(b) P is increasing from 1990 ($t = 0$) to 1995 ($t \approx 5.7$), and from 2001 ($t \approx 11.8$) to 2004. P is decreasing from 1995 to 2001.

(c) The maximum population was about 1,821,000 in 1995 ($t \approx 5.7$).

93. False. The domain of $f(x) = \sqrt{x^2}$ is the set of all real numbers.

95. c   **97.** b   **99.** a

101.
$$f(x) = a_{2n+1}x^{2n+1} + a_{2n-1}x^{2n-1} + \cdots + a_3x^3 + a_1x$$
$$f(-x) = a_{2n+1}(-x)^{2n+1} + a_{2n-1}(-x)^{2n-1} + \cdots + a_3(-x)^3 + a_1(-x)$$
$$= -a_{2n+1}x^{2n+1} - a_{2n-1}x^{2n-1} - \cdots - a_3x^3 - a_1x = -f(x)$$

Therefore, $f(x)$ is odd.

103. f is an even function.

(a) $g(x) = -f(x)$ is even because
$g(-x) = -f(-x) = -f(x) = g(x)$.

(b) $g(x) = f(-x)$ is even because
$g(-x) = f(-(-x)) = f(x) = f(-x) = g(x)$.

(c) $g(x) = f(x) - 2$ is even because
$g(-x) = f(-x) - 2 = f(x) - 2 = g(x)$.

(d) $g(x) = -f(x - 2)$ is neither even nor odd because
$g(-x) = -f(-x - 2) = -f(x + 2) \neq g(x)$ nor $-g(x)$.

105. No, $x^2 + y^2 = 25$ does not represent x as a function of y. For instance, $(-3, 4)$ and $(3, 4)$ both lie on the graph.

107. $-2x^2 + 8x$

Terms: $-2x^2, 8x$

Coefficients: $-2, 8$

109. $\dfrac{x}{3} - 5x^2 + x^3$

Terms: $\dfrac{x}{3}, -5x^2, x^3$

Coefficients: $\dfrac{1}{3}, -5, 1$

111. (a) $d = \sqrt{(6 - (-2))^2 + (3 - 7)^2}$
$= \sqrt{64 + 16} = \sqrt{80} = 4\sqrt{5}$

(b) midpoint $= \left(\dfrac{-2 + 6}{2}, \dfrac{7 + 3}{2}\right) = (2, 5)$

113. (a) $d = \sqrt{\left(-\dfrac{3}{2} - \dfrac{5}{2}\right)^2 + (4 - (-1))^2}$
$= \sqrt{16 + 25} = \sqrt{41}$

(b) midpoint $= \left(\dfrac{\frac{5}{2} - \frac{3}{2}}{2}, \dfrac{-1 + 4}{2}\right) = \left(\dfrac{1}{2}, \dfrac{3}{2}\right)$

115. $f(x) = 5x - 1$

(a) $f(6) = 5(6) - 1 = 29$

(b) $f(-1) = 5(-1) - 1 = -6$

(c) $f(x - 3) = 5(x - 3) - 1 = 5x - 16$

117. $f(x) = x\sqrt{x - 3}$

(a) $f(3) = 3\sqrt{3 - 3} = 0$

(b) $f(12) = 12\sqrt{12 - 3}$
$= 12\sqrt{9} = 12(3) = 36$

(c) $f(6) = 6\sqrt{6 - 3} = 6\sqrt{3}$

119. $f(x) = x^2 - 2x + 9$

$$f(3 + h) = (3 + h)^2 - 2(3 + h) + 9 = 9 + 6h + h^2 - 6 - 2h + 9$$
$$= h^2 + 4h + 12$$

$$f(3) = 3^2 - 2(3) + 9 = 12$$

$$\frac{f(3 + h) - f(3)}{h} = \frac{(h^2 + 4h + 12) - 12}{h} = \frac{h(h + 4)}{h} = h + 4, h \neq 0$$

Section 1.4   Shifting, Reflecting, and Stretching Graphs

- ■ You should know the graphs of the most commonly used functions in algebra, and be able to reproduce them on your graphing utility.

    (a) Constant function: $f(x) = c$ (b) Identity function: $f(x) = x$

    (c) Absolute value function: $f(x) = |x|$ (d) Square root function: $f(x) = \sqrt{x}$

    (e) Squaring function: $f(x) = x^2$ (f) Cubing function: $f(x) = x^3$

- ■ You should know how the graph of a function is changed by vertical and horizontal shifts.
- ■ You should know how the graph of a function is changed by reflection.
- ■ You should know how the graph of a function is changed by nonrigid transformations, like stretches and shrinks.
- ■ You should know how the graph of a function is changed by a sequence of transformations.

Vocabulary Check

1. quadratic function

2. absolute value function

3. rigid transformations

4. $-f(x), f(-x)$

5. $c > 1, 0 < c < 1$

6. (a) ii   (b) iv   (c) iii   (d) i

1.

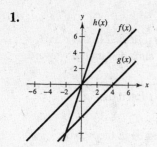

3.

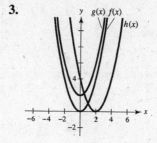

5.

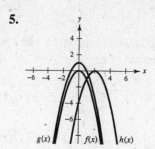

7.

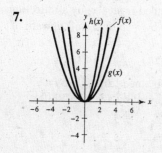

9.

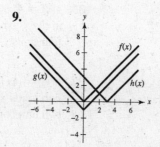

11.

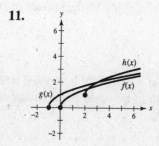

13. (a) $y = f(x) + 2$

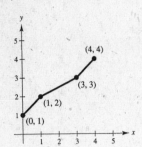

(b) $y = -f(x)$

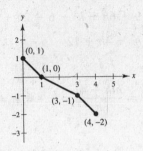

(c) $y = f(x - 2)$

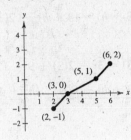

(d) $y = f(x + 3)$

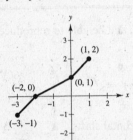

(e) $y = 2f(x)$

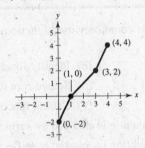

(f) $y = f(-x)$

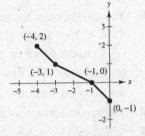

(g) Let $g(x) = f\left(\frac{1}{2}x\right)$. Then from the graph,

$$g(0) = f\left(\frac{1}{2}(0)\right) = f(0) = -1$$

$$g(2) = f\left(\frac{1}{2}(2)\right) = f(1) = 0$$

$$g(6) = f\left(\frac{1}{2}(6)\right) = f(3) = 1$$

$$g(8) = f\left(\frac{1}{2}(8)\right) = f(4) = 2$$

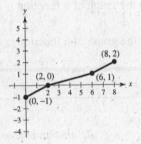

15. Horizontal shift three units to left of $y = x$:
$y = x + 3$ (or vertical shift three units upward)

17. Vertical shift one unit downward of $y = x^2$

$y = x^2 - 1$

19. Reflection in the x-axis and a vertical shift one unit upward of $y = \sqrt{x}$: $y = 1 - \sqrt{x}$

21. $y = -\sqrt{x} - 1$ is $f(x)$ reflected in the x-axis, followed by a vertical shift one unit downward.

23. $y = \sqrt{x - 2}$ is $f(x)$ shifted right two units.

25. $y = 2\sqrt{x}$ is a vertical stretch of $f(x) = \sqrt{x}$.

27. $y = |x + 5|$ is $f(x)$ shifted left five units.

29. $y = -|x|$ is $f(x)$ reflected in the x-axis.

31. $y = 4|x|$ is a vertical stretch of $f(x)$.

33. $g(x) = 4 - x^3$ is obtained from $f(x)$ by a reflection in the x-axis followed by a vertical shift upward of four units.

35. $h(x) = \frac{1}{4}(x + 2)^3$ is obtained from $f(x)$ by a left shift of two units and a vertical shrink by a factor of $\frac{1}{4}$.

37. $p(x) = \left(\frac{1}{3}x\right)^3 + 2$ is obtained from $f(x)$ by a horizontal stretch followed by a vertical shift two units upward.

39. $f(x) = x^3 - 3x^2$

$g(x) = f(x + 2) = (x + 2)^3 - 3(x + 2)^2$ is a horizontal shift two units to left.

$h(x) = \frac{1}{2}f(x) = \frac{1}{2}(x^3 - 3x^2)$ is a vertical shrink.

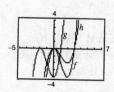

41. $f(x) = x^3 - 3x^2$

$g(x) = -\frac{1}{3}f(x) = -\frac{1}{3}(x^3 - 3x^2)$ reflection in the *x*-axis and vertical shrink

$h(x) = f(-x) = (-x)^3 - 3(-x)^2$ reflection in the *y*-axis

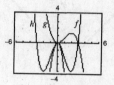

43. (a) $f(x) = x^2$

(b) $g(x) = 2 - (x + 5)^2$ is obtained from f by a horizontal shift to the left five units, a reflection in the *x*-axis, and a vertical shift upward two units.

(c)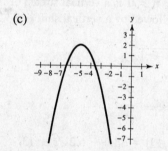

(d) $g(x) = 2 - f(x + 5)$

45. (a) $f(x) = x^2$

(b) $g(x) = 3 + 2(x - 4)^2$ is obtained from f by a horizontal shift four units to the right, a vertical stretch of 2, and a vertical shift upward three units.

(c)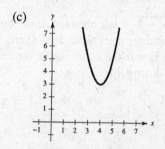

(d) $g(x) = 3 + 2f(x - 4)$

47. (a) $f(x) = x^3$

(b) $g(x) = 3(x - 2)^3$ is obtained from f by a horizontal shift two units to the right followed by a vertical stretch of 3.

(c)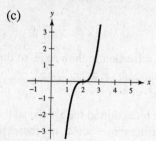

(d) $g(x) = 3f(x - 2)$

49. (a) $f(x) = x^3$

(b) $g(x) = (x - 1)^3 + 2$ is obtained from f by a horizontal shift one unit to the right, and a vertical shift upward two units.

(c)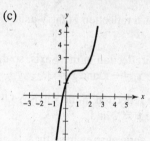

(d) $g(x) = f(x - 1) + 2$

51. (a) $f(x) = |x|$

(b) $g(x) = |x + 4| + 8$ is obtained from f by a horizontal shift four units to the left, followed by a vertical shift eight units upward.

(c)

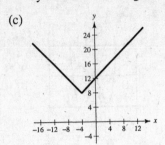

(d) $g(x) = f(x + 4) + 8$

55. (a) $f(x) = \sqrt{x}$

(b) $g(x) = -\frac{1}{2}\sqrt{x + 3} - 1$ is obtained from f by a horizontal shift three units to the left, a vertical shrink, a reflection in the x-axis, and a vertical shift one unit downward.

(c)

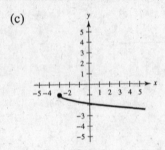

(d) $g(x) = -\frac{1}{2}f(x + 3) - 1$

59. False. $y = f(-x)$ is a reflection in the y-axis.

61. (a) $y = f(-x)$ is a reflection in the y-axis, so the x-intercepts are $x = -2$ and $x = 3$.

(b) $y = -f(x)$ is a reflection in the x-axis, so the x-intercepts are $x = 2$ and $x = -3$.

(c) $y = 2f(x)$ is a vertical stretch, so the x-intercepts are the same: $x = 2, -3$.

(d) $y = f(x) + 2$ is a vertical shift, so you cannot determine the x-intercepts.

(e) $y = f(x - 3)$ is a horizontal shift 3 units to the right, so the x-intercepts are $x = 5$ and $x = 0$.

53. (a) $f(x) = |x|$

(b) $g(x) = -2|x - 1| - 4$ is obtained from f by a horizontal shift one unit to the right, a vertical stretch of 2, a reflection in the x-axis, and a vertical shift downward four units.

(c)

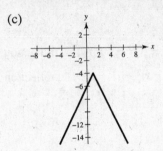

(d) $g(x) = -2f(x - 1) - 4$

57. (a) $F(t) = 33.0 + 6.2\sqrt{t}$ is a vertical stretch of $f(t) = \sqrt{t}$, followed by a vertical shift of 33.0.

(b)

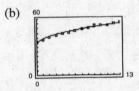

(c) $G(t) = F(t + 13) = 33.0 + 6.2\sqrt{t + 13}$,

$-13 \le t \le 0$.

$G(-13) = F(0)$ corresponds to 1990.

$G(0) = F(13)$ corresponds to 2003.

63. (a) $y = f(-x)$ is a reflection in the y-axis, so the graph is increasing on $(-\infty, -2)$ and decreasing on $(-2, \infty)$.

(b) $y = -f(x)$ is a reflection in the x-axis, so the graph is decreasing on $(-\infty, 2)$ and increasing on $(2, \infty)$.

(c) $y = 2f(x)$ is a vertical stretch, so the graph is increasing on $(-\infty, 2)$ and decreasing on $(2, \infty)$.

(d) $y = f(x) - 3$ is a vertical shift, so the graph is increasing on $(-\infty, 2)$ and decreasing on $(2, \infty)$.

(e) $y = f(x + 1)$ is a horizontal shift one unit to the left, so the graph is increasing on $(-\infty, 1)$ and decreasing on $(1, \infty)$.

65. The vertex is approximately at $(2, 1)$ and the graph opens upward. Matches (c).

67. The vertex is approximately $(2, -4)$ and the graph opens upward. Matches (c).

69. Slope L_1: $\dfrac{10 + 2}{2 + 2} = 3$

71. Domain: All $x \neq 9$

Slope L_2: $\dfrac{9 - 3}{3 + 1} = \dfrac{3}{2}$

Neither parallel nor perpendicular

73. Domain:

$100 - x^2 \geq 0 \implies x^2 \leq 100 \implies -10 \leq x \leq 10$

Section 1.5   Combinations of Functions

> ■ Given two functions, f and g, you should be able to form the following functions (if defined):
> 1. Sum: $(f + g)(x) = f(x) + g(x)$
> 2. Difference: $(f - g)(x) = f(x) - g(x)$
> 3. Product: $(fg)(x) = f(x)g(x)$
> 4. Quotient: $(f/g)(x) = f(x)/g(x),\ g(x) \neq 0$
> 5. Composition of f with g: $(f \circ g)(x) = f(g(x))$
> 6. Composition of g with f: $(g \circ f)(x) = g(f(x))$

Vocabulary Check

1. addition, subtraction, multiplication, division

2. composition

3. $g(x)$

4. inner, outer

1.

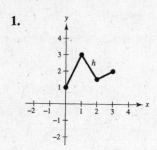

3.

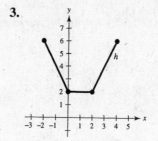

5. $f(x) = x + 3$, $\quad g(x) = x - 3$

    (a) $(f + g)(x) = f(x) + g(x) = (x + 3) + (x - 3) = 2x$

    (b) $(f - g)(x) = f(x) - g(x) = (x + 3) - (x - 3) = 6$

    (c) $(fg)(x) = f(x)g(x) = (x + 3)(x - 3) = x^2 - 9$

    (d) $\left(\dfrac{f}{g}\right)(x) = \dfrac{f(x)}{g(x)} = \dfrac{x + 3}{x - 3}, \quad x \neq 3$

    Domain: all $x \neq 3$

7. $f(x) = x^2$, $g(x) = 1 - x$

    (a) $(f + g)(x) = f(x) + g(x) = x^2 + (1 - x) = x^2 - x + 1$

    (b) $(f - g)(x) = f(x) - g(x) = x^2 - (1 - x) = x^2 + x - 1$

    (c) $(fg)(x) = f(x) \cdot g(x) = x^2(1 - x) = x^2 - x^3$

    (d) $\left(\dfrac{f}{g}\right)(x) = \dfrac{f(x)}{g(x)} = \dfrac{x^2}{1 - x}, x \neq 1$

    Domain: all $x \neq 1$.

9. $f(x) = x^2 + 5$, $g(x) = \sqrt{1 - x}$

    (a) $(f + g)(x) = x^2 + 5 + \sqrt{1 - x}$

    (b) $(f - g)(x) = x^2 + 5 - \sqrt{1 - x}$

    (c) $(fg)(x) = (x^2 + 5)\sqrt{1 - x}$

    (d) $\left(\dfrac{f}{g}\right)(x) = \dfrac{x^2 + 5}{\sqrt{1 - x}}$

    Domain: $x < 1$

11. $f(x) = \dfrac{1}{x}$, $g(x) = \dfrac{1}{x^2}$

    (a) $(f + g)(x) = \dfrac{1}{x} + \dfrac{1}{x^2} = \dfrac{x + 1}{x^2}$

    (b) $(f - g)(x) = \dfrac{1}{x} - \dfrac{1}{x^2} = \dfrac{x - 1}{x^2}$

    (c) $(fg)(x) = \dfrac{1}{x} \cdot \dfrac{1}{x^2} = \dfrac{1}{x^3}$

    (d) $\left(\dfrac{f}{g}\right)(x) = \dfrac{1/x}{1/x^2} = x, x \neq 0$

    Domain: $x \neq 0$

13. $(f + g)(3) = f(3) + g(3)$

              $= (3^2 - 1) + (3 - 2)$

              $= 8 + 1 = 9$

15. $(f - g)(0) = f(0) - g(0)$

              $= (0 - 1) - (0 - 2)$

              $= 1$

17. $(fg)(4) = f(4)g(4)$

           $= (4^2 - 1)(4 - 2)$

           $= 15(2)$

           $= 30$

19. $\left(\dfrac{f}{g}\right)(-5) = \dfrac{f(-5)}{g(-5)}$

             $= \dfrac{(-5)^2 - 1}{-5 - 2}$

             $= \dfrac{24}{-7}$

             $= -\dfrac{24}{7}$

21. $(f - g)(2t) = f(2t) - g(2t)$
$$= ((2t)^2 - 1) - (2t - 2)$$
$$= 4t^2 - 2t + 1$$

23. $(fg)(-5t) = f(-5t)g(-5t)$
$$= ((-5t)^2 - 1)(-5t - 2)$$
$$= (25t^2 - 1)(-5t - 2)$$
$$= -125t^3 - 50t^2 + 5t + 2$$

25. $\left(\dfrac{f}{g}\right)(-t) = \dfrac{f(-t)}{g(-t)}$
$$= \dfrac{(-t)^2 - 1}{-t - 2}$$
$$= \dfrac{t^2 - 1}{-t - 2}$$
$$= \dfrac{1 - t^2}{t + 2}, \quad t \neq -2$$

27.

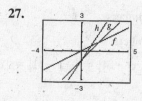

29.

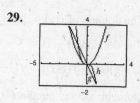

31. $f(x) = 3x,\ g(x) = -\dfrac{x^3}{10},\ (f + g)(x) = 3x - \dfrac{x^3}{10}$

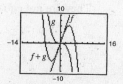

For $0 \leq x \leq 2$, $f(x)$ contributes more to the magnitude.

For $x > 6$, $g(x)$ contributes more to the magnitude.

33. $f(x) = 3x + 2,\ g(x) = -\sqrt{x + 5}$,
$(f + g)(x) = 3x + 2 - \sqrt{x + 5}$

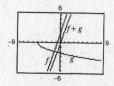

$f(x) = 3x + 2$ contributes more to the magnitude in both intervals.

35. $f(x) = x^2,\ g(x) = x - 1$

(a) $(f \circ g)(x) = f(g(x)) = f(x - 1) = (x - 1)^2$

(b) $(g \circ f)(x) = g(f(x)) = g(x^2) = x^2 - 1$

(c) $(f \circ g)(0) = (0 - 1)^2 = 1$

37. $f(x) = 3x + 5,\ g(x) = 5 - x$

(a) $(f \circ g)(x) = f(g(x)) = f(5 - x) = 3(5 - x) + 5 = 20 - 3x$

(b) $(g \circ f)(x) = g(f(x)) = g(3x + 5) = 5 - (3x + 5) = -3x$

(c) $(f \circ g)(0) = 20$

39. (a) The domain of $f(x) = \sqrt{x + 4}$ is $x + 4 \geq 0$ or $x \geq -4$.

(b) The domain of $g(x) = x^2$ is all real numbers.

(c) $(f \circ g)(x) = f(g(x)) = f(x^2) = \sqrt{x^2 + 4}$.

The domain of $(f \circ g)$ is all real numbers.

41. (a) The domain of $f(x) = x^2 + 1$ is all real numbers.

(b) The domain of $g(x) = \sqrt{x}$ is all $x \geq 0$.

(c) $(f \circ g)(x) = f(g(x)) = f(\sqrt{x})$
$$= (\sqrt{x})^2 + 1 = x + 1, \quad x \geq 0$$

The domain of $f \circ g$ is $x \geq 0$.

43. (a) The domain of $f(x) = \dfrac{1}{x}$ is all $x \neq 0$.

(b) The domain of $g(x) = x + 3$ is all real numbers.

(c) The domain of $(f \circ g)(x) = f(x + 3) = \dfrac{1}{x + 3}$ is all $x \neq -3$.

45. (a) The domain of $f(x) = |x - 4|$ is all real numbers.

(b) The domain of $g(x) = 3 - x$ is all real numbers.

(c) $(f \circ g)(x) = f(g(x)) = f(3 - x) = |(3 - x) - 4| = |-x - 1| = |x + 1|$

Domain: all real numbers

47. (a) The domain of $f(x) = x + 2$ is all real numbers.

(b) The domain of $g(x) = \dfrac{1}{x^2 - 4}$ is all $x \neq \pm 2$

(c) $(f \circ g)(x) = f(g(x)) = f\left(\dfrac{1}{x^2 - 4}\right) = \dfrac{1}{x^2 - 4} + 2$

Domain: $x \neq \pm 2$

49. (a) $(f \circ g)(x) = f(g(x)) = f(x^2) = \sqrt{x^2 + 4}$

Domain: all x

$(g \circ f)(x) = g(f(x)) = g\left(\sqrt{x + 4}\right) = \left(\sqrt{x + 4}\right)^2$

$= x + 4, \; x \geq -4$

(b) 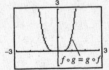   They are not equal.

51. (a) $(f \circ g)(x) = f(g(x)) = f(3x + 9)$

$= \tfrac{1}{3}(3x + 9) - 3 = x$

Domain: all x

$(g \circ f)(x) = g(f(x)) = g\left(\tfrac{1}{3}x - 3\right)$

$= 3\left(\tfrac{1}{3}x - 3\right) + 9 = x$

The domain of $f \circ g$ is all real numbers.

(b) 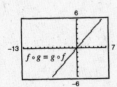   They are equal.

53. (a) $(f \circ g)(x) = f(g(x)) = f(x^6) = (x^6)^{2/3} = x^4$

Domain: all x

$(g \circ f)(x) = g(f(x)) = g(x^{2/3}) = (x^{2/3})^6 = x^4$

(b)   They are equal.

55. (a) $(f \circ g)(x) = f(g(x)) = f(4 - x) = 5(4 - x) + 4 = 24 - 5x$

$(g \circ f)(x) = g(f(x)) = g(5x + 4) = 4 - (5x + 4) = -5x$

(b) No, $(f \circ g)(x) \neq (g \circ f)(x)$ because $24 - 5x \neq -5x$.

(c)

x	$f(g(x))$	$g(f(x))$
0	24	0
1	19	-5
2	14	-10
3	9	-15

57. (a) $(f \circ g)(x) = f(g(x)) = f(x^2 - 5) = \sqrt{(x^2 - 5) + 6} = \sqrt{x^2 + 1}$

$(g \circ f)(x) = g(f(x)) = g(\sqrt{x + 6}) = (\sqrt{x + 6})^2 - 5$

$= (x + 6) - 5 = x + 1, x \geq -6$

(b) No, $(f \circ g)(x) \neq (g \circ f)(x)$ because $\sqrt{x^2 + 1} \neq x + 1$.

(c)

x	$f(g(x))$	$g(f(x))$
0	1	1
-2	$\sqrt{5}$	-1
3	$\sqrt{10}$	4

59. (a) $(f \circ g)(x) = f(g(x)) = f(2x - 1) = |(2x - 1) + 3|$

$= |2x + 2| = 2|x + 1|$

$(g \circ f)(x) = g(f(x)) = g(|x + 3|) = 2|x + 3| - 1$

(b) No, $(f \circ g)(x) \neq (g \circ f)(x)$ because $2|x + 1| \neq 2|x + 3| - 1$.

(c)

x	$f(g(x))$	$g(f(x))$
-1	0	3
0	2	5
1	4	7

61. (a) $(f + g)(3) = f(3) + g(3) = 2 + 1 = 3$

(b) $\left(\dfrac{f}{g}\right)(2) = \dfrac{f(2)}{g(2)} = \dfrac{0}{2} = 0$

63. (a) $(f \circ g)(2) = f(g(2)) = f(2) = 0$

(b) $(g \circ f)(2) = g(f(2)) = g(0) = 4$

65. Let $f(x) = x^2$ and $g(x) = 2x + 1$, then $(f \circ g)(x) = h(x)$. This is not a unique solution. For example, if $f(x) = (x + 1)^2$ and $g(x) = 2x$, then $(f \circ g)(x) = h(x)$ as well.

67. Let $f(x) = \sqrt[3]{x}$ and $g(x) = x^2 - 4$, then $(f \circ g)(x) = h(x)$. This answer is not unique. Other possibilities may be:

$f(x) = \sqrt[3]{x - 4}$ and $g(x) = x^2$ or

$f(x) = \sqrt[3]{-x}$ and $g(x) = 4 - x^2$ or

$f(x) = \sqrt[9]{x}$ and $g(x) = (x^2 - 4)^3$

69. Let $f(x) = 1/x$ and $g(x) = x + 2$, then $(f \circ g)(x) = h(x)$. Again, this is not a unique solution. Other possibilities may be:

$f(x) = \dfrac{1}{x + 2}$ and $g(x) = x$

or $f(x) = \dfrac{1}{x + 1}$ and $g(x) = x + 1$

71. Let $f(x) = x^2 + 2x$ and $g(x) = x + 4$. Then $(f \circ g)(x) = h(x)$. (Answer is not unique.)

73. (a) $T(x) = R(x) + B(x) = \frac{3}{4}x + \frac{1}{15}x^2$

(b)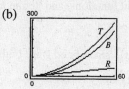

(c) $B(x)$ contributes more to $T(x)$ at higher speeds.

75. $t = 5$ corresponds to 1995.

Year	1995	1996	1997	1998	1999	2000	2001	2002	2003	2004	2005
y_1	140	151.4	162.8	174.2	185.6	197	208.4	219.8	231.2	242.6	254
y_2	325.8	342.8	364.4	390.6	421.5	457	497.1	541.8	591.2	645.2	703.8
y_3	458.8	475.3	497.9	526.5	561.2	602	648.8	701.7	760.7	825.7	896.8

77. $(A \circ r)(t)$ gives the area of the circle as a function of time.

$$(A \circ r)(t) = A(r(t))$$

$$= A(0.6t)$$

$$= \pi(0.6t)^2 = 0.36\pi t^2$$

79. $C(x) = 60x + 750$

$x(t) = 50t$

(a) $C(x(t)) = C(50t)$

$$= 60(50t) + 750$$

$$= 3000t + 750$$

$C(x(t))$ represents the cost after t hours.

(b) $x(4) = 50(4) = 200$ units

(c)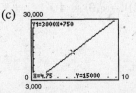

$t = 4.75$, or 4 hours 45 minutes

81. (a) $(N \circ T)(t) = N(T(t))$

$$= N(2t + 1)$$

$$= 10(2t + 1)^2 - 20(2t + 1) + 600$$

$$= 40t^2 + 590$$

$N \circ T$ represents the number of bacteria as a function of time.

(b) $(N \circ T)(6) = 10(13^2) - 20(13) + 600 = 2030$

At time $t = 6$, there are 2030 bacteria.

(c) $N = 800$ when $t \approx 2.3$ hours.

83. $g(f(x)) = g(x - 500,000) = 0.03(x - 500,000)$ represents 3 percent of the amount over $500,000.

85. False. $(f \circ g)(x) = f(6x) = 6x + 1$, but $(g \circ f)(x) = g(x + 1) = 6(x + 1)$.

87. Let A, B, and C be the three siblings, in decreasing age. Then $A = 2B$ and $B = \frac{1}{2}C + 6$.

(a) $A = 2B = 2\left(\frac{1}{2}C + 6\right) = C + 12$

(b) If $A = 16$, then $B = 8$ and $C = 4$.

89. Let $f(x)$ and $g(x)$ be odd functions, and define $h(x) = f(x)g(x)$. Then,

$$h(-x) = f(-x)g(-x)$$

$$= [-f(x)][-g(x)] \text{ since } f \text{ and } g \text{ are both odd}$$

$$= f(x)g(x) = h(x).$$

Thus, h is even.

Let $f(x)$ and $g(x)$ be even functions, and define $h(x) = f(x)g(x)$. Then,

$$h(-x) = f(-x)g(-x)$$

$$= f(x)g(x) \text{ since } f \text{ and } g \text{ are both even}$$

$$= h(x).$$

Thus, h is even.

91. $g(-x) = \frac{1}{2}[f(-x) + f(-(-x))] = \frac{1}{2}[f(-x) + f(x)] = g(x)$, which shows that g is even.

$h(-x) = \frac{1}{2}[f(-x) - f(-(-x))] = \frac{1}{2}[f(-x) - f(x)]$

$= -\frac{1}{2}[f(x) - f(-x)] = -h(x)$,

which shows that h is odd.

93. $(0, -5), (1, -5), (2, -7)$ (other answers possible)

95. $\left(\sqrt{24}, 0\right), \left(-\sqrt{24}, 0\right), \left(0, \sqrt{24}\right)$

     (other answers possible)

97. $y - (-2) = \dfrac{8 - (-2)}{-3 - (-4)}(x - (-4))$

$y + 2 = 10(x + 4)$

$y - 10x - 38 = 0$

99. $y - (-1) = \dfrac{4 - (-1)}{-(1/3) - (3/2)}\left(x - \dfrac{3}{2}\right)$

$y + 1 = \dfrac{5}{-11/6}\left(x - \dfrac{3}{2}\right) = -\dfrac{30}{11}\left(x - \dfrac{3}{2}\right)$

$11y + 11 = -30x + 45$

$30x + 11y - 34 = 0$

Section 1.6   Inverse Functions

- ■   Two functions f and g are inverses of each other if $f(g(x)) = x$ for every x in the domain of g and $g(f(x)) = x$ for every x in the domain of f.
- ■   Be able to find the inverse of a function, if it exists.
  1. Replace $f(x)$ with y.
  2. Interchange x and y.
  3. Solve for y. If this equation represents y as a function of x, then you have found $f^{-1}(x)$. If this equation does not represent y as a function of x, then f does not have an inverse function.
- ■   A function f has an inverse function if and only if no **horizontal** line crosses the graph of f at more than one point.
- ■   A function f has an inverse function if and only if f is one-to-one.

Vocabulary Check

1. inverse, f^{-1}

2. range, domain

3. $y = x$

4. one-to-one

5. Horizontal

1. $f(x) = 6x$

$f^{-1}(x) = \frac{1}{6}x$

$f(f^{-1}(x)) = f\left(\frac{1}{6}x\right) = 6\left(\frac{1}{6}x\right) = x$

$f^{-1}(f(x)) = f^{-1}(6x) = \frac{1}{6}(6x) = x$

3. $f(x) = x + 7$

$f^{-1}(x) = x - 7$

$f(f^{-1}(x)) = f(x - 7) = (x - 7) + 7 = x$

$f^{-1}(f(x)) = f^{-1}(x + 7) = (x + 7) - 7 = x$

5. $f^{-1}(x) = \dfrac{x-1}{2}$

$f(f^{-1}(x)) = f\left(\dfrac{x-1}{2}\right) = 2\left(\dfrac{x-1}{2}\right) + 1 = (x-1) + 1 = x$

$f^{-1}(f(x)) = f^{-1}(2x+1) = \dfrac{(2x+1)-1}{2} = \dfrac{2x}{2} = x$

7. $f^{-1}(x) = x^3$

$f(f^{-1}(x)) = f(x^3) = \sqrt[3]{x^3} = x$

$f^{-1}(f(x)) = f^{-1}(\sqrt[3]{x}) = (\sqrt[3]{x})^3 = x$

9. (a) $f(g(x)) = f\left(-\dfrac{2x+6}{7}\right) = -\dfrac{7}{2}\left(-\dfrac{2x+6}{7}\right) - 3 = \dfrac{2x+6}{2} - 3 = (x+3) - 3 = x$

$g(f(x)) = g\left(-\dfrac{7}{2}x - 3\right) = -\dfrac{2(-\frac{7}{2}x - 3) + 6}{7} = -\dfrac{-7x - 6 + 6}{7} = \dfrac{7x}{7} = x$

(b)

x	2	0	-2	-4	-6
$f(x)$	-10	-3	4	11	18

x	-10	-3	4	11	18
$g(x)$	2	0	-2	-4	-6

Note that the entries in the tables are the same except that the rows are interchanged.

11. (a) $f(g(x)) = f(\sqrt[3]{x-5}) = [\sqrt[3]{x-5}]^3 + 5 = (x-5) + 5 = x$

$g(f(x)) = g(x^3 + 5) = \sqrt[3]{(x^3 + 5) - 5} = \sqrt[3]{x^3} = x$

(b)

x	-3	-2	-1	0	1
$f(x)$	-22	-3	4	5	6

x	-22	-3	4	5	6
$g(x)$	-3	-2	-1	0	1

Note that the entries in the tables are the same except that the rows are interchanged.

13. (a) $f(g(x)) = f(8 + x^2) = -\sqrt{(8 + x^2) - 8} = -\sqrt{x^2} = -(-x) = x \quad x \le 0$

   [Since $x \le 0$, $\sqrt{x^2} = -x$]

$g(f(x)) = g(-\sqrt{x-8}) = 8 + [-\sqrt{x-8}]^2 = 8 + (x-8) = x$

(b)

x	8	9	12	17	24
$f(x)$	0	-1	-2	-3	-4

x	0	-1	-2	-3	-4
$g(x)$	8	9	12	17	24

Note that the entries in the tables are the same except that the rows are interchanged.

15. $f(g(x)) = f(\sqrt[3]{x}) = (\sqrt[3]{x})^3 = x$

$g(f(x)) = g(x^3) = \sqrt[3]{x^3} = x$

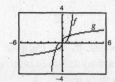

Reflections in the line $y = x$

17. $f(g(x)) = f(x^2 + 4),\ x \geq 0$

$= \sqrt{(x^2 + 4) - 4} = x$

$g(f(x)) = g(\sqrt{x - 4})$

$= (\sqrt{x - 4})^2 + 4 = x$

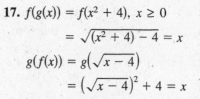

Reflections in the line $y = x$

19. $f(g(x)) = f(\sqrt[3]{1 - x}) = 1 - (\sqrt[3]{1 - x})^3 = 1 - (1 - x) = x$

$g(f(x)) = g(1 - x^3) = \sqrt[3]{1 - (1 - x^3)} = \sqrt[3]{x^3} = x$

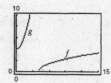

Reflections in the line $y = x$

21. The inverse is a line through $(-1, 0)$.

Matches graph (c).

23. The inverse is half a parabola starting at $(1, 0)$.

Matches graph (a).

25. $f(x) = 2x,\quad g(x) = \dfrac{x}{2}$

(a)

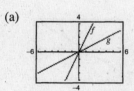

Reflection in the line $y = x$

(b)

x	-2	-1	0	1	2
$f(x)$	-4	-2	0	2	4

x	-4	-2	0	2	4
$g(x)$	-2	-1	0	1	2

The entries in the tables are the same, except that the rows are interchanged.

27. $f(x) = \dfrac{x - 1}{x + 5},\quad g(x) = -\dfrac{5x + 1}{x - 1} = \dfrac{5x + 1}{1 - x}$

(a)

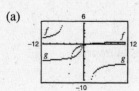

Reflection in the line $y = x$

(b)

x	-2	-1	0	3	5
$f(x)$	-1	$-\frac{1}{2}$	$-\frac{1}{5}$	$\frac{1}{4}$	$\frac{2}{5}$

x	-1	$-\frac{1}{2}$	$-\frac{1}{5}$	$\frac{1}{4}$	$\frac{2}{5}$
$g(x)$	-2	-1	0	3	5

The entries in the tables are the same, except that the rows are interchanged.

29. Not a function

31. It is the graph of a one-to-one function.

33. It is the graph of a one-to-one function.

35. $f(x) = 3 - \dfrac{1}{2}x$

f is one-to-one because a horizontal line will intersect the graph at most once.

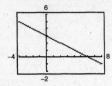

37. $h(x) = \dfrac{x^2}{x^2 + 1}$

h is not one-to-one because some horizontal lines intersect the graph twice.

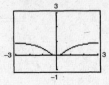

39. $h(x) = \sqrt{16 - x^2}$

h is not one-to-one because some horizontal lines intersect the graph twice.

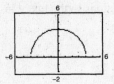

41. $f(x) = 10$

f is not one-to-one because the horizontal line $y = 10$ intersects the graph at every point on the graph.

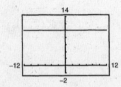

43. $g(x) = (x + 5)^3$

g is one-to-one because a horizontal line will intersect the graph at most once.

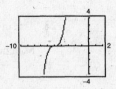

45. $h(x) = |x + 4| - |x - 4|$

h is not one-to-one because some horizontal lines intersect the graph more than once.

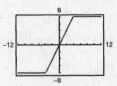

47. $f(x) = x^4$

$y = x^4$

$x = y^4$

$y = \pm\sqrt[4]{x}$

f is not one-to-one.

This does not represent y as a function of x. f does not have an inverse.

49. $f(x) = \dfrac{3x + 4}{5}$

$y = \dfrac{3x + 4}{5}$

$x = \dfrac{3y + 4}{5}$

$5x = 3y + 4$

$5x - 4 = 3y$

$\dfrac{5x - 4}{3} = y$

$f^{-1}(x) = \dfrac{5x - 4}{3}$

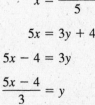

f is one-to-one and has an inverse.

51. $f(x) = \dfrac{1}{x^2}$ is not one-to-one, and does not have an inverse. For example, $f(1) = f(-1) = 1$.

53. $f(x) = (x + 3)^2,\ x \geq -3,\ y \geq 0$

$\quad y = (x + 3)^2,\ x \geq -3,\ y \geq 0$

$\quad x = (y + 3)^2,\ y \geq -3,\ x \geq 0$

$\quad \sqrt{x} = y + 3,\ y \geq -3,\ x \geq 0$

$\quad y = \sqrt{x} - 3,\ x \geq 0,\ y \geq -3$

f is one-to-one.

This is a function of x, so f has an inverse.

$f^{-1}(x) = \sqrt{x} - 3,\ x \geq 0$

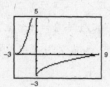

55. $f(x) = \sqrt{2x + 3}\ \Rightarrow\ x \geq -\dfrac{3}{2},\ y \geq 0$

$\quad y = \sqrt{2x + 3},\ x \geq -\dfrac{3}{2},\ y \geq 0$

$\quad x = \sqrt{2y + 3},\ y \geq -\dfrac{3}{2},\ x \geq 0$

$\quad x^2 = 2y + 3,\ x \geq 0,\ y \geq -\dfrac{3}{2}$

$\quad y = \dfrac{x^2 - 3}{2},\ x \geq 0,\ y \geq -\dfrac{3}{2}$

f is one to one.

This is a function of x, so f has an inverse.

$f^{-1}(x) = \dfrac{x^2 - 3}{2},\ x \geq 0$

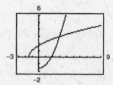

57. $f(x) = |x - 2|,\ x \leq 2,\ y \geq 0$

$\quad y = |x - 2|$

$\quad x = |y - 2|,\quad y \leq 2,\quad x \geq 0$

$\quad x = -(y - 2)$ since $y - 2 \leq 0$.

$\quad x = -y + 2$

$\quad y = -x + 2,\quad x \geq 0,\quad y \leq 2$

$\quad f^{-1}(x) = -x + 2,\quad x \geq 0$

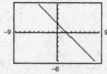

59. $f(x) = 2x - 3$

$\quad y = 2x - 3$

$\quad x = 2y - 3$

$\quad y = \dfrac{x + 3}{2}$

$\quad f^{-1}(x) = \dfrac{x + 3}{2}$

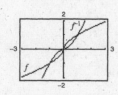

Reflections in the line $y = x$

61. $f(x) = x^5$

$\quad y = x^5$

$\quad x = y^5$

$\quad y = \sqrt[5]{x}$

$\quad f^{-1}(x) = \sqrt[5]{x}$

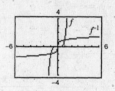

Reflections in the line $y = x$

63. $f(x) = x^{3/5}$

$\quad y = x^{3/5}$

$\quad x = y^{3/5}$

$\quad y = x^{5/3}$

$\quad f^{-1}(x) = x^{5/3}$

Reflections in the line $y = x$

65. $f(x) = \sqrt{4 - x^2},\ 0 \leq x \leq 2$

$\quad y = \sqrt{4 - x^2}$

$\quad x = \sqrt{4 - y^2}$

$\quad x^2 = 4 - y^2$

$\quad y^2 = 4 - x^2$

$\quad y = \sqrt{4 - x^2}$

$\quad f^{-1}(x) = \sqrt{4 - x^2},\ 0 \leq x \leq 2$

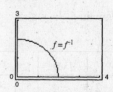

Reflections in the line $y = x$

67. $f(x) = \dfrac{4}{x}$

$y = \dfrac{4}{x}$

$x = \dfrac{4}{y}$

$xy = 4$

$y = \dfrac{4}{x}$

$f^{-1}(x) = \dfrac{4}{x}$

Reflections in the line $y = x$

69. If we let $f(x) = (x - 2)^2$, $x \geq 2$, then f has an inverse. [**Note:** We could also let $x \leq 2$.]

$f(x) = (x - 2)^2$, $x \geq 2$, $y \geq 0$

$y = (x - 2)^2$, $x \geq 2$, $y \geq 0$

$x = (y - 2)^2$, $x \geq 0$, $y \geq 2$

$\sqrt{x} = y - 2$, $x \geq 0$, $y \geq 2$

$\sqrt{x} + 2 = y$, $x \geq 0$, $y \geq 2$

Thus, $f^{-1}(x) = \sqrt{x} + 2$, $x \geq 0$.

71. If we let $f(x) = |x + 2|$, $x \geq -2$, then f has an inverse. [**Note:** We could also let $x \leq -2$.]

$f(x) = |x + 2|$, $x \geq -2$

$f(x) = x + 2$ when $x \geq -2$

$y = x + 2$, $x \geq -2$, $y \geq 0$

$x = y + 2$, $x \geq 0$, $y \geq -2$

$x - 2 = y$, $x \geq 0$, $y \geq -2$

Thus, $f^{-1}(x) = x - 2$, $x \geq 0$.

73. Let $f(x) = (x + 3)^2$, $x \geq -3$.

$y = (x + 3)^2$

$x = (y + 3)^2$

$\sqrt{x} = y + 3$

$y = \sqrt{x} - 3$

$f^{-1}(x) = \sqrt{x} - 3$

Domain f: $x \geq -3$ Range f: $y \geq 0$

Domain f^{-1}: $x \geq 0$ Range f^{-1}: $y \geq -3$

75. Let $f(x) = -2x^2 + 5$, $x \geq 0$.

$y = -2x^2 + 5$

$x = -2y^2 + 5$

$x - 5 = -2y^2$

$y^2 = \dfrac{x - 5}{-2} = \dfrac{5 - x}{2}$

$y = \sqrt{(5 - x)/2}$

$f^{-1}(x) = \sqrt{\dfrac{5 - x}{2}}$

Domain f: $x \geq 0$ Range f: $y \leq 5$

Domain f^{-1}: $x \leq 5$ Range f^{-1}: $y \geq 0$

77. Let $f(x) = |x - 4| + 1$, $x \geq 4$ and $y \geq 1$.

$y = |x - 4| + 1$

$y = x - 3$ because $x \geq 4$.

$x = y - 3$

$y = x + 3$

$f^{-1}(x) = x + 3$, $x \geq 1$

Domain f: $x \geq 4$ Range f: $y \geq 1$

Domain f^{-1}: $x \geq 1$ Range f^{-1}: $y \geq 4$

79.

x	$f(x)$
-2	-4
-1	-2
1	2
3	3

x	$f^{-1}(x)$
-4	-2
-2	-1
2	1
3	3

81. $f^{-1}(0) = \frac{1}{2}$ because $f\left(\frac{1}{2}\right) = 0$.

83. $(f \circ g)(2) = f(3) = -2$

85. $f^{-1}(g(0)) = f^{-1}(2) = 0$

87. $(g \circ f^{-1})(2) = g(0) = 2$

89. $f(x) = x^3 + x + 1$

The graph of the inverse relation is an inverse function since it satisfies the Vertical Line Test.

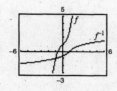

91. $g(x) = \dfrac{3x^2}{x^2 + 1}$

The graph of the inverse relation is not an inverse function since it does not satisfy the Vertical Line Test.

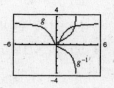

In Exercises 93–97, $f(x) = \frac{1}{8}x - 3$, $f^{-1}(x) = 8(x + 3)$, $g(x) = x^3$, $g^{-1}(x) = \sqrt[3]{x}$.

93. $(f^{-1} \circ g^{-1})(1) = f^{-1}(g^{-1}(1)) = f^{-1}(\sqrt[3]{1}) = 8(\sqrt[3]{1} + 3) = 8(1 + 3) = 32$

95. $(f^{-1} \circ f^{-1})(6) = f^{-1}(f^{-1}(6)) = f^{-1}(8[6 + 3]) = f^{-1}(72) = 8(72 + 3) = 600$

97. $(fg)(x) = f(g(x)) = f(x^3) = \frac{1}{8}x^3 - 3$ Now find the inverse of $(f \circ g)(x) = \frac{1}{8}x^3 - 3$:

$$y = \tfrac{1}{8}x^3 - 3$$
$$x = \tfrac{1}{8}y^3 - 3$$
$$x + 3 = \tfrac{1}{8}y^3$$
$$8(x + 3) = y^3$$
$$\sqrt[3]{8(x + 3)} = y$$
$$(f \circ g)^{-1}(x) = 2\sqrt[3]{x + 3}$$

Note: $(f \circ g)^{-1} = g^{-1} \circ f^{-1}$

In Exercises 99 and 101, $f(x) = x + 4$, $f^{-1}(x) = x - 4$, $g(x) = 2x - 5$, $g^{-1}(x) = \dfrac{x + 5}{2}$.

99. $(g^{-1} \circ f^{-1})(x) = g^{-1}(f^{-1}(x))$

$$= g^{-1}(x - 4)$$
$$= \frac{(x - 4) + 5}{2}$$
$$= \frac{x + 1}{2}$$

101. $(f \circ g)(x) = f(g(x)) = f(2x - 5) = (2x - 5) + 4 = 2x - 1$. Now find the inverse of $(f \circ g)(x) = 2x - 1$:

$$y = 2x - 1$$
$$x = 2y - 1$$
$$x + 1 = 2y$$
$$y = \frac{x + 1}{2}$$

$$(f \circ g)^{-1}(x) = \frac{x + 1}{2}$$

Note that $(f \circ g)^{-1}(x) = (g^{-1} \circ f^{-1})(x)$; see Exercise 99.

103. (a) Yes, f is one-to-one. For each European shoe size, there is exactly one U.S. shoe size.

(b) $f(11) = 45$

(c) $f^{-1}(43) = 10$ because $f(10) = 43$.

(d) $f(f^{-1}(41)) = f(8) = 41$

(e) $f^{-1}(f(13)) = f^{-1}(47) = 13$

105. (a) Yes, f is one-to-one, so f^{-1} exists.

(b) f^{-1} gives the year corresponding to the 10 values in the second column.

(c) $f^{-1}(650.3) = 10$ because $f(10) = 650.3$.

(d) No, because $f(11) = f(15) = 690.4$.

107. False. $f(x) = x^2$ is even, but f^{-1} does not exist.

109. We will show that $(f \circ g)^{-1}(x) = (g^{-1} \circ f^{-1})(x)$ for all x in their domains.
Let $y = (f \circ g)^{-1}(x) \implies (f \circ g)(y) = x$ then $f(g(y)) = x \implies f^{-1}(x) = g(y)$.
Hence, $(g^{-1} \circ f^{-1})(x) = g^{-1}(f^{-1}(x)) = g^{-1}(g(y)) = y = (f \circ g)^{-1}(x)$.
Thus, $g^{-1} \circ f^{-1} = (f \circ g)^{-1}$.

111. No, the graphs are not reflections of each other in the line $y = x$.

113. Yes, the graphs are reflections of each other in the line $y = x$.

115. Yes. The inverse would give the time it took to complete n miles.

117. No. The function oscillates.

119. $\dfrac{27x^3}{3x^2} = 9x, \ x \neq 0$

121. $\dfrac{x^2 - 36}{6 - x} = \dfrac{(x-6)(x+6)}{-(x-6)} = \dfrac{x+6}{-1} = -x - 6, \ x \neq 6$

123. $4x - y = 3$

$y = 4x - 3$

Yes, y is a function of x.

125. $x^2 + y^2 = 9$

$y = \pm\sqrt{9 - x^2}$

No, y is not a function of x.

127. $y = \sqrt{x + 2}$

Yes, y is a function of x.

Section 1.7 Linear Models and Scatter Plots

- ■ You should know how to construct a scatter plot for a set of data
- ■ You should recognize if a set of data has a positive correlation, negative correlation, or neither.
- ■ You should be able to fit a line to data using the point-slope formula.
- ■ You should be able to use the regression feature of a graphing utility to find a linear model for a set of data.
- ■ You should be able to find and interpret the correlation coefficient of a linear model.

Vocabulary Check

1. positive　　　　**2.** negative　　　　**3.** fitting a line to data　　　**4.** $-1, 1$

1. (a)

Monthly sales (in thousands of dollars) vs. Years of experience scatter plot.

(b) Yes, the data appears somewhat linear. The more experience, *x*, corresponds to higher sales, *y*.

3. Negative correlation—*y* decreases as *x* increases.

5. No correlation

7. (a)

$y = \frac{2}{3}x + \frac{5}{3}$

(b) $y = 0.46x + 1.62$

Correlation coefficient: 0.95095

(c)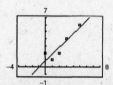

(d) Yes, the model appears valid.

9. (a)

$y = \frac{3}{2}x - \frac{1}{2}$

(b) $y = 0.95x + 0.92$

Correlation coefficient: 0.90978

(c)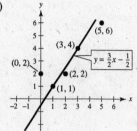

(d) Yes, the model appears valid.

11. (a)

Elongation vs. Force scatter plot.

(b) $d = 0.07F - 0.3$

(c) $d = 0.066F$ or $F = 15.13d + 0.096$

(d) If $F = 55$, $d = 0.066(55) \approx 3.63$ cm.

13. (a)

(b) $y = 136.1t + 836$

(c)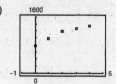

Yes, the model is a good fit.

(d) For 2005, $t = 5$ and $y \approx 1516.5$, or \$1,516,500.

For 2010, $t = 10$ and $y \approx 2197$, or \$2,197,000.

Yes, the answers seem reasonable.

(e) The slope is 136.1. It says that the mean salary increases by \$136,100 per year.

15. (a)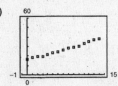

(b) $C = 1.552t + 15.70$

Correlation coefficient: 0.99544

(c)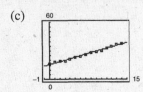

(d) The model is a good fit.

(e) For 2005, $t = 15$, $y_1 \approx \$38.98$.

For 2010, $t = 20$, $y_1 \approx \$46.74$.

(f) Answers will vary.

17. (a)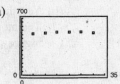

(b) $P = 0.6t + 512$

(c)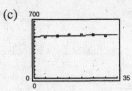

The model is not a good fit.

(d) For 2050, $t = 50$ and $P = 542$, or 542,000 people. Answers will vary.

19. (a) $T = 36.7t + 926$

Correlation coefficient: 0.79495

(b)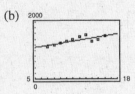

(c) The slope indicated the number of new stores opened per year.

(d) $T = 36.7t + 926 > 1800$

$36.7t > 874$

$t > 23.8$

The number of stores will exceed 1800 near the end of 2013.

(e)

Year	1997	1998	1999	2000	2001	2002	2003	2004	2005	2006
Data	1130	1182	1243	1307	1381	1475	1553	1308	1400	1505
Model	1183	1220	1256	1293	1330	1366	1403	1440	1477	1513

The model is not a good fit, especially around $t = 14$.

21. True. To have positive correlation, the y-values tend to increase as x increases.

23. Answers will vary.

25. $f(x) = 2x^2 - 3x + 5$

(a) $f(-1) = 2 + 3 + 5 = 10$

(b) $f(w + 2) = 2(w + 2)^2 - 3(w + 2) + 5$

$= 2w^2 + 5w + 7$

27. $h(x) = \begin{cases} 1 - x^2, & x \le 0 \\ 2x + 3, & x > 0 \end{cases}$

(a) $h(1) = 2(1) + 3 = 5$ (b) $h(0) = 1 - 0 = 1$

29. $6x + 1 = -9x - 8$

$15x = -9$

$x = -\frac{9}{15} = -\frac{3}{5}$

31. $8x^2 - 10x - 3 = 0$

$(4x + 1)(2x - 3) = 0$

$x = -\frac{1}{4}, \frac{3}{2}$

33. $2x^2 - 7x + 4 = 0$

$x = \dfrac{7 \pm \sqrt{49 - 4(4)(2)}}{4}$

$= \dfrac{7 \pm \sqrt{17}}{4}$

Review Exercises for Chapter 1

1.

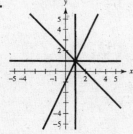

3. $m = \dfrac{2-2}{8-(-3)} = \dfrac{0}{11} = 0$

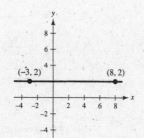

5. $m = \dfrac{(5/2)-1}{5-(3/2)} = \dfrac{3/2}{7/2} = \dfrac{3}{7}$

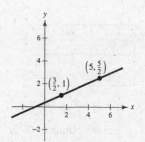

7. $(-4.5, 6), (2.1, 3)$

$m = \dfrac{3-6}{2.1-(-4.5)} = \dfrac{-3}{6.6} = -\dfrac{30}{66} = -\dfrac{5}{11}$

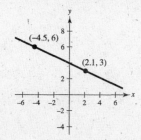

9. (a) $\qquad y + 1 = \dfrac{1}{4}(x - 2)$

$4y + 4 = x - 2$

$-x + 4y + 6 = 0$

(b) Three additional points:

$(2 + 4, -1 + 1) = (6, 0)$

$(6 + 4, 0 + 1) = (10, 1)$

$(10 + 4, 1 + 1) = (14, 2)$

(other answers possible)

11. (a) $\qquad y + 5 = \frac{3}{2}(x - 0)$

$2y + 10 = 3x$

$-3x + 2y + 10 = 0$

(b) Three additional points:

$(0 + 2, -5 + 3) = (2, -2)$

$(2 + 2, -2 + 3) = (4, 1)$

$(4 + 2, 1 + 3) = (6, 4)$

(other answers possible)

13. (a) $\qquad y + 5 = -1\left(x - \frac{1}{5}\right)$

$y + 5 = -x + \frac{1}{5}$

$5y + 25 = -5x + 1$

$5x + 5y + 24 = 0$

(b) Three additional points:

$\left(\frac{1}{5} + 1, -5 - 1\right) = \left(\frac{6}{5}, -6\right)$

$\left(\frac{6}{5} + 1, -6 - 1\right) = \left(\frac{11}{5}, -7\right)$

$\left(\frac{11}{5} + 1, -7 - 1\right) = \left(\frac{16}{5}, -8\right)$

(other answers possible)

15. (a) $y - 6 = 0(x + 2)$

$y - 6 = 0$

(b) Three additional points:

$(0, 6), (1, 6), (2, 6)$

(other answers possible)

17. (a) m is undefined means that the line is vertical.

$x - 10 = 0$

(b) Three additional points: $(10, 0), (10, 1), (10, 2)$

(other answers possible)

19. $y + 1 = \dfrac{-1 + 1}{4 - 2}(x - 2)$

$\qquad = 0(x - 2) = 0 \implies y = -1$

(Slope = 0)

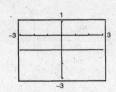

21. $y - 0 = \dfrac{2 - 0}{6 - (-1)}(x + 1)$

$\qquad = \dfrac{2}{7}(x + 1) = \dfrac{2}{7}x + \dfrac{2}{7} \implies y = \dfrac{2}{7}x + \dfrac{2}{7}$

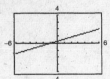

23. $t = 8$ corresponds to 2008.

Point: $(8, 12{,}500)$, slope: 850

$V - 12{,}500 = 850(t - 8)$

$\qquad V = 850t + 5700$

25. $m = 42.70$

Point: $(8, 625.50)$

$V - 625.50 = 42.70(t - 8)$

$\qquad V = 42.70t + 283.90$

27. $(2, 160{,}000), (3, 185{,}000)$

$m = \dfrac{185{,}000 - 160{,}000}{3 - 2} = 25{,}000$

$S - 160{,}000 = 25{,}000(t - 2)$

$\qquad S = 25{,}000t + 110{,}000$

For the fourth quarter let $t = 4$. Then we have $S = 25{,}000(4) + 110{,}000 = \$210{,}000$.

29. $5x - 4y = 8 \implies y = \frac{5}{4}x - 2$ and $m = \frac{5}{4}$

(a) Parallel slope: $m = \frac{5}{4}$

$\qquad y - (-2) = \frac{5}{4}(x - 3)$

$\qquad\quad 4y + 8 = 5x - 15$

$\qquad\qquad\quad 0 = 5x - 4y - 23$

$\qquad\qquad\quad y = \frac{5}{4}x - \frac{23}{4}$

(b) Perpendicular slope: $m = -\frac{4}{5}$

$\qquad y - (-2) = -\frac{4}{5}(x - 3)$

$\qquad\quad 5y + 10 = -4x + 12$

$\qquad 4x + 5y - 2 = 0$

$\qquad\qquad\quad y = -\frac{4}{5}x + \frac{2}{5}$

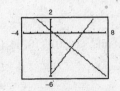

31. $x = 4$ is a vertical line; the slope is not defined.

(a) Parallel line: $x = -6$

(b) Perpendicular slope: $m = 0$

Perpendicular line: $y - 2 = 0(x + 6)$

$\qquad\qquad\qquad\qquad = 0 \implies y = 2$

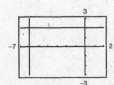

33. (a) Not a function. 20 is assigned two different values.

(b) Function

(c) Function

(d) Not a function. No value is assigned to 30.

35. No, y is not a function of x. Some x-values correspond to two y-values. For example, $x = 1$ corresponds to $y = 4$ and $y = -4$.

37. $y = \sqrt{1 - x}$

Each x value, $x \leq 1$, corresponds to only one y-value so y is a function of x.

39. $f(x) = x^2 + 1$

(a) $f(1) = 1^2 + 1 = 2$

(b) $f(-3) = (-3)^2 + 1 = 10$

(c) $f(b^3) = (b^3)^2 + 1 = b^6 + 1$

(d) $f(x - 1) = (x - 1)^2 + 1 = x^2 - 2x + 2$

41. $h(x) = \begin{cases} 2x + 1, & x \leq -1 \\ x^2 + 2, & x > -1 \end{cases}$

(a) $h(-2) = 2(-2) + 1 = -3$

(b) $h(-1) = 2(-1) + 1 = -1$

(c) $h(0) = 0^2 + 2 = 2$

(d) $h(2) = 2^2 + 2 = 6$

43. The domain of $f(x) = \dfrac{x - 1}{x + 2}$ is all real numbers $x \neq -2$.

45. $f(x) = \sqrt{25 - x^2}$

Domain: $25 - x^2 \geq 0$

$(5 + x)(5 - x) \geq 0$

Domain: $[-5, 5]$

47. The domain of $g(5) = \dfrac{5s + 5}{3s - 9}$ is all real numbers $s \neq 3$.

49. (a) $C(x) = 16{,}000 + 5.35x$

(b) $P(x) = R(x) - C(x)$

$= 8.20x - (16{,}000 + 5.35x)$

$= 2.85x - 16{,}000$

51.
$$f(x) = 2x^2 + 3x - 1$$
$$f(x + h) = 2(x + h)^2 + 3(x + h) - 1$$
$$= 2x^2 + 4xh + 2h^2 + 3x + 3h - 1$$
$$\frac{f(x + h) - f(x)}{h} = \frac{(2x^2 + 4xh + 2h^2 + 3x + 3h - 1) - (2x^2 + 3x - 1)}{h}$$
$$= \frac{4xh + 2h^2 + 3h}{h}$$
$$= 4x + 2h + 3, \quad h \neq 0$$

53. Domain: All real numbers

Range: $y \leq 3$

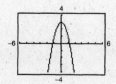

55. Domain: $36 - x^2 \geq 0 \Rightarrow x^2 \leq 36 \Rightarrow -6 \leq x \leq 6$

Range: $0 \leq y \leq 6$

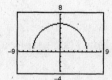

57. (a) $y = \dfrac{x^2 + 3x}{6}$

(b) y is a function of x.

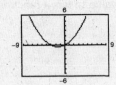

59. (a) $3x + y^2 = 2$

$$y^2 = 2 - 3x$$

$$y = \pm\sqrt{2 - 3x}$$

(b) y is not a function of x.

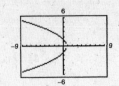

61. $f(x) = x^3 - 3x$

(a)

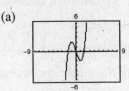

(b) Increasing on $(-\infty, -1)$ and $(1, \infty)$

Decreasing on $(-1, 1)$

63. $f(x) = x\sqrt{x - 6}$

(a)

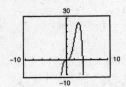

(b) Increasing on $(6, \infty)$

65. $f(x) = (x^2 - 4)^2$

Relative minima: $(-2, 0)$ and $(2, 0)$

Relative maximum: $(0, 16)$

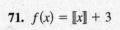

67. $h(x) = 4x^3 - x^4$

Relative maximum: $(3, 27)$

69. $f(x) = \begin{cases} 3x + 5, & x < 0 \\ x - 4, & x \ge 0 \end{cases}$

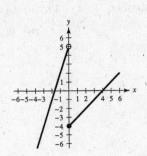

71. $f(x) = [\![x]\!] + 3$

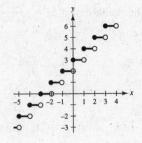

73. $f(-x) = (-x)^2 + 6$

$\quad = x^2 + 6$

$\quad = f(x)$

Even

75. $f(-x) = ((-x)^2 - 8)^2$

$\quad = (x^2 - 8)^2$

$\quad = f(x)$

f is even.

77. $f(-x) = 3(-x)^{5/2} \ne f(x)$ and $f(-x) \ne -f(x)$

Neither even nor odd

(Note that the domain of f is $x \ge 0$.)

79. $f(x) = -2$ is a constant function.

81. $g(x) = x^2$ is the parent function. f is obtained from g by a horizontal shift two units to the right, followed by a vertical shift one unit upward.

$$f(x) = (x - 2)^2 + 1 = g(x - 2) + 1$$

83. $g(x) = |x| + 3$ is obtained from $f(x) = |x|$ by a vertical shift three units upward.

$$g(x) = f(x) + 3$$

85.

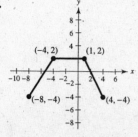

$y = f(-x)$ is a reflection in the y-axis.

87.

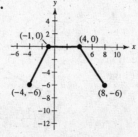

$y = f(x) - 2$ is a vertical shift two units downward.

89. (a) $f(x) = x^2$

(b) h is a vertical shift six units downward.

(c)

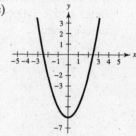

(d) $h(x) = f(x) - 6$

91. $h(x) = (x - 2)^3 + 5$

(a) $f(x) = x^3$

(b) The graph of h is a horizontal shift of f two units to the right, followed by a vertical shift five units upward.

(c)

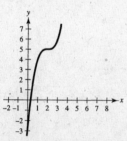

(d) $h(x) = (x - 2)^3 + 5 = f(x - 2) + 5$

93. (a) $f(x) = x^2$

(b) h is a horizontal shift two units to the right, a reflection in the x-axis, followed by a vertical shift eight units downward.

(c)

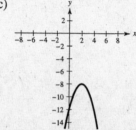

(d) $h(x) = -f(x - 2) - 8$

95. $h(x) = -\sqrt{x} + 5$

(a) $f(x) = \sqrt{x}$

(b) The graph of h is a reflection of f in the x-axis, followed by a vertical shift five units upward.

(c)

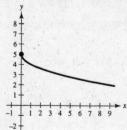

(d) $h(x) = -\sqrt{x} + 5$

$= -f(x) + 5$

97. $h(x) = \sqrt{x - 1} + 3$

(a) $f(x) = \sqrt{x}$

(b) The graph of h is a horizontal shift of one unit to the right, followed by a vertical shift three units upward.

(c)

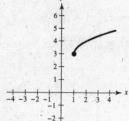

(d) $h(x) = f(x - 1) + 3$

99. $h(x) = -\frac{1}{2}|x| + 9$

  (a) $f(x) = |x|$

  (b) h is a vertical shrink, followed by a reflection in the x-axis, followed by a vertical shift nine units upward.

(c)

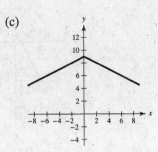

(d) $h(x) = -\frac{1}{2}f(x) + 9$

101. $(f - g)(4) = f(4) - g(4)$
$$= [3 - 2(4)] - \sqrt{4}$$
$$= -5 - 2 = -7$$

103. $(f + g)(25) = f(25) + g(25)$
$$= -47 + 5$$
$$= -42$$

105. $(fh)(1) = f(1)h(1)$
$$= (3 - 2(1))(3(1)^2 + 2)$$
$$= (1)(5) = 5$$

107. $(h \circ g)(7) = h(g(7))$
$$= h(\sqrt{7})$$
$$= 3(\sqrt{7})^2 + 2$$
$$= 23$$

109. $(f \circ h)(-4) = f(h(-4))$
$$= f(50)$$
$$= -97$$

111. $f(x) = x^2, g(x) = x + 3$
$(f \circ g)(x) = f(x + 3)$
$$= (x + 3)^2 = h(x)$$

113. $f(x) = \sqrt{x}, g(x) = 4x + 2$
$(f \circ g)(x) = f(4x + 2) = \sqrt{4x + 2} = h(x)$

115. $f(x) = \dfrac{4}{x}, g(x) = x + 2$

$(f \circ g)(x) = f(x + 2) = \dfrac{4}{x + 2} = h(x)$

117.

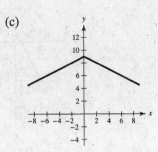

119. $f(x) = 6x$
$f^{-1}(x) = \frac{1}{6}x$
$f(f^{-1}(x)) = f\left(\frac{1}{6}x\right) = 6\left(\frac{1}{6}x\right) = x$
$f^{-1}(f(x)) = f^{-1}(6x) = \frac{1}{6}(6x) = x$

121. $f(x) = \dfrac{1}{2}x + 3 \Rightarrow f^{-1}(x) = 2(x - 3) = 2x - 6$

$f(f^{-1}(x)) = f(2(x - 3))$

$$= \frac{1}{2}(2(x - 3)) + 3 = x - 3 + 3 = x$$

$f^{-1}(f(x)) = f^{-1}\left(\dfrac{1}{2}x + 3\right)$

$$= 2\left(\frac{1}{2}x + 3 - 3\right) = 2\left(\frac{1}{2}x\right) = x$$

123. (a)

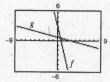

Reflection in the line $y = x$

(b)

x	-5	-1	0	1	3
$f(x)$	23	7	3	-1	-9

x	23	7	3	-1	-9
$g(x)$	-5	-1	0	1	3

The entries in the table are the same except that their rows are interchanged.

125.

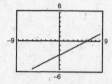

$f(x) = \frac{1}{2}x - 3$ passes the Horizontal Line Test, and hence is one-to-one and has an inverse $(f^{-1}(x) = 2(x + 3))$.

127.

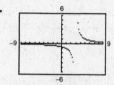

$h(t) = \dfrac{2}{t - 3}$ passes the Horizontal Line Test, and hence is one-to-one.

129.
$$y = \frac{1}{2}x - 5$$
$$x = \frac{1}{2}y - 5$$
$$x + 5 = \frac{1}{2}y$$
$$y = 2(x + 5)$$
$$f^{-1}(x) = 2x + 10$$

131. $f(x) = 4x^3 - 3$
$$y = 4x^3 - 3$$
$$x = 4y^3 - 3$$
$$x + 3 = 4y^3$$
$$\frac{x + 3}{4} = y^3$$
$$f^{-1}(x) = \sqrt[3]{\frac{x + 3}{4}}$$

133.
$$f(x) = \sqrt{x + 10}$$
$$y = \sqrt{x + 10}, x \geq -10, y \geq 0$$
$$x = \sqrt{y + 10}, y \geq -10, x \geq 0$$
$$x^2 = y + 10$$
$$x^2 - 10 = y$$
$$f^{-1}(x) = x^2 - 10, x \geq 0$$

135. Negative correlation

137. (a)

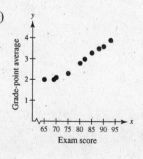

(b) Yes, the relationship is approximately linear. Higher entrance exam scores, x, are associated with higher grade-point averages, y.

139. (a)

(b) $s \approx 10t$ (Approximations will vary.)

(c) $s = 9.7t + 0.4$; 0.99933

(d) For $t = 2.5$, $S \approx 24.7$ m/sec.

141. $y = 95.174x - 458.423$

143. The model does not fit well.

145. False. $g(x) = -[(x - 6)^2 + 3] = -(x - 6)^2 - 3$ and $g(-1) = -52 \neq 28$

147. False. $f(x) = \dfrac{1}{x}$ or $f(x) = x$ satisfies $f = f^{-1}$.

Chapter 1   Practice Test

1. Find the slope of the line passing through the points $(-2, 2)$ and $(1, 3)$.

2. Find an equation for the line passing through the points $(3, -2)$ and $(4, -5)$. Use a graphing utility to sketch a graph of the line.

3. Find an equation of the line that passes through the point $(-1, 5)$ and has slope -3. Use a graphing utility to sketch a graph of the line.

4. Find the slope-intercept form of the line that passes through the point $(-3, 2)$ and is perpendicular to $3x + 5y = 7$.

5. Does the equation $x^4 + y^4 = 16$ represent y as a function of x?

6. Evaluate the function $f(x) = |x - 2|/(x - 2)$ at the points $x = 0$, $x = 2$, and $x = 4$.

7. Find the domain of the function $f(x) = 5/(x^2 - 16)$.

8. Find the domain of the function $g(t) = \sqrt{4 - t}$.

9. Use a graphing utility to sketch the graph of the function $f(x) = 3 - x^6$ and determine if the function is even, odd, or neither.

10. Determine the open interval(s) on which the function $f(x) = 12x - x^3$ is increasing.

11. Use a graphing utility to approximate any relative minimum or maximum values of the function $y = 4 - x + x^3$.

12. Compare the graph of $f(x) = x^3 - 3$ with the graph of $y = x^3$.

13. Compare the graph of $f(x) = \sqrt{x - 6}$ with the graph of $y = \sqrt{x}$.

14. Find $g \circ f$ if $f(x) = \sqrt{x}$ and $g(x) = x^2 - 2$. What is the domain of $g \circ f$?

15. Find f/g if $f(x) = 3x^2$ and $g(x) = 16 - x^4$. What is the domain of f/g?

16. Show that $f(x) = 3x + 1$ and $g(x) = \dfrac{x - 1}{3}$ are inverse functions algebraically and graphically.

17. Find the inverse of $f(x) = \sqrt{9 - x^2}$, $0 \le x \le 3$. Graph f and f^{-1} in the same viewing rectangle.

18. Use a graphing utility to find the least squares regression line for the points $(-1, 0)$, $(0, 1)$, $(3, 3)$, $(4, 5)$. Graph the points and the line.

CHAPTER 2
Polynomial and Rational Functions

Section 2.1   Quadratic Functions **49**

Section 2.2   Polynomial Functions of Higher Degree **54**

Section 2.3   Real Zeros of Polynomial Functions **62**

Section 2.4   Complex Numbers **70**

Section 2.5   The Fundamental Theorem of Algebra **73**

Section 2.6   Rational Functions and Asymptotes **79**

Section 2.7   Graphs of Rational Functions **83**

Section 2.8   Quadratic Models **91**

Review Exercises **94**

Practice Test **105**

C H A P T E R   2
Polynomial and Rational Functions

Section 2.1   Quadratic Functions

> You should know the following facts about parabolas.
>
> ■ $f(x) = ax^2 + bx + c$, $a \neq 0$, is a quadratic function, and its graph is a parabola.
>
> ■   If $a > 0$, the parabola opens upward and the vertex is the minimum point. If $a < 0$, the parabola opens downward and the vertex is the maximum point.
>
> ■   The vertex is $(-b/2a, f(-b/2a))$.
>
> ■   To find the x-intercepts (if any), solve
>
> $$ax^2 + bx + c = 0.$$
>
> ■   The standard form of the equation of a parabola is
>
> $$f(x) = a(x - h)^2 + k$$
>
> where $a \neq 0$.
>
> (a) The vertex is (h, k).
>
> (b) The axis is the vertical line $x = h$.

Vocabulary Check

1. nonnegative integer, real   **2.** quadratic, parabola   **3.** axis

4. positive, minimum   **5.** negative, maximum

---

1. $f(x) = (x - 2)^2$ opens upward and has vertex $(2, 0)$. Matches graph (c).

3. $f(x) = x^2 + 3$ opens upward and has vertex $(0, 3)$. Matches graph (b).

5.

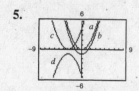

(a) $y = \frac{1}{2}x^2$, vertical shrink

(b) $y = \frac{1}{2}x^2 - 1$, vertical shrink and vertical shift one unit downward

(c) $y = \frac{1}{2}(x + 3)^2$, vertical shrink and horizontal shift three units to the left

(d) $y = -\frac{1}{2}(x + 3)^2 - 1$, horizontal shift three units to the left, vertical shrink, reflection in x-axis, and vertical shift one unit downward

7. $f(x) = 25 - x^2$

Vertex: $(0, 25)$

x-intercepts: $(-5, 0), (5, 0)$

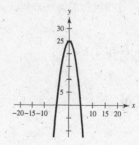

9. $f(x) = \frac{1}{2}x^2 - 4$

Vertex: $(0, -4)$

x-intercepts: $(\pm 2\sqrt{2}, 0)$

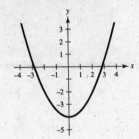

11. $f(x) = (x + 4)^2 - 3$

Vertex: $(-4, -3)$

x-intercepts: $(-4 \pm \sqrt{3}, 0)$

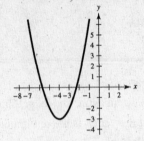

13. $h(x) = x^2 - 8x + 16 = (x - 4)^2$

Vertex: $(4, 0)$

x-intercepts: $(4, 0)$

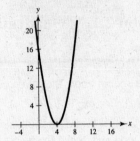

15. $f(x) = x^2 - x + \frac{5}{4} = \left(x - \frac{1}{2}\right)^2 + 1$

Vertex: $\left(\frac{1}{2}, 1\right)$

x-intercepts: None

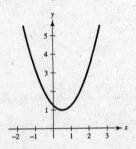

17. $f(x) = -x^2 + 2x + 5 = -(x - 1)^2 + 6$

Vertex: $(1, 6)$

x-intercepts: $\left(1 - \sqrt{6}, 0\right), \left(1 + \sqrt{6}, 0\right)$

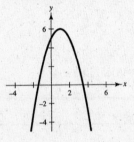

19. $h(x) = 4x^2 - 4x + 21 = 4\left(x - \frac{1}{2}\right)^2 + 20$

Vertex: $\left(\frac{1}{2}, 20\right)$

x-intercept: None

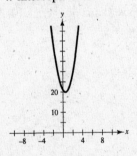

21. $f(x) = -(x^2 + 2x - 3) = -(x + 1)^2 + 4$

Vertex: $(-1, 4)$

x-intercepts: $(-3, 0), (1, 0)$

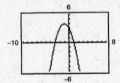

23. $g(x) = x^2 + 8x + 11 = (x + 4)^2 - 5$

Vertex: $(-4, -5)$

x-intercepts: $(-4 \pm \sqrt{5}, 0)$

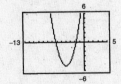

25. $f(x) = -2x^2 + 16x - 31$

$\qquad = -2\left(x^2 - 8x + \frac{31}{2}\right)$

$\qquad = -2\left(x^2 - 8x + 16 - \frac{1}{2}\right)$

$\qquad = -2(x - 4)^2 + 1$

Vertex: $(4, 1)$

x-intercept: $\left(4 \pm \frac{1}{2}\sqrt{2}, 0\right)$

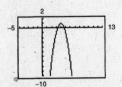

27. $(-1, 4)$ is the vertex.

$f(x) = a(x + 1)^2 + 4$

Since the graph passes through the point $(1, 0)$, we have:

$0 = a(1 + 1)^2 + 4$

$0 = 4a + 4$

$-1 = a$

Thus, $f(x) = -(x + 1)^2 + 4$. Note that $(-3, 0)$ is on the parabola.

29. $(-2, 5)$ is the vertex.

$f(x) = a(x + 2)^2 + 5$

Since the graph passes through the point $(0, 9)$, we have:

$9 = a(0 + 2)^2 + 5$

$4 = 4a$

$1 = a$

$f(x) = 1(x + 2)^2 + 5 = (x + 2)^2 + 5$

31. $(1, -2)$ is the vertex.

$f(x) = a(x - 1)^2 - 2$

Since the graph passes through the point $(-1, 14)$, we have:

$14 = a(-1 - 1)^2 - 2$

$14 = 4a - 2$

$16 = 4a$

$4 = a$

$f(x) = 4(x - 1)^2 - 2$

33. $\left(\frac{1}{2}, 1\right)$ is the vertex.

$f(x) = a\left(x - \frac{1}{2}\right)^2 + 1$

Since the graph passes through the point $\left(-2, -\frac{21}{5}\right)$, we have:

$-\frac{21}{5} = a\left(-2 - \frac{1}{2}\right)^2 + 1$

$-\frac{21}{5} = \frac{25}{4}a + 1$

$-\frac{26}{5} = \frac{25}{4}a$

$-\frac{104}{125} = a$

$f(x) = -\frac{104}{125}\left(x - \frac{1}{2}\right)^2 + 1$

35. $y = x^2 - 4x - 5$

x-intercepts: $(5, 0), (-1, 0)$

$0 = x^2 - 4x - 5$

$0 = (x - 5)(x + 1)$

$x = 5$ or $x = -1$

37. $y = x^2 + 8x + 16$

x-intercept: $(-4, 0)$

$0 = x^2 + 8x + 16$

$0 = (x + 4)^2$

$x = -4$

39. $y = x^2 - 4x$

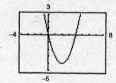

$0 = x^2 - 4x$

$0 = x(x - 4)$

$x = 0$ or $x = 4$

x-intercepts: $(0, 0), (4, 0)$

41. $y = 2x^2 - 7x - 30$

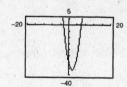

$0 = 2x^2 - 7x - 30$

$0 = (2x + 5)(x - 6)$

$x = -\frac{5}{2}$ or $x = 6$

x-intercepts: $\left(-\frac{5}{2}, 0\right), (6, 0)$

43. $y = -\frac{1}{2}(x^2 - 6x - 7)$

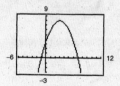

$0 = -\frac{1}{2}(x^2 - 6x - 7)$

$0 = x^2 - 6x - 7$

$0 = (x + 1)(x - 7)$

$x = -1, 7$

x-intercepts: $(-1, 0), (7, 0)$

45. $f(x) = [x - (-1)](x - 3)$, opens upward

$\quad = (x + 1)(x - 3)$

$\quad = x^2 - 2x - 3$

$g(x) = -[x - (-1)](x - 3)$, opens downward

$\quad = -(x + 1)(x - 3)$

$\quad = -(x^2 - 2x - 3)$

$\quad = -x^2 + 2x + 3$

Note: $f(x) = a(x + 1)(x - 3)$ has x-intercepts $(-1, 0)$ and $(3, 0)$ for all real numbers $a \neq 0$.

47. $f(x) = [x - (-3)]\left[x - \left(-\frac{1}{2}\right)\right](2)$, opens upward

$\quad = (x + 3)\left(x + \frac{1}{2}\right)(2)$

$\quad = (x + 3)(2x + 1)$

$\quad = 2x^2 + 7x + 3$

$g(x) = -(2x^2 + 7x + 3)$, opens downward

$\quad = -2x^2 - 7x - 3$

Note: $f(x) = a(x + 3)(2x + 1)$ has x-intercepts $(-3, 0)$ and $\left(-\frac{1}{2}, 0\right)$ for all real numbers $a \neq 0$.

49. Let $x =$ the first number and $y =$ the second number. Then the sum is

$x + y = 110 \implies y = 110 - x$.

The product is

$P(x) = xy = x(110 - x) = 110x - x^2$.

$P(x) = -x^2 + 110x$

$\quad = -(x^2 - 110x + 3025 - 3025)$

$\quad = -[(x - 55)^2 - 3025]$

$\quad = -(x - 55)^2 + 3025$

The maximum value of the product occurs at the vertex of $P(x)$ and is 3025. This happens when $x = y = 55$.

51. Let x be the first number and y be the second number. Then $x + 2y = 24 \implies x = 24 - 2y$.

The product is $P = xy = (24 - 2y)y = 24y - 2y^2$.

Completing the square,

$P = -2y^2 + 24y$

$\quad = -2(y^2 - 12y + 36) + 72$

$\quad = -2(y - 6)^2 + 72$.

The maximum value of the product P occurs at the vertex of the parabola and equals 72. This happens when $y = 6$ and $x = 24 - 2(6) = 12$.

53. (a)

(b) Radius of semicircular ends of track: $r = \frac{1}{2}y$

Distance around two semicircular parts of track:

$$d = 2\pi r = 2\pi\left(\frac{1}{2}y\right) = \pi y$$

(c) Distance traveled around track in one lap:

$d = \pi y + 2x = 200$

$\pi y = 200 - 2x$

$y = \dfrac{200 - 2x}{\pi}$

(e)

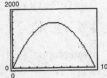

The area is maximum when $x = 50$ and

$y = \dfrac{200 - 2(50)}{\pi} = \dfrac{100}{\pi}$.

(d) Area of rectangular region:

$A = xy = x\left(\dfrac{200 - 2x}{\pi}\right)$

$\quad = \dfrac{1}{\pi}(200x - 2x^2)$

$\quad = -\dfrac{2}{\pi}(x^2 - 100x)$

$\quad = -\dfrac{2}{\pi}(x^2 - 100x + 2500 - 2500)$

$\quad = -\dfrac{2}{\pi}(x - 50)^2 + \dfrac{5000}{\pi}$

The area is maximum when $x = 50$ and

$y = \dfrac{200 - 2(50)}{\pi} = \dfrac{100}{\pi}$.

55. (a)

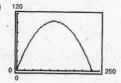

(b) When $x = 0$, $y = \dfrac{3}{2}$ feet.

(c) The vertex occurs at

$$x = \frac{-b}{2a} = \frac{-9/5}{2(-16/2025)} = \frac{3645}{32} \approx 113.9.$$

The maximum height is

$$y = \frac{-16}{2025}\left(\frac{3645}{32}\right)^2 + \frac{9}{5}\left(\frac{3645}{32}\right) + \frac{3}{2}$$

$$\approx 104.0 \text{ feet.}$$

(d) Using a graphing utility, the zero of y occurs at $x \approx 228.6$, or 228.6 feet from the punter.

57. $C = 800 - 10x + 0.25x^2$

x	10	15	20	25	30
C	725	706.25	700	706.25	725

From the table, the minimum cost seems to be at $x = 20$.

The minimum cost occurs at the vertex.

$$x = \frac{-b}{2a} = -\frac{(-10)}{2(0.25)} = \frac{10}{0.5} = 20$$

$C(20) = 700$ is the minimum cost.

Graphically, you could graph $C = 800 - 10x + 0.25x^2$ in the window $[0, 40] \times [0, 1000]$ and find the vertex $(20, 700)$.

59. (a) $R(20) = -25(20)^2 + 1200(20)$

$\qquad\qquad = \$14{,}000$ thousand

$\qquad R(25) = -25(25)^2 + 1200(25)$

$\qquad\qquad = \$14{,}375$ thousand

$\qquad R(30) = -25(30)^2 + 1200(30)$

$\qquad\qquad = \$13{,}500$ thousand

(b) The vertex occurs at

$$p = \frac{-b}{2a} = \frac{-1200}{2(-25)} = \$24.$$

(c) $R(24) = -25(24)^2 + 1200(24)$

$\qquad\qquad = \$14{,}400$ thousand

(d) Answers will vary.

61. $C(t) = 4306 - 3.4t - 1.32t^2$, $0 \le t \le 44$

($t = 0$ corresponds to 1960.)

(a)

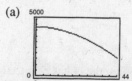

(b) The maximum consumption per year of 4306 cigarettes per person per year occurred in 1960 ($t = 0$). Answers will vary.

(c) For 2000, $C(40) = 2058$.

$$(2058)\frac{209{,}117{,}000}{48{,}306{,}000} \approx 8909 \text{ cigarettes per}$$

smoker per year, $\dfrac{8909}{365} \approx 24$ cigarettes per smoker per day

63. True

$$-12x^2 - 1 = 0$$

$$12x^2 = -1, \text{ impossible}$$

65. The parabola opens downward and the vertex is $(-2, -4)$. Matches (c) and (d).

67. For $a < 0$, $f(x) = a\left(x + \dfrac{b}{2a}\right)^2 + \left(c - \dfrac{b^2}{4a}\right)$ is a

maximum when $x = \dfrac{-b}{2a}$. In this case, the

maximum value is $c - \dfrac{b^2}{4a}$. Hence,

$$25 = -75 - \frac{b^2}{4(-1)}$$

$$-100 = 300 - b^2$$

$$400 = b^2$$

$$b = \pm 20.$$

69. For $a > 0$, $f(x) = a\left(x + \dfrac{b}{2a}\right)^2 + \left(c - \dfrac{b^2}{4a}\right)$ is a

minimum when $x = \dfrac{-b}{2a}$. In this case, the minimum

value is $c - \dfrac{b^2}{4a}$. Hence,

$$10 = 26 - \frac{b^2}{4}$$

$$40 = 104 - b^2$$

$$b^2 = 64$$

$$b = \pm 8.$$

71. Model (a) is preferable. $a > 0$ means the parabola opens upward and profits are increasing for t to the right of the vertex,

$$t \geq -\frac{b}{(2a)}.$$

73. $x + y = 8 \implies y = 8 - x$

Then $-\frac{2}{3}x + y = -\frac{2}{3}x + (8 - x) = 6$

$\implies -\frac{5}{3}x = -2 \implies x = \frac{6}{5}$ and $y = 8 - \frac{6}{5} = \frac{34}{5}.$

$(1.2, 6.8)$

75. $\quad y = x + 3 = 9 - x^2$

$$x^2 + x - 6 = 0$$

$$(x + 3)(x - 2) = 0$$

$$x = -3, x = 2$$

Thus, $(-3, 0)$ and $(2, 5)$ are the points of intersection.

77. Answers will vary. (Make a Decision.)

Section 2.2   Polynomial Functions of Higher Degree

■ You should know the following basic principles about polynomials.

■ $f(x) = a_n x^n + a_{n-1}x^{n-1} + \cdots + a_2 x^2 + a_1 x + a_0$, $a_n \neq 0$, is a polynomial function of degree n.

■ If f is of odd degree and

    (a) $a_n > 0$, then

        1. $f(x) \to \infty$ as $x \to \infty$.

        2. $f(x) \to -\infty$ as $x \to -\infty$.

    (b) $a_n < 0$, then

        1. $f(x) \to -\infty$ as $x \to \infty$.

        2. $f(x) \to \infty$ as $x \to -\infty$.

■ If f is of even degree and

    (a) $a_n > 0$, then

        1. $f(x) \to \infty$ as $x \to \infty$.

        2. $f(x) \to \infty$ as $x \to -\infty$.

    (b) $a_n < 0$, then

        1. $f(x) \to -\infty$ as $x \to \infty$.

        2. $f(x) \to -\infty$ as $x \to -\infty$.

—CONTINUED—

Section 2.2   —CONTINUED—

■ The following are equivalent for a polynomial function.

(a) $x = a$ is a zero of a function.

(b) $x = a$ is a solution of the polynomial equation $f(x) = 0$.

(c) $(x - a)$ is a factor of the polynomial.

(d) $(a, 0)$ is an x-intercept of the graph of f.

■ A polynomial of degree n has at most n distinct zeros.

■ If f is a polynomial function such that $a < b$ and $f(a) \neq f(b)$, then f takes on every value between $f(a)$ and $f(b)$ in the interval $[a, b]$.

■ If you can find a value where a polynomial is positive and another value where it is negative, then there is at least one real zero between the values.

Vocabulary Check

1. continuous

2. Leading Coefficient Test

3. $n, n - 1$, relative extrema

4. solution, $(x - a)$, x-intercept

5. touches, crosses

6. Intermediate Value

---

1. $f(x) = -2x + 3$ is a line with y-intercept $(0, 3)$. Matches graph (f).

3. $f(x) = -2x^2 - 5x$ is a parabola with x-intercepts $(0, 0)$ and $\left(-\frac{5}{2}, 0\right)$ and opens downward. Matches graph (c).

5. $f(x) = -\frac{1}{4}x^4 + 3x^2$ has intercepts $(0, 0)$ and $\left(\pm 2\sqrt{3}, 0\right)$. Matches graph (e).

7. $f(x) = x^4 + 2x^3$ has intercepts $(0, 0)$ and $(-2, 0)$. Matches graph (g).

9. $y = x^3$

(a) $f(x) = (x - 2)^3$

Horizontal shift two units to the right

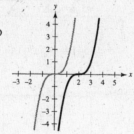

(b) $f(x) = x^3 - 2$

Vertical shift two units downward

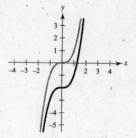

(c) $f(x) = -\frac{1}{2}x^3$

Reflection in the x-axis and a vertical shrink

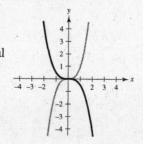

(d) $f(x) = (x - 2)^3 - 2$

Horizontal shift two units to the right and a vertical shift two units downward

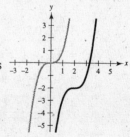

11. $f(x) = 3x^3 - 9x + 1$; $g(x) = 3x^3$

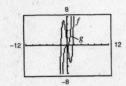

13. $f(x) = -(x^4 - 4x^3 + 16x)$; $g(x) = -x^4$

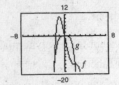

15. $f(x) = 2x^4 - 3x + 1$

Degree: 4

Leading coefficient: 2

The degree is even and the leading coefficient is positive. The graph rises to the left and right.

17. $g(x) = 5 - \frac{7}{2}x - 3x^2$

Degree: 2

Leading coefficient: -3

The degree is even and the leading coefficient is negative. The graph falls to the left and right.

19. Degree: 5 (odd)

Leading coefficient: $\frac{6}{3} = 2 > 0$

Falls to the left and rises to the right

21. $h(t) = -\frac{2}{3}(t^2 - 5t + 3)$

Degree: 2

Leading coefficient: $-\frac{2}{3}$

The degree is even and the leading coefficient is negative. The graph falls to the left and right.

23. $f(x) = x^2 - 25$

$\quad = (x + 5)(x - 5)$

$\quad x = \pm 5$

25. $h(t) = t^2 - 6t + 9$

$\quad = (t - 3)^2$

$\quad t = 3$ (multiplicity 2)

27. $f(x) = x^2 + x - 2$

$\quad = (x + 2)(x - 1)$

$\quad x = -2, 1$

29. $f(t) = t^3 - 4t^2 + 4t$

$\quad = t(t - 2)^2$

$\quad t = 0, 2$ (multiplicity 2)

31. $f(x) = \frac{1}{2}x^2 + \frac{5}{2}x - \frac{3}{2}$

$\quad = \frac{1}{2}(x^2 + 5x - 3)$

$\quad x = \dfrac{-5 \pm \sqrt{25 - 4(-3)}}{2} = -\dfrac{5}{2} \pm \dfrac{\sqrt{37}}{2}$

$\quad \approx 0.5414, -5.5414$

33. (a)

-7   2   11   -10

(b) $x \approx 3.732, 0.268$

(c) $f(x) = 3x^2 - 12x + 3$

$\quad = 3(x^2 - 4x + 1)$

$\quad x = \dfrac{4 \pm \sqrt{16 - 4}}{2} = 2 \pm \sqrt{3}$

35. (a)

-6   6   6   -2

(b) $t = \pm 1$

(c) $g(t) = \frac{1}{2}t^4 - \frac{1}{2}$

$\quad = \frac{1}{2}(t + 1)(t - 1)(t^2 + 1)$

$\quad t = \pm 1$

37. (a)

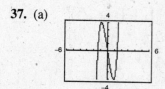

(b) $x = 0, 1.414, -1.414$

(c) $f(x) = x^5 + x^3 - 6x$

$\qquad = x(x^4 + x^2 - 6)$

$\qquad = x(x^2 + 3)(x^2 - 2)$

$\qquad x = 0, \pm\sqrt{2}$

39. (a)

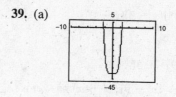

(b) $2.236, -2.236$

(c) $f(x) = 2x^4 - 2x^2 - 40$

$\qquad = 2(x^4 - x^2 - 20)$

$\qquad = 2(x^2 + 4)(x + \sqrt{5})(x - \sqrt{5})$

$\qquad x = \pm\sqrt{5}$

41. (a)

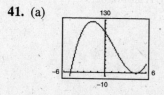

(b) $x = 4, 5, -5$

(c) $f(x) = x^3 - 4x^2 - 25x + 100$

$\qquad = x^2(x - 4) - 25(x - 4)$

$\qquad = (x^2 - 25)(x - 4)$

$\qquad = (x - 5)(x + 5)(x - 4)$

$\qquad x = \pm 5, 4$

43. (a)

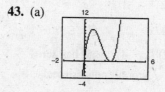

(b) $x = 0, \frac{5}{2}$

(c) $y = 4x^3 - 20x^2 + 25x$

$\qquad 0 = 4x^3 - 20x^2 + 25x$

$\qquad 0 = x(2x - 5)^2$

$\qquad x = 0$ or $x = \frac{5}{2}$ (multiplicity 2)

45. $f(x) = 2x^4 - 6x^2 + 1$

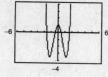

Zeros: $x \approx \pm 0.421, \pm 1.680$

Relative maximum: $(0, 1)$

Relative minimums:
$(1.225, -3.5), (-1.225, -3.5)$

47. $f(x) = x^5 + 3x^3 - x + 6$

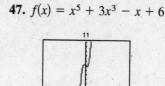

Zeros: $x \approx -1.178$

Relative maximum: $(-0.324, 6.218)$

Relative minimum: $(0.324, 5.782)$

49. $f(x) = (x - 0)(x - 4) = x^2 - 4x$

Note: $f(x) = a(x - 0)(x - 4) = ax(x - 4)$ has zeros 0 and 4 for all nonzero real numbers a.

51. $f(x) = (x - 0)(x + 2)(x + 3) = x^3 + 5x^2 + 6x$

Note: $f(x) = ax(x + 2)(x + 3)$ has zeros 0, -2, and -3 for all nonzero real numbers a.

53. $f(x) = (x - 4)(x + 3)(x - 3)(x - 0)$

$\quad = (x - 4)(x^2 - 9)x$

$\quad = x^4 - 4x^3 - 9x^2 + 36x$

Note: $f(x) = a(x^4 - 4x^3 - 9x^2 + 36x)$ has zeros 4, −3, 3, and 0 for all nonzero real numbers a.

55. $f(x) = \left[x - \left(1 + \sqrt{3}\right)\right]\left[x - \left(1 - \sqrt{3}\right)\right]$

$\quad = \left[(x - 1) - \sqrt{3}\right]\left[(x - 1) + \sqrt{3}\right]$

$\quad = (x - 1)^2 - \left(\sqrt{3}\right)^2$

$\quad = x^2 - 2x + 1 - 3$

$\quad = x^2 - 2x - 2$

Note: $f(x) = a(x^2 - 2x - 2)$ has zeros $1 + \sqrt{3}$ and $1 - \sqrt{3}$ for all nonzero real numbers a.

57. $f(x) = (x - 2)\left[x - \left(4 + \sqrt{5}\right)\right]\left[x - \left(4 - \sqrt{5}\right)\right]$

$\quad = (x - 2)\left[(x - 4) - \sqrt{5}\right]\left[(x - 4) + \sqrt{5}\right]$

$\quad = (x - 2)\left[(x - 4)^2 - 5\right]$

$\quad = x^3 - 10x^2 + 27x - 22$

Note: $f(x) = a(x - 2)\left[(x - 4)^2 - 5\right]$ has zeros 2, $4 + \sqrt{5}$, and $4 - \sqrt{5}$ for all nonzero real numbers a.

59. $f(x) = (x + 2)^2(x + 1) = x^3 + 5x^2 + 8x + 4$

Note: $f(x) = a(x + 2)^2(x + 1)$ has zeros −2, −2, and −1 for all nonzero real numbers a.

61. $f(x) = (x + 4)^2(x - 3)^2$

$\quad = x^4 + 2x^3 - 23x^2 - 24x + 144$

Note: $f(x) = a(x + 4)^2(x - 3)^2$ has zeros −4, −4, 3, 3 for all nonzero real numbers a.

63. $f(x) = -(x + 1)^2(x + 2)$

$\quad = -x^3 - 4x^2 - 5x - 2$

Note: $f(x) = a(x + 1)^2(x + 2)^2$, $a < 0$, has zeros −1, −1, −2, rises to the left, and falls to the right.

65.

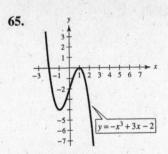

For example,

$\quad f(x) = -(x + 2)(x - 1)^2 = -x^3 + 3x - 2.$

67.

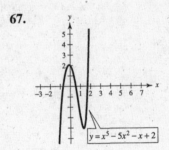

69. (a) The degree of f is odd and the leading coefficient is 1. The graph falls to the left and rises to the right.

(b) $f(x) = x^3 - 9x = x(x^2 - 9) = x(x - 3)(x + 3)$

   Zeros: 0, 3, −3

(c) and (d)

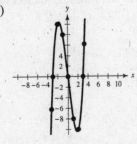

71. (a) The degree of f is odd and the leading coefficient is 1. The graph falls to the left and rises to the right.

(b) $f(x) = x^3 - 3x^2 = x^2(x - 3)$

   Zeros: 0, 3

(c) and (d)

73. (a) The degree of f is even and the leading coefficient is -1. The graph falls to the left and falls to the right.

(b) $f(x) = -x^4 + 9x^2 - 20 = -(x^2 - 4)(x^2 - 5)$

Zeros: $\pm 2, \pm \sqrt{5}$: $(\pm 2, 0), (\pm \sqrt{5}, 0)$

(c) and (d)

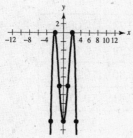

75. (a) The degree is odd and the leading coefficient is 1. The graph falls to the left and rises to the right.

(b) $x^3 + 3x^2 - 9x - 27 = x^2(x + 3) - 9(x + 3)$

$$= (x^2 - 9)(x + 3)$$

$$= (x - 3)(x + 3)^2$$

Zeros: $3, -3$: $(3, 0), (-3, 0)$

(c) and (d)

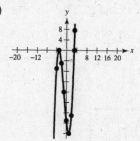

77. $g(t) = -\frac{1}{4}t^4 + 2t^2 - 4$

(a) Falls to left and falls to right; $\left(-\frac{1}{4} < 0\right)$

(b) $g(t) = -\frac{1}{4}(t^4 - 8t^2 + 16) = -\frac{1}{4}(t^2 - 4)^2$

$t = -2, -2, 2, 2 \Rightarrow (-2, 0), (2, 0)$; zeros

(c) and (d)

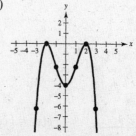

79. $f(x) = x^3 - 3x^2 + 3$

(a)

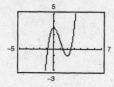

The function has three zeros. They are in the intervals $(-1, 0)$, $(1, 2)$ and $(2, 3)$.

(b) Zeros: $-0.879, 1.347, 2.532$

(c)

x	y_1	x	y_1	x	y_1
-0.9	-0.159	1.3	0.127	2.5	-0.125
-0.89	-0.0813	1.31	0.09979	2.51	-0.087
-0.88	-0.0047	1.32	0.07277	2.52	-0.0482
-0.87	0.0708	1.33	0.04594	2.53	-0.0084
-0.86	0.14514	1.34	0.0193	2.54	0.03226
-0.85	0.21838	1.35	-0.0071	2.55	0.07388
-0.84	0.2905	1.36	-0.0333	2.56	0.11642

81. $g(x) = 3x^4 + 4x^3 - 3$

(a)

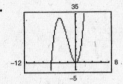

The function has two zeros.
They are in the intervals
$(-2, -1)$ and $(0, 1)$.

(b) Zeros: $-1.585, 0.779$

(c)

x	y_1		x	y_1
-1.6	0.2768		0.75	-0.3633
-1.59	0.09515		0.76	-0.2432
-1.58	-0.0812		0.77	-0.1193
-1.57	-0.2524		0.78	0.00866
-1.56	-0.4184		0.79	0.14066
-1.55	-0.5795		0.80	0.2768
-1.54	-0.7356		0.81	0.41717

83.

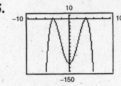

$f(x) = x^2(x + 6)$

No symmetry

Two x-intercepts

85.

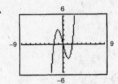

$g(t) = -\frac{1}{2}(t - 4)^2(t + 4)^2$

Symmetric about the y-axis

Two x-intercepts

87.

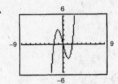

$f(x) = x^3 - 4x$

$\quad = x(x + 2)(x - 2)$

Symmetric to origin

Three x-intercepts

89.

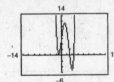

$g(x) = \frac{1}{5}(x + 1)^2(x - 3)(2x - 9)$

Three x-intercepts

No symmetry

91. (a) Volume = length × width × height

Because the box is made from a square, length = width.

Thus:

Volume = (length)^2 × height = $(36 - 2x)^2 x$

(b) Domain: $\quad 0 < 36 - 2x < 36$

$\quad\quad -36 < -2x < 0$

$\quad\quad 18 > x > 0$

(c)

Height, x	Length and Width	Volume, V
1	$36 - 2(1)$	$1[36 - 2(1)]^2 = 1156$
2	$36 - 2(2)$	$2[36 - 2(2)]^2 = 2048$
3	$36 - 2(3)$	$3[36 - 2(3)]^2 = 2700$
4	$36 - 2(4)$	$4[36 - 2(4)]^2 = 3136$
5	$36 - 2(5)$	$5[36 - 2(5)]^2 = 3380$
6	$36 - 2(6)$	$6[36 - 2(6)]^2 = 3456$
7	$36 - 2(7)$	$7[36 - 2(7)]^2 = 3388$

Maximum volume 3456 for $x = 6$

(d)

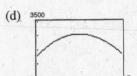

$x = 6$ when $V(x)$ is maximum.

93. The point of diminishing returns (where the graph changes from curving upward to curving downward) occurs when $x = 200$. The point is $(200, 160)$ which corresponds to spending $2,000,000 on advertising to obtain a revenue of $160 million.

95.

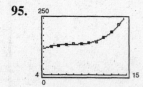

The model is a good fit.

97. For 2010, $t = 20$, and

$y_1 \approx \$730.2$ thousand

$y_2 \approx \$285.0$ thousand.

Answers will vary.

99. True. $f(x) = x^6$ has only one zero, 0.

101. False. The graph touches at $x = 1$, but does not cross the x-axis there.

103. True. The exponent of $(x + 2)$ is odd (3).

105. The zeros are 0, 1, 1, and the graph rises to the right. Matches (b).

107. The zeros are 1, 1, -2, -2, and the graph rises to the right. Matches (a).

109. $(g - f)(3) = g(3) - f(3) = 8(3)^2 - [14(3) - 3]$

$\qquad = 72 - 39 = 33$

111. $\left(\dfrac{f}{g}\right)(-1.5) = \dfrac{f(-1.5)}{g(-1.5)} = \dfrac{-24}{18} = -\dfrac{4}{3}$

113. $(g \circ f)(0) = g(f(0)) = g(-3) = 8(-3)^2 = 72$

115. $\qquad 2x^2 - x \geq 1$

$\qquad 2x^2 - x - 1 \geq 0$

$(2x + 1)(x - 1) \geq 0$

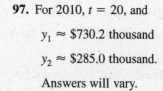

$[2x + 1 \geq 0 \text{ and } x - 1 \geq 0]$ or $[2x + 1 \leq 0 \text{ and } x - 1 \leq 0]$

$\qquad [x \geq -\tfrac{1}{2} \text{ and } x \geq 1]$ or $[x \leq -\tfrac{1}{2} \text{ and } x \leq 1]$

$\qquad\qquad x \geq 1$ or $x \leq -\tfrac{1}{2}$

117. $|x + 8| - 1 \geq 15$

$\qquad |x + 8| \geq 16$

$\qquad x + 8 \geq 16$ or $x + 8 \leq -16$

$\qquad\qquad x \geq 8$ or $x \leq -24$

Section 2.3   Real Zeros of Polynomial Functions

You should know the following basic techniques and principles of polynomial division.

■ The Division Algorithm (Long Division of Polynomials)

■ Synthetic Division

■ $f(k)$ is equal to the remainder of $f(x)$ divided by $(x - k)$.

■ $f(k) = 0$ if and only if $(x - k)$ is a factor of $f(x)$.

■ The Rational Zero Test

■ The Upper and Lower Bound Rule

Vocabulary Check

1. $f(x)$ is the dividend, $d(x)$ is the divisor, $q(x)$ is the quotient, and $r(x)$ is the remainder.

2. improper, proper   **3.** synthetic division   **4.** Rational Zero

5. Descartes's Rule, Signs   **6.** Remainder Theorem   **7.** upper bound, lower bound

---

1.
$$
\begin{array}{r}
2x + 4 \\
x + 3 \overline{\smash{)}\ 2x^2 + 10x + 12} \\
\underline{2x^2 + 6x} \\
4x + 12 \\
\underline{4x + 12} \\
0
\end{array}
$$

$$\frac{2x^2 + 10x + 12}{x + 3} = 2x + 4,\ x \neq -3$$

3.
$$
\begin{array}{r}
x^3 + 3x^2 - 1 \\
x + 2 \overline{\smash{)}\ x^4 + 5x^3 + 6x^2 - x - 2} \\
\underline{x^4 + 2x^3} \\
3x^3 + 6x^2 \\
\underline{3x^3 + 6x^2} \\
-x - 2 \\
\underline{-x - 2} \\
0
\end{array}
$$

$$\frac{x^4 + 5x^3 + 6x^2 - x - 2}{x + 2} = x^3 + 3x^2 - 1,\ x \neq -2$$

5.
$$
\begin{array}{r}
x^2 - 3x + 1 \\
4x + 5 \overline{\smash{)}\ 4x^3 - 7x^2 - 11x + 5} \\
\underline{-(4x^3 + 5x^2)} \\
-12x^2 - 11x \\
\underline{-(-12x^2 - 15x)} \\
4x + 5 \\
\underline{-(4x + 5)} \\
0
\end{array}
$$

$$\frac{4x^3 - 7x^2 - 11x + 5}{4x + 5} = x^2 - 3x + 1,\quad x \neq -\frac{5}{4}$$

7.
$$
\begin{array}{r}
7x^2 - 14x + 28 \\
x + 3 \overline{\smash{)}\ 7x^3 + 0x^2 + 0x + 3} \\
\underline{7x^3 + 14x^2} \\
-14x^2 \\
\underline{-14x^2 - 28x} \\
28x + 3 \\
\underline{28x + 56} \\
-53
\end{array}
$$

$$\frac{7x^3 + 3}{x + 2} = 7x^2 - 14x + 28 - \frac{53}{x + 2}$$

9.

$$
\begin{array}{r}
3x + 5 \\
2x^2 + 0x + 1 \overline{)\ 6x^3 + 10x^2 + x + 8} \\
-\ (6x^3 + 0x^2 + 3x) \\
\hline
10x^2 - 2x + 8 \\
-\ (10x^2 + 0x + 5) \\
\hline
-2x + 3
\end{array}
$$

$$\frac{6x^3 + 10x^2 + x + 8}{2x^2 + 1} = 3x + 5 - \frac{2x - 3}{2x^2 + 1}$$

11.

$$
\begin{array}{r}
x \\
x^2 + 1 \overline{)\ x^3 + 0x^2 + 0x - 9} \\
x^3 + x \\
\hline
-x - 9
\end{array}
$$

$$\frac{x^3 - 9}{x^2 + 1} = x - \frac{x + 9}{x^2 + 1}$$

13.

$$
\begin{array}{r}
2x \\
x^2 - 2x + 1 \overline{)\ 2x^3 - 4x^2 - 15x + 5} \\
2x^3 - 4x^2 + 2x \\
\hline
-17x + 5
\end{array}
$$

$$\frac{2x^3 - 4x^2 - 15x + 5}{(x - 1)^2} = 2x - \frac{17x - 5}{(x - 1)^2}$$

15.

$$
\begin{array}{r|rrrr}
5 & 3 & -17 & 15 & -25 \\
  &   & 15 & -10 & 25 \\
\hline
  & 3 & -2 & 5 & 0
\end{array}
$$

$$\frac{3x^3 - 17x^2 + 15x - 25}{x - 5} = 3x^2 - 2x + 5, \ x \neq 5$$

17.

$$
\begin{array}{r|rrrr}
3 & 6 & 7 & -1 & 26 \\
  &   & 18 & 75 & 222 \\
\hline
  & 6 & 25 & 74 & 248
\end{array}
$$

$$\frac{6x^3 + 7x^2 - x + 26}{x - 3} = 6x^2 + 25x + 74 + \frac{248}{x - 3}$$

19.

$$
\begin{array}{r|rrrr}
2 & 9 & -18 & -16 & 32 \\
  &   & 18 & 0 & -32 \\
\hline
  & 9 & 0 & -16 & 0
\end{array}
$$

$$\frac{9x^3 - 18x^2 - 16x + 32}{x - 2} = 9x^2 - 16, \ x \neq 2$$

21.

$$
\begin{array}{r|rrrr}
-8 & 1 & 0 & 0 & 512 \\
   &   & -8 & 64 & -512 \\
\hline
   & 1 & -8 & 64 & 0
\end{array}
$$

$$\frac{x^3 + 512}{x + 8} = x^2 - 8x + 64, \ x \neq -8$$

23.

$$
\begin{array}{r|rrrr}
-\frac{1}{2} & 4 & 16 & -23 & -15 \\
   &   & -2 & -7 & 15 \\
\hline
   & 4 & 14 & -30 & 0
\end{array}
$$

$$\frac{4x^3 + 16x^2 - 23x - 15}{x + \frac{1}{2}} = 4x^2 + 14x - 30, \ x \neq -\frac{1}{2}$$

25. $y_2 = x - 2 + \dfrac{4}{x + 2}$

$$= \frac{(x - 2)(x + 2) + 4}{x + 2}$$

$$= \frac{x^2 - 4 + 4}{x + 2}$$

$$= \frac{x^2}{x + 2}$$

$$= y_1$$

27. $y_2 = x^2 - 8 + \dfrac{39}{x^2 + 5}$

$$= \frac{(x^2 - 8)(x^2 + 5) + 39}{x^2 + 5}$$

$$= \frac{x^4 - 8x^2 + 5x^2 - 40 + 39}{x^2 + 5}$$

$$= \frac{x^4 - 3x^2 - 1}{x^2 + 5}$$

$$= y_1$$

29. $f(x) = x^3 - x^2 - 14x + 11, \quad k = 4$

$$
\begin{array}{r|rrrr}
4 & 1 & -1 & -14 & 11 \\
  &   & 4 & 12 & -8 \\
\hline
  & 1 & 3 & -2 & 3
\end{array}
$$

$f(x) = (x - 4)(x^2 + 3x - 2) + 3$

$f(4) = (0)(26) + 3 = 3$

31.

$$
\begin{array}{r|rrrr}
\sqrt{2} & 1 & 3 & -2 & -14 \\
         &   & \sqrt{2} & 2 + 3\sqrt{2} & 6 \\
\hline
         & 1 & 3 + \sqrt{2} & 3\sqrt{2} & -8
\end{array}
$$

$f(x) = (x - \sqrt{2})(x^2 + (3 + \sqrt{2})x + 3\sqrt{2}) - 8$

$f(\sqrt{2}) = 0(4 + 6\sqrt{2}) - 8 = -8$

33.

$$
\begin{array}{r|rrrr}
1 - \sqrt{3} & 4 & -6 & -12 & -4 \\
             &   & 4 - 4\sqrt{3} & 10 - 2\sqrt{3} & 4 \\
\hline
             & 4 & -2 - 4\sqrt{3} & -2 - 2\sqrt{3} & 0
\end{array}
$$

$f(x) = (x - 1 + \sqrt{3})[4x^2 - (2 + 4\sqrt{3})x - (2 + 2\sqrt{3})]$

$f(1 - \sqrt{3}) = 0$

35. $f(x) = 2x^3 - 7x + 3$

(a)
$$
\begin{array}{r|rrrr}
1 & 2 & 0 & -7 & 3 \\
  &   & 2 & 2 & -5 \\
\hline
  & 2 & 2 & -5 & -2 \quad = f(1)
\end{array}
$$

(b)
$$
\begin{array}{r|rrrr}
-2 & 2 & 0 & -7 & 3 \\
   &   & -4 & 8 & -2 \\
\hline
   & 2 & -4 & 1 & 1 \quad = f(-2)
\end{array}
$$

(c)
$$
\begin{array}{r|rrrr}
\frac{1}{2} & 2 & 0 & -7 & 3 \\
            &   & 1 & \frac{1}{2} & -\frac{13}{4} \\
\hline
            & 2 & 1 & -\frac{13}{2} & -\frac{1}{4} \quad = f\left(\frac{1}{2}\right)
\end{array}
$$

(d)
$$
\begin{array}{r|rrrr}
2 & 2 & 0 & -7 & 3 \\
  &   & 4 & 8 & 2 \\
\hline
  & 2 & 4 & 1 & 5 \quad = f(2)
\end{array}
$$

37. $h(x) = x^3 - 5x^2 - 7x + 4$

(a)
$$
\begin{array}{r|rrrr}
3 & 1 & -5 & -7 & 4 \\
  &   & 3 & -6 & -39 \\
\hline
  & 1 & -2 & -13 & -35 \quad = h(3)
\end{array}
$$

(b)
$$
\begin{array}{r|rrrr}
2 & 1 & -5 & -7 & 4 \\
  &   & 2 & -6 & -26 \\
\hline
  & 1 & -3 & -13 & -22 \quad = h(2)
\end{array}
$$

(c)
$$
\begin{array}{r|rrrr}
-2 & 1 & -5 & -7 & 4 \\
   &   & -2 & 14 & -14 \\
\hline
   & 1 & -7 & 7 & -10 \quad = h(-2)
\end{array}
$$

(d)
$$
\begin{array}{r|rrrr}
-5 & 1 & -5 & -7 & 4 \\
   &   & -5 & 50 & -215 \\
\hline
   & 1 & -10 & 43 & -211 \quad = h(-5)
\end{array}
$$

39.
$$
\begin{array}{r|rrrr}
2 & 1 & 0 & -7 & 6 \\
  &   & 2 & 4 & -6 \\
\hline
  & 1 & 2 & -3 & 0
\end{array}
$$

$x^3 - 7x + 6 = (x - 2)(x^2 + 2x - 3)$

$ = (x - 2)(x + 3)(x - 1)$

Zeros: $2, -3, 1$

41.
$$
\begin{array}{r|rrrr}
\frac{1}{2} & 2 & -15 & 27 & -10 \\
            &   & 1 & -7 & 10 \\
\hline
            & 2 & -14 & 20 & 0
\end{array}
$$

$2x^3 - 15x^2 + 27x - 10$

$ = \left(x - \frac{1}{2}\right)(2x^2 - 14x + 20)$

$ = (2x - 1)(x - 2)(x - 5)$

Zeros: $\frac{1}{2}, 2, 5$

43. (a)

$$-2 \begin{array}{|rrrr} 2 & 1 & -5 & 2 \\ & -4 & 6 & -2 \\ \hline 2 & -3 & 1 & 0 \end{array}$$

(b) $2x^2 - 3x + 1 = (2x - 1)(x - 1)$

Remaining factors: $(2x - 1)$, $(x - 1)$

(c) $f(x) = (x + 2)(2x - 1)(x - 1)$

45. (a)

$$5 \begin{array}{|rrrrr} 1 & -4 & -15 & 58 & -40 \\ & 5 & 5 & -50 & 40 \\ \hline 1 & 1 & -10 & 8 & 0 \end{array}$$

$$-4 \begin{array}{|rrrr} 1 & 1 & -10 & 8 \\ & -4 & 12 & -8 \\ \hline 1 & -3 & 2 & 0 \end{array}$$

47. (a)

$$-\tfrac{1}{2} \begin{array}{|rrrr} 6 & 41 & -9 & -14 \\ & -3 & -19 & 14 \\ \hline 6 & 38 & -28 & 0 \end{array}$$

(b) $6x^2 + 38x - 28 = (3x - 2)(2x + 14)$

Remaining factors: $(3x - 2)$, $(x + 7)$

(c) $f(x) = (2x + 1)(3x - 2)(x + 7)$

(d) Real zeros: $-\tfrac{1}{2}, \tfrac{2}{3}, -7$

(e)

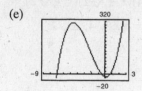

(d) Real zeros: $-2, \tfrac{1}{2}, 1$

(e)

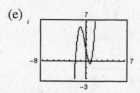

(b) $x^2 - 3x + 2 = (x - 2)(x - 1)$

Remaining factors: $(x - 2)$, $(x - 1)$

(c) $f(x) = (x - 5)(x + 4)(x - 2)(x - 1)$

(d) Real zeros: $5, -4, 2, 1$

(e)

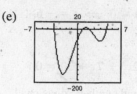

49. $f(x) = x^3 + 3x^2 - x - 3$

p = factor of -3

q = factor of 1

Possible rational zeros: $\pm 1, \pm 3$

$f(x) = x^2(x + 3) - (x + 3) = (x + 3)(x^2 - 1)$

Rational zeros: $\pm 1, -3$

51. $f(x) = 2x^4 - 17x^3 + 35x^2 + 9x - 45$

p = factor of -45

q = factor of 2

Possible rational zeros: $\pm 1, \pm 3, \pm 5, \pm 9, \pm 15, \pm 45, \pm\tfrac{1}{2}, \pm\tfrac{3}{2}, \pm\tfrac{5}{2}, \pm\tfrac{9}{2}, \pm\tfrac{15}{2}, \pm\tfrac{45}{2}$

Using synthetic division, -1, 3, and 5 are zeros.

$f(x) = (x + 1)(x - 3)(x - 5)(2x - 3)$

Rational zeros: $-1, 3, 5, \tfrac{3}{2}$

53. $z^4 - z^3 - 2z - 4 = 0$

Possible rational zeros: $\pm 1, \pm 2, \pm 4$

$$
\begin{array}{r|rrrrr}
-1 & 1 & -1 & 0 & -2 & -4 \\
   &   & -1 & 2 & -2 & 4 \\
\hline
   & 1 & -2 & 2 & -4 & 0
\end{array}
\qquad
\begin{array}{r|rrrr}
2 & 1 & -2 & 2 & -4 \\
  &   & 2 & 0 & 4 \\
\hline
  & 1 & 0 & 2 & 0
\end{array}
$$

$z^4 - z^3 - 2z - 4 = (z + 1)(z - 2)(z^2 + 2) = 0$

The only real zeros are -1 and 2. You can verify this by graphing the function $f(z) = z^4 - z^3 - 2z - 4$.

55. $2y^4 + 7y^3 - 26y^2 + 23y - 6 = 0$

Using a graphing utility and synthetic division, $1/2$, 1, and -6 are rational zeros. Hence,

$(y + 6)(y - 1)^2(2y - 1) = 0 \implies y = -6, 1, \frac{1}{2}.$

57. $4x^4 - 55x^2 - 45x + 36 = 0$

Using a graphing utility and synthetic division, $4, -3, \frac{1}{2}, -\frac{3}{2}$ are rational zeros. Hence,
$(x - 4)(x + 3)(2x - 1)(2x + 3) = 0 \implies$

$x = 4, -3, \frac{1}{2}, -\frac{3}{2}.$

59. $4x^5 + 12x^4 - 11x^3 - 42x^2 + 7x + 30 = 0$

Using a graphing utility and synthetic division, 1, -1, -2, $\frac{3}{2}$, and $-\frac{5}{2}$ are rational zeros. Hence,
$(x - 1)(x + 1)(x + 2)(2x - 3)(2x + 5) = 0 \implies$
$x = 1, -1, -2, \frac{3}{2}, -\frac{5}{2}.$

61. $h(t) = t^3 - 2t^2 - 7t + 2$

(a) Zeros: $-2, 3.732, 0.268$

(b)
$$
\begin{array}{r|rrrr}
-2 & 1 & -2 & -7 & 2 \\
   &   & -2 & 8 & -2 \\
\hline
   & 1 & -4 & 1 & 0
\end{array}
\quad t = -2 \text{ is a zero.}
$$

(c) $h(t) = (t + 2)(t^2 - 4t + 1)$

$= (t + 2)\left[t - \left(\sqrt{3} + 2\right)\right]\left[t + \left(\sqrt{3} - 2\right)\right]$

63. $h(x) = x^5 - 7x^4 + 10x^3 + 14x^2 - 24x$

(a) $h(x) = x(x^4 - 7x^3 + 10x^2 + 14x - 24)$

From the calculator we have $x = 0, 3, 4$ and $x \approx \pm 1.414.$

(b)
$$
\begin{array}{r|rrrrr}
3 & 1 & -7 & 10 & 14 & -24 \\
  &   & 3 & -12 & -6 & 24 \\
\hline
  & 1 & -4 & -2 & 8 & 0
\end{array}
$$
$$
\begin{array}{r|rrrr}
4 & 1 & -4 & -2 & 8 \\
  &   & 4 & 0 & -8 \\
\hline
  & 1 & 0 & -2 & 0
\end{array}
$$

(c) $h(x) = x(x - 3)(x - 4)(x^2 - 2)$

$= x(x - 3)(x - 4)\left(x - \sqrt{2}\right)\left(x + \sqrt{2}\right)$

The exact roots are $x = 0, 3, 4, \pm\sqrt{2}.$

65. $f(x) = 2x^4 - x^3 + 6x^2 - x + 5$

4 variations in sign $\implies$ 4, 2 or 0 positive real zeros

$f(-x) = 2x^4 + x^3 + 6x^2 + x + 5$

0 variations in sign $\implies$ 0 negative real zeros

67. $g(x) = 4x^3 - 5x + 8$

2 variations in sign $\implies$ 2 or 0 positive real zeros

$g(-x) = -4x^3 + 5x + 8$

1 variation in sign $\implies$ 1 negative real zero

69. $f(x) = x^3 + x^2 - 4x - 4$

(a) $f(x)$ has 1 variation in sign $\implies$ 1 positive real zero.

$f(-x) = -x^3 + x^2 + 4x - 4$ has 2 variations in sign $\implies$ 2 or 0 negative real zeros.

(b) Possible rational zeros: $\pm 1, \pm 2, \pm 4$

(c)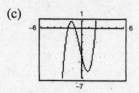

(d) Real zeros: $-2, -1, 2$

71. $f(x) = -2x^4 + 13x^3 - 21x^2 + 2x + 8$

(a) $f(x)$ has 3 variations in sign $\implies$ 3 or 1 positive real zeros.

$f(-x) = -2x^4 - 13x^3 - 21x^2 - 2x + 8$ has 1 variation in sign $\implies$ 1 negative real zero.

(b) Possible rational zeros: $\pm\frac{1}{2}, \pm 1, \pm 2, \pm 4, \pm 8$

(c)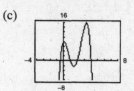

(d) Real zeros: $-\frac{1}{2}, 1, 2, 4$

73. $f(x) = 32x^3 - 52x^2 + 17x + 3$

(a) $f(x)$ has 2 variations in sign $\implies$ 2 or 0 positive real zeros.

$f(-x) = -32x^3 - 52x^2 - 17x + 3$ has 1 variation in sign $\implies$ 1 negative real zero.

(b) Possible rational zeros: $\pm\frac{1}{32}, \pm\frac{1}{16}, \pm\frac{1}{8}, \pm\frac{1}{4},$

$\pm\frac{1}{2}, \pm 1, \pm\frac{3}{32}, \pm\frac{3}{16}, \pm\frac{3}{8}, \pm\frac{3}{4}, \pm\frac{3}{2}, \pm 3$

(c)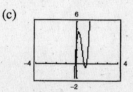

(d) Real zeros: $1, \frac{3}{4}, -\frac{1}{8}$

75. $f(x) = x^4 - 4x^3 + 15$

4	1	−4	0	0	15
		4	0	0	0
	1	0	0	0	15

4 is an upper bound.

−1	1	−4	0	0	15
		−1	5	−5	5
	1	−5	5	−5	20

−1 is a lower bound.

Real zeros: 1.937, 3.705

77. $f(x) = x^4 - 4x^3 + 16x - 16$

5	1	−4	0	16	−16
		25	105	525	2705
	5	21	105	541	2689

5 is an upper bound.

−3	1	−4	0	16	−16
		−3	21	−63	141
	1	−7	21	−47	125

−3 is a lower bound.

Real zeros: $-2, 2$

79. $P(x) = x^4 - \frac{25}{4}x^2 + 9$

$= \frac{1}{4}(4x^4 - 25x^2 + 36)$

$= \frac{1}{4}(4x^2 - 9)(x^2 - 4)$

$= \frac{1}{4}(2x + 3)(2x - 3)(x + 2)(x - 2)$

The rational zeros are $\pm\frac{3}{2}$ and ± 2.

81. $f(x) = x^3 - \frac{1}{4}x^2 - x + \frac{1}{4}$

$\quad = \frac{1}{4}(4x^3 - x^2 - 4x + 1)$

$\quad = \frac{1}{4}[x^2(4x - 1) - 1(4x - 1)]$

$\quad = \frac{1}{4}(4x - 1)(x^2 - 1)$

$\quad = \frac{1}{4}(4x - 1)(x + 1)(x - 1)$

The rational zeros are $\frac{1}{4}$ and ± 1.

85. $f(x) = x^3 - x = x(x + 1)(x - 1)$

Rational zeros: $3\ (x = 0, \pm 1)$

Irrational zeros: 0

Matches (b).

87. $y = 2x^4 - 9x^3 + 5x^2 + 3x - 1$

Using the graph and synthetic division, $-1/2$ is a zero:

$$
\begin{array}{r|rrrrr}
-\frac{1}{2} & 2 & -9 & 5 & 3 & -1 \\
 & & -1 & 5 & -5 & 1 \\
\hline
 & 2 & -10 & 10 & -2 & 0
\end{array}
$$

$y = \left(x + \frac{1}{2}\right)(2x^3 - 10x^2 + 10x - 2)$

$x = 1$ is a zero of the cubic, so

$y = (2x + 1)(x - 1)(x^2 - 4x + 1).$

For the quadratic term, use the Quadratic Formula.

$x = \dfrac{4 \pm \sqrt{16 - 4}}{2} = 2 \pm \sqrt{3}$

The real zeros are $-\frac{1}{2}, 1, 2 \pm \sqrt{3}$.

91. (a) $P(t) = 0.0058t^3 + 0.500t^2 + 1.38t + 4.6$

(b)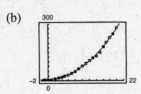

(c) The model fits the data well.

83. $f(x) = x^3 - 1$

$\quad = (x - 1)(x^2 + x + 1)$

Rational zeros: $1\ (x = 1)$

Irrational zeros: 0

Matches (d).

89. $y = -2x^4 + 17x^3 - 3x^2 - 25x - 3$

Using the graph and synthetic division, -1 and $3/2$ are zeros:

$y = -(x + 1)(2x - 3)(x^2 - 8x - 1)$

Using the Quadratic Formula:

$x = \dfrac{8 \pm \sqrt{64 + 4}}{2} = 4 \pm \sqrt{17}$

The real zeros are $-1, 3/2, 4 \pm \sqrt{17}$.

(d) For 2010, $t = 21$ and:

$$
\begin{array}{r|rrrr}
21 & 0.0058 & 0.5 & 1.38 & 4.6 \\
 & & 0.1218 & 13.0578 & 303.1938 \\
\hline
 & 0.0058 & 0.6218 & 14.4378 & 307.7938
\end{array}
$$

Hence, the population will be about 307.8 million, which seems reasonable.

93. (a) Combined length and width:

$$4x + y = 120 \implies y = 120 - 4x$$

Volume $= l \cdot w \cdot h = x^2 y$

$$= x^2(120 - 4x)$$

$$= 4x^2(30 - x)$$

(b)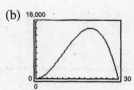

Dimensions with maximum volume:
$20 \times 20 \times 40$

(c)
$$13,500 = 4x^2(30 - x)$$

$$4x^3 - 120x^2 + 13,500 = 0$$

$$x^3 - 30x^2 + 3375 = 0$$

$$
\begin{array}{r|rrrr}
15 & 1 & -30 & 0 & 3375 \\
   &   & 15 & -225 & -3375 \\
\hline
   & 1 & -15 & -225 & 0
\end{array}
$$

$$(x - 15)(x^2 - 15x - 225) = 0$$

Using the Quadratic Formula, $x = 15$ or $\dfrac{15 \pm 15\sqrt{5}}{2}$.

The value of $\dfrac{15 - 15\sqrt{5}}{2}$ is not possible because it is negative.

95. False, $-\frac{4}{7}$ is a zero of f.

97. The zeros are 1, 1, and -2. The graph falls to the right.

$$y = a(x - 1)^2(x + 2) \quad a < 0$$

Since $f(0) = -4$, $a = -2$.

$$y = -2(x - 1)^2(x + 2) = -2x^3 + 6x - 4$$

99. $f(x) = -(x + 1)(x - 1)(x + 2)(x - 2)$

101.
$$
\begin{array}{r|rrrr}
4 & 1 & -k & 2k & -8 \\
  &   & 4 & 16 - 4k & 64 - 8k \\
\hline
  & 1 & 4 - k & 16 - 2k & 56 - 8k
\end{array}
$$

Hence, $56 - 8k = 0 \implies k = 7$.

103. (a) $\dfrac{x^2 - 1}{x - 1} = x + 1, \quad x \neq 1$

(b) $\dfrac{x^3 - 1}{x - 1} = x^2 + x + 1, \quad x \neq 1$

(c) $\dfrac{x^4 - 1}{x - 1} = x^3 + x^2 + x + 1, \quad x \neq 1$

In general,

$$\dfrac{x^n - 1}{x - 1} = x^{n-1} + x^{n-2} + \cdots + x + 1, \quad x \neq 1.$$

105.
$$9x^2 - 25 = 0$$

$$(3x + 5)(3x - 5) = 0$$

$$x = -\frac{5}{3}, \frac{5}{3}$$

107. $2x^2 + 6x + 3 = 0$

$$x = \frac{-6 \pm \sqrt{6^2 - 4(2)(3)}}{2(2)}$$

$$= \frac{-6 \pm \sqrt{12}}{4}$$

$$= \frac{-3 \pm \sqrt{3}}{2}$$

$$x = -\frac{3}{2} + \frac{\sqrt{3}}{2}, \quad -\frac{3}{2} - \frac{\sqrt{3}}{2}$$

109. $f(x) = (x - 0)(x + 12) = x^2 + 12x$

[Answer not unique]

111. $f(x) = (x - 0)(x + 1)(x - 2)(x - 5)$

$= (x^2 + x)(x^2 - 7x + 10)$

$= x^4 - 6x^3 + 3x^2 + 10x$

[Answer not unique]

Section 2.4   Complex Numbers

■ You should know how to work with complex numbers.

■ Operations on complex numbers

(a) Addition: $(a + bi) + (c + di) = (a + c) + (b + d)i$

(b) Subtraction: $(a + bi) - (c + di) = (a - c) + (b - d)i$

(c) Multiplication: $(a + bi)(c + di) = (ac - bd) + (ad + bc)i$

(d) Division: $\dfrac{a + bi}{c + di} = \dfrac{a + bi}{c + di} \cdot \dfrac{c - di}{c - di} = \dfrac{ac + bd}{c^2 + d^2} + \dfrac{bc - ad}{c^2 + d^2}i$

■ The complex conjugate of $a + bi$ is $a - bi$:

$(a + bi)(a - bi) = a^2 + b^2$

■ The additive inverse of $a + bi$ is $-a - bi$.

■ The multiplicative inverse of $a + bi$ is

$\dfrac{a - bi}{a^2 + b^2}.$

■ $\sqrt{-a} = \sqrt{a}\,i$ for $a > 0$.

Vocabulary Check

1. (a) ii   (b) iii   (c) i

2. $\sqrt{-1}, -1$

3. complex, $a + bi$

4. real, imaginary

5. Mandelbrot Set

1. $a + bi = -9 + 4i$

$a = -9$

$b = 4$

3. $(a - 1) + (b + 3)i = 5 + 8i$

$a - 1 = 5 \implies a = 6$

$b + 3 = 8 \implies b = 5$

5. $5 + \sqrt{-16} = 5 + \sqrt{16(-1)}$

$= 5 + 4i$

7. $-6 = -6 + 0i$

9. $-5i + i^2 = -5i - 1 = -1 - 5i$

11. $\left(\sqrt{-75}\right)^2 = -75$

13. $\sqrt{-0.09} = \sqrt{0.09}\,i = 0.3i$

15. $(4 + i) - (7 - 2i) = (4 - 7) + (1 + 2)i$

$= -3 + 3i$

17. $\left(-1 + \sqrt{-8}\right) + \left(8 - \sqrt{-50}\right) = 7 + 2\sqrt{2}i - 5\sqrt{2}i = 7 - 3\sqrt{2}i$

19. $13i - (14 - 7i) = 13i - 14 + 7i = -14 + 20i$

21. $\left(\dfrac{3}{2} + \dfrac{5}{2}i\right) + \left(\dfrac{5}{3} + \dfrac{11}{3}i\right) = \left(\dfrac{3}{2} + \dfrac{5}{3}\right) + \left(\dfrac{5}{2} + \dfrac{11}{3}\right)i$

$\qquad = \dfrac{9 + 10}{6} + \dfrac{15 + 22}{6}i$

$\qquad = \dfrac{19}{6} + \dfrac{37}{6}i$

23. $(1.6 + 3.2i) + (-5.8 + 4.3i) = -4.2 + 7.5i$

25. $\sqrt{-6} \cdot \sqrt{-2} = \left(\sqrt{6}i\right)\left(\sqrt{2}i\right)$

$\qquad = \sqrt{12}i^2 = \left(2\sqrt{3}\right)(-1) = -2\sqrt{3}$

27. $\left(\sqrt{-10}\right)^2 = \left(\sqrt{10}i\right)^2 = 10i^2 = -10$

29. $(1 + i)(3 - 2i) = 3 - 2i + 3i - 2i^2$

$\qquad = 3 + i + 2$

$\qquad = 5 + i$

31. $4i(8 + 5i) = 32i + 20i^2$

$\qquad = 32i + 20(-1)$

$\qquad = -20 + 32i$

33. $\left(\sqrt{14} + \sqrt{10}i\right)\left(\sqrt{14} - \sqrt{10}i\right) = 14 - 10i^2 = 14 + 10 = 24$

35. $(4 + 5i)^2 - (4 - 5i)^2 = [(4 + 5i) + (4 - 5i)][(4 + 5i) - (4 - 5i)] = 8(10i) = 80i$

37. $4 - 3i$ is the complex conjugate of $4 + 3i$.

$(4 + 3i)(4 - 3i) = 16 + 9 = 25$

39. $-6 + \sqrt{5}i$ is the complex conjugate of $-6 - \sqrt{5}i$.

$\left(-6 - \sqrt{5}i\right)\left(-6 + \sqrt{5}i\right) = 36 + 5 = 41$

41. $-\sqrt{20}i$ is the complex conjugate of $\sqrt{-20} = \sqrt{20}i$.

$\left(\sqrt{20}i\right)\left(-\sqrt{20}i\right) = 20$

43. $3 + \sqrt{2}i$ is the complex conjugate of $3 - \sqrt{-2} = 3 - \sqrt{2}i$.

$\left(3 - \sqrt{2}i\right)\left(3 + \sqrt{2}i\right) = 9 + 2 = 11$

45. $\dfrac{6}{i} = \dfrac{6}{i} \cdot \dfrac{-i}{-i} = \dfrac{-6i}{-i^2} = \dfrac{-6i}{1} = -6i$

47. $\dfrac{2}{4 - 5i} = \dfrac{2}{4 - 5i} \cdot \dfrac{4 + 5i}{4 + 5i} = \dfrac{8 + 10i}{16 + 25} = \dfrac{8}{41} + \dfrac{10}{41}i$

49. $\dfrac{2 + i}{2 - i} = \dfrac{2 + i}{2 - i} \cdot \dfrac{2 + i}{2 + i}$

$\qquad = \dfrac{4 + 4i + i^2}{4 + 1}$

$\qquad = \dfrac{3 + 4i}{5} = \dfrac{3}{5} + \dfrac{4}{5}i$

51. $\dfrac{i}{(4 - 5i)^2} = \dfrac{i}{16 - 25 - 40i}$

$\qquad = \dfrac{i}{-9 - 40i} \cdot \dfrac{-9 + 40i}{-9 + 40i}$

$\qquad = \dfrac{-40 - 9i}{81 + 40^2}$

$\qquad = \dfrac{-40}{1681} - \dfrac{9}{1681}i$

53. $\dfrac{2}{1+i} - \dfrac{3}{1-i} = \dfrac{2(1-i) - 3(1+i)}{(1+i)(1-i)}$

$\qquad\qquad = \dfrac{2 - 2i - 3 - 3i}{1 + 1}$

$\qquad\qquad = \dfrac{-1 - 5i}{2}$

$\qquad\qquad = -\dfrac{1}{2} - \dfrac{5}{2}i$

55. $\dfrac{i}{3-2i} + \dfrac{2i}{3+8i} = \dfrac{3i + 8i^2 + 6i - 4i^2}{(3-2i)(3+8i)}$

$\qquad\qquad = \dfrac{-4 + 9i}{9 + 18i + 16}$

$\qquad\qquad = \dfrac{-4 + 9i}{25 + 18i} \cdot \dfrac{25 - 18i}{25 - 18i}$

$\qquad\qquad = \dfrac{-100 + 72i + 225i + 162}{25^2 + 18^2}$

$\qquad\qquad = \dfrac{62 + 297i}{949}$

$\qquad\qquad = \dfrac{62}{949} + \dfrac{297}{949}i$

57. $-6i^3 + i^2 = -6i^2 i + i^2$

$\qquad\qquad = -6(-1)i + (-1)$

$\qquad\qquad = 6i - 1 = -1 + 6i$

59. $\left(\sqrt{-75}\right)^3 = \left(5\sqrt{3}i\right)^3 = 5^3\left(\sqrt{3}\right)^3 i^3 = 125\left(3\sqrt{3}\right)(-i)$

$\qquad\qquad = -375\sqrt{3}i$

61. $\dfrac{1}{i^3} = \dfrac{1}{i^3} \cdot \dfrac{i}{i} = \dfrac{i}{i^4} = \dfrac{i}{1} = i$

63. $(2)^3 = 8$

$\left(-1 + \sqrt{3}i\right)^3 = (-1)^3 + 3(-1)^2\left(\sqrt{3}i\right) + 3(-1)\left(\sqrt{3}i\right)^2 + \left(\sqrt{3}i\right)^3$

$\qquad\qquad = -1 + 3\sqrt{3}i - 9i^2 + 3\sqrt{3}i^3$

$\qquad\qquad = -1 + 3\sqrt{3}i + 9 - 3\sqrt{3}i$

$\qquad\qquad = 8$

$\left(-1 - \sqrt{3}i\right)^3 = (-1)^3 + 3(-1)^2\left(-\sqrt{3}i\right) + 3(-1)\left(-\sqrt{3}i\right)^2 + \left(-\sqrt{3}i\right)^3$

$\qquad\qquad = -1 - 3\sqrt{3}i - 9i^2 - 3\sqrt{3}i^3$

$\qquad\qquad = -1 - 3\sqrt{3}i + 9 + 3\sqrt{3}i$

$\qquad\qquad = 8$

The three numbers are cube roots of 8.

65. $4 + 3i$

67. $5i$

69. 2

71. $4 - 5i$

73. $3i$

75. 1

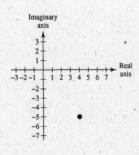

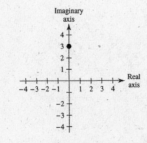

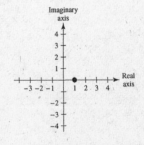

77. The complex number $\frac{1}{2}i$, is in the Mandelbrot Set since for $c = \frac{1}{2}i$, the corresponding Mandelbrot sequence is

$$\frac{1}{2}i, \; -\frac{1}{4} + \frac{1}{2}i, \; -\frac{3}{16} + \frac{1}{4}i, \; -\frac{7}{256} + \frac{13}{32}i,$$

$$-\frac{10{,}767}{65{,}536} + \frac{1957}{4096}i,$$

$$-\frac{864{,}513{,}055}{4{,}294{,}967{,}296} + \frac{46{,}037{,}845}{134{,}217{,}728}i$$

which is bounded. Or in decimal form

$0.5i, \; -0.25 + 0.5i, \; -0.1875 + 0.25i,$

$-0.02734 + 0.40625i, \; -0.164291 + 0.477783i,$

$-0.201285 + 0.343009i.$

79. $z_1 = 5 + 2i$

$z_2 = 3 - 4i$

$$\frac{1}{z} = \frac{1}{z_1} + \frac{1}{z_2} = \frac{1}{5 + 2i} + \frac{1}{3 - 4i}$$

$$= \frac{(3 - 4i) + (5 + 2i)}{(5 + 2i)(3 - 4i)}$$

$$= \frac{8 - 2i}{23 - 14i}$$

$$z = \frac{23 - 14i}{8 - 2i}\left(\frac{8 + 2i}{8 + 2i}\right)$$

$$= \frac{212 - 66i}{68} \approx 3.118 - 0.971i$$

81. False. A real number $a + 0i = a$ is equal to its conjugate.

83. False. For example, $(1 + 2i) + (1 - 2i) = 2$, which is not an imaginary number.

85. True. Let $z_1 = a_1 + b_1i$ and $z_2 = a_2 + b_2i$. Then

$$\overline{z_1 z_2} = \overline{(a_1 + b_1i)(a_2 + b_2i)}$$

$$= \overline{(a_1a_2 - b_1b_2) + (a_1b_2 + b_1a_2)i}$$

$$= (a_1a_2 - b_1b_2) - (a_1b_2 + b_1a_2)i$$

$$= (a_1 - b_1i)(a_2 - b_2i)$$

$$= \overline{a_1 + b_1i} \; \overline{a_2 + b_2i}$$

$$= \overline{z_1} \; \overline{z_2}.$$

87. $(4x - 5)(4x + 5) = 16x^2 - 20x + 20x - 25$

$ = 16x^2 - 25$

89. $\left(3x - \frac{1}{2}\right)(x + 4) = 3x^2 - \frac{1}{2}x + 12x - 2$

$\phantom{\left(3x - \frac{1}{2}\right)(x + 4)} = 3x^2 + \frac{23}{2}x - 2$

Section 2.5   The Fundamental Theorem of Algebra

- ■ You should know that if f is a polynomial of degree $n > 0$, then f has at least one zero in the complex number system. (Fundamental Theorem of Algebra)
- ■ You should know that if $a + bi$ is a complex zero of a polynomial f, with real coefficients, then $a - bi$ is also a complex zero of f.
- ■ You should know the difference between a factor that is irreducible over the rationals (such as $x^2 - 7$) and a factor that is irreducible over the reals (such as $x^2 + 9$).

Vocabulary Check

1. Fundamental Theorem, Algebra

2. Linear Factorization Theorem

3. irreducible, reals

4. complex conjugate

---

1. $f(x) = x^2(x + 3)$

The three zeros are $x = 0$, $x = 0$ and $x = -3$.

3. $f(x) = (x + 9)(x + 4i)(x - 4i)$

Zeros: $-9, \pm 4i$

5. $f(x) = x^3 - 4x^2 + x - 4 = x^2(x - 4) + 1(x - 4) = (x - 4)(x^2 + 1)$

Zeros: $4, \pm i$

The only real zero of $f(x)$ is $x = 4$. This corresponds to the x-intercept of $(4, 0)$ on the graph.

7. $f(x) = x^4 + 4x^2 + 4 = (x^2 + 2)^2$

Zeros: $\pm\sqrt{2}i, \pm\sqrt{2}i$

$f(x)$ has no real zeros and the graph of $f(x)$ has no x-intercepts.

9. $h(x) = x^2 - 4x + 1$

h has no rational zeros. By the Quadratic Formula, the zeros are

$$x = \frac{4 \pm \sqrt{16 - 4}}{2} = 2 \pm \sqrt{3}.$$

$$h(x) = \left[x - \left(2 + \sqrt{3}\right)\right]\left[x - \left(2 - \sqrt{3}\right)\right]$$
$$= \left(x - 2 - \sqrt{3}\right)\left(x - 2 + \sqrt{3}\right)$$

11. $f(x) = x^2 - 12x + 26$

f has no rational zeros. By the Quadratic Formula, the zeros are

$$x = \frac{12 \pm \sqrt{(-12)^2 - 4(26)}}{2} = 6 \pm \sqrt{10}.$$

$$f(x) = \left[x - \left(6 + \sqrt{10}\right)\right]\left[x - \left(6 - \sqrt{10}\right)\right]$$
$$= \left(x - 6 - \sqrt{10}\right)\left(x - 6 + \sqrt{10}\right)$$

13. $f(x) = x^2 + 25$

$\qquad = (x + 5i)(x - 5i)$

The zeros of $f(x)$ are $x = \pm 5i$.

15. $f(x) = 16x^4 - 81$

$\qquad = (4x^2 - 9)(4x^2 + 9)$

$\qquad = (2x - 3)(2x + 3)(2x + 3i)(2x - 3i)$

Zeros: $\pm\frac{3}{2}, \pm\frac{3}{2}i$

17. $f(z) = z^2 - z + 56$

$$z = \frac{1 \pm \sqrt{1 - 4(56)}}{2}$$

$$= \frac{1 \pm \sqrt{-223}}{2}$$

$$= \frac{1}{2} \pm \frac{\sqrt{223}}{2}i$$

$$f(z) = \left(z - \frac{1}{2} + \frac{\sqrt{223}i}{2}\right)\left(z - \frac{1}{2} - \frac{\sqrt{223}i}{2}\right)$$

19. $f(x) = x^4 + 10x^2 + 9$

$\qquad = (x^2 + 1)(x^2 + 9)$

$\qquad = (x + i)(x - i)(x + 3i)(x - 3i)$

The zeros of $f(x)$ are $x = \pm i$ and $x = \pm 3i$.

21. $f(x) = 3x^3 - 5x^2 + 48x - 80$

Using synthetic division, $\frac{5}{3}$ is a zero:

$$\frac{5}{3} \begin{array}{|rrrr} 3 & -5 & 48 & -80 \\ & 5 & 0 & 80 \\ \hline 3 & 0 & 48 & 0 \end{array}$$

$$
\begin{aligned}
f(x) &= \left(x - \tfrac{5}{3}\right)(3x^2 + 48) \\
&= (3x - 5)(x^2 + 16) \\
&= (3x - 5)(x + 4i)(x - 4i)
\end{aligned}
$$

The zeros are $\frac{5}{3}$, $4i$, $-4i$.

23. $f(t) = t^3 - 3t^2 - 15t + 125$

Possible rational zeros: $\pm 1, \pm 5, \pm 25, \pm 125$

$$-5 \begin{array}{|rrrr} 1 & -3 & -15 & 125 \\ & -5 & 40 & -125 \\ \hline 1 & -8 & 25 & 0 \end{array}$$

By the Quadratic Formula, the zeros of $t^2 - 8t + 25$ are

$$t = \frac{8 \pm \sqrt{64 - 100}}{2} = 4 \pm 3i.$$

The zeros of $f(t)$ are $t = -5$ and $t = 4 \pm 3i$.

$$
\begin{aligned}
f(t) &= [t - (-5)][t - (4 + 3i)][t - (4 - 3i)] \\
&= (t + 5)(t - 4 - 3i)(t - 4 + 3i)
\end{aligned}
$$

25. $f(x) = 5x^3 - 9x^2 + 28x + 6$

Possible rational zeros: $\pm 6, \pm \dfrac{6}{5}, \pm 3, \pm \dfrac{3}{5}, \pm 2, \pm \dfrac{2}{5}, \pm 1, \pm \dfrac{1}{5}$

$$-\tfrac{1}{5} \begin{array}{|rrrr} 5 & -9 & 28 & 6 \\ & -1 & 2 & -6 \\ \hline 5 & -10 & 30 & 0 \end{array}$$

By the Quadratic Formula, the zeros of $5x^2 - 10x + 30$ are those of $x^2 - 2x + 6$:

$$x = \frac{2 \pm \sqrt{4 - 4(6)}}{2} = 1 \pm \sqrt{5}\,i$$

Zeros: $-\dfrac{1}{5}, 1 \pm \sqrt{5}\,i$

$$f(x) = 5\left(x + \tfrac{1}{5}\right)\left(x - \left(1 + \sqrt{5}\,i\right)\right)\left(x - \left(1 - \sqrt{5}\,i\right)\right) = (5x + 1)\left(x - 1 - \sqrt{5}\,i\right)\left(x - 1 + \sqrt{5}\,i\right)$$

27. $g(x) = x^4 - 4x^3 + 8x^2 - 16x + 16$

Possible rational zeros: $\pm 1, \pm 2, \pm 4, \pm 8, \pm 16$

$$2 \begin{array}{|rrrrr} 1 & -4 & 8 & -16 & 16 \\ & 2 & -4 & 8 & -16 \\ \hline 1 & -2 & 4 & -8 & 0 \end{array}$$

$$2 \begin{array}{|rrrr} 1 & -2 & 4 & -8 \\ & 2 & 0 & 8 \\ \hline 1 & 0 & 4 & 0 \end{array}$$

$$
\begin{aligned}
g(x) &= (x - 2)(x - 2)(x^2 + 4) \\
&= (x - 2)^2(x + 2i)(x - 2i)
\end{aligned}
$$

The zeros of g are 2, 2, and $\pm 2i$.

29. (a) $f(x) = x^2 - 14x + 46$.

By the Quadratic Formula,

$$x = \frac{14 \pm \sqrt{(-14)^2 - 4(46)}}{2} = 7 \pm \sqrt{3}.$$

The zeros are $7 + \sqrt{3}$ and $7 - \sqrt{3}$.

(b) $f(x) = \left[x - \left(7 + \sqrt{3}\right)\right]\left[x - \left(7 - \sqrt{3}\right)\right]$

$$= \left(x - 7 - \sqrt{3}\right)\left(x - 7 + \sqrt{3}\right)$$

(c) x-intercepts: $\left(7 + \sqrt{3}, 0\right)$ and $\left(7 - \sqrt{3}, 0\right)$

(d)

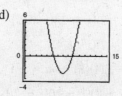

31. (a) $f(x) = 2x^3 - 3x^2 + 8x - 12$

$= (2x - 3)(x^2 + 4)$

The zeros are $\dfrac{3}{2}$, $\pm 2i$.

(b) $f(x) = (2x - 3)(x + 2i)(x - 2i)$

(c) x-intercept: $\left(\dfrac{3}{2}, 0\right)$

(d)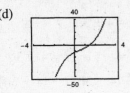

33. (a) $f(x) = x^3 - 11x + 150$

$= (x + 6)(x^2 - 6x + 25)$

Use the Quadratic Formula to find the zeros of $x^2 - 6x + 25$.

$$x = \frac{6 \pm \sqrt{(-6)^2 - 4(25)}}{2} = 3 \pm 4i.$$

The zeros are $-6, 3 + 4i$, and $3 - 4i$.

(b) $f(x) = (x + 6)(x - 3 + 4i)(x - 3 - 4i)$

(c) x-intercept: $(-6, 0)$

(d)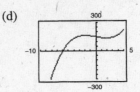

35. (a) $f(x) = x^4 + 25x^2 + 144$

$= (x^2 + 9)(x^2 + 16)$

The zeros are $\pm 3i$, $\pm 4i$.

(b) $f(x) = (x^2 + 9)(x^2 + 16)$

$= (x + 3i)(x - 3i)(x + 4i)(x - 4i)$

(c) No x-intercepts

(d)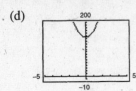

37. $f(x) = (x - 2)(x - i)(x + i)$

$= (x - 2)(x^2 + 1)$

$= (x^3 - 2x^2 + x - 2)$

Note that $f(x) = a(x^3 - 2x^2 + x - 2)$, where a is any nonzero real number, has zeros $2, \pm i$.

39. $f(x) = (x - 2)^2(x - 4 - i)(x - 4 + i)$

$= (x - 2)^2(x - 8x + 16 + 1)$

$= (x^2 - 4x + 4)(x^2 - 8x + 17)$

$= x^4 - 12x^3 + 53x^2 - 100x + 68$

Note that $f(x) = a(x^4 - 12x^3 + 53x^2 - 100x + 68)$, where a is any nonzero real number, has zeros $2, 2, 4 \pm i$.

41. Because $1 + \sqrt{2}i$ is a zero, so is $1 - \sqrt{2}i$.

$f(x) = (x - 0)(x + 5)\left(x - 1 - \sqrt{2}i\right)\left(x - 1 + \sqrt{2}i\right)$

$= (x^2 + 5x)(x^2 - 2x + 1 + 2)$

$= (x^2 + 5x)(x^2 - 2x + 3)$

$= x^4 + 3x^3 - 7x^2 + 15x$

Note that $f(x) = a(x^4 + 3x^3 - 7x^2 + 15x)$, where a is any nonzero real number, has zeros $0, -5, 1 \pm \sqrt{2}i$.

43. (a) $f(x) = a(x - 1)(x + 2)(x - 2i)(x + 2i)$

$= a(x - 1)(x + 2)(x^2 + 4)$

$f(-1) = 10 = a(-2)(1)(5) \Rightarrow a = -1$

$f(x) = -(x - 1)(x + 2)(x - 2i)(x + 2i)$

(b) $f(x) = -(x - 1)(x + 2)(x^2 + 4)$

$= -(x^2 + x - 2)(x^2 + 4)$

$= -x^4 - x^3 - 2x^2 - 4x + 8$

45. (a) $f(x) = a(x + 1)(x - 2 - \sqrt{5}i)(x - 2 + \sqrt{5}i)$

$= a(x + 1)(x^2 - 4x + 4 + 5)$

$= a(x + 1)(x^2 - 4x + 9)$

$f(-2) = 42 = a(-1)(4 + 8 + 9) \Rightarrow a = -2$

$f(x) = -2(x + 1)(x - 2 - \sqrt{5}i)(x - 2 + \sqrt{5}i)$

(b) $f(x) = -2(x + 1)(x^2 - 4x + 9)$

$= -2x^3 + 6x^2 - 10x - 18$

47. $f(x) = x^4 - 6x^2 - 7$

(a) $f(x) = (x^2 - 7)(x^2 + 1)$

(b) $f(x) = (x - \sqrt{7})(x + \sqrt{7})(x^2 + 1)$

(c) $f(x) = (x - \sqrt{7})(x + \sqrt{7})(x + i)(x - i)$

49. $f(x) = x^4 - 2x^3 - 3x^2 + 12x - 18$

(a) $f(x) = (x^2 - 6)(x^2 - 2x + 3)$

(b) $f(x) = (x + \sqrt{6})(x - \sqrt{6})(x^2 - 2x + 3)$

(c) $f(x) = (x + \sqrt{6})(x - \sqrt{6})(x - 1 - \sqrt{2}i)(x - 1 + \sqrt{2}i)$

51. $f(x) = 2x^3 + 3x^2 + 50x + 75$

Since $5i$ is a zero, so is $-5i$.

$$
\begin{array}{r|rrrr}
5i & 2 & 3 & 50 & 75 \\
   &   & 10i & -50 + 15i & -75 \\
\hline
   & 2 & 3 + 10i & 15i & 0 \\
\end{array}
$$

$$
\begin{array}{r|rrr}
-5i & 2 & 3 + 10i & 15i \\
    &   & -10i & -15i \\
\hline
    & 2 & 3 & 0 \\
\end{array}
$$

The zero of $2x + 3$ is $x = -\frac{3}{2}$. The zeros of f are $x = -\frac{3}{2}$ and $x = \pm 5i$.

Alternate Solution

Since $x = \pm 5i$ are zeros of $f(x)$, $(x + 5i)(x - 5i) = x^2 + 25$ is a factor of $f(x)$. By long division we have:

$$
\begin{array}{r}
2x + 3 \\
x^2 + 0x + 25 \overline{)\, 2x^3 + 3x^2 + 50x + 75} \\
\underline{2x^3 + 0x^2 + 50x} \\
3x^2 + 0x + 75 \\
\underline{3x^2 + 0x + 75} \\
0
\end{array}
$$

Thus, $f(x) = (x^2 + 25)(2x + 3)$ and the zeros of f are $x = \pm 5i$ and $x = -\frac{3}{2}$.

53. $g(x) = x^3 - 7x^2 - x + 87$. Since $5 + 2i$ is a zero, so is $5 - 2i$.

$$
\begin{array}{r|rrrr}
5 + 2i & 1 & -7 & -1 & 87 \\
       &   & 5 + 2i & -14 + 6i & -87 \\
\hline
       & 1 & -2 + 2i & -15 + 6i & 0 \\
\end{array}
$$

$$
\begin{array}{r|rrr}
5 - 2i & 1 & -2 + 2i & -15 + 6i \\
       &   & 5 - 2i & 15 - 6i \\
\hline
       & 1 & 3 & 0 \\
\end{array}
$$

The zero of $x + 3$ is $x = -3$.

The zeros of f are $-3, 5 \pm 2i$.

55. $h(x) = 3x^3 - 4x^2 + 8x + 8$. Since $1 - \sqrt{3}i$ is a zero, so is $1 + \sqrt{3}i$.

$$
\begin{array}{r|rrrr}
1 - \sqrt{3}i & 3 & -4 & 8 & 8 \\
& & 3 - 3\sqrt{3}i & -10 - 2\sqrt{3}i & -8 \\
\hline
& 3 & -1 - 3\sqrt{3}i & -2 - 2\sqrt{3}i & 0
\end{array}
$$

$$
\begin{array}{r|rrr}
1 + \sqrt{3}i & 3 & -1 - 3\sqrt{3}i & -2 - 2\sqrt{3}i \\
& & 3 + 3\sqrt{3}i & 2 + 2\sqrt{3}i \\
\hline
& 3 & 2 & 0
\end{array}
$$

The zero of $3x + 2$ is $x = -\frac{2}{3}$. The zeros of h are $x = -\frac{2}{3}, 1 \pm \sqrt{3}i$.

57. $h(x) = 8x^3 - 14x^2 + 18x - 9$. Since $\frac{1}{2}\left(1 - \sqrt{5}i\right)$ is a zero, so is $\frac{1}{2}\left(1 + \sqrt{5}i\right)$.

$$
\begin{array}{r|rrrr}
\frac{1}{2}\left(1 - \sqrt{5}i\right) & 8 & -14 & 18 & -9 \\
& & 4 - 4\sqrt{5}i & -15 + 3\sqrt{5}i & 9 \\
\hline
& 8 & -10 - 4\sqrt{5}i & 3 + 3\sqrt{5}i & 0
\end{array}
$$

$$
\begin{array}{r|rrr}
\frac{1}{2}\left(1 + \sqrt{5}i\right) & 8 & -10 - 4\sqrt{5}i & 3 + 3\sqrt{5}i \\
& & 4 + 4\sqrt{5}i & -3 - 3\sqrt{5}i \\
\hline
& 8 & -6 & 0
\end{array}
$$

The zero of $8x - 6$ is $x = \frac{3}{4}$. The zeros of h are $x = \frac{3}{4}, \frac{1}{2}\left(1 \pm \sqrt{5}i\right)$.

59. $f(x) = x^4 + 3x^3 - 5x^2 - 21x + 22$

(a) The root feature yields the real roots 1 and 2, and the complex roots $-3 \pm 1.414i$.

(b) By synthetic division:

$$
\begin{array}{r|rrrrr}
1 & 1 & 3 & -5 & -21 & 22 \\
& & 1 & 4 & -1 & -22 \\
\hline
& 1 & 4 & -1 & -22 & 0
\end{array}
$$

$$
\begin{array}{r|rrrr}
2 & 1 & 4 & -1 & -22 \\
& & 2 & 12 & 22 \\
\hline
& 1 & 6 & 11 & 0
\end{array}
$$

The complex roots of $x^2 + 6x + 11$ are $x = \dfrac{-6 \pm \sqrt{6^2 - 4(11)}}{2} = -3 \pm \sqrt{2}i$.

61. $h(x) = 8x^3 - 14x^2 + 18x - 9$

(a) The root feature yields the real root 0.75, and the complex roots $0.5 \pm 1.118i$.

(b) By synthetic division:

$$
\begin{array}{r|rrrr}
\frac{3}{4} & 8 & -14 & 18 & -9 \\
& & 6 & -6 & 9 \\
\hline
& 8 & -8 & 12 & 0
\end{array}
$$

The complex roots of $8x^2 - 8x + 12$ are

$$
x = \frac{8 \pm \sqrt{64 - 4(8)(12)}}{2(8)} = \frac{1}{2} \pm \frac{\sqrt{5}}{2}i.
$$

63. $-16t^2 + 48t = 64, \quad 0 \le t \le 3$

$$-16t^2 + 48t - 64 = 0$$

$$t = \frac{-48 \pm \sqrt{1792}i}{-32}$$

Since the roots are imaginary, the ball never will reach a height of 64 feet. You can verify this graphically by observing that $y_1 = -16t^2 + 48t$ and $y_2 = 64$ do not intersect.

65. False, a third degree polynomial must have at least one real zero.

67. $f(x) = x^4 - 4x^2 + k$

(a) f has two real zeros each of multiplicity 2 for $k = 4$: $f(x) = x^4 - 4x^2 + 4 = (x^2 - 2)^2$.

(b) f has two real zeros and two complex zeros if $k < 0$.

69. $f(x) = x^2 - 7x - 8 = \left(x^2 - 7x + \frac{49}{4}\right) - 8 - \frac{49}{4}$

$\qquad = \left(x - \frac{7}{2}\right)^2 - \frac{81}{4}$

Vertex: $\left(\frac{7}{2}, -\frac{81}{4}\right)$

$f(x) = (x - 8)(x + 1)$

Intercepts: $(8, 0), (-1, 0), (0, -8)$

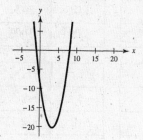

71. $f(x) = 6x^2 + 5x - 6 = (3x - 2)(2x + 3)$

Intercepts: $\left(\frac{2}{3}, 0\right), \left(-\frac{3}{2}, 0\right), (0, -6)$

$f(x) = 6x^2 + 5x - 6$

$\qquad = 6\left(x^2 + \frac{5}{6}x + \frac{25}{144}\right) - 6 - \frac{25}{24}$

$\qquad = 6\left(x + \frac{5}{12}\right)^2 - \frac{169}{24}$

Vertex: $\left(-\frac{5}{12}, -\frac{169}{24}\right)$

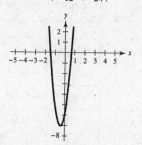

Section 2.6   Rational Functions and Asymptotes

- You should know the following basic facts about rational functions.
  (a) A function of the form $f(x) = P(x)/Q(x)$, $Q(x) \neq 0$, where $P(x)$ and $Q(x)$ are polynomials, is called a rational function.
  (b) The domain of a rational function is the set of all real numbers except those which make the denominator zero.
  (c) If $f(x) = P(x)/Q(x)$ is in reduced form, and a is a value such that $Q(a) = 0$, then the line $x = a$ is a vertical asymptote of the graph of f. $f(x) \rightarrow \infty$ or $f(x) \rightarrow -\infty$ as $x \rightarrow a$.
  (d) The line $y = b$ is a horizontal asymptote of the graph of f if $f(x) \rightarrow b$ as $x \rightarrow \infty$ or $x \rightarrow -\infty$.
  (e) Let $f(x) = \dfrac{P(x)}{Q(x)} = \dfrac{a_n x^n + a_{n-1} x^{n-1} + \cdots + a_1 x + a_0}{b_m x^m + b'_{m-1} x^{m-1} + \cdots + b_1 x + b_0}$ where $P(x)$ and $Q(x)$ have no common factors.

  1. If $n < m$, then the x-axis $(y = 0)$ is a horizontal asymptote.

  2. If $n = m$, then $y = \dfrac{a_n}{b_m}$ is a horizontal asymptote.

  3. If $n > m$, then there are no horizontal asymptotes.

Vocabulary Check

1. rational functions   **2.** vertical asymptote   **3.** horizontal asymptote

1. $f(x) = \dfrac{1}{x-1}$

  (a) Domain: all $x \neq 1$

  (b)

x	$f(x)$
0.5	-2
0.9	-10
0.99	-100
0.999	-1000

x	$f(x)$
1.5	2
1.1	10
1.01	100
1.001	1000

x	$f(x)$
5	0.25
10	$0.\overline{1}$
100	$0.\overline{01}$
1000	$0.\overline{001}$

x	$f(x)$
-5	$-0.\overline{16}$
-10	$-0.\overline{09}$
-100	$-0.\overline{0099}$
-1000	$-0.\overline{00099}$

  (c) f approaches $-\infty$ from the left of 1 and ∞ from the right of 1.

3. $f(x) = \dfrac{3x}{|x-1|}$

  (a) Domain: all $x \neq 1$

  (b)

x	$f(x)$
0.5	3
0.9	27
0.99	297
0.999	2997

x	$f(x)$
1.5	9
1.1	33
1.01	303
1.001	3003

x	$f(x)$
5	3.75
10	$3.\overline{33}$
100	$3.\overline{03}$
1000	$3.\overline{003}$

x	$f(x)$
-5	-2.5
-10	-2.727
-100	-2.970
-1000	-2.997

  (c) f approaches ∞ from both the left and the right of 1.

5. $f(x) = \dfrac{3x^2}{x^2 - 1}$

  (a) Domain: all $x \neq \pm 1$

  (b)

x	$f(x)$
0.5	-1
0.9	-12.79
0.99	-147.8
0.999	-1498

x	$f(x)$
1.5	5.4
1.1	17.29
1.01	152.3
1.001	1502.3

x	$f(x)$
5	3.125
10	$3.\overline{03}$
100	$3.\overline{0003}$
1000	3

x	$f(x)$
-5	3.125
-10	$3.\overline{03}$
-100	$3.\overline{0003}$
-1000	3

  (c) f approaches $-\infty$ from the left of 1, and ∞ from the right of 1. f approaches ∞ from the left of -1, and $-\infty$ from the right of -1.

7. $f(x) = \dfrac{2}{x + 2}$

Vertical asymptote: $x = -2$

Horizontal asymptote: $y = 0$

Matches graph (a).

9. $f(x) = \dfrac{4x + 1}{x}$

Vertical asymptote: $x = 0$

Horizontal asymptote: $y = 4$

Matches graph (c).

11. $f(x) = \dfrac{x - 2}{x - 4}$

Vertical asymptote: $x = 4$

Horizontal asymptote: $y = 1$

Matches graph (b).

13. $f(x) = \dfrac{1}{x^2}$

(a) Vertical asymptote: $x = 0$

Horizontal asymptote: $y = 0$

(b) Holes: none

15. $f(x) = \dfrac{x(2 + x)}{2x - x^2} = \dfrac{2 + x}{2 - x}, \; x \neq 0$

(a) Vertical asymptote: $x = 2$

Horizontal asymptote: $y = -1$

(b) Hole at $x = 0$: $(0, 1)$

17. $f(x) = \dfrac{x^2 - 25}{x^2 + 5x}$

$\quad = \dfrac{(x - 5)(x + 5)}{x(x + 5)}$

$\quad = \dfrac{x - 5}{x}, \; x \neq -5$

(a) Vertical asymptote: $x = 0$

Horizontal asymptote: $y = 1$

(b) Hole at $x = -5$: $(-5, 2)$

19. $f(x) = \dfrac{3x^2 + x - 5}{x^2 + 1}$

(a) Domain: all real numbers

(b) Vertical asymptote: none

Horizontal asymptote: $y = 3$

(c)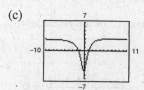

21. $f(x) = \dfrac{x - 3}{|x|}$

(a) Domain: all real numbers except $x = 0$

(b) Vertical asymptote: $x = 0$

Horizontal asymptote:

$y = 1$ to the right

$y = -1$ (to the left)

(c)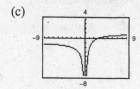

23. $f(x) = \dfrac{x^2 - 16}{x - 4}, \; g(x) = x + 4$

(a) Domain of f: all real numbers except 4

Domain of g: all real numbers

(b) $f(x) = \dfrac{(x - 4)(x + 4)}{x - 4} = x + 4, \; x \neq 4$

f has no vertical asymptotes.

(c) Hole at $x = 4$

(d)

x	1	2	3	4	5	6	7
$f(x)$	5	6	7	Undef.	9	10	11
$g(x)$	5	6	7	8	9	20	11

(e) f and g differ at $x = 4$, where f is undefined.

25. $f(x) = \dfrac{x^2 - 1}{x^2 - 2x - 3} = \dfrac{(x-1)(x+1)}{(x+1)(x-3)}, \quad g(x) = \dfrac{x-1}{x-3}$

(a) Domain of f: all real numbers except $-1, 3$

Domain of g: all real numbers except 3

(b) $f(x) = \dfrac{(x-1)(x+1)}{(x+1)(x-3)} = \dfrac{x-1}{x-3}, \quad x \neq -1$

f has a vertical asymptote at $x = 3$.

(c) The graph has a hole at $x = -1$.

(d)

x	-2	-1	0	1	2	3	4
$f(x)$	$\frac{3}{5}$	Undef.	$\frac{1}{3}$	0	-1	Undef.	3
$g(x)$	$\frac{3}{5}$	$\frac{1}{2}$	$\frac{1}{3}$	0	-1	Undef.	3

(e) f and g differ at $x = -1$, where f is undefined.

27. $f(x) = 4 - \dfrac{1}{x}$

(a) As $x \to \pm\infty, f(x) \to 4$

(b) As $x \to \infty, f(x) \to 4$ but is less than 4.

(c) As $x \to -\infty, f(x) \to 4$ but is greater than 4.

29. $f(x) = \dfrac{2x - 1}{x - 3}$

(a) As $x \to \pm\infty, f(x) \to 2$.

(b) As $x \to \infty, f(x) \to 2$ but is greater than 2.

(c) As $x \to -\infty, f(x) \to 2$ but is less than 2.

31. $g(x) = \dfrac{x^2 - 4}{x + 3} = \dfrac{(x-2)(x+2)}{x+3}$

The zeros of g are the zeros of the numerator:
$x = \pm 2$

33. $f(x) = 1 - \dfrac{2}{x - 5} = \dfrac{x - 7}{x - 5}$

The zero of f corresponds to the zero of the numerator and is $x = 7$.

35. $g(x) = \dfrac{x^2 - 2x - 3}{x^2 + 1} = \dfrac{(x-3)(x+1)}{x^2+1} = 0$

Zeros: $x = -1, 3$

37. $f(x) = \dfrac{2x^2 - 5x + 2}{2x^2 - 7x + 3} = \dfrac{(2x-1)(x-2)}{(2x-1)(x-3)} = \dfrac{x-2}{x-3},$

$x \neq \dfrac{1}{2}$

Zero: $x = 2 \left(x = \dfrac{1}{2} \text{ is not in the domain.} \right)$

39. $C = \dfrac{255p}{100 - p}, \quad 0 \le p < 100$

(a) $C(10) = \dfrac{255(10)}{100 - 10} \approx 28.33$ million dollars

(b) $C(40) = \dfrac{255(40)}{100 - 40} = 170$ million dollars

(c) $C(75) = \dfrac{255(75)}{100 - 75} = 765$ million dollars

(d)

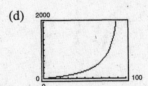

(e) $C \to \infty$ as $x \to 100$. No, it would not be possible to remove 100% of the pollutants.

41. (a) Use data $\left(16, \dfrac{1}{3}\right), \left(32, \dfrac{1}{4.7}\right), \left(44, \dfrac{1}{9.8}\right),$

$\left(50, \dfrac{1}{19.7}\right), \left(60, \dfrac{1}{39.4}\right).$

$\dfrac{1}{y} = -0.007x + 0.445$

$y = \dfrac{1}{0.445 - 0.007x}$

(b)

x	16	32	44	50	60
y	3.0	4.5	7.3	10.5	40

(Answers will vary.)

(c) No, the function is negative for $x = 70$.

43. $N = \dfrac{20(5 + 3t)}{1 + 0.04t}$, $0 \le t$

(a)

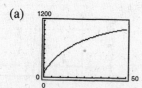

(b) $N(5) \approx 333$ deer

$N(10) = 500$ deer

$N(25) = 800$ deer

(c) The herd is limited by the horizontal asymptote:

$$N = \dfrac{60}{0.04} = 1500 \text{ deer}$$

45. False. A rational function can have at most n vertical asymptotes, where n is the degree of the denominator.

47. There are vertical asymptotes at $x = \pm 3$, and zeros at $x = \pm 2$. Matches (b).

49. $f(x) = \dfrac{x - 1}{x^3 - 8}$

51. $f(x) = \dfrac{2(x + 3)(x - 3)}{(x + 2)(x - 1)} = \dfrac{2x^2 - 18}{x^2 + x - 2}$

53. $y - 2 = \dfrac{-1 - 2}{0 - 3}(x - 3) = 1(x - 3)$

$y = x - 1$

$y - x + 1 = 0$

55. $y - 7 = \dfrac{10 - 7}{3 - 2}(x - 2) = 3(x - 2)$

$y = 3x + 1$

$3x - y + 1 = 0$

57.

$$
\begin{array}{r}
x + 9 \\
x - 4 \overline{)\, x^2 + 5x + 6} \\
\underline{x^2 - 4x } \\
9x + 6 \\
\underline{9x - 36} \\
42
\end{array}
$$

$$\dfrac{x^2 + 5x + 6}{x - 4} = x + 9 + \dfrac{42}{x - 4}$$

59.

$$
\begin{array}{r}
2x^2 - 9 \\
x^2 + 5 \overline{)\, 2x^4 + 0x^3 + x^2 + 0x - 11} \\
\underline{2x^4 + 10x^2 } \\
-9x^2 - 11 \\
\underline{-9x^2 - 45} \\
34
\end{array}
$$

$$\dfrac{2x^4 + x^2 - 11}{x^2 + 5} = 2x^2 - 9 + \dfrac{34}{x^2 + 5}$$

Section 2.7   Graphs of Rational Functions

■ You should be able to graph $f(x) = \dfrac{p(x)}{q(x)}$.

(a) Find the x- and y-intercepts.

(b) Find any vertical or horizontal asymptotes.

(c) Plot additional points.

(d) If the degree of the numerator is one more than the degree of the denominator, use long division to find the slant asymptote.

Vocabulary Check

1. slant, asymptote

2. vertical

1. $g(x) = \dfrac{2}{x} + 1$

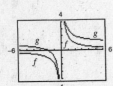

Vertical shift one unit upward

3. $g(x) = -\dfrac{2}{x}$

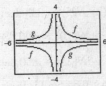

Reflection in the x-axis

5. $g(x) = \dfrac{2}{x^2} - 2$

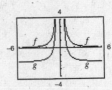

Vertical shift two units downward

7. $g(x) = \dfrac{2}{(x-2)^2}$

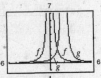

Horizontal shift two units to the right

9. $f(x) = \dfrac{1}{x+2}$

y-intercept: $\left(0, \dfrac{1}{2}\right)$

Vertical asymptote: $x = -2$

Horizontal asymptote: $y = 0$

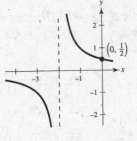

x	-4	-3	-1	0	1
y	$-\frac{1}{2}$	-1	1	$\frac{1}{2}$	$\frac{1}{3}$

11. $C(x) = \dfrac{5 + 2x}{1 + x} = \dfrac{2x + 5}{x + 1}$

x-intercept: $\left(-\dfrac{5}{2}, 0\right)$

y-intercept: $(0, 5)$

Vertical asymptote: $x = -1$

Horizontal asymptote: $y = 2$

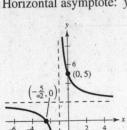

x	-4	-3	-2	0	1	2
C(x)	1	$\frac{1}{2}$	-1	5	$\frac{7}{2}$	3

13. $f(t) = \dfrac{1 - 2t}{t} = -\dfrac{2t - 1}{t}$

t-intercept: $\left(\dfrac{1}{2}, 0\right)$

Vertical asymptote: $t = 0$

Horizontal asymptote: $y = -2$

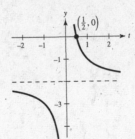

x	-2	-1	$\frac{1}{2}$	1	2
y	$-\frac{5}{2}$	-3	0	-1	$-\frac{3}{2}$

15. $f(x) = \dfrac{x^2}{x^2 - 4}$

Intercept: $(0, 0)$

Vertical asymptotes: $x = 2$, $x = -2$

Horizontal asymptote: $y = 1$

y-axis symmetry

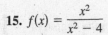

x	-4	-1	0	-1	4
y	$\frac{4}{3}$	$-\frac{1}{3}$	0	$-\frac{1}{3}$	$\frac{4}{3}$

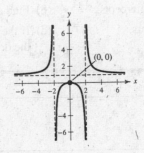

17. $f(x) = \dfrac{x}{x^2 - 1} = \dfrac{x}{(x + 1)(x - 1)}$

Intercept: $(0, 0)$

Vertical asymptotes: $x = 1$ and $x = -1$

Horizontal asymptote: $y = 0$

Origin symmetry

x	-3	-2	$-\frac{1}{2}$	0	$\frac{1}{2}$	2	3	4
y	$-\frac{3}{8}$	$-\frac{2}{3}$	$\frac{2}{3}$	0	$-\frac{2}{3}$	$\frac{2}{3}$	$\frac{3}{8}$	$\frac{4}{15}$

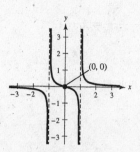

19. $g(x) = \dfrac{4(x + 1)}{x(x - 4)}$

Intercept: $(-1, 0)$

Vertical asymptotes: $x = 0$ and $x = 4$

Horizontal asymptote: $y = 0$

x	-2	-1	1	2	3	5	6
y	$-\frac{1}{3}$	0	$-\frac{8}{3}$	-3	$-\frac{16}{3}$	$\frac{24}{5}$	$\frac{7}{3}$

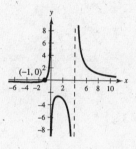

21. $f(x) = \dfrac{3x}{x^2 - x - 2} = \dfrac{3x}{(x + 1)(x - 2)}$

Intercept: $(0, 0)$

Vertical asymptotes: $x = -1, 2$

Horizontal asymptote: $y = 0$

x	-3	0	1	3	4
y	$-\frac{9}{10}$	0	$-\frac{3}{2}$	$\frac{9}{4}$	$\frac{6}{5}$

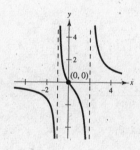

23. $f(x) = \dfrac{x^2 + 3x}{x^2 + x - 6} = \dfrac{x(x + 3)}{(x - 2)(x + 3)} = \dfrac{x}{x - 2},$

$x \neq -3$

Intercept: $(0, 0)$

Vertical asymptote: $x = 2$

(There is a hole at $x = -3$.)

Horizontal asymptote: $y = 1$

x	-2	-1	0	1	2	3
y	$\frac{1}{2}$	$\frac{1}{3}$	0	-1	Undef.	3

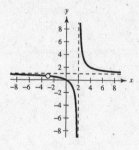

25. $f(x) = \dfrac{x^2 - 1}{x + 1} = \dfrac{(x + 1)(x - 1)}{x + 1} = x - 1$,

$x \neq -1$

The graph is a line, with a hole at $x = -1$.

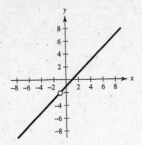

27. $f(x) = \dfrac{2 + x}{1 - x} = -\dfrac{x + 2}{x - 1}$

Vertical asymptote: $x = 1$

Horizontal asymptote: $y = -1$

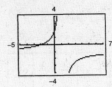

Domain: $x \neq 1$ or $(-\infty, 1) \cup (1, \infty)$

29. $f(t) = \dfrac{3t + 1}{t}$

Vertical asymptote: $t = 0$

Horizontal asymptote: $y = 3$

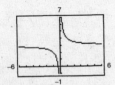

Domain: $t \neq 0$ or $(-\infty, 0) \cup (0, \infty)$

31. $h(t) = \dfrac{4}{t^2 + 1}$

Domain: all real numbers OR $(-\infty, \infty)$

Horizontal asymptote: $y = 0$

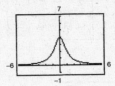

33. $f(x) = \dfrac{x + 1}{x^2 - x - 6} = \dfrac{x + 1}{(x - 3)(x + 2)}$

Domain: all real numbers except $x = 3, -2$

Vertical asymptotes: $x = 3$, $x = -2$

Horizontal asymptote: $y = 0$

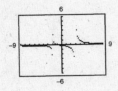

35. $f(x) = \dfrac{20x}{x^2 + 1} - \dfrac{1}{x} = \dfrac{19x^2 - 1}{x(x^2 + 1)}$

Domain: all real numbers except 0,
OR $(-\infty, 0) \cup (0, \infty)$

Vertical asymptote: $x = 0$

Horizontal asymptote: $y = 0$

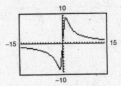

37. $h(x) = \dfrac{6x}{\sqrt{x^2 + 1}}$

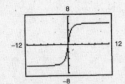

There are two horizontal
asymptotes, $y = \pm 6$.

39. $g(x) = \dfrac{4|x - 2|}{x + 1}$

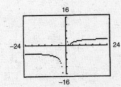

There are two horizontal
asymptotes, $y = \pm 4$.

One vertical asymptote: $x = -1$

41. $f(x) = \dfrac{4(x - 1)^2}{x^2 - 4x + 5}$

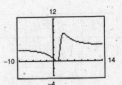

The graph crosses its horizontal
asymptote, $y = 4$.

43. $f(x) = \dfrac{2x^2 + 1}{x} = 2x + \dfrac{1}{x}$

Vertical asymptote: $x = 0$

Slant asymptote: $y = 2x$

Origin symmetry

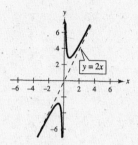

45. $h(x) = \dfrac{x^2}{x - 1} = x + 1 + \dfrac{1}{x - 1}$

Intercept: $(0, 0)$

Vertical asymptote: $x = 1$

Slant asymptote: $y = x + 1$

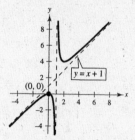

47. $g(x) = \dfrac{x^3}{2x^2 - 8} = \dfrac{1}{2}x + \dfrac{4x}{2x^2 - 8}$

Intercept: $(0, 0)$

Vertical asymptotes: $x = \pm 2$

Slant asymptote: $y = \dfrac{1}{2}x$

Origin symmetry

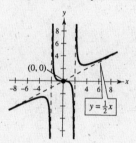

49. $f(x) = \dfrac{x^3 + 2x^2 + 4}{2x^2 + 1} = \dfrac{x}{2} + 1 + \dfrac{3 - \dfrac{x}{2}}{2x^2 + 1}$

Intercepts: $(-2.594, 0), (0, 4)$

Slant asymptote: $y = \dfrac{x}{2} + 1$

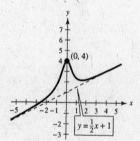

51. $y = \dfrac{x + 1}{x - 3}$

(a) x-intercept: $(-1, 0)$

(b) $0 = \dfrac{x + 1}{x - 3}$

$0 = x + 1$

$-1 = x$

53. $y = \dfrac{1}{x} - x$

(a) x-intercepts: $(\pm 1, 0)$

(b) $0 = \dfrac{1}{x} - x$

$x = \dfrac{1}{x}$

$x^2 = 1$

$x = \pm 1$

55. $y = \dfrac{2x^2 + x}{x + 1} = 2x - 1 + \dfrac{1}{x + 1}$

Domain: all real numbers except $x = -1$

Vertical asymptote: $x = -1$

Slant asymptote: $y = 2x - 1$

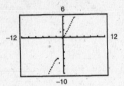

57. $y = \dfrac{1 + 3x^2 - x^3}{x^2} = \dfrac{1}{x^2} + 3 - x = -x + 3 + \dfrac{1}{x^2}$

Domain: all real numbers except 0

or $(-\infty, 0) \cup (0, \infty)$

Vertical asymptote: $x = 0$

Slant asymptote: $y = -x + 3$

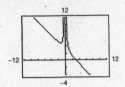

59. $f(x) = \dfrac{x^2 - 5x + 4}{x^2 - 4} = \dfrac{(x - 4)(x - 1)}{(x - 2)(x + 2)}$

Vertical asymptotes: $x = 2$, $x = -2$

Horizontal asymptote: $y = 1$

No slant asymptotes, no holes

61. $f(x) = \dfrac{2x^2 - 5x + 2}{2x^2 - x - 6}$

$\quad = \dfrac{(2x - 1)(x - 2)}{(2x + 3)(x - 2)}$

$\quad = \dfrac{2x - 1}{2x + 3}, \; x \neq 2$

Vertical asymptote: $x = -\dfrac{3}{2}$

Horizontal asymptote: $y = 1$

No slant asymptotes

Hole at $x = 2$, $\left(2, \dfrac{3}{7}\right)$

63. $f(x) = \dfrac{2x^3 - x^2 - 2x + 1}{x^2 + 3x + 2}$

$\quad = \dfrac{(x - 1)(x + 1)(2x - 1)}{(x + 1)(x + 2)}$

$\quad = \dfrac{(x - 1)(2x - 1)}{x + 2}, \; x \neq -1$

Long division gives

$f(x) = \dfrac{2x^2 - 3x + 1}{x + 2} = 2x - 7 + \dfrac{15}{x + 2}.$

Vertical asymptote: $x = -2$

No horizontal asymptote

Slant asymptote: $y = 2x - 7$

Hole at $x = -1$, $(-1, 6)$

65. $y = \dfrac{1}{x + 5} + \dfrac{4}{x}$

(a)

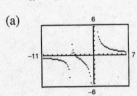

x-intercept: $(-4, 0)$

(b) $\qquad 0 = \dfrac{1}{x + 5} + \dfrac{4}{x}$

$-\dfrac{4}{x} = \dfrac{1}{x + 5}$

$-4(x + 5) = x$

$-4x - 20 = x$

$-5x = 20$

$x = -4$

67. $y = \dfrac{1}{x+2} + \dfrac{2}{x+4}$

(a)

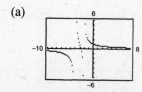

x-intercept: $\left(-\dfrac{8}{3}, 0\right)$

(b) $\dfrac{1}{x+2} + \dfrac{2}{x+4} = 0$

$$\dfrac{1}{x+2} = \dfrac{-2}{x+4}$$

$$x + 4 = -2x - 4$$

$$3x = -8$$

$$x = -\dfrac{8}{3}$$

69. $y = x - \dfrac{6}{x-1}$

(a)

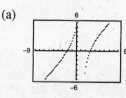

x-intercept: $(-2, 0), (3, 0)$

(b) $0 = x - \dfrac{6}{x-1}$

$$\dfrac{6}{x-1} = x$$

$$6 = x(x-1)$$

$$0 = x^2 - x - 6$$

$$0 = (x+2)(x-3)$$

$$x = -2, \quad x = 3$$

71. $y = x + 2 - \dfrac{1}{x+1}$

(a)

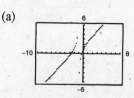

x-intercepts:
$(-2.618, 0), (-0.382, 0)$

(b) $x + 2 = \dfrac{1}{x+2}$

$$x^2 + 3x + 2 = 1$$

$$x^2 + 3x + 1 = 0$$

$$x = \dfrac{-3 \pm \sqrt{9-4}}{2}$$

$$= \dfrac{-3}{2} \pm \dfrac{\sqrt{5}}{2}$$

$$\approx -2.618, -0.382$$

73. $y = x + 1 + \dfrac{2}{x-1}$

(a)

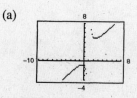

No x-intercepts

(b) $x + 1 + \dfrac{2}{x-1} = 0$

$$\dfrac{2}{x-1} = -x - 1$$

$$2 = -x^2 + 1$$

$x^2 + 1 = 0$

No real zeros

75. $y = x + 3 - \dfrac{2}{2x-1}$

(a)

x-intercepts: $(0.766, 0), (-3.266, 0)$

(b) $x + 3 - \dfrac{2}{2x-1} = 0$

$$x + 3 = \dfrac{2}{2x-1}$$

$$2x^2 + 5x - 3 = 2$$

$$2x^2 + 5x - 5 = 0$$

$$x = \dfrac{-5 \pm \sqrt{25 - 4(2)(-5)}}{4}$$

$$= \dfrac{-5 \pm \sqrt{65}}{4}$$

$$\approx 0.766, -3.266$$

77. (a) $0.25(50) + 0.75(x) = C(50 + x)$

$$\frac{12.5 + 0.75x}{50 + x} = C$$

$$\frac{50 + 3x}{200 + 4x} = C$$

$$C = \frac{3x + 50}{4(x + 50)}$$

(b) Domain: $x \geq 0$ and $x \leq 1000 - 50 = 950$

Thus, $0 \leq x \leq 950$.

(c)

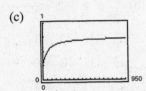

As the tank fills, the rate that the concentration is increasing slows down. It approaches the horizontal asymptote $C = \frac{3}{4} = 0.75$. When the tank is full ($x = 950$), the concentration is $C \doteq 0.725$.

81. $C = 100\left(\dfrac{200}{x^2} + \dfrac{x}{x + 30}\right),\ 1 \leq x$

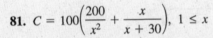

The minimum occurs when $x \approx 40.4 \approx 40$.

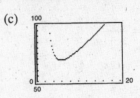

83. $C = \dfrac{3t^2 + t}{t^3 + 50},\ 0 \leq t$

(a) The horizontal asymptote is the t-axis, or $C = 0$. This indicates that the chemical eventually dissipates.

(b)

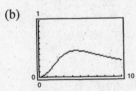

The maximum occurs when $t \approx 4.5$.

(c) Graph C together with $y = 0.345$. The graphs intersect at $t \approx 2.65$ and $t \approx 8.32$. $C < 0.345$ when $0 \leq t < 2.65$ hours and when $t > 8.32$ hours.

79. (a) $A = xy$ and

$$(x - 2)(y - 4) = 30$$

$$y - 4 = \frac{30}{x - 2}$$

$$y = 4 + \frac{30}{x - 2} = \frac{4x + 22}{x - 2}$$

Thus, $A = xy = x\left(\dfrac{4x + 22}{x - 2}\right) = \dfrac{2x(2x + 11)}{x - 2}$.

(b) Domain: Since the margins on the left and right are each 1 inch, $x > 2$, or $(2, \infty)$.

(c)

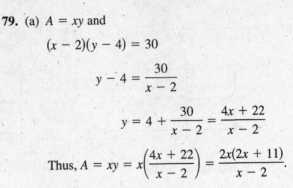

The area is minimum when $x \approx 5.87$ in. and $y \approx 11.75$ in.

85. (a) $y_1 = 583.8t + 2414$ ($t = 0$ corresponds to 1990)

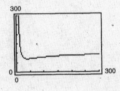

(b) Using the data $\left(t = \dfrac{1}{A}\right)$, we obtain:

$$y_2 = -0.00001855t + 0.0003150$$

$$y_3 = \frac{1}{-0.00001855t + 0.000315} = \frac{1}{A}$$

(c)

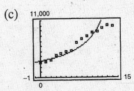

87. False, you will have to lift your pencil to cross the vertical asymptote.

89. $h(x) = \dfrac{6 - 2x}{3 - x} = \dfrac{2(3 - x)}{3 - x} = 2, \quad x \neq 3$

Since $h(x)$ is not reduced and $(3 - x)$ is a factor of both the numerator and the denominator, $x = 3$ is not a horizontal asymptote.

There is a hole in the graph at $x = 3$.

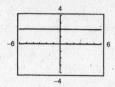

91. $y = x + 1 + \dfrac{a}{x + 2}$ has a slant asymptote

$y = x + 1$ and a vertical asymptote $x = -2$.

$0 = 2 + 1 + \dfrac{a}{2 + 2}$

$0 = 3 + \dfrac{a}{4}$

$\dfrac{a}{4} = -3$

$a = -12$

Hence, $y = x + 1 - \dfrac{12}{x + 2} = \dfrac{x^2 + 3x - 10}{x + 2}$.

93. $\left(\dfrac{x}{8}\right)^{-3} = \left(\dfrac{8}{x}\right)^3 = \dfrac{512}{x^3}$

95. $\dfrac{3^{7/6}}{3^{1/6}} = 3^{6/6} = 3$

97.

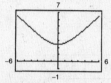

Domain: all x

Range: $y \geq \sqrt{6}$

99.

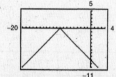

Domain: all x

Range: $y \leq 0$

101. Answers will vary. (Make a Decision)

Section 2.8   Quadratic Models

> You should know how to
> ■ Construct and classify scatter plots.
> ■ Fit a quadratic model to data.
> ■ Choose an appropriate model given a set of data.

Vocabulary Check

1. linear

2. quadratic

1. A quadratic model is better.

3. A linear model is better.

5. Neither linear nor quadratic

7. (a)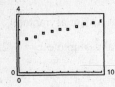

(b) Linear model is better.

(d)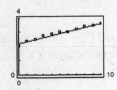

(c) $y = 0.14x + 2.2$, linear

$[y = -0.00478x^2 + 0.1887x + 2.1692,$ quadratic$]$

(e)

x	0	1	2	3	4	5	6	7	8	9	10
y	2.1	2.4	2.5	2.8	2.9	3.0	3.0	3.2	3.4	3.5	3.6
Model	2.2	2.4	2.5	2.7	2.8	2.9	3.1	3.2	3.4	3.5	3.6

9. (a)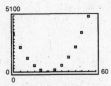

(b) Quadratic model is better.

(d)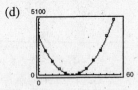

(c) $y = 5.55x^2 - 277.5x + 3478$

(e)

x	0	5	10	15	20	25	30	35	40	45	50	55
y	3480	2235	1250	565	150	12	145	575	1275	2225	3500	5010
Model	3478	2229	1258	564	148	9	148	564	1258	2229	3478	5004

11. (a) $y = 2.48x + 1.1$, linear

$y = 0.071x^2 + 1.69x + 2.7$, quadratic

(b) 0.98995 for linear model

0.99519 for quadratic model

(c) Quadratic fits better.

13. (a) $y = -0.89x + 5.3$, linear

$y = 0.001x^2 - 0.90x + 5.3$, quadratic

(b) 0.99982 for the linear model

0.99987 for the quadratic model

(c) The quadratic model is slightly better.

15. (a)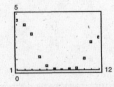

(b) $P = 0.1322t^2 - 1.901t + 6.87$

(c)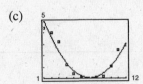

(d) The model's minimum is $H \approx 0.03$ at $t = 7.2$.
This corresponds to July.

17. (a)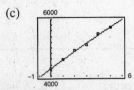

(b) $y = -2.630t^2 + 301.74t + 4270.2$

(c)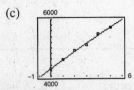

(d) According to the model, $y > 10,000$ when
$t \approx 24$, or 2024.

(e) Answers will vary.

19. (a)

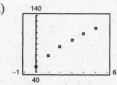

(b)

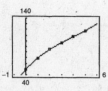

(c) $y = -1.1357t^2 + 18.999t + 50.32,$
quadratic model,   (0.99859)

(d)

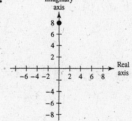

(e) The cubic model is a better fit.

(f)

Year	2006 ($t = 6$)	2007 ($t = 7$)	2008 ($t = 8$)
A^*	127.76	140.15	154.29
Cubic	129.91	145.13	164.96
Quadratic	123.40	127.64	129.60

21. True

23. The model is above all data points.

25. (a) $f(g(x)) = f(2x^2 - 1) = 5(2x^2 - 1) + 8 = 10x^2 + 3$

(b) $g(f(x)) = g(5x + 8) = 2(5x + 8)^2 - 1 = 50x^2 + 160x + 127$

27. (a) $f(g(x)) = f(x^3 - 5) = \sqrt[3]{x^3 - 5 + 5} = x$

(b) $g(f(x)) = g(\sqrt[3]{x + 5}) = [\sqrt[3]{x + 5}]^3 - 5 = x$

29. f is one-to-one.

$y = \dfrac{x - 4}{5}$

$x = \dfrac{y - 4}{5}$

$5x + 4 = y \Rightarrow f^{-1}(x) = 5x + 4$

31. f is one-to-one.

$y = 2x^2 - 3, \quad x \geq 0$

$x = 2y^2 - 3, \quad y \geq 0$

$y^2 = \dfrac{(x + 3)}{2}$

$y = \sqrt{\dfrac{x + 3}{2}} \Rightarrow f^{-1}(x) = \sqrt{\dfrac{x + 3}{2}}$

$= \dfrac{\sqrt{2x + 6}}{2}, \quad x \geq -3$

33.

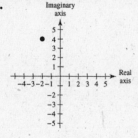

35.

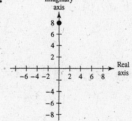

Review Exercises for Chapter 2

1.

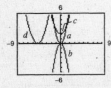

(a) $y = 2x^2$ is a vertical stretch.

(b) $y = -2x^2$ is a vertical stretch and reflection in the x-axis.

(c) $y = x^2 + 2$ is a vertical shift two units upward.

(d) $y = (x + 5)^2$ is a horizontal shift five units to the left.

3. $f(x) = \left(x + \frac{3}{2}\right)^2 + 1$

Vertex: $\left(-\frac{3}{2}, 1\right)$

y-intercept: $\left(0, \frac{13}{4}\right)$

No x-intercepts

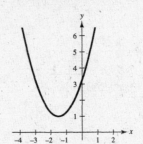

5. $f(x) = \frac{1}{3}(x^2 + 5x - 4)$

$\qquad = \frac{1}{3}\left(x^2 + 5x + \frac{25}{4} - \frac{25}{4} - 4\right)$

$\qquad = \frac{1}{3}\left[\left(x + \frac{5}{2}\right)^2 - \frac{41}{4}\right]$

$\qquad = \frac{1}{3}\left(x + \frac{5}{2}\right)^2 - \frac{41}{12}$

Vertex: $\left(-\frac{5}{2}, -\frac{41}{12}\right)$

y-intercept: $\left(0, -\frac{4}{3}\right)$

x-intercepts: $0 = \frac{1}{3}(x^2 + 5x - 4)$

$\qquad\qquad\quad 0 = x^2 + 5x - 4$

$\qquad\qquad\quad x = \dfrac{-5 \pm \sqrt{41}}{2}$ Use the Quadratic Formula.

$\qquad\qquad\quad \left(\dfrac{-5 \pm \sqrt{41}}{2}, 0\right)$

7. $f(x) = 3 - x^2 - 4x$

$\qquad = 3 - (x^2 + 4x + 4) + 4$

$\qquad = 7 - (x + 2)^2$

Vertex: $(-2, 7)$

Intercepts: $(0, 3)$, $\left(-2 \pm \sqrt{7}, 0\right)$

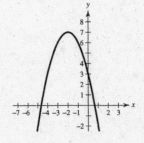

9. Vertex: $(1, -4) \Rightarrow f(x) = a(x - 1)^2 - 4$

Point: $(2, -3) \Rightarrow -3 = a(2 - 1)^2 - 4$

$\qquad\qquad\qquad\qquad 1 = a$

Thus, $f(x) = (x - 1)^2 - 4$.

11. Vertex: $(-2, -2) \Rightarrow f(x) = a(x + 2)^2 - 2$

Point: $(-1, 0) \Rightarrow 0 = a(-1 + 2)^2 - 2$

$\qquad\qquad\qquad\qquad a = 2$

Thus, $f(x) = 2(x + 2)^2 - 2$.

13. (a) $A = xy = x\left(\dfrac{8-x}{2}\right)$, since $x + 2y - 8 = 0 \implies y = \dfrac{8-x}{2}$.

Since the figure is in the first quadrant and x and y must be positive, the domain of

$A = x\left(\dfrac{8-x}{2}\right)$ is $0 < x < 8$.

(b)

x	y	Area
1	$4 - \frac{1}{2}(1)$	$(1)\left[4 - \frac{1}{2}(1)\right] = \frac{7}{2}$
2	$4 - \frac{1}{2}(2)$	$(2)\left[4 - \frac{1}{2}(2)\right] = 6$
3	$4 - \frac{1}{2}(3)$	$(3)\left[4 - \frac{1}{2}(3)\right] = \frac{15}{2}$
4	$4 - \frac{1}{2}(4)$	$(4)\left[4 - \frac{1}{2}(4)\right] = 8$
5	$4 - \frac{1}{2}(5)$	$(5)\left[4 - \frac{1}{2}(5)\right] = \frac{15}{2}$
6	$4 - \frac{1}{2}(6)$	$(6)\left[4 - \frac{1}{2}(6)\right] = 6$

The dimensions that will produce a maximum area seem to be $x = 4$ and $y = 2$.

(c)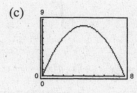

The maximum area of 8 occurs at the vertex

when $x = 4$ and $y = \dfrac{8-4}{2} = 2$.

(d) $A = x\left(\dfrac{8-x}{2}\right)$

$= \dfrac{1}{2}(8x - x^2)$

$= -\dfrac{1}{2}(x^2 - 8x)$

$= -\dfrac{1}{2}(x^2 - 8x + 16 - 16)$

$= -\dfrac{1}{2}[(x - 4)^2 - 16]$

$= -\dfrac{1}{2}(x - 4)^2 + 8$

(e) The answers are the same.

The maximum area of 8 occurs when $x = 4$ and $y = \dfrac{8-4}{2} = 2$.

15.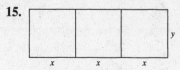

$6x + 4y = 1500$ Total amount of fencing

$A = 3xy$ Area enclosed

Because $y = \dfrac{1}{4}(1500 - 6x)$,

$A = 3x\left(\dfrac{1}{4}\right)(1500 - 6x) = -\dfrac{9}{2}x^2 + 1125x.$

The vertex is at $x = \dfrac{-b}{2a} = \dfrac{-1125}{2(-9/2)} = 125$. Thus $x = 125$ feet, $y = \dfrac{1}{4}(1500 - 6(125)) = 187.5$

and the dimensions are 375 feet by 187.5 feet.

17. (a) 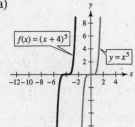 (b) 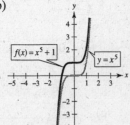 (c) 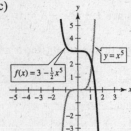 (d)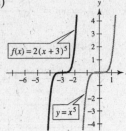

19. $f(x) = \frac{1}{2}x^3 - 2x + 1$; $g(x) = \frac{1}{2}x^3$

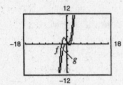

21. $f(x) = -x^2 + 6x + 9$

The degree is even and the leading coefficient is negative. The graph falls to the left and right.

23. $f(x) = \frac{3}{4}(x^4 + 3x^2 + 2)$

The degree is even and the leading coefficient is positive. The graph rises to the left and right.

25. (a) $x^4 - x^3 - 2x^2 = x^2(x^2 - x - 2)$
$$= x^2(x - 2)(x + 1) = 0$$

Zeros: $x = -1, 0, 2$

(b)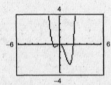

(c) Zeros: $x = -1, 0, 2$; the same

27. (a) $t^3 - 3t = t(t^2 - 3) = t(t + \sqrt{3})(t - \sqrt{3}) = 0$

Zeros: $t = 0, \pm\sqrt{3}$

(b)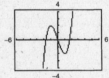

(c) Zeros: $t = 0, \pm 1.732$, the same

29. (a) $x(x + 3)^2 = 0$

Zeros: $x = 0, -3$

(b)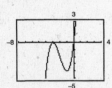

(c) Zeros: $x = -3, 0$, the same

31. $f(x) = (x + 2)(x - 1)^2(x - 5)$
$$= x^4 - 5x^3 - 3x^2 + 17x - 10$$

33. $f(x) = (x - 3)(x - 2 + \sqrt{3})(x - 2 - \sqrt{3})$
$$= x^3 - 7x^2 + 13x - 3$$

35. (a) Degree is even and leading coefficient is $1 > 0$. Rises to the left and rises to the right.

(b) $x^4 - 2x^3 - 12x^2 + 18x + 27 = (x - 3)^2(x + 1)(x + 3)$

Zeros: $\pm 3, -1$

(c) and (d)

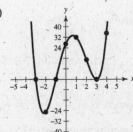

37. $f(x) = x^3 + 2x^2 - x - 1$

    (a) $f(-3) < 0, f(-2) > 0 \implies$ zero in $[-3, -2]$

        $f(-1) > 0, f(0) < 0 \implies$ zero in $[-1, 0]$

        $f(0) < 0, f(1) > 0 \implies$ zero in $[0, 1]$

    (b) Zeros: $-2.247, -0.555, 0.802$

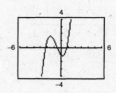

39. $f(x) = x^4 - 6x^2 - 4$

    (a) $f(-3) > 0, f(-2) < 0 \implies$ zero in $[-3, -2]$

        $f(2) < 0, f(3) > 0 \implies$ zero in $[2, 3]$

    (b) Zeros: ± 2.570

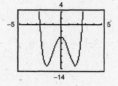

41. $y_1 = \dfrac{x^2}{x - 2}$

$y_2 = x + 2 + \dfrac{4}{x - 2}$

$ = \dfrac{(x + 2)(x - 2)}{x - 2} + \dfrac{4}{x - 2}$

$ = \dfrac{x^2 - 4}{x - 2} + \dfrac{4}{x - 2}$

$ = \dfrac{x^2}{x - 2} = y_1$

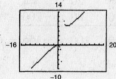

43. $y_1 = \dfrac{x^4 + 1}{x^2 + 2}$

$y_2 = x^2 - 2 + \dfrac{5}{x^2 + 2}$

$ = \dfrac{x^2(x^2 + 2)}{x^2 + 2} - \dfrac{2(x^2 + 2)}{x^2 + 2} + \dfrac{5}{x^2 + 2}$

$ = \dfrac{x^4 + 2x^2 - 2x^2 - 4 + 5}{x^2 + 2}$

$ = \dfrac{x^4 + 1}{x^2 + 2} = y_1$

45.

$$
\begin{array}{r}
8x + 5 \\
3x - 2 \overline{\smash{)}\, 24x^2 - x - 8} \\
\underline{24x^2 - 16x} \\
15x - 8 \\
\underline{15x - 10} \\
2
\end{array}
$$

Thus, $\dfrac{24x^2 - x - 8}{3x - 2} = 8x + 5 + \dfrac{2}{3x - 2}.$

47.

$$
\begin{array}{r}
x^2 - 2 \\
x^2 - 1 \overline{\smash{)}\, x^4 - 3x^2 + 2} \\
\underline{x^4 - x^2} \\
-2x^2 + 2 \\
\underline{-2x^2 + 2} \\
0
\end{array}
$$

Thus, $\dfrac{x^4 - 3x^2 + 2}{x^2 - 1} = x^2 - 2, (x \neq \pm 1).$

49.

$$
\begin{array}{r}
5x + 2 \\
x^2 - 3x + 1 \overline{\smash{)}\, 5x^3 - 13x^2 - x + 2} \\
\underline{5x^3 - 15x^2 + 5x} \\
2x^2 - 6x + 2 \\
\underline{2x^2 - 6x + 2} \\
0
\end{array}
$$

Thus, $\dfrac{5x^3 - 13x^2 - x + 2}{x^2 - 3x + 1} = 5x + 2, x \neq \dfrac{1}{2}\left(3 \pm \sqrt{5}\right).$

51.

$$
\begin{array}{r}
3x^2 + 5x + 8 \\
2x^2 + 0x - 1 \overline{)\, 6x^4 + 10x^3 + 13x^2 - 5x + 2} \\
\underline{6x^4 + 0x^3 - 3x^2} \\
10x^3 + 16x^2 - 5x \\
\underline{10x^3 + 0x^2 - 5x} \\
16x^2 - 0 + 2 \\
\underline{16x^2 + 0 - 8} \\
10
\end{array}
$$

$$\frac{6x^4 + 10x^3 + 13x^2 - 5x + 2}{2x^2 - 1} = 3x^2 + 5x + 8 + \frac{10}{2x^2 - 1}$$

53.

$$
\begin{array}{r|rrrrr}
-2 & 0.25 & -4 & 0 & 0 & 0 \\
   &   & -\frac{1}{2} & 9 & -18 & 36 \\
\hline
   & \frac{1}{4} & -\frac{9}{2} & 9 & -18 & 36
\end{array}
$$

Hence,

$$\frac{0.25x^4 - 4x^3}{x + 2} = \frac{1}{4}x^3 - \frac{9}{2}x^2 + 9x - 18 + \frac{36}{x + 2}.$$

55.

$$
\begin{array}{r|rrrrr}
\frac{2}{3} & 6 & -4 & -27 & 18 & 0 \\
            &   & 4 & 0   & -18 & 0 \\
\hline
            & 6 & 0 & -27 & 0 & 0
\end{array}
$$

Thus,

$$\frac{6x^4 - 4x^3 - 27x^2 + 18x}{x - (2/3)} = 6x^3 - 27x, \; x \neq \frac{2}{3}.$$

57.

$$
\begin{array}{r|rrrr}
4 & 3 & -10 & 12 & -22 \\
  &   & 12 & 8 & 80 \\
\hline
  & 3 & 2   & 20 & 58
\end{array}
$$

Thus,

$$\frac{3x^3 - 10x^2 + 12x - 22}{x - 4} = 3x^2 + 2x + 20 + \frac{58}{x - 4}.$$

59. (a)

$$
\begin{array}{r|rrrrr}
-3 & 1 & 10 & -24 & 20 & 44 \\
   &   & -3 & -21 & 135 & -465 \\
\hline
   & 1 & 7 & -45 & 155 & -421 = f(-3)
\end{array}
$$

(b)

$$
\begin{array}{r|rrrrr}
-2 & 1 & 10 & -24 & 20 & 44 \\
   &   & -2 & -16 & 80 & -200 \\
\hline
   & 1 & 8 & -40 & 100 & -156 = f(-2)
\end{array}
$$

61. $f(x) = x^3 + 4x^2 - 25x - 28$

(a)
$$
\begin{array}{r|rrrr}
4 & 1 & 4 & -25 & -28 \\
  &   & 4 & 32 & 28 \\
\hline
  & 1 & 8 & 7   & 0
\end{array}
$$

$(x - 4)$ is a factor.

(b) $x^2 + 8x + 7 = (x + 1)(x + 7)$

Remaining factors: $(x + 1), (x + 7)$

(c) $f(x) = (x - 4)(x + 1)(x + 7)$

(d) Zeros: $4, -1, -7$

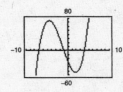

63. $f(x) = x^4 - 4x^3 - 7x^2 + 22x + 24$

(a)
$$
\begin{array}{r|rrrrr}
-2 & 1 & -4 & -7 & 22 & 24 \\
   &   & -2 & 12 & -10 & -24 \\
\hline
   & 1 & -6 & 5 & 12 & 0
\end{array}
$$

$(x + 2)$ is a factor.

$$
\begin{array}{r|rrrr}
3 & 1 & -6 & 5 & 12 \\
  &   & 3 & -9 & -12 \\
\hline
  & 1 & -3 & -4 & 0
\end{array}
$$

$(x - 3)$ is a factor.

(b) $x^2 - 3x - 4 = (x - 4)(x + 1)$

Remaining factors: $(x - 4), (x + 1)$

(c) $f(x) = (x + 2)(x - 3)(x - 4)(x + 1)$

(d) Zeros: $-2, 3, 4, -1$

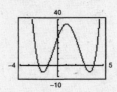

65. $f(x) = 4x^3 - 11x^2 + 10x - 3$

Possible rational zeros: $\pm 3, \pm\frac{3}{2}, \pm\frac{3}{4}, \pm 1, \pm\frac{1}{2}, \pm\frac{1}{4}$

Zeros: $1, 1, \frac{3}{4}$

67. $f(x) = 6x^3 - 5x^2 + 24x - 20$

$= (6x - 5)(x^2 + 4)$

Real zero: $\frac{5}{6}$

69. $f(x) = 6x^4 - 25x^3 + 14x^2 + 27x - 18$

Possible rational zeros: $\pm 1, \pm 2, \pm 3, \pm 6, \pm 9, \pm 18, \pm\frac{1}{2}, \pm\frac{3}{2}, \pm\frac{9}{2}, \pm\frac{1}{3}, \pm\frac{2}{3}, \pm\frac{1}{6}$

Use a graphing utility to see that $x = -1$ and $x = 3$ are probably zeros.

$$
\begin{array}{r|rrrrr}
-1 & 6 & -25 & 14 & 27 & -18 \\
   &   & -6 & 31 & -45 & 18 \\
\hline
   & 6 & -31 & 45 & -18 & 0
\end{array}
\qquad
\begin{array}{r|rrrr}
3 & 6 & -31 & 45 & -18 \\
  &   & 18 & -39 & 18 \\
\hline
  & 6 & -13 & 6 & 0
\end{array}
$$

$6x^4 - 25x^3 + 14x^2 + 27x - 18 = (x + 1)(x - 3)(6x^2 - 13x + 6)$

$\qquad\qquad\qquad\qquad\qquad\quad = (x + 1)(x - 3)(3x - 2)(2x - 3)$

Thus, the zeros of f are $x = -1$, $x = 3$, $x = \frac{2}{3}$, and $x = \frac{3}{2}$.

71. $g(x) = 5x^3 - 6x + 9$ has two variations in sign
$\Rightarrow$ 0 or 2 positive real zeros.

$g(-x) = -5x^3 + 6x + 9$ has one variation in sign
$\Rightarrow$ 1 negative real zero.

73.
$$
\begin{array}{r|rrrr}
1 & 4 & -3 & 4 & -3 \\
  &   & 4 & 1 & 5 \\
\hline
  & 4 & 1 & 5 & 2
\end{array}
$$

All entries positive; $x = 1$ is upper bound.

$$
\begin{array}{r|rrrr}
-\frac{1}{4} & 4 & -3 & 4 & -3 \\
             &   & -1 & 1 & -\frac{5}{4} \\
\hline
             & 4 & -4 & 5 & -\frac{17}{4}
\end{array}
$$

Alternating signs; $x = -\frac{1}{4}$ is lower bound.

75. $6 + \sqrt{-25} = 6 + 5i$

77. $-2i^2 + 7i = 2 + 7i$

79. $(7 + 5i) + (-4 + 2i) = (7 - 4) + (5i + 2i)$

$\qquad\qquad\qquad\qquad\quad = 3 + 7i$

81. $5i(13 - 8i) = 65i - 40i^2 = 40 + 65i$

83. $\left(\sqrt{-16} + 3\right)\left(\sqrt{-25} - 2\right) = (4i + 3)(5i - 2)$

$\qquad\qquad\qquad\qquad\qquad = -20 - 8i + 15i - 6$

$\qquad\qquad\qquad\qquad\qquad = -26 + 7i$

85. $\sqrt{-9} + 3 + \sqrt{-36} = 3i + 3 + 6i$

$\qquad\qquad\qquad\qquad\quad = 3 + 9i$

87. $(10 - 8i)(2 - 3i) = 20 - 30i - 16i + 24i^2 = -4 - 46i$

89. $(3 + 7i)^2 + (3 - 7i)^2 = (9 + 42i - 49) + (9 - 42i - 49) = -80$

91. $\dfrac{6 + i}{i} = \dfrac{6 + i}{i} \cdot \dfrac{-i}{-i} = \dfrac{-6i - i^2}{-i^2}$

$\qquad = \dfrac{-6i + 1}{1} = 1 - 6i$

93. $\dfrac{3 + 2i}{5 + i} \cdot \dfrac{5 - i}{5 - i} = \dfrac{15 + 10i - 3i + 2}{25 + 1}$

$\qquad\qquad\qquad = \dfrac{17}{26} + \dfrac{7}{26}i$

95. $-3 - 2i$

97. $2 - 5i$

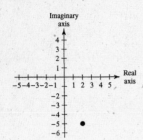

99. $-6i$

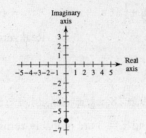

101. 3

103. $f(x) = 3x(x - 2)^2$

Zeros: $0, 2, 2$

105. $f(x) = 2x^4 - 5x^3 + 10x - 12$

$$
\begin{array}{r|rrrrr}
2 & 2 & -5 & 0 & 10 & -12 \\
  &   & 4 & -2 & -4 & 12 \\
\hline
  & 2 & -1 & -2 & 6 & 0
\end{array}
$$

$x = 2$ is a zero.

$$
\begin{array}{r|rrrr}
-\frac{3}{2} & 2 & -1 & -2 & 6 \\
  &   & -3 & 6 & -6 \\
\hline
  & 2 & -4 & 4 & 0
\end{array}
$$

$x = -\dfrac{3}{2}$ is a zero.

$$f(x) = (x - 2)\left(x + \frac{3}{2}\right)(2x^2 - 4x + 4)$$

$$= (x - 2)(2x + 3)(x^2 - 2x + 2)$$

By the Quadratic Formula, applied to $x^2 - 2x + 2$,

$$x = \frac{2 \pm \sqrt{4 - 4(2)}}{2} = 1 \pm i.$$

Zeros: $2, -\dfrac{3}{2}, 1 \pm i$

$$f(x) = (x - 2)(2x + 3)(x - 1 + i)(x - 1 - i)$$

107. $h(x) = x^3 - 7x^2 + 18x - 24$

$$
\begin{array}{r|rrrr}
4 & 1 & -7 & 18 & -24 \\
  &   & 4 & -12 & 24 \\
\hline
  & 1 & -3 & 6 & 0
\end{array}
$$

$x = 4$ is a zero. Applying the Quadratic Formula on $x^2 - 3x + 6$,

$$x = \frac{3 \pm \sqrt{9 - 4(6)}}{2} = \frac{3}{2} \pm \frac{\sqrt{15}}{2}i.$$

Zeros: $4, \dfrac{3}{2} + \dfrac{\sqrt{15}}{2}i, \dfrac{3}{2} - \dfrac{\sqrt{15}}{2}i$

$$h(x) = (x - 4)\left(x - \frac{3 + \sqrt{15}i}{2}\right)\left(x - \frac{3 - \sqrt{15}i}{2}\right)$$

109. $f(x) = x^5 + x^4 + 5x^3 + 5x^2$

$$= x^2(x^3 + x^2 + 5x + 5)$$

$$= x^2[x^2(x + 1) + 5(x + 1)]$$

$$= x^2(x + 1)(x^2 + 5)$$

$$= x^2(x + 1)\left(x + \sqrt{5}i\right)\left(x - \sqrt{5}i\right)$$

Zeros: $0, 0, -1, \pm\sqrt{5}i$

111. $f(x) = x^3 - 4x^2 + 6x - 4$

(a) $x^3 - 4x^2 + 6x - 4 = (x - 2)(x^2 - 2x + 2)$

By the Quadratic Formula, for $x^2 - 2x + 2$,

$$x = \frac{2 \pm \sqrt{(-2)^2 - 4(2)}}{2} = 1 \pm i.$$

Zeros: $2, 1 + i, 1 - i$

(b) $f(x) = (x - 2)(x - 1 - i)(x - 1 + i)$

(c) x-intercept: $(2, 0)$

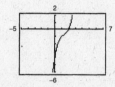

113. (a) $f(x) = -3x^3 - 19x^2 - 4x + 12$

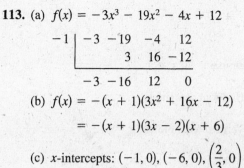

(b) $f(x) = -(x + 1)(3x^2 + 16x - 12)$

$= -(x + 1)(3x - 2)(x + 6)$

(c) x-intercepts: $(-1, 0), (-6, 0), \left(\frac{2}{3}, 0\right)$

115. $f(x) = x^4 + 34x^2 + 225$

(a) $x^4 + 34x^2 + 225 = (x^2 + 9)(x^2 + 25)$

Zeros: $\pm 3i, \pm 5i$

(b) $(x + 3i)(x - 3i)(x + 5i)(x - 5i)$

(c) No x-intercepts

(d)

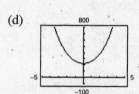

117. Since $5i$ is a zero, so is $-5i$.

$f(x) = (x - 4)(x + 2)(x - 5i)(x + 5i)$

$= (x^2 - 2x - 8)(x^2 + 25)$

$= x^4 - 2x^3 + 17x^2 - 50x - 200$

119. $f(x) = (x - 1)(x + 4)(x + 3 - 5i)(x + 3 + 5i)$

$= (x^2 + 3x - 4)((x + 3)^2 + 25)$

$= (x^2 + 3x - 4)(x^2 + 6x + 34)$

$= x^4 + 9x^3 + 48x^2 + 78x - 136$

121. $f(x) = x^4 - 2x^3 + 8x^2 - 18x - 9$

(a) $f(x) = (x^2 + 9)(x^2 - 2x - 1)$

For the quadratic $x^2 - 2x - 1$, $x = \frac{2 \pm \sqrt{(-2)^2 - 4(-1)}}{2} = 1 \pm \sqrt{2}$.

(b) $f(x) = (x^2 + 9)\left(x - 1 + \sqrt{2}\right)\left(x - 1 - \sqrt{2}\right)$

(c) $f(x) = (x + 3i)(x - 3i)\left(x - 1 + \sqrt{2}\right)\left(x - 1 - \sqrt{2}\right)$

123. Zeros: $-2i, 2i$

$(x + 2i)(x - 2i) = x^2 + 4$ is a factor.

$f(x) = (x^2 + 4)(x + 3)$

Zeros: $\pm 2i, -3$

125. (a) Domain: all $x \neq -3$

(b) Horizontal asymptote: $y = -1$

Vertical asymptote: $x = -3$

127. $f(x) = \frac{2}{x^2 - 3x - 18} = \frac{2}{(x - 6)(x + 3)}$

(a) Domain: all $x \neq 6, -3$

(b) Horizontal asymptote: $y = 0$

Vertical asymptotes: $x = 6, x = -3$

129. $f(x) = \frac{7 + x}{7 - x}$

(a) Domain: all $x \neq 7$

(b) Horizontal asymptote: $y = -1$

Vertical asymptote: $x = 7$

131. $f(x) = \dfrac{4x^2}{2x^2 - 3}$

   (a) Domain: all $x \neq \pm\sqrt{\dfrac{3}{2}} = \pm\dfrac{\sqrt{6}}{2}$

   (b) Horizontal asymptote: $y = 2$

       Vertical asymptotes: $x = \pm\sqrt{\dfrac{3}{2}} = \pm\dfrac{\sqrt{6}}{2}$

133. $f(x) = \dfrac{2x - 10}{x^2 - 2x - 15} = \dfrac{2(x - 5)}{(x - 5)(x + 3)} = \dfrac{2}{x + 3}$,

   $x \neq 5$

   (a) Domain: all $x \neq 5, -3$

   (b) Vertical asymptote: $x = -3$

       (There is a hole at $x = 5$.)

       Horizontal asymptote: $y = 0$

135. $f(x) = \dfrac{x - 2}{|x| + 2}$

   (a) Domain: all real numbers

   (b) No vertical asymptotes

       Horizontal asymptotes: $y = 1, y = -1$

137. $C = \dfrac{528p}{100 - p}$, $0 \le p < 100$

   (a) When $p = 25$, $C = \dfrac{528(25)}{100 - 25} = 176$ million.

       When $p = 50$, $C = \dfrac{528(50)}{100 - 50} = 528$ million.

       When $p = 75$, $C = \dfrac{528(75)}{100 - 75} = 1584$ million.

   (b)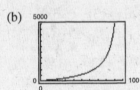

   (c) No. As $p \to 100$, C tends to infinity..

139. $f(x) = \dfrac{x^2 - 5x + 4}{x^2 - 1}$

        $= \dfrac{(x - 4)(x - 1)}{(x - 1)(x + 1)}$

        $= \dfrac{x - 4}{x + 1}$, $x \neq 1$

    Vertical asymptotes: $x = -1$

    Horizontal asymptote: $y = 1$

    No slant asymptotes

    Hole at $x = 1$: $\left(1, -\dfrac{3}{2}\right)$

141. $f(x) = \dfrac{2x^2 - 7x + 3}{2x^2 - 3x - 9}$

        $= \dfrac{(x - 3)(2x - 1)}{(x - 3)(2x + 3)}$

        $= \dfrac{2x - 1}{2x + 3}, x \neq 3$

    Vertical asymptote: $x = -\dfrac{3}{2}$

    Horizontal asymptote: $y = 1$

    No slant asymptotes

    Hole at $x = 3$: $\left(3, \dfrac{5}{9}\right)$

143. $f(x) = \dfrac{3x^3 - x^2 - 12x + 4}{x^2 + 3x + 2}$

        $= \dfrac{(x - 2)(x + 2)(3x - 1)}{(x + 1)(x + 2)}$

        $= \dfrac{(x - 2)(3x - 1)}{x + 1}$, $x \neq -2$

        $= 3x - 10 + \dfrac{12}{x + 1}$, $x \neq -2$

    Vertical asymptote: $x = -1$

    No horizontal asymptotes

    Slant asymptote: $y = 3x - 10$

    Hole at $x = -2$: $(-2, -28)$

145. $f(x) = \dfrac{2x - 1}{x - 5}$

Intercepts: $\left(0, \dfrac{1}{5}\right), \left(\dfrac{1}{2}, 0\right)$

Vertical asymptote: $x = 5$

Horizontal asymptote: $y = 2$

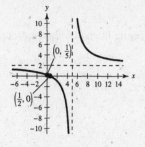

147. $f(x) = \dfrac{2x}{x^2 + 4}$

Intercept: $(0, 0)$

Origin symmetry

Horizontal asymptote: $y = 0$

x	-2	-1	0	1	2
y	$-\dfrac{1}{2}$	$-\dfrac{2}{5}$	0	$\dfrac{2}{5}$	$\dfrac{1}{2}$

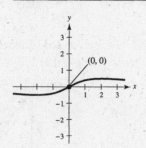

149. $f(x) = \dfrac{x^2}{x^2 + 1}$

Intercept: $(0, 0)$

y-axis symmetry

Horizontal asymptote: $y = 1$

x	± 3	± 2	± 1	0
y	$\dfrac{9}{10}$	$\dfrac{4}{5}$	$\dfrac{1}{2}$	0

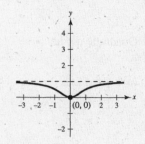

151. $f(x) = \dfrac{2}{(x + 1)^2}$

Intercept: $(0, 2)$

Horizontal asymptote: $y = 0$

Vertical asymptote: $x = -1$

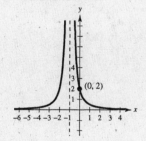

153. $f(x) = \dfrac{2x^3}{x^2 + 1} = 2x - \dfrac{2x}{x^2 + 1}$

Intercept: $(0, 0)$

Origin symmetry

Slant asymptote: $y = 2x$

x	-2	-1	0	1	2
y	$-\dfrac{16}{5}$	-1	0	1	$\dfrac{16}{5}$

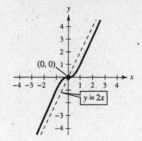

155. $f(x) = \dfrac{x^2 - x + 1}{x - 3} = x + 2 + \dfrac{7}{x - 3}$

Intercept: $\left(0, -\dfrac{1}{3}\right)$

Vertical asymptote: $x = 3$

Slant asymptote: $y = x + 2$

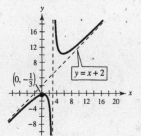

157. $N = \dfrac{20(4 + 3t)}{1 + 0.05t}$, $t \geq 0$

(a)

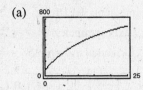

(b) $N(5)$ = 304,000 fish

  $N(10) \approx 453,333$ fish

  $N(25) \approx 702,222$ fish

(c) The limit is

  $\dfrac{60}{0.05} = 1{,}200{,}000$ fish, the horizontal asymptote.

159. Quadratic model

161. Linear model

163. (a)

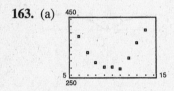

(b) $y = 8.03t^2 - 157.1t + 1041$; 0.98348

(c)

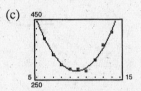

Yes, the model is a good fit.

(d) From the model, $y \geq 500$ when $t \approx 15.1$, or 2005.

(e) Answers will vary.

165. False. The degree of the numerator is two more than the degree of the denominator.

167. False. $(1 + i) + (1 - i) = 2$, a real number

169. Not every rational function has a vertical asymptote. For example,

  $y = \dfrac{x}{x^2 + 1}$.

171. The error is $\sqrt{-4} \neq 4i$. In fact,

  $-i\left(\sqrt{-4} - 1\right) = -i(2i - 1) = 2 + i$.

Chapter 2   Practice Test

1. Sketch the graph of $f(x) = x^2 - 6x + 5$ by hand and identify the vertex and the intercepts.

2. Find the number of units x that produce a minimum cost C if $C = 0.01x^2 - 90x + 15,000$.

3. Find the quadratic function that has a maximum at $(1, 7)$ and passes through the point $(2, 5)$.

4. Find two quadratic functions that have x-intercepts $(2, 0)$ and $\left(\frac{4}{3}, 0\right)$.

5. Use the leading Coefficient Test to determine the right-hand and left-hand behavior of the graph of the polynomial function $f(x) = -3x^5 + 2x^3 - 17$.

6. Find all the real zeros of $f(x) = x^5 - 5x^3 + 4x$. Verify your answer with a graphing utility.

7. Find a polynomial function with 0, 3, and -2 as zeros.

8. Sketch $f(x) = x^3 - 12x$ by hand.

9. Divide $3x^4 - 7x^2 + 2x - 10$ by $x - 3$ using long division.

10. Divide $x^3 - 11$ by $x^2 + 2x - 1$.

11. Use synthetic division to divide $3x^5 + 13x^4 + 12x - 1$ by $x + 5$.

12. Use synthetic division to find $f(-6)$ when $f(x) = 7x^3 + 40x^2 - 12x + 15$.

13. Find the real zeros of $f(x) = x^3 - 19x - 30$.

14. Find the real zeros of $f(x) = x^4 + x^3 - 8x^2 - 9x - 9$.

15. List all possible rational zeros of the function $f(x) = 6x^3 - 5x^2 + 4x - 15$.

16. Find the rational zeros of the polynomial $f(x) = x^3 - \frac{20}{3}x^2 + 9x - \frac{10}{3}$.

17. Write $f(x) = x^4 + x^3 + 3x^2 + 5x - 10$ as a product of linear factors.

18. Write $\dfrac{2}{1 + i}$ in standard form.

19. Write $\dfrac{3 + i}{2} - \dfrac{i + 1}{4}$ in standard form.

20. Find a polynomial with real coefficients that has 2, $3 + i$, and $3 - 2i$ as zeros.

21. Use synthetic division to show that $3i$ is a zero of $f(x) = x^3 + 4x^2 + 9x + 36$.

22. Find a mathematical model for the statement, "z varies directly as the square of x and inversely as the square root of y".

23. Sketch the graph of $f(x) = \dfrac{x - 1}{2x}$ and label all intercepts and asymptotes.

24. Sketch the graph of $f(x) = \dfrac{3x^2 - 4}{x}$ and label all intercepts and asymptotes.

25. Find all the asymptotes of $f(x) = \dfrac{8x^2 - 9}{x^2 + 1}$.

26. Find all the asymptotes of $f(x) = \dfrac{4x^2 - 2x + 7}{x - 1}$.

27. Sketch the graph of $f(x) = \dfrac{x - 5}{(x - 5)^2}$.

C H A P T E R   3
Exponential and Logarithmic Functions

Section 3.1   Exponential Functions and Their Graphs **107**

Section 3.2   Logarithmic Functions and Their Graphs **113**

Section 3.3   Properties of Logarithms **118**

Section 3.4   Solving Exponential and Logarithmic Equations **124**

Section 3.5   Exponential and Logarithmic Models **132**

Section 3.6   Nonlinear Models **137**

Review Exercises **141**

Practice Test **148**

CHAPTER 3
Exponential and Logarithmic Functions

Section 3.1   Exponential Functions and Their Graphs

- You should know that a function of the form $y = a^x$, where $a > 0$, $a \neq 1$, is called an exponential function with base a.
- You should be able to graph exponential functions.
- You should be familiar with the number e and the natural exponential function $f(x) = e^x$.
- You should know formulas for compound interest.

    (a) For n compoundings per year: $A = P\left(1 + \dfrac{r}{n}\right)^{nt}$.

    (b) For continuous compoundings: $A = Pe^{rt}$.

Vocabulary Check

1. algebraic

2. transcendental

3. natural exponential, natural

4. $A = P\left(1 + \dfrac{r}{n}\right)^{nt}$

5. $A = Pe^{rt}$

---

1. $(3.4)^{6.8} \approx 4112.033$

3. $5^{-\pi} \approx 0.006$

5. $g(x) = 5^x$

x	-2	-1	0	1	2
y	$\frac{1}{25}$	$\frac{1}{5}$	1	5	25

Asymptote: $y = 0$

Intercept: $(0, 1)$

Increasing

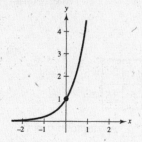

7. $f(x) = \left(\frac{1}{5}\right)^x = 5^{-x}$

x	-2	-1	0	1	2
y	25	5	1	$\frac{1}{5}$	$\frac{1}{25}$

Asymptote: $y = 0$

Intercept: $(0, 1)$

Decreasing

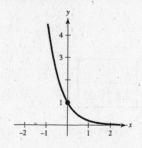

9. $h(x) = 5^{x-2}$

x	-1	0	1	2	3
y	$\frac{1}{125}$	$\frac{1}{25}$	$\frac{1}{5}$	1	5

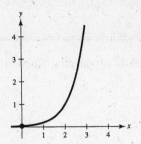

Asymptote: $y = 0$

Intercept: $\left(0, \frac{1}{25}\right)$

Increasing

11. $g(x) = 5^{-x} - 3$

x	-1	0	1	2
y	2	-2	$-2\frac{4}{5}$	$-2\frac{24}{25}$

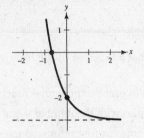

Asymptote: $y = -3$

Intercepts:

$(0, -2), (-0.683, 0)$

Decreasing

13. $f(x) = 2^{x-2}$ rises to the right.

Asymptote: $y = 0$

Intercept: $\left(0, \frac{1}{4}\right)$

Matches graph (d).

15. $f(x) = 2^x - 4$ rises to the right.

Asymptote: $y = -4$

Intercept: $(0, -3)$

Matches graph (c).

17. $f(x) = 3^x$

$g(x) = 3^{x-5} = f(x - 5)$

Horizontal shift five units to the right

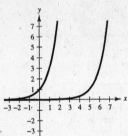

19. $f(x) = \left(\frac{3}{5}\right)^x$

$g(x) = -\left(\frac{3}{5}\right)^{x+4} = -f(x + 4)$

Horizontal shift four units to the left, followed by reflection in x-axis

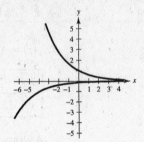

21. $f(x) = 4^x$

$g(x) = 4^{x-2} - 3$

$= f(x - 2) - 3$

Horizontal shift two units to the right followed by vertical shift three units downward

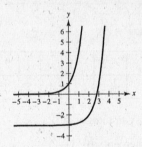

23. $e^{9.2} \approx 9897.129$

25. $50e^{4(0.02)} \approx 54.164$

27. $f(x) = \left(\frac{5}{2}\right)^x$

x	-2	-1	0	1	2
$f(x)$	0.16	0.4	1	2.5	6.25

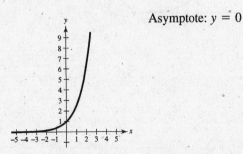

Asymptote: $y = 0$

29. $f(x) = 6^x$

x	-2	-1	0	1	2
$f(x)$	0.03	0.17	1	6	36

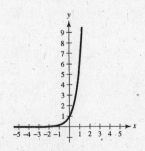

Asymptote: $y = 0$

31. $f(x) = 3^{x+2}$

x	-3	-2	-1	0	1
$f(x)$	0.33	1	3	9	27

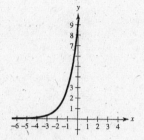

Asymptote: $y = 0$

33. $y = 2^{-x^2}$

x	-2	-1	0	1	2
y	0.06	0.5	1	0.5	0.06

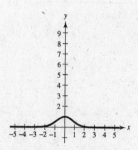

Asymptote: $y = 0$

35. $y = 3^{x-2} + 1$

x	-1	0	1	2	3	4
y	1.04	1.11	1.33	2	4	10

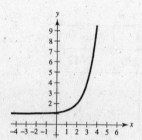

Asymptote: $y = 1$

37. $f(x) = e^{-x}$

x	-2	-1	0	1	2
$f(x)$	7.39	2.72	1	0.37	0.14

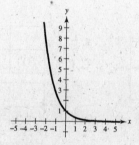

Asymptote: $y = 0$

39. $f(x) = 3e^{x+4}$

x	-6	-5	-4	-3	-2
$f(x)$	0.41	1.10	3	8.15	22.17

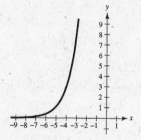

Asymptote: $y = 0$

41. $f(x) = 2 + e^{x-5}$

x	3	4	5	6	7
$f(x)$	2.14	2.37	3	4.72	9.39

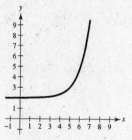

Asymptote: $y = 2$

43. $s(t) = 2e^{0.12t}$

t	-2	-1	0	1	2
$s(t)$	1.57	1.77	2	2.26	2.54

Asymptote: $y = 0$

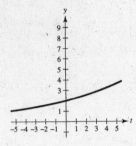

45. $f(x) = \dfrac{8}{1 + e^{-0.5x}}$

(a)

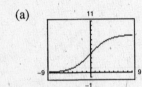

(b)

x	-30	-20	-10	0	10	20	30
$f(x)$	≈ 0	≈ 0	0.05	4	7.95	≈ 8	≈ 8

Horizontal asymptotes: $y = 0$, $y = 8$

47. $f(x) = \dfrac{-6}{2 - e^{0.2x}}$

(a)

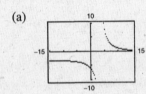

(b)

x	-20	-10	0	3	3.4	3.46
$f(x)$	-3.03	-3.22	-6	-34	-230	-2617

x	3.47	4	5	10	20
$f(x)$	3516	26.6	8.4	1.11	0.11

Horizontal asymptotes: $y = -3$, $y = 0$

Vertical asymptote: $x \approx 3.47$

49.

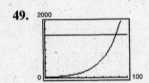

Intersection: $(86.350, 1500)$

51. $f(x) = x^2 e^{-x}$

(a)

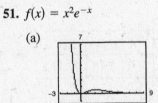

(b) Decreasing: $(-\infty, 0)$, $(2, \infty)$

   Increasing: $(0, 2)$

(c) Relative maximum: $(2, 4e^{-2}) \approx (2, 0.541)$

   Relative minimum: $(0, 0)$

53. $P = 2500, r = 2.5\% = 0.025, t = 10$

Compounded n times per year: $A = P\left(1 + \dfrac{r}{n}\right)^{nt} = 2500\left(1 + \dfrac{0.025}{n}\right)^{10n}$

Compounded continuously: $A = Pe^{rt} = 2500e^{(0.025)(10)}$

n	1	2	4	12	365	Continuous
A	3200.21	3205.09	3207.57	3209.23	3210.04	3210.06

55. $P = 2500, r = 4\% = 0.04, t = 20$

Compounded n times per year: $A = P\left(1 + \dfrac{r}{n}\right)^{nt} = 2500\left(1 + \dfrac{0.04}{n}\right)^{20n}$

Compounded continuously: $A = Pe^{rt} = 2500e^{(0.04)(20)}$

n	1	2	4	12	365	Continuous
A	5477.81	5520.10	5541.79	5556.46	5563.61	5563.85

57. $P = 12,000, r = 4\% = 0.04$

$A = Pe^{rt} = 12000e^{0.04t}$

t	1	10	20	30	40	50
A	12,489.73	17,901.90	26,706.49	39,841.40	59,436.39	88,668.67

59. $P = 12,000, r = 3.5\% = 0.035$

$A = Pe^{rt} = 12,000e^{0.035t}$

t	1	10	20	30	40	50
A	12,427.44	17,028.81	24,165.03	34,291.81	48,662.40	69,055.23

61. $A = 25\left[\dfrac{(1 + 0.12/12)^{48} - 1}{0.12/12}\right]$

$= 25\left[\dfrac{1.01^{48} - 1}{0.01}\right]$

$= \$1530.57$

63. $A = 200\left[\dfrac{(1 + 0.06/12)^{72} - 1}{0.06/12}\right]$

$= \$17,281.77$

65. $p = 5000\left(1 - \dfrac{4}{4 + e^{-0.002x}}\right)$

(a)
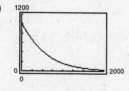

(b) If $x = 500, p \approx \$421.12$.

(c) For $x = 600, p \approx \$350.13$.

x	100	200	300	400	500	600	700
p	849.53	717.64	603.25	504.94	421.12	350.13	290.35

67. $Q = 25\left(\frac{1}{2}\right)^{t/1599}$

(a) When $t = 0$,

$$Q = 25\left(\frac{1}{2}\right)^{0/1599} = 25(1) = 25 \text{ grams.}$$

(b) When $t = 1000$,

$$Q = 25\left(\frac{1}{2}\right)^{1000/1599} \approx 16.21 \text{ grams.}$$

(c)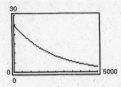

(d) Never. The graph has a horizontal asymptote $Q = 0$.

69. $P(t) = 100e^{0.2197t}$

(a)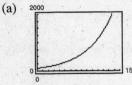

(b) $P(0) = 100$

$P(5) \approx 300$

$P(10) \approx 900$

(c) $P(0) = 100e^{0.2197(0)} = 100$

$P(5) = 100e^{0.2197(5)} = 299.966 \approx 300$

$P(10) = 100e^{0.2197(10)} = 899.798 \approx 900$

71. $C(t) = P(1.04)^t$

(a)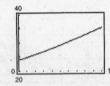

(b) $C(10) \approx 35.45$

(c) $C(10) = 23.95(1.04)^{10} \approx 35.45$

73. True. $f(x) = 1^x$ is not an exponential function.

75. The graph decreases for all x and has positive y-intercept. Matches (d).

77. $f(x) = \left(1 + \dfrac{0.5}{x}\right)^x$ and $g(x) = e^{0.5} \approx 1.6487$ (Horizontal line)

As $x \to \infty, f(x) \to g(x)$.

79. $e^\pi \approx 23.14, \ \pi^e \approx 22.46$

$e^\pi > \pi^e$

81. $5^{-3} = 0.008, \ 3^{-5} \approx 0.0041$

$5^{-3} > 3^{-5}$

83. f has an inverse because f is one-to-one.

$$y = 5x - 7$$
$$x = 5y - 7$$
$$x + 7 = 5y$$
$$f^{-1}(x) = \tfrac{1}{5}(x + 7)$$

85. f has an inverse because f is one-to-one.

$$y = \sqrt[3]{x + 8}$$
$$x = \sqrt[3]{y + 8}$$
$$x^3 = y + 8$$
$$x^3 - 8 = y$$
$$f^{-1}(x) = x^3 - 8$$

87. $f(x) = \dfrac{2x}{x-7}$

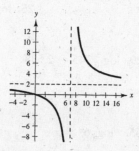

Vertical asymptote: $x = 7$

Horizontal asymptote: $y = 2$

Intercept: $(0, 0)$

89. Answers will vary.

Section 3.2   Logarithmic Functions and Their Graphs

■ You should know that a function of the form $y = \log_a x$, where $a > 0$, $a \neq 1$, and $x > 0$, is called a logarithm of x to base a.

■ You should be able to convert from logarithmic form to exponential form and vice versa.

$$y = \log_a x \iff a^y = x$$

■ You should know the following properties of logarithms.

(a) $\log_a 1 = 0$ since $a^0 = 1$.   (c) $\log_a a^x = x$ since $a^x = a^x$.

(b) $\log_a a = 1$ since $a^1 = a$.   (d) If $\log_a x = \log_a y$, then $x = y$.

■ You should know the definition of the natural logarithmic function.

$$\log_e x = \ln x, \, x > 0$$

■ You should know the properties of the natural logarithmic function.

(a) $\ln 1 = 0$ since $e^0 = 1$.   (c) $\ln e^x = x$ since $e^x = e^x$.

(b) $\ln e = 1$ since $e^1 = e$.   (d) If $\ln x = \ln y$, then $x = y$.

■ You should be able to graph logarithmic functions.

Vocabulary Check

1. logarithmic function   **2.** 10   **3.** natural logarithmic

4. $a^{\log_a x} = x$ **5.** $x = y$

---

1. $\log_4 64 = 3 \implies 4^3 = 64$ **3.** $\log_7 \frac{1}{49} = -2 \implies 7^{-2} = \frac{1}{49}$ **5.** $\log_{32} 4 = \frac{2}{5} \implies 32^{2/5} = 4$

7. $\ln 1 = 0 \implies e^0 = 1$ **9.** $\ln e = 1 \implies e^1 = e$ **11.** $\ln \sqrt{e} = \frac{1}{2} \implies e^{1/2} = \sqrt{e}$

13. $5^3 = 125 \implies \log_5 125 = 3$ **15.** $81^{1/4} = 3 \implies \log_{81} 3 = \frac{1}{4}$ **17.** $6^{-2} = \frac{1}{36} \implies \log_6 \frac{1}{36} = -2$

19. $e^3 = 20.0855\ldots \implies \ln 20.0855\ldots = 3$ **21.** $e^{1.3} = 3.6692\ldots \implies \ln 3.6692\ldots = 1.3$

23. $\sqrt[3]{e} = 1.3956\ldots \Rightarrow \ln(1.3956\ldots) = \frac{1}{3}$

25. $\log_2 16 = \log_2 2^4 = 4$

27. $g\left(\dfrac{1}{1000}\right) = \log_{10}\left(\dfrac{1}{1000}\right)$
$= \log_{10}(10^{-3}) = -3$

29. $\log_{10} 345 \approx 2.538$

31. $6 \log_{10} 14.8 \approx 7.022$

33. $\log_7 x = \log_7 9$
$x = 9$

35. $\log_6 6^2 = x$
$2 \log_6 6 = x$
$2 = x$

37. $\log_8 x = \log_8 10^{-1}$
$x = 10^{-1} = \frac{1}{10}$

39. $\log_4 4^{3x} = (3x)\log_4 4 = 3x$

41. $3 \log_2\left(\frac{1}{2}\right) = 3 \log_2(2^{-1}) = 3(-1) = -3$

43. $f(x) = 3^x$ and $g(x) = \log_3 x$ are inverses of each other.

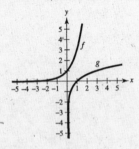

45. $f(x) = e^{2x}$ and $g(x) = \frac{1}{2}\ln x$ are inverses of each other.

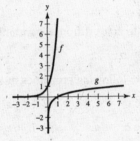

47. $y = \log_2(x + 2)$

Domain: $x + 2 > 0 \Rightarrow x > -2$
Vertical asymptote: $x = -2$
$\log_2(x + 2) = 0$
$\quad x + 2 = 1$
$\qquad\quad x = -1$
x-intercept: $(-1, 0)$

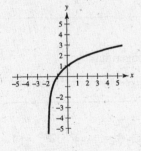

49. $y = 1 + \log_2 x$

Domain: $x > 0$
Vertical asymptote: $x = 0$
$1 + \log_2 x = 0$
$\quad \log_2 x = -1$
$\qquad\quad x = 2^{-1} = \frac{1}{2}$
x-intercept: $\left(\frac{1}{2}, 0\right)$

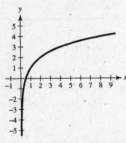

51. $y = 1 + \log_2(x - 2)$

Domain: $x - 2 > 0 \Rightarrow x > 2$
Vertical asymptote: $x = 2$
$1 + \log_2(x - 2) = 0$
$\quad \log_2(x - 2) = -1$
$\qquad x - 2 = 2^{-1} = \frac{1}{2}$
$\qquad\qquad x = \frac{5}{2}$
x-intercept: $\left(\frac{5}{2}, 0\right)$

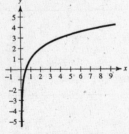

53. $f(x) = \log_3 x + 2$

Asymptote: $x = 0$
Point on graph: $(1, 2)$
Matches graph (b).

55. $f(x) = -\log_3(x + 2)$

Asymptote: $x = -2$
Point on graph: $(-1, 0)$
Matches graph (d).

57. $f(x) = \log_{10} x$

$g(x) = -\log_{10} x$ is a reflection in the x-axis of the graph of f.

59. $f(x) = \log_2 x$

$g(x) = 4 - \log_2 x$ is obtained from f by a reflection in the x-axis followed by a vertical shift four units upward.

61. Horizontal shift three units to the left and a vertical shift two units downward

63. $\ln \sqrt{42} \approx 1.869$

65. $-\ln\left(\frac{1}{2}\right) \approx 0.693$

67. $\ln e^2 = 2$

(Inverse Property)

69. $e^{\ln 1.8} = 1.8$

(Inverse Property)

71. $f(x) = \ln(x - 1)$

Domain: $x > 1$

Vertical asymptote: $x = 1$

x-intercept: $(2, 0)$

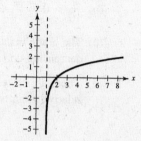

73. $g(x) = \ln(-x)$

Domain: $-x > 0 \implies x < 0$

The domain is $(-\infty, 0)$.

Vertical asymptote: $-x = 0 \implies x = 0$

x-intercept: $0 = \ln(-x)$

$e^0 = -x$

$-1 = x$

The x-intercept is $(-1, 0)$.

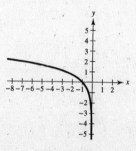

75. $g(x) = \ln(x + 3)$ is a horizontal shift three units to the left.

77. $g(x) = \ln x - 5$ is a vertical shift five units downward.

79. $g(x) = \ln(x - 1) + 2$ is a horizontal shift one unit to the right and a vertical shift two units upward.

81. $f(x) = \dfrac{x}{2} - \ln \dfrac{x}{4}$

(a)
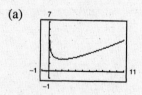

(b) Domain: $(0, \infty)$

(c) Increasing on $(2, \infty)$

Decreasing on $(0, 2)$

(d) Relative minimum: $(2, 1.693)$

83. $h(x) = 4x \ln x$

(a)
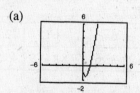

(b) Domain: $(0, \infty)$

(c) Increasing on $(0.368, \infty)$

Decreasing on $(0, 0.368)$

(d) Relative minimum: $(0.368, -1.472)$

85. $f(x) = \ln\left(\dfrac{x+2}{x-1}\right)$

(a)
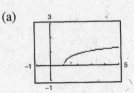

(b) $\dfrac{x+2}{x-1} > 0$; Critical numbers: $1, -2$

Test intervals: $(-\infty, -2), (-2, 1), (1, \infty)$

Testing these three intervals, we see that the domain is $(-\infty, -2) \cup (1, \infty)$.

(c) The graph is decreasing on $(-\infty, -2)$ and decreasing on $(1, \infty)$.

(d) There are no relative maximum or minimum values.

87. $f(x) = \ln\left(\dfrac{x^2}{10}\right)$

(a)
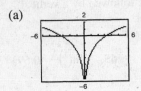

(b) $\dfrac{x^2}{10} > 0 \implies x \neq 0$; Domain: all $x \neq 0$

(c) The graph is increasing on $(0, \infty)$ and decreasing on $(-\infty, 0)$.

(d) There are no relative maximum or relative minimum values.

89. $f(x) = \sqrt{\ln x}$

(a)
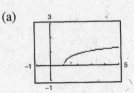

(b) $\ln x \geq 0 \implies x \geq 1$; Domain: $x \geq 1$

(c) The graph is increasing on $(1, \infty)$.

(d) There are no relative maximum or relative minimum values.

91. $f(t) = 80 - 17 \log_{10}(t+1), \quad 0 \leq t \leq 12$

(a) $f(0) = 80 - 17 \log_{10}(0+1) = 80$

(b) $f(4) = 80 - 17 \log_{10}(4+1) \approx 68.1$

(c) $f(10) = 80 - 17 \log_{10}(10+1) \approx 62.3$

(d)
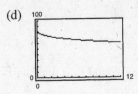

93. $t = \dfrac{\ln K}{0.055}$

(a)

K	1	2	4	6	8	10	12
t	0	12.6	25.2	32.6	37.8	41.9	45.2

As the amount increases, the time increases, but at a lesser rate.

(b)
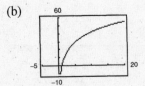

95. $\beta = 10 \log_{10}\left(\dfrac{I}{10^{-12}}\right)$

(a) $I = 1$: $\beta = 10 \log_{10}\left(\dfrac{1}{10^{-12}}\right) = 10 \cdot \log_{10}(10^{12}) = 10(12) = 120$ decibels

(b) $I = 10^{-2}$: $\beta = 10 \log_{10}\left(\dfrac{10^{-2}}{10^{-12}}\right) = 10 \log_{10}(10^{10}) = 10(10) = 100$ decibels

(c) No, this is a logarithmic scale.

97. $y = 80.4 - 11 \ln x$

$y(300) = 80.4 - 11 \ln 300 \approx 17.66 \text{ ft}^3/\text{min}$

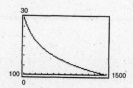

99. False. You would reflect $y = 6^x$ in the line $y = x$.

101. $5 = \log_b 32$

$b^5 = 32 = 2^5$

$b = 2$

103. $2 = \log_b\left(\frac{1}{16}\right)$

$b^2 = \frac{1}{16} = \left(\frac{1}{4}\right)^2$

$b = \frac{1}{4}$

105. The vertical asymptote is to the right of the y-axis, and the graph increases. Matches (b).

107. $f(x) = \log_a x$ is the inverse of $g(x) = a^x$, where $a > 0, a \neq 1$.

109. (a) False, y is not an exponential function of x. (y can never be 0.)

(b) True, y could be $\log_2 x$.

(c) True, x could be 2^y.

(d) False, y is not linear. (The points are not collinear.)

111. $f(x) = \dfrac{\ln x}{x}$

(a)

x	1	5	10	10^2	10^4	10^6
$f(x)$	0	0.322	0.230	0.046	0.00092	0.0000138

(b) As x increases without bound, $f(x)$ approaches 0.

(c)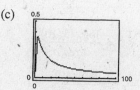

113. $x^2 + 2x - 3 = (x + 3)(x - 1)$

115. $12x^2 + 5x - 3 = (4x + 3)(3x - 1)$

117. $16x^2 - 25 = (4x + 5)(4x - 5)$

119. $2x^3 + x^2 - 45x = x(2x^2 + x - 45)$

$= x(2x - 9)(x + 5)$

121. $(f + g)(2) = f(2) + g(2)$

$= [3(2) + 2] + [2^3 - 1]$

$= 8 + 7 = 15$

123. $(fg)(6) = f(6)g(6)$

$= [3(6) + 2][6^3 - 1]$

$= [20][215] = 4300$

125. $5x - 7 = x + 4$

The graphs of $y = 5x - 7$ and $y_2 = x + 4$ intersect when $x = 2.75$ or $\frac{11}{4}$.

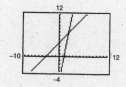

127. $\sqrt{3x - 2} = 9$

The graphs of $y_1 = \sqrt{3x - 2}$ and $y_2 = 9$ intersect when $x \approx 27.667$ or $\frac{83}{3}$.

Section 3.3   Properties of Logarithms

- ■ You should know the following properties of logarithms.

    (a) $\log_a x = \dfrac{\log_b x}{\log_b a}$

    (b) $\log_a(uv) = \log_a u + \log_a v$ $\qquad \ln(uv) = \ln u + \ln v$

    (c) $\log_a(u/v) = \log_a u - \log_a v$ $\qquad \ln(u/v) = \ln u - \ln v$

    (d) $\log_a u^n = n \log_a u$ $\qquad \ln u^n = n \ln u$

- ■ You should be able to rewrite logarithmic expressions using these properties.

Vocabulary Check

1. change-of-base

2. $\dfrac{\ln x}{\ln a}$

3. $\log_a u^n$

4. $\ln u + \ln v$

---

1. (a) $\log_5 x = \dfrac{\log_{10} x}{\log_{10} 5}$

  (b) $\log_5 x = \dfrac{\ln x}{\ln 5}$

3. (a) $\log_{1/5} x = \dfrac{\log_{10} x}{\log_{10} 1/5} = \dfrac{\log_{10} x}{-\log_{10} 5}$

  (b) $\log_{1/5} x = \dfrac{\ln x}{\ln 1/5} = \dfrac{\ln x}{-\ln 5}$

5. (a) $\log_a\left(\dfrac{3}{10}\right) = \dfrac{\log_{10}(3/10)}{\log_{10} a}$

  (b) $\log_a\left(\dfrac{3}{10}\right) = \dfrac{\ln(3/10)}{\ln a}$

7. (a) $\log_{2.6} x = \dfrac{\log_{10} x}{\log_{10} 2.6}$

  (b) $\log_{2.6} x = \dfrac{\ln x}{\ln 2.6}$

9. $\log_3 7 = \dfrac{\ln 7}{\ln 3} \approx 1.771$

11. $\log_{1/2} 4 = \dfrac{\ln 4}{\ln (1/2)} = -2$

13. $\log_9(0.8) = \dfrac{\ln(0.8)}{\ln 9} \approx -0.102$

15. $\log_{15} 1460 = \dfrac{\ln 1460}{\ln 15} \approx 2.691$

17. $\ln 20 = \ln(4 \cdot 5)$
$\qquad = \ln 4 + \ln 5$

19. $\ln \frac{5}{64} = \ln 5 - \ln 64$
$\qquad = \ln 5 - \ln 4^3$
$\qquad = \ln 5 - 3 \ln 4$

21. $\log_b 25 = \log_b 5^2$
$\qquad = 2 \log_b 5$
$\qquad \approx 2(0.8271) \approx 1.6542$

23. $\log_b \sqrt{3} = \frac{1}{2} \log_b 3 \approx \frac{1}{2}(0.5646) \approx 0.2823$

25. $f(x) = \log_3(x + 2) = \dfrac{\ln(x + 2)}{\ln 3}$

27. $f(x) = \log_{1/2}(x - 2) = \dfrac{\ln(x - 2)}{\ln(1/2)} = \dfrac{\ln(x - 2)}{-\ln 2}$

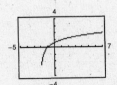

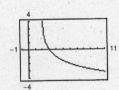

29. $f(x) = \log_{1/4}(x^2) = \dfrac{\ln x^2}{\ln(1/4)} = \dfrac{\ln x^2}{-\ln 4}$

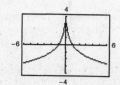

31. $\log_4 8 = \log_4 2^3 = 3 \log_4 2$

$\qquad = 3 \log_4 4^{1/2} = 3\left(\tfrac{1}{2}\right) \log_4 4$

$\qquad = \tfrac{3}{2}$

33. $\ln(5e^6) = \ln 5 + \ln e^6 = \ln 5 + 6 = 6 + \ln 5$

35. $\log_5 \tfrac{1}{250} = \log_5 1 - \log_5 250 = 0 - \log_5(125 \cdot 2)$

$\qquad = -\log_5(5^3 \cdot 2) = -[\log_5 5^3 + \log_5 2]$

$\qquad = -[3 \log_5 5 + \log_5 2] = -3 - \log_5 2$

37. $\log_{10} 5x = \log_{10} 5 + \log_{10} x$

39. $\log_{10} \dfrac{5}{x} = \log_{10} 5 - \log_{10} x$

41. $\log_8 x^4 = 4 \log_8 x$

43. $\ln \sqrt{z} = \ln z^{1/2} = \tfrac{1}{2} \ln z$

45. $\ln xyz = \ln x + \ln y + \ln z$

47. $\log_3(a^2bc^3) = \log_3 a^2 + \log_3 b + \log_3 c^3$

$\qquad = 2 \log_3 a + \log_3 b + 3 \log_3 c$

49. $\ln\left(a^2 \sqrt{a-1}\right) = \ln a^2 + \ln(a-1)^{1/2}$

$\qquad = 2 \ln a + \tfrac{1}{2} \ln(a-1),\, a > 1$

51. $\ln \sqrt[3]{\dfrac{x}{y}} = \tfrac{1}{3} \ln \dfrac{x}{y}$

$\qquad = \tfrac{1}{3}[\ln x - \ln y]$

$\qquad = \tfrac{1}{3} \ln x - \tfrac{1}{3} \ln y$

53. $\ln\left(\dfrac{x^2 - 1}{x^3}\right) = \ln(x^2 - 1) - \ln x^3$

$\qquad = \ln[(x-1)(x+1)] - 3 \ln x$

$\qquad = \ln(x-1) + \ln(x+1) - 3 \ln x,\, x > 1$

55. $\ln\left(\dfrac{x^4 \sqrt{y}}{z^5}\right) = \ln x^4 \sqrt{y} - \ln z^5$

$\qquad = \ln x^4 + \ln \sqrt{y} - \ln z^5$

$\qquad = 4 \ln x + \tfrac{1}{2} \ln y - 5 \ln z$

57. $y_1 = \ln[x^3(x+4)]$

$\quad y_2 = 3 \ln x + \ln(x+4)$

(a)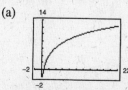

(b)

x	0.5	1	1.5	2	3	10
y_1	−0.5754	1.6094	2.9211	3.8712	5.2417	9.5468
y_2	−0.5754	1.6094	2.9211	3.8712	5.2417	9.5468

(c) The graphs and table suggest that

$\quad y_1 = y_2$ for $x > 0$. In fact,

$\quad y_1 = \ln[x^3(x+4)] = \ln x^3 + \ln(x+4)$

$\qquad = 3 \ln x + \ln(x+4) = y_2.$

59. $\ln x + \ln 4 = \ln 4x$

61. $\log_4 z - \log_4 y = \log_4 \dfrac{z}{y}$

63. $2 \log_2(x + 3) = \log_2(x + 3)^2$

65. $\frac{1}{2} \ln(x^2 + 4) = \ln(x^2 + 4)^{1/2}$
$$= \ln \sqrt{x^2 + 4}$$

67. $\ln x - 3 \ln(x + 1) = \ln x - \ln(x + 1)^3$
$$= \ln \frac{x}{(x + 1)^3}$$

69. $\ln(x - 2) - \ln(x + 2) = \ln\left(\dfrac{x - 2}{x + 2}\right)$

71. $\ln x - 2[\ln(x + 2) + \ln(x - 2)] = \ln x - 2 \ln[(x + 2)(x - 2)]$
$$= \ln x - 2 \ln(x^2 - 4)$$
$$= \ln x - \ln(x^2 - 4)^2$$
$$= \ln \frac{x}{(x^2 - 4)^2}$$

73. $\frac{1}{3}[2 \ln(x + 3) + \ln x - \ln(x^2 - 1)] = \frac{1}{3}[\ln(x + 3)^2 + \ln x - \ln(x^2 - 1)]$
$$= \frac{1}{3}[\ln[x(x + 3)^2] - \ln(x^2 - 1)]$$
$$= \frac{1}{3} \ln \frac{x(x + 3)^2}{x^2 - 1} = \ln \sqrt[3]{\frac{x(x + 3)^2}{x^2 - 1}}$$

75. $\frac{1}{3}[\ln y + 2 \ln(y + 4)] - \ln(y - 1) = \frac{1}{3}[\ln y + \ln(y + 4)^2] - \ln(y - 1)$
$$= \frac{1}{3} \ln[y(y + 4)^2] - \ln(y - 1)$$
$$= \ln \sqrt[3]{y(y + 4)^2} - \ln(y - 1)$$
$$= \ln \frac{\sqrt[3]{y(y + 4)^2}}{y - 1}$$

77. $y_1 = 2[\ln 8 - \ln(x^2 + 1)]$

$y_2 = \ln\left[\dfrac{64}{(x^2 + 1)^2}\right]$

(a)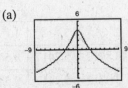

(c) The graphs and table suggest that $y_1 = y_2$. In fact,
$$y_1 = 2[\ln 8 - \ln(x^2 + 1)]$$
$$= 2 \ln \frac{8}{x^2 + 1} = \ln \frac{64}{(x^2 + 1)^2} = y_2.$$

(b)

x	-8	-4	-2	0	2	4	8
y_1	-4.1899	-1.5075	0.9400	4.1589	0.9400	-1.5075	-4.1899
y_2	-4.1899	-1.5075	0.9400	4.1589	0.9400	-1.5075	-4.1899

79. $y_1 = \ln x^2$

$y_2 = 2 \ln x$

(a)

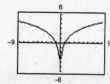

(The domain of y_2 is $x > 0$.)

(b)

x	-8	-4	1	2	4
y_1	4.1589	2.7726	0	1.3863	2.7726
y_2	undefined	undefined	0	1.3863	2.7726

(c) The graphs and table suggest that $y_1 = y_2$ for $x > 0$. The functions are not equivalent because the domains are different.

81. $\log_3 9 = 2 \log_3 3 = 2$

83. $\log_4 16^{3.4} = 3.4 \log_4(4^2) = 6.8 \log_4 4 = 6.8$

85. $\log_2(-4)$ is undefined. -4 is not in the domain of $f(x) = \log_2 x$.

87. $\log_5 75 - \log_5 3 = \log_5 \frac{75}{3} = \log_5 25 = \log_5 5^2 = 2$

89. $\ln e^3 - \ln e^7 = 3 - 7 = -4$

91. $2 \ln e^4 = 2(4) \ln e = 8$

93. $\ln\left(\frac{1}{\sqrt{e}}\right) = \ln(1) - \ln e^{1/2} = 0 - \frac{1}{2} \ln e = -\frac{1}{2}$

95. (a) $\beta = 10 \cdot \log_{10}\left(\frac{I}{10^{-12}}\right) = 10[\log_{10} I - \log_{10} 10^{-12}]$

$= 10[\log_{10} I - (-12) \log_{10} 10]$

$= 10[\log_{10} I + 12] = 120 + 10 \cdot \log_{10} I$

(b)

I	10^{-4}	10^{-6}	10^{-8}	10^{-10}	10^{-12}	10^{-14}
β	80	60	40	20	0	-20

(c) $\beta(10^{-4}) = 120 + 10 \cdot \log_{10} 10^{-4} = 120 - 40 = 80$

$\beta(10^{-6}) = 120 + 10 \cdot \log_{10} 10^{-6} = 120 - 60 = 60$

$\beta(10^{-8}) = 120 + 10 \cdot \log_{10} 10^{-8} = 120 - 80 = 40$

$\beta(10^{-10}) = 120 + 10 \cdot \log_{10} 10^{-10} = 120 - 100 = 20$

$\beta(10^{-12}) = 120 + 10 \cdot \log_{10} 10^{-12} = 120 - 120 = 0$

$\beta(10^{-14}) = 120 + 10 \cdot \log_{10} 10^{-14} = 120 - 140 = -20$

97. (a)

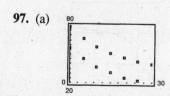

(b) $T - 21 = 54.4(0.964)^t$

$T = 21 + 54.4(0.964)^t$

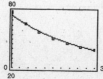

The data $(t, T - 21)$ fits the model $T - 21 = 54.4(0.964)^t$.

The model $T = 21 + 54.4(0.964)^t$ fits the original data.

—CONTINUED—

97. **—CONTINUED—**

(c) $\ln(T - 21) = -0.0372t + 3.9971$, linear model

$$T - 21 = e^{-0.0372t + 3.9971}$$

$$T = 21 + 54.4e^{-0.0372t}$$

$$= 21 + 54.4(0.964)^t$$

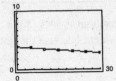

(d)

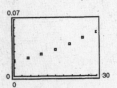

$$\frac{1}{T - 21} = 0.00121t + 0.01615, \quad \text{linear model}$$

$$T - 21 = \frac{1}{0.00121t + 0.01615}$$

$$T = 21 + \frac{1}{0.00121t + 0.01615}$$

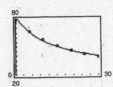

99. True

101. False. For example, let $x = 1$ and $a = 2$.

Then $f\left(\dfrac{x}{a}\right) = \ln\left(\dfrac{1}{2}\right)$. But $\dfrac{f(x)}{f(a)} = \dfrac{\ln 1}{\ln 2} = 0$.

103. False. $\sqrt{\ln x} \neq \frac{1}{2}\ln x$

In fact, $\ln x^{1/2} = \frac{1}{2}\ln x$.

105. True. In fact, if $\ln x < 0$, then $0 < x < 1$.

107. Let $y = \log_a x$ and $z = \log_{a/b} x$, then $a^y = x = \left(\dfrac{a}{b}\right)^z$ and

$$\left(\frac{1}{b}\right)^z = a^{y-z}$$

$$\frac{1}{b} = a^{(y-z)/z}$$

$$\log_a\left(\frac{1}{b}\right) = \frac{y - z}{z} = \frac{y}{z} - 1 \implies 1 + \log_a\left(\frac{1}{b}\right) = \frac{\log_a x}{\log_{a/b} x}.$$

109. $f(x) = \log_2 x = \dfrac{\ln x}{\ln 2}$

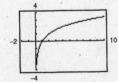

111. $f(x) = \log_3 \sqrt{x} = \dfrac{1}{2}\dfrac{\ln x}{\ln 3}$

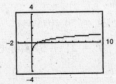

113. $f(x) = \log_5\left(\dfrac{x}{3}\right) = \dfrac{\ln(x/3)}{\ln 5}$

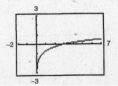

115. $\ln 1 = 0, \ln 2 \approx 0.6931, \ln 3 \approx 1.0986, \ln 5 \approx 1.6094$

$\ln 2 \approx 0.6931$

$\ln 3 \approx 1.0986$

$\ln 4 = \ln 2 + \ln 2 \approx 0.6931 + 0.6931 = 1.3862$

$\ln 5 \approx 1.6094$

$\ln 6 = \ln 2 + \ln 3 \approx 0.6931 + 1.0986 = 1.7917$

$\ln 8 = \ln 2^3 = 3 \ln 2 \approx 3(0.6931) = 2.0793$

$\ln 9 = \ln 3^2 = 2 \ln 3 \approx 2(1.0986) = 2.1972$

$\ln 10 = \ln 5 + \ln 2 \approx 1.6094 + 0.6931 = 2.3025$

$\ln 12 = \ln 2^2 + \ln 3 = 2 \ln 2 + \ln 3 \approx 2(0.6931) + 1.0986 = 2.4848$

$\ln 15 = \ln 5 + \ln 3 \approx 1.6094 + 1.0986 = 2.7080$

$\ln 16 = \ln 2^4 = 4 \ln 2 \approx 4(0.6931) = 2.7724$

$\ln 18 = \ln 3^2 + \ln 2 = 2 \ln 3 + \ln 2 \approx 2(1.0986) + 0.6931 = 2.8903$

$\ln 20 = \ln 5 + \ln 2^2 = \ln 5 + 2 \ln 2 \approx 1.6094 + 2(0.6931) = 2.9956$

117. $\left(\dfrac{2x^2}{3y}\right)^{-3} = \left(\dfrac{3y}{2x^2}\right)^3$

$= \dfrac{(3y)^3}{(2x^2)^3}$

$= \dfrac{27y^3}{8x^6}$

119. $xy(x^{-1} + y^{-1})^{-1} = \dfrac{xy}{x^{-1} + y^{-1}}$

$= \dfrac{xy}{\dfrac{1}{x} + \dfrac{1}{y}}$

$= \dfrac{xy}{\dfrac{y + x}{xy}}$

$= \dfrac{(xy)^2}{x + y}, x \neq 0, y \neq 0$

121. $2x^3 + 20x^2 + 50x = 0$

$2x(x^2 + 10x + 25) = 0$

$2x(x + 5)^2 = 0$

$x = 0, -5, -5$

123. $9x^4 - 37x^2 + 4 = 0$

$(x^2 - 4)(9x^2 - 1) = 0$

$(x - 2)(x + 2)(3x - 1)(3x + 1) = 0$

$x = \pm 2, \pm\frac{1}{3}$

125. $9x^4 - 226x^2 + 25 = 0$

$(x^2 - 25)(9x^2 - 1) = 0$

$(x - 5)(x + 5)(3x + 1)(3x - 1) = 0$

$x = \pm 5, \pm\frac{1}{3}$

Section 3.4   Solving Exponential and Logarithmic Equations

■   To solve an exponential equation, isolate the exponential expression, then take the logarithm of both sides. Then solve for the variable.

   1. $\log_a a^x = x$

   2. $\ln e^x = x$

■   To solve a logarithmic equation, rewrite it in exponential form. Then solve for the variable.

   1. $a^{\log_a x} = x$

   2. $e^{\ln x} = x$

■   If $a > 0$ and $a \neq 1$ we have the following:

   1. $\log_a x = \log_a y \implies x = y$

   2. $a^x = a^y \implies x = y$

■   Use your graphing utility to approximate solutions.

Vocabulary Check

1. solve

2. (a) $x = y$ (b) $x = y$ (c) x (d) x

3. extraneous

---

1. $4^{2x-7} = 64$

   (a) $x = 5$

     $4^{2(5)-7} = 4^3 = 64$

     Yes, $x = 5$ is a solution.

   (b) $x = 2$

     $4^{2(2)-7} = 4^{-3} = \frac{1}{64} \neq 64$

     No, $x = 2$ is not a solution.

3. $3e^{x+2} = 75$

   (a) $x = -2 + e^{25}$

     $3e^{(-2+e^{25})+2} = 3e^{e^{25}} \neq 75$

     No, $x = -2 + e^{25}$ is not a solution.

   (b) $x = -2 + \ln 25$

     $3e^{(-2+\ln 25)+2} = 3e^{\ln 25} = 3(25) = 75$

     Yes, $x = -2 + \ln 25$ is a solution.

   (c) $x \approx 1.2189$

     $3e^{1.2189+2} = 3e^{3.2189} \approx 75$

     Yes, $x \approx 1.2189$ is a solution.

5. $\log_4(3x) = 3$

     $4^3 = 3x$

     $x = \frac{64}{3} \approx 21.333$

   (a) $x = 21.3560$ is an approximate solution.

   (b) No, $x = -4$ is not a solution.

   (c) Yes, $x = \frac{64}{3}$ is a solution.

7. $\ln(x - 1) = 3.8$

    (a) $x = 1 + e^{3.8}$

       $\ln(1 + e^{3.8} - 1) = \ln e^{3.8} = 3.8$

       Yes, $x = 1 + e^{3.8}$ is a solution.

    (b) $x \approx 45.7012$

       $\ln(45.7012 - 1) = \ln(44.7012) \approx 3.8$

       Yes, $x \approx 45.7012$ is a solution.

    (c) $x = 1 + \ln 3.8$

       $\ln(1 + \ln 3.8 - 1) = \ln(\ln 3.8) \approx 0.289$

       No, $x = 1 + \ln 3.8$ is not a solution.

9.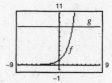

Point of intersection: $(3, 8)$

Algebraically: $2^x = 8$

             $2^x = 2^3$

           $x = 3 \Rightarrow y = 8 \Rightarrow (3, 8)$

11. Point of intersection: $(4, 10)$

Algebraically: $5^{x-2} - 15 = 10$

             $5^{x-2} = 25 = 5^2$

             $x - 2 = 2$

                 $x = 4$

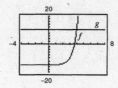

$(4, 10)$

13.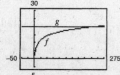

Point of intersection: $(243, 20)$

Algebraically: $4 \log_3 x = 20$

             $\log_3 x = 5$

               $x = 3^5 = 243$

$(243, 20)$

15.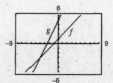

Point of intersection: $(-4, -3)$

Algebraically: $\ln e^{x+1} = 2x + 5$

              $x + 1 = 2x + 5$

                 $-4 = x$

$(-4, -3)$

17. $4^x = 16$

    $4^x = 4^2$

     $x = 2$

19. $5^x = \dfrac{1}{625}$

    $5^x = \dfrac{1}{5^4} = 5^{-4}$

      $x = -4$

21. $\left(\dfrac{1}{8}\right)^x = 64$

    $8^{-x} = 8^2$

    $-x = 2$

      $x = -2$

23. $\left(\dfrac{2}{3}\right)^x = \dfrac{81}{16}$

    $\left(\dfrac{3}{2}\right)^{-x} = \left(\dfrac{3}{2}\right)^4$

      $-x = 4$

        $x = -4$

25. $6(10^x) = 216$

       $10^x = 36$

  $\log_{10} 10^x = \log_{10} 36$

         $x = \log_{10} 36 \approx 1.5563$

27. $2^{x+3} = 256$

    $2^x \cdot 2^3 = 256$

       $2^x = 32$

        $x = 5$

Alternate solution:

    $2^{x+3} = 2^8$

    $x + 3 = 8$

       $x = 5$

29. $\ln x - \ln 5 = 0$

$\ln x = \ln 5$

$x = 5$

31. $\ln x = -7$

$x = e^{-7}$

33. $\log_x 625 = 4$

$x^4 = 625$

$x^4 = 5^4$

$x = 5$

35. $\log_{10} x = -1$

$x = 10^{-1}$

$x = \frac{1}{10}$

37. $\ln(2x - 1) = 5$

$2x - 1 = e^5$

$x = \dfrac{1 + e^5}{2} \approx 74.707$

39. $\ln e^{x^2} = x^2 \ln e^x = x^2$

41. $e^{\ln(5x+2)} = 5x + 2$

43. $-1 + \ln e^{2x} = -1 + 2x = 2x - 1$

45. $8^{3x} = 360$

$\ln 8^{3x} = \ln 360$

$3x \ln 8 = \ln 360$

$3x = \dfrac{\ln 360}{\ln 8}$

$x = \dfrac{1}{3}\dfrac{\ln 360}{\ln 8}$

$x \approx 0.944$

47. $5^{-t/2} = 0.20 = \dfrac{1}{5}$

$-\dfrac{t}{2} \ln 5 = \ln\left(\dfrac{1}{5}\right)$

$-\dfrac{t}{2} \ln 5 = -\ln 5$

$\dfrac{t}{2} = 1$

$t = 2$

49. $5(2^{3-x}) - 13 = 100$

$5(2^{3-x}) = 113$

$2^{3-x} = \dfrac{113}{5}$

$\ln 2^{3-x} = \ln\left(\dfrac{113}{5}\right)$

$3 - x = \dfrac{\ln(113/5)}{\ln 2}$

$x = 3 - \dfrac{\ln(113/5)}{\ln 2}$

$x \approx -1.498$

51. $\left(1 + \dfrac{0.10}{12}\right)^{12t} = 2$

$\left(\dfrac{12.1}{12}\right)^{12t} = 2$

$(12t) \ln\left(\dfrac{12.1}{12}\right) = \ln 2$

$t = \dfrac{1}{12}\dfrac{\ln 2}{\ln(12.1/12)}$

$t \approx 6.960$

53. $5000\left[\dfrac{(1 + 0.005)^x}{0.005}\right] = 250{,}000$

$5000(1.005)^x = 1250$

$1.005^x = 0.25$

$x \ln(1.005) = \ln 0.25$

$x = \dfrac{\ln 0.25}{\ln(1.005)}$

$x \approx -277.951$

55. $2e^{5x} = 18$

$e^{5x} = 9$

$5x = \ln 9$

$x = \tfrac{1}{5} \ln 9$

$x \approx 0.439$

57. $500e^{-x} = 300$

$e^{-x} = \tfrac{3}{5}$

$-x = \ln \tfrac{3}{5}$

$x = -\ln \tfrac{3}{5} = \ln \tfrac{5}{3} \approx 0.511$

59. $7 - 2e^x = 5$

$-2e^x = -2$

$e^x = 1$

$x = \ln 1 = 0$

61. $e^{2x} - 4e^x - 5 = 0$

$(e^x - 5)(e^x + 1) = 0$

$e^x = 5 \text{ or } e^x = -1$

$x = \ln 5 \approx 1.609$

$(e^x = -1 \text{ is impossible.})$

63. $250e^{0.02x} = 10{,}000$

$e^{0.02x} = 40$

$0.02x = \ln 40$

$x = \dfrac{\ln 40}{0.02}$

$x \approx 184.444$

65. $e^x = e^{x^2 - 2}$

$x = x^2 - 2$

$x^2 - x - 2 = 0$

$(x - 2)(x + 1) = 0$

$x = 2, -1$

67. $e^{x^2 - 3x} = e^{x - 2}$

$x^2 - 3x = x - 2$

$x^2 - 4x + 2 = 0$

$x = \dfrac{4 \pm \sqrt{16 - 8}}{2}$

$x = 2 \pm \sqrt{2}$

$x \approx 3.414, 0.586$

69. $\dfrac{400}{1 + e^{-x}} = 350$

$1 + e^{-x} = \dfrac{400}{350} = \dfrac{8}{7}$

$e^{-x} = \dfrac{1}{7}$

$-x = \ln\left(\dfrac{1}{7}\right) = -\ln 7$

$x = \ln 7 \approx 1.946$

71. $\dfrac{40}{1 - 5e^{-0.01x}} = 200$

$1 - 5e^{-0.01x} = \dfrac{40}{200} = \dfrac{1}{5}$

$5e^{-0.01x} = \dfrac{4}{5}$

$e^{-0.01x} = \dfrac{4}{25}$

$-0.01x = \ln\left(\dfrac{4}{25}\right)$

$x = \dfrac{\ln(4/25)}{-0.01}$

$x \approx 183.258$

73. $e^{3x} = 12$

x	0.6	0.7	0.8	0.9	1.0
$f(x)$	6.05	8.17	11.02	14.88	20.09

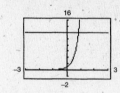

$x \approx 0.828$

75. $20(100 - e^{x/2}) = 500$

x	5	6	7	8	9
$f(x)$	1756	1598	1338	908	200

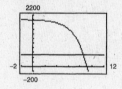

$x \approx 8.635$

77. $\left(1 + \dfrac{0.065}{365}\right)^{365t} = 4 \Rightarrow t = 21.330$

79. $\dfrac{3000}{2 + e^{2x}} = 2$

The zero of $y = \dfrac{3000}{2 + e^{2x}} - 2$ is $x \approx 3.656$.

81. $g(x) = 6e^{1-x} - 25$

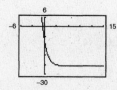

Zero at $x = -0.427$

83. $g(t) = e^{0.09t} - 3$

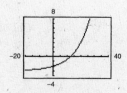

Zero at $t = 12.207$

85. $\ln x = -3$

$x = e^{-3} \approx 0.050$

87. $\ln 4x = 2.1$

$\quad 4x = e^{2.1}$

$\quad\quad x = \frac{1}{4}e^{2.1}$

$\quad\quad\quad \approx 2.042$

89. $-2 + 2\ln 3x = 17$

$\quad\quad 2\ln 3x = 19$

$\quad\quad\quad \ln 3x = \frac{19}{2}$

$\quad\quad\quad\quad 3x = e^{19/2}$

$\quad\quad\quad\quad\quad x = \frac{1}{3}e^{19/2}$

$\quad\quad\quad\quad\quad x \approx 4453.242$

91. $\log_5(3x + 2) = \log_5(6 - x)$

$\quad\quad 3x + 2 = 6 - x$

$\quad\quad\quad 4x = 4$

$\quad\quad\quad\quad x = 1$

93. $\log_{10}(z - 3) = 2$

$\quad\quad z - 3 = 10^2$

$\quad\quad\quad z = 10^2 + 3$

$\quad\quad\quad\quad = 103$

95. $7\log_4(0.6x) = 12$

$\quad\quad \log_4(0.6x) = \frac{12}{7}$

$\quad\quad 4^{12/7} = 0.6x = \frac{3}{5}x$

$\quad\quad\quad\quad x = \frac{5}{3}4^{12/7}$

$\quad\quad\quad\quad\quad \approx 17.945$

97. $\ln\sqrt{x + 2} = 1$

$\quad\quad \sqrt{x + 2} = e^1$

$\quad\quad\quad x + 2 = e^2$

$\quad\quad\quad\quad x = e^2 - 2 \approx 5.389$

99. $\ln(x + 1)^2 = 2$

$\quad e^{\ln(x+1)^2} = e^2$

$\quad\quad (x + 1)^2 = e^2$

$\quad x + 1 = e$ or $x + 1 = -e$

$\quad x = e - 1 \approx 1.718$

or

$\quad x = -e - 1 \approx -3.718$

101. $\log_4 x - \log_4(x - 1) = \dfrac{1}{2}$

$\quad\quad \log_4\left(\dfrac{x}{x - 1}\right) = \dfrac{1}{2}$

$\quad\quad 4^{\log_4(x/x-1)} = 4^{1/2}$

$\quad\quad\quad \dfrac{x}{x - 1} = 2$

$\quad\quad\quad\quad x = 2(x - 1)$

$\quad\quad\quad\quad x = 2x - 2$

$\quad\quad\quad\quad 2 = x$

103. $\quad\ln(x + 5) = \ln(x - 1) - \ln(x + 1)$

$\quad\ln(x + 5) = \ln\left(\dfrac{x - 1}{x + 1}\right)$

$\quad\quad x + 5 = \dfrac{x - 1}{x + 1}$

$\quad (x + 5)(x + 1) = x - 1$

$\quad x^2 + 6x + 5 = x - 1$

$\quad x^2 + 5x + 6 = 0$

$\quad (x + 2)(x + 3) = 0$

$\quad x = -2$ or $x = -3$

Both of these solutions are extraneous, so the equation has no solution.

105. $\log_{10} 8x - \log_{10}\left(1 + \sqrt{x}\right) = 2$

$\quad\quad \log_{10}\dfrac{8x}{1 + \sqrt{x}} = 2$

$\quad\quad\quad \dfrac{8x}{1 + \sqrt{x}} = 10^2$

$\quad\quad\quad\quad 8x = 100 + 100\sqrt{x}$

$\quad 8x - 100\sqrt{x} - 100 = 0$

$\quad 2x - 25\sqrt{x} - 25 = 0$

$\quad \sqrt{x} = \dfrac{25 \pm \sqrt{25^2 - 4(2)(-25)}}{4}$

$\quad\quad = \dfrac{25 \pm 5\sqrt{33}}{4}$

Choosing the positive value, we have
$\sqrt{x} \approx 13.431$ and $x \approx 180.384$.

107. $\ln 2x = 2.4$

x	2	3	4	5	6
$f(x)$	1.39	1.79	2.08	2.30	2.48

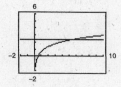

$x \approx 5.512$

109. $6 \log_3(0.5x) = 11$

x	12	13	14	15	16
$f(x)$	9.79	10.22	10.63	11.00	11.36

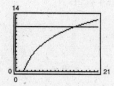

$x \approx 14.988$

111. $\log_{10} x = x^3 - 3$

Graphing $y = \log_{10} x - x^3 + 3$, you obtain two zeros, $x \approx 1.469$ and $x \approx 0.001$.

113. $\ln x + \ln(x - 2) = 1$

Graphing $y = \ln x + \ln(x - 2) - 1$, you obtain one zero, $x \approx 2.928$.

115. $\ln(x - 3) + \ln(x + 3) = 1$

Graphing $y = \ln(x - 3) + \ln(x + 3) - 1$, you obtain $x \approx 3.423$.

117. $y_1 = 7$

$y_2 = 2^{x-1} - 5$

Intersection: $(4.585, 7)$

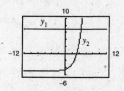

119. $y_1 = 80$

$y_2 = 4e^{-0.2x}$

Intersection: $(-14.979, 80)$

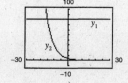

121. $y_1 = 3.25$

$y_2 = \frac{1}{2} \ln(x + 2)$

Intersection: $(663.142, 3.25)$

123. $2x^2 e^{2x} + 2xe^{2x} = 0$

$(2x^2 + 2x)e^{2x} = 0$

$2x^2 + 2x = 0 \quad$ (since $e^{2x} \neq 0$)

$2x(x + 1) = 0$

$x = 0, -1$

125. $-xe^{-x} + e^{-x} = 0$

$(-x + 1)e^{-x} = 0$

$-x + 1 = 0 \quad$ (since $e^{-x} \neq 0$)

$x = 1$

127. $2x \ln x + x = 0$

$x(2 \ln x + 1) = 0$

$2 \ln x + 1 = 0 \quad$ (since $x > 0$)

$\ln x = -\frac{1}{2}$

$x = e^{-1/2} \approx 0.607$

129. $\dfrac{1 + \ln x}{2} = 0$

$1 + \ln x = 0$

$\ln x = -1$

$x = e^{-1} = \dfrac{1}{e} \approx 0.368$

131. (a) $2000 = 1000e^{0.075t}$

$2 = e^{0.075t}$

$\ln 2 = 0.075t$

$t = \dfrac{\ln 2}{0.075} \approx 9.24$ years

(b) $3000 = 1000e^{0.075t}$

$3 = e^{0.075t}$

$\ln 3 = 0.075t$

$t = \dfrac{\ln 3}{0.075} \approx 14.65$ years

133. (a) $2000 = 1000e^{0.025t}$

$2 = e^{0.025t}$

$\ln 2 = 0.025t$

$t = \dfrac{\ln 2}{0.025} \approx 27.73$ years

(b) $3000 = 1000e^{0.025t}$

$3 = e^{0.025t}$

$\ln 3 = 0.025t$

$t = \dfrac{\ln 3}{0.025} \approx 43.94$ years

135. $p = 500 - 0.5(e^{0.004x})$

(a) $p = 350$

$350 = 500 - 0.5(e^{0.004x})$

$300 = e^{0.004x}$

$0.004x = \ln 300$

$x \approx 1426$ units

(b) $p = 300$

$300 = 500 - 0.5(e^{0.004x})$

$400 = e^{0.004x}$

$0.004x = \ln 400$

$x \approx 1498$ units

137. $7247 - 596.5 \ln t = 5800$

$-596.5 \ln t = -1447$

$\ln t \approx 2.4258$

$t \approx 11.3$, or 2001

139. (a)

(b) From the graph we see horizontal asymptotes at $y = 0$ and $y = 100$. These represent the lower and upper percent bounds.

(c) Males: $50 = \dfrac{100}{1 + e^{-0.6114(x - 69.71)}}$

$1 + e^{-0.6114(x - 69.71)} = 2$

$e^{-0.6114(x - 69.71)} = 1$

$-0.6114(x - 69.71) = \ln 1$

$-0.6114(x - 69.71) = 0$

$x = 69.71$ inches

Females: $50 = \dfrac{100}{1 + e^{-0.66607(x - 64.51)}}$

$1 + e^{-0.66607(x - 64.51)} = 2$

$e^{-0.66607(x - 64.51)} = 1$

$-0.66607(x - 64.51) = \ln 1$

$-0.66607(x - 64.51) = 0$

$x = 64.51$ inches

141. $T = 20[1 + 7(2^{-h})]$

(a)

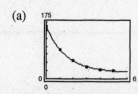

(b) We see a horizontal asymptote at $y = 20$.
This represents the room temperature.

(c) $$100 = 20[1 + 7(2^{-h})]$$
$$5 = 1 + 7(2^{-h})$$
$$4 = 7(2^{-h})$$
$$\frac{4}{7} = 2^{-h}$$
$$\ln\left(\frac{4}{7}\right) = \ln 2^{-h}$$
$$\ln\left(\frac{4}{7}\right) = -h\ln 2$$
$$\frac{\ln(4/7)}{-\ln 2} = h$$
$$h \approx 0.81 \text{ hour}$$

143. False. The equation $e^x = 0$ has no solutions.

145. Answers will vary.

147. Yes. The doubling time is given by
$$2P = Pe^{rt}$$
$$2 = e^{rt}$$
$$\ln 2 = rt$$
$$t = \frac{\ln 2}{r}.$$

The time to quadruple is given by
$$4P = Pe^{rt}$$
$$4 = e^{rt}$$
$$\ln 4 = rt$$
$$t = \frac{\ln 4}{r} = \frac{\ln 2^2}{r} = \frac{2\ln 2}{r} = 2\left[\frac{\ln 2}{r}\right]$$

which is twice as long.

149. $f(x) = 3x^3 - 4$

151. $f(x) = |x| + 9$

153. $f(x) = \begin{cases} 2x, & x < 0 \\ -x^2 + 4, & x \geq 0 \end{cases}$

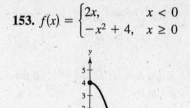

Section 3.5 Exponential and Logarithmic Models

■ You should be able to solve compound interest problems.

1. $A = P\left(1 + \dfrac{r}{n}\right)^{nt}$

2. $A = Pe^{rt}$

■ You should be able to solve growth and decay problems.

(a) Exponential growth if $b > 0$ and $y = ae^{bx}$.

(b) Exponential decay if $b > 0$ and $y = ae^{-bx}$.

■ You should be able to use the Gaussian model

$y = ae^{-(x-b)^2/c}$.

■ You should be able to use the logistics growth model

$y = \dfrac{a}{1 + be^{-(x-c)/d}}$.

■ You should be able to use the logarithmic models

$y = \ln(ax + b)$ and $y = \log_{10}(ax + b)$.

Vocabulary Check

1. (a) iv   (b) i   (c) vi   (d) iii   (e) vii   (f) ii   (g) v

2. Normally

3. Sigmoidal

4. Bell-shaped, mean

---

1. $y = 2e^{x/4}$

This is an exponential growth model.

Matches graph (c).

3. $y = 6 + \log_{10}(x + 2)$

This is a logarithmic model, and contains $(-1, 6)$.

Matches graph (b).

5. $y = \ln(x + 1)$

This is a logarithmic model.

Matches graph (d).

7. Since $A = 10,000e^{0.035t}$, the time to double is given by

$20,000 = 10,000e^{0.035t}$

$2 = e^{0.035t}$

$\ln 2 = 0.035t$

$t = \dfrac{\ln 2}{0.035} \approx 19.8$ years.

Amount after 10 years:

$A = 10,000e^{0.035(10)} \approx \$14,190.68$

9. Since $A = 7500e^{rt}$ and $A = 15,000$ when $t = 21$, we have the following.

$15,000 = 7500e^{21r}$

$2 = e^{21r}$

$\ln 2 = 21r$

$r = \dfrac{\ln 2}{21} \approx 0.033 = 3.3\%$

Amount after 10 years:

$A = 7500e^{0.033(10)} \approx \$10,432.26$

11. Since $A = 5000e^{rt}$ and $A = 5665.74$ when $t = 10$, we have the following.

$$5665.74 = 5000e^{10r}$$

$$\frac{5665.74}{5000} = e^{10r}$$

$$\ln\left(\frac{5665.74}{5000}\right) = 10r$$

$$r = \frac{1}{10}\ln\left(\frac{5665.74}{5000}\right)$$

$$\approx 0.0125 = 1.25\%$$

The time to double is given by

$$10,000 = 5000e^{0.0125t}$$

$$2 = e^{0.0125t}$$

$$\ln 2 = 0.0125t$$

$$t = \frac{\ln 2}{0.0125} \approx 55.5 \text{ years.}$$

13. Since $A = Pe^{0.045t}$ and $A = 100,000$ when $t = 10$, we have the following.

$$100,000 = Pe^{0.045(10)}$$

$$\frac{100,000}{e^{0.45}} = P \approx 63,762.82$$

The time to double is given by

$$127,525.64 = 63,762.82e^{0.045t}$$

$$2 = e^{0.045t}$$

$$\ln 2 = 0.045t$$

$$t = \frac{\ln 2}{0.045} \approx 15.4 \text{ years.}$$

15.

$$3P = Pe^{rt}$$
$$3 = e^{rt}$$
$$\ln 3 = rt$$
$$\frac{\ln 3}{r} = t$$

r	2%	4%	6%	8%	10%	12%
$t = \dfrac{\ln 3}{r}$	54.93	27.47	18.31	13.73	10.99	9.16

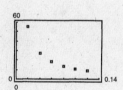

17.

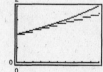

Continuous compounding results in faster growth.

$$A = 1 + 0.075[\![t]\!]$$

and $A = e^{0.07t}$

19.

$$\frac{1}{2}C = Ce^{k(1599)}$$

$$\frac{1}{2} = e^{1599k}$$

$$k = \frac{\ln(1/2)}{1599}$$

$$y = Ce^{kt}$$

$$= 10e^{[\ln(1/2)/1599]1000}$$

$$\approx 6.48 \text{ g}$$

21.

$$\frac{1}{2}C = Ce^{k(5715)}$$

$$\frac{1}{2} = e^{5715k}$$

$$k = \frac{\ln(1/2)}{5715}$$

$$y = Ce^{kt}$$

$$= 3e^{[\ln(1/2)/5715]1000}$$

$$\approx 2.66 \text{ g}$$

23.
$$y = ae^{bx}$$
$$1 = ae^{b(0)} \implies 1 = a$$
$$10 = e^{b(3)}$$
$$\ln 10 = 3b$$
$$\frac{\ln 10}{3} = b \implies b \approx 0.7675$$

Thus, $y = e^{0.7675x}$.

25. $(0, 4) \implies a = 4$

$$(5, 1) \implies 1 = 4e^{b(5)} \implies b = \frac{1}{5} \ln\left(\frac{1}{4}\right)$$
$$= -\frac{1}{5} \ln 4 \approx -0.2773$$
$$y = 4e^{-0.2773x}$$

27. (a) Australia: $(0, 19.2)$, $(10, 20.9)$

$\quad a = 19.2$ and $20.9 = 19.2e^{b(10)} \implies b = 0.008484$

$\quad y = 19.2e^{0.008484t}$

$\quad$ For 2030, $y \approx 24.8$ million.

Canada: $(0, 31.3)$, $(10, 34.3)$

$\quad a = 31.3$ and $34.3 = 31.3e^{b(10)} \implies b = 0.009153$

$\quad y = 31.3e^{0.009153t}$

$\quad$ For 2030, $y \approx 41.2$ million.

Philippines: $(0, 79.7)$, $(10, 95.9)$

$\quad a = 79.7$ and $95.9 = 79.7e^{b(10)} \implies b = 0.0185$

$\quad y = 79.7e^{0.0185t}$

$\quad$ For 2030, $y \approx 138.8$ million.

South Africa: $(0, 44.1)$, $(10, 43.3)$

$\quad a = 44.1$ and $43.3 = 44.1e^{b(10)} \implies b = -0.00183$

$\quad y = 44.1e^{-0.00183t}$

$\quad$ For 2030, $y \approx 41.7$ million.

Turkey: $(0, 65.7)$, $(10, 73.3)$

$\quad a = 65.7$ and $73.3 = 65.7e^{b(10)} \implies b = 0.01095$

$\quad y = 65.7e^{0.01095t}$

$\quad$ For 2030, $y \approx 91.2$ million.

(b) The constant b gives the growth rates.

(c) The constant b is negative for South Africa.

29. (a) $180 = 134.0e^{k(10)}$

$$10k = \ln \frac{180}{134.0}$$
$$k \approx 0.0295$$

(b) For 2010, $t = 20$ and

$$P = 134.0e^{0.0295(20)} \approx 241{,}734 \text{ people.}$$

31.
$$y = Ce^{kt}$$
$$\frac{1}{2}C = Ce^{(1599)k}$$
$$\ln\left(\frac{1}{2}\right) = 1599k$$
$$k = \frac{\ln(1/2)}{1599}$$

When $t = 100$, we have

$$y = Ce^{[(\ln(1/2)\cdot 100)/1599]} \approx 0.958C, \text{ or } 95.8\%.$$

33. (a) $V = mt + b$, $V(0) = 30,788 \Rightarrow b = 30,788$

$$V(2) = 24,000 \Rightarrow 24,000 = 2m + 30,788$$

$$\Rightarrow m = -3394$$

$$V(t) = -3394t + 30,788$$

(b) $V = ae^{kt}$, $V(0) = 30,788 \Rightarrow b = 30,788$

$$V(2) = 24,000 \Rightarrow 24,000 = 30,788e^{2k}$$

$$\Rightarrow k = \frac{1}{2}\ln\left(\frac{24,000}{30,788}\right) \approx -0.1245$$

$$V = 30,788e^{-0.1245t}$$

(c)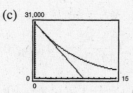

(d) The exponential model depreciates faster in the first year.

(e) Answers will vary.

35. $S(t) = 100(1 - e^{kt})$

(a) $\quad 15 = 100(1 - e^{k(1)})$

$$-85 = -100e^k$$

$$k = \ln 0.85$$

$$k \approx -0.1625$$

$$S(t) = 100(1 - e^{-0.1625t})$$

(b)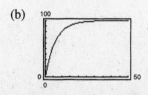

(c) $S(5) = 100(1 - e^{-0.1625(5)})$

$$\approx 55.625 = 55,625 \text{ units}$$

37. $y = 0.0266e^{-(x-100)^2/450}$, $70 \le x \le 115$

(a)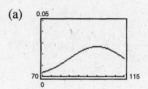

(b) Maximum point is $x = 100$, the average IQ score.

39. $p(t) = \dfrac{1000}{1 + 9e^{-0.1656t}}$

(a) $p(5) = \dfrac{1000}{1 + 9e^{-0.1656(5)}} \approx 203 \text{ animals}$

(b) $\qquad 500 = \dfrac{1000}{1 + 9e^{-0.1656t}}$

$$1 + 9e^{-0.1656t} = 2$$

$$9e^{-0.1656t} = 1$$

$$e^{-0.1656t} = \frac{1}{9}$$

$$t = \frac{-\ln(1/9)}{0.1656} \approx 13 \text{ months}$$

(c)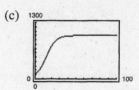

The horizontal asymptotes are $p = 0$ and $p = 1000$. The population will approach 1000 as time increases.

41. $R = \log_{10}\left(\dfrac{I}{I_0}\right) = \log_{10}(I) \Rightarrow I = 10^R$

(a) $I = 10^{6.1} \approx 1,258,925$

(b) $I = 10^{7.6} \approx 39,810,717$

(c) $I = 10^{9.0} \approx 1,000,000,000$

43. $\beta(I) = 10 \log_{10}(I/I_0)$, where $I_0 = 10^{-12}$ watt per square meter.

    (a) $\beta(10^{-10}) = 10 \cdot \log_{10}\left(\dfrac{10^{-10}}{10^{-12}}\right) = 10 \log_{10} 10^2 = 20$ decibels

    (b) $\beta(10^{-5}) = 10 \cdot \log_{10}\left(\dfrac{10^{-5}}{10^{-12}}\right) = 10 \log_{10} 10^7 = 70$ decibels

    (c) $\beta(10^0) = 10 \cdot \log_{10}\left(\dfrac{10^0}{10^{-12}}\right) = 10 \log_{10} 10^{12} = 120$ decibels

45.
$$\beta = 10 \log_{10}\left(\frac{I}{I_0}\right)$$
$$10^{\beta/10} = \frac{I}{I_0}$$
$$I = I_0 10^{\beta/10}$$
$$\% \text{ decrease} = \frac{I_0 10^{8.8} - I_0 10^{7.2}}{I_0 10^{8.8}} \times 100 = 97.5\%$$

47. $\text{pH} = -\log_{10}[\text{H}^+] = -\log_{10}[2.3 \times 10^{-5}] \approx 4.64$

49.
$$\text{pH} = -\log_{10}[\text{H}^+]$$
$$-\text{pH} = \log_{10}[\text{H}^+]$$
$$10^{-\text{pH}} = [\text{H}^+]$$
$$\frac{\text{Hydrogen ion concentration of grape}}{\text{Hydrogen ion concentration of milk of magnesia}} = \frac{10^{-3.5}}{10^{-10.5}} = 10^7$$

51. (a) $P = 120,000$, $r = 0.075$, $M = 839.06$
$$u = M - \left(M - \frac{Pr}{12}\right)\left(1 + \frac{r}{12}\right)^{12t}$$
$$= 839.06 - (839.06 - 750)(1 + 0.00625)^{12t}$$
$$v = (839.06 - 750)(1.00625)^{12t}$$

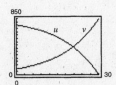

(b) In the early years, the majority of the monthly payment goes toward interest. The interest and principle are equal when $t \approx 20.729 \approx 21$ years.

(c) $P = 120,000$, $r = 0.075$, $M = 966.71$
$$u = 966.71 - (966.71 - 750)(1.00625)^{12t}$$
$$v = (966.71 - 750)(1.00625)^{12t}$$

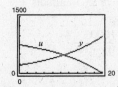

$u = v$ when $t \approx 10.73$ years.

53. $t = -10 \ln\left(\dfrac{T - 70}{98.6 - 70}\right)$

At 9:00 A.M. we have $t = -10 \ln\left[(85.7 - 70)/(98.6 - 70)\right] \approx 6$ hours.

Thus, we can conclude that the person died 6 hours before 9 A.M., or 3:00 A.M.

55. False. The domain could be all real numbers.

57. True. For the Gaussian model, $y > 0$.

59. $4x - 3y - 9 = 0 \implies y = \frac{1}{3}(4x - 9)$

Slope: $\frac{4}{3}$

Matches (a).

Intercepts: $(0, -3), \left(\frac{9}{4}, 0\right)$

61. $y = 25 - 2.25x$

Slope: -2.25

Matches (d).

Intercepts: $(0, 25), \left(\frac{100}{9}, 0\right)$

63. $f(x) = 2x^3 - 3x^2 + x - 1$

The graph falls to the left and rises to the right.

65. $g(x) = -1.6x^5 + 4x^2 - 2$

The graph rises to the left and falls to the right.

67.

$$
\begin{array}{r|rrrr}
4 & 2 & -8 & 3 & -9 \\
  &   & 8 & 0 & 12 \\
\hline
  & 2 & 0 & 3 & 3
\end{array}
$$

$$\frac{2x^3 - 8x^2 + 3x - 9}{x - 4} = 2x^2 + 3 + \frac{3}{x - 4}$$

69. Answers will vary.

Section 3.6   Nonlinear Models

■ You should be able to use a graphing utility to find nonlinear models, including:

(a) Quadratic models

(b) Exponential models

(c) Power models

(d) Logarithmic models

(e) Logistic models

■ You should be able to use a scatter plot to determine which model is best.

■ You should be able to determine the sum of squared differences for a model.

Vocabulary Check

1. $y = ax + b$

2. quadratic

3. $y = ax^b$

4. sum, squared differences

5. $y = ab^x, ae^{cx}$

1. Logarithmic model   **3.** Quadratic model   **5.** Exponential model   **7.** Quadratic model

9.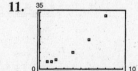

Logarithmic model

11.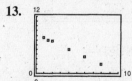

Exponential model

13.

Linear model

15. $y = 4.752(1.2607)^x$

Coefficient of determination: 0.96773

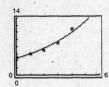

17. $y = 8.463(0.7775)^x$

Coefficient of determination: 0.86639

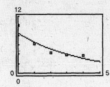

19. $y = 2.083 + 1.257 \ln x$

Coefficient of determination: 0.98672

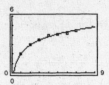

21. $y = 9.826 - 4.097 \ln x$

Coefficient of determination: 0.93704

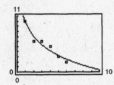

23. $y = 1.985x^{0.760}$

Coefficient of determination: 0.99686

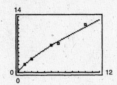

25. $y = 16.103x^{-3.174}$

Coefficient of determination: 0.88161

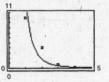

27. (a) Quadratic model: $R = 0.031t^2 + 1.13t + 97.1$

Exponential model: $R = 94.435(1.0174)^t$

Power model: $R = 77.837t^{0.1918}$

(b)

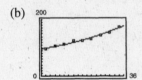
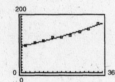
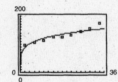

(c) The exponential model fits best. Answers will vary.

(d) For 2008, $t = 38$ and $R \approx 181.9$ million.

For 2012, $t = 42$ and $R \approx 194.9$ million.

Answers will vary.

29. (a) Linear model: $P = 3.11t + 250.9$

Coefficient of determination: 0.99942

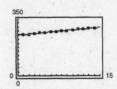

(b) Power model: $P = 246.52t^{0.0587}$

Coefficient of determination: 0.90955

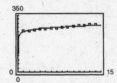

—CONTINUED—

29. **—CONTINUED—**

(c) Exponential model:
$$P = 251.57(1.0114)^t$$

Coefficient of determination:
0.99811

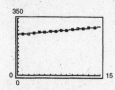

(d) Quadratic model:
$$P = -0.020t^2 + 3.41t + 250.1$$

Coefficient of determination:
0.99994

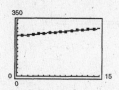

(e) The quadratic model is best because its coefficient of determination is closest to 1.

(f) Linear model:

Year	2005	2006	2007	2008	2009	2010
Population (in millions)	297.6	300.7	303.8	306.9	310.0	313.1

Power model:

Year	2005	2006	2007	2008	2009	2010
Population (in millions)	289.0	290.1	291.1	292.1	293.0	293.9

Exponential model:

Year	2005	2006	2007	2008	2009	2010
Population (in millions)	298.2	301.6	305.0	308.5	312.0	315.6

Quadratic model:

Year	2005	2006	2007	2008	2009	2010
Population (in millions)	296.8	299.5	302.3	305.0	307.7	310.3

(g) and (h) Answers will vary.

31. (a) $T = -1.239t + 73.02$

No, the data does not appear linear.

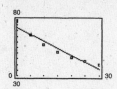

(c) Subtracting 21 from the T-values, the exponential model is $y = 54.438(0.9635)^t$. Adding back 21, $T = 54.438(0.9635)^t + 21$.

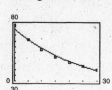

(b) $T = 0.034t^2 - 2.26t + 77.3$

Yes, the data appears quadratic. But, for $t = 60$, the graph is increasing, which is incorrect.

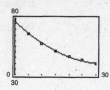

(d) Answers will vary.

33. (a) $P = \dfrac{162.4}{1 + 0.34e^{0.5609x}}$

(b) 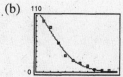   The model is a good fit.

35. (a) Linear model: $y = 15.71t + 51.0$

Logarithmic model: $y = 134.67 \ln t - 97.5$

Quadratic model: $y = -1.292t^2 + 38.96t - 45.0$

Exponential model: $y = 85.97(1.091)^t$

Power model: $y = 37.27t^{0.7506}$

(b)

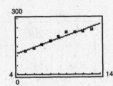

Linear model:

Logarithmic model:

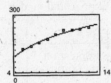

Quadratic model:

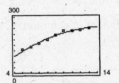

Exponential model:

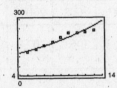

Power model:

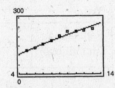

(c) Linear: 803.9

Logarithmic: 411.7

Quadratic: 289.8 (Best)

Exponential: 1611.4

Power: 667.1

(d) Linear: 0.9485

Logarithmic: 0.9736

Quadratic: 0.9814 (Best)

Exponential: 0.9274

Power: 0.9720

(e) Quadratic model is best.

37. True

39. $2x + 5y = 10$

$5y = -2x + 10$

$y = -\frac{2}{5}x + 2$

Slope: $-\frac{2}{5}$

y-intercept: $(0, 2)$

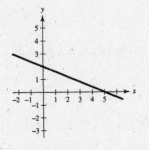

41. $1.2x + 3.5y = 10.5$

$35y = -12x + 105$

$y = -\frac{12}{35}x + \frac{105}{35} = -\frac{12}{35}x + 3$

Slope: $-\frac{12}{35}$

y-intercept: $(0, 3)$

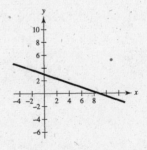

Review Exercises for Chapter 3

1. $(1.45)^{2\pi} \approx 10.3254$

3. $60^{2(-1.1)} = 60^{-2.2}$
$\approx 0.0001225 \approx 0.0$

5. $e^8 \approx 2980.958$

7. $e^{-(-2.1)} \approx e^{2.1} \approx 8.1662$

9. $f(x) = 4^x$

    Intercept: $(0, 1)$

    Horizontal asymptote: x-axis

    Increasing on: $(-\infty, \infty)$

    Matches graph (c).

11. $f(x) = -4^x$

    Intercept: $(0, -1)$

    Horizontal asymptote: x-axis

    Decreasing on: $(-\infty, \infty)$

    Matches graph (b).

13. $f(x) = 6^x$

    Intercept: $(0, 1)$

    Horizontal asymptote: x-axis

    Increasing on: $(-\infty, \infty)$

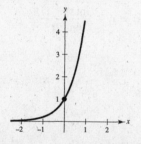

15. $g(x) = 1 + 6^{-x}$

    Intercept: $(0, 2)$

    Horizontal asymptote: $y = 1$

    Decreasing on: $(-\infty, \infty)$

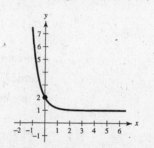

17.

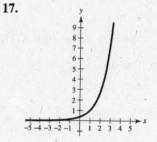

$h(x) = e^{x-1}$

Horizontal asymptote: $y = 0$

19.

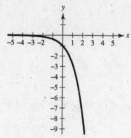

$h(x) = -e^x$

Horizontal asymptote: $y = 0$

21.

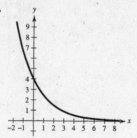

$f(x) = 4e^{-0.5x}$

Horizontal asymptote: $y = 0$

23. $f(x) = \dfrac{10}{1 + 2e^{-0.05x}}$

(a)

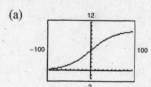

(b) Horizontal asymptotes:
    $y = 0,\ y = 10$

25. $A = Pe^{rt} = 10,000e^{0.08t}$

t	1	10	20	30	40	50
A	10,832.87	22,255.41	49,530.32	110,231.76	245,325.30	545,981.50

27. $V(t) = 26{,}000\left(\frac{3}{4}\right)^t$

(a)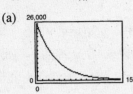

(b) For $t = 2$, $V(2) = \$14{,}625.$

(c) The car depreciates most rapidly at the beginning, which is realistic.

29. $\log_5 125 = 3$
    $5^3 = 125$

31. $\log_{64} 2 = \frac{1}{6}$
    $64^{1/6} = 2$

33. $\ln e^4 = 4$
    $e^4 = e^4$

35. $4^3 = 64$
    $\log_4 64 = 3$

37. $25^{3/2} = 125$
    $\log_{25} 125 = \frac{3}{2}$

39. $\left(\frac{1}{2}\right)^{-3} = 8$
    $\log_{1/2} 8 = -3$

41. $e^7 = 1096.6331\ldots$
    $\ln 1096.6331\ldots = 7$

43. $\log_6 216 = \log_6 6^3$
    $= 3\log_6 6 = 3$

45. $\log_4\left(\frac{1}{4}\right) = \log_4(4^{-1})$
    $= -\log_4 4 = -1$

47. $g(x) = -\log_2 x + 5 = 5 - \dfrac{\ln x}{\ln 2}$

Domain: $x > 0$

Vertical asymptote: $x = 0$

x-intercept: $(32, 0)$

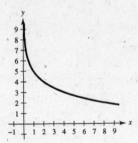

49. $f(x) = \log_2(x - 1) + 6 = 6 + \dfrac{\ln(x - 1)}{\ln (2)}$

Domain: $x > 1$

Vertical asymptote: $x = 1$

x-intercept: $(1.016, 0)$

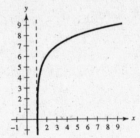

51. $\ln(21.5) \approx 3.068$

53. $\ln\sqrt{6} \approx 0.896$

55. $\log_5 3 = \log_5 x$
    $3 = x$

57. $\log_9 x = \log_9 3^{-2}$
    $x = 3^{-2} = \frac{1}{9}$

59. $f(x) = \ln x + 3$

Domain: $(0, \infty)$

Vertical asymptote: $x = 0$

x-intercept: $(0.05, 0)$

61. $h(x) = \frac{1}{2}\ln x$

Domain: $x > 0$

Vertical asymptote: $x = 0$

x-intercept: $(1, 0)$

63. $t = 50 \log_{10} \dfrac{18{,}000}{18{,}000 - h}$

(a) $0 \le h < 18{,}000$

(c) The plane climbs at a faster rate as it approaches its absolute ceiling.

(d) If $h = 4000$, $t = 50 \log_{10} \dfrac{18{,}000}{18{,}000 - 4000} \approx 5.46$ minutes.

(b)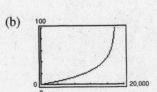

Vertical asymptote: $h = 18{,}000$

65. $\log_4 9 = \dfrac{\log_{10} 9}{\log_{10} 4} \approx 1.585$

$\log_4 9 = \dfrac{\ln 9}{\ln 4} \approx 1.585$

67. $\log_{12} 200 = \dfrac{\log_{10} 200}{\log_{10} 12} \approx 2.132$

$\log_{12} 200 = \dfrac{\ln 200}{\ln 12} \approx 2.132$

69. $f(x) = \log_2(x - 1) = \dfrac{\ln(x - 1)}{\ln 2}$

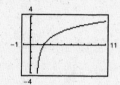

71. $f(x) = -\log_{1/2}(x + 2) = -\dfrac{\ln(x + 2)}{\ln(1/2)} = \dfrac{\ln(x + 2)}{\ln 2}$

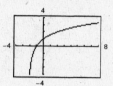

73. $\log_b 9 = \log_b 3^2$

$= 2 \log_b 3$

$= 2(0.5646)$

$= 1.1292$

75. $\log_b \sqrt{5} = \log_b 5^{1/2}$

$= \tfrac{1}{2} \log_b 5$

$= \tfrac{1}{2}(0.8271)$

$= 0.41355$

77. $\ln(5e^{-2}) = \ln 5 + \ln e^{-2}$

$= \ln 5 - 2 \ln e$

$= \ln 5 - 2$

79. $\log_{10} 200 = \log_{10}(2 \cdot 100)$

$= \log_{10} 2 + \log_{10} 10^2$

$= \log_{10} 2 + 2$

81. $\log_5 5x^2 = \log_5 5 + \log_5 x^2 = 1 + 2 \log_5 x$

83. $\log_{10} \dfrac{5\sqrt{y}}{x^2} = \log_{10} 5\sqrt{y} - \log_{10} x^2$

$= \log_{10} 5 + \log_{10} \sqrt{y} - \log_{10} x^2$

$= \log_{10} 5 + \dfrac{1}{2} \log_{10} y - 2 \log_{10} x$

85. $\ln\left(\dfrac{x + 3}{xy}\right) = \ln(x + 3) - \ln(xy)$

$= \ln(x + 3) - \ln x - \ln y$

87. $\log_2 5 + \log_2 x = \log_2 5x$

89. $\dfrac{1}{2} \ln(2x - 1) - 2\ln(x + 1) = \ln\sqrt{2x - 1} - \ln(x + 1)^2 = \ln\dfrac{\sqrt{2x - 1}}{(x + 1)^2}$

91. $\ln 3 + \dfrac{1}{3}\ln(4 - x^2) - \ln x = \ln\left[\dfrac{3(4 - x^2)^{1/3}}{x}\right] = \ln\left[\dfrac{3\sqrt[3]{4 - x^2}}{x}\right]$

93. $s = 25 - \dfrac{13 \ln(h/12)}{\ln 3}$

(a)

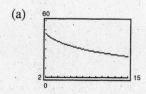

(b)

h	4	6	8	10	12	14
s	38	33.2	29.8	27.2	25	23.2

(c) As the depth increases, the number of miles of roads cleared decreases.

95. $8^x = 512 = 8^3 \implies x = 3$

97. $6^x = \dfrac{1}{216} = \dfrac{1}{6^3} = 6^{-3} \implies x = -3$

99. $2^{x+1} = \dfrac{1}{16}$

$2^{x+1} = 2^{-4}$

$x + 1 = -4$

$x = -5$

101. $\log_7 x = 4 \implies x = 7^4 = 2401$

103. $\log_2(x - 1) = 3$

$2^3 = x - 1$

$x = 9$

105. $\ln x = 4$

$x = e^4 \approx 54.598$

107. $\ln(x - 1) = 2$

$e^2 = x - 1$

$x = 1 + e^2$

109. $3e^{-5x} = 132$

$e^{-5x} = 44$

$-5x = \ln 44$

$x = -\dfrac{\ln 44}{5} \approx -0.757$

111. $2^x + 13 = 35$

$2^x = 22$

$x \ln 2 = \ln 22$

$x = \dfrac{\ln 22}{\ln 2} \approx 4.459$

113. $-4(5^x) = -68$

$5^x = 17$

$x \ln 5 = \ln 17$

$x = \dfrac{\ln 17}{\ln 5} \approx 1.760$

115. $2e^{x-3} - 1 = 4$

$2e^{x-3} = 5$

$e^{x-3} = \dfrac{5}{2}$

$x - 3 = \ln\left(\dfrac{5}{2}\right)$

$x = 3 + \ln\left(\dfrac{5}{2}\right) \approx 3.916$

117. $e^{2x} - 7e^x + 10 = 0$

$(e^x - 5)(e^x - 2) = 0$

$e^x = 5 \implies x = \ln 5 \approx 1.609$

$e^x = 2 \implies x = \ln 2 \approx 0.693$

119. $\ln 3x = 8.2$

$3x = e^{8.2}$

$x = \dfrac{e^{8.2}}{3} \approx 1213.650$

121. $\ln x - \ln 3 = 2$

$\ln \dfrac{x}{3} = 2$

$\dfrac{x}{3} = e^2$

$x = 3e^2 \approx 22.167$

123. $\ln\sqrt{x + 1} = 2$

$\dfrac{1}{2}\ln(x + 1) = 2$

$\ln(x + 1) = 4$

$x + 1 = e^4$

$x = e^4 - 1$

≈ 53.598

125. $\log_4(x - 1) = \log_4(x - 2) - \log_4(x + 2)$

$$\log_4(x - 1) = \log_4\left(\frac{x - 2}{x + 2}\right)$$

$$x - 1 = \frac{x - 2}{x + 2}$$

$$(x - 1)(x + 2) = x - 2$$

$$x^2 + x - 2 = x - 2$$

$$x^2 = 0$$

$$x = 0 \quad \text{(extraneous)}$$

No solution

127. $\log_{10}(1 - x) = -1$

$$10^{-1} = 1 - x$$

$$x = 1 - 10^{-1} = 0.9$$

129. $xe^x + e^x = 0$

$$(x + 1)e^x = 0$$

$$x + 1 = 0 \quad (\text{since } e^x \neq 0)$$

$$x = -1$$

131. $x \ln x + x = 0$

$$x(\ln x + 1) = 0$$

$$\ln x + 1 = 0 \quad (\text{since } x > 0)$$

$$\ln x = -1$$

$$x = e^{-1} = \frac{1}{e} \approx 0.368$$

133. $3(7550) = 7550e^{0.0725t}$

$$3 = e^{0.0725t}$$

$$\ln 3 = 0.0725t$$

$$t = \frac{\ln 3}{0.0725} \approx 15.2 \text{ years}$$

135. $y = 3e^{-2x/3}$

Decreasing exponential

Matches graph (e).

137. $y = \ln(x + 3)$

Logarithmic function shifted to left

Matches graph (f).

139. $y = 2e^{-(x+4)^2/3}$

Gaussian model

Matches graph (a).

141. $y = ae^{bx}$

$$2 = ae^{b(0)} \implies a = 2$$

$$3 = 2e^{b(4)}$$

$$1.5 = e^{4b}$$

$\ln 1.5 = 4b \implies b \approx 0.1014$

Thus, $y = 2e^{0.1014x}$.

143. $y = ae^{bx}$

$$\frac{1}{2} = ae^{b(0)} \implies a = \frac{1}{2}$$

$$5 = \frac{1}{2}e^{b(5)}$$

$$10 = e^{5b}$$

$\ln 10 = 5b \implies b \approx 0.4605$

Thus, $y = \frac{1}{2}e^{0.4605x}$.

145. $P = 361e^{kt}$

$t = 0$ corresponds to 2000.

$(-20, 215)$:

$215 = 361e^{k(-20)}$

$\dfrac{215}{361} = e^{-20k}$

$-20k = \ln\left(\dfrac{215}{361}\right)$

$k = -\dfrac{1}{20}\ln\left(\dfrac{215}{361}\right) = \dfrac{1}{20}\ln\left(\dfrac{361}{215}\right) \approx 0.02591$

$P = 361e^{0.02591t}$

For 2020, $P(20) = 361e^{0.02591(20)} \approx 606.1$ or

606,100 population in 2020.

147. (a) $20,000 = 10,000e^{r(12)}$

$2 = e^{12r}$

$\ln 2 = 12r$

$r = \dfrac{\ln 2}{12} \approx 0.0578$ or 5.78%

(b) $10,000e^{0.0578(1)} \approx \$10,595.03$

149. (a)　　　$50 = \dfrac{158}{1 + 5.4e^{-0.12t}}$

$1 + 5.4e^{-0.12t} = \dfrac{158}{50}$

$5.4e^{-0.12t} = \dfrac{108}{50}$

$e^{-0.12t} = \dfrac{108}{50(5.4)}$

$-0.12t = \ln\dfrac{108}{270}$

$t = \dfrac{\ln(108/270)}{-0.12} \approx 7.6$ weeks

(b) Similarly:

$75 = \dfrac{158}{1 + 5.4e^{-0.12t}}$

$1 + 5.4e^{-0.12t} = \dfrac{158}{75}$

$e^{-0.12t} = 0.20494$

$-0.12t = \ln(0.20494)$

$t \approx 13.2$ weeks

151. Logistic model

153. Logarithmic model

155. (a) Linear model: $y = 297.8t + 739$; 0.97653

Quadratic model: $y = 11.79t^2 + 38.5t + 2118$; 0.98112

Exponential model: $y = 1751.5(1.077)^t$; 0.98225

Logarithmic model: $y = 3169.8 \ln t - 3532$; 0.95779

Power model: $y = 598.1t^{0.7950}$; 0.97118

—CONTINUED—

155. —CONTINUED—

(b)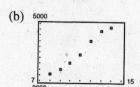

Linear model:

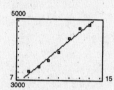

Quadratic model:

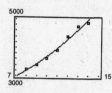

Exponential model:

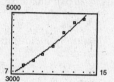

Logarithmic model:

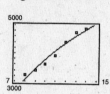

Power model:

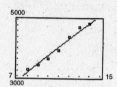

(c) The exponential model is best because its coefficient of determination is closest to 1. Answers will vary.

(d) For 2010, $t = 20$ and $y \approx \$7722$ million.

(e) $y = 5250$ when $t \approx 14.8$, or 2004.

157. (a) $P = \dfrac{9999.887}{1 + 19.0e^{-0.2x}}$

(b)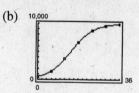

(c) The model is a good fit.

(d) The limiting size is $\dfrac{9999.887}{1 + 0} \approx 10,000$ fish.

159. True; by the Inverse Properties, $\log_b b^{2x} = 2x$.

161. False; $\ln x + \ln y = \ln(xy) \neq \ln(x + y)$

163. False. The domain of $f(x) = \ln(x)$ is $x > 0$.

165. Since $1 < \sqrt{2} < 2, 2^1 < 2^{\sqrt{2}} < 2^2 \Rightarrow 2 < 2^{\sqrt{2}} < 4.$

Chapter 3   Practice Test

1. Solve for x: $x^{3/5} = 8$

2. Solve for x: $3^{x-1} = \frac{1}{81}$

3. Graph $f(x) = 2^{-x}$ by hand.

4. Graph $g(x) = e^x + 1$ by hand.

5. If \$5000 is invested at 9% interest, find the amount after three years if the interest is compounded

   (a) monthly.   (b) quarterly.   (c) continuously.

6. Write the equation in logarithmic form: $7^{-2} = \frac{1}{49}$

7. Solve for x: $x - 4 = \log_2 \frac{1}{64}$

8. Given $\log_b 2 = 0.3562$ and $\log_b 5 = 0.8271$, evaluate $\log_b \sqrt[4]{8/25}$.

9. Write $5 \ln x - \frac{1}{2} \ln y + 6 \ln z$ as a single logarithm.

10. Using your calculator and the change of base formula, evaluate $\log_9 28$.

11. Use your calculator to solve for N: $\log_{10} N = 0.6646$

12. Graph $y = \log_4 x$ by hand.

13. Determine the domain of $f(x) = \log_3(x^2 - 9)$.

14. Graph $y = \ln(x - 2)$ by hand.

15. True or false: $\dfrac{\ln x}{\ln y} = \ln(x - y)$

16. Solve for x: $5^x = 41$

17. Solve for x: $x - x^2 = \log_5 \frac{1}{25}$

18. Solve for x: $\log_2 x + \log_2(x - 3) = 2$

19. Solve for x: $\dfrac{e^x + e^{-x}}{3} = 4$

20. Six thousand dollars is deposited into a fund at an annual percentage rate of 13%.
Find the time required for the investment to double if the interest is compounded continuously.

21. Use a graphing utility to find the points of intersection of the graphs of $y = \ln(3x)$ and $y = e^x - 4$.

22. Use a graphing utility to find the power model $y = ax^b$ for the data $(1, 1)$, $(2, 5)$, $(3, 8)$, and $(4, 17)$.

C H A P T E R 4
Trigonometric Functions

Section 4.1   Radian and Degree Measure **150**

Section 4.2   Trigonometric Functions: The Unit Circle **155**

Section 4.3   Right Triangle Trigonometry **159**

Section 4.4   Trigonometric Functions of Any Angle **165**

Section 4.5   Graphs of Sine and Cosine Functions **173**

Section 4.6   Graphs of Other Trigonometric Functions **179**

Section 4.7   Inverse Trigonometric Functions **184**

Section 4.8   Applications and Models **190**

Review Exercises **195**

Practice Test **204**

CHAPTER 4
Trigonometric Functions

Section 4.1   Radian and Degree Measure

You should know the following basic facts about angles, their measurement, and their applications.

- ■ Types of Angles:
  - (a) Acute: Measure between 0° and 90°.
  - (b) Right: Measure 90°.
  - (c) Obtuse: Measure between 90° and 180°.
  - (d) Straight: Measure 180°.
- ■ α and β are complementary if $\alpha + \beta = 90°$. They are supplementary if $\alpha + \beta = 180°$.
- ■ Two angles in standard position that have the same terminal side are called coterminal angles.
- ■ To convert degrees to radians, use $1° = \pi/180$ radians.
- ■ To convert radians to degrees, use 1 radian $= (180/\pi)°$.
- ■ $1' =$ one minute $= 1/60$ of $1°$
- ■ $1'' =$ one second $= 1/60$ of $1' = 1/3600$ of $1°$
- ■ The length of a circular arc is $s = r\theta$ where θ is measured in radians.
- ■ Speed $=$ distance/time
- ■ Angular speed $= \theta/t = s/rt$

Vocabulary Check

1. Trigonometry   **2.** angle   **3.** standard position   **4.** coterminal

5. radian   **6.** complementary   **7.** supplementary   **8.** degree

9. linear   **10.** angular

---

1. 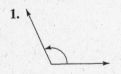 The angle shown is approximately 2 radians.

3. (a) Since $\dfrac{3\pi}{2} < \dfrac{7\pi}{4} < 2\pi, \dfrac{7\pi}{4}$ lies in Quadrant IV.

    (b) Since $\dfrac{5\pi}{2} < \dfrac{11\pi}{4} < 3\pi, \dfrac{11\pi}{4}$ lies in Quadrant II.

5. (a) Since $-\dfrac{\pi}{2} < -1 < 0; -1$ lies in Quadrant IV.

    (b) Since $-\pi < -2 < -\dfrac{\pi}{2}; -2$ lies in Quadrant III.

7. (a) $\dfrac{13\pi}{4}$

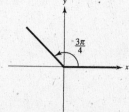

(b) $\dfrac{4\pi}{3}$

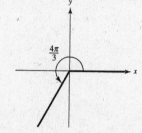

9. (a) $\dfrac{11\pi}{6}$

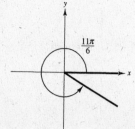

(b) $\dfrac{2\pi}{3}$

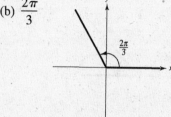

11. (a) Coterminal angles for $\dfrac{\pi}{6}$:

$$\frac{\pi}{6} + 2\pi = \frac{13\pi}{6}$$

$$\frac{\pi}{6} - 2\pi = -\frac{11\pi}{6}$$

(b) Coterminal angles for $\dfrac{2\pi}{3}$:

$$\frac{2\pi}{3} + 2\pi = \frac{8\pi}{3}$$

$$\frac{2\pi}{3} - 2\pi = -\frac{4\pi}{3}$$

13. (a) Coterminal angles for $-\dfrac{9\pi}{4}$:

$$-\frac{9\pi}{4} + 2\pi = -\frac{\pi}{4}$$

$$-\frac{9\pi}{4} + 4\pi = \frac{7\pi}{4}$$

(b) Coterminal angles for $-\dfrac{2\pi}{15}$:

$$-\frac{2\pi}{15} + 2\pi = \frac{28\pi}{15}$$

$$-\frac{2\pi}{15} - 2\pi = -\frac{32\pi}{15}$$

15. Complement: $\dfrac{\pi}{2} - \dfrac{\pi}{3} = \dfrac{\pi}{6}$

Supplement: $\pi - \dfrac{\pi}{3} = \dfrac{2\pi}{3}$

17. Complement: $\dfrac{\pi}{2} - \dfrac{\pi}{6} = \dfrac{\pi}{3}$

Supplement: $\pi - \dfrac{\pi}{6} = \dfrac{5\pi}{6}$

19. Complement: $\dfrac{\pi}{2} - 1 \approx 0.57$

Supplement: $\pi - 1 \approx 2.14$

21.

The angle shown is approximately $210°$.

23. (a) Since $90° < 150° < 180°$, $150°$ lies in Quadrant II.

(b) Since $270° < 282° < 360°$, $282°$ lies in Quadrant IV.

25. (a) Since $-180° < -132° \, 50' < -90°$, $-132° \, 50'$ lies in Quadrant III.

(b) Since $-360° < -336° \, 30' < -270°$, $-336° \, 30'$ lies in Quadrant I.

27. (a) 30°   (b) 150°

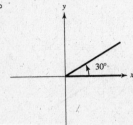

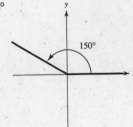

29. (a) 405°   (b) 780°

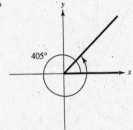

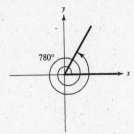

31. (a) Coterminal angles for 52°:

$$52° + 360° = 412°$$
$$52° - 360° = -308°$$

(b) Coterminal angles for $-36°$:

$$-36° + 360° = 324°$$
$$-36° - 360° = -396°$$

33. (a) Coterminal angles for 300°:

$$300° + 360° = 660°$$
$$300° - 360° = -60°$$

(b) Coterminal angles for 230°:

$$230° + 360° = 590°$$
$$230° - 360° = -130°$$

35. Complement: $90° - 24° = 66°$

Supplement: $180° - 24° = 156°$

37. Complement: $90° - 87° = 3°$

Supplement: $180° - 87° = 93°$

39. (a) $30° = 30°\left(\dfrac{\pi}{180°}\right) = \dfrac{\pi}{6}$

(b) $150° = 150°\left(\dfrac{\pi}{180°}\right) = \dfrac{5\pi}{6}$

41. (a) $-20° = -20°\left(\dfrac{\pi}{180°}\right) = -\dfrac{\pi}{9}$

(b) $-240° = -240°\left(\dfrac{\pi}{180°}\right) = -\dfrac{4\pi}{3}$

43. (a) $\dfrac{3\pi}{2} = \dfrac{3\pi}{2}\left(\dfrac{180°}{\pi}\right) = 270°$

(b) $-\dfrac{7\pi}{6} = -\dfrac{7\pi}{6}\left(\dfrac{180°}{\pi}\right) = -210°$

45. (a) $\dfrac{7\pi}{3} = \dfrac{7\pi}{3}\left(\dfrac{180°}{\pi}\right) = 420°$

(b) $-\dfrac{13\pi}{60} = -\dfrac{13\pi}{60}\left(\dfrac{180°}{\pi}\right) = -39°$

47. $115° = 115\left(\dfrac{\pi}{180°}\right) \approx 2.007$ radians

49. $-216.35° = -216.35\left(\dfrac{\pi}{180°}\right) \approx -3.776$ radians

51. $-0.78° = -0.78\left(\dfrac{\pi}{180°}\right) \approx -0.014$ radians

53. $\dfrac{\pi}{7} = \dfrac{\pi}{7}\left(\dfrac{180°}{\pi}\right) \approx 25.714°$

55. $6.5\pi = 6.5\pi\left(\dfrac{180°}{\pi}\right) = 1170°$

57. $-2 = -2\left(\dfrac{180°}{\pi}\right) \approx -114.592°$

59. $64° \, 45' = 64° + \left(\dfrac{45}{60}\right)° = 64.75°$

61. $85° \, 18' \, 30'' = 85° + \left(\dfrac{18}{60}\right)° + \left(\dfrac{30}{3600}\right)° \approx 85.308°$

63. $-125° \, 36'' = -125° - \left(\dfrac{36}{3600}\right)° = -125.01°$

65. $280.6° = 280° + 0.6(60)' = 280° \, 36'$

67. $-345.12° = -345° \, 7' \, 12''$

69. $-0.355 = -0.355\left(\dfrac{180°}{\pi}\right)$

$\approx -20.34° = -20° \, 20' \, 24''$

71. $s = r\theta$

$6 = 5\theta$

$\theta = \dfrac{6}{5}$ radians

73. $s = r\theta$

$32 = 7\theta$

$\theta = \dfrac{32}{7} = 4\dfrac{4}{7}$ radians

75. The angles in radians are:

$0° = 0$

$30° = \dfrac{\pi}{6}$

$45° = \dfrac{\pi}{4}$

$60° = \dfrac{\pi}{3}$

$90° = \dfrac{\pi}{2}$

$135° = \dfrac{3\pi}{4}$

$180° = \pi$

$210° = \dfrac{7\pi}{6}$

$270° = \dfrac{3\pi}{2}$

$330° = \dfrac{11\pi}{6}$

77. $s = r\theta$

$8 = 15\theta$

$\theta = \dfrac{8}{15}$ radians

79. $s = r\theta$

$35 = 14.5\theta$

$\theta = \dfrac{70}{29} \approx 2.414$ radians

81. $s = r\theta$, θ in radians

$s = 14(180)\left(\dfrac{\pi}{180}\right) = 14\pi \approx 43.982$ inches

83. $s = r\theta$, θ in radians

$s = 27\left(\dfrac{2\pi}{3}\right) = 18\pi$ meters ≈ 56.55 meters

85. $r = \dfrac{s}{\theta} = \dfrac{36}{\pi/2} = \dfrac{72}{\pi}$ feet ≈ 22.92 feet

87. $r = \dfrac{s}{\theta} = \dfrac{82}{135°(\pi/180°)} = \dfrac{328}{3\pi}$ miles ≈ 34.80 miles

89. $\theta = 42° \, 7' \, 33'' - 25° \, 46' \, 32''$

$= 16° \, 21' \, 1'' \approx 0.28537$ radian

$s = r\theta = 4000(0.28537) \approx 1141.48$ miles

91. $\theta = \dfrac{s}{r} = \dfrac{450}{6378} \approx 0.07056$ radian $\approx 4.04°$

$\approx 4° \, 2' \, 33.02''$

93. $\theta = \dfrac{s}{r} = \dfrac{2.5}{6} = \dfrac{25}{60} = \dfrac{5}{12}$ radian $\approx 23.87°$

95. (a) single axel: $1\frac{1}{2}$ revolutions $= 360° + 180° = 540°$

$$= 2\pi + \pi = 3\pi \text{ radians}$$

(b) double axel: $2\frac{1}{2}$ revolutions $= 720° + 180° = 900°$

$$= 4\pi + \pi = 5\pi \text{ radians}$$

(c) triple axel: $3\frac{1}{2}$ revolutions $= 1260°$

$$= 7\pi \text{ radians}$$

97. (a) $\dfrac{\text{Revolutions}}{\text{Second}} = \dfrac{2400}{60} = 40 \text{ rev/sec}$

Angular speed $= (2\pi)(40) = 80\pi \text{ rad/sec}$

(b) Radius of saw blade $= \dfrac{7.5}{2} = 3.75$ in.

Radius in feet $= \dfrac{3.75}{12} = 0.3125$ ft

Speed $= \dfrac{s}{t} = \dfrac{r\theta}{t} = r\dfrac{\theta}{t} = r(\text{angular speed})$

$$= 0.3125(80\pi) = 78.54 \text{ ft/sec}$$

99. (a) $\dfrac{\text{Revolutions}}{\text{Hour}} = \dfrac{480}{(1/60)} = 28{,}800 \text{ rev/hr}$

Angular speed $= 2\pi(28{,}800) = 57{,}600\pi \text{ rad/hr}$

Radius of wheel $= \dfrac{25/2}{(12 \text{ in./ft})(5280 \text{ ft/mi})} = \dfrac{5}{25{,}344} \text{ miles}$

Speed $= \dfrac{s}{t} = \dfrac{r\theta}{t} = r\dfrac{\theta}{t} = r(\text{angular speed})$

$$= \dfrac{5}{25{,}344} \cdot 57{,}600\pi = \dfrac{125\pi}{11} \approx 35.70 \text{ miles/hr}$$

(b) Let $x =$ spin balance machine rate.

$$70 = r(\text{angular speed}) = r\left(2\pi \cdot \dfrac{x}{(1/60)}\right) = \dfrac{5}{25{,}344}120\pi x \Rightarrow x \approx 941.18 \text{ rev/min}$$

101. False, 1 radian $= \left(\dfrac{180}{\pi}\right)° \approx 57.3°$, so one radian is much larger than one degree.

103. True: $\dfrac{2\pi}{3} + \dfrac{\pi}{4} + \dfrac{\pi}{12} = \dfrac{8\pi + 3\pi + \pi}{12} = \pi = 180°$

105. If θ is constant, the length of the arc is proportional to the radius ($s = r\theta$), and hence increasing.

107. $A = \dfrac{1}{2}r^2\theta = \dfrac{1}{2}(10)^2 \cdot \dfrac{\pi}{3} = \dfrac{50}{3}\pi$ square meters

109. $A = \frac{1}{2}r^2\theta,\ s = r\theta$

(a) $\theta = 0.8 \implies A = \frac{1}{2}r^2(0.8) = 0.4r^2$ Domain: $r > 0$

$s = r\theta = r(0.8)$ Domain: $r > 0$

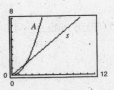

The area function changes more rapidly for $r > 1$ because it is quadratic and the arc length function is linear.

—CONTINUED—

109. **—CONTINUED—**

(b) $r = 10 \implies A = \frac{1}{2}(10^2)\theta = 50\theta$ Domain: $0 < \theta < 2\pi$

$\qquad\qquad\quad s = r\theta = 10\theta$ Domain: $0 < \theta < 2\pi$

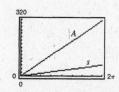

111. Answers will vary.

113.

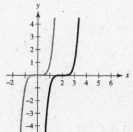

115.

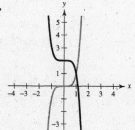

117.

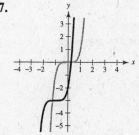

Section 4.2   Trigonometric Functions: The Unit Circle

- ■ You should know how to evaluate trigonometric functions using the unit circle.
- ■ You should know the definition of a *periodic* function.
- ■ You should be able to recognize even and odd trigonometric functions.
- ■ You should be able to evaluate trigonometric functions with a calculator in both radian and degree mode.

Vocabulary Check

1. unit circle   **2.** periodic   **3.** odd, even

1. $\sin\theta = y = \dfrac{15}{17}$

$\cos\theta = x = -\dfrac{8}{17}$

$\tan\theta = \dfrac{y}{x} = -\dfrac{15}{8}$

$\cot\theta = \dfrac{x}{y} = -\dfrac{8}{15}$

$\sec\theta = \dfrac{1}{x} = -\dfrac{17}{8}$

$\csc\theta = \dfrac{1}{y} = \dfrac{17}{15}$

3. $\sin\theta = y = -\dfrac{5}{13}$

$\cos\theta = x = \dfrac{12}{13}$

$\tan\theta = \dfrac{y}{x} = -\dfrac{5}{12}$

$\cot\theta = \dfrac{x}{y} = -\dfrac{12}{5}$

$\sec\theta = \dfrac{1}{x} = \dfrac{13}{12}$

$\csc\theta = \dfrac{1}{y} = -\dfrac{13}{5}$

5. $t = \dfrac{\pi}{4}$ corresponds to $\left(\dfrac{\sqrt{2}}{2}, \dfrac{\sqrt{2}}{2}\right)$.

7. $t = \dfrac{7\pi}{6}$ corresponds to $\left(-\dfrac{\sqrt{3}}{2}, -\dfrac{1}{2}\right)$.

9. $t = \dfrac{2\pi}{3}$ corresponds to $\left(-\dfrac{1}{2}, \dfrac{\sqrt{3}}{2}\right)$.

11. $t = \dfrac{3\pi}{2}$ corresponds to $(0, -1)$.

13. $t = -\dfrac{7\pi}{4}$ corresponds to $\left(\dfrac{\sqrt{2}}{2}, \dfrac{\sqrt{2}}{2}\right)$.

15. $t = -\dfrac{3\pi}{2}$ corresponds to $(0, 1)$.

17. $t = \dfrac{\pi}{4}$ corresponds to $\left(\dfrac{\sqrt{2}}{2}, \dfrac{\sqrt{2}}{2}\right)$.

$$\sin t = y = \frac{\sqrt{2}}{2}$$

$$\cos t = x = \frac{\sqrt{2}}{2}$$

$$\tan t = \frac{y}{x} = 1$$

19. $t = \dfrac{7\pi}{6}$ corresponds to $\left(-\dfrac{\sqrt{3}}{2}, -\dfrac{1}{2}\right)$.

$$\sin t = y = -\frac{1}{2}$$

$$\cos t = x = -\frac{\sqrt{3}}{2}$$

$$\tan t = \frac{y}{x} = \frac{1}{\sqrt{3}} = \frac{\sqrt{3}}{3}$$

21. $t = \dfrac{2\pi}{3}$ corresponds to $\left(-\dfrac{1}{2}, \dfrac{\sqrt{3}}{2}\right)$.

$$\sin t = y = \frac{\sqrt{3}}{2}$$

$$\cos t = x = -\frac{1}{2}$$

$$\tan t = \frac{y}{x} = -\sqrt{3}$$

23. $t = -\dfrac{5\pi}{3}$ corresponds to $\left(\dfrac{1}{2}, \dfrac{\sqrt{3}}{2}\right)$.

$$\sin t = y = \frac{\sqrt{3}}{2}$$

$$\cos t = x = \frac{1}{2}$$

$$\tan t = \frac{y}{x} = \sqrt{3}$$

25. $t = -\dfrac{\pi}{6}$ corresponds to $\left(\dfrac{\sqrt{3}}{2}, -\dfrac{1}{2}\right)$.

$$\sin t = y = -\frac{1}{2}$$

$$\cos t = x = \frac{\sqrt{3}}{2}$$

$$\tan t = \frac{y}{x} = -\frac{\sqrt{3}}{3}$$

27. $t = -\dfrac{7\pi}{4}$ corresponds to $\left(\dfrac{\sqrt{2}}{2}, \dfrac{\sqrt{2}}{2}\right)$.

$$\sin t = y = \frac{\sqrt{2}}{2}$$

$$\cos t = x = \frac{\sqrt{2}}{2}$$

$$\tan t = \frac{y}{x} = 1$$

29. $t = -\dfrac{3\pi}{2}$ corresponds to $(0, 1)$.

$$\sin t = y = 1$$
$$\cos t = x = 0$$

$$\tan t = \frac{y}{x} \text{ is undefined.}$$

31. $t = \dfrac{3\pi}{4}$ corresponds to $\left(-\dfrac{\sqrt{2}}{2}, \dfrac{\sqrt{2}}{2}\right)$.

$$\sin t = y = \frac{\sqrt{2}}{2} \qquad \csc t = \frac{1}{y} = \sqrt{2}$$

$$\cos t = x = -\frac{\sqrt{2}}{2} \qquad \sec t = \frac{1}{x} = -\sqrt{2}$$

$$\tan t = \frac{y}{x} = -1 \qquad \cot t = \frac{x}{y} = -1$$

33. $t = \dfrac{\pi}{2}$ corresponds to $(0, 1)$.

$\sin t = y = 1$ $\csc t = \dfrac{1}{y} = 1$

$\cos t = x = 0$ $\sec t = \dfrac{1}{x}$ is undefined.

$\tan t = \dfrac{y}{x}$ is undefined. $\cot t = \dfrac{x}{y} = 0$

35. $t = -\dfrac{2\pi}{3}$ corresponds to $\left(-\dfrac{1}{2}, -\dfrac{\sqrt{3}}{2}\right)$.

$\sin t = y = -\dfrac{\sqrt{3}}{2}$ $\csc t = -\dfrac{2\sqrt{3}}{3}$

$\cos t = x = -\dfrac{1}{2}$ $\sec \theta = -2$

$\tan t = \dfrac{y}{x} = \sqrt{3}$ $\cot \theta = \dfrac{\sqrt{3}}{3}$

37. $\sin 5\pi = \sin \pi = 0$

39. $\cos \dfrac{8\pi}{3} = \cos \dfrac{2\pi}{3} = -\dfrac{1}{2}$

41. $\cos\left(-\dfrac{13\pi}{6}\right) = \cos\left(-\dfrac{\pi}{6}\right) = \cos\left(\dfrac{11\pi}{6}\right) = \dfrac{\sqrt{3}}{2}$

43. $\sin\left(-\dfrac{9\pi}{4}\right) = \sin\left(-\dfrac{\pi}{4}\right) = -\dfrac{\sqrt{2}}{2}$

45. $\sin t = \dfrac{1}{3}$

  (a) $\sin(-t) = -\sin t = -\dfrac{1}{3}$

  (b) $\csc(-t) = -\csc t = -3$

47. $\cos(-t) = -\dfrac{1}{5}$

  (a) $\cos t = \cos(-t) = -\dfrac{1}{5}$

  (b) $\sec(-t) = \dfrac{1}{\cos(-t)} = -5$

49. $\sin t = \dfrac{4}{5}$

  (a) $\sin(\pi - t) = \sin t = \dfrac{4}{5}$

  (b) $\sin(t + \pi) = -\sin t = -\dfrac{4}{5}$

51. $\sin \dfrac{7\pi}{9} \approx 0.6428$

53. $\cos \dfrac{11\pi}{5} \approx 0.8090$

55. $\csc 1.3 \approx 1.0378$

57. $\cos(-1.7) \approx -0.1288$

59. $\csc 0.8 = \dfrac{1}{\sin 0.8} \approx 1.3940$

61. $\sec 22.8 = \dfrac{1}{\cos 22.8} \approx -1.4486$

63. $\cot 2.5 = \dfrac{1}{\tan 2.5} \approx -1.3386$

65. $\csc(-1.5) = \dfrac{1}{\sin(-1.5)} \approx -1.0025$

67. $\sec(-4.5) = \dfrac{1}{\cos(-4.5)} \approx -4.7439$

69. (a) $\sin 5 \approx -1$

    (b) $\cos 2 \approx -0.4$

71. (a) $\sin t = 0.25$

        $t \approx 0.25$ or 2.89

    (b) $\cos t = -0.25$

        $t \approx 1.82$ or 4.46

73. $I = 5e^{-2t} \sin t$

      $I(0.7) = 5e^{-1.4} \sin 0.7$

        ≈ 0.79 amperes

75. $y(t) = \frac{1}{4}\cos 6t$

(a) $y(0) = \frac{1}{4}\cos 0 = 0.2500$ ft

(b) $y\left(\frac{1}{4}\right) = \frac{1}{4}\cos\frac{3}{2} \approx 0.0177$ ft

(c) $y\left(\frac{1}{2}\right) = \frac{1}{4}\cos 3 \approx -0.2475$ ft

77. False. $\sin\left(\frac{-4\pi}{3}\right) = \frac{\sqrt{3}}{2} > 0$

79. False. 0 corresponds to $(1, 0)$.

81. (a) The points have y-axis symmetry.

(b) $\sin t_1 = \sin(\pi - t_1)$ since they have the same y-value.

(c) $-\cos t_1 = \cos(\pi - t_1)$ since the x-values have opposite signs.

83. $\cos 1.5 \approx 0.0707$, $2\cos 0.75 \approx 1.4634$

Thus, $\cos 2t \neq 2\cos t$.

85. $\cos\theta = x = \cos(-\theta)$

$\sec\theta = \frac{1}{\cos\theta} = \frac{1}{\cos(-\theta)} = \sec(-\theta)$

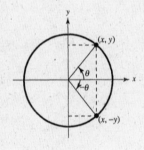

87. $h(t) = f(t)g(t)$ is odd.

$h(-t) = f(-t)g(-t) = -f(t)g(t) = -h(t)$

89. $f(x) = \frac{1}{2}(3x - 2)$

$y = \frac{1}{2}(3x - 2)$

$x = \frac{1}{2}(3y - 2)$

$2x = 3y - 2$

$2x + 2 = 3y$

$\frac{2}{3}(x + 1) = y$

$f^{-1}(x) = \frac{2}{3}(x + 1)$

91. $f(x) = \sqrt{x^2 - 4}$, $x \geq 2$, $y \geq 0$

$y = \sqrt{x^2 - 4}$

$x = \sqrt{y^2 - 4}$

$x^2 = y^2 - 4$

$x^2 + 4 = y^2$

$\sqrt{x^2 + 4} = y$, $x \geq 0$

$f^{-1}(x) = \sqrt{x^2 + 4}$, $x \geq 0$

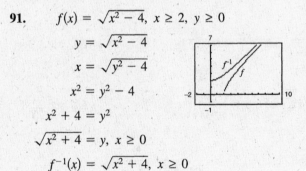

93. $f(x) = \frac{2x}{x - 3}$

Asymptotes: $x = 3$, $y = 2$

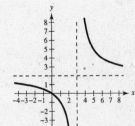

95. $f(x) = \dfrac{x^2 + 3x - 10}{2x^2 - 8} = \dfrac{(x-2)(x+5)}{2(x-2)(x+2)}$

$\qquad = \dfrac{x+5}{2(x+2)}, \quad x \neq 2$

Asymptotes: $x = -2, \; y = \dfrac{1}{2}$

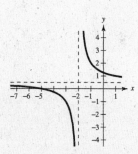

97. $y = x^2 + 3x - 4$

Domain: all real numbers

$0 = x^2 + 3x - 4 = (x+4)(x-1) \Rightarrow x = 1, -4$

Intercepts: $(0, -4), (1, 0), (-4, 0)$

No asymptotes

99. $f(x) = 3^{x+1} + 2$

Domain: all real numbers

Intercept: $(0, 5)$

Asymptote: $y = 2$

Section 4.3   Right Triangle Trigonometry

■ You should know the right triangle definition of trigonometric functions.

    (a) $\sin\theta = \dfrac{\text{opp}}{\text{hyp}}$ (b) $\cos\theta = \dfrac{\text{adj}}{\text{hyp}}$ (c) $\tan\theta = \dfrac{\text{opp}}{\text{adj}}$

    (d) $\csc\theta = \dfrac{\text{hyp}}{\text{opp}}$ (e) $\sec\theta = \dfrac{\text{hyp}}{\text{adj}}$ (f) $\cot\theta = \dfrac{\text{adj}}{\text{opp}}$

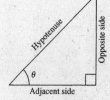

■ You should know the following identities.

    (a) $\sin\theta = \dfrac{1}{\csc\theta}$ (b) $\csc\theta = \dfrac{1}{\sin\theta}$ (c) $\cos\theta = \dfrac{1}{\sec\theta}$

    (d) $\sec\theta = \dfrac{1}{\cos\theta}$ (e) $\tan\theta = \dfrac{1}{\cot\theta}$ (f) $\cot\theta = \dfrac{1}{\tan\theta}$

    (g) $\tan\theta = \dfrac{\sin\theta}{\cos\theta}$ (h) $\cot\theta = \dfrac{\cos\theta}{\sin\theta}$ (i) $\sin^2\theta + \cos^2\theta = 1$

    (j) $1 + \tan^2\theta = \sec^2\theta$ (k) $1 + \cot^2\theta = \csc^2\theta$

■ You should know that two acute angles α and β are complementary if $\alpha + \beta = 90°$, and cofunctions of complementary angles are equal.

■ You should know the trigonometric function values of $30°, 45°$, and $60°$, or be able to construct triangles from which you can determine them.

Vocabulary Check

1. (a) iii   (b) vi   (c) ii   (d) v   (e) i   (f) iv

2. hypotenuse, opposite, adjacent   **3.** elevation, depression

1.

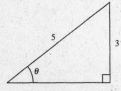

$$\text{adj} = \sqrt{5^2 - 3^2} = \sqrt{16} = 4$$

$$\sin \theta = \frac{\text{opp}}{\text{hyp}} = \frac{3}{5}$$

$$\cos \theta = \frac{\text{adj}}{\text{hyp}} = \frac{4}{5}$$

$$\tan \theta = \frac{\text{opp}}{\text{adj}} = \frac{3}{4}$$

$$\csc \theta = \frac{\text{hyp}}{\text{opp}} = \frac{5}{3}$$

$$\sec \theta = \frac{\text{hyp}}{\text{adj}} = \frac{5}{4}$$

$$\cot \theta = \frac{\text{adj}}{\text{opp}} = \frac{4}{3}$$

3.

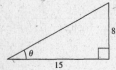

$$\text{hyp} = \sqrt{8^2 + 15^2} = 17$$

$$\sin \theta = \frac{\text{opp}}{\text{hyp}} = \frac{8}{17}$$

$$\cos \theta = \frac{\text{adj}}{\text{hyp}} = \frac{15}{17}$$

$$\tan \theta = \frac{\text{opp}}{\text{adj}} = \frac{8}{15}$$

$$\csc \theta = \frac{\text{hyp}}{\text{opp}} = \frac{17}{8}$$

$$\sec \theta = \frac{\text{hyp}}{\text{adj}} = \frac{17}{15}$$

$$\cot \theta = \frac{\text{adj}}{\text{opp}} = \frac{15}{8}$$

5. $\text{opp} = \sqrt{10^2 - 8^2} = 6$

$$\sin \theta = \frac{\text{opp}}{\text{hyp}} = \frac{6}{10} = \frac{3}{5}$$

$$\cos \theta = \frac{\text{adj}}{\text{hyp}} = \frac{8}{10} = \frac{4}{5}$$

$$\tan \theta = \frac{\text{opp}}{\text{adj}} = \frac{6}{8} = \frac{3}{4}$$

$$\csc \theta = \frac{\text{hyp}}{\text{opp}} = \frac{10}{6} = \frac{5}{3}$$

$$\sec \theta = \frac{\text{hyp}}{\text{adj}} = \frac{10}{8} = \frac{5}{4}$$

$$\cot \theta = \frac{\text{adj}}{\text{opp}} = \frac{8}{6} = \frac{4}{3}$$

$\text{opp} = \sqrt{2.5^2 - 2^2} = 1.5$

$$\sin \theta = \frac{\text{opp}}{\text{hyp}} = \frac{1.5}{2.5} = \frac{3}{5}$$

$$\cos \theta = \frac{\text{adj}}{\text{hyp}} = \frac{2}{2.5} = \frac{4}{5}$$

$$\tan \theta = \frac{\text{opp}}{\text{adj}} = \frac{1.5}{2} = \frac{3}{4}$$

$$\csc \theta = \frac{\text{hyp}}{\text{opp}} = \frac{2.5}{1.5} = \frac{5}{3}$$

$$\sec \theta = \frac{\text{hyp}}{\text{adj}} = \frac{2.5}{2} = \frac{5}{4}$$

$$\cot \theta = \frac{\text{adj}}{\text{opp}} = \frac{2}{1.5} = \frac{4}{3}$$

The function values are the same since the triangles are similar and the corresponding sides are proportional.

7. $\text{adj} = \sqrt{3^2 - 1^2} = \sqrt{8} = 2\sqrt{2}$

$\sin \theta = \dfrac{\text{opp}}{\text{hyp}} = \dfrac{1}{3}$

$\cos \theta = \dfrac{\text{adj}}{\text{hyp}} = \dfrac{2\sqrt{2}}{3}$

$\tan \theta = \dfrac{\text{opp}}{\text{adj}} = \dfrac{1}{2\sqrt{2}} = \dfrac{\sqrt{2}}{4}$

$\csc \theta = \dfrac{\text{hyp}}{\text{opp}} = 3$

$\sec \theta = \dfrac{\text{hyp}}{\text{adj}} = \dfrac{3}{2\sqrt{2}} = \dfrac{3\sqrt{2}}{4}$

$\cot \theta = \dfrac{\text{adj}}{\text{opp}} = 2\sqrt{2}$

$\text{adj} = \sqrt{6^2 - 2^2} = \sqrt{32} = 4\sqrt{2}$

$\sin \theta = \dfrac{\text{opp}}{\text{hyp}} = \dfrac{2}{6} = \dfrac{1}{3}$

$\cos \theta = \dfrac{\text{adj}}{\text{hyp}} = \dfrac{4\sqrt{2}}{6} = \dfrac{2\sqrt{2}}{3}$

$\tan \theta = \dfrac{\text{opp}}{\text{adj}} = \dfrac{2}{4\sqrt{2}} = \dfrac{1}{2\sqrt{2}} = \dfrac{\sqrt{2}}{4}$

$\csc \theta = \dfrac{\text{hyp}}{\text{opp}} = \dfrac{6}{2} = 3$

$\sec \theta = \dfrac{\text{hyp}}{\text{adj}} = \dfrac{6}{4\sqrt{2}} = \dfrac{3}{2\sqrt{2}} = \dfrac{3\sqrt{2}}{4}$

$\cot \theta = \dfrac{\text{adj}}{\text{opp}} = \dfrac{4\sqrt{2}}{2} = 2\sqrt{2}$

The function values are the same since the triangles are similar and the corresponding sides are proportional.

9. Given: $\sin \theta = \dfrac{5}{6} = \dfrac{\text{opp}}{\text{hyp}}$

$5^2 + (\text{adj})^2 = 6^2$

$\text{adj} = \sqrt{11}$

$\cos \theta = \dfrac{\text{adj}}{\text{hyp}} = \dfrac{\sqrt{11}}{6}$

$\tan \theta = \dfrac{\text{opp}}{\text{adj}} = \dfrac{5}{\sqrt{11}} = \dfrac{5\sqrt{11}}{11}$

$\cot \theta = \dfrac{\text{adj}}{\text{opp}} = \dfrac{\sqrt{11}}{5}$

$\sec \theta = \dfrac{\text{hyp}}{\text{adj}} = \dfrac{6}{\sqrt{11}} = \dfrac{6\sqrt{11}}{11}$

$\csc \theta = \dfrac{\text{hyp}}{\text{opp}} = \dfrac{6}{5}$

11. Given: $\sec \theta = 4 = \dfrac{4}{1} = \dfrac{\text{hyp}}{\text{adj}}$

$(\text{opp})^2 + 1^2 = 4^2$

$\text{opp} = \sqrt{15}$

$\sin \theta = \dfrac{\sqrt{15}}{4}$

$\cos \theta = \dfrac{1}{4}$

$\tan \theta = \sqrt{15}$

$\cot \theta = \dfrac{1}{\sqrt{15}} = \dfrac{\sqrt{15}}{15}$

$\csc \theta = \dfrac{4}{\sqrt{15}} = \dfrac{4\sqrt{15}}{15}$

13. Given: $\tan \theta = 3 = \dfrac{3}{1} = \dfrac{\text{opp}}{\text{adj}}$

$3^2 + 1^2 = (\text{hyp})^2$

$\text{hyp} = \sqrt{10}$

$\sin \theta = \dfrac{3\sqrt{10}}{10}$

$\cos \theta = \dfrac{\sqrt{10}}{10}$

$\sec \theta = \sqrt{10}$

$\cot \theta = \dfrac{1}{3}$

$\csc \theta = \dfrac{\sqrt{10}}{3}$

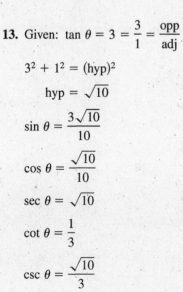

15. Given: $\cot \theta = \dfrac{9}{4} = \dfrac{\text{adj}}{\text{opp}}$

$4^2 + 9^2 = (\text{hyp})^2$

$\text{hyp} = \sqrt{97}$

$\sin \theta = \dfrac{4}{\sqrt{97}} = \dfrac{4\sqrt{97}}{97}$

$\cos \theta = \dfrac{9}{\sqrt{97}} = \dfrac{9\sqrt{97}}{97}$

$\tan \theta = \dfrac{4}{9}$

$\sec \theta = \dfrac{\sqrt{97}}{9}$

$\csc \theta = \dfrac{\sqrt{97}}{4}$

Function	θ (deg)	θ (rad)	Function Value
17. sin	30°	$\dfrac{\pi}{6}$	$\dfrac{1}{2}$
19. tan	60°	$\dfrac{\pi}{3}$	$\sqrt{3}$
21. cot	60°	$\dfrac{\pi}{3}$	$\dfrac{\sqrt{3}}{3}$
23. cos	30°	$\dfrac{\pi}{6}$	$\dfrac{\sqrt{3}}{2}$
25. cot	45°	$\dfrac{\pi}{4}$	1

27. $\sin \theta = \dfrac{1}{\csc \theta}$

29. $\tan \theta = \dfrac{1}{\cot \theta}$

31. $\sec \theta = \dfrac{1}{\cos \theta}$

33. $\tan \theta = \dfrac{\sin \theta}{\cos \theta}$

35. $\sin^2 \theta + \cos^2 \theta = 1$

37. $\sin(90° - \theta) = \cos \theta$

39. $\tan(90° - \theta) = \cot \theta$

41. $\sec(90° - \theta) = \csc \theta$

43. $\sin 60° = \dfrac{\sqrt{3}}{2}$, $\cos 60° = \dfrac{1}{2}$

(a) $\tan 60° = \dfrac{\sin 60°}{\cos 60°} = \sqrt{3}$

(b) $\sin 30° = \cos 60° = \dfrac{1}{2}$

(c) $\cos 30° = \sin 60° = \dfrac{\sqrt{3}}{2}$

(d) $\cot 60° = \dfrac{\cos 60°}{\sin 60°} = \dfrac{1}{\sqrt{3}} = \dfrac{\sqrt{3}}{3}$

45. $\csc \theta = 3$, $\sec \theta = \dfrac{3\sqrt{2}}{4}$

(a) $\sin \theta = \dfrac{1}{\csc \theta} = \dfrac{1}{3}$

(b) $\cos \theta = \dfrac{1}{\sec \theta} = \dfrac{2\sqrt{2}}{3}$

(c) $\tan \theta = \dfrac{\sin \theta}{\cos \theta} = \dfrac{1/3}{(2\sqrt{2})/3} = \dfrac{\sqrt{2}}{4}$

(d) $\sec(90° - \theta) = \csc \theta = 3$

47. $\cos \alpha = \dfrac{1}{4}$

(a) $\sec \alpha = \dfrac{1}{\cos \alpha} = 4$

(b) $\sin^2 \alpha + \cos^2 \alpha = 1$

$\sin^2 \alpha + \left(\dfrac{1}{4}\right)^2 = 1$

$\sin^2 \alpha = \dfrac{15}{16}$

$\sin \alpha = \dfrac{\sqrt{15}}{4}$

(c) $\cot \alpha = \dfrac{\cos \alpha}{\sin \alpha}$

$= \dfrac{1/4}{\sqrt{15}/4}$

$= \dfrac{1}{\sqrt{15}} = \dfrac{\sqrt{15}}{15}$

(d) $\sin(90° - \alpha) = \cos \alpha = \dfrac{1}{4}$

49. $\tan \theta \cot \theta = \tan \theta \left(\dfrac{1}{\tan \theta} \right) = 1$

51. $\tan \theta \cos \theta = \left(\dfrac{\sin \theta}{\cos \theta} \right) \cos \theta = \sin \theta$

53. $(1 + \cos \theta)(1 - \cos \theta) = 1 - \cos^2 \theta$

$\qquad\qquad\qquad = (\sin^2 \theta + \cos^2 \theta) - \cos^2 \theta$

$\qquad\qquad\qquad = \sin^2 \theta$

55. $\dfrac{\sin \theta}{\cos \theta} + \dfrac{\cos \theta}{\sin \theta} = \dfrac{\sin^2 \theta + \cos^2 \theta}{\sin \theta \cos \theta}$

$\qquad\qquad\qquad = \dfrac{1}{\sin \theta \cos \theta}$

$\qquad\qquad\qquad = \dfrac{1}{\sin \theta} \cdot \dfrac{1}{\cos \theta} = \csc \theta \sec \theta$

57. (a) $\sin 41° \approx 0.6561$

(b) $\cos 87° \approx 0.0523$

59. (a) $\sec 42° \, 12' = \sec 42.2° = \dfrac{1}{\cos 42.2°} \approx 1.3499$

(b) $\csc 48° \, 7' = \dfrac{1}{\sin\left(48 + \frac{7}{60}\right)°} \approx 1.3432$

61. Make sure that your calculator is in radian mode.

(a) $\cot \dfrac{\pi}{16} = \dfrac{1}{\tan(\pi/16)} \approx 5.0273$

(b) $\tan \dfrac{\pi}{8} \approx 0.4142$

63. (a) $\sin \theta = \dfrac{1}{2} \implies \theta = 30° = \dfrac{\pi}{6}$

(b) $\csc \theta = 2 \implies \theta = 30° = \dfrac{\pi}{6}$

65. (a) $\sec \theta = 2 \implies \theta = 60° = \dfrac{\pi}{3}$

(b) $\cot \theta = 1 \implies \theta = 45° = \dfrac{\pi}{4}$

67. (a) $\csc \theta = \dfrac{2\sqrt{3}}{3} \implies \theta = 60° = \dfrac{\pi}{3}$

(b) $\sin \theta = \dfrac{\sqrt{2}}{2} \implies \theta = 45° = \dfrac{\pi}{4}$

69. $\tan 30° = \dfrac{y}{105} \implies y = 105 \tan 30° = 105 \cdot \dfrac{\sqrt{3}}{3} = 35\sqrt{3}$

$\cos 30° = \dfrac{105}{r} \implies r = \dfrac{105}{\cos 30°} = \dfrac{105}{\sqrt{3}/2} = \dfrac{210}{\sqrt{3}} = 70\sqrt{3}$

71. $\cos 60° = \dfrac{x}{16} \implies x = 16 \cos 60° = 16\left(\dfrac{1}{2}\right) = 8$

$\sin 60° = \dfrac{y}{16} \implies y = 16 \sin 60° = 16\left(\dfrac{\sqrt{3}}{2}\right) = 8\sqrt{3}$

73. $\tan 45° = \dfrac{20}{x} \implies 1 = \dfrac{20}{x} \implies x = 20$

$r^2 = 20^2 + 20^2 \implies r = 20\sqrt{2}$

75. $\tan 45° = \dfrac{2\sqrt{5}}{x} \implies 1 = \dfrac{2\sqrt{5}}{x} \implies x = 2\sqrt{5}$

$r^2 = \left(2\sqrt{5}\right)^2 + \left(2\sqrt{5}\right)^2 = 20 + 20 = 40 \implies r = 2\sqrt{10}$

77. (a)

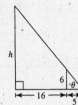

(b) $\tan \theta = \dfrac{6}{5}$ and $\tan \theta = \dfrac{h}{21}$ Thus, $\dfrac{6}{5} = \dfrac{h}{21}$.

(c) $h = \dfrac{6(21)}{5} = 25.2$ feet

79. $\tan \theta = \dfrac{\text{opp}}{\text{adj}}$

$\tan 58° = \dfrac{w}{100}$

$w = 100 \tan 58° \approx 160$ feet

81. (a) $\tan \theta = \dfrac{50}{50} = 1 \Longrightarrow \theta = 45°$

(b) $L^2 = 50^2 + 50^2 = 2 \cdot 50^2 \Longrightarrow L = 50\sqrt{2}$ feet

(c) $\dfrac{50\sqrt{2}}{6} = \dfrac{25\sqrt{2}}{3}$ ft/sec   rate down the zip line

$\dfrac{50}{6} = \dfrac{25}{3}$ ft/sec   vertical rate

83.

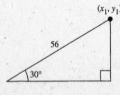

$\sin 30° = \dfrac{y_1}{56}$

$y_1 = (\sin 30°)(56) = \left(\dfrac{1}{2}\right)(56) = 28$

$\cos 30° = \dfrac{x_1}{56}$

$x_1 = \cos 30°(56) = \dfrac{\sqrt{3}}{2}(56) = 28\sqrt{3}$

$(x_1, y_1) = \left(28\sqrt{3}, 28\right)$

$\sin 60° = \dfrac{y_2}{56}$

$y_2 = \sin 60°(56) = \left(\dfrac{\sqrt{3}}{2}\right)(56) = 28\sqrt{3}$

$\cos 60° = \dfrac{x_2}{56}$

$x_2 = (\cos 60°)(56) = \left(\dfrac{1}{2}\right)(56) = 28$

$(x_2, y_2) = \left(28, 28\sqrt{3}\right)$

85. True

$\sin 60° \csc 60° = \sin 60° \dfrac{1}{\sin 60°} = 1$

87. True

$1 + \cot^2 \theta = \csc^2 \theta$ for all θ

89. (a)

θ	0°	20°	40°	60°	80°
$\sin \theta$	0	0.3420	0.6428	0.8660	0.9848
$\cos \theta$	1	0.9397	0.7660	0.5000	0.1736
$\tan \theta$	0	0.3640	0.8391	1.7321	5.6713

(b) Sine and tangent are increasing, cosine is decreasing.

(c) In each case, $\tan \theta = \dfrac{\sin \theta}{\cos \theta}$.

91. 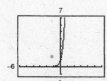 **93.** 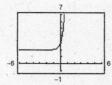 **95.** 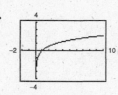 **97.**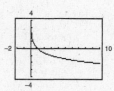

Section 4.4   Trigonometric Functions of Any Angle

- ■ Know the Definitions of Trigonometric Functions of Any Angle.

  If θ is in standard position, (x, y) a point on the terminal side and $r = \sqrt{x^2 + y^2} \neq 0$, then:

  $$\sin \theta = \frac{y}{r} \qquad\qquad \csc \theta = \frac{r}{y}, \ y \neq 0$$

  $$\cos \theta = \frac{x}{r} \qquad\qquad \sec \theta = \frac{r}{x}, \ x \neq 0$$

  $$\tan \theta = \frac{y}{x}, \ x \neq 0 \qquad \cot \theta = \frac{x}{y}, \ y \neq 0$$

- ■ You should know the signs of the trigonometric functions in each quadrant.

- ■ You should know the trigonometric function values of the quadrant angles $0, \dfrac{\pi}{2}, \pi,$ and $\dfrac{3\pi}{2}$.

- ■ You should be able to find reference angles.
- ■ You should be able to evaluate trigonometric functions of any angle. (Use reference angles.)
- ■ You should know that the period of sine and cosine is 2π.
- ■ You should know which trigonometric functions are odd and even.

  Even: $\cos x$ and $\sec x$

  Odd: $\sin x, \tan x, \cot x, \csc x$

Vocabulary Check

1. $\dfrac{y}{r}$ **2.** $\csc \theta$ **3.** $\dfrac{y}{x}$ **4.** $\dfrac{r}{x}$

5. $\cos \theta$ **6.** $\cot \theta$ **7.** reference

1. (a) $(x, y) = (4, 3)$

  $r = \sqrt{16 + 9} = 5$

  $$\sin \theta = \frac{y}{r} = \frac{3}{5} \qquad \csc \theta = \frac{r}{y} = \frac{5}{3}$$

  $$\cos \theta = \frac{x}{r} = \frac{4}{5} \qquad \sec \theta = \frac{r}{x} = \frac{5}{4}$$

  $$\tan \theta = \frac{y}{x} = \frac{3}{4} \qquad \cot \theta = \frac{x}{y} = \frac{4}{3}$$

(b) $(x, y) = (-8, -15)$

  $r = \sqrt{64 + 225} = 17$

  $$\sin \theta = \frac{y}{r} = -\frac{15}{17} \qquad \csc \theta = \frac{r}{y} = -\frac{17}{15}$$

  $$\cos \theta = \frac{x}{r} = -\frac{8}{17} \qquad \sec \theta = \frac{r}{x} = -\frac{17}{8}$$

  $$\tan \theta = \frac{y}{x} = \frac{15}{8} \qquad \cot \theta = \frac{x}{y} = \frac{8}{15}$$

3. (a) $(x, y) = \left(-\sqrt{3}, -1\right)$

$r = \sqrt{3 + 1} = 2$

$\sin \theta = \dfrac{y}{r} = -\dfrac{1}{2}$ $\qquad$ $\csc \theta = \dfrac{r}{y} = -2$

$\cos \theta = \dfrac{x}{r} = \dfrac{-\sqrt{3}}{2}$ $\qquad$ $\sec \theta = \dfrac{r}{x} = \dfrac{-2\sqrt{3}}{3}$

$\tan \theta = \dfrac{y}{x} = \dfrac{\sqrt{3}}{3}$ $\qquad$ $\cot \theta = \dfrac{x}{y} = \sqrt{3}$

(b) $(x, y) = (-2, 2)$

$r = \sqrt{4 + 4} = 2\sqrt{2}$

$\sin \theta = \dfrac{y}{r} = \dfrac{\sqrt{2}}{2}$ $\qquad$ $\csc \theta = \dfrac{r}{y} = \sqrt{2}$

$\cos \theta = \dfrac{x}{r} = -\dfrac{\sqrt{2}}{2}$ $\qquad$ $\sec \theta = \dfrac{r}{x} = -\sqrt{2}$

$\tan \theta = \dfrac{y}{x} = -1$ $\qquad$ $\cot \theta = \dfrac{x}{y} = -1$

5. $(x, y) = (7, 24)$

$r = \sqrt{49 + 576} = 25$

$\sin \theta = \dfrac{y}{r} = \dfrac{24}{25}$ $\qquad$ $\csc \theta = \dfrac{r}{y} = \dfrac{25}{24}$

$\cos \theta = \dfrac{x}{r} = \dfrac{7}{25}$ $\qquad$ $\sec \theta = \dfrac{r}{x} = \dfrac{25}{7}$

$\tan \theta = \dfrac{y}{x} = \dfrac{24}{7}$ $\qquad$ $\cot \theta = \dfrac{x}{y} = \dfrac{7}{24}$

7. $(x, y) = (5, -12)$

$r = \sqrt{5^2 + (-12)^2} = \sqrt{25 + 144} = \sqrt{169} = 13$

$\sin \theta = \dfrac{y}{r} = -\dfrac{12}{13}$

$\cos \theta = \dfrac{x}{r} = \dfrac{5}{13}$

$\tan \theta = \dfrac{y}{x} = -\dfrac{12}{5}$

$\csc \theta = \dfrac{r}{y} = -\dfrac{13}{12}$

$\sec \theta = \dfrac{r}{x} = \dfrac{13}{5}$

$\cot \theta = \dfrac{x}{y} = -\dfrac{5}{12}$

9. $(x, y) = (-4, 10)$

$r = \sqrt{16 + 100} = 2\sqrt{29}$

$\sin \theta = \dfrac{y}{r} = \dfrac{5\sqrt{29}}{29}$

$\cos \theta = \dfrac{x}{r} = -\dfrac{2\sqrt{29}}{29}$

$\tan \theta = \dfrac{y}{x} = -\dfrac{5}{2}$

$\csc \theta = \dfrac{r}{y} = \dfrac{\sqrt{29}}{5}$

$\sec \theta = \dfrac{r}{x} = -\dfrac{\sqrt{29}}{2}$

$\cot \theta = \dfrac{x}{y} = -\dfrac{2}{5}$

11. $(x, y) = (-10, 8)$

$r = \sqrt{(-10)^2 + 8^2} = \sqrt{164} = 2\sqrt{41}$

$\sin \theta = \dfrac{y}{r} = \dfrac{8}{2\sqrt{41}} = \dfrac{4\sqrt{41}}{41}$

$\cos \theta = \dfrac{x}{r} = \dfrac{-10}{2\sqrt{41}} = -\dfrac{5\sqrt{41}}{41}$

$\tan \theta = \dfrac{y}{x} = \dfrac{8}{-10} = -\dfrac{4}{5}$

$\csc \theta = \dfrac{r}{y} = \dfrac{\sqrt{41}}{4}$

$\sec \theta = \dfrac{r}{x} = -\dfrac{\sqrt{41}}{5}$

$\cot \theta = \dfrac{x}{y} = -\dfrac{5}{4}$

13. $\sin \theta < 0 \implies \theta$ lies in Quadrant III or Quadrant IV.

$\cos \theta < 0 \implies \theta$ lies in Quadrant II or Quadrant III.

$\sin \theta < 0$ *and* $\cos \theta < 0 \implies \theta$ lies in Quadrant III.

15. $\cot \theta > 0 \implies \theta$ lies in Quadrant I or Quadrant III.

$\cos \theta > 0 \implies \theta$ lies in Quadrant I or Quadrant IV.

$\cot \theta > 0$ and $\cos \theta > 0 \implies \theta$ lies in Quadrant I.

17. $\sin \theta = \dfrac{y}{r} = \dfrac{3}{5} \implies x^2 = 25 - 9 = 16$

θ in Quadrant II $\implies x = -4$

$$\sin \theta = \frac{y}{r} = \frac{3}{5} \qquad \csc \theta = \frac{r}{y} = \frac{5}{3}$$

$$\cos \theta = \frac{x}{r} = -\frac{4}{5} \qquad \sec \theta = \frac{r}{x} = -\frac{5}{4}$$

$$\tan \theta = \frac{y}{x} = -\frac{3}{4} \qquad \cot \theta = \frac{x}{y} = -\frac{4}{3}$$

19. $\sin \theta < 0 \implies y < 0$

$\tan \theta = \dfrac{y}{x} = \dfrac{-15}{8} \implies r = 17$

$$\sin \theta = \frac{y}{r} = -\frac{15}{17} \qquad \csc \theta = \frac{r}{y} = -\frac{17}{15}$$

$$\cos \theta = \frac{x}{r} = \frac{8}{17} \qquad \sec \theta = \frac{r}{x} = \frac{17}{8}$$

$$\cot \theta = \frac{x}{y} = -\frac{8}{15}$$

21. $\sec \theta = \dfrac{r}{x} = \dfrac{2}{-1} \implies y^2 = 4 - 1 = 3$

$\sin \theta \geq 0 \implies y = \sqrt{3}$

$$\sin \theta = \frac{y}{r} = \frac{\sqrt{3}}{2} \qquad \csc \theta = \frac{r}{y} = \frac{2\sqrt{3}}{3}$$

$$\cos \theta = \frac{x}{r} = -\frac{1}{2} \qquad \sec \theta = \frac{r}{x} = -2$$

$$\tan \theta = \frac{y}{x} = -\sqrt{3} \qquad \cot \theta = \frac{x}{y} = -\frac{\sqrt{3}}{3}$$

23. $\cot \theta$ is undefined $\implies \theta = n\pi$.

$\dfrac{\pi}{2} \leq \theta \leq \dfrac{3\pi}{2} \implies \theta = \pi, y = 0, x = -r$

$$\sin \theta = \frac{y}{r} = 0 \qquad \csc \theta = \frac{r}{y} \text{ is undefined.}$$

$$\cos \theta = \frac{x}{r} = \frac{-r}{r} = -1 \qquad \sec \theta = \frac{r}{x} = -1$$

$$\tan \theta = \frac{y}{x} = \frac{0}{x} = 0 \qquad \cot \theta \text{ is undefined.}$$

25. To find a point on the terminal side of θ, use any point on the line $y = -x$ that lies in Quadrant II. $(-1, 1)$ is one such point.

$x = -1, y = 1, r = \sqrt{2}$

$$\sin \theta = \frac{1}{\sqrt{2}} = \frac{\sqrt{2}}{2} \qquad \csc \theta = \sqrt{2}$$

$$\cos \theta = -\frac{1}{\sqrt{2}} = -\frac{\sqrt{2}}{2} \qquad \sec \theta = -\sqrt{2}$$

$$\tan \theta = -1 \qquad \cot \theta = -1$$

27. To find a point on the terminal side of θ, use any point on the line $y = 2x$ that lies in Quadrant III. $(-1, -2)$ is one such point.

$x = -1, y = -2, r = \sqrt{5}$

$$\sin \theta = -\frac{2}{\sqrt{5}} = -\frac{2\sqrt{5}}{5}$$

$$\cos \theta = -\frac{1}{\sqrt{5}} = -\frac{\sqrt{5}}{5}$$

$$\tan \theta = \frac{-2}{-1} = 2$$

$$\csc \theta = \frac{\sqrt{5}}{-2} = -\frac{\sqrt{5}}{2}$$

$$\sec \theta = \frac{\sqrt{5}}{-1} = -\sqrt{5}$$

$$\cot \theta = \frac{-1}{-2} = \frac{1}{2}$$

29. $(x, y) = (-1, 0)$

$$\sec \pi = \frac{r}{x} = \frac{1}{-1} = -1$$

31. $(x, y) = (0, -1)$

$$\cot\left(\frac{3\pi}{2}\right) = \frac{x}{y} = \frac{0}{-1} = 0$$

33. $(x, y) = (1, 0)$

$$\sec 0 = \frac{r}{x} = \frac{1}{1} = 1$$

35. $(x, y) = (-1, 0)$

$$\cot \pi = \frac{x}{y} = -\frac{1}{0} \implies \text{undefined}$$

37. $\theta = 120°$

$\theta' = 180° - 120° = 60°$

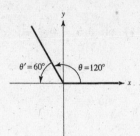

39. $\theta = -135°$ is coterminal with $225°$.

$\theta' = 225° - 180° = 45°$

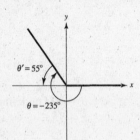

41. $\theta = \frac{5\pi}{3}$, $\theta' = 2\pi - \frac{5\pi}{3} = \frac{\pi}{3}$

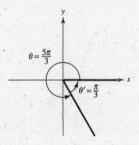

43. $\theta = -\frac{5\pi}{6}$ is coterminal with $\frac{7\pi}{6}$.

$$\theta' = \frac{7\pi}{6} - \pi = \frac{\pi}{6}$$

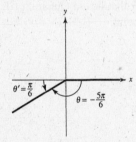

45. $\theta = 208°$

$\theta' = 208° - 180° = 28°$

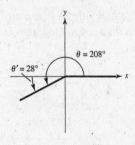

47. $\theta = -292°$

$\theta' = 360° - 292° = 68°$

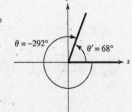

49. $\theta = \frac{11\pi}{5}$ is coterminal with $\frac{\pi}{5}$.

$$\theta' = \frac{\pi}{5}$$

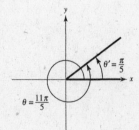

51. $\theta = -1.8$ lies in Quadrant III.

Reference angle: $\pi - 1.8 \approx 1.342$

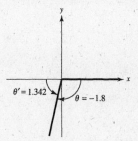

53. $\theta' = 45°$, Quadrant III

$$\sin 225° = -\sin 45° = -\frac{\sqrt{2}}{2}$$

$$\cos 225° = -\cos 45° = -\frac{\sqrt{2}}{2}$$

$$\tan 225° = \tan 45° = 1$$

55. $\theta = -750°$ is coterminal with $330°$, Quadrant IV.

$$\theta' = 360° - 330° = 30°$$

$$\sin(-750°) = -\sin 30° = -\frac{1}{2}$$

$$\cos(-750°) = \cos 30° = \frac{\sqrt{3}}{2}$$

$$\tan(-750°) = -\tan 30° = -\frac{\sqrt{3}}{3}$$

57. $\theta = \dfrac{5\pi}{3}$, Quadrant IV

$$\theta' = 2\pi - \frac{5\pi}{3} = \frac{\pi}{3}$$

$$\sin\left(\frac{5\pi}{3}\right) = -\sin\left(\frac{\pi}{3}\right) = -\frac{\sqrt{3}}{2}$$

$$\cos\left(\frac{5\pi}{3}\right) = \cos\left(\frac{\pi}{3}\right) = \frac{1}{2}$$

$$\tan\left(\frac{5\pi}{3}\right) = -\tan\left(\frac{\pi}{3}\right) = -\sqrt{3}$$

59. $\theta' = \dfrac{\pi}{6}$, Quadrant IV

$$\sin\left(-\frac{\pi}{6}\right) = -\sin\frac{\pi}{6} = -\frac{1}{2}$$

$$\cos\left(-\frac{\pi}{6}\right) = \cos\frac{\pi}{6} = \frac{\sqrt{3}}{2}$$

$$\tan\left(-\frac{\pi}{6}\right) = -\tan\frac{\pi}{6} = -\frac{\sqrt{3}}{3}$$

61. $\theta' = \dfrac{\pi}{4}$, Quadrant II

$$\sin\frac{11\pi}{4} = \sin\frac{\pi}{4} = \frac{\sqrt{2}}{2}$$

$$\cos\frac{11\pi}{4} = -\cos\frac{\pi}{4} = -\frac{\sqrt{2}}{2}$$

$$\tan\frac{11\pi}{4} = -\tan\frac{\pi}{4} = -1$$

63. $\theta = -\dfrac{17\pi}{6}$ is coterminal with $\dfrac{7\pi}{6}$.

$$\theta' = \frac{7\pi}{6} - \pi = \frac{\pi}{6}, \text{ Quadrant III}$$

$$\sin\left(-\frac{17\pi}{6}\right) = -\sin\left(\frac{\pi}{6}\right) = -\frac{1}{2}$$

$$\cos\left(-\frac{17\pi}{6}\right) = -\cos\left(\frac{\pi}{6}\right) = -\frac{\sqrt{3}}{2}$$

$$\tan\left(-\frac{17\pi}{6}\right) = \tan\left(\frac{\pi}{6}\right) = \frac{\sqrt{3}}{3}$$

65.
$$\sin\theta = -\frac{3}{5}$$

$$\sin^2\theta + \cos^2\theta = 1$$

$$\cos^2\theta = 1 - \sin^2\theta$$

$$\cos^2\theta = 1 - \left(-\frac{3}{5}\right)^2$$

$$\cos^2\theta = 1 - \frac{9}{25}$$

$$\cos^2\theta = \frac{16}{25}$$

$\cos\theta > 0$ in Quadrant IV.

$$\cos\theta = \frac{4}{5}$$

67. $\csc \theta = -2$

$1 + \cot^2 \theta = \csc^2 \theta$

$\cot^2 \theta = \csc^2 \theta - 1$

$\cot^2 \theta = (-2)^2 - 1$

$\cot^2 \theta = 3$

$\cot \theta < 0$ in Quadrant IV.

$\cot \theta = -\sqrt{3}$

69. $\sec \theta = -\dfrac{9}{4}$

$1 + \tan^2 \theta = \sec^2 \theta$

$\tan^2 \theta = \sec^2 \theta - 1$

$\tan^2 \theta = \left(-\dfrac{9}{4}\right)^2 - 1$

$\tan^2 \theta = \dfrac{65}{16}$

$\tan \theta > 0$ in Quadrant III.

$\tan \theta = \dfrac{\sqrt{65}}{4}$

71. $\sin \theta = \dfrac{2}{5}$ and $\cos \theta < 0 \Rightarrow \theta$ is in Quadrant II.

$\cos \theta = -\sqrt{1 - \sin^2 \theta} = -\sqrt{1 - \dfrac{4}{25}} = -\dfrac{\sqrt{21}}{5}$

$\tan \theta = \dfrac{\sin \theta}{\cos \theta} = \dfrac{2/5}{-\sqrt{21}/5} = \dfrac{-2}{\sqrt{21}} = \dfrac{-2\sqrt{21}}{21}$

$\csc \theta = \dfrac{1}{\sin \theta} = \dfrac{5}{2}$

$\sec \theta = \dfrac{1}{\cos \theta} = \dfrac{-5}{\sqrt{21}} = \dfrac{-5\sqrt{21}}{21}$

$\cot \theta = \dfrac{1}{\tan \theta} = \dfrac{-\sqrt{21}}{2}$

73. $\tan \theta = -4$ and $\cos \theta < 0 \Rightarrow \theta$ is in Quadrant II.

$\sec \theta = -\sqrt{1 + \tan^2 \theta} = -\sqrt{1 + 16} = -\sqrt{17}$

$\cos \theta = \dfrac{1}{\sec \theta} = \dfrac{-1}{\sqrt{17}} = \dfrac{-\sqrt{17}}{17}$

$\sin \theta = \tan \theta \cos \theta = (-4)\left(-\dfrac{\sqrt{17}}{17}\right) = \dfrac{4\sqrt{17}}{17}$

$\csc \theta = \dfrac{1}{\sin \theta} = \dfrac{17}{4\sqrt{17}} = \dfrac{\sqrt{17}}{4}$

$\cot \theta = \dfrac{1}{\tan \theta} = -\dfrac{1}{4}$

75. $\csc \theta = -\dfrac{3}{2}$ and $\tan \theta < 0 \Rightarrow \theta$ is in Quadrant IV.

$\sin \theta = \dfrac{1}{\csc \theta} = -\dfrac{2}{3}$

$\cos \theta = \sqrt{1 - \sin^2 \theta} = \sqrt{1 - \dfrac{4}{9}} = \dfrac{\sqrt{5}}{3}$

$\sec \theta = \dfrac{1}{\cos \theta} = \dfrac{3}{\sqrt{5}} = \dfrac{3\sqrt{5}}{5}$

$\tan \theta = \dfrac{\sin \theta}{\cos \theta} = \dfrac{-2/3}{\sqrt{5}/3} = \dfrac{-2}{\sqrt{5}} = \dfrac{-2\sqrt{5}}{5}$

$\cot \theta = \dfrac{1}{\tan \theta} = \dfrac{-\sqrt{5}}{2}$

77. $\sin 10° \approx 0.1736$

79. $\tan 245° \approx 2.1445$

81. $\cos(-110°) \approx -0.3420$

83. $\sec(-280°) = \dfrac{1}{\cos(-280°)}$

≈ 5.7588

85. $\tan\left(\dfrac{2\pi}{9}\right) \approx 0.8391$

87. $\csc\left(-\dfrac{8\pi}{9}\right) = \dfrac{1}{\sin\left(-\dfrac{8\pi}{9}\right)}$

≈ -2.9238

89. (a) $\sin \theta = \dfrac{1}{2} \implies$ reference angle is 30° or

$\dfrac{\pi}{6}$ and θ is in Quadrant I or Quadrant II.

Values in degrees: 30°, 150°

Values in radian: $\dfrac{\pi}{6}, \dfrac{5\pi}{6}$

(b) $\sin \theta = -\dfrac{1}{2} \implies$ reference angle is 30° or

$\dfrac{\pi}{6}$ and θ is in Quadrant III or Quadrant IV.

Values in degrees: 210°, 330°

Values in radians: $\dfrac{7\pi}{6}, \dfrac{11\pi}{6}$

91. (a) $\csc \theta = \dfrac{2\sqrt{3}}{3} \implies$ reference angle is 60° or

$\dfrac{\pi}{3}$ and θ is in Quadrant I or Quadrant II.

Values in degrees: 60°, 120°

Values in radians: $\dfrac{\pi}{3}, \dfrac{2\pi}{3}$

(b) $\cot \theta = -1 \implies$ reference angle is 45° or

$\dfrac{\pi}{4}$ and θ is in Quadrant II or Quadrant IV.

Values in degrees: 135°, 315°

Values in radians: $\dfrac{3\pi}{4}, \dfrac{7\pi}{4}$

93. (a) $\sec \theta = -\dfrac{2\sqrt{3}}{3} \implies$ reference angle is $\dfrac{\pi}{6}$ or 30°, and θ is in Quadrant II or Quadrant III.

Values in degrees: 150°, 210°

Values in radians: $\dfrac{5\pi}{6}, \dfrac{7\pi}{6}$

(b) $\cos \theta = -\dfrac{1}{2} \implies$ reference angle is $\dfrac{\pi}{3}$ or 60°, and θ is in Quadrant II or Quadrant III.

Values in degrees: 120°, 240°

Values in radians: $\dfrac{2\pi}{3}, \dfrac{4\pi}{3}$

95. (a) $f(\theta) + g(\theta) = \sin 30° + \cos 30° = \dfrac{1}{2} + \dfrac{\sqrt{3}}{2} = \dfrac{1 + \sqrt{3}}{2}$

(b) $\cos 30° - \sin 30° = \dfrac{\sqrt{3} - 1}{2}$

(c) $[\cos 30°]^2 = \left(\dfrac{\sqrt{3}}{2} \right)^2 = \dfrac{3}{4}$

(d) $\sin 30° \cos 30° = \left(\dfrac{1}{2} \right)\left(\dfrac{\sqrt{3}}{2} \right) = \dfrac{\sqrt{3}}{4}$

(e) $2 \sin 30° = 2\left(\dfrac{1}{2} \right) = 1$

(f) $\cos(-30°) = \cos 30° = \dfrac{\sqrt{3}}{2}$

97. (a) $f(\theta) + g(\theta) = \sin 315° + \cos 315° = -\dfrac{\sqrt{2}}{2} + \dfrac{\sqrt{2}}{2} = 0$

(b) $\cos 315° - \sin 315° = \dfrac{\sqrt{2}}{2} - \left(\dfrac{-\sqrt{2}}{2} \right) = \sqrt{2}$

(c) $[\cos 315°]^2 = \left(\dfrac{\sqrt{2}}{2} \right)^2 = \dfrac{1}{2}$

(d) $\sin 315° \cos 315° = \left(\dfrac{-\sqrt{2}}{2} \right)\left(\dfrac{\sqrt{2}}{2} \right) = -\dfrac{1}{2}$

(e) $2 \sin 315° = 2\left(\dfrac{-\sqrt{2}}{2} \right) = -\sqrt{2}$

(f) $\cos(-315°) = \cos(315°) = \dfrac{\sqrt{2}}{2}$

99. (a) $f(\theta) + g(\theta) = \sin 150° + \cos 150° = \dfrac{1}{2} + \dfrac{-\sqrt{3}}{2} = \dfrac{1 - \sqrt{3}}{2}$

(b) $\cos 150° - \sin 150° = \dfrac{-\sqrt{3}}{2} - \dfrac{1}{2} = \dfrac{-1 - \sqrt{3}}{2}$

(c) $[\cos 150°]^2 = \left(\dfrac{-\sqrt{3}}{2} \right)^2 = \dfrac{3}{4}$

(d) $\sin 150° \cos 150° = \dfrac{1}{2} \cdot \dfrac{-\sqrt{3}}{2} = \dfrac{-\sqrt{3}}{4}$

(e) $2 \sin 150° = 2\left(\dfrac{1}{2} \right) = 1$

(f) $\cos(-150°) = \cos(150°) = \dfrac{-\sqrt{3}}{2}$

101. (a) $f(\theta) + g(\theta) = \sin\dfrac{7\pi}{6} + \cos\dfrac{7\pi}{6} = -\dfrac{1}{2} - \dfrac{\sqrt{3}}{2} = \dfrac{-1-\sqrt{3}}{2}$

(b) $\cos\dfrac{7\pi}{6} - \sin\dfrac{7\pi}{6} = \dfrac{-\sqrt{3}}{2} - \left(-\dfrac{1}{2}\right) = \dfrac{1-\sqrt{3}}{2}$

(c) $\left[\cos\dfrac{7\pi}{6}\right]^2 = \left(\dfrac{-\sqrt{3}}{2}\right)^2 = \dfrac{3}{4}$

(e) $2\sin\dfrac{7\pi}{6} = 2\left(-\dfrac{1}{2}\right) = -1$

(d) $\sin\dfrac{7\pi}{6}\cos\dfrac{7\pi}{6} = \left(-\dfrac{1}{2}\right)\left(-\dfrac{\sqrt{3}}{2}\right) = \dfrac{\sqrt{3}}{4}$

(f) $\cos\left(\dfrac{-7\pi}{6}\right) = \cos\left(\dfrac{7\pi}{6}\right) = \dfrac{-\sqrt{3}}{2}$

103. (a) $f(\theta) + g(\theta) = \sin\dfrac{4\pi}{3} + \cos\dfrac{4\pi}{3} = \dfrac{-\sqrt{3}}{2} - \dfrac{1}{2} = \dfrac{-1-\sqrt{3}}{2}$

(b) $\cos\dfrac{4\pi}{3} - \sin\dfrac{4\pi}{3} = -\dfrac{1}{2} - \left(\dfrac{-\sqrt{3}}{2}\right) = \dfrac{\sqrt{3}-1}{2}$

(c) $\left[\cos\dfrac{4\pi}{3}\right]^2 = \left(-\dfrac{1}{2}\right)^2 = \dfrac{1}{4}$

(e) $2\sin\dfrac{4\pi}{3} = 2\left(\dfrac{-\sqrt{3}}{2}\right) = -\sqrt{3}$

(d) $\sin\dfrac{4\pi}{3}\cos\dfrac{4\pi}{3} = \left(\dfrac{-\sqrt{3}}{2}\right)\left(-\dfrac{1}{2}\right) = \dfrac{\sqrt{3}}{4}$

(f) $\cos\left(\dfrac{-4\pi}{3}\right) = \cos\left(\dfrac{4\pi}{3}\right) = -\dfrac{1}{2}$

105. (a) $f(\theta) + g(\theta) = \sin 270° + \cos 270° = -1 + 0 = -1$

(d) $\sin 270° \cos 270° = (-1)(0) = 0$

(b) $\cos 270° - \sin 270° = 0 - (-1) = 1$

(e) $2\sin 270° = 2(-1) = -2$

(c) $[\cos 270°]^2 = 0^2 = 0$

(f) $\cos(-270°) = \cos(270°) = 0$

107. (a) $f(\theta) + g(\theta) = \sin\dfrac{7\pi}{2} + \cos\dfrac{7\pi}{2} = -1 + 0 = -1$

(d) $\sin\dfrac{7\pi}{2}\cos\dfrac{7\pi}{2} = (-1)(0) = 0$

(b) $\cos\dfrac{7\pi}{2} - \sin\dfrac{7\pi}{2} = 0 - (-1) = 1$

(e) $2\sin\dfrac{7\pi}{2} = 2(-1) = -2$

(c) $\left[\cos\dfrac{7\pi}{2}\right]^2 = 0^2 = 0$

(f) $\cos\left(\dfrac{-7\pi}{2}\right) = \cos\left(\dfrac{7\pi}{2}\right) = 0$

109. $T = 49.5 + 20.5\cos\left(\dfrac{\pi t}{6} - \dfrac{7\pi}{6}\right)$

(a) January: $t = 1 \Rightarrow T = 49.5 + 20.5\cos\left(\dfrac{\pi(1)}{6} - \dfrac{7\pi}{6}\right) = 29°$

(b) July: $t = 7 \Rightarrow T = 70°$

(c) December: $t = 12 \Rightarrow T \approx 31.75°$

111. $\sin\theta = \dfrac{6}{d} \Rightarrow d = \dfrac{6}{\sin\theta}$

(a) $\theta = 30°$

$d = \dfrac{6}{\sin 30°} = \dfrac{6}{(1/2)} = 12$ miles

(c) $\theta = 120°$

$d = \dfrac{6}{\sin 120°} \approx 6.9$ miles

(b) $\theta = 90°$

$d = \dfrac{6}{\sin 90°} = \dfrac{6}{1} = 6$ mile

113. True. The reference angle for $\theta = 151°$ is
$\theta' = 180° - 151° = 29°$, and sine is positive
in Quadrants I and II.

115. (a)

θ	0°	20°	40°	60°	80°
$\sin \theta$	0	0.3420	0.6428	0.8660	0.9848
$\sin(180° - \theta)$	0	0.3420	0.6428	0.8660	0.9848

(b) It appears that $\sin \theta = \sin(180° - \theta)$.

117. $3x - 7 = 14$

$3x = 21$

$x = 7$

119. $x^2 - 2x - 5 = 0$

$x = \dfrac{2 \pm \sqrt{4 + 20}}{2} = 1 \pm \sqrt{6}$

$x \approx 3.449, x \approx -1.449$

121. $\dfrac{3}{x - 1} = \dfrac{x + 2}{9}$

$27 = (x - 1)(x + 2)$

$x^2 + x - 29 = 0$

$x = \dfrac{-1 \pm \sqrt{1 + 4(29)}}{2} = \dfrac{-1 \pm \sqrt{117}}{2}$

$x \approx -5.908, 4.908$

123. $4^{3-x} = 726$

$3 - x = \log_4 726$

$x = 3 - \log_4 726$

$= 3 - \dfrac{\ln 726}{\ln 4}$

≈ -1.752

125. $\ln x = -6$

$x = e^{-6} \approx 0.002479 \approx 0.002$

Section 4.5   Graphs of Sine and Cosine Functions

- ■ You should be able to graph $y = a \sin(bx - c)$ and $y = a \cos(bx - c)$.
- ■ Amplitude: $|a|$
- ■ Period: $\dfrac{2\pi}{|b|}$
- ■ Shift: Solve $bx - c = 0$ and $bx - c = 2\pi$.
- ■ Key increments: $\dfrac{1}{4}$ (period)

Vocabulary Check

1. amplitude   **2.** one cycle   **3.** $\dfrac{2\pi}{b}$ **4.** phase shift

1. $f(x) = \sin x$

  (a) x-intercepts: $(-2\pi, 0), (-\pi, 0), (0, 0), (\pi, 0), (2\pi, 0)$

  (b) y-intercept: $(0, 0)$

  (c) Increasing on: $\left(-2\pi, -\dfrac{3\pi}{2}\right), \left(-\dfrac{\pi}{2}, \dfrac{\pi}{2}\right), \left(\dfrac{3\pi}{2}, 2\pi\right)$

  Decreasing on: $\left(-\dfrac{3\pi}{2}, -\dfrac{\pi}{2}\right), \left(\dfrac{\pi}{2}, \dfrac{3\pi}{2}\right)$

  (d) Relative maxima: $\left(-\dfrac{3\pi}{2}, 1\right), \left(\dfrac{\pi}{2}, 1\right)$

  Relative minima: $\left(-\dfrac{\pi}{2}, -1\right), \left(\dfrac{3\pi}{2}, -1\right)$

3. $y = 3 \sin 2x$

  Period: $\dfrac{2\pi}{2} = \pi$

  Amplitude: $|3| = 3$

  Xmin = -2π
  Xmax = 2π
  Xscl = $\pi/2$
  Ymin = -4
  Ymax = 4
  Yscl = 1

5. $y = \dfrac{5}{2} \cos \dfrac{x}{2}$

  Period: $\dfrac{2\pi}{1/2} = 4\pi$

  Amplitude: $\left|\dfrac{5}{2}\right| = \dfrac{5}{2}$

  Xmin = -4π
  Xmax = 4π
  Xscl = π
  Ymin = -3
  Ymax = 3
  Yscl = 1

7. $y = \dfrac{2}{3} \sin \pi x$

  Period: $\dfrac{2\pi}{\pi} = 2$

  Amplitude: $\left|\dfrac{2}{3}\right| = \dfrac{2}{3}$

9. $y = -2 \sin x$

  Period: $\dfrac{2\pi}{1} = 2\pi$

  Amplitude: $|-2| = 2$

11. $y = \dfrac{1}{4} \cos \dfrac{2x}{3}$

  Period: $\dfrac{2\pi}{2/3} = 3\pi$

  Amplitude: $\left|\dfrac{1}{4}\right| = \dfrac{1}{4}$

13. $y = \dfrac{1}{3} \sin 4\pi x$

  Period: $\dfrac{2\pi}{4\pi} = \dfrac{1}{2}$

  Amplitude: $\left|\dfrac{1}{3}\right| = \dfrac{1}{3}$

15. $f(x) = \sin x$

  $g(x) = \sin(x - \pi)$

  The graph of g is a horizontal shift to the right π units of the graph of f (a phase shift).

17. $f(x) = \cos 2x$

  $g(x) = -\cos 2x$

  The graph of g is a reflection in the x-axis of the graph of f.

19. $f(x) = \cos x$

  $g(x) = -5 \cos x$

  The graph of g has five times the amplitude of f, and reflected in the x-axis.

21. $f(x) = \sin 2x$

  $g(x) = 5 + \sin 2x$

  The graph of g is a vertical shift upward of five units of the graph of f.

23. The graph of g has twice the amplitude as the graph of f. The period is the same.

25. The graph of g is a horizontal shift π units to the right of the graph of f.

27. $f(x) = \sin x$

Period: 2π

Amplitude: 1

$g(x) = -4 \sin x$

Period: 2π

Amplitude: $|-4| = 4$

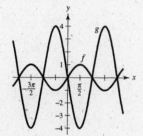

29. $f(x) = \cos x$

Period: 2π

Amplitude: 1

$g(x) = 4 + \cos x$ is a vertical shift of the graph of $f(x)$ four units upward.

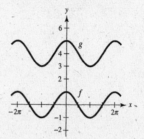

31. $f(x) = -\dfrac{1}{2} \sin \dfrac{x}{2}$

Period: 4π

Amplitude: $\dfrac{1}{2}$

$g(x) = 3 - \dfrac{1}{2} \sin \dfrac{x}{2}$ is the graph of $f(x)$

shifted vertically three units upward.

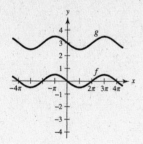

33. $f(x) = 2 \cos x$

Period: 2π

Amplitude: 2

$g(x) = 2 \cos(x + \pi)$ is the graph of $f(x)$ shifted π units to the left.

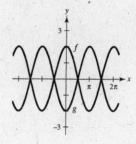

35. $f(x) = \sin x,\ g(x) = \cos\left(x - \dfrac{\pi}{2}\right)$

$\sin x = \cos\left(x - \dfrac{\pi}{2}\right)$

Period: 2π

Amplitude: 1

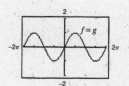

37. $f(x) = \cos x$

$g(x) = -\sin\left(x - \dfrac{\pi}{2}\right) = \sin\left(\dfrac{\pi}{2} - x\right) = \cos x$

Thus, $f(x) = g(x)$.

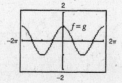

39. $y = 3 \sin x$

Period: 2π

Amplitude: 3

Key points: $(0, 0), \left(\dfrac{\pi}{2}, 3\right), (\pi, 0), \left(\dfrac{3\pi}{2}, -3\right), (2\pi, 0)$

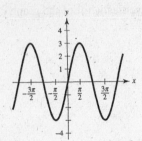

41. $y = \cos \dfrac{x}{2}$

Period: 4π

Amplitude: 1

Key points: $(0, 1), (\pi, 0), (2\pi, -1), (3\pi, 0), (4\pi, 1)$

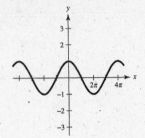

43. $y = \sin\left(x - \dfrac{\pi}{4}\right);\ a = 1,\ b = 1,\ c = \dfrac{\pi}{4}$

Period: 2π

Amplitude: 1

Shift: Set $x - \dfrac{\pi}{4} = 0$ and $x - \dfrac{\pi}{4} = 2\pi$

$\qquad x = \dfrac{\pi}{4} \qquad\qquad x = \dfrac{9\pi}{4}$

Key points: $\left(\dfrac{\pi}{4}, 0\right), \left(\dfrac{3\pi}{4}, 1\right), \left(\dfrac{5\pi}{4}, 0\right), \left(\dfrac{7\pi}{4}, -1\right), \left(\dfrac{9\pi}{4}, 0\right)$

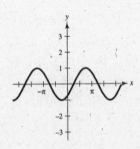

45. $y = -8 \cos(x + \pi)$

Period: 2π

Amplitude: 8

Key points: $(-\pi, -8), \left(-\dfrac{\pi}{2}, 0\right), (0, 8), \left(\dfrac{\pi}{2}, 0\right), (\pi, -8)$

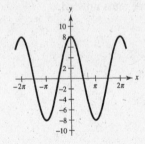

47. $y = -2 \sin \dfrac{2\pi x}{3}$

Amplitude: 2

Period: $\dfrac{2\pi}{2\pi/3} = 3$

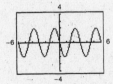

49. $y = -4 + 5 \cos \dfrac{\pi t}{12}$

Amplitude: 5

Period: $\dfrac{2\pi}{\pi/12} = 24$

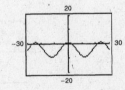

51. $y = \dfrac{2}{3} \cos\left(\dfrac{x}{2} - \dfrac{\pi}{4}\right)$

Amplitude: $\dfrac{2}{3}$

Period: $\dfrac{2\pi}{1/2} = 4\pi$

53. $y = -2 \sin(4x + \pi)$

Amplitude: 2

Period: $\dfrac{\pi}{2}$

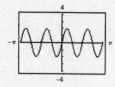

55. $y = \cos\left(2\pi x - \dfrac{\pi}{2}\right) + 1$

Amplitude: 1

Period: 1

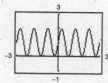

57. $y = 5 \sin(\pi - 2x) + 10$

Amplitude: 5

Period: π

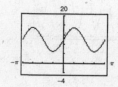

59. $y = \dfrac{1}{100} \sin 120\pi t$

Amplitude: $\dfrac{1}{100}$

Period: $\dfrac{1}{60}$

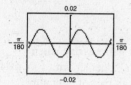

61. $f(x) = a \cos x + d$

Amplitude: $\dfrac{1}{2}[8 - 0] = 4$

Since $f(x)$ is the graph of $g(x) = 4 \cos x$ reflected about the x-axis and shifted vertically four units upward, we have $a = -4$ and $d = 4$. Thus,

$f(x) = -4 \cos x + 4$

$\quad\ = 4 - 4 \cos x.$

63. $f(x) = a \cos x + d$

Amplitude: $\dfrac{1}{2}[7 - (-5)] = 6$

Graph of f is the graph of $g(x) = 6 \cos x$ reflected about the x-axis and shifted vertically one unit upward. Thus, $f(x) = -6 \cos x + 1.$

65. $f(x) = a \sin(bx - c)$

Amplitude: $|a| = 3$

Since the graph is reflected about the x-axis, we have $a = -3$.

Period: $\dfrac{2\pi}{b} = \pi \implies b = 2$

Phase shift: $c = 0$

Thus, $f(x) = -3 \sin 2x.$

67. $f(x) = a \sin(bx - c)$

Amplitude: $a = 1$

Period: $2\pi \implies b = 1$

Phase shift: $bx - c = 0$ when $x = \dfrac{\pi}{4}$.

$(1)\left(\dfrac{\pi}{4}\right) - c = 0 \implies c = \dfrac{\pi}{4}$

Thus, $f(x) = \sin\left(x - \dfrac{\pi}{4}\right).$

69. $y_1 = \sin x$

$y_2 = -\dfrac{1}{2}$

In the interval $[-2\pi, 2\pi]$, $\sin x = -\dfrac{1}{2}$ when

$x = -\dfrac{5\pi}{6}, -\dfrac{\pi}{6}, \dfrac{7\pi}{6}, \dfrac{11\pi}{6}.$

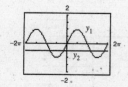

71. $v = 0.85 \sin \dfrac{\pi t}{3}$

(a)

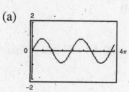

(b) Time for one cycle = one period = $\dfrac{2\pi}{\pi/3} = 6$ sec

(c) Cycles per min = $\dfrac{60}{6} = 10$ cycles per min

(d) The period would change.

73. $h = 25 \sin \frac{\pi}{15}(t - 75) + 30$

(a)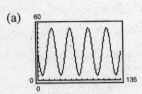

(b) Minimum: $30 - 25 = 5$ feet

   Maximum: $30 + 25 = 55$ feet

75. $C = 30.3 + 21.6 \sin\left(\frac{2\pi t}{365} + 10.9\right)$

(a) Period: $\dfrac{2\pi}{b} = \dfrac{2\pi}{(2\pi/365)} = 365$ days

   This is to be expected: 365 days = 1 year

(b) The constant 30.3 gallons is the average daily fuel consumption.

(c)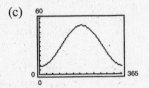

Consumption exceeds 40 gallons/day when $124 \le x \le 252$. (Graph C together with $y = 40$.) (Beginning of May through part of September)

77. (a)

(b) $y = 0.506 \sin(0.209x - 1.336) + 0.526$

(c)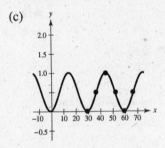

The model is a good fit.

(d) The period is $\dfrac{2\pi}{0.209} \approx 30.06$.

(e) June 29, 2007 is day 545. Using the model, $y \approx 0.2709$ or 27.09%.

79. True. The period is $\dfrac{2\pi}{3/10} = \dfrac{20\pi}{3}$.

81. True

83. The graph passes through $(0, 0)$ and has period π. Matches (e).

85. The period is 4π and the amplitude is 1. Since $(0, 1)$ and $(\pi, 0)$ are on the graph, matches (c).

87. (a) $h(x) = \cos^2 x$ is even.

   (b) $h(x) = \sin^2 x$ is even.

   (c) $h(x) = \sin x \cos x$ is odd.

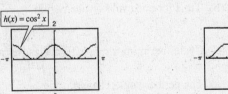

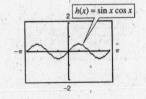

89. (a)

x	-1	-0.1	-0.01	-0.001
$\dfrac{\sin x}{x}$	0.8415	0.9983	1.0	1.0

x	0	0.001	0.01	0.1	1
$\dfrac{\sin x}{x}$	Undef.	1.0	1.0	0.9983	0.8415

(b)

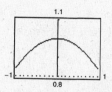

As $x \to 0$, $f(x) = \dfrac{\sin x}{x}$ approaches 1.

(c) As x approaches 0, $\dfrac{\sin x}{x}$ approaches 1.

91. (a)

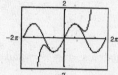

(b)

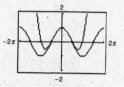

(c) Next term for sine approximation: $-\dfrac{x^7}{7!}$

Next term for cosine approximation: $-\dfrac{x^6}{6!}$

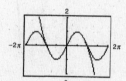

93.

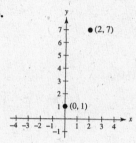

Slope $= \dfrac{7 - 1}{2 - 0} = 3$

95. $8.5 = 8.5\left(\dfrac{180°}{\pi}\right) \approx 487.014°$

97. Answers will vary. (Make a Decision)

Section 4.6   Graphs of Other Trigonometric Functions

- You should be able to graph:

  $y = a \tan(bx - c)$ $y = a \cot(bx - c)$

  $y = a \sec(bx - c)$ $y = a \csc(bx - c)$

- When graphing $y = a \sec(bx - c)$ or $y = a \csc(bx - c)$ you should know to first graph $y = a \cos(bx - c)$ or $y = a \sin(bx - c)$ since

  (a) The intercepts of sine and cosine are vertical asymptotes of cosecant and secant.

  (b) The maximums of sine and cosine are local minimums of cosecant and secant.

  (c) The minimums of sine and cosine are local maximums of cosecant and secant.

- You should be able to graph using a damping factor.

Vocabulary Check

1. vertical   **2.** reciprocal   **3.** damping

1. $f(x) = \tan x$

  (a) x-intercepts: $(-2\pi, 0), (-\pi, 0), (0, 0), (\pi, 0), (2\pi, 0)$

  (b) y-intercept: $(0, 0)$

  (c) Increasing on: $\left(-2\pi, -\frac{3\pi}{2}\right), \left(-\frac{3\pi}{2}, -\frac{\pi}{2}\right), \left(-\frac{\pi}{2}, \frac{\pi}{2}\right), \left(\frac{\pi}{2}, \frac{3\pi}{2}\right), \left(\frac{3\pi}{2}, 2\pi\right)$

  (d) No relative extrema

    Never decreasing

  (e) Vertical asymptotes: $x = -\frac{3\pi}{2}, -\frac{\pi}{2}, \frac{\pi}{2}, \frac{3\pi}{2}$

3. $f(x) = \sec x$

  (a) No x-intercepts

  (b) y-intercept: $(0, 1)$

  (c) Increasing on intervals:

$$\left(-2\pi, -\frac{3\pi}{2}\right), \left(-\frac{3\pi}{2}, -\pi\right), \left(0, \frac{\pi}{2}\right), \left(\frac{\pi}{2}, \pi\right)$$

    Decreasing on intervals:

$$\left(-\pi, -\frac{\pi}{2}\right), \left(-\frac{\pi}{2}, 0\right), \left(\pi, \frac{3\pi}{2}\right), \left(\frac{3\pi}{2}, 2\pi\right)$$

  (d) Relative minima: $(-2\pi, 1), (0, 1), (2\pi, 1)$,

    Relative maxima: $(-\pi, -1), (\pi, -1)$

  (e) Vertical asymptotes: $x = -\frac{3\pi}{2}, -\frac{\pi}{2}, \frac{\pi}{2}, \frac{3\pi}{2}$

5. $y = \frac{1}{2} \tan x$

Period: π

Two consecutive asymptotes: $x = -\frac{\pi}{2}$ and $x = \frac{\pi}{2}$

x	$-\frac{\pi}{4}$	0	$\frac{\pi}{4}$
y	$-\frac{1}{2}$	0	$\frac{1}{2}$

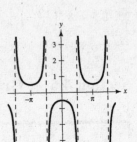

7. $y = -2 \tan 2x$

Period: $\frac{\pi}{2}$

Two consecutive asymptotes:

$$2x = -\frac{\pi}{2} \Rightarrow x = -\frac{\pi}{4}$$

$$2x = \frac{\pi}{2} \Rightarrow x = \frac{\pi}{4}$$

x	$-\frac{\pi}{8}$	0	$\frac{\pi}{8}$
y	2	0	-2

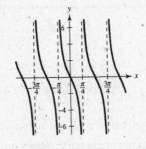

9. $y = -\frac{1}{2} \sec x$

Graph $y = -\frac{1}{2} \cos x$ first.

Period: 2π

One cycle: 0 to 2π

11. $y = \sec \pi x - 3$

Period: 2

Shift graph of $\sec \pi x$ down three units.

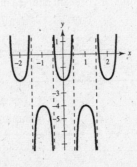

13. $y = 3 \csc \frac{x}{2}$

Graph $y = 3 \sin \frac{x}{2}$ first.

Period: $\frac{2\pi}{1/2} = 4\pi$

One cycle: 0 to 4π

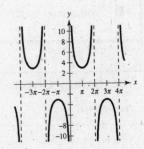

15. $y = \dfrac{1}{2} \cot \dfrac{x}{2}$

Period: $\dfrac{\pi}{1/2} = 2\pi$

Two consecutive
asymptotes:

$\dfrac{x}{2} = 0 \Rightarrow x = 0$

$\dfrac{x}{2} = \pi \Rightarrow x = 2\pi$

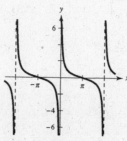

x	$\dfrac{\pi}{2}$	π	$\dfrac{3\pi}{2}$
y	$\dfrac{1}{2}$	0	$-\dfrac{1}{2}$

17. $y = 2 \tan \dfrac{\pi x}{4}$

Period: $\dfrac{\pi}{\pi/4} = 4$

Two consecutive
asymptotes:

$\dfrac{\pi x}{4} = -\dfrac{\pi}{2} \Rightarrow x = -2$

$\dfrac{\pi x}{4} = \dfrac{\pi}{2} \Rightarrow x = 2$

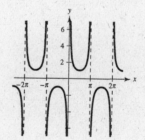

x	-1	0	1
y	-2	0	2

19. $y = \dfrac{1}{2} \sec(2x - \pi)$

Period: $\dfrac{2\pi}{2} = \pi$

Asymptotes: $x = \pm \dfrac{\pi}{4}$

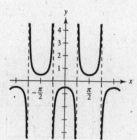

21. $y = \csc(\pi - x)$

Graph $y = \sin(\pi - x)$ first.

Period: 2π

Asymptotes: Set $\pi - x = 0$ and $\pi - x = 2\pi$

$\qquad\qquad\qquad x = \pi \qquad\qquad x = -\pi$

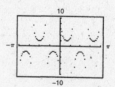

23. $y = 2 \cot\left(x - \dfrac{\pi}{2}\right)$

Period: π

Two consecutive
asymptotes:

$x - \dfrac{\pi}{2} = 0 \Rightarrow x = \dfrac{\pi}{2}$

$x - \dfrac{\pi}{2} = \pi \Rightarrow x = \dfrac{3\pi}{2}$

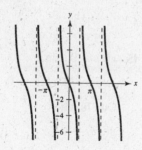

x	$\dfrac{3\pi}{4}$	π	$\dfrac{5\pi}{4}$
y	2	0	-2

25. $y = 2 \csc 3x = \dfrac{2}{\sin(3x)}$

Period: $\dfrac{2\pi}{3}$

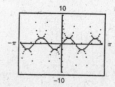

27. $y = -2 \sec 4x$

$= \dfrac{-2}{\cos 4x}$

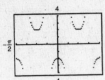

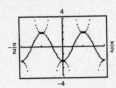

29. $y = \dfrac{1}{3} \sec\left(\dfrac{\pi x}{2} + \dfrac{\pi}{2}\right)$

$= \dfrac{1}{3 \cos\left(\dfrac{\pi x}{2} + \dfrac{\pi}{2}\right)}$

Period: 4

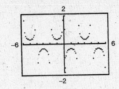

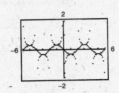

31. $\tan x = 1$

$x = -\dfrac{7\pi}{4},\ -\dfrac{3\pi}{4},\ \dfrac{\pi}{4},\ \dfrac{5\pi}{4}$

33. $\sec x = -2$

$x = \pm\dfrac{2\pi}{3},\ \pm\dfrac{4\pi}{3}$

35. The graph of $f(x) = \sec x$ has y-axis symmetry. Thus, the function is even.

37. The function

$f(x) = \csc 2x = \dfrac{1}{\sin 2x}$

has origin symmetry. Thus, the function is odd.

39. $y_1 = \sin x \csc x$ and $y_2 = 1$

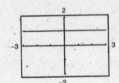

Not equivalent because y_1 is not defined at 0.

$\sin x \csc x = \sin x\left(\dfrac{1}{\sin x}\right) = 1, \quad \sin x \neq 0$

41. $y_1 = \dfrac{\cos x}{\sin x}$ and $y_2 = \cot x = \dfrac{1}{\tan x}$

Equivalent

$\cot x = \dfrac{\cos x}{\sin x}$

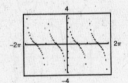

43. $f(x) = x \cos x$

As $x \to 0$, $f(x) \to 0$.

Odd function

$f\left(\dfrac{3\pi}{2}\right) = 0$

Matches graph (d).

45. $g(x) = |x| \sin x$

As $x \to 0$, $g(x) \to 0$.

Odd function

$g(2\pi) = 0$

Matches graph (b).

47. $f(x) = \sin x + \cos\left(x + \dfrac{\pi}{2}\right)$, $g(x) = 0$

$f(x) = g(x)$

The graph is the line $y = 0$.

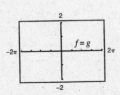

49. $f(x) = \sin^2 x$, $g(x) = \dfrac{1}{2}(1 - \cos 2x)$

$f(x) = g(x)$

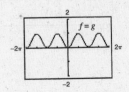

51. $f(x) = e^{-x} \cos x$

Damping factor: e^{-x}

$x \to \infty, f(x) \to 0$

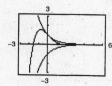

53. $h(x) = e^{-x^2/4} \cos x$

Damping factor: $e^{-x^2/4}$

$h \to 0$ as $x \to \infty$

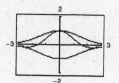

55. (a) As $x \to \dfrac{\pi^+}{2}, f(x) \to -\infty$.

(b) As $x \to \dfrac{\pi^-}{2}, f(x) \to \infty$.

(c) As $x \to -\dfrac{\pi^+}{2}, f(x) \to -\infty$.

(d) As $x \to -\dfrac{\pi^-}{2}, f(x) \to \infty$.

57. $f(x) = \cot x$

(a) As $x \to 0^+, f(x) \to \infty$.

(b) As $x \to 0^-, f(x) \to -\infty$.

(c) As $x \to \pi^+, f(x) \to \infty$.

(d) As $x \to \pi^-, f(x) \to -\infty$.

59. As the predator population increases, the number of prey decreases. When the number of prey is small, the number of predators decreases.

61. $\tan x = \dfrac{5}{d}$

$d = \dfrac{5}{\tan x} = 5 \cot x$

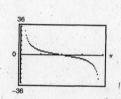

63. (a)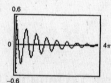

(b) The displacement function is not periodic, but damped. It approaches 0 as t increases.

65. True. $-\dfrac{3\pi}{2} + \pi = -\dfrac{\pi}{2}$ and $x = -\dfrac{\pi}{2}$ is a vertical asymptote for the tangent function.

67. $f(x) = 2 \sin x, g(x) = \dfrac{1}{2} \csc x$

(a)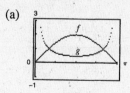

(b) $f(x) > g(x)$ for $\dfrac{\pi}{6} < x < \dfrac{5\pi}{6}$

(c) As $x \to \pi$, $2 \sin x \to 0$ and $\dfrac{1}{2} \csc x \to \infty$, since $g(x)$ is the reciprocal of $f(x)$.

69. (a)

x	-1	-0.1	-0.01	-0.001
$\dfrac{\tan x}{x}$	1.5574	1.0033	1.0	1.0

x	0	0.001	0.01	0.1	1
$\dfrac{\tan x}{x}$	Undef.	1.0	1.0	1.0033	1.5574

(b)

As $x \to 0, f(x) = \dfrac{\tan x}{x} \to 1$.

(c) The ratio approaches 1 as x approaches 0.

71. Period is $\dfrac{\pi}{2}$ and graph is increasing on $\left(-\dfrac{\pi}{4}, \dfrac{\pi}{4}\right)$.
Matches (a).

73.

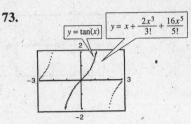

The graphs of $y_1 = \tan x$ and $y_2 = x + \dfrac{2x^3}{3!} + \dfrac{16x^5}{5!}$

are similar on the interval $\left(-\dfrac{\pi}{2}, \dfrac{\pi}{2}\right)$.

75. Distributive Property

77. Additive Identity Property

79. Not one-to-one

81. $y = \sqrt{3x - 14}, \ x \geq \frac{14}{3}, \ y \geq 0$

$x = \sqrt{3y - 14}, \ y \geq \frac{14}{3}, \ x \geq 0$

$x^2 = 3y - 14$

$y = \frac{1}{3}(x^2 + 14)$

$f^{-1}(x) = \frac{1}{3}(x^2 + 14), \ x \geq 0$

Section 4.7   Inverse Trigonometric Functions

■ You should know the definitions, domains, and ranges of $y = \arcsin x$, $y = \arccos x$, and $y = \arctan x$.

Function	Domain	Range
$y = \arcsin x \implies x = \sin y$	$-1 \leq x \leq 1$	$-\dfrac{\pi}{2} \leq y \leq \dfrac{\pi}{2}$
$y = \arccos x \implies x = \cos y$	$-1 \leq x \leq 1$	$0 \leq y \leq \pi$
$y = \arctan x \implies x = \tan y$	$-\infty < x < \infty$	$-\dfrac{\pi}{2} < y < \dfrac{\pi}{2}$

■ You should know the inverse properties of the inverse trigonometric functions.

$\sin(\arcsin x) = x, \ -1 \leq x \leq 1$ and $\arcsin(\sin y) = y, \ -\dfrac{\pi}{2} \leq y \leq \dfrac{\pi}{2}$

$\cos(\arccos x) = x, \ -1 \leq x \leq 1$ and $\arccos(\cos y) = y, \ 0 \leq y \leq \pi$

$\tan(\arctan x) = x$ and $\arctan(\tan y) = y, \ -\dfrac{\pi}{2} < y < \dfrac{\pi}{2}$

■ You should be able to use the triangle technique to convert trigonometric functions or inverse trigonometric functions into algebraic expressions.

Vocabulary Check

1. $y = \sin^{-1} x, -1 \leq x \leq 1$

2. $y = \arccos x, 0 \leq y \leq \pi$

3. $y = \tan^{-1} x, -\infty < x < \infty, -\dfrac{\pi}{2} < y < \dfrac{\pi}{2}$

---

1. (a) $\arcsin \dfrac{1}{2} = \dfrac{\pi}{6}$ (b) $\arcsin 0 = 0$

3. (a) $\arcsin 1 = \dfrac{\pi}{2}$ because $\sin \dfrac{\pi}{2} = 1$ and $-\dfrac{\pi}{2} \leq \dfrac{\pi}{2} \leq \dfrac{\pi}{2}$.

   (b) $\arccos 1 = 0$ because $\cos 0 = 1$ and $0 \leq 1 \leq \pi$.

5. (a) $\arctan \dfrac{\sqrt{3}}{3} = \dfrac{\pi}{6}$ (b) $\arctan(-1) = -\dfrac{\pi}{4}$

7. (a) $y = \arctan(-\sqrt{3}) \Rightarrow \tan y = -\sqrt{3}$ for $-\dfrac{\pi}{2} < y < \dfrac{\pi}{2} \Rightarrow y = -\dfrac{\pi}{3}$

   (b) $y = \arctan \sqrt{3} \Rightarrow \tan y = \sqrt{3} \Rightarrow y = \dfrac{\pi}{3}$

9. (a) $y = \sin^{-1} \dfrac{\sqrt{3}}{2} \Rightarrow \sin y = \dfrac{\sqrt{3}}{2}$ for $-\dfrac{\pi}{2} \leq y \leq \dfrac{\pi}{2} \Rightarrow y = \dfrac{\pi}{3}$

   (b) $y = \tan^{-1}\left(\dfrac{-\sqrt{3}}{3}\right) \Rightarrow \tan y = \dfrac{-\sqrt{3}}{3} \Rightarrow y = -\dfrac{\pi}{6}$

11. $y = \arccos x$

(a)

x	-1	-0.8	-0.6	-0.4	-0.2	
y	3.1416	2.4981	2.2143	1.9823	1.7722	

x	0	0.2	0.4	0.6	0.8	1.0
y	1.5708	1.3694	1.1593	0.9273	0.6435	0

(b)

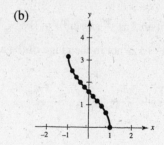

(c)

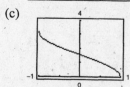

(d) Intercepts: $\left(0, \dfrac{\pi}{2}\right)$, $(1, 0)$, no symmetry

13. $y = \arctan x \leftrightarrow \tan y = x$

$\left(-\sqrt{3}, -\dfrac{\pi}{3}\right), \left(-\dfrac{\sqrt{3}}{3}, -\dfrac{\pi}{6}\right), \left(1, \dfrac{\pi}{4}\right)$

15. $\cos^{-1}(0.75) \approx 0.72$ **17.** $\arcsin(-0.75) \approx -0.85$ **19.** $\arctan(-6) \approx -1.41$

21. $\tan \theta = \dfrac{x}{8}$

$\theta = \arctan \dfrac{x}{8}$

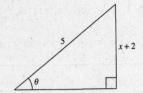

23. $\sin \theta = \dfrac{x+2}{5}$

$\theta = \arcsin\left(\dfrac{x+2}{5}\right)$

25. Let y be the third side. Then

$y^2 = 2^2 - x^2 = 4 - x^2 \implies y = \sqrt{4-x^2}$

$\sin \theta = \dfrac{x}{2} \implies \theta = \arcsin \dfrac{x}{2}$

$\cos \theta = \dfrac{\sqrt{4-x^2}}{2} \implies \theta = \arccos \dfrac{\sqrt{4-x^2}}{2}$

$\tan \theta = \dfrac{x}{\sqrt{4-x^2}} \implies \theta = \arctan \dfrac{x}{\sqrt{4-x^2}}$

27. Let y be the hypotenuse. Then $y^2 = (x+1)^2 + 2^2 = x^2 + 2x + 5 \implies y = \sqrt{x^2 + 2x + 5}$.

$\sin \theta = \dfrac{x+1}{\sqrt{x^2+2x+5}} \implies \theta = \arcsin \dfrac{x+1}{\sqrt{x^2+2x+5}}$

$\cos \theta = \dfrac{2}{\sqrt{x^2+2x+5}} \implies \theta = \arccos \dfrac{2}{\sqrt{x^2+2x+5}}$

$\tan \theta = \dfrac{x+1}{2} \implies \theta = \arctan \dfrac{x+1}{2}$

29. $\sin(\arcsin 0.7) = 0.7$ **31.** $\cos[\arccos(-0.3)] = -0.3$

33. $\arcsin(\sin 3\pi) = \arcsin(0) = 0$ **35.** $\arctan\left(\tan \dfrac{11\pi}{6}\right) = \arctan\left(-\dfrac{\sqrt{3}}{3}\right) = -\dfrac{\pi}{6}$

Note: 3π is not in the range of the arcsine function.

37. $\sin^{-1}\left(\sin \dfrac{5\pi}{2}\right) = \sin^{-1} 1 = \dfrac{\pi}{2}$ **39.** $\sin^{-1}\left(\tan \dfrac{5\pi}{4}\right) = \sin^{-1} 1 = \dfrac{\pi}{2}$

41. $\tan(\arcsin 0) = \tan 0 = 0$ **43.** $\sin(\arctan 1) = \sin\left(\dfrac{\pi}{4}\right) = \dfrac{\sqrt{2}}{2}$

45. $\arcsin\left[\cos\left(-\dfrac{\pi}{6}\right)\right] = \arcsin\left(\dfrac{\sqrt{3}}{2}\right) = \dfrac{\pi}{3}$ **47.** Let $y = \arctan \dfrac{4}{3}$. Then

$\tan y = \dfrac{4}{3}, 0 < y < \dfrac{\pi}{2}$, and

$\sin y = \dfrac{4}{5}$.

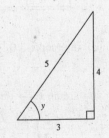

49. Let $y = \arcsin \dfrac{24}{25}$. Then

$\sin y = \dfrac{24}{25}$, and $\cos y = \dfrac{7}{25}$.

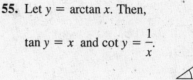

51. Let $y = \arctan\left(-\dfrac{3}{5}\right)$. Then,

$\tan y = -\dfrac{3}{5}$, $-\dfrac{\pi}{2} < y < 0$ and $\sec y = \dfrac{\sqrt{34}}{5}$.

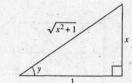

53. Let $y = \arccos\left(-\dfrac{2}{3}\right)$. Then,

$\cos y = -\dfrac{2}{3}$, $\dfrac{\pi}{2} < y < \pi$ and $\sin y = \dfrac{\sqrt{5}}{3}$.

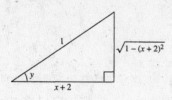

55. Let $y = \arctan x$. Then,

$\tan y = x$ and $\cot y = \dfrac{1}{x}$.

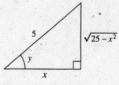

57. Let $y = \arccos(x + 2)$, $\cos y = x + 2$.

Opposite side: $\sqrt{1 - (x + 2)^2}$

$\sin y = \dfrac{\sqrt{1 - (x + 2)^2}}{1} = \sqrt{-x^2 - 4x - 3}$

59. Let $y = \arccos \dfrac{x}{5}$. Then $\cos y = \dfrac{x}{5}$, and

$\tan y = \dfrac{\sqrt{25 - x^2}}{x}$.

61. Let $y = \arctan \dfrac{x}{\sqrt{7}}$. Then $\tan y = \dfrac{x}{\sqrt{7}}$ and

$\csc y = \dfrac{\sqrt{7 + x^2}}{x}$.

63. Let $y = \arctan \dfrac{14}{x}$. Then $\tan y = \dfrac{14}{x}$ and

$\sin y = \dfrac{14}{\sqrt{196 + x^2}}$. Thus,

$y = \arcsin\left(\dfrac{14}{\sqrt{196 + x^2}}\right)$.

65. Let $y = \arccos \dfrac{3}{\sqrt{x^2 - 2x + 10}}$. Then,

$$\cos y = \frac{3}{\sqrt{x^2 - 2x + 10}} = \frac{3}{\sqrt{(x - 1)^2 + 9}}$$

and $\sin y = \dfrac{|x - 1|}{\sqrt{(x - 1)^2 + 9}}$. Thus,

$$y = \arcsin \frac{|x - 1|}{\sqrt{(x - 1)^2 + 9}} = \arcsin \frac{|x - 1|}{\sqrt{x^2 - 2x + 10}}.$$

67. $y = 2 \arccos x$

Domain: $-1 \le x \le 1$

Range: $0 \le y \le 2\pi$

Vertical stretch of $f(x) = \arccos x$

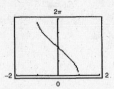

69. The graph of $f(x) = \arcsin(x - 2)$ is a horizontal translation of the graph of $y = \arcsin x$ by two units.

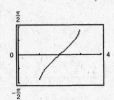

71. $f(x) = \arctan 2x$

Domain: all real numbers

Range: $-\dfrac{\pi}{2} < y < \dfrac{\pi}{2}$

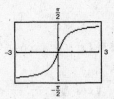

73. $f(t) = 3 \cos 2t + 3 \sin 2t$

$$= \sqrt{3^2 + 3^2} \sin\left(2t + \arctan \frac{3}{3}\right)$$

$$= 3\sqrt{2} \sin(2t + \arctan 1)$$

$$= 3\sqrt{2} \sin\left(2t + \frac{\pi}{4}\right)$$

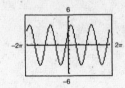

The graphs are the same

75. As $x \to 1^-$, $\arcsin x \to \dfrac{\pi}{2}$.

77. As $x \to \infty$, $\arctan x \to \dfrac{\pi}{2}$.

79. As $x \to -1^+$, $\arccos x \to \pi$.

81. (a) $\sin \theta = \dfrac{10}{s} \implies \theta = \arcsin\left(\dfrac{10}{s}\right)$

(b) $s = 52$: $\theta = \arcsin\left(\dfrac{10}{52}\right) \approx 0.1935$, $(\approx 11.1°)$

$s = 26$: $\theta = \arcsin\left(\dfrac{10}{26}\right) \approx 0.3948$, $(\approx 22.6°)$

83. (a) $\tan \theta = \dfrac{s}{750}$

$$\theta = \arctan\left(\frac{s}{750}\right)$$

(b) When $s = 400$,

$$\theta = \arctan\left(\frac{400}{750}\right) \approx 0.49 \text{ radian}, \quad (\approx 28°).$$

When $s = 1600$, $\theta \approx 1.13$ radians, $(\approx 65°)$.

85. (a) $\tan \theta = \dfrac{6}{x}$

$$\theta = \arctan\left(\dfrac{6}{x}\right)$$

(b) When $x = 10$,

$$\theta = \arctan\left(\dfrac{6}{10}\right) \approx 0.54 \text{ radian}, \ (\approx 31°).$$

When $x = 3$,

$$\theta = \arctan\left(\dfrac{6}{3}\right) \approx 1.11 \text{ radians}, \ (\approx 63°).$$

87. False. $\arcsin \dfrac{1}{2} = \dfrac{\pi}{6}$

89. $y = \operatorname{arccot} x$ if and only if $\cot y = x$, $-\infty < x < \infty$ and $0 < y < \pi$.

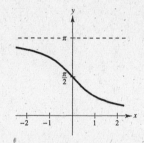

91. $y = \operatorname{arccsc} x$ if and only if $\csc y = x$.

Domain: $(-\infty, -1] \cup [1, \infty)$

Range: $\left[-\dfrac{\pi}{2}, 0\right) \cup \left(0, \dfrac{\pi}{2}\right]$

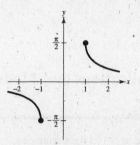

93. $y = \operatorname{arcsec} \sqrt{2} \implies \sec y = \sqrt{2}$ and $0 \le y < \dfrac{\pi}{2} \cup \dfrac{\pi}{2} < y \le \pi \implies y = \dfrac{\pi}{4}$

95. $y = \operatorname{arccot}\left(-\sqrt{3}\right) \implies \cot y = -\sqrt{3}$ and $0 < y < \pi \implies y = \dfrac{5\pi}{6}$

97. Let $y = \arcsin(-x)$. Then,

$$\sin y = -x$$
$$-\sin y = x$$
$$\sin(-y) = x$$
$$-y = \arcsin x$$
$$y = -\arcsin x.$$

Therefore, $\arcsin(-x) = -\arcsin x$.

99. $\arcsin x + \arccos x = \dfrac{\pi}{2}$

Let $\alpha = \arcsin x \implies \sin \alpha = x$.

Let $\beta = \arccos x \implies \cos \beta = x$.

Hence, $\sin a = \cos \beta \implies a$ and β are

complementary angles $\implies a + \beta = \dfrac{\pi}{2} \implies \arcsin x + \arccos x = \dfrac{\pi}{2}$.

101. $\dfrac{4}{4\sqrt{2}} = \dfrac{1}{\sqrt{2}} = \dfrac{\sqrt{2}}{2}$

103. $\dfrac{2\sqrt{3}}{6} = \dfrac{\sqrt{3}}{3}$

105. $\sin \theta = \dfrac{5}{6}$

Adjacent side: $\sqrt{6^2 - 5^2} = \sqrt{11}$

$\cos \theta = \dfrac{\sqrt{11}}{6}$

$\tan \theta = \dfrac{5}{\sqrt{11}} = \dfrac{5\sqrt{11}}{11}$

$\csc \theta = \dfrac{6}{5}$

$\sec \theta = \dfrac{6}{\sqrt{11}} = \dfrac{6\sqrt{11}}{11}$

$\cot \theta = \dfrac{\sqrt{11}}{5}$

107. $\sin \theta = \dfrac{3}{4}$

Adjacent side: $\sqrt{16 - 9} = \sqrt{7}$

$\cos \theta = \dfrac{\sqrt{7}}{4}$

$\tan \theta = \dfrac{3}{\sqrt{7}} = \dfrac{3\sqrt{7}}{7}$

$\csc \theta = \dfrac{4}{3}$

$\sec \theta = \dfrac{4}{\sqrt{7}} = \dfrac{4\sqrt{7}}{7}$

$\cot \theta = \dfrac{\sqrt{7}}{3}$

Section 4.8   Applications and Models

- ■ You should be able to solve right triangles.
- ■ You should be able to solve right triangle applications.
- ■ You should be able to solve applications of simple harmonic motion: $d = a \sin wt$ or $d = a \cos wt$.

Vocabulary Check

1. elevation, depression   2. bearing   3. harmonic motion

1. Given: $A = 30°$, $b = 10$

$B = 90° - 30° = 60°$

$\tan A = \dfrac{a}{b} \implies a = b \tan A = 10 \tan 30° \approx 5.77$

$\cos A = \dfrac{b}{c} \implies c = \dfrac{b}{\cos A} = \dfrac{10}{\cos 30°} \approx 11.55$

3. Given: $B = 71°$, $b = 14$

$\tan B = \dfrac{b}{a} \implies a = \dfrac{b}{\tan B} = \dfrac{14}{\tan 71°} \approx 4.82$

$\sin B = \dfrac{b}{c} \implies c = \dfrac{b}{\sin B} = \dfrac{14}{\sin 71°} \approx 14.81$

$A = 90° - 71° = 19°$

5. Given: $a = 6$, $b = 12$

$c^2 = a^2 + b^2 \implies c = \sqrt{36 + 144} \approx 13.42$

$\tan A = \dfrac{a}{b} = \dfrac{6}{12} = \dfrac{1}{2} \implies A = \arctan \dfrac{1}{2} \approx 26.57°$

$B = 90° - 26.57° = 63.43°$

7. Given: $b = 16$, $c = 54$

$a = \sqrt{c^2 - b^2} = \sqrt{2660} \approx 51.58$

$\cos A = \dfrac{b}{c} = \dfrac{16}{54} \implies A = \arccos\left(\dfrac{16}{54}\right) \approx 72.76°$

$B = 90° - 72.76° = 17.24°$

9. $A = 12° \, 15'$, $c = 430.5$

$B = 90° - 12° \, 15' = 77° \, 45'$

$\sin 12° \, 15' = \dfrac{a}{430.5}$

$a = 430.5 \sin 12° \, 15' \approx 91.34$

$\cos 12° \, 15' = \dfrac{b}{430.5}$

$b = 430.5 \cos 12° \, 15' \approx 420.70$

11. $\tan \theta = \dfrac{h}{b/2}$

$h = \dfrac{1}{2} b \tan \theta$

$h = \dfrac{1}{2}(8) \tan 52° \approx 5.12$ in.

13. $\tan \theta = \dfrac{h}{b/2}$

$h = \dfrac{1}{2} b \tan \theta$

$h = \dfrac{1}{2}(18.5) \tan 41.6° \approx 8.21$ ft

15. (a)

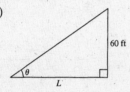

(b) $\tan \theta = \dfrac{60}{L}$

$L = \dfrac{60}{\tan \theta}$

$= 60 \cot \theta$

(c)

θ	10°	20°	30°	40°	50°
L	340	165	104	72	50

(d) No, the shadow lengths do not increase in equal increments. The cotangent function is not linear.

17. $\sin 80° = \dfrac{h}{20}$

$h = 20 \sin 80°$

≈ 19.70 feet

19. $\sin 50° = \dfrac{h}{100}$

$h = 100 \sin 50°$

≈ 76.6 feet

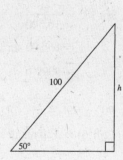

21. (a)

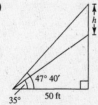

(b) Let the height of the church $= x$ and the height of the church and steeple $= y$. Then:

$\tan 35° = \dfrac{x}{50}$ and $\tan 47° \, 40' = \dfrac{y}{50}$

$x = 50 \tan 35°$ and $y = 50 \tan 47° \, 40'$

$h = y - x = 50(\tan 47° \, 40' - \tan 35°)$

(c) $h \approx 19.9$ feet

23. (a) $l^2 = 100^2 + (20 - 3 + h)^2$

$l = \sqrt{100^2 + (17 + h)^2}$

$ = \sqrt{h^2 + 34h + 10{,}289}$

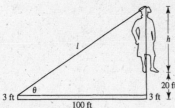

(b) $\cos\theta = \dfrac{100}{l} \implies \theta = \arccos\left(\dfrac{100}{l}\right)$

(c) If $\theta = 35°$, then $l = \dfrac{100}{\cos 35°} \approx 122.077$. Then

$$(17 + h)^2 + 100^2 = l^2$$

$$(17 + h)^2 = l^2 - 100^2 \approx 4902.906$$

$$17 + h \approx 70.02$$

$$h \approx 53.02 \text{ feet}$$

$$\approx 53 \text{ feet}, \frac{1}{4} \text{ inch.}$$

25. $\tan\theta = \dfrac{75}{95}$

$\theta = \arctan\dfrac{15}{19} \approx 38.29°$

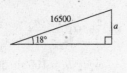

27. $\sin\alpha = \dfrac{4000}{16{,}500} \implies \alpha \approx 14.03°$

$\theta \approx 90° - \alpha \approx 75.97°$

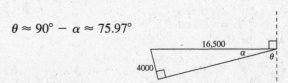

29. Since the airplane speed is

$$\left(275 \frac{\text{ft}}{\text{sec}}\right)\left(60 \frac{\text{sec}}{\text{min}}\right) = 16{,}500 \frac{\text{ft}}{\text{min}},$$

after one minute its distance travelled is 16,500 feet.

$\sin 18° = \dfrac{a}{16{,}500}$

$a = 16{,}500 \sin 18°$

$ \approx 5099 \text{ ft}$

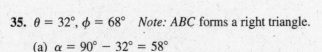

31. $\sin 9.5° = \dfrac{x}{4}$

$x = 4 \sin 9.5°$

$ \approx 0.66 \text{ mile}$

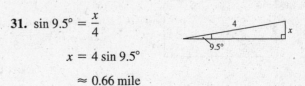

33. $90° - 29° = 61°$

$(20)(6) = 120$ nautical miles

$\sin 61° = \dfrac{a}{120} \implies a = 120 \sin 61° \approx 104.95$ nautical miles

$\cos 61° = \dfrac{b}{120} \implies b = 120 \cos 61° \approx 58.18$ nautical miles

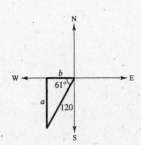

35. $\theta = 32°$, $\phi = 68°$ *Note: ABC forms a right triangle.*

(a) $\alpha = 90° - 32° = 58°$

Bearing from A to C: N $58°$ E

(b) $\beta = \theta = 32°$

$\gamma = 90° - \phi = 22°$

$C = \beta + \gamma = 54°$

$\tan C = \dfrac{d}{50} \implies \tan 54° = \dfrac{d}{50} \implies d \approx 68.82$ m

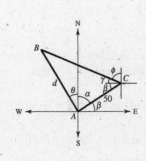

37. $\tan \theta = \dfrac{45}{30} = \dfrac{3}{2} \implies \theta \approx 56.31°$

Bearing: N 56.31° W

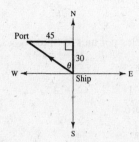

39. $\tan 6.5° = \dfrac{350}{d} \implies d \approx 3071.91$ ft

$\tan 4° = \dfrac{350}{D} \implies D \approx 5005.23$ ft

Distance between ships: $D - d \approx 1933.3$ ft

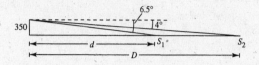

41. $\tan 57° = \dfrac{a}{x} \implies x = a \cot 57°$

$\tan 16° = \dfrac{a}{x + (55/6)}$

$\tan 16° = \dfrac{a}{a \cot 57° + (55/6)}$

$\cot 16° = \dfrac{a \cot 57° + (55/6)}{a}$

$a \cot 16° - a \cot 57° = \dfrac{55}{6}$

$\implies a \approx 3.23$ miles $\approx 17,054$ feet

43. (a)

$\theta = \arctan \dfrac{56}{40} \approx 54.5°$

Similarly for the back row,

$\theta = \arctan \dfrac{56}{150} \approx 20.5°$.

(b) For 45°, you need to be 56 feet away since

$\arctan \dfrac{56}{56} = 45°$.

45. $L_1 : 3x - 2y = 5 \implies y = \dfrac{3}{2}x - \dfrac{5}{2} \implies m_1 = \dfrac{3}{2}$

$L_2 : x + y = 1 \implies y = -x + 1 \implies m_2 = -1$

$\tan \alpha = \left| \dfrac{-1 - (3/2)}{1 + (-1)(3/2)} \right| = \left| \dfrac{-5/2}{-1/2} \right| = 5$

$\alpha = \arctan 5 \approx 78.7°$

47. The diagonal of the base has a length of $\sqrt{a^2 + a^2} = \sqrt{2}\,a$. Now, we have:

$\tan \theta = \dfrac{a}{\sqrt{2}\,a} = \dfrac{1}{\sqrt{2}}$

$\theta = \arctan \dfrac{1}{\sqrt{2}} \approx 35.3°$

49. $\cos 30° = \dfrac{b}{r}$

$b = \cos 30° \, r$

$b = \dfrac{\sqrt{3}\,r}{2}$

$y = 2b$

$\quad = 2\left(\dfrac{\sqrt{3}\,r}{2} \right) = \sqrt{3}\,r$

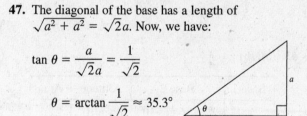

51. $d = 0$ when $t = 0$, $a = 8$, period $= 2$

Use $d = a \sin \omega t$ since $d = 0$ when $t = 0$.

$\dfrac{2\pi}{\omega} = 2 \implies \omega = \pi$

Thus, $d = 8 \sin \pi t$.

53. $d = 3$ when $t = 0$, $a = 3$, period $= 1.5$

Use $d = a \cos \omega t$ since $d = 3$ when $t = 0$.

$$\frac{2\pi}{\omega} = 1.5 \implies \omega = \frac{4}{3}\pi$$

Thus, $d = 3 \cos\left(\frac{4}{3}\pi t\right)$.

55. $d = 4 \cos 8\pi t$

(a) Maximum displacement = amplitude = 4

(b) Frequency $= \dfrac{\omega}{2\pi} = \dfrac{8\pi}{2\pi}$

$\qquad\qquad\quad = 4$ cycles per unit of time

(c) $d = 4 \cos(8\pi(5)) = 4$

(d) $8\pi t = \dfrac{\pi}{2} \implies t = \dfrac{1}{16}$

57. $d = \dfrac{1}{16} \sin 140\pi t$

(a) Maximum displacement = amplitude $= \dfrac{1}{16}$

(b) Frequency $= \dfrac{\omega}{2\pi} = \dfrac{140\pi}{2\pi}$

$\qquad\qquad\quad = 70$ cycles per unit of time

(c) $d = 0$

(d) $140\pi t = \pi \implies t = \dfrac{1}{140}$

59. $d = a \sin \omega t$

$$\text{Period} = \frac{2\pi}{\omega} = \frac{1}{\text{frequency}}$$

$$\frac{2\pi}{\omega} = \frac{1}{264}$$

$$\omega = 2\pi(264) = 528\pi$$

61. $y = \dfrac{1}{4} \cos 16t,\ t > 0$

(a)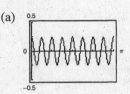

(b) Period: $\dfrac{2\pi}{16} = \dfrac{\pi}{8}$ seconds

(c) $\dfrac{1}{4} \cos 16t = 0$ when

$$16t = \frac{\pi}{2} \implies t = \frac{\pi}{32} \text{ seconds.}$$

63. (a), (b)

Base 1	Base 2	Altitude	Area
8	$8 + 16 \cos 10°$	$8 \sin 10°$	22.1
8	$8 + 16 \cos 20°$	$8 \sin 20°$	42.5
8	$8 + 16 \cos 30°$	$8 \sin 30°$	59.7
8	$8 + 16 \cos 40°$	$8 \sin 40°$	72.7
8	$8 + 16 \cos 50°$	$8 \sin 50°$	80.5
8	$8 + 16 \cos 60°$	$8 \sin 60°$	83.1
8	$8 + 16 \cos 70°$	$8 \sin 70°$	80.7

Maximum ≈ 83.1 square feet

(c) $A = \frac{1}{2}(b_1 + b_2)h$

$\quad = \frac{1}{2}[8 + 8 + 16 \cos \theta]8 \sin \theta$

$\quad = 64(1 + \cos \theta) \sin \theta$

(d) Maximum area is approximately 83.1 square feet for $\theta = 60°$.

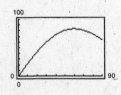

65. $S(t) = 18.09 + 1.41 \sin\left(\dfrac{\pi t}{6} + 4.60\right)$

(a)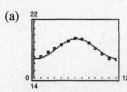

(b) The period is $\dfrac{2\pi}{(\pi/6)} = 12$ months, which is 1 year.

(c) The amplitude is 1.41. This gives the maximum change in time from the average time (18.09) of sunset.

67. False. The other acute angle is $90° - 48.1° = 41.9°$. Then

$$\tan(41.9°) = \frac{\text{opp}}{\text{adj}}$$

$$= \frac{a}{22.56} \implies a = 22.56 \cdot \tan(41.9°).$$

69. $y - 2 = 4(x + 1)$

$4x - y + 6 = 0$

71. Slope $= \dfrac{6 - 2}{-2 - 3} = -\dfrac{4}{5}$

$$y - 2 = -\frac{4}{5}(x - 3)$$

$$5y - 10 = -4x + 12$$

$$4x + 5y - 22 = 0$$

73. Domain: $(-\infty, \infty)$

75. Domain: $(-\infty, \infty)$

Review Exercises for Chapter 4

1. $40°$ or 0.7 radian

3. (a)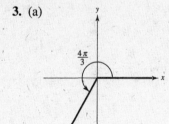

(b) Quadrant III

(c) $\dfrac{4\pi}{3} + 2\pi = \dfrac{10\pi}{3}$

$\dfrac{4\pi}{3} - 2\pi = -\dfrac{2\pi}{3}$

5. (a)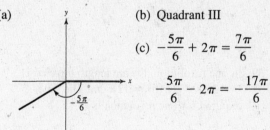

(b) Quadrant III

(c) $-\dfrac{5\pi}{6} + 2\pi = \dfrac{7\pi}{6}$

$-\dfrac{5\pi}{6} - 2\pi = -\dfrac{17\pi}{6}$

7. Complement: $\dfrac{\pi}{2} - \dfrac{\pi}{8} = \dfrac{3\pi}{8}$

Supplement: $\pi - \dfrac{\pi}{8} = \dfrac{7\pi}{8}$

9. Complement: $\dfrac{\pi}{2} - \dfrac{3\pi}{10} = \dfrac{\pi}{5}$

Supplement: $\pi - \dfrac{3\pi}{10} = \dfrac{7\pi}{10}$

11. (a)

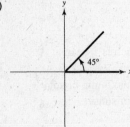

(b) Quadrant I

(c) $45° + 360° = 405°$

$45° - 360° = -315°$

15. Complement of 5°: $90° - 5° = 85°$

Supplement of 5°: $180° - 5° = 175°$

19. $135° 16' 45'' = \left(135 + \dfrac{16}{60} + \dfrac{45}{3600}\right)^° \approx 135.279°$

23. $135.29° = 135° + (0.29)(60)' = 135° 17' 24''$

27. $415° = 415° \cdot \dfrac{\pi \, \text{rad}}{180°} = \dfrac{83\pi}{36} \, \text{rad} \approx 7.243 \, \text{rad}$

31. $\dfrac{5\pi}{7} = \dfrac{5\pi}{7}\left(\dfrac{180°}{\pi}\right) \approx 128.571°$

35. $s = r\theta$

$25 = 12\theta$

$\theta = \dfrac{25}{12} \approx 2.083$

39. In one revolution, the arc length traveled is $s = 2\pi r = 2\pi(6) = 12\pi$ cm. The time required for one revolution is

$t = \dfrac{1}{500}$ minutes $= \dfrac{1}{500}(60) = \dfrac{3}{25}$ seconds.

Linear speed $= \dfrac{s}{t} = \dfrac{12\pi}{3/25} = 100\pi$ cm/sec

43. $\cos \dfrac{5\pi}{6} = -\dfrac{\sqrt{3}}{2}$

$\sin \dfrac{5\pi}{6} = \dfrac{1}{2}$

$(x, y) = \left(-\dfrac{\sqrt{3}}{2}, \dfrac{1}{2}\right)$

13. (a)

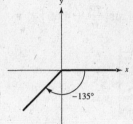

(b) Quadrant III

(c) $-135° + 360° = 225°$

$-135° - 360° = -495°$

17. Complement: not possible

Supplement: $180° - 171° = 9°$

21. $5° 22' 53'' = \left(5 + \dfrac{22}{60} + \dfrac{53}{3600}\right)^° \approx 5.381°$

25. $-85.36° = -[85 + 0.36(60')] = -85° 21' 36''$

29. $-72° = -72° \cdot \dfrac{\pi \, \text{rad}}{180°} = -\dfrac{2\pi}{5} \, \text{rad} \approx -1.257 \, \text{rad}$

33. $-3.5 = -3.5\left(\dfrac{180°}{\pi}\right) \approx -200.535°$

37. $s = r\theta$

$s = 20(138°)\dfrac{\pi}{180°}$

$s \approx 48.171$ m

41. $t = \dfrac{7\pi}{4}$ corresponds to $\left(\dfrac{\sqrt{2}}{2}, -\dfrac{\sqrt{2}}{2}\right)$.

45. $t = -\dfrac{2\pi}{3}$ corresponds to $\left(-\dfrac{1}{2}, -\dfrac{\sqrt{3}}{2}\right)$.

47. $t = -\dfrac{5\pi}{4}$ corresponds to $\left(-\dfrac{\sqrt{2}}{2}, \dfrac{\sqrt{2}}{2}\right)$.

49. $\sin\dfrac{7\pi}{6} = -\dfrac{1}{2}$ $\cot\dfrac{7\pi}{6} = \sqrt{3}$

$\cos\dfrac{7\pi}{6} = -\dfrac{\sqrt{3}}{2}$ $\sec\dfrac{7\pi}{6} = -\dfrac{2}{\sqrt{3}} = -\dfrac{2\sqrt{3}}{3}$

$\tan\dfrac{7\pi}{6} = \dfrac{1}{\sqrt{3}} = \dfrac{\sqrt{3}}{3}$ $\csc\dfrac{7\pi}{6} = -2$

51. $t = 2\pi$ corresponds to $(1, 0)$.

$\sin 2\pi = y = 0$ $\cot 2\pi = \dfrac{x}{y}$, undefined

$\cos 2\pi = x = 1$ $\sec 2\pi = \dfrac{1}{x} = 1$

$\tan 2\pi = \dfrac{y}{x} = 0$ $\csc 2\pi = \dfrac{1}{y}$, undefined

53. $t = -\dfrac{11\pi}{6}$ corresponds to $\left(\dfrac{\sqrt{3}}{2}, \dfrac{1}{2}\right)$.

$\sin\left(-\dfrac{11\pi}{6}\right) = y = \dfrac{1}{2}$

$\cos\left(-\dfrac{11\pi}{6}\right) = x = \dfrac{\sqrt{3}}{2}$

$\tan\left(-\dfrac{11\pi}{6}\right) = \dfrac{y}{x} = \dfrac{\sqrt{3}}{3}$

$\cot\left(-\dfrac{11\pi}{6}\right) = \dfrac{x}{y} = \sqrt{3}$

$\sec\left(-\dfrac{11\pi}{6}\right) = \dfrac{1}{x} = \dfrac{2\sqrt{3}}{3}$

$\csc\left(-\dfrac{11\pi}{6}\right) = \dfrac{1}{y} = 2$

55. $t = -\dfrac{\pi}{2}$ corresponds to $(0, -1)$.

$\sin\left(-\dfrac{\pi}{2}\right) = y = -1$ $\cot\left(-\dfrac{\pi}{2}\right) = \dfrac{x}{y} = 0$

$\cos\left(-\dfrac{\pi}{2}\right) = x = 0$ $\sec\left(-\dfrac{\pi}{2}\right) = \dfrac{1}{x}$, undefined

$\tan\left(-\dfrac{\pi}{2}\right) = \dfrac{y}{x}$, undefined $\csc\left(-\dfrac{\pi}{2}\right) = \dfrac{1}{y} = -1$

57. $\sin\left(\dfrac{11\pi}{4}\right) = \sin\left(\dfrac{3\pi}{4}\right) = \dfrac{\sqrt{2}}{2}$

59. $\sin\left(-\dfrac{17\pi}{6}\right) = \sin\left(\dfrac{7\pi}{6}\right) = -\dfrac{1}{2}$

61. $\sin t = \dfrac{3}{5}$

(a) $\sin(-t) = -\sin t = -\dfrac{3}{5}$

(b) $\csc(-t) = \dfrac{1}{\sin(-t)} = -\dfrac{5}{3}$

63. $\sin(-t) = -\dfrac{2}{3}$

(a) $\sin t = -\sin(-t) = -\left(-\dfrac{2}{3}\right) = \dfrac{2}{3}$

(b) $\csc t = \dfrac{1}{\sin t} = \dfrac{3}{2}$

65. $\cot 2.3 = \dfrac{1}{\tan 2.3} \approx -0.8935$

67. $\cos\dfrac{5\pi}{3} = \dfrac{1}{2}$

69. The opposite side is $\sqrt{9^2 - 4^2} = \sqrt{81 - 16} = \sqrt{65}$.

$\sin\theta = \dfrac{\text{opp}}{\text{hyp}} = \dfrac{\sqrt{65}}{9}$ $\csc\theta = \dfrac{1}{\sin\theta} = \dfrac{9}{\sqrt{65}} = \dfrac{9\sqrt{65}}{65}$

$\cos\theta = \dfrac{\text{adj}}{\text{hyp}} = \dfrac{4}{9}$ $\sec\theta = \dfrac{1}{\cos\theta} = \dfrac{9}{4}$

$\tan\theta = \dfrac{\text{opp}}{\text{adj}} = \dfrac{\sqrt{65}}{4}$ $\cot\theta = \dfrac{1}{\tan\theta} = \dfrac{4}{\sqrt{65}} = \dfrac{4\sqrt{65}}{65}$

71. The hypotenuse is $\sqrt{12^2 + 10^2} = \sqrt{244} = 2\sqrt{61}$.

$\sin \theta = \dfrac{\text{opp}}{\text{hyp}} = \dfrac{10}{2\sqrt{61}} = \dfrac{5}{\sqrt{61}} = \dfrac{5\sqrt{61}}{61}$

$\cos \theta = \dfrac{\text{adj}}{\text{hyp}} = \dfrac{12}{2\sqrt{61}} = \dfrac{6}{\sqrt{61}} = \dfrac{6\sqrt{61}}{61}$

$\tan \theta = \dfrac{\text{opp}}{\text{adj}} = \dfrac{10}{12} = \dfrac{5}{6}$

$\csc \theta = \dfrac{1}{\sin \theta} = \dfrac{\sqrt{61}}{5}$

$\sec \theta = \dfrac{1}{\cos \theta} = \dfrac{\sqrt{61}}{6}$

$\cot \theta = \dfrac{1}{\tan \theta} = \dfrac{6}{5}$

73. $\csc \theta \tan \theta = \dfrac{1}{\sin \theta} \cdot \dfrac{\sin \theta}{\cos \theta} = \dfrac{1}{\cos \theta} = \sec \theta$

75. (a) $\cos 84° \approx 0.1045$

(b) $\sin 6° \approx 0.1045$

77. (a) $\cos \dfrac{\pi}{4} \approx 0.7071$

(b) $\sec \dfrac{\pi}{4} \approx 1.4142$

79. $\tan 62° = \dfrac{\omega}{125}$

$x = 125 \tan 62°$

≈ 235 feet

81. $x = 12, y = 16, r = \sqrt{144 + 256} = \sqrt{400} = 20$

$\sin \theta = \dfrac{y}{r} = \dfrac{4}{5}$ $\qquad$ $\csc \theta = \dfrac{r}{y} = \dfrac{5}{4}$

$\cos \theta = \dfrac{x}{r} = \dfrac{3}{5}$ $\qquad$ $\sec \theta = \dfrac{r}{x} = \dfrac{5}{3}$

$\tan \theta = \dfrac{y}{x} = \dfrac{4}{3}$ $\qquad$ $\cot \theta = \dfrac{x}{y} = \dfrac{3}{4}$

83. $x = -7, y = 2, r = \sqrt{49 + 4} = \sqrt{53}$

$\sin \theta = \dfrac{y}{r} = \dfrac{2}{\sqrt{53}} = \dfrac{2\sqrt{53}}{53}$

$\cos \theta = \dfrac{x}{r} = -\dfrac{7}{\sqrt{53}} = -\dfrac{7\sqrt{53}}{53}$

$\tan \theta = \dfrac{y}{x} = -\dfrac{2}{7}$

$\csc \theta = \dfrac{\sqrt{53}}{2}$

$\sec \theta = -\dfrac{\sqrt{53}}{7}$

$\cot \theta = -\dfrac{7}{2}$

85. $x = \dfrac{2}{3}$ and $y = \dfrac{5}{8}, r = \sqrt{\left(\dfrac{2}{3}\right)^2 + \left(\dfrac{5}{8}\right)^2} = \dfrac{\sqrt{481}}{24}$

$\sin \theta = \dfrac{y}{r} = \dfrac{5/8}{\sqrt{481}/24} = \dfrac{15\sqrt{481}}{481}$

$\cos \theta = \dfrac{x}{r} = \dfrac{2/3}{\sqrt{481}/24} = \dfrac{16\sqrt{481}}{481}$

$\tan \theta = \dfrac{y}{x} = \dfrac{5/8}{2/3} = \dfrac{15}{16}$

$\csc \theta = \dfrac{r}{y} = \dfrac{\sqrt{481}}{15}$

$\sec \theta = \dfrac{r}{x} = \dfrac{\sqrt{481}}{16}$

$\cot \theta = \dfrac{x}{y} = \dfrac{16}{15}$

87. $\sec\theta = \dfrac{6}{5}$, $\tan\theta < 0 \implies \theta$ is in Quadrant IV.

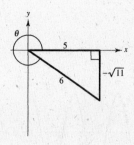

$$r = 6, x = 5, y = -\sqrt{36 - 25} = -\sqrt{11}$$

$$\sin\theta = \frac{y}{r} = -\frac{\sqrt{11}}{6} \qquad \csc\theta = -\frac{6\sqrt{11}}{11}$$

$$\cos\theta = \frac{x}{r} = \frac{5}{6} \qquad \sec\theta = \frac{6}{5}$$

$$\tan\theta = \frac{y}{x} = -\frac{\sqrt{11}}{5} \qquad \cot\theta = -\frac{5\sqrt{11}}{11}$$

89. $\sin\theta = \dfrac{3}{8}$, $\cos\theta < 0 \implies \theta$ is in Quadrant II.

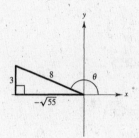

$$y = 3, r = 8, x = -\sqrt{55}$$

$$\sin\theta = \frac{y}{r} = \frac{3}{8} \qquad\qquad \csc\theta = \frac{8}{3}$$

$$\cos\theta = \frac{x}{r} = -\frac{\sqrt{55}}{8} \qquad\qquad \sec\theta = -\frac{8}{\sqrt{55}} = -\frac{8\sqrt{55}}{55}$$

$$\tan\theta = \frac{y}{x} = -\frac{3}{\sqrt{55}} = -\frac{3\sqrt{55}}{55} \qquad \cot\theta = -\frac{\sqrt{55}}{3}$$

91. Reference angle:

$$264° - 180° = 84°$$

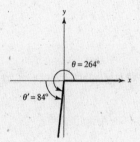

93. Coterminal angle:

$$2\pi - \frac{6\pi}{5} = \frac{4\pi}{5}$$

Reference angle:

$$\pi - \frac{4\pi}{5} = \frac{\pi}{5}$$

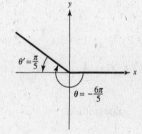

95. $240°$ is in Quadrant III with reference angle $60°$.

$$\sin 240° = -\sin 60° = -\frac{\sqrt{3}}{2}$$

$$\cos 240° = -\cos 60° = -\frac{1}{2}$$

$$\tan 240° = \frac{-\sqrt{3}/2}{-1/2} = \sqrt{3}$$

97. $-210°$ is coterminal with $150°$ in Quadrant II with reference angle $30°$.

$$\sin(-210°) = \sin(30°) = \frac{1}{2}$$

$$\cos(-210°) = -\cos(30°) = -\frac{\sqrt{3}}{2}$$

$$\tan(-210°) = \frac{1/2}{-\sqrt{3}/2} = -\frac{1}{\sqrt{3}} = -\frac{\sqrt{3}}{3}$$

99. $-9\pi/4$ is coterminal with $7\pi/4$ in Quadrant IV with reference angle $\pi/4$.

$$\sin\left(-\frac{9\pi}{4}\right) = -\sin\left(\frac{\pi}{4}\right) = -\frac{\sqrt{2}}{2}$$

$$\cos\left(-\frac{9\pi}{4}\right) = \cos\left(\frac{\pi}{4}\right) = \frac{\sqrt{2}}{2}$$

$$\tan\left(-\frac{9\pi}{4}\right) = \frac{-\sqrt{2}/2}{\sqrt{2}/2} = -1$$

101. $\sin(4\pi) = \sin(0) = 0$

$$\cos(4\pi) = 1$$

$$\tan(4\pi) = 0$$

103. $\tan 33° \approx 0.6494$

105. $\sec \dfrac{12\pi}{5} = \dfrac{1}{\cos(12\pi/5)} \approx 3.2361$

107. $f(x) = 3 \sin x$

Amplitude: 3

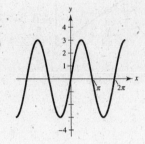

109. $f(x) = \frac{1}{4} \cos x$

Amplitude: $\frac{1}{4}$

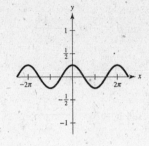

111. Period: $\dfrac{2\pi}{\pi} = 2$

Amplitude: 5

113. Period: $\dfrac{2\pi}{2} = \pi$

Amplitude: 3.4

115. $y = 3 \cos 2\pi x$

Amplitude: 3

Period: $\dfrac{2\pi}{2\pi} = 1$

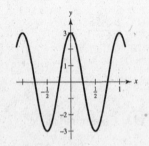

117. $f(x) = 5 \sin \dfrac{2x}{5}$

Amplitude: 5

Period: $\dfrac{2\pi}{2/5} = 5\pi$

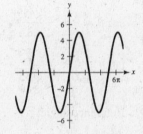

119. $f(x) = -\dfrac{5}{2} \cos\left(\dfrac{x}{4}\right)$

Amplitude: $\dfrac{5}{2}$

Period: $\dfrac{2\pi}{1/4} = 8\pi$

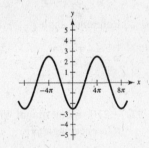

121. $f(x) = \dfrac{5}{2} \sin(x - \pi)$

Amplitude: $\dfrac{5}{2}$

Period: 2π

Shift:

$x - \pi = 0$ and $x - \pi = 2\pi$

$x = \pi \qquad\qquad x = 3\pi$

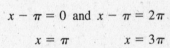

123. $f(x) = 2 - \cos \dfrac{\pi x}{2}$

125. $f(x) = -3 \cos\left(\dfrac{x}{2} - \dfrac{\pi}{4}\right)$

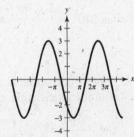

127. $f(x) = -2 \cos\left(x - \dfrac{\pi}{4}\right)$

129. $f(x) = -4 \cos\left(2x - \dfrac{\pi}{2}\right)$

131. $S = 48.4 - 6.1 \cos \dfrac{\pi t}{6}$

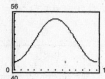

Maximum sales: $t = 6$
(June)

Minimum sales: $t = 12$
(December)

133. $f(x) = -\tan \dfrac{\pi x}{4}$

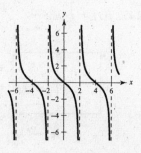

Period: $\dfrac{\pi}{(\pi/4)} = 4$

Asymptotes:
$x = -2, x = 2$

Reflected in x-axis

x	-1	0	1
y	1	0	-1

135. $f(x) = \dfrac{1}{4} \tan\left(x - \dfrac{\pi}{2}\right)$

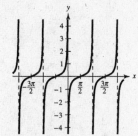

137. $f(x) = 3 \cot \dfrac{x}{2}$

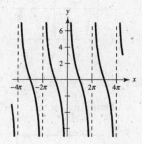

Period: $\dfrac{\pi}{1/2} = 2\pi$

Two consecutive
asymptotes:

$\dfrac{x}{2} = 0 \implies x = 0$

$\dfrac{x}{2} = \pi \implies x = 2\pi$

139. $f(x) = \dfrac{1}{2} \cot\left(x - \dfrac{\pi}{2}\right)$

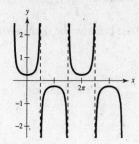

Period: π

Two consecutive
asymptotes:

$x - \dfrac{\pi}{2} = 0 \implies x = \dfrac{\pi}{2}$

$x - \dfrac{\pi}{2} = \pi \implies x = \dfrac{3\pi}{2}$

141. $f(x) = \dfrac{1}{4} \sec x$

Period: 2π

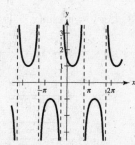

143. $f(x) = \dfrac{1}{4} \csc 2x$

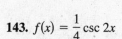

Period: π

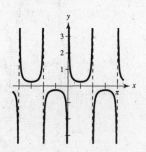

145. $f(x) = \sec\left(x - \dfrac{\pi}{4}\right)$

Secant function shifted
$\dfrac{\pi}{4}$ to right

147. $f(x) = \dfrac{1}{4} \tan \dfrac{\pi x}{2}$

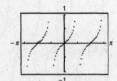

149. $f(x) = 4 \cot(2x - \pi) = \dfrac{4}{\tan(2x - \pi)}$

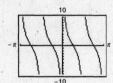

151. $f(x) = 2 \sec(x - \pi) = \dfrac{2}{\cos(x - \pi)}$

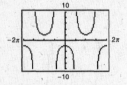

153. $f(x) = \csc\left(3x - \dfrac{\pi}{2}\right) = \dfrac{1}{\sin(3x - (\pi/2))}$

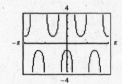

155. $f(x) = e^x \sin 2x$

Damping factor: $y = e^x$

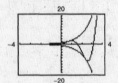

157. $f(x) = 2x \cos x$

Damping factor: $g(x) = 2x$

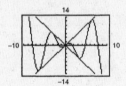

As $x \to \infty$, f oscillates between $2x$ and $-2x$.

159. (a) $\arcsin(-1) = -\dfrac{\pi}{2}$ because $\sin\left(-\dfrac{\pi}{2}\right) = -1$.

(b) $\arcsin 4$ does not exist because the domain of $\arcsin$ is $[-1, 1]$.

161. (a) $\arccos\left(\dfrac{\sqrt{2}}{2}\right) = \dfrac{\pi}{4}$ because $\cos \dfrac{\pi}{4} = \dfrac{\sqrt{2}}{2}$.

(b) $\arccos\left(-\dfrac{\sqrt{3}}{2}\right) = \dfrac{5\pi}{6}$ because $\cos \dfrac{5\pi}{6} = -\dfrac{\sqrt{3}}{2}$.

163. $\arccos(0.42) \approx 1.14$

165. $\sin^{-1}(-0.94) \approx -1.22$

167. $\arctan(-12) \approx -1.49$

169. $\tan^{-1}(0.81) \approx 0.68$

171. $\sin \theta = \dfrac{x + 3}{16} \implies \theta = \arcsin\left(\dfrac{x + 3}{16}\right)$

173. Let $y = \arcsin(x - 1)$. Then,

$\sin y = (x - 1) = \dfrac{x - 1}{1}$ and

$\sec y = \dfrac{1}{\sqrt{-x^2 + 2x}}$

$= \dfrac{\sqrt{-x^2 + 2x}}{-x^2 + 2x}$.

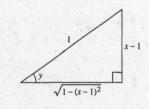

175. Let $y = \arccos \dfrac{x^2}{4 - x^2}$. Then $\cos y = \dfrac{x^2}{4 - x^2}$ and

$\sin y = \dfrac{\sqrt{(4 - x^2)^2 - (x^2)^2}}{4 - x^2}$.

$= \dfrac{\sqrt{16 - 8x^2}}{4 - x^2}$

$= \dfrac{2\sqrt{4 - 2x^2}}{4 - x^2}$.

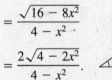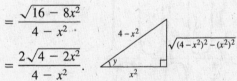

177. $\sin(1° \, 10') = \dfrac{a}{3.5}$

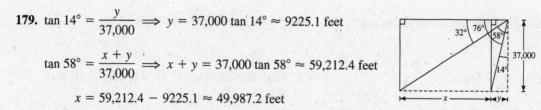

$a = 3.5 \sin(1° \, 10') = 3.5 \sin\!\left(\dfrac{7°}{6}\right) \approx 0.0713$ or 71 meters

not drawn to scale

179. $\tan 14° = \dfrac{y}{37{,}000} \implies y = 37{,}000 \tan 14° \approx 9225.1$ feet

$\tan 58° = \dfrac{x+y}{37{,}000} \implies x + y = 37{,}000 \tan 58° \approx 59{,}212.4$ feet

$x = 59{,}212.4 - 9225.1 \approx 49{,}987.2$ feet

The towns are approximately 50,000 feet apart or 9.47 miles.

181. Use cosine model with amplitude 3 feet.

Period: 15 seconds

$y = 3 \cos\!\left(\dfrac{2\pi}{15}t\right)$

183. False. $y = \sin \theta$ is a function, but it is not one-to-one.

185. $\tan \theta = \dfrac{0.672s^2}{3000}$

(a)

s	10	20	30	40	50	60
θ	1.28°	5.12°	11.40°	19.72°	29.25°	38.88°

(b) θ increases at an increasing rate. The function is not linear.

Chapter 4   Practice Test

1. Express $350°$ in radian measure.

2. Express $(5\pi)/9$ in degree measure.

3. Convert $135°\,14'\,12''$ to decimal form.

4. Convert $-22.569°$ to $D°\,M'\,S''$ form.

5. If $\cos \theta = \frac{2}{3}$, use the trigonometric identities to find $\tan \theta$.

6. Find θ given $\sin \theta = 0.9063$.

7. Solve for x in the figure below.

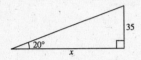

8. Find the magnitude of the reference angle for $\theta = (6\pi)/5$.

9. Evaluate $\csc 3.92$.

10. Find $\sec \theta$ given that θ lies in Quadrant III and $\tan \theta = 6$.

11. Graph $y = 3 \sin \dfrac{x}{2}$.

12. Graph $y = -2 \cos(x - \pi)$.

13. Graph $y = \tan 2x$.

14. Graph $y = -\csc\left(x + \dfrac{\pi}{4}\right)$.

15. Graph $y = 2x + \sin x$, using a graphing calculator.

16. Graph $y = 3x \cos x$, using a graphing calculator.

17. Evaluate $\arcsin 1$.

18. Evaluate $\arctan(-3)$.

19. Evaluate $\sin\left(\arccos \dfrac{4}{\sqrt{35}}\right)$.

20. Write an algebraic expression for $\cos\left(\arcsin \dfrac{x}{4}\right)$.

For Exercises 21–23, solve the right triangle.

21. $A = 40°,\ c = 12$

22. $B = 6.84°,\ a = 21.3$

23. $a = 5,\ b = 9$

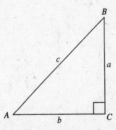

24. A 20-foot ladder leans against the side of a barn. Find the height of the top of the ladder if the angle of elevation of the ladder is $67°$.

25. An observer in a lighthouse 250 feet above sea level spots a ship off the shore. If the angle of depression to the ship is $5°$, how far out is the ship?

C H A P T E R 5
Analytic Trigonometry

Section 5.1   Using Fundamental Identities **206**

Section 5.2   Verifying Trigonometric Identities **212**

Section 5.3   Solving Trigonometric Equations **218**

Section 5.4   Sum and Difference Formulas **224**

Section 5.5   Multiple-Angle and Product-to-Sum Formulas **232**

Review Exercises **243**

Practice Test **251**

CHAPTER 5
Analytic Trigonometry

Section 5.1 Using Fundamental Identities

■ You should know the fundamental trigonometric identities.

(a) Reciprocal Identities

$$\sin u = \frac{1}{\csc u} \qquad\qquad \csc u = \frac{1}{\sin u}$$

$$\cos u = \frac{1}{\sec u} \qquad\qquad \sec u = \frac{1}{\cos u}$$

$$\tan u = \frac{1}{\cot u} = \frac{\sin u}{\cos u} \qquad \cot u = \frac{1}{\tan u} = \frac{\cos u}{\sin u}$$

(b) Pythagorean Identities

$$\sin^2 u + \cos^2 u = 1$$
$$1 + \tan^2 u = \sec^2 u$$
$$1 + \cot^2 u = \csc^2 u$$

(c) Cofunction Identities

$$\sin\left(\frac{\pi}{2} - u\right) = \cos u \qquad\qquad \cos\left(\frac{\pi}{2} - u\right) = \sin u$$

$$\tan\left(\frac{\pi}{2} - u\right) = \cot u \qquad\qquad \cot\left(\frac{\pi}{2} - u\right) = \tan u$$

$$\sec\left(\frac{\pi}{2} - u\right) = \csc u \qquad\qquad \csc\left(\frac{\pi}{2} - u\right) = \sec u$$

(d) Negative Angle Identities

$$\sin(-x) = -\sin x \qquad \csc(-x) = -\csc x$$
$$\cos(-x) = \cos x \qquad\; \sec(-x) = \sec x$$
$$\tan(-x) = -\tan x \qquad \cot(-x) = -\cot x$$

■ You should be able to use these fundamental identities to find function values.

■ You should be able to convert trigonometric expressions to equivalent forms by using the fundamental identities.

■ You should be able to check your answers with a graphing utility.

Vocabulary Check

1. $\sec u$ **2.** $\tan u$ **3.** $\cot u$ **4.** $\csc u$

5. $\tan^2 u$ **6.** $\csc^2 u$ **7.** $\sin u$ **8.** $\sec u$

9. $-\tan u$ **10.** $\cos u$

1. $\sin x = \dfrac{1}{2}$, $\cos x = \dfrac{\sqrt{3}}{2}$, x is in Quadrant I.

$\tan x = \dfrac{1/2}{\sqrt{3}/2} = \dfrac{1}{\sqrt{3}} = \dfrac{\sqrt{3}}{3}$

$\cot x = \sqrt{3}$

$\csc x = 2$

$\sec x = \dfrac{2}{\sqrt{3}} = \dfrac{2\sqrt{3}}{3}$

3. $\sec \theta = \sqrt{2}$, $\sin \theta = -\dfrac{\sqrt{2}}{2} \implies \theta$ is in Quadrant IV.

$\cos \theta = \dfrac{1}{\sec \theta} = \dfrac{1}{\sqrt{2}} = \dfrac{\sqrt{2}}{2}$

$\tan \theta = \dfrac{\sin \theta}{\cos \theta} = \dfrac{-\sqrt{2}/2}{\sqrt{2}/2} = -1$

$\cot \theta = \dfrac{1}{\tan \theta} = -1$

$\csc \theta = -\sqrt{2}$

5. $\tan x = \dfrac{7}{24}$, $\sec x = \dfrac{-25}{24} \implies x$ is in Quadrant III.

$\cot x = \dfrac{24}{7}$

$\cos x = -\dfrac{24}{25}$

$\sin x = -\sqrt{1 - \cos^2 x} = -\dfrac{7}{25}$

$\csc x = \dfrac{1}{\sin x} = -\dfrac{25}{7}$

7. $\sec \phi = -\dfrac{17}{15}$, $\sin \phi = \dfrac{8}{17}$, ϕ is in Quadrant II.

$\cos \phi = -\dfrac{15}{17}$

$\csc \phi = \dfrac{17}{8}$

$\tan \phi = \dfrac{8/17}{-15/17} = -\dfrac{8}{15}$

$\cot \phi = -\dfrac{15}{8}$

9. $\sin(-x) = -\sin x = -\dfrac{2}{3} \implies \sin x = \dfrac{2}{3}$

$\sin x = \dfrac{2}{3}$, $\tan x = -\dfrac{2\sqrt{5}}{5} \implies x$ is in Quadrant II.

$\cos x = -\sqrt{1 - \sin^2 x} = -\sqrt{1 - \dfrac{4}{9}} = -\dfrac{\sqrt{5}}{3}$

$\cot x = \dfrac{1}{\tan x} = -\dfrac{\sqrt{5}}{2}$

$\sec x = \dfrac{1}{\cos x} = -\dfrac{3\sqrt{5}}{5}$

$\csc x = \dfrac{1}{\sin x} = \dfrac{3}{2}$

11. $\tan \theta = 2$, $\sin \theta < 0 \implies \theta$ is in Quadrant III.

$\sec \theta = -\sqrt{\tan^2 \theta + 1} = -\sqrt{5}$

$\cos \theta = \dfrac{1}{-\sqrt{5}} = -\dfrac{\sqrt{5}}{5}$

$\cot \theta = \dfrac{1}{2}$

$\sin \theta = -\sqrt{1 - \cos^2 \theta}$

$\quad = -\sqrt{1 - \dfrac{1}{5}} = -\sqrt{\dfrac{4}{5}} = \dfrac{-2}{\sqrt{5}} = \dfrac{-2\sqrt{5}}{5}$

$\csc \theta = -\dfrac{\sqrt{5}}{2}$

13. $\csc \theta$ is undefined and $\cos \theta < 0 \implies \theta = \pi$.

$\sin \theta = 0$

$\cos \theta = -1$

$\tan \theta = 0$

$\cot \theta$ is undefined.

$\sec \theta = -1$

15. $\sec x \cos x = \dfrac{1}{\cos x} \cdot \cos x = 1$

Matches (d).

17. $\cot^2 x - \csc^2 x = \cot^2 x - (1 + \cot^2 x) = -1$

Matches (b).

19. $\dfrac{\sin(-x)}{\cos(-x)} = \dfrac{-\sin x}{\cos x} = -\tan x$

Matches (e).

21. $\sin x \sec x = \sin x \left(\dfrac{1}{\cos x}\right) = \tan x$

Matches (b).

23. $\sec^4 x - \tan^4 x = (\sec^2 x + \tan^2 x)(\sec^2 x - \tan^2 x)$
$$= (\sec^2 x + \tan^2 x)(1)$$
$$= \sec^2 x + \tan^2 x$$

Matches (f).

25. $\dfrac{\sec^2 x - 1}{\sin^2 x} = \dfrac{\tan^2 x}{\sin^2 x} = \dfrac{\sin^2 x}{\cos^2 x} \cdot \dfrac{1}{\sin^2 x} = \sec^2 x$

Matches (e).

27. $\cot x \sin x = \dfrac{\cos x}{\sin x} \sin x = \cos x$

29. $\sin \phi (\csc \phi - \sin \phi) = \sin \phi \csc \phi - \sin^2 \phi$
$$= \sin \phi \cdot \dfrac{1}{\sin \phi} - \sin^2 \phi$$
$$= 1 - \sin^2 \phi$$
$$= \cos^2 \phi$$

31. $\dfrac{\csc x}{\cot x} = \dfrac{1}{\sin x} \cdot \dfrac{\sin x}{\cos x} = \dfrac{1}{\cos x} = \sec x$

33. $\sec \alpha \dfrac{\sin \alpha}{\tan \alpha} = \dfrac{1}{\cos \alpha}(\sin \alpha) \cot \alpha$
$$= \dfrac{1}{\cos \alpha}(\sin \alpha)\left(\dfrac{\cos \alpha}{\sin \alpha}\right) = 1$$

35. $\sin\left(\dfrac{\pi}{2} - x\right) \csc x = \cos x \cdot \dfrac{1}{\sin x} = \cot x$

37. $\dfrac{\cos^2 y}{1 - \sin y} = \dfrac{1 - \sin^2 y}{1 - \sin y}$
$$= \dfrac{(1 + \sin y)(1 - \sin y)}{1 - \sin y}$$
$$= 1 + \sin y$$

39. $\sin \theta + \cos \theta \cot \theta = \sin \theta + \cos \theta \dfrac{\cos \theta}{\sin \theta}$
$$= \dfrac{\sin^2 \theta + \cos^2 \theta}{\sin \theta}$$
$$= \dfrac{1}{\sin \theta} = \csc \theta$$

41. $\dfrac{\cos \theta}{1 - \sin \theta} = \dfrac{\cos \theta}{1 - \sin \theta} \cdot \dfrac{1 + \sin \theta}{1 + \sin \theta}$
$$= \dfrac{\cos \theta (1 + \sin \theta)}{1 - \sin^2 \theta}$$
$$= \dfrac{\cos \theta (1 + \sin \theta)}{\cos^2 \theta}$$
$$= \dfrac{1 + \sin \theta}{\cos \theta} = \sec \theta + \tan \theta$$

43. $\dfrac{1 + \cos\theta}{\sin\theta} + \dfrac{\sin\theta}{1 + \cos\theta} = \dfrac{1 + 2\cos\theta + \cos^2\theta + \sin^2\theta}{\sin\theta(1 + \cos\theta)}$

$$= \dfrac{2 + 2\cos\theta}{\sin\theta(1 + \cos\theta)}$$

$$= \dfrac{2(1 + \cos\theta)}{\sin\theta(1 + \cos\theta)}$$

$$= \dfrac{2}{\sin\theta} = 2\csc\theta$$

45. $\csc\theta\tan\theta = \dfrac{1}{\sin\theta}\cdot\dfrac{\sin\theta}{\cos\theta} = \dfrac{1}{\cos\theta} = \sec\theta$

47. $1 - \dfrac{\sin^2\theta}{1 - \cos\theta} = \dfrac{1 - \cos\theta - \sin^2\theta}{1 - \cos\theta}$

$$= \dfrac{\cos^2\theta - \cos\theta}{1 - \cos\theta}$$

$$= \dfrac{\cos\theta(\cos\theta - 1)}{1 - \cos\theta}$$

$$= -\cos\theta$$

49. $\dfrac{\cot(-\theta)}{\csc\theta} = \dfrac{\cos(-\theta)}{\sin(-\theta)}\sin\theta$

$$= \dfrac{\cos\theta}{-\sin\theta}\sin\theta = -\cos\theta$$

51. $\cot^2 x - \cot^2 x\cos^2 x = \cot^2 x(1 - \cos^2 x)$

$$= \dfrac{\cos^2 x}{\sin^2 x}\sin^2 x = \cos^2 x$$

53. $\dfrac{\cos^2 x - 4}{\cos x - 2} = \dfrac{(\cos x + 2)(\cos x - 2)}{\cos x - 2} = \cos x + 2$

55. $\tan^4 x + 2\tan^2 x + 1 = (\tan^2 x + 1)^2$

$$= (\sec^2 x)^2 = \sec^4 x$$

57. $\sin^4 x - \cos^4 x = (\sin^2 x + \cos^2 x)(\sin^2 x - \cos^2 x)$

$$= (1)(\sin^2 x - \cos^2 x) = \sin^2 x - \cos^2 x$$

59. $\csc^3 x - \csc^2 x - \csc x + 1 = \csc^2 x(\csc x - 1) - (\csc x - 1)$

$$= (\csc^2 x - 1)(\csc x - 1)$$

$$= \cot^2 x(\csc x - 1)$$

61. $(\sin x + \cos x)^2 = \sin^2 x + 2\sin x\cos x + \cos^2 x$

$$= (\sin^2 x + \cos^2 x) + 2\sin x\cos x$$

$$= 1 + 2\sin x\cos x$$

63. $(\csc x + 1)(\csc x - 1) = \csc^2 x - 1 = \cot^2 x$

65. $\dfrac{1}{1 + \cos x} + \dfrac{1}{1 - \cos x} = \dfrac{1 - \cos x + 1 + \cos x}{(1 + \cos x)(1 - \cos x)}$

$$= \dfrac{2}{1 - \cos^2 x}$$

$$= \dfrac{2}{\sin^2 x}$$

$$= 2\csc^2 x$$

67. $\tan x - \dfrac{\sec^2 x}{\tan x} = \dfrac{\tan^2 x - \sec^2 x}{\tan x} = \dfrac{-1}{\tan x} = -\cot x$

69. $\dfrac{\sin^2 y}{1 - \cos y} = \dfrac{1 - \cos^2 y}{1 - \cos y}$

$\qquad = \dfrac{(1 + \cos y)(1 - \cos y)}{1 - \cos y}$

$\qquad = 1 + \cos y$

71. $\dfrac{3}{\sec x - \tan x} \cdot \dfrac{\sec x + \tan x}{\sec x + \tan x} = \dfrac{3(\sec x + \tan x)}{\sec^2 x - \tan^2 x}$

$\qquad\qquad\qquad = \dfrac{3(\sec x + \tan x)}{1}$

$\qquad\qquad\qquad = 3(\sec x + \tan x)$

73. $y_1 = \cos\!\left(\dfrac{\pi}{2} - x\right),\ y_2 = \sin x$

x	0.2	0.4	0.6	0.8	1.0	1.2	1.4
y_1	0.1987	0.3894	0.5646	0.7174	0.8415	0.9320	0.9854
y_2	0.1987	0.3894	0.5646	0.7174	0.8415	0.9320	0.9854

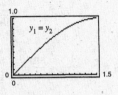

Conjecture: $y_1 = y_2$

75. $y_1 = \dfrac{\cos x}{1 - \sin x},\ y_2 = \dfrac{1 + \sin x}{\cos x}$

x	0.2	0.4	0.6	0.8	1.0	1.2	1.4
y_1	1.2230	1.5085	1.8958	2.4650	3.4082	5.3319	11.6814
y_2	1.2230	1.5085	1.8958	2.4650	3.4082	5.3319	11.6814

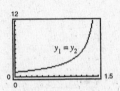

Conjecture: $y_1 = y_2$

77. $y_1 = \cos x \cot x + \sin x = \csc x$

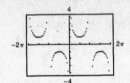

79. $y_1 = \sec x - \dfrac{\cos x}{1 + \sin x} = \tan x$

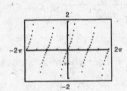

81. $\sqrt{25 - x^2} = \sqrt{25 - (5 \sin \theta)^2},\ x = 5 \sin \theta$

$\qquad = \sqrt{25 - 25 \sin^2 \theta}$

$\qquad = \sqrt{25(1 - \sin^2 \theta)}$

$\qquad = \sqrt{25 \cos^2 \theta}$

$\qquad = 5 \cos \theta$

83. $\sqrt{x^2 - 9} = \sqrt{(3 \sec \theta)^2 - 9},\ x = 3 \sec \theta$

$\qquad = \sqrt{9 \sec^2 \theta - 9}$

$\qquad = \sqrt{9(\sec^2 \theta - 1)}$

$\qquad = \sqrt{9 \tan^2 \theta}$

$\qquad = 3 \tan \theta$

85. $x = 3 \sin \theta,\ 0 < \theta < \dfrac{\pi}{2}$

$\qquad \sqrt{9 - x^2} = \sqrt{9 - 9 \sin^2 \theta}$

$\qquad\qquad = \sqrt{9 \cos^2 \theta} = 3 \cos \theta$

87. $2x = 3 \tan \theta,\ 0 < \theta < \dfrac{\pi}{2}$

$\qquad \sqrt{4x^2 + 9} = \sqrt{9 \tan^2 \theta + 9}$

$\qquad\qquad = \sqrt{9 \sec^2 \theta} = 3 \sec \theta$

89. $4x = 3 \sec \theta,\ 0 < \theta < \dfrac{\pi}{2}$

$\qquad \sqrt{16x^2 - 9} = \sqrt{9 \sec^2 \theta - 9}$

$\qquad\qquad = \sqrt{9 \tan^2 \theta} = 3 \tan \theta$

91. $x = \sqrt{2} \sin \theta,\ 0 < \theta < \dfrac{\pi}{2}$

$\qquad \sqrt{2 - x^2} = \sqrt{2 - 2 \sin^2 \theta}$

$\qquad\qquad = \sqrt{2 \cos^2 \theta} = \sqrt{2} \cos \theta$

93. $\sin \theta = \sqrt{1 - \cos^2 \theta}$

Let $y_1 = \sin x$ and $y_2 = \sqrt{1 - \cos^2 x}$, $0 \le x < 2\pi$.

$y_1 = y_2$ for $0 \le x \le \pi$, so we have

$\sin \theta = \sqrt{1 - \cos^2 \theta}$ for $0 \le \theta \le \pi$.

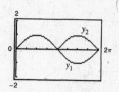

95. $\sec \theta = \sqrt{1 + \tan^2 \theta}$

Let $y_1 = \dfrac{1}{\cos x}$ and $y_2 = \sqrt{1 + \tan^2 x}$, $0 \le x < 2\pi$.

$y_1 = y_2$ for $0 \le x < \dfrac{\pi}{2}$ and $\dfrac{3\pi}{2} < x < 2\pi$, so we

have $\sec \theta = \sqrt{1 + \tan^2 \theta}$ for $0 \le \theta < \dfrac{\pi}{2}$ and $\dfrac{3\pi}{2} < \theta < 2\pi$.

97. $\ln|\cos \theta| - \ln|\sin \theta| = \ln \dfrac{|\cos \theta|}{|\sin \theta|} = \ln|\cot \theta|$

99. $\ln(1 + \sin x) - \ln|\sec x| = \ln \left| \dfrac{1 + \sin x}{\sec x} \right|$

$= \ln|\cos x(1 + \sin x)|$

101. Let $\theta = \dfrac{7\pi}{6}$. Then

$\cos \theta = \cos \dfrac{7\pi}{6} = -\dfrac{\sqrt{3}}{2} \neq \sqrt{1 - \sin^2 \theta} = \dfrac{\sqrt{3}}{2}$.

103. Let $\theta = \dfrac{5\pi}{3}$. Then

$\sin \theta = \sin \dfrac{5\pi}{3} = -\dfrac{\sqrt{3}}{2} \neq \sqrt{1 - \cos^2 \theta} = \dfrac{\sqrt{3}}{2}$.

105. Let $\theta = \dfrac{7\pi}{4}$. Then

$\csc \theta = \csc \dfrac{7\pi}{4} = -\sqrt{2} \neq \sqrt{1 + \cot^2 \theta} = \sqrt{2}$.

107. (a) $\csc^2 132° - \cot^2 132° \approx 1.8107 - 0.8107 = 1$

(b) $\csc^2 \dfrac{2\pi}{7} - \cot^2 \dfrac{2\pi}{7} \approx 1.6360 - 0.6360 = 1$

109. $\cos\left(\dfrac{\pi}{2} - \theta\right) = \sin \theta$

(a) $\theta = 80°$

$\cos(90° - 80°) = \sin 80°$

$0.9848 = 0.9848$

(b) $\theta = 0.8$

$\cos\left(\dfrac{\pi}{2} - 0.8\right) = \sin 0.8$

$0.7174 = 0.7174$

111. $\csc x \cot x - \cos x = \dfrac{1}{\sin x} \cdot \dfrac{\cos x}{\sin x} - \cos x$

$= \cos x(\csc^2 x - 1)$

$= \cos x \cdot \cot^2 x$

113. True for all $\theta \neq n\pi$

$\sin \theta \cdot \csc \theta = \sin \theta \left(\dfrac{1}{\sin \theta}\right) = 1$

115. As $x \to \dfrac{\pi^-}{2}$, $\sin x \to 1$ and $\csc x \to 1$.

117. As $x \to \dfrac{\pi^-}{2}$, $\tan x \to \infty$ and $\cot x \to 0$.

119. $\sin \theta$

$\cos \theta = \pm \sqrt{1 - \sin^2 \theta}$

$\tan \theta = \dfrac{\sin \theta}{\cos \theta} = \pm \dfrac{\sin \theta}{\sqrt{1 - \sin^2 \theta}}$

$\csc \theta = \dfrac{1}{\sin \theta}$

$\sec \theta = \dfrac{\pm 1}{\sqrt{1 - \sin^2 \theta}}$

$\cot \theta = \pm \dfrac{\sqrt{1 - \sin^2 \theta}}{\sin \theta}$

The sign + or − depends on the choice of θ.

121. $\sin \theta = \dfrac{\text{opp}}{\text{hyp}},\ \cos \theta = \dfrac{\text{adj}}{\text{hyp}}$

From the Pythagorean Theorem,

$(\text{opp})^2 + (\text{adj})^2 = (\text{hyp})^2$

$\sin^2 \theta + \cos^2 \theta = 1.$

123. $f(x) = \dfrac{1}{2} \sin \pi x$

Period: $\dfrac{2\pi}{\pi} = 2$

Amplitude: $\dfrac{1}{2}$

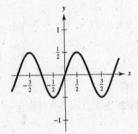

125. $f(x) = \dfrac{1}{2} \cot\left(x + \dfrac{\pi}{4}\right)$

Period: π

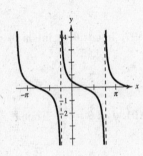

Section 5.2   Verifying Trigonometric Identities

> ■ You should know the difference between an expression, a conditional equation, and an identity.
>
> ■ You should be able to solve trigonometric identities, using the following techniques.
>
> (a) Work with *one* side at a time. Do not "cross" the equal sign.
>
> (b) Use algebraic techniques such as combining fractions, factoring expressions, rationalizing denominators, and squaring binomials.
>
> (c) Use the fundamental identities.
>
> (d) Convert all the terms into sines and cosines.

Vocabulary Check

1. conditional   **2.** identity   **3.** $\cot u$ **4.** $\sin u$

5. $\tan u$ **6.** $\cos u$ **7.** $\cos^2 u$ **8.** $\cot u$

9. $-\sin u$ **10.** $\sec u$

1. $\sin t \csc t = \sin t \left(\dfrac{1}{\sin t}\right) = 1$

3. $\dfrac{\csc^2 x}{\cot x} = \dfrac{1}{\sin^2 x} \cdot \dfrac{\sin x}{\cos x} = \dfrac{1}{\sin x \cdot \cos x}$

$= \csc x \cdot \sec x$

5. $\cos^2 \beta - \sin^2 \beta = (1 - \sin^2 \beta) - \sin^2 \beta$

$\qquad\qquad = 1 - 2 \sin^2 \beta$

7. $\tan^2 \theta + 6 = (\tan^2 \theta + 1) + 5$

$\qquad\qquad = \sec^2 \theta + 5$

9. $(1 + \sin x)(1 - \sin x) = 1 - \sin^2 x = \cos^2 x$

11. $\dfrac{1}{\sec x \tan x} = \cos x \cdot \dfrac{\cos x}{\sin x}$

$\qquad = \dfrac{\cos^2 x}{\sin x}$

$\qquad = \dfrac{1 - \sin^2 x}{\sin x}$

$\qquad = \dfrac{1}{\sin x} - \sin x$

$\qquad = \csc x - \sin x$

x	0.2	0.4	0.6	0.8	1.0	1.2	1.4
y_1	4.8348	2.1785	1.2064	0.6767	0.3469	0.1409	0.0293
y_2	4.8348	2.1785	1.2064	0.6767	0.3469	0.1409	0.0293

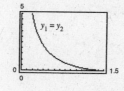

13. $\csc x - \sin x = \dfrac{1}{\sin x} - \sin x$

$\qquad = \dfrac{1 - \sin^2 x}{\sin x}$

$\qquad = \dfrac{\cos^2 x}{\sin x}$

$\qquad = \cos x \cdot \dfrac{\cos x}{\sin x}$

$\qquad = \cos x \cdot \cot x$

x	0.2	0.4	0.6	0.8	1.0	1.2	1.4
y_1	4.8348	2.1785	1.2064	0.6767	0.3469	0.1409	0.0293
y_2	4.8348	2.1785	1.2064	0.6767	0.3469	0.1409	0.0293

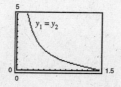

15. $\sin x + \cos x \cot x = \sin x + \cos x \dfrac{\cos x}{\sin x}$

$\qquad = \dfrac{\sin^2 x + \cos^2 x}{\sin x}$

$\qquad = \dfrac{1}{\sin x}$

$\qquad = \csc x$

x	0.2	0.4	0.6	0.8	1.0	1.2	1.4
y_1	5.0335	2.5679	1.7710	1.3940	1.1884	1.0729	1.0148
y_2	5.0335	2.5679	1.7710	1.3940	1.1884	1.0729	1.0148

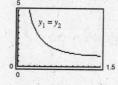

17. $\dfrac{1}{\tan x} + \dfrac{1}{\cot x} = \dfrac{\cot x + \tan x}{\tan x \cdot \cot x}$

$\qquad\qquad = \cot x + \tan x$

x	0.2	0.4	0.6	0.8	1.0	1.2	1.4
y_1	5.1359	2.7880	2.1458	2.0009	2.1995	2.9609	5.9704
y_2	5.1359	2.7880	2.1458	2.0009	2.1995	2.9609	5.9704

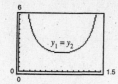

19. The error is in line 1: $\cot(-x) \neq \cot x$.

21. $\sin^{1/2} x \cos x - \sin^{5/2} x \cos x = \sin^{1/2} x \cos x(1 - \sin^2 x) = \sin^{1/2} x \cos x \cdot \cos^2 x = \cos^3 x\sqrt{\sin x}$

23. $\cot\left(\dfrac{\pi}{2} - x\right) \csc x = \tan x \csc x$

$$= \frac{\sin x}{\cos x} \cdot \frac{1}{\sin x}$$

$$= \frac{1}{\cos x} = \sec x$$

25. $\dfrac{\csc(-x)}{\sec(-x)} = \dfrac{1/\sin(-x)}{1/\cos(-x)}$

$$= \frac{\cos(-x)}{\sin(-x)}$$

$$= \frac{\cos x}{-\sin x}$$

$$= -\cot x$$

27. $\dfrac{\cos(-\theta)}{1 + \sin(-\theta)} = \dfrac{\cos \theta}{1 - \sin \theta} \cdot \dfrac{1 + \sin \theta}{1 + \sin \theta}$

$$= \frac{\cos \theta(1 + \sin \theta)}{1 - \sin^2 \theta}$$

$$= \frac{\cos \theta(1 + \sin \theta)}{\cos^2 \theta}$$

$$= \frac{1 + \sin \theta}{\cos \theta}$$

$$= \frac{1}{\cos \theta} + \frac{\sin \theta}{\cos \theta}$$

$$= \sec \theta + \tan \theta$$

29. $\dfrac{\sin x \cos y + \cos x \sin y}{\cos x \cos y - \sin x \sin y} = \dfrac{\dfrac{\sin x \cos y}{\cos x \cos y} + \dfrac{\cos x \sin y}{\cos x \cos y}}{\dfrac{\cos x \cos y}{\cos x \cos y} - \dfrac{\sin x \sin y}{\cos x \cos y}} = \dfrac{\tan x + \tan y}{1 - \tan x \tan y}$

31. $\dfrac{\cos x - \cos y}{\sin x + \sin y} + \dfrac{\sin x - \sin y}{\cos x + \cos y} = \dfrac{(\cos x + \cos y)(\cos x - \cos y) + (\sin x + \sin y)(\sin x - \sin y)}{(\sin x + \sin y)(\cos x + \cos y)}$

$$= \frac{\cos^2 x - \cos^2 y + \sin^2 x - \sin^2 y}{(\sin x + \sin y)(\cos x + \cos y)}$$

$$= \frac{1 - 1}{(\sin x + \sin y)(\cos x + \cos y)}$$

$$= 0$$

33. $\sqrt{\dfrac{1 + \sin \theta}{1 - \sin \theta}} = \sqrt{\dfrac{1 + \sin \theta}{1 - \sin \theta} \cdot \dfrac{1 + \sin \theta}{1 + \sin \theta}}$

$$= \sqrt{\frac{(1 + \sin \theta)^2}{1 - \sin^2 \theta}}$$

$$= \sqrt{\frac{(1 + \sin \theta)^2}{\cos^2 \theta}}$$

$$= \frac{1 + \sin \theta}{|\cos \theta|}$$

Note: Check your answer with a graphing utility. What happens if you leave off the absolute value?

35. $\sin^2\left(\dfrac{\pi}{2} - x\right) + \sin^2 x = \cos^2 x + \sin^2 x = 1$

37. $\sin x \csc\left(\dfrac{\pi}{2} - x\right) = \sin x \sec x$

$$= \sin x\left(\dfrac{1}{\cos x}\right)$$

$$= \tan x$$

39. $2\sec^2 x - 2\sec^2 x \sin^2 x - \sin^2 x - \cos^2 x = 2\sec^2 x(1 - \sin^2 x) - (\sin^2 x + \cos^2 x)$

$$= 2\sec^2 x(\cos^2 x) - 1$$

$$= 2 \cdot \dfrac{1}{\cos^2 x} \cdot \cos^2 x - 1$$

$$= 2 - 1 = 1$$

41. $\dfrac{\cot x \tan x}{\sin x} = \dfrac{1}{\sin x} = \csc x$

43. $\csc^4 x - 2\csc^2 x + 1 = (\csc^2 x - 1)^2$

$$= (\cot^2 x)^2 = \cot^4 x$$

45. $\sec^4 \theta - \tan^4 \theta = (\sec^2 \theta + \tan^2 \theta)(\sec^2 \theta - \tan^2 \theta)$

$$= (1 + \tan^2 \theta + \tan^2 \theta)(1)$$

$$= 1 + 2\tan^2 \theta$$

47. $\dfrac{\sin \beta}{1 - \cos \beta} \cdot \dfrac{1 + \cos \beta}{1 + \cos \beta} = \dfrac{\sin \beta(1 + \cos \beta)}{1 - \cos^2 \beta}$

$$= \dfrac{\sin \beta(1 + \cos \beta)}{\sin^2 \beta}$$

$$= \dfrac{1 + \cos \beta}{\sin \beta}$$

49. $\dfrac{\tan^3 \alpha - 1}{\tan \alpha - 1} = \dfrac{(\tan \alpha - 1)(\tan^2 \alpha + \tan \alpha + 1)}{\tan \alpha - 1} = \tan^2 \alpha + \tan \alpha + 1$

51. It appears that $y_1 = 1$. Analytically,

$$\dfrac{1}{\cot x + 1} + \dfrac{1}{\tan x + 1} = \dfrac{\tan x + 1 + \cot x + 1}{(\cot x + 1)(\tan x + 1)}$$

$$= \dfrac{\tan x + \cot x + 2}{\cot x \tan x + \cot x + \tan x + 1}$$

$$= \dfrac{\tan x + \cot x + 2}{\tan x + \cot x + 2}$$

$$= 1.$$

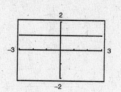

53. It appears that $y_1 = \sin x$. Analytically,

$$\dfrac{1}{\sin x} - \dfrac{\cos^2 x}{\sin x} = \dfrac{1 - \cos^2 x}{\sin x} = \dfrac{\sin^2 x}{\sin x} = \sin x.$$

55. $\ln|\cot \theta| = \ln\left|\dfrac{\cos \theta}{\sin \theta}\right|$

$$= \ln\dfrac{|\cos \theta|}{|\sin \theta|}$$

$$= \ln|\cos \theta| - \ln|\sin \theta|$$

57. $-\ln(1 + \cos \theta) = \ln(1 + \cos \theta)^{-1}$

$$= \ln\left[\frac{1}{1 + \cos \theta} \cdot \frac{1 - \cos \theta}{1 - \cos \theta}\right]$$

$$= \ln \frac{1 - \cos \theta}{1 - \cos^2 \theta}$$

$$= \ln \frac{1 - \cos \theta}{\sin^2 \theta}$$

$$= \ln(1 - \cos \theta) - \ln \sin^2 \theta$$

$$= \ln(1 - \cos \theta) - 2 \ln|\sin \theta|$$

59. $\sin^2 35° + \sin^2 55° = \cos^2(90° - 35°) + \sin^2 55°$

$$= \cos^2 55° + \sin^2 55° = 1$$

61. $\cos^2 20° + \cos^2 52° + \cos^2 38° + \cos^2 70° = \cos^2 20° + \cos^2 52° + \sin^2(90° - 38°) + \sin^2(90° - 70°)$

$$= \cos^2 20° + \cos^2 52° + \sin^2 52° + \sin^2 20°$$

$$= (\cos^2 20° + \sin^2 20°) + (\cos^2 52° + \sin^2 52°)$$

$$= 1 + 1 = 2$$

63. $\tan^5 x = \tan^3 x \cdot \tan^2 x$

$$= \tan^3 x(\sec^2 x - 1)$$

$$= \tan^3 x \sec^2 x - \tan^3 x$$

65. $(\sin^2 x - \sin^4 x) \cos x = \sin^2 x(1 - \sin^2 x) \cos x$

$$= \sin^2 x \cdot \cos^2 x \cdot \cos x$$

$$= \cos^3 x \sin^2 x$$

67. Let $\theta = \sin^{-1} x \implies \sin \theta = x = \dfrac{x}{1}$.

From the diagram,

$$\tan(\sin^{-1} x) = \tan \theta = \frac{x}{\sqrt{1 - x^2}}.$$

69. Let $\theta = \sin^{-1} \dfrac{x - 1}{4} \implies \sin \theta = \dfrac{x - 1}{4}$.

From the diagram,

$$\tan\left(\sin^{-1} \frac{x - 1}{4}\right) = \tan \theta = \frac{x - 1}{\sqrt{16 - (x - 1)^2}}.$$

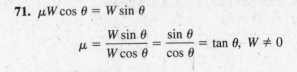

71. $\mu W \cos \theta = W \sin \theta$

$$\mu = \frac{W \sin \theta}{W \cos \theta} = \frac{\sin \theta}{\cos \theta} = \tan \theta, \ W \neq 0$$

73. True

75. False. Just because the equation is true for one value of θ, you cannot conclude that the equation is an identity. For example,

$$\sin^2 \frac{\pi}{4} + \cos^2 \frac{\pi}{4} = 1 \neq 1 + \tan^2 \frac{\pi}{4}.$$

77. (a) $\dfrac{\sin x}{1 + \cos x} = \dfrac{\sin x}{1 + \cos x} \cdot \dfrac{1 - \cos x}{1 - \cos x}$

$= \dfrac{\sin x(1 - \cos x)}{1 - \cos^2 x}$

$= \dfrac{\sin x(1 - \cos x)}{\sin^2 x}$

$= \dfrac{1 - \cos x}{\sin x}$

(b) Not true for $x = 0$ because $(1 - \cos x)/\sin x$ is not defined for $x = 0$.

79. (a) $\dfrac{\sin x}{1 + \cos x} = \dfrac{\sin x}{1 + \cos x} \cdot \dfrac{1 - \cos x}{1 - \cos x}$

$= \dfrac{\sin x(1 - \cos x)}{1 - \cos^2 x}$

$= \dfrac{\sin x(1 - \cos x)}{\sin^2 x}$

$= \dfrac{1 - \cos x}{\sin x} = \csc x - \cot x$

(b) The identity is true for $x = \dfrac{\pi}{2}$:

$\dfrac{\sin(\pi/2)}{1 + \cos(\pi/2)} = \dfrac{1}{1 + 0}$

$= 1 = \csc \dfrac{\pi}{2} - \cot \dfrac{\pi}{2}$

81. $\sqrt{a^2 - u^2} = \sqrt{a^2 - a^2 \sin^2 \theta}$

$= \sqrt{a^2(1 - \sin^2 \theta)}$

$= \sqrt{a^2 \cos^2 \theta}$

$= a \cos \theta$

83. $\sqrt{a^2 + u^2} = \sqrt{a^2 + a^2 \tan^2 \theta}$

$= \sqrt{a^2(1 + \tan^2 \theta)}$

$= \sqrt{a^2 \sec^2 \theta}$

$= a \sec \theta$

85. $\sqrt{\tan^2 x} = |\tan x|$

Let $x = \dfrac{3\pi}{4}$. Then, $\sqrt{\tan^2 x} = \sqrt{(-1)^2} = 1 \neq \tan\left(\dfrac{3\pi}{4}\right) = -1.$

87. When n is even, $\cos\left[\dfrac{(2n + 1)\pi}{2}\right] = \cos \dfrac{\pi}{2} = 0.$

When n is odd, $\cos\left[\dfrac{(2n + 1)\pi}{2}\right] = \cos \dfrac{3\pi}{2} = 0.$

Thus, $\cos\left[\dfrac{(2n + 1)\pi}{2}\right] = 0$ for all n.

89. $(x - 1)(x - 8i)(x + 8i) = (x - 1)(x^2 + 64)$

$= x^3 - x^2 + 64x - 64$

Answers will vary.

91. $(x - 4)(x - 6 - i)(x - 6 + i) = (x - 4)((x - 6)^2 + 1)$

$= (x - 4)(x^2 - 12x + 37)$

$= x^3 - 16x^2 + 85x - 148$

Answers will vary.

93. $f(x) = 2^x + 3$

95. $f(x) = 2^{-x} + 1$

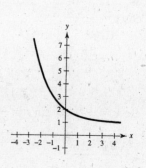

97. $\csc\theta > 0$ and $\tan\theta < 0 \implies$ Quadrant II

99. $\sec\theta > 0$ and $\sin\theta < 0 \implies$ Quadrant IV

Section 5.3   Solving Trigonometric Equations

- You should be able to identify and solve trigonometric equations.
- A trigonometric equation is a conditional equation. It is true for a specific set of values.
- To solve trigonometric equations, use algebraic techniques such as collecting like terms, taking square roots, factoring, squaring, converting to quadratic form, using formulas, and using inverse functions. Study the examples in this section.
- Use your graphing utility to calculate solutions and verify results.

Vocabulary Check

1. general   **2.** quadratic   **3.** extraneous

1. $2\cos x - 1 = 0$

(a) $x = \dfrac{\pi}{3}$: $2\cos\dfrac{\pi}{3} - 1 = 2\left(\dfrac{1}{2}\right) - 1 = 0$ (b) $x = \dfrac{5\pi}{3}$: $2\cos\dfrac{5\pi}{3} - 1 = 2\left(\dfrac{1}{2}\right) - 1 = 0$

3. $3\tan^2 2x - 1 = 0$

(a) $x = \dfrac{\pi}{12}$: $3\left[\tan\left(\dfrac{2\pi}{12}\right)\right]^2 - 1 = 3\tan^2\dfrac{\pi}{6} - 1 = 3\left(\dfrac{1}{\sqrt{3}}\right)^2 - 1 = 0$

(b) $x = \dfrac{5\pi}{12}$: $3\left[\tan\left(\dfrac{10\pi}{12}\right)\right]^2 - 1 = 3\tan^2\dfrac{5\pi}{6} - 1 = 3\left(-\dfrac{1}{\sqrt{3}}\right)^2 - 1 = 0$

5. $2\sin^2 x - \sin x - 1 = 0$

(a) $x = \dfrac{\pi}{2}$: $2\sin^2\left(\dfrac{\pi}{2}\right) - \sin\left(\dfrac{\pi}{2}\right) - 1 = 2 - 1 - 1 = 0$

(b) $x = \dfrac{7\pi}{6}$: $2\sin^2\left(\dfrac{7\pi}{6}\right) - \sin\left(\dfrac{7\pi}{6}\right) - 1 = 2\left(\dfrac{1}{4}\right) - \left(-\dfrac{1}{2}\right) - 1 = 0$

7. $\sin x = \dfrac{1}{2}$

$x = 30°, 150°$

9. $\cos x = -\dfrac{1}{2}$

$x = 120°, 240°$

11. $\tan x = 1$

$x = 45°, 225°$

13. $\cos x = -\dfrac{\sqrt{3}}{2}$

$x = \dfrac{5\pi}{6}, \dfrac{7\pi}{6}$

15. $\cot x = -1$

$x = \dfrac{3\pi}{4}, \dfrac{7\pi}{4}$

17. $\tan x = -\dfrac{\sqrt{3}}{3}$

$x = \dfrac{5\pi}{6}, \dfrac{11\pi}{6}$

19. $\csc x = -2 \implies \sin x = -\dfrac{1}{2}$

$x = \dfrac{7\pi}{6}, \dfrac{11\pi}{6}$

21. $\cot x = \sqrt{3} \implies \tan x = \dfrac{\sqrt{3}}{3}$

$x = \dfrac{\pi}{6}, \dfrac{7\pi}{6}$

23. $\tan x = -1$

$x = \dfrac{3\pi}{4}, \dfrac{7\pi}{4}$

25. $2 \cos x + 1 = 0$

$$2 \cos x = -1$$

$$\cos x = -\frac{1}{2}$$

$$x = \frac{2\pi}{3} + 2n\pi$$

$$\text{or } x = \frac{4\pi}{3} + 2n\pi$$

27. $\sqrt{3} \sec x - 2 = 0$

$$\sqrt{3} \sec x = 2$$

$$\sec x = \frac{2}{\sqrt{3}}$$

$$\cos x = \frac{\sqrt{3}}{2}$$

$$x = \frac{\pi}{6} + 2n\pi$$

$$\text{or } x = \frac{11\pi}{6} + 2n\pi$$

29. $3 \csc^2 x - 4 = 0$

$$\csc^2 x = \frac{4}{3}$$

$$\csc x = \pm \frac{2}{\sqrt{3}}$$

$$\sin x = \pm \frac{\sqrt{3}}{2}$$

$$x = \frac{\pi}{3} + n\pi \ \text{ or } \ x = \frac{2\pi}{3} + n\pi$$

31. $4 \cos^2 x - 1 = 0$

$$\cos^2 x = \frac{1}{4}$$

$$\cos x = \pm \frac{1}{2}$$

$$x = \frac{\pi}{3} + n\pi \text{ or } x = \frac{2\pi}{3} + n\pi$$

33.

$$\sin^2 x = 3 \cos^2 x$$

$$\sin^2 x - 3(1 - \sin^2 x) = 0$$

$$4 \sin^2 x = 3$$

$$\sin x = \pm \frac{\sqrt{3}}{2}$$

$$x = \frac{\pi}{3} + n\pi$$

$$\text{or } x = \frac{2\pi}{3} + n\pi$$

35. $\tan x + \sqrt{3} = 0$

$$\tan x = -\sqrt{3}$$

$$x = \frac{2\pi}{3}, \frac{5\pi}{3}$$

37. $\csc^2 x - 2 = 0$

$$\csc^2 x = 2$$

$$\csc x = \pm \sqrt{2}$$

$$\sin x = \pm \frac{1}{\sqrt{2}}$$

$$x = \frac{\pi}{4}, \frac{3\pi}{4}, \frac{5\pi}{4}, \frac{7\pi}{4}$$

39. $3 \tan^3 x - \tan x = 0$

$$\tan x(3 \tan^2 x - 1) = 0$$

$$\tan x = 0 \quad \text{or} \quad 3 \tan^2 x - 1 = 0$$

$$x = 0, \pi \qquad \qquad \tan x = \pm \frac{\sqrt{3}}{3}$$

$$x = \frac{\pi}{6}, \frac{5\pi}{6}, \frac{7\pi}{6}, \frac{11\pi}{6}$$

41. $\sec^2 x - \sec x - 2 = 0$

$(\sec x - 2)(\sec x + 1) = 0$

$\sec x - 2 = 0 \quad$ or $\quad \sec x + 1 = 0$

$\sec x = 2 \qquad\qquad \sec x = -1$

$x = \dfrac{\pi}{3}, \dfrac{5\pi}{3} \qquad\qquad x = \pi$

43. $2 \sin x + \csc x = 0$

$2 \sin x + \dfrac{1}{\sin x} = 0$

$2 \sin^2 x + 1 = 0$

Since $2 \sin^2 x + 1 > 0$, there are no solutions.

45. $\cos x + \sin x \tan x = 2$

$\cos x + \dfrac{\sin^2 x}{\cos x} = 2$

$\dfrac{\cos^2 x + \sin^2 x}{\cos x} = 2$

$\dfrac{1}{\cos x} = 2$

$\cos x = \dfrac{1}{2}$

$x = \dfrac{\pi}{3}, \dfrac{5\pi}{3}$

47. $\sec^2 x + \tan x = 3$

$(1 + \tan^2 x) + \tan x = 3$

$\tan^2 x + \tan x - 2 = 0$

$(\tan x + 2)(\tan x - 1) = 0$

$\tan x = -2 \qquad$ or $\quad \tan x = 1$

$x \approx 2.0344, 5.1760 \qquad x = \dfrac{\pi}{4}, \dfrac{5\pi}{4}$

49. $2 \sin^2 x + 3 \sin x + 1 = 0$

$y = 2 \sin^2 x + 3 \sin x + 1$

$x \approx 3.6652, 5.7596, 4.7124$

51. $y = 4 \sin^2 x - 2 \cos x - 1$

$x \approx 0.8614, 5.4218$

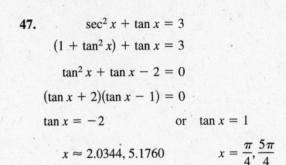

53. $y = \csc x + \cot x - 1 = \dfrac{1}{\sin x} + \dfrac{\cos x}{\sin x} - 1$

$x \approx 1.5708, \qquad \left(\dfrac{\pi}{2} \right)$

55. $\dfrac{\cos x \cot x}{1 - \sin x} = 3$

Graph $y = \dfrac{\cos x}{(1 - \sin x) \tan x} - 3.$

The solutions are approximately
$x \approx 0.5236, x \approx 2.6180.$

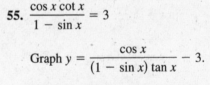

57. (a) 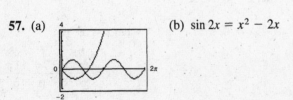   (b) $\sin 2x = x^2 - 2x$

(c) Points of intersection: $(0, 0)$, $(1.7757, -0.3984)$

59. (a) 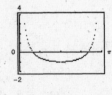   (b) $\sin^2 x = e^x - 4x$

(c) Points of intersection:

$(0.3194, 0.0986), (2.2680, 0.5878)$

61. $\cos \dfrac{x}{4} = 0$

$\dfrac{x}{4} = \dfrac{\pi}{2} + 2n\pi$ or $\dfrac{x}{4} = \dfrac{3\pi}{2} + 2n\pi$

$x = 2\pi + 8n\pi$ or $x = 6\pi + 8n\pi$

Combining, $x = 2\pi + 4n\pi$.

63. $\sin 4x = 1$

$4x = \dfrac{\pi}{2} + 2n\pi$

$x = \dfrac{\pi}{8} + \dfrac{n\pi}{2}$

65. $\sin 2x = -\dfrac{\sqrt{3}}{2}$

$2x = \dfrac{4\pi}{3} + 2n\pi$ or $2x = \dfrac{5\pi}{3} + 2n\pi$

$x = \dfrac{2\pi}{3} + n\pi$ \qquad $x = \dfrac{5\pi}{6} + n\pi$

67. $2 \sin^2 2x = 1$

$\sin^2 2x = \dfrac{1}{2}$

$\sin 2x = \pm \dfrac{\sqrt{2}}{2}$

$2x = \dfrac{\pi}{4} + \dfrac{n\pi}{2}$

$x = \dfrac{\pi}{8} + \dfrac{n\pi}{4}$

69. $\tan 3x(\tan x - 1) = 0$

$\tan 3x = 0$ or $\tan x - 1 = 0$

$3x = n\pi$ \quad or \quad $x = \dfrac{\pi}{4} + n\pi$

$x = \dfrac{n\pi}{3}$ \qquad $x = \dfrac{\pi}{4} + n\pi$

71. $\cos \dfrac{x}{2} = \dfrac{\sqrt{2}}{2}$

$\dfrac{x}{2} = \dfrac{\pi}{4} + 2n\pi$ or $\dfrac{x}{2} = \dfrac{7\pi}{4} + 2n\pi$

$x = \dfrac{\pi}{2} + 4n\pi$ \qquad $x = \dfrac{7\pi}{2} + 4n\pi$

73. $y = \sin \dfrac{\pi x}{2} + 1$

From the graph in the textbook we see that the curve has x-intercepts at $x = -1$ and at $x = 3$.

75. $y = \tan^2 \left(\dfrac{\pi x}{6} \right) - 3$

From the graph in the textbook, we see that the curve has x-intercepts at $x = \pm 2$.

77. $2 \cos x - \sin x = 0$

Graph $y_1 = 2 \cos x - \sin x$ and estimate the zeros.

$x \approx 1.1071, 4.2487$

79. $x \tan x - 1 = 0$

Graph $y_1 = x \tan x - 1$ and estimate the zeros.

$x \approx 0.8603, 3.4256$

81. $\sec^2 x + 0.5 \tan x - 1 = 0$

Graph $y_1 = \dfrac{1}{(\cos x)^2} + 0.5 \tan x - 1$.

$x = 0, x \approx 2.6779, 3.1416, 5.8195$

83. $12 \sin^2 x - 13 \sin x + 3 = 0$

Graph $y_1 = 12 \sin^2 x - 13 \sin x + 3$.

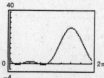

$x \approx 0.3398, 0.8481, 2.2935, 2.8018$

85. $3 \tan^2 x + 5 \tan x - 4 = 0$, $\left[-\dfrac{\pi}{2}, \dfrac{\pi}{2} \right]$

$$x \approx -1.154, 0.534$$

87. $4 \cos^2 x - 2 \sin x + 1 = 0$, $\left[-\dfrac{\pi}{2}, \dfrac{\pi}{2} \right]$

$$x \approx 1.110$$

89. $f(x) = \sin 2x$

(a)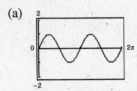

Maxima: $(0.7854, 1)$, $(3.9270, 1)$

Minima: $(2.3562, -1)$, $(5.4978, -1)$

(b) $2 \cos 2x = 0$

$\cos 2x = 0$

$2x = \dfrac{\pi}{2} + n\pi$

$x = \dfrac{\pi}{4} + \dfrac{n\pi}{2}$

The zeros are 0.7854, 2.3562, 3.9270, and 5.4978.

91. $f(x) = \sin^2 x + \cos x$

(a)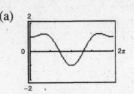

Maxima: $(1.0472, 1.25)$, $(5.2360, 1.25)$

Minima: $(3.1416, -1)$

(b) $2 \sin x \cos x - \sin x = 0$

$\sin x(2 \cos x - 1) = 0$

$\sin x = 0 \implies x = n\pi$

$\cos x = \dfrac{1}{2} \implies x = \dfrac{\pi}{3} + 2n\pi, \dfrac{5\pi}{3} + 2n\pi$

The zeros are 1.0472, 3.1416, and 5.2360.

93. (a) $f(x) = \sin x + \cos x$

Maximum:
$(0.7854, 1.4142)$

Minimum:
$(3.9270, -1.4142)$

(b) $\cos x - \sin x = 0$

$\cos x = \sin x$

$1 = \dfrac{\sin x}{\cos x}$

$\tan x = 1$

$x = \dfrac{\pi}{4}, \dfrac{5\pi}{4}$

$f\left(\dfrac{\pi}{4} \right) = \sin \dfrac{\pi}{4} + \cos \dfrac{\pi}{4} = \dfrac{\sqrt{2}}{2} + \dfrac{\sqrt{2}}{2} = \sqrt{2}$

$f\left(\dfrac{5\pi}{4} \right) = \sin \dfrac{5\pi}{4} + \cos \dfrac{5\pi}{4}$

$= -\sin \dfrac{\pi}{4} + \left(-\cos \dfrac{\pi}{4} \right)$

$= -\dfrac{\sqrt{2}}{2} - \dfrac{\sqrt{2}}{2} = -\sqrt{2}$

Therefore, the maximum point in the interval $[0, 2\pi)$ is $\left(\pi/4, \sqrt{2} \right)$ and the minimum point is $\left(5\pi/4, -\sqrt{2} \right)$.

95. $f(x) = \tan \dfrac{\pi x}{4}$

$\tan 0 = 0$, but 0 is not positive. By graphing

$$y = \tan \dfrac{\pi x}{4} - x,$$

you see that the smallest positive fixed point is $x = 1$.

97. $f(x) = \cos \dfrac{1}{x}$

(a) The domain of $f(x)$ is all real numbers except 0.

(b) The graph has y-axis symmetry and a horizontal asymptote at $y = 1$.

(c) As $x \to 0$, $f(x)$ oscillates between -1 and 1.

(d) There are an infinite number of solutions in the interval $[-1, 1]$.

$$\frac{1}{x} = \frac{\pi}{2} + n\pi = \frac{\pi + 2n\pi}{2} \Rightarrow x = \frac{2}{\pi(2n+1)}$$

(e) The greatest solution appears to occur at $x \approx 0.6366$.

99. $S = 74.50 - 43.75 \cos \dfrac{\pi t}{6}$

t	1	2	3	4	5	6	7	8	9	10	11	12
S	36.6	52.6	74.5	96.4	112.4	118.3	112.4	96.4	74.5	52.6	36.6	30.8

$S > 100$ for $t = 5, 6, 7$ (May, June, July)

101.
$$y = \frac{1}{12}(\cos 8t - 3 \sin 8t)$$

$$\frac{1}{12}(\cos 8t - 3 \sin 8t) = 0$$

$$\cos 8t = 3 \sin 8t$$

$$\frac{1}{3} = \tan 8t$$

$$8t = 0.32175 + n\pi$$

$$t = 0.04 + \frac{n\pi}{8}$$

In the interval $0 \le t \le 1$, $t = 0.04, 0.43,$ and 0.83 second.

103.
$$r = \frac{1}{32}v_0^2 \sin 2\theta$$

$$300 = \frac{1}{32}(100)^2 \sin 2\theta$$

$$\sin 2\theta = 0.96$$

$2\theta \approx 1.287 \qquad$ or $\quad 2\theta \approx \pi - 1.287 \approx 1.855$

$\theta \approx 0.6435 \approx 37°$ or $\quad \theta \approx 0.9275 \approx 53°$

105. (a)

Year (0 ↔ 1990)

(b) Models 1 and 2 are both good fits, but model 1 seems better.

(c) The constant term 5.45 gives the average unemployment rate, 5.45%.

(d) The length is approximately one period

$$\frac{2\pi}{0.47} \approx 13.37 \text{ years.}$$

(e) $r = 1.24 \sin(0.47t + 0.40) + 5.45 = 5.0$

Using a graphing utility, $t \approx 19.99 \approx 20$, or 2010.

107. False. $\sin x - x = 0$ has one solution, $x = 0$.

109. False. The equation has no solution because $-1 \le \sin x \le 1$.

111. $124° = 124°\left(\dfrac{\pi}{180°}\right) \approx 2.164$ radians

113. $-0.41° = -0.41°\left(\dfrac{\pi}{180°}\right) \approx -0.007$ radian

115. $\tan 30° = \dfrac{14}{x} \implies x = \dfrac{14}{\tan 30°} = \dfrac{14}{\sqrt{3}/3} \approx 24.249$

117. $\tan 87.5° = \dfrac{x}{100}$

$x = 100 \tan 87.5°$

≈ 2290.4 feet ≈ 0.43 mile

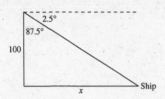

Section 5.4   Sum and Difference Formulas

- ■ You should memorize the sum and difference formulas.

  $\sin(u \pm v) = \sin u \cos v \pm \cos u \sin v$

  $\cos(u \pm v) = \cos u \cos v \mp \sin u \sin v$

  $\tan(u \pm v) = \dfrac{\tan u \pm \tan v}{1 \mp \tan u \tan v}$

- ■ You should be able to use these formulas to find the values of the trigonometric functions of angles whose sums or differences are special angles.

- ■ You should be able to use these formulas to solve trigonometric equations.

Vocabulary Check

1. $\sin u \cos v - \cos u \sin v$

2. $\cos u \cos v - \sin u \sin v$

3. $\dfrac{\tan u + \tan v}{1 - \tan u \tan v}$

4. $\sin u \cos v + \cos u \sin v$

5. $\cos u \cos v + \sin u \sin v$

6. $\dfrac{\tan u - \tan v}{1 + \tan u \tan v}$

1. (a) $\cos(240° - 0°) = \cos(240°) = -\dfrac{1}{2}$

  (b) $\cos(240°) - \cos 0° = -\dfrac{1}{2} - 1 = -\dfrac{3}{2}$

3. (a) $\cos\left(\dfrac{\pi}{4} + \dfrac{\pi}{3}\right) = \cos\dfrac{\pi}{4}\cos\dfrac{\pi}{3} - \sin\dfrac{\pi}{4}\sin\dfrac{\pi}{3}$

  (b) $\cos\dfrac{\pi}{4} + \cos\dfrac{\pi}{3} = \dfrac{\sqrt{2}}{2} + \dfrac{1}{2} = \dfrac{\sqrt{2}+1}{2}$

$= \dfrac{\sqrt{2}}{2}\left(\dfrac{1}{2}\right) - \dfrac{\sqrt{2}}{2}\left(\dfrac{\sqrt{3}}{2}\right)$

$= \dfrac{\sqrt{2} - \sqrt{6}}{4}$

5. (a) $\sin(315° - 60°) = \sin 315° \cos 60° - \cos 315° \sin 60° = -\dfrac{\sqrt{2}}{2} \cdot \dfrac{1}{2} - \dfrac{\sqrt{2}}{2} \cdot \dfrac{\sqrt{3}}{2} = \dfrac{-\sqrt{2} - \sqrt{6}}{4}$

  (b) $\sin 315° - \sin 60° = -\dfrac{\sqrt{2}}{2} - \dfrac{\sqrt{3}}{2} = -\dfrac{\sqrt{2} + \sqrt{3}}{2}$

7. $\sin 105° = \sin(60° + 45°)$

$= \sin 60° \cos 45° + \sin 45° \cos 60°$

$= \dfrac{\sqrt{3}}{2} \cdot \dfrac{\sqrt{2}}{2} + \dfrac{\sqrt{2}}{2} \cdot \dfrac{1}{2}$

$= \dfrac{\sqrt{2}}{4}\left(\sqrt{3} + 1\right)$

$\cos 105° = \cos(60° + 45°)$

$= \cos 60° \cos 45° - \sin 60° \sin 45°$

$= \dfrac{1}{2} \cdot \dfrac{\sqrt{2}}{2} - \dfrac{\sqrt{3}}{2} \cdot \dfrac{\sqrt{2}}{2}$

$= \dfrac{\sqrt{2}}{4}\left(1 - \sqrt{3}\right)$

$\tan 105° = \tan(60° + 45°)$

$= \dfrac{\tan 60° + \tan 45°}{1 - \tan 60° \tan 45°}$

$= \dfrac{\sqrt{3} + 1}{1 - \sqrt{3}} = \dfrac{\sqrt{3} + 1}{1 - \sqrt{3}} \cdot \dfrac{1 + \sqrt{3}}{1 + \sqrt{3}}$

$= \dfrac{4 + 2\sqrt{3}}{-2} = -2 - \sqrt{3}$

9. $\sin 195° = \sin(225° - 30°)$

$= \sin 225° \cos 30° - \sin 30° \cos 225°$

$= -\sin 45° \cos 30° + \sin 30° \cos 45°$

$= -\dfrac{\sqrt{2}}{2} \cdot \dfrac{\sqrt{3}}{2} + \dfrac{1}{2} \cdot \dfrac{\sqrt{2}}{2} = \dfrac{\sqrt{2}}{4}\left(1 - \sqrt{3}\right)$

$\cos 195° = \cos(225° - 30°)$

$= \cos 225° \cos 30° + \sin 225° \sin 30°$

$= -\cos 45° \cos 30° - \sin 45° \sin 30°$

$= -\dfrac{\sqrt{2}}{2} \cdot \dfrac{\sqrt{3}}{2} - \dfrac{\sqrt{2}}{2} \cdot \dfrac{1}{2} = -\dfrac{\sqrt{2}}{4}\left(\sqrt{3} + 1\right)$

$\tan 195° = \tan(225° - 30°)$

$= \dfrac{\tan 225° - \tan 30°}{1 + \tan 225° \tan 30°}$

$= \dfrac{\tan 45° - \tan 30°}{1 + \tan 45° \tan 30°}$

$= \dfrac{1 - \left(\sqrt{3}/3\right)}{1 + \left(\sqrt{3}/3\right)} = \dfrac{3 - \sqrt{3}}{3 + \sqrt{3}} \cdot \dfrac{3 - \sqrt{3}}{3 - \sqrt{3}}$

$= \dfrac{12 - 6\sqrt{3}}{6} = 2 - \sqrt{3}$

11. $\sin \dfrac{11\pi}{12} = \sin\left(\dfrac{3\pi}{4} + \dfrac{\pi}{6}\right)$

$= \sin \dfrac{3\pi}{4} \cos \dfrac{\pi}{6} + \sin \dfrac{\pi}{6} \cos \dfrac{3\pi}{4}$

$= \dfrac{\sqrt{2}}{2} \cdot \dfrac{\sqrt{3}}{2} + \dfrac{1}{2}\left(-\dfrac{\sqrt{2}}{2}\right) = \dfrac{\sqrt{2}}{4}\left(\sqrt{3} - 1\right)$

$\cos \dfrac{11\pi}{12} = \cos\left(\dfrac{3\pi}{4} + \dfrac{\pi}{6}\right)$

$= \cos \dfrac{3\pi}{4} \cos \dfrac{\pi}{6} - \sin \dfrac{3\pi}{4} \sin \dfrac{\pi}{6}$

$= -\dfrac{\sqrt{2}}{2} \cdot \dfrac{\sqrt{3}}{2} - \dfrac{\sqrt{2}}{2} \cdot \dfrac{1}{2}$

$= -\dfrac{\sqrt{2}}{4}\left(\sqrt{3} + 1\right)$

$\tan \dfrac{11\pi}{12} = \tan\left(\dfrac{3\pi}{4} + \dfrac{\pi}{6}\right)$

$= \dfrac{\tan(3\pi/4) + \tan(\pi/6)}{1 - \tan(3\pi/4) \tan(\pi/6)}$

$= \dfrac{-1 + \left(\sqrt{3}/3\right)}{1 - (-1)\left(\sqrt{3}/3\right)}$

$= \dfrac{-3 + \sqrt{3}}{3 + \sqrt{3}} \cdot \dfrac{3 - \sqrt{3}}{3 - \sqrt{3}}$

$= \dfrac{-12 + 6\sqrt{3}}{6} = -2 + \sqrt{3}$

13. $-\dfrac{\pi}{12} = \dfrac{\pi}{6} - \dfrac{\pi}{4}$

$$\sin\left(-\frac{\pi}{12}\right) = \sin\left(\frac{\pi}{6} - \frac{\pi}{4}\right)$$

$$= \sin\frac{\pi}{6}\cos\frac{\pi}{4} - \sin\frac{\pi}{4}\cos\frac{\pi}{6}$$

$$= \frac{1}{2}\cdot\frac{\sqrt{2}}{2} - \frac{\sqrt{2}}{2}\cdot\frac{\sqrt{3}}{2} = \frac{\sqrt{2}}{4}\left(1 - \sqrt{3}\right)$$

$$\cos\left(-\frac{\pi}{12}\right) = \cos\left(\frac{\pi}{6} - \frac{\pi}{4}\right)$$

$$= \cos\frac{\pi}{6}\cos\frac{\pi}{4} + \sin\frac{\pi}{6}\sin\frac{\pi}{4}$$

$$= \frac{\sqrt{3}}{2}\cdot\frac{\sqrt{2}}{2} + \frac{1}{2}\cdot\frac{\sqrt{2}}{2} = \frac{\sqrt{2}}{4}\left(\sqrt{3} + 1\right)$$

$$\tan\left(-\frac{\pi}{12}\right) = \tan\left(\frac{\pi}{6} - \frac{\pi}{4}\right)$$

$$= \frac{\tan(\pi/6) - \tan(\pi/4)}{1 + \tan(\pi/6)\tan(\pi/4)}$$

$$= \frac{(\sqrt{3}/3) - 1}{1 + (\sqrt{3}/3)} = \frac{\sqrt{3} - 3}{\sqrt{3} + 3}\cdot\frac{\sqrt{3} - 3}{\sqrt{3} - 3}$$

$$= \frac{12 - 6\sqrt{3}}{-6} = -2 + \sqrt{3}$$

15. $\sin 75° = \sin(30° + 45°)$

$$= \sin 30°\cos 45° + \sin 45°\cos 30°$$

$$= \frac{1}{2}\cdot\frac{\sqrt{2}}{2} + \frac{\sqrt{2}}{2}\cdot\frac{\sqrt{3}}{2} = \frac{\sqrt{2}}{4}\left(1 + \sqrt{3}\right)$$

$$\cos 75° = \cos(30° + 45°)$$

$$= \cos 30°\cos 45° - \sin 30°\sin 45°$$

$$= \frac{\sqrt{3}}{2}\cdot\frac{\sqrt{2}}{2} - \frac{1}{2}\cdot\frac{\sqrt{2}}{2} = \frac{\sqrt{2}}{4}\left(\sqrt{3} - 1\right)$$

$$\tan 75° = \tan(30° + 45°)$$

$$= \frac{\tan 30° + \tan 45°}{1 - \tan 30°\tan 45°}$$

$$= \frac{(\sqrt{3}/3) + 1}{1 - (\sqrt{3}/3)} = \frac{\sqrt{3} + 3}{3 - \sqrt{3}}\cdot\frac{3 + \sqrt{3}}{3 + \sqrt{3}}$$

$$= \frac{6\sqrt{3} + 12}{6} = \sqrt{3} + 2$$

17. $-225°$ is coterminal with $135°$, and lies in Quadrant II.

$$\sin(-225°) = \frac{\sqrt{2}}{2}$$

$$\cos(-225°) = -\frac{\sqrt{2}}{2}$$

$$\tan(-225°) = -1$$

19. $\dfrac{13\pi}{12} = \dfrac{3\pi}{4} + \dfrac{\pi}{3}$

$$\sin\frac{13\pi}{12} = \sin\left(\frac{3\pi}{4} + \frac{\pi}{3}\right) = \sin\frac{3\pi}{4}\cos\frac{\pi}{3} + \sin\frac{\pi}{3}\cos\frac{3\pi}{4}$$

$$= \frac{\sqrt{2}}{2}\cdot\frac{1}{2} + \frac{\sqrt{3}}{2}\left(-\frac{\sqrt{2}}{2}\right) = \frac{\sqrt{2} - \sqrt{6}}{4}$$

$$\cos\frac{13\pi}{12} = \cos\left(\frac{3\pi}{4} + \frac{\pi}{3}\right) = \cos\frac{3\pi}{4}\cos\frac{\pi}{3} - \sin\frac{3\pi}{4}\sin\frac{\pi}{3}$$

$$= \left(-\frac{\sqrt{2}}{2}\right)\left(\frac{1}{2}\right) - \left(\frac{\sqrt{2}}{2}\right)\left(\frac{\sqrt{3}}{2}\right) = -\frac{\sqrt{6} + \sqrt{2}}{4}$$

$$\tan\frac{13\pi}{12} = \tan\left(\frac{3\pi}{4} + \frac{\pi}{3}\right) = \frac{\tan(3\pi/4) + \tan(\pi/3)}{1 - \tan(3\pi/4)\tan(\pi/3)}$$

$$= \frac{(-1) + \sqrt{3}}{1 - (-1)\sqrt{3}} = \frac{\sqrt{3} - 1}{\sqrt{3} + 1} = 2 - \sqrt{3}$$

21. $-\dfrac{7\pi}{12} = \dfrac{\pi}{6} - \dfrac{3\pi}{4}$

$$\sin\left(-\dfrac{7\pi}{12}\right) = \sin\left(\dfrac{\pi}{6} - \dfrac{3\pi}{4}\right) = \sin\dfrac{\pi}{6}\cos\dfrac{3\pi}{4} - \sin\dfrac{3\pi}{4}\cos\dfrac{\pi}{6}$$

$$= \dfrac{1}{2}\left(-\dfrac{\sqrt{2}}{2}\right) - \dfrac{\sqrt{2}}{2}\left(\dfrac{\sqrt{3}}{2}\right) = -\dfrac{\sqrt{2} + \sqrt{6}}{4}$$

$$\cos\left(-\dfrac{7\pi}{12}\right) = \cos\left(\dfrac{\pi}{6} - \dfrac{3\pi}{4}\right) = \cos\dfrac{\pi}{6}\cos\dfrac{3\pi}{4} + \sin\dfrac{\pi}{6}\sin\dfrac{3\pi}{4}$$

$$= \dfrac{\sqrt{3}}{2}\left(-\dfrac{\sqrt{2}}{2}\right) + \dfrac{1}{2}\cdot\dfrac{\sqrt{2}}{2} = \dfrac{\sqrt{2} - \sqrt{6}}{4}$$

$$\tan\left(-\dfrac{7\pi}{12}\right) = \tan\left(\dfrac{\pi}{6} - \dfrac{3\pi}{4}\right) = \dfrac{\tan(\pi/6) - \tan(3\pi/4)}{1 + \tan(\pi/6)\tan(3\pi/4)}$$

$$= \dfrac{\left(\sqrt{3}/3\right) - (-1)}{1 + \left(\sqrt{3}/3\right)(-1)} = \dfrac{3 + \sqrt{3}}{3 - \sqrt{3}} = 2 + \sqrt{3}$$

23. $\cos 60° \cos 20° - \sin 60° \sin 20° = \cos(60° + 20°) = \cos 80°$

25. $\dfrac{\tan 325° - \tan 86°}{1 + \tan 325° \tan 86°} = \tan(325° - 86°) = \tan 239°$

27. $\sin 3.5 \cos 1.2 - \cos 3.5 \sin 1.2 = \sin(3.5 - 1.2) = \sin 2.3$

29. $\cos\dfrac{\pi}{9}\cos\dfrac{\pi}{7} - \sin\dfrac{\pi}{9}\sin\dfrac{\pi}{7} = \cos\left(\dfrac{\pi}{9} + \dfrac{\pi}{7}\right) = \cos\left(\dfrac{16\pi}{63}\right)$

31. $y_1 = \sin\left(\dfrac{\pi}{6} + x\right)$

$$= \sin\dfrac{\pi}{6}\cos x + \sin x \cdot \cos\dfrac{\pi}{6}$$

$$= \dfrac{1}{2}\cos x + \dfrac{\sqrt{3}}{2}\sin x$$

$$= \dfrac{1}{2}\left(\cos x + \sqrt{3}\sin x\right) = y_2$$

x	0.2	0.4	0.6	0.8	1.0	1.2	1.4
y_1	0.6621	0.7978	0.9017	0.9696	0.9989	0.9883	0.9384
y_2	0.6621	0.7978	0.9017	0.9696	0.9989	0.9883	0.9384

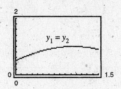

33. $y_1 = \cos(x + \pi)\cos(x - \pi)$

$$= (\cos x \cdot \cos \pi - \sin x \cdot \sin \pi)[\cos x \cos \pi + \sin x \sin \pi]$$

$$= [-\cos x][-\cos x] = \cos^2 x = y_2$$

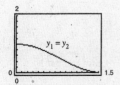

x	0.2	0.4	0.6	0.8	1.0	1.2	1.4
y_1	0.9605	0.8484	0.6812	0.4854	0.2919	0.1313	0.0289
y_2	0.9605	0.8484	0.6812	0.4854	0.2919	0.1313	0.0289

For Exercises 35 and 37,

$\sin u = \frac{5}{13}$ and u in **Quadrant II** $\Rightarrow \cos u = -\frac{12}{13}$

$\cos v = -\frac{3}{5}$ and v in **Quadrant II** $\Rightarrow \sin v = \frac{4}{5}$

$\tan u = -\frac{5}{12}$ and $\tan v = -\frac{4}{3}$.

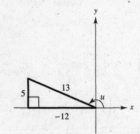 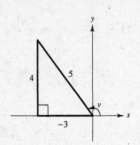

35. $\sin(u + v) = \sin u \cos v + \sin v \cos u$

$$= \frac{5}{13}\left(\frac{-3}{5}\right) + \frac{4}{5}\left(\frac{-12}{13}\right) = -\frac{63}{65}$$

37. $\tan(u + v) = \dfrac{\tan u + \tan v}{1 - \tan u \tan v}$

$$= \frac{(-5/12) - (4/3)}{1 - (-5/12)(-4/3)}$$

$$= \frac{-63/36}{16/36} = -\frac{63}{16}$$

For Exercises 39 and 41, $\sin u = -\frac{8}{17}$, $\cos u = -\frac{15}{17}$, $\sin v = -\frac{3}{5}$, $\cos v = -\frac{4}{5}$, $\tan u = \frac{8}{15}$, $\tan v = \frac{3}{4}$.

39. $\cos(u + v) = \cos u \cos v - \sin u \sin v$

$$= \left(-\tfrac{15}{17}\right)\left(-\tfrac{4}{5}\right) - \left(-\tfrac{8}{17}\right)\left(-\tfrac{3}{5}\right) = \tfrac{36}{85}$$

41. $\sin(v - u) = \sin v \cos u - \cos v \sin u$

$$= \left(-\tfrac{3}{5}\right)\left(-\tfrac{15}{17}\right) - \left(-\tfrac{4}{5}\right)\left(-\tfrac{8}{17}\right) = \tfrac{13}{85}$$

43. $\sin(\arcsin x + \arccos x) = \sin(\arcsin x)\cos(\arccos x) + \sin(\arccos x)\cos(\arcsin x)$

$$= x \cdot x + \sqrt{1 - x^2} \cdot \sqrt{1 - x^2}$$

$$= x^2 + 1 - x^2$$

$$= 1$$

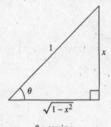 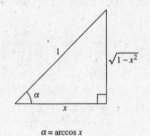

$\theta = \arcsin x$ $\qquad$ $\alpha = \arccos x$

45. Let: $\quad u = \arctan 2x \quad$ and $\quad v = \arccos x$

$\tan u = 2x \qquad\qquad \cos v = x$

$\sin(\arctan 2x - \arccos x) = \sin(u - v)$

$$= \sin u \cos v - \cos u \sin v$$

$$= \frac{2x}{\sqrt{4x^2 + 1}}(x) - \frac{1}{\sqrt{4x^2 + 1}}\left(\sqrt{1 - x^2}\right)$$

$$= \frac{2x^2 - \sqrt{1 - x^2}}{\sqrt{4x^2 + 1}}$$

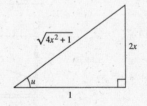 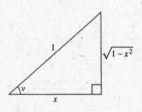

47. $\sin^{-1} 1 = \dfrac{\pi}{2}$ because $\sin \dfrac{\pi}{2} = 1$.

$\cos^{-1} 1 = 0$ because $\cos 0 = 1$.

$\sin(\sin^{-1} 1 + \cos^{-1} 1) = \sin\left(\dfrac{\pi}{2} + 0\right) = 1$

49. $\sin^{-1} 1 = \dfrac{\pi}{2}$ and $\cos^{-1}(-1) = \pi$

$\sin(\sin^{-1} 1 - \cos^{-1}(-1)) = \sin\left(\dfrac{\pi}{2} - \pi\right)$

$$= \sin\left(-\dfrac{\pi}{2}\right) = -1$$

51. $\sin^{-1} \dfrac{1}{2} = \dfrac{\pi}{6}$ and $\cos^{-1} \dfrac{1}{2} = \dfrac{\pi}{3}$

$$\sin\left(\sin^{-1} \dfrac{1}{2} - \cos^{-1} \dfrac{1}{2}\right) = \sin\left(\dfrac{\pi}{6} - \dfrac{\pi}{3}\right)$$

$$= \sin\left(-\dfrac{\pi}{6}\right) = -\dfrac{1}{2}$$

53. $\sin^{-1} 0 = 0$ and $\sin^{-1} \dfrac{1}{2} = \dfrac{\pi}{6}$

$$\tan\left(\sin^{-1} 0 + \sin^{-1} \dfrac{1}{2}\right) = \tan\left(0 + \dfrac{\pi}{6}\right) = \dfrac{\sqrt{3}}{3}$$

55. $\sin^{-1}(-1) = -\dfrac{\pi}{2}$

$$\sin\left(\dfrac{\pi}{2} + \sin^{-1}(-1)\right) = \sin\left(\dfrac{\pi}{2} - \dfrac{\pi}{2}\right) = \sin 0 = 0$$

57. $\sin^{-1} 1 = \dfrac{\pi}{2}$

$$\cos(\pi + \sin^{-1} 1) = \cos\left(\pi + \dfrac{\pi}{2}\right) = \cos \dfrac{3\pi}{2} = 0$$

59. Let $\theta = \cos^{-1} \dfrac{3}{5} \implies \cos \theta = \dfrac{3}{5}$.

Let $\phi = \sin^{-1} \dfrac{5}{13} \implies \sin \phi = \dfrac{5}{13}$.

$$\sin\left(\cos^{-1} \dfrac{3}{5} - \sin^{-1} \dfrac{5}{13}\right) = \sin(\theta - \phi)$$

$$= \sin \theta \cos \phi - \cos \theta \sin \phi$$

$$= \left(\dfrac{4}{5}\right)\left(\dfrac{12}{13}\right) - \left(\dfrac{3}{5}\right)\left(\dfrac{5}{13}\right) = \dfrac{33}{65}$$

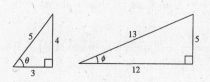

61. Let $\theta = \tan^{-1} \dfrac{3}{4} \implies \tan \theta = \dfrac{3}{4}$.

Let $\phi = \sin^{-1} \dfrac{3}{5} \implies \sin \phi = \dfrac{3}{5}$.

$$\sin\left(\tan^{-1} \dfrac{3}{4} + \sin^{-1} \dfrac{3}{5}\right) = \sin(\theta + \phi)$$

$$= \sin \theta \cos \phi + \sin \phi \cos \theta$$

$$= \dfrac{3}{5}\left(\dfrac{4}{5}\right) + \dfrac{3}{5}\left(\dfrac{4}{5}\right) = \dfrac{24}{25}$$

Note: $\theta = \phi$

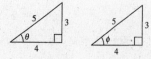

63. $\sin\left(\dfrac{\pi}{2} + x\right) = \sin \dfrac{\pi}{2} \cos x + \sin x \cos \dfrac{\pi}{2}$

$$= (1) \cos x + 0 = \cos x$$

65. $\tan(x + \pi) - \tan(\pi - x) = \dfrac{\tan x + \tan \pi}{1 - \tan x \cdot \tan \pi} - \dfrac{\tan \pi - \tan x}{1 + \tan \pi \tan x}$

$$= \dfrac{\tan x}{1} - \left(-\dfrac{\tan x}{1}\right) = 2 \tan x$$

67. $\sin(x + y) + \sin(x - y) = \sin x \cos y + \sin y \cos x + \sin x \cos y - \sin y \cos x = 2 \sin x \cos y$

69. $\cos(x + y) \cos(x - y) = [\cos x \cos y - \sin x \sin y][\cos x \cos y + \sin x \sin y]$

$$= \cos^2 x \cos^2 y - \sin^2 x \sin^2 y \ = \cos^2 x(1 - \sin^2 y) - \sin^2 x \sin^2 y$$

$$= \cos^2 x - \sin^2 y(\cos^2 x + \sin^2 x) = \cos^2 x - \sin^2 y$$

71.
$$\sin\left(x + \frac{\pi}{3}\right) + \sin\left(x - \frac{\pi}{3}\right) = 1$$

$$\sin x \cos \frac{\pi}{3} + \cos x \sin \frac{\pi}{3} + \sin x \cos \frac{\pi}{3} - \cos x \sin \frac{\pi}{3} = 1$$

$$2 \sin x(0.5) = 1$$

$$\sin x = 1$$

$$x = \frac{\pi}{2}$$

73.
$$\tan(x + \pi) + 2\sin(x + \pi) = 0$$

$$\frac{\tan x + \tan \pi}{1 - \tan x \tan \pi} + 2(\sin x \cos \pi + \cos x \sin \pi) = 0$$

$$\frac{\tan x + 0}{1 - \tan x(0)} + 2[\sin x(-1) + \cos x(0)] = 0$$

$$\frac{\tan x}{1} - 2\sin x = 0$$

$$\frac{\sin x}{\cos x} = 2\sin x$$

$$\sin x = 2 \sin x \cos x$$

$$\sin x(1 - 2\cos x) = 0$$

$$\sin x = 0 \qquad \text{or} \qquad \cos x = \frac{1}{2}$$

$$x = 0, \pi \qquad\qquad x = \frac{\pi}{3}, \frac{5\pi}{3}$$

75. Graph $y_1 = \cos\left(x + \frac{\pi}{4}\right) + \cos\left(x - \frac{\pi}{4}\right)$ and $y_2 = 1$.

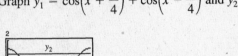

$x \approx 0.7854, 5.4978$

77. $\tan(x + \pi) - \cos\left(x + \frac{\pi}{2}\right) = 0$

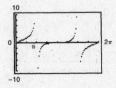

Answers: 0.0, 3.1416 ($x = 0, \pi$)

79. $y_1 + y_2 = A \cos 2\pi\left(\frac{t}{T} - \frac{x}{\lambda}\right) + A \cos 2\pi\left(\frac{t}{T} + \frac{x}{\lambda}\right)$

$$= A\left[\cos\left(\frac{2\pi t}{T}\right)\cos\left(\frac{2\pi x}{\lambda}\right) + \sin\left(\frac{2\pi t}{T}\right)\sin\left(\frac{2\pi x}{\lambda}\right)\right] + A\left[\cos\left(\frac{2\pi t}{T}\right)\cos\left(\frac{2\pi x}{\lambda}\right) - \sin\left(\frac{2\pi t}{T}\right)\sin\left(\frac{2\pi x}{\lambda}\right)\right]$$

$$= 2A \cos\left(\frac{2\pi t}{T}\right)\cos\left(\frac{2\pi x}{\lambda}\right)$$

81. False. See page 380.

83. $\cos(n\pi + \theta) = \cos n\pi \cos \theta - \sin n\pi \sin \theta$

$\qquad = (-1)^n(\cos \theta) - (0)(\sin \theta)$

$\qquad = (-1)^n(\cos \theta)$, where n is an integer.

85. $C = \arctan \dfrac{b}{a} \implies \tan C = \dfrac{b}{a} \implies \sin C = \dfrac{b}{\sqrt{a^2 + b^2}}, \cos C = \dfrac{a}{\sqrt{a^2 + b^2}}$

$\sqrt{a^2 + b^2}\,\sin(B\theta + C) = \sqrt{a^2 + b^2}\left(\sin B\theta \cdot \dfrac{a}{\sqrt{a^2 + b^2}} + \dfrac{b}{\sqrt{a^2 + b^2}} \cdot \cos B\theta\right) = a \sin B\theta + b \cos B\theta$

87. $\sin \theta + \cos \theta$

$a = 1, \; b = 1, \; B = 1$

(a) $C = \arctan \dfrac{b}{a} = \arctan 1 = \dfrac{\pi}{4}$

$\sin \theta + \cos \theta = \sqrt{a^2 + b^2}\,\sin(B\theta + C)$

$\qquad = \sqrt{2} \sin\left(\theta + \dfrac{\pi}{4}\right)$

(b) $C = \arctan \dfrac{a}{b} = \arctan 1 = \dfrac{\pi}{4}$

$\sin \theta + \cos \theta = \sqrt{a^2 + b^2}\,\cos(B\theta - C)$

$\qquad = \sqrt{2} \cos\left(\theta - \dfrac{\pi}{4}\right)$

89. $12 \sin 3\theta + 5 \cos 3\theta; \; a = 12, \; b = 5, \; B = 3$

(a) $C = \arctan \dfrac{b}{a} = \arctan \dfrac{5}{12} \approx 0.3948$

$12 \sin 3\theta + 5 \cos 3\theta = \sqrt{a^2 + b^2}\,\sin(B\theta + C)$

$\qquad \approx 13 \sin(3\theta + 0.3948)$

(b) $C = \arctan \dfrac{a}{b} = \arctan \dfrac{12}{5} \approx 1.1760$

$12 \sin 3\theta + 5 \cos 3\theta = \sqrt{a^2 + b^2}\,\cos(B\theta - C)$

$\qquad \approx 13 \cos(3\theta - 1.1760)$

91. $C = \arctan \dfrac{b}{a} = \dfrac{\pi}{2} \implies a = 0$

$\sqrt{a^2 + b^2} = 2 \implies b = 2$

$B = 1$

$2 \sin\left(\theta + \dfrac{\pi}{2}\right) = (0)(\sin \theta) + (2)(\cos \theta) = 2 \cos \theta$

93. $\dfrac{\sin(x + h) - \sin x}{h} = \dfrac{\sin x \cos h + \cos x \sin h - \sin x}{h}$

$\qquad = \dfrac{\sin x(\cos h - 1) + \cos x \sin h}{h}$

$\qquad = \dfrac{\cos x \sin h}{h} - \dfrac{\sin x(1 - \cos h)}{h}$

95. From the figure, it appears that $u + v = w$. Assume that u, v, and w are all in Quadrant I. From the figure:

$\tan u = \dfrac{s}{3s} = \dfrac{1}{3}$

$\tan v = \dfrac{s}{2s} = \dfrac{1}{2}$

$\tan w = \dfrac{s}{s} = 1$

$\tan(u + v) = \dfrac{\tan u + \tan v}{1 - \tan u \tan v} = \dfrac{(1/3) + (1/2)}{1 - (1/3)(1/2)} = \dfrac{5/6}{1 - (1/6)} = 1 = \tan w.$

Thus, $\tan(u + v) = \tan w$. Because u, v, and w are all in Quadrant I, we have

$\arctan[\tan(u + v)] = \arctan[\tan w]$

$\qquad u + v = w.$

97. $x = 0$: $y = -\frac{1}{2}(0 - 10) + 14 = 5 + 14 = 19$

y-intercept: $(0, 19)$

$y = 0$: $0 = -\frac{1}{2}(x - 10) + 14$

$\quad\quad\quad = -\frac{1}{2}x + 19 \implies x = 38$

x-intercept: $(38, 0)$

99. $x = 0$: $|2(0) - 9| - 5 = 9 - 5 = 4$

y-intercept: $(0, 4)$

$y = 0$: $|2x - 9| = 5 \implies x = 7, 2$

x-intercepts: $(2, 0), (7, 0)$

101. $\arccos\left(\dfrac{\sqrt{3}}{2}\right) = \dfrac{\pi}{6}$ because $\cos\dfrac{\pi}{6} = \dfrac{\sqrt{3}}{2}$.

103. $\arcsin 1 = \dfrac{\pi}{2}$ because $\sin\dfrac{\pi}{2} = 1$.

Section 5.5   Multiple-Angle and Product-to-Sum Formulas

■ You should know the following double-angle formulas.

(a) $\sin 2u = 2 \sin u \cos u$

(b) $\cos 2u = \cos^2 u - \sin^2 u$

$\quad\quad\quad\quad = 2 \cos^2 u - 1$

$\quad\quad\quad\quad = 1 - 2 \sin^2 u$

(c) $\tan 2u = \dfrac{2 \tan u}{1 - \tan^2 u}$

■ You should be able to reduce the power of a trigonometric function.

(a) $\sin^2 u = \dfrac{1 - \cos 2u}{2}$

(b) $\cos^2 u = \dfrac{1 + \cos 2u}{2}$

(c) $\tan^2 u = \dfrac{1 - \cos 2u}{1 + \cos 2u}$

■ You should be able to use the half-angle formulas.

(a) $\sin\dfrac{u}{2} = \pm\sqrt{\dfrac{1 - \cos u}{2}}$

(b) $\cos\dfrac{u}{2} = \pm\sqrt{\dfrac{1 + \cos u}{2}}$

(c) $\tan\dfrac{u}{2} = \dfrac{1 - \cos u}{\sin u} = \dfrac{\sin u}{1 + \cos u}$

■ You should be able to use the product-sum formulas.

(a) $\sin u \sin v = \dfrac{1}{2}[\cos(u - v) - \cos(u + v)]$

(b) $\cos u \cos v = \dfrac{1}{2}[\cos(u - v) + \cos(u + v)]$

(c) $\sin u \cos v = \dfrac{1}{2}[\sin(u + v) + \sin(u - v)]$

(d) $\cos u \sin v = \dfrac{1}{2}[\sin(u + v) - \sin(u - v)]$

■ You should be able to use the sum-product formulas.

(a) $\sin x + \sin y = 2 \sin\left(\dfrac{x + y}{2}\right)\cos\left(\dfrac{x - y}{2}\right)$

(b) $\sin x - \sin y = 2 \cos\left(\dfrac{x + y}{2}\right)\sin\left(\dfrac{x - y}{2}\right)$

(c) $\cos x + \cos y = 2 \cos\left(\dfrac{x + y}{2}\right)\cos\left(\dfrac{x - y}{2}\right)$

(d) $\cos x - \cos y = -2 \sin\left(\dfrac{x + y}{2}\right)\sin\left(\dfrac{x - y}{2}\right)$

Vocabulary Check

1. $2 \sin u \cos u$

2. $\dfrac{1 + \cos 2u}{2}$

3. $\cos 2u$

4. $\tan\dfrac{u}{2}$

5. $\dfrac{2 \tan u}{1 - \tan^2 u}$

6. $\dfrac{1}{2}[\cos(u - v) + \cos(u + v)]$

7. $\sin^2 u$

8. $\cos\dfrac{u}{2}$

9. $\dfrac{1}{2}[\sin(u + v) + \sin(u - v)]$

10. $2 \sin\left(\dfrac{u + v}{2}\right)\cos\left(\dfrac{u - v}{2}\right)$

1. (a) $\sin\theta = \dfrac{3}{5}$

(b) $\cos\theta = \dfrac{4}{5}$

(c) $\cos 2\theta = \cos^2\theta - \sin^2\theta$

$$= \dfrac{16}{25} - \dfrac{9}{25} = \dfrac{7}{25}$$

(d) $\sin 2\theta = 2\sin\theta\cos\theta$

$$= 2\left(\dfrac{3}{5}\right)\left(\dfrac{4}{5}\right) = \dfrac{24}{25}$$

(e) $\tan 2\theta = \dfrac{\sin 2\theta}{\cos 2\theta} = \dfrac{24}{7}$

(f) $\sec 2\theta = \dfrac{1}{\cos 2\theta} = \dfrac{25}{7}$

(g) $\csc 2\theta = \dfrac{1}{\sin 2\theta} = \dfrac{25}{24}$

(h) $\cot 2\theta = \dfrac{1}{\tan 2\theta} = \dfrac{7}{24}$

3. $\sin 2x - \sin x = 0$

Solutions: 0, 1.047, 3.142, 5.236

Analytically:

$$\sin 2x - \sin x = 0$$
$$2\sin x\cos x - \sin x = 0$$
$$\sin x(2\cos x - 1) = 0$$

$\sin x = 0$ or $2\cos x - 1 = 0$

$x = 0,\ \pi$ $\cos x = \dfrac{1}{2}$

$x = 0,\ \dfrac{\pi}{3},\ \pi,\ \dfrac{5\pi}{3}$ $x = \dfrac{\pi}{3},\ \dfrac{5\pi}{3}$

5. $4\sin x\cos x = 1$

$x \approx 0.2618,\ 1.3090,\ 3.4034,\ 4.4506$

Analytically:

$$4\sin x\cos x = 1$$
$$2\sin(2x) = 1$$
$$\sin(2x) = \dfrac{1}{2}$$
$$2x = \dfrac{\pi}{6},\ \dfrac{5\pi}{6},\ \dfrac{13\pi}{6},\ \dfrac{17\pi}{6}$$
$$x = \dfrac{\pi}{12},\ \dfrac{5\pi}{12},\ \dfrac{13\pi}{12},\ \dfrac{17\pi}{12}$$

7. $\cos 2x - \cos x = 0$

$x \approx 0,\ 2.094,\ 4.189,\ (6.283\text{ not in interval})$

Analytically:

$$\cos 2x - \cos x = 0$$
$$2\cos^2 x - 1 - \cos x = 0$$
$$(2\cos x + 1)(\cos x - 1) = 0$$
$$\cos x = -\dfrac{1}{2}, \qquad \cos x = 1$$
$$x = \dfrac{2\pi}{3},\ \dfrac{4\pi}{3},\ 0,\ (2\pi\text{ not in interval})$$

9. Solutions: 0, 1.571, 3.142, 4.712

$$\sin 4x = -2\sin 2x$$
$$\sin 4x + 2\sin 2x = 0$$
$$2\sin 2x\cos 2x + 2\sin 2x = 0$$
$$2\sin 2x(\cos 2x + 1) = 0$$

$2\sin 2x = 0$ or $\cos 2x + 1 = 0$

$\sin 2x = 0$ $\cos 2x = -1$

$2x = n\pi$ $2x = \pi + 2n\pi$

$x = \dfrac{n}{2}\pi$ $x = \dfrac{\pi}{2} + n\pi$

$x = 0,\ \dfrac{\pi}{2},\ \pi,\ \dfrac{3\pi}{2}$ $x = \dfrac{\pi}{2},\ \dfrac{3\pi}{2}$

11. $\cos 2x + \sin x = 0$

Using a graphing utility,
$x \approx 1.5708,\ 3.6652,\ 5.7596.$

Algebraically:

$$\cos 2x + \sin x = 0$$
$$1 - 2\sin^2 x + \sin x = 0$$
$$2\sin^2 x - \sin x - 1 = 0$$
$$(2\sin x + 1)(\sin x - 1) = 0$$
$$\sin x = -\dfrac{1}{2} \Rightarrow x = \dfrac{7\pi}{6},\ \dfrac{11\pi}{6}$$
$$\sin x = 1 \ \Rightarrow x = \dfrac{\pi}{2}$$

13. $\sin u = \dfrac{3}{5}$, $0 < u < \dfrac{\pi}{2}$ $\Rightarrow$ $\cos u = \dfrac{4}{5}$

$\sin 2u = 2 \sin u \cos u = 2 \cdot \dfrac{3}{5} \cdot \dfrac{4}{5} = \dfrac{24}{25}$

$\cos 2u = \cos^2 u - \sin^2 u = \dfrac{16}{25} - \dfrac{9}{25} = \dfrac{7}{25}$

$\tan 2u = \dfrac{2 \tan u}{1 - \tan^2 u} = \dfrac{2(3/4)}{1 - (9/16)} = \dfrac{24}{7}$

15. $\tan u = \dfrac{1}{2}$, $\pi < u < \dfrac{3\pi}{2}$ $\Rightarrow$ $\sin u = -\dfrac{1}{\sqrt{5}}$

and $\cos u = -\dfrac{2}{\sqrt{5}}$

$\sin 2u = 2 \sin u \cos u = 2\left(-\dfrac{1}{\sqrt{5}}\right)\left(-\dfrac{2}{\sqrt{5}}\right) = \dfrac{4}{5}$

$\cos 2u = \cos^2 u - \sin^2 u = \left(-\dfrac{2}{\sqrt{5}}\right)^2 - \left(-\dfrac{1}{\sqrt{5}}\right)^2 = \dfrac{3}{5}$

$\tan 2u = \dfrac{2 \tan u}{1 - \tan^2 u} = \dfrac{2(1/2)}{1 - (1/4)} = \dfrac{4}{3}$

17. $\sec u = -\dfrac{5}{2}$, $\dfrac{\pi}{2} < u < \pi$

$\cos u = -\dfrac{2}{5}$ $\Rightarrow$ $\sin u = \dfrac{\sqrt{21}}{5}$

$\sin 2u = 2 \sin u \cos u = 2\left(\dfrac{\sqrt{21}}{5}\right)\left(-\dfrac{2}{5}\right) = \dfrac{-4\sqrt{21}}{25}$

$\cos 2u = \cos^2 u - \sin^2 u = \dfrac{4}{25} - \dfrac{21}{25} = -\dfrac{17}{25}$

$\tan 2u = \dfrac{2 \tan u}{1 - \tan^2 u}$

$= \dfrac{2\left(\sqrt{21}/-2\right)}{1 - (21/4)} = \dfrac{-\sqrt{21}}{-17/4} = \dfrac{4\sqrt{21}}{17}$

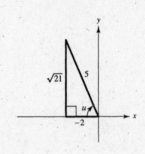

19. $8 \sin x \cos x = 4(2 \sin x \cos x) = 4 \sin 2x$

21. $6 - 12 \sin^2 x = 6(1 - 2 \sin^2 x) = 6 \cos 2x$

23. $\cos^4 x = (\cos^2 x)(\cos^2 x) = \left(\dfrac{1 + \cos 2x}{2}\right)\left(\dfrac{1 + \cos 2x}{2}\right) = \dfrac{1 + 2 \cos 2x + \cos^2 2x}{4}$

$= \dfrac{1 + 2 \cos 2x + (1 + \cos 4x)/2}{4} = \dfrac{2 + 4 \cos 2x + 1 + \cos 4x}{8}$

$= \dfrac{3 + 4 \cos 2x + \cos 4x}{8} = \dfrac{1}{8}(3 + 4 \cos 2x + \cos 4x)$

25. $(\sin^2 x)(\cos^2 x) = \left(\dfrac{1 - \cos 2x}{2}\right)\left(\dfrac{1 + \cos 2x}{2}\right)$

$= \dfrac{1 - \cos^2 2x}{4}$

$= \dfrac{1}{4}\left(1 - \dfrac{1 + \cos 4x}{2}\right)$

$= \dfrac{1}{8}(2 - 1 - \cos 4x)$

$= \dfrac{1}{8}(1 - \cos 4x)$

27. $\sin^2 x \cos^4 x = \sin^2 x \cos^2 x \cos^2 x = \left(\dfrac{1 - \cos 2x}{2}\right)\left(\dfrac{1 + \cos 2x}{2}\right)\left(\dfrac{1 + \cos 2x}{2}\right)$

$$= \frac{1}{8}(1 - \cos 2x)(1 + \cos 2x)(1 + \cos 2x)$$

$$= \frac{1}{8}(1 - \cos^2 2x)(1 + \cos 2x)$$

$$= \frac{1}{8}(1 + \cos 2x - \cos^2 2x - \cos^3 2x)$$

$$= \frac{1}{8}\left[1 + \cos 2x - \left(\frac{1 + \cos 4x}{2}\right) - \cos 2x\left(\frac{1 + \cos 4x}{2}\right)\right]$$

$$= \frac{1}{16}[2 + 2\cos 2x - 1 - \cos 4x - \cos 2x - \cos 2x \cos 4x]$$

$$= \frac{1}{16}\left[1 + \cos 2x - \cos 4x - \left(\frac{1}{2}\cos 2x + \frac{1}{2}\cos 6x\right)\right]$$

$$= \frac{1}{32}(2 + 2\cos 2x - 2\cos 4x - \cos 2x - \cos 6x)$$

$$= \frac{1}{32}(2 + \cos 2x - 2\cos 4x - \cos 6x)$$

29. $\sin^2 2x = \dfrac{1 - \cos 4x}{2}$

$$= \frac{1}{2} - \frac{1}{2}\cos 4x$$

$$= \frac{1}{2}(1 - \cos 4x)$$

31. $\cos^2 \dfrac{x}{2} = \dfrac{1 + \cos x}{2}$

$$= \frac{1}{2} + \frac{1}{2}\cos x$$

$$= \frac{1}{2}(1 + \cos x)$$

33. $\sin^2 2x \cos^2 2x = \left(\dfrac{1 - \cos 4x}{2}\right)\left(\dfrac{1 + \cos 4x}{2}\right)$

$$= \frac{1}{4}(1 - \cos 4x)(1 + \cos 4x)$$

$$= \frac{1}{4}(1 - \cos^2 4x)$$

$$= \frac{1}{4}\left(1 - \frac{1 + \cos 8x}{2}\right)$$

$$= \frac{1 - \cos 8x}{8}$$

35. $\sin^4 \dfrac{x}{2} = \left(\sin^2 \dfrac{x}{2}\right)\left(\sin^2 \dfrac{x}{2}\right)$

$$= \left(\frac{1 - \cos x}{2}\right)\left(\frac{1 - \cos x}{2}\right)$$

$$= \frac{1}{4}[1 - 2\cos x + \cos^2 x]$$

$$= \frac{1}{4}\left[1 - 2\cos x + \frac{1 + \cos 2x}{2}\right]$$

$$= \frac{1}{8}[2 - 4\cos x + 1 + \cos 2x]$$

$$= \frac{1}{8}[3 - 4\cos x + \cos 2x]$$

37. (a) $\cos\dfrac{\theta}{2} = \sqrt{\dfrac{1 + \cos\theta}{2}} = \sqrt{\dfrac{1 + (15/17)}{2}} = \sqrt{\dfrac{16}{17}} = \dfrac{4}{\sqrt{17}} = \dfrac{4\sqrt{17}}{17}$

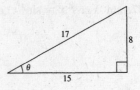

(b) $\sin\dfrac{\theta}{2} = \sqrt{\dfrac{1 - \cos\theta}{2}} = \sqrt{\dfrac{1 - (15/17)}{2}} = \dfrac{\sqrt{17}}{17}$

$\sin\theta = \frac{8}{17}$

$\cos\theta = \frac{15}{17}$

(c) $\tan\dfrac{\theta}{2} = \dfrac{\sin\theta}{1 + \cos\theta} = \dfrac{8/17}{1 + (15/17)} = \dfrac{8}{32} = \dfrac{1}{4}$

(d) $\sec\dfrac{\theta}{2} = \dfrac{1}{\cos(\theta/2)} = \dfrac{1}{\sqrt{(1 + \cos\theta)/2}} = \dfrac{\sqrt{2}}{\sqrt{1 + (15/17)}} = \dfrac{\sqrt{17}}{4}$

(e) $\csc\dfrac{\theta}{2} = \dfrac{1}{\sin(\theta/2)} = \dfrac{1}{\sqrt{(1 - \cos\theta)/2}} = \dfrac{1}{\sqrt{[1 - (15/17)]/2}} = \dfrac{1}{1/\sqrt{17}} = \sqrt{17}$

(f) $\cot\dfrac{\theta}{2} = \dfrac{1 + \cos\theta}{\sin\theta} = \dfrac{1 + (15/17)}{8/17} = 4$

(g) $2\sin\dfrac{\theta}{2}\cos\dfrac{\theta}{2} = 2\left(\dfrac{1}{\sqrt{17}}\right)\left(\dfrac{4\sqrt{17}}{17}\right) = \dfrac{8}{17}, \quad (= \sin\theta)$

(h) $2\cos\dfrac{\theta}{2}\tan\dfrac{\theta}{2} = 2\sin\dfrac{\theta}{2} = \dfrac{2\sqrt{17}}{17}$

39. $\sin 15° = \sin\left(\dfrac{1}{2} \cdot 30°\right) = \sqrt{\dfrac{1 - \cos 30°}{2}} = \sqrt{\dfrac{1 - \left(\sqrt{3}/2\right)}{2}} = \dfrac{1}{2}\sqrt{2 - \sqrt{3}}$

$\cos 15° = \cos\left(\dfrac{1}{2} \cdot 30°\right) = \sqrt{\dfrac{1 + \cos 30°}{2}} = \sqrt{\dfrac{1 + \left(\sqrt{3}/2\right)}{2}} = \dfrac{1}{2}\sqrt{2 + \sqrt{3}}$

$\tan 15° = \tan\left(\dfrac{1}{2} \cdot 30°\right) = \dfrac{\sin 30°}{1 + \cos 30°} = \dfrac{1/2}{1 + \left(\sqrt{3}/2\right)} = \dfrac{1}{2 + \sqrt{3}} = 2 - \sqrt{3}$

41. $\sin 112° 30' = \sin\left(\dfrac{1}{2} \cdot 225°\right) = \sqrt{\dfrac{1 - \cos 225°}{2}} = \sqrt{\dfrac{1 + \left(\sqrt{2}/2\right)}{2}} = \dfrac{1}{2}\sqrt{2 + \sqrt{2}}$

$\cos 112° 30' = \cos\left(\dfrac{1}{2} \cdot 225°\right) = -\sqrt{\dfrac{1 + \cos 225°}{2}} = -\sqrt{\dfrac{1 - \left(\sqrt{2}/2\right)}{2}} = -\dfrac{1}{2}\sqrt{2 - \sqrt{2}}$

$\tan 112° 30' = \tan\left(\dfrac{1}{2} \cdot 225°\right) = \dfrac{\sin 225°}{1 + \cos 225°} = \dfrac{-\sqrt{2}/2}{1 - \left(\sqrt{2}/2\right)} = -1 - \sqrt{2}$

43. $\sin\dfrac{\pi}{8} = \sin\left[\dfrac{1}{2}\left(\dfrac{\pi}{4}\right)\right] = \sqrt{\dfrac{1 - \cos(\pi/4)}{2}} = \dfrac{1}{2}\sqrt{2 - \sqrt{2}}$

$\cos\dfrac{\pi}{8} = \cos\left[\dfrac{1}{2}\left(\dfrac{\pi}{4}\right)\right] = \sqrt{\dfrac{1 + \cos(\pi/4)}{2}} = \dfrac{1}{2}\sqrt{2 + \sqrt{2}}$

$\tan\dfrac{\pi}{8} = \tan\left[\dfrac{1}{2}\left(\dfrac{\pi}{4}\right)\right] = \dfrac{\sin(\pi/4)}{1 + \cos(\pi/4)} = \dfrac{\sqrt{2}/2}{1 + \left(\sqrt{2}/2\right)} = \sqrt{2} - 1$

45. $\sin\dfrac{3\pi}{8} = \sin\left(\dfrac{1}{2} \cdot \dfrac{3\pi}{4}\right) = \sqrt{\dfrac{1 - \cos(3\pi/4)}{2}} = \sqrt{\dfrac{1 + \left(\sqrt{2}/2\right)}{2}} = \dfrac{1}{2}\sqrt{2 + \sqrt{2}}$

$\cos\dfrac{3\pi}{8} = \cos\left(\dfrac{1}{2} \cdot \dfrac{3\pi}{4}\right) = \sqrt{\dfrac{1 + \cos(3\pi/4)}{2}} = \sqrt{\dfrac{1 - \left(\sqrt{2}/2\right)}{2}} = \dfrac{1}{2}\sqrt{2 - \sqrt{2}}$

$\tan\dfrac{3\pi}{8} = \tan\left(\dfrac{1}{2} \cdot \dfrac{3\pi}{4}\right) = \dfrac{\sin(3\pi/4)}{1 + \cos(3\pi/4)} = \dfrac{\sqrt{2}/2}{1 - \left(\sqrt{2}/2\right)} = \dfrac{\sqrt{2}}{2 - \sqrt{2}} = \sqrt{2} + 1$

47. $\sin u = \dfrac{5}{13}, \dfrac{\pi}{2} < u < \pi \implies \cos u = -\dfrac{12}{13}$

$$\sin\!\left(\dfrac{u}{2}\right) = \sqrt{\dfrac{1 - \cos u}{2}}$$

$$= \sqrt{\dfrac{1 + (12/13)}{2}} = \dfrac{5\sqrt{26}}{26}$$

$$\cos\!\left(\dfrac{u}{2}\right) = \sqrt{\dfrac{1 + \cos u}{2}}$$

$$= \sqrt{\dfrac{1 - (12/13)}{2}} = \dfrac{\sqrt{26}}{26}$$

$$\tan\!\left(\dfrac{u}{2}\right) = \dfrac{\sin u}{1 + \cos u} = \dfrac{5/13}{1 - (12/13)} = \dfrac{5}{1} = 5$$

49. $\tan u = -\dfrac{8}{5}, \dfrac{3\pi}{2} < u < 2\pi,$ Quadrant IV

$$\sin u = -\dfrac{8}{\sqrt{89}}, \cos u = \dfrac{5}{\sqrt{89}}$$

$$\sin\!\left(\dfrac{u}{2}\right) = \sqrt{\dfrac{1 - \cos u}{2}} = \sqrt{\dfrac{1 - \left(5/\sqrt{89}\right)}{2}} = \sqrt{\dfrac{\sqrt{89} - 5}{2\sqrt{89}}} = \sqrt{\dfrac{89 - 5\sqrt{89}}{178}}$$

$$\cos\!\left(\dfrac{u}{2}\right) = -\sqrt{\dfrac{1 + \cos u}{2}} = -\sqrt{\dfrac{1 + \left(5/\sqrt{89}\right)}{2}} = -\sqrt{\dfrac{\sqrt{89} + 5}{2\sqrt{89}}} = -\sqrt{\dfrac{89 + 5\sqrt{89}}{178}}$$

$$\tan\!\left(\dfrac{u}{2}\right) = \dfrac{1 - \cos u}{\sin u} = \dfrac{1 - \left(5/\sqrt{89}\right)}{-8/\sqrt{89}} = \dfrac{5 - \sqrt{89}}{8}$$

51. $\csc u = -\dfrac{5}{3}, \pi < u < \dfrac{3\pi}{2},$ Quadrant III

$$\sin u = -\dfrac{3}{5}, \cos u = -\dfrac{4}{5}$$

$$\sin\!\left(\dfrac{u}{2}\right) = \sqrt{\dfrac{1 - \cos u}{2}} = \sqrt{\dfrac{1 + (4/5)}{2}} = \dfrac{3}{\sqrt{10}} = \dfrac{3\sqrt{10}}{10}$$

$$\cos\!\left(\dfrac{u}{2}\right) = -\sqrt{\dfrac{1 + \cos u}{2}} = -\sqrt{\dfrac{1 - (4/5)}{2}} = \dfrac{-1}{\sqrt{10}} = -\dfrac{\sqrt{10}}{10}$$

$$\tan\!\left(\dfrac{u}{2}\right) = \dfrac{1 - \cos u}{\sin u} = \dfrac{1 + (4/5)}{-3/5} = -3$$

53. $\sqrt{\dfrac{1 - \cos 6x}{2}} = |\sin 3x|$

55. $-\sqrt{\dfrac{1 - \cos 8x}{1 + \cos 8x}} = -\dfrac{\sqrt{(1 - \cos 8x)/2}}{\sqrt{(1 + \cos 8x)/2}}$

$$= -\left|\dfrac{\sin 4x}{\cos 4x}\right| = -|\tan 4x|$$

57. $\sin \dfrac{x}{2} - \cos x = 0$

$$\pm \sqrt{\dfrac{1 - \cos x}{2}} = \cos x$$

$$\dfrac{1 - \cos x}{2} = \cos^2 x$$

$$0 = 2\cos^2 x + \cos x - 1$$

$$= (2\cos x - 1)(\cos x + 1)$$

$$\cos x = \dfrac{1}{2} \quad \text{or} \quad \cos x = -1$$

$$x = \dfrac{\pi}{3}, \dfrac{5\pi}{3} \qquad x = \pi$$

By checking these values in the original equations, we see that $x = \pi/3$ and $x = 5\pi/3$ are the only solutions. $x = \pi$ is extraneous.

59. $\cos \dfrac{x}{2} - \sin x = 0$

$$\pm \sqrt{\dfrac{1 + \cos x}{2}} = \sin x$$

$$\dfrac{1 + \cos x}{2} = \sin^2 x$$

$$1 + \cos x = 2\sin^2 x$$

$$1 + \cos x = 2 - 2\cos^2 x$$

$$2\cos^2 x + \cos x - 1 = 0$$

$$(2\cos x - 1)(\cos x + 1) = 0$$

$$2\cos x - 1 = 0 \quad \text{or} \quad \cos x + 1 = 0$$

$$\cos x = \dfrac{1}{2} \qquad\qquad \cos x = -1$$

$$x = \dfrac{\pi}{3}, \dfrac{5\pi}{3} \qquad\qquad x = \pi$$

$$x = \dfrac{\pi}{3}, \pi, \dfrac{5\pi}{3}$$

$\pi/3$, π, and $5\pi/3$ are all solutions to the equation.

61. $6 \sin \dfrac{\pi}{3} \cos \dfrac{\pi}{3} = 6 \cdot \dfrac{1}{2}\left[\sin\left(\dfrac{\pi}{3} + \dfrac{\pi}{3}\right) + \sin\left(\dfrac{\pi}{3} - \dfrac{\pi}{3}\right) \right] = 3\left[\sin \dfrac{2\pi}{3} + \sin 0 \right] = 3 \sin \dfrac{2\pi}{3}$

63. $\sin 5\theta \cos 3\theta = \dfrac{1}{2}[\sin(5\theta + 3\theta) + \sin(5\theta - 3\theta)] = \dfrac{1}{2}(\sin 8\theta + \sin 2\theta)$

65. $10 \cos 75° \cos 15° = 5[\cos(75° - 15°) + \cos(75° + 15°)] = 5[\cos 60° + \cos 90°]$

67. $5 \cos(-5\beta) \cos 3\beta = 5 \cdot \dfrac{1}{2}[\cos(-5\beta - 3\beta) + \cos(-5\beta + 3\beta)]$

$$= \dfrac{5}{2}[\cos(-8\beta) + \cos(-2\beta)] = \dfrac{5}{2}(\cos 8\beta + \cos 2\beta)$$

69. $\sin(x + y) \sin(x - y) = \dfrac{1}{2}[\cos((x + y) - (x - y)) - \cos((x + y) + (x - y))] = \dfrac{1}{2}[\cos 2y - \cos 2x]$

71. $\cos(\theta - \pi) \sin(\theta + \pi) = \dfrac{1}{2}[\sin((\theta - \pi) + (\theta + \pi)) - \sin((\theta - \pi) - (\theta + \pi))]$

$$= \dfrac{1}{2}[\sin 2\theta - \sin(-2\pi)] = \dfrac{1}{2}[\sin 2\theta + \sin 2\pi]$$

73. $\sin 5\theta - \sin \theta = 2 \cos\left(\dfrac{5\theta + \theta}{2}\right) \sin\left(\dfrac{5\theta - \theta}{2}\right)$

$$= 2 \cos 3\theta \cdot \sin 2\theta$$

75. $\cos 6x + \cos 2x = 2 \cos\left(\dfrac{6x + 2x}{2}\right) \cos\left(\dfrac{6x - 2x}{2}\right)$

$$= 2 \cos 4x \cos 2x$$

77. $\sin(\alpha + \beta) - \sin(\alpha - \beta) = 2 \cos\left(\dfrac{\alpha + \beta + \alpha - \beta}{2}\right) \sin\left(\dfrac{\alpha + \beta - \alpha + \beta}{2}\right) = 2 \cos \alpha \sin \beta$

79. $\cos\left(\theta + \dfrac{\pi}{2}\right) - \cos\left(\theta - \dfrac{\pi}{2}\right) = -2\sin\left(\dfrac{\theta + (\pi/2) + \theta - (\pi/2)}{2}\right)\sin\left(\dfrac{\theta + (\pi/2) - \theta + (\pi/2)}{2}\right)$

$$= -2\sin\theta\sin\dfrac{\pi}{2} = -2\sin\theta$$

81. $\sin 195° + \sin 105° = 2\sin\left(\dfrac{195° + 105°}{2}\right)\cos\left(\dfrac{195° - 105°}{2}\right) = 2\sin(150°)\cos(45°) = 2\left(\dfrac{1}{2}\right)\left(\dfrac{\sqrt{2}}{2}\right) = \dfrac{\sqrt{2}}{2}$

83. $\cos\dfrac{5\pi}{12} + \cos\dfrac{\pi}{12} = 2\cos\left(\dfrac{(5\pi/12) + (\pi/12)}{2}\right)\cos\left(\dfrac{(5\pi/12) - (\pi/12)}{2}\right)$

$$= 2\cos\left(\dfrac{\pi}{4}\right)\cos\left(\dfrac{\pi}{6}\right) = 2\left(\dfrac{\sqrt{2}}{2}\right)\left(\dfrac{\sqrt{3}}{2}\right) = \dfrac{2\sqrt{6}}{4} = \dfrac{\sqrt{6}}{2}$$

85.
$$\sin 6x + \sin 2x = 0$$
$$2\sin\left(\dfrac{6x + 2x}{2}\right)\cos\left(\dfrac{6x - 2x}{2}\right) = 0$$
$$\sin 4x\cos 2x = 0$$
$$\sin 4x = 0 \quad \text{or} \quad \cos 2x = 0$$
$$4x = n\pi \qquad\qquad 2x = \dfrac{\pi}{2} + n\pi$$
$$x = \dfrac{n\pi}{4} \qquad\qquad x = \dfrac{\pi}{4} + \dfrac{n\pi}{2}$$

In the interval we have

$$x = 0, \dfrac{\pi}{4}, \dfrac{\pi}{2}, \dfrac{3\pi}{4}, \pi, \dfrac{5\pi}{4}, \dfrac{3\pi}{2}, \dfrac{7\pi}{4}.$$

87. $\dfrac{\cos 2x}{\sin 3x - \sin x} - 1 = 0$

$$\dfrac{\cos 2x}{\sin 3x - \sin x} = 1$$

$$\dfrac{\cos 2x}{2\cos 2x\sin x} = 1$$

$$2\sin x = 1$$

$$\sin x = \dfrac{1}{2}$$

$$x = \dfrac{\pi}{6}, \dfrac{5\pi}{6}$$

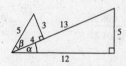

Figure for Exercises 89 and 91

89. $\sin^2\alpha = \left(\dfrac{5}{13}\right)^2 = \dfrac{25}{169}$

$$\sin^2\alpha = 1 - \cos^2\alpha = 1 - \left(\dfrac{12}{13}\right)^2$$

$$= 1 - \dfrac{144}{169} = \dfrac{25}{169}$$

91. $\sin\alpha\cos\beta = \left(\dfrac{5}{13}\right)\left(\dfrac{4}{5}\right) = \dfrac{4}{13}$

$$\sin\alpha\cos\beta = \cos\left(\dfrac{\pi}{2} - \alpha\right)\sin\left(\dfrac{\pi}{2} - \beta\right)$$

$$= \left(\dfrac{5}{13}\right)\left(\dfrac{4}{5}\right) = \dfrac{4}{13}$$

93. $\csc 2\theta = \dfrac{1}{\sin 2\theta}$

$$= \dfrac{1}{2\sin\theta\cos\theta}$$

$$= \dfrac{1}{\sin\theta}\cdot\dfrac{1}{2\cos\theta}$$

$$= \dfrac{\csc\theta}{2\cos\theta}$$

95. $\cos^2 2\alpha - \sin^2 2\alpha = \cos[2(2\alpha)]$

$$= \cos 4\alpha$$

97. $(\sin x + \cos x)^2 = \sin^2 x + 2\sin x\cos x + \cos^2 x$

$$= (\sin^2 x + \cos^2 x) + 2\sin x\cos x$$

$$= 1 + \sin 2x$$

99. $1 + \cos 10y = 1 + \cos^2 5y - \sin^2 5y$

$$= 1 + \cos^2 5y - (1 - \cos^2 5y)$$

$$= 2\cos^2 5y$$

101. $\sec\dfrac{u}{2} = \dfrac{1}{\cos(u/2)}$

$$= \pm\sqrt{\dfrac{2}{1+\cos u}}$$

$$= \pm\sqrt{\dfrac{2\sin u}{\sin u(1+\cos u)}}$$

$$= \pm\sqrt{\dfrac{2\sin u}{\sin u + \sin u\cos u}}$$

$$= \pm\sqrt{\dfrac{(2\sin u)/(\cos u)}{(\sin u)/(\cos u) + (\sin u\cos u)/(\cos u)}}$$

$$= \pm\sqrt{\dfrac{2\tan u}{\tan u + \sin u}}$$

103. $\cos 3\beta = \cos(2\beta + \beta)$

$$= \cos 2\beta\cos\beta - \sin 2\beta\sin\beta$$

$$= (\cos^2\beta - \sin^2\beta)\cos\beta - 2\sin\beta\cos\beta\sin\beta$$

$$= \cos^3\beta - \sin^2\beta\cos\beta - 2\sin^2\beta\cos\beta$$

$$= \cos^3\beta - 3\sin^2\beta\cos\beta$$

105. $\dfrac{\cos 4x - \cos 2x}{2\sin 3x} = \dfrac{-2\sin\left(\dfrac{4x+2x}{2}\right)\sin\left(\dfrac{4x-2x}{2}\right)}{2\sin 3x}$

$$= \dfrac{-2\sin 3x\sin x}{2\sin 3x}$$

$$= -\sin x$$

107. $\dfrac{\cos 4x + \cos 2x}{\sin 4x + \sin 2x} = \dfrac{2\cos(3x)\cos x}{2\sin(3x)\cos x}$

$$= \cot 3x$$

109. $\sin\left(\dfrac{\pi}{6}+x\right) + \sin\left(\dfrac{\pi}{6}-x\right) = 2\sin\left(\dfrac{\dfrac{\pi}{6}+x+\dfrac{\pi}{6}-x}{2}\right)\cos\left(\dfrac{\dfrac{\pi}{6}+x-\dfrac{\pi}{6}+x}{2}\right) = 2\sin\dfrac{\pi}{6}\cos x = \cos x$

111. $\sin^2 x = \dfrac{1-\cos 2x}{2} = \dfrac{1}{2} - \dfrac{\cos 2x}{2}$

113. $f(x) = \cos^4 x = \dfrac{1}{8}(3 + 4\cos 2x + \cos 4x)$

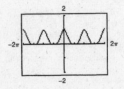

115. $\sin(2 \arcsin x) = 2 \sin(\arcsin x) \cos(\arcsin x)$

$$= 2x\sqrt{1 - x^2}$$

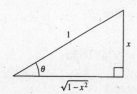

117. $\cos(2 \arcsin x) = 1 - 2 \sin^2(\arcsin x)$

$$= 1 - 2x^2$$

119. $\cos(2 \arctan x) = 1 - 2 \sin^2(\arctan x)$

$$= 1 - 2\left(\frac{x}{\sqrt{1 + x^2}}\right)^2$$

$$= 1 - \frac{2x^2}{1 + x^2}$$

$$= \frac{1 - x^2}{1 + x^2}$$

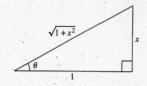

121. (a) $y = 4 \sin \dfrac{x}{2} + \cos x$

Maximum: $(\pi, 3)$

(b) $\quad 2 \cos \dfrac{x}{2} - \sin x = 0$

$$2\left(\pm \sqrt{\frac{1 + \cos x}{2}}\right) = \sin x$$

$$4\left(\frac{1 + \cos x}{2}\right) = \sin^2 x$$

$$2(1 + \cos x) = 1 - \cos^2 x$$

$$\cos^2 x + 2 \cos x + 1 = 0$$

$$(\cos x + 1)^2 = 0$$

$$\cos x = -1$$

$$x = \pi$$

123. (a) $y = 2 \cos \dfrac{x}{2} + \sin 2x$

Maximum: $(0.699, 2.864)$

Minimum: $(5.584, -2.864)$

(b) $2 \cos 2x - \sin(x/2) = 0$ has four zeros on $[0, 2\pi)$. Two of the zeros are $x \approx 0.699$ and $x \approx 5.584$. (The other two are $2.608, 3.675$.)

125. (a) $\qquad f(x) = \sin 2x - \sin x = 0$

$$2 \sin x \cos x - \sin x = 0$$

$$\sin x(2 \cos x - 1) = 0$$

$$\sin x = 0 \implies x = 0, \pi, 2\pi$$

$$\cos x = \frac{1}{2} \implies x = \frac{\pi}{3}, \frac{5\pi}{3}$$

(b) $\qquad 2 \cos 2x - \cos x = 0$

$$2(2 \cos^2 x - 1) - \cos x = 0$$

$$4 \cos^2 x - \cos x - 2 = 0$$

$$\cos x = \frac{1 \pm \sqrt{1 + 32}}{8} = \frac{1 \pm \sqrt{33}}{8}$$

$$x = \arccos\left(\frac{1 \pm \sqrt{33}}{8}\right)$$

$$x = 2\pi - \arccos\left(\frac{1 \pm \sqrt{33}}{8}\right)$$

$$x \approx 0.5678, 2.2057, 4.0775, 5.7154$$

127. (a) $r = \dfrac{1}{32}v_0^2 \sin 2\theta$

$= \dfrac{1}{32}v_0^2(2\sin\theta\cos\theta)$

$= \dfrac{1}{16}v_0^2 \sin\theta\cos\theta$

(b) $r = \dfrac{1}{16}v_0^2 \sin\theta\cos\theta$

$= \dfrac{1}{16}(80)^2 \sin 42°\cos 42° \approx 198.90$ feet

(c) $r = \dfrac{1}{16}v_0^2 \sin\theta\cos\theta$

$300 = \dfrac{1}{16}v_0^2 \sin 40°\cos 40°$

$v_0^2 = \dfrac{300(16)}{\sin 40°\cos 40°} \approx 9748.0955$

$v_0 \approx 98.73$ feet per second

(d) $\sin 2\theta$ is greatest when $\theta = 45°$.

129. $\dfrac{x}{2} = 2r\sin^2\dfrac{\theta}{2}$, $\quad x = 4r\left[\sin\dfrac{\theta}{2}\right]^2 = 4r\,\dfrac{1-\cos\theta}{2} = 2r(1-\cos\theta)$

131. False. If $x = \pi$, $\sin\dfrac{x}{2} = \sin\dfrac{\pi}{2} = 1$, whereas

$-\sqrt{\dfrac{1-\cos\pi}{2}} = -1.$

133. $f(x) = 2\sin x\left[2\cos^2\left(\dfrac{x}{2}\right) - 1\right]$

(a)

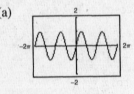

(b) The graph appears to be that of $y = \sin 2x$.

(c) $2\sin x\left[2\cos^2\left(\dfrac{x}{2}\right) - 1\right] = 2\sin x\left[2\,\dfrac{1+\cos x}{2} - 1\right]$

$= 2\sin x[\cos x] = \sin 2x$

135. Answers will vary.

137. (a)

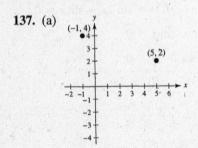

(b) Distance:

$\sqrt{(5+1)^2 + (2-4)^2} = \sqrt{40} = 2\sqrt{10}$

(c) Midpoint: $\left(\dfrac{-1+5}{2}, \dfrac{4+2}{2}\right) = (2, 3)$

139. (a)

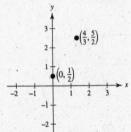

(b) Distance:

$\sqrt{\left(\dfrac{4}{3}\right)^2 + \left(\dfrac{5}{2} - \dfrac{1}{2}\right)^2} = \sqrt{\dfrac{16}{9} + 4}$

$= \dfrac{\sqrt{52}}{3} = \dfrac{2\sqrt{13}}{3}$

(c) Midpoint:

$\left(\dfrac{0 + (4/3)}{2}, \dfrac{(1/2) + (5/2)}{2}\right) = \left(\dfrac{2}{3}, \dfrac{3}{2}\right)$

141. (a) Complement: $90° - 55° = 35°$

Supplement: $180° - 55° = 125°$

(b) Complement: None

Supplement: $180° - 162° = 18°$

143. (a) Complement: $\dfrac{\pi}{2} - \dfrac{\pi}{18} = \dfrac{8\pi}{18} = \dfrac{4\pi}{9}$

Supplement: $\pi - \dfrac{\pi}{18} = \dfrac{17\pi}{18}$

(b) Complement: $\dfrac{\pi}{2} - \dfrac{9\pi}{20} = \dfrac{\pi}{20}$

Supplement: $\pi - \dfrac{9\pi}{20} = \dfrac{11\pi}{20}$

145. $s = r\theta \implies \theta = \dfrac{s}{r} = \dfrac{7}{15} \approx 0.467$ rad

147. $f(x) = \dfrac{3}{2}\cos(2x)$

Period: $\dfrac{2\pi}{2} = \pi$

Amplitude: $\dfrac{3}{2}$

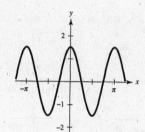

149. $f(x) = \dfrac{1}{2}\tan(2\pi x)$

Period: $\dfrac{\pi}{2\pi} = \dfrac{1}{2}$

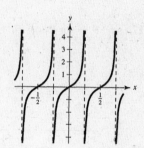

Review Exercises for Chapter 5

1. $\dfrac{1}{\cos x} = \sec x$

3. $\dfrac{1}{\sec x} = \cos x$

5. $\sqrt{1 - \cos^2 x} = |\sin x|$

7. $\csc\left(\dfrac{\pi}{2} - x\right) = \sec x$

9. $\sec(-x) = \sec x$

11. $\sin x = \dfrac{4}{5}, \cos x = \dfrac{3}{5},$ Quadrant I

$\tan x = \dfrac{\sin x}{\cos x} = \dfrac{4}{3}$

$\cot x = \dfrac{3}{4}$

$\sec x = \dfrac{5}{3}$

$\csc x = \dfrac{5}{4}$

13. $\sin\left(\dfrac{\pi}{2} - x\right) = \cos x = \dfrac{1}{\sqrt{2}} = \dfrac{\sqrt{2}}{2},$

$\sin x = -\dfrac{1}{\sqrt{2}} = -\dfrac{\sqrt{2}}{2},$ Quadrant IV

$\tan x = -1$

$\cot x = -1$

$\sec x = \sqrt{2}$

$\csc x = -\sqrt{2}$

15. $\dfrac{1}{\tan^2 x + 1} = \dfrac{1}{\sec^2 x} = \cos^2 x$

17. $\dfrac{\sin^2 \alpha - \cos^2 \alpha}{\sin^2 \alpha - \sin \alpha \cos \alpha} = \dfrac{(\sin \alpha + \cos \alpha)(\sin \alpha - \cos \alpha)}{\sin \alpha(\sin \alpha - \cos \alpha)} = \dfrac{\sin \alpha + \cos \alpha}{\sin \alpha} = 1 + \cot \alpha$

19. $\tan^2 \theta (\csc^2 \theta - 1) = \tan^2 \theta (\cot^2 \theta)$

$$= \tan^2 \theta \left(\frac{1}{\tan^2 \theta} \right) = 1$$

21. $\tan\left(\dfrac{\pi}{2} - x \right) \sec x = \cot x \sec x$

$$= \frac{\cos x}{\sin x} \cdot \frac{1}{\cos x} = \frac{1}{\sin x} = \csc x$$

23. $\sin^{-1/2} x \cos x = \dfrac{\cos x}{\sin^{1/2} x}$

$$= \frac{\cos x}{\sqrt{\sin x}} \cdot \frac{\sqrt{\sin x}}{\sqrt{\sin x}}$$

$$= \frac{\cos x}{\sin x} \sqrt{\sin x} = \cot x \sqrt{\sin x}$$

25. $\cos x (\tan^2 x + 1) = \cos x \sec^2 x$

$$= \frac{1}{\sec x} \sec^2 x = \sec x$$

27. $\sin^3 \theta + \sin \theta \cos^2 \theta = \sin \theta (\sin^2 \theta + \cos^2 \theta)$

$$= \sin \theta$$

29. $\sin^5 x \cos^2 x = \sin^4 x \cos^2 x \sin x$

$$= (1 - \cos^2 x)^2 \cos^2 x \sin x$$

$$= (1 - 2\cos^2 x + \cos^4 x) \cos^2 x \sin x$$

$$= (\cos^2 x - 2\cos^4 x + \cos^6 x) \sin x$$

31. $\sqrt{\dfrac{1 - \sin \theta}{1 + \sin \theta}} = \sqrt{\dfrac{1 - \sin \theta}{1 + \sin \theta} \cdot \dfrac{1 - \sin \theta}{1 - \sin \theta}} = \sqrt{\dfrac{(1 - \sin \theta)^2}{1 - \sin^2 \theta}} = \sqrt{\dfrac{(1 - \sin \theta)^2}{\cos^2 \theta}} = \dfrac{|1 - \sin \theta|}{|\cos \theta|} = \dfrac{1 - \sin \theta}{|\cos \theta|}$

Note: We can drop the absolute value on $1 - \sin \theta$ since it is always nonnegative.

33. $\dfrac{\csc(-x)}{\sec(-x)} = -\dfrac{\csc x}{\sec x} = -\dfrac{\cos x}{\sin x} = -\cot x$

35. $\csc^2\left(\dfrac{\pi}{2} - x \right) - 1 = \sec^2 x - 1 = \tan^2 x$

37. $2 \sin x - 1 = 0$

$$\sin x = \frac{1}{2}$$

$$x = \frac{\pi}{6} + 2n\pi$$

$$x = \frac{5\pi}{6} + 2n\pi$$

39. $\sin x = \sqrt{3} - \sin x$

$$2 \sin x = \sqrt{3}$$

$$\sin x = \frac{\sqrt{3}}{2}$$

$$x = \frac{\pi}{3} + 2n\pi$$

$$x = \frac{2\pi}{3} + 2n\pi$$

41. $3\sqrt{3} \tan x = 3$

$$\tan x = \frac{1}{\sqrt{3}}$$

$$x = \frac{\pi}{6} + n\pi$$

43. $3 \csc^2 x = 4$

$$\csc^2 x = \frac{4}{3}$$

$$\sin^2 x = \frac{3}{4}$$

$$\sin x = \pm \frac{\sqrt{3}}{2}$$

$$x = \frac{\pi}{3} + n\pi$$

$$x = \frac{2\pi}{3} + n\pi$$

45. $4 \cos^2 x - 3 = 0$

$$\cos^2 x = \frac{3}{4}$$

$$\cos x = \pm \frac{\sqrt{3}}{2}$$

$$x = \frac{\pi}{6} + n\pi$$

$$x = \frac{5\pi}{6} + n\pi$$

47. $\sin x - \tan x = 0$

$$\sin x - \frac{\sin x}{\cos x} = 0$$

$$\sin x \cos x - \sin x = 0$$

$$\sin x (\cos x - 1) = 0$$

$$\sin x = 0 \quad \text{or} \quad \cos x - 1 = 0$$

$$x = n\pi \qquad\qquad \cos x = 1$$

49. $2 \cos^2 x - \cos x - 1 = 0$

$(2 \cos x + 1)(\cos x - 1) = 0$

$2 \cos x + 1 = 0 \qquad \text{or} \quad \cos x - 1 = 0$

$\cos x = -\dfrac{1}{2} \qquad\qquad \cos x = 1$

$x = \dfrac{2\pi}{3}, \dfrac{4\pi}{3} \qquad\qquad x = 0$

51. $\cos^2 x + \sin x = 1$

$1 - \sin^2 x + \sin x = 1$

$\sin x(\sin x - 1) = 0$

$\sin x = 0 \qquad \text{or} \quad \sin x = 1$

$x = 0, \pi \qquad\qquad x = \dfrac{\pi}{2}$

53. $2 \sin 2x = \sqrt{2}$

$\sin 2x = \dfrac{\sqrt{2}}{2}$

$2x = \dfrac{\pi}{4}, \dfrac{3\pi}{4}, \dfrac{9\pi}{4}, \dfrac{11\pi}{4}$

$x = \dfrac{\pi}{8}, \dfrac{3\pi}{8}, \dfrac{9\pi}{8}, \dfrac{11\pi}{8}$

55. $\cos 4x(\cos x - 1) = 0$

$\cos 4x = 0 \text{ or } \cos x - 1 = 0$

$4x = \dfrac{\pi}{2}, \dfrac{3\pi}{2}, \dfrac{5\pi}{2}, \dfrac{7\pi}{2}, \dfrac{9\pi}{2}, \dfrac{11\pi}{2}, \dfrac{13\pi}{2}, \dfrac{15\pi}{2}$

or $\cos x = 1$

$x = \dfrac{\pi}{8}, \dfrac{3\pi}{8}, \dfrac{5\pi}{8}, \dfrac{7\pi}{8}, \dfrac{9\pi}{8}, \dfrac{11\pi}{8}, \dfrac{13\pi}{8}, \dfrac{15\pi}{8}, 0$

57. $\cos 4x - 7 \cos 2x = 8$

$2 \cos^2 2x - 1 - 7 \cos 2x = 8$

$2 \cos^2 2x - 7 \cos 2x - 9 = 0$

$(2 \cos 2x - 9)(\cos 2x + 1) = 0$

$2 \cos 2x - 9 = 0 \quad \text{or} \quad \cos 2x + 1 = 0$

$\cos 2x = \dfrac{9}{2} \qquad\qquad \cos 2x = -1$

No solution $\qquad\qquad 2x = \pi + 2n\pi$

$\qquad\qquad\qquad\qquad x = \dfrac{\pi}{2} + n\pi$

$\qquad\qquad\qquad\qquad x = \dfrac{\pi}{2}, \dfrac{3\pi}{2}$

59. $2 \sin 2x - 1 = 0$

$\sin 2x = \dfrac{1}{2}$

$2x = \dfrac{\pi}{6} + 2n\pi \text{ or } 2x = \dfrac{5\pi}{6} + 2n\pi$

$x = \dfrac{\pi}{12} + n\pi \text{ or } x = \dfrac{5\pi}{12} + n\pi$

61. $2 \sin^2 3x - 1 = 0$

$\sin^2 3x = \dfrac{1}{2}$

$\sin 3x = \pm \dfrac{\sqrt{2}}{2}$

$3x = \dfrac{\pi}{4} + \dfrac{n\pi}{2}$

$x = \dfrac{\pi}{12} + \dfrac{n\pi}{6}$

63. $\sin^2 x - 2 \sin x = 0$

$\sin x(\sin x - 2) = 0$

$\sin x = 0 \text{ or } \sin x = 2 \text{ (impossible)}$

$x = 0, \pi$

65. $\tan^2 \theta + 3 \tan \theta - 10 = 0$

$(\tan \theta + 5)(\tan \theta - 2) = 0$

$\tan \theta = -5 \implies$

$\qquad \theta = \arctan(-5) + \pi, \arctan(-5) + 2\pi$

$\tan \theta = 2 \implies \theta = \arctan(2), \arctan(2) + \pi$

$\theta \approx 1.1071, 1.7682, 4.2487, 4.9098$

67. $\sin 285° = \sin(315° - 30°)$

$\qquad = \sin 315° \cos 30° - \cos 315° \sin 30°$

$\qquad = \left(-\dfrac{\sqrt{2}}{2}\right)\left(\dfrac{\sqrt{3}}{2}\right) - \left(\dfrac{\sqrt{2}}{2}\right)\left(\dfrac{1}{2}\right) = -\dfrac{\sqrt{6} + \sqrt{2}}{4}$

$\cos 285° = \cos(315° - 30°) \quad = \cos 315° \cos 30° + \sin 315° \sin 30°$

$\qquad = \left(\dfrac{\sqrt{2}}{2}\right)\left(\dfrac{\sqrt{3}}{2}\right) + \left(-\dfrac{\sqrt{2}}{2}\right)\left(\dfrac{1}{2}\right) = \dfrac{\sqrt{6} - \sqrt{2}}{4}$

$\tan 285° = -\dfrac{\sqrt{6} + \sqrt{2}}{\sqrt{6} - \sqrt{2}} = -2 - \sqrt{3}$

69. $\sin \dfrac{31\pi}{12} = \sin\left(\dfrac{11\pi}{6} + \dfrac{3\pi}{4}\right) = \sin \dfrac{11\pi}{6} \cos \dfrac{3\pi}{4} + \sin \dfrac{3\pi}{4} \cos \dfrac{11\pi}{6}$

$\qquad = \left(-\dfrac{1}{2}\right)\left(-\dfrac{\sqrt{2}}{2}\right) + \left(\dfrac{\sqrt{2}}{2}\right)\left(\dfrac{\sqrt{3}}{2}\right) = \dfrac{\sqrt{2} + \sqrt{6}}{4}$

$\cos \dfrac{31\pi}{12} = \cos\left(\dfrac{11\pi}{6} + \dfrac{3\pi}{4}\right) = \cos \dfrac{11\pi}{6} \cos \dfrac{3\pi}{4} - \sin \dfrac{11\pi}{6} \sin \dfrac{3\pi}{4}$

$\qquad = \left(\dfrac{\sqrt{3}}{2}\right)\left(-\dfrac{\sqrt{2}}{2}\right) - \left(-\dfrac{1}{2}\right)\left(\dfrac{\sqrt{2}}{2}\right) = \dfrac{\sqrt{2} - \sqrt{6}}{4}$

$\tan \dfrac{31\pi}{12} = \dfrac{\sin(31\pi/12)}{\cos(31\pi/12)} = \dfrac{\sqrt{2} + \sqrt{6}}{\sqrt{2} - \sqrt{6}} = -2 - \sqrt{3}$

71. $\sin 130° \cos 50° + \cos 130° \sin 50° = \sin(130° + 50°) = \sin 180° = 0$

73. $\dfrac{\tan 25° + \tan 10°}{1 - \tan 25° \tan 10°} = \tan(25° + 10°) = \tan 35°$

For Exercises 75–79, $\sin u = \frac{3}{5}, \cos v = -\frac{7}{25}, \cos u = -\frac{4}{5}, \sin v = \frac{24}{25}.$

75. $\sin(u + v) = \sin u \cos v + \sin v \cos u$

$\qquad = \dfrac{3}{5}\left(-\dfrac{7}{25}\right) + \dfrac{24}{25}\left(-\dfrac{4}{5}\right) = \dfrac{-117}{125}$

77. $\tan(u - v) = \dfrac{\tan u - \tan v}{1 + \tan u \tan v}$

$\qquad = \dfrac{(-3/4) - (-24/7)}{1 + (-3/4)(-24/7)} = \dfrac{3}{4}$

79. $\cos(u + v) = \cos u \cos v - \sin u \sin v$

$\qquad = \left(-\dfrac{4}{5}\right)\left(-\dfrac{7}{25}\right) - \left(\dfrac{3}{5}\right)\left(\dfrac{24}{25}\right) = -\dfrac{44}{125}$

81. $\sin^{-1} 0 = 0$ and $\cos^{-1}(-1) = \pi$

$\sin(\sin^{-1} 0 + \cos^{-1}(-1)) = \sin(0 + \pi) = 0$

83. $\cos^{-1} 1 = 0$ and $\sin^{-1}(-1) = -\dfrac{\pi}{2}$

$\cos(\cos^{-1} 1 - \sin^{-1}(-1)) = \cos\left(0 + \dfrac{\pi}{2}\right) = 0$

85. $\cos\left(x + \dfrac{\pi}{2}\right) = \cos x \cos\dfrac{\pi}{2} - \sin x \sin\dfrac{\pi}{2}$

$\qquad = (\cos x)(0) - (\sin x)(1) = -\sin x$

87. $\cot\left(\dfrac{\pi}{2} - x\right) = \dfrac{\cos[(\pi/2) - x]}{\sin[(\pi/2) - x]}$

$\qquad = \dfrac{\cos(\pi/2)\cos x + \sin(\pi/2)\sin x}{\sin(\pi/2)\cos x - \sin x \cos(\pi/2)}$

$\qquad = \dfrac{\sin x}{\cos x} = \tan x$

89. $\cos 3x = \cos(2x + x)$

$\qquad = \cos 2x \cos x - \sin 2x \sin x$

$\qquad = (\cos^2 x - \sin^2 x)\cos x - 2\sin x \cos x \sin x$

$\qquad = \cos^3 x - 3\sin^2 x \cos x$

$\qquad = \cos^3 x - 3\cos x(1 - \cos^2 x)$

$\qquad = \cos^3 x - 3\cos x + 3\cos^3 x$

$\qquad = 4\cos^3 x - 3\cos x$

91. $\sin\left(x + \dfrac{\pi}{2}\right) - \sin\left(x - \dfrac{\pi}{2}\right) = \sqrt{2}$

$\qquad 2\cos x \sin\dfrac{\pi}{2} = \sqrt{2}$ (Sum-to-Product)

$\qquad\qquad \cos x = \dfrac{\sqrt{2}}{2}$

$\qquad\qquad\quad x = \dfrac{\pi}{4}, \dfrac{7\pi}{4}$

93. $\sin u = -\dfrac{5}{7}, \ \pi < u < \dfrac{3\pi}{2},$ Quadrant III

$\cos^2 u = 1 - \left(-\dfrac{5}{7}\right)^2 = \dfrac{24}{49} \Rightarrow \cos u = -\dfrac{2\sqrt{6}}{7}$

$\sin 2u = 2\sin u \cos u = 2\left(-\dfrac{5}{7}\right)\left(-\dfrac{2\sqrt{6}}{7}\right) = \dfrac{20\sqrt{6}}{49}$

$\cos 2u = 1 - 2\sin^2 u$

$\qquad = 1 - 2\left(-\dfrac{5}{7}\right)^2 = 1 - \dfrac{50}{49} = -\dfrac{1}{49}$

$\tan 2u = \dfrac{\sin 2u}{\cos 2u} = \dfrac{20\sqrt{6}}{-1} = -20\sqrt{6}$

95. $\tan u = -\dfrac{2}{9}, \dfrac{\pi}{2} < u < \pi,$ Quadrant II

$\sec^2 u = \tan^2 u + 1 = \dfrac{4}{81} + 1 = \dfrac{85}{81} \Rightarrow$

$\quad \sec u = -\dfrac{\sqrt{85}}{9}$

$\cos u = \dfrac{-9\sqrt{85}}{85}, \ \sin u = (\tan u)(\cos u) = \dfrac{2\sqrt{85}}{85}$

$\sin 2u = 2\sin u \cos u$

$\qquad = 2\left(\dfrac{2\sqrt{85}}{85}\right)\left(\dfrac{-9\sqrt{85}}{85}\right) = -\dfrac{36}{85}$

$\cos 2u = 1 - 2\sin^2 u = 1 - 2\left(\dfrac{4}{85}\right) = \dfrac{77}{85}$

$\tan 2u = \dfrac{\sin 2u}{\cos 2u} = -\dfrac{36}{77}$

97. $6\sin x \cos x = 3[2\sin x \cos x] = 3\sin 2x$

99. $1 - 4\sin^2 x \cos^2 x = 1 - (2\sin x \cos x)^2$

$\qquad = 1 - \sin^2 2x = \cos^2 2x$

101. $\qquad r = \frac{1}{32}v_0{}^2 \sin 2\theta$

$\qquad\qquad 100 = \frac{1}{32}(80)^2 \sin 2\theta$

$\qquad\quad \sin 2\theta = 0.5$

$\qquad\qquad 2\theta = 30° \quad \text{or} \quad 2\theta = 180° - 30° = 150°$

$\qquad\qquad\quad \theta = 15° \qquad\qquad \theta = 75°$

103. $\sin^6 x = \left(\dfrac{1 - \cos 2x}{2}\right)^3 = \dfrac{1}{8}(1 - 3\cos 2x + 3\cos^2 2x - \cos^3 2x)$

$$= \frac{1}{8}\left[1 - 3\cos 2x + 3\left(\frac{1 + \cos 4x}{2}\right) - \cos 2x\left(\frac{1 + \cos 4x}{2}\right)\right]$$

$$= \frac{1}{8}\left(1 - 3\cos 2x + \frac{3}{2} + \frac{3}{2}\cos 4x - \frac{1}{2}\cos 2x - \frac{1}{2}\cos 2x \cos 4x\right)$$

$$= \frac{1}{16}\left(5 - 7\cos 2x + 3\cos 4x - \frac{1}{2}[\cos 2x + \cos 6x]\right)$$

$$= \frac{1}{32}(10 - 15\cos 2x + 6\cos 4x - \cos 6x)$$

105. $\cos^4 2x = \left(\dfrac{1 + \cos 4x}{2}\right)^2$

$$= \frac{1}{4}(1 + 2\cos 4x + \cos^2 4x)$$

$$= \frac{1}{4}\left(1 + 2\cos 4x + \frac{1 + \cos 8x}{2}\right)$$

$$= \frac{1}{8}(2 + 4\cos 4x + 1 + \cos 8x)$$

$$= \frac{1}{8}(3 + 4\cos 4x + \cos 8x)$$

107. $\sin 105° = \sin\left(\dfrac{1}{2} \cdot 210°\right) = \sqrt{\dfrac{1 - \cos 210°}{2}} = \sqrt{\dfrac{1 + (\sqrt{3}/2)}{2}} = \dfrac{\sqrt{2 + \sqrt{3}}}{2}$

$\cos 105° = \cos\left(\dfrac{1}{2} \cdot 210°\right) = -\sqrt{\dfrac{1 + \cos 210°}{2}} = -\sqrt{\dfrac{1 - (\sqrt{3}/2)}{2}} = \dfrac{-\sqrt{2 - \sqrt{3}}}{2}$

$\tan 105° = \tan\left(\dfrac{1}{2} \cdot 210°\right) = \dfrac{\sin 210°}{1 + \cos 210°} = \dfrac{-1/2}{1 - (\sqrt{3}/2)} = \dfrac{1}{\sqrt{3} - 2} = -2 - \sqrt{3}$

109. $\sin\left(\dfrac{7\pi}{8}\right) = \sin\left(\dfrac{1}{2} \cdot \dfrac{7\pi}{4}\right) = \sqrt{\dfrac{1 - \cos(7\pi/4)}{2}} = \sqrt{\dfrac{1 - (\sqrt{2}/2)}{2}} = \dfrac{\sqrt{2 - \sqrt{2}}}{2}$

$\cos\left(\dfrac{7\pi}{8}\right) = \cos\left(\dfrac{1}{2} \cdot \dfrac{7\pi}{4}\right) = -\sqrt{\dfrac{1 + \cos(7\pi/4)}{2}} = -\sqrt{\dfrac{1 + (\sqrt{2}/2)}{2}} = \dfrac{-\sqrt{2 + \sqrt{2}}}{2}$

$\tan\left(\dfrac{7\pi}{8}\right) = \tan\left(\dfrac{1}{2} \cdot \dfrac{7\pi}{4}\right) = \dfrac{\sin(7\pi/4)}{1 + \cos(7\pi/4)} = \dfrac{-\sqrt{2}/2}{1 + (\sqrt{2}/2)} = \dfrac{-\sqrt{2}}{2 + \sqrt{2}} = 1 - \sqrt{2}$

111. $\sin u = \dfrac{3}{5}, 0 < u < \dfrac{\pi}{2} \implies \cos u = \dfrac{4}{5}$

$\sin\left(\dfrac{u}{2}\right) = \sqrt{\dfrac{1 - \cos u}{2}} = \sqrt{\dfrac{1 - (4/5)}{2}} = \dfrac{1}{\sqrt{10}} = \dfrac{\sqrt{10}}{10}$

$\cos\left(\dfrac{u}{2}\right) = \sqrt{\dfrac{1 + \cos u}{2}} = \sqrt{\dfrac{1 + (4/5)}{2}} = \dfrac{3}{\sqrt{10}} = \dfrac{3\sqrt{10}}{10}$

$\tan\left(\dfrac{u}{2}\right) = \dfrac{1 - \cos u}{\sin u} = \dfrac{1 - (4/5)}{3/5} = \dfrac{1}{3}$

113. $\cos u = -\dfrac{2}{7}, \dfrac{\pi}{2} < u < \pi \implies \sin u = \sqrt{1 - \dfrac{4}{49}} = \dfrac{\sqrt{45}}{7}$

$\sin\left(\dfrac{u}{2}\right) = \sqrt{\dfrac{1 - \cos u}{2}} = \sqrt{\dfrac{1 + (2/7)}{2}} = \sqrt{\dfrac{9}{14}} = \dfrac{3\sqrt{14}}{14}$

$\cos\left(\dfrac{u}{2}\right) = \sqrt{\dfrac{1 + \cos u}{2}} = \sqrt{\dfrac{1 - (2/7)}{2}} = \sqrt{\dfrac{5}{14}} = \dfrac{\sqrt{70}}{14}$

$\tan\left(\dfrac{u}{2}\right) = \dfrac{1 - \cos u}{\sin u} = \dfrac{1 + (2/7)}{\sqrt{45}/7} = \dfrac{9}{\sqrt{45}} = \dfrac{9\sqrt{45}}{45} = \dfrac{\sqrt{45}}{5} = \dfrac{3\sqrt{5}}{5}$

115. $-\sqrt{\dfrac{1 + \cos 8x}{2}} = -|\cos 4x|$

117. Volume V of the trough will be the area A of the isosceles triangle times the length l of the trough.

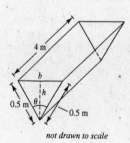

not drawn to scale

$V = A \cdot l$

$A = \dfrac{1}{2}bh$

$\cos\dfrac{\theta}{2} = \dfrac{h}{0.5} \implies h = 0.5 \cos\dfrac{\theta}{2}$

$\sin\dfrac{\theta}{2} = \dfrac{b/2}{0.5} \implies \dfrac{b}{2} = 0.5 \sin\dfrac{\theta}{2}$

$A = 0.5 \sin\dfrac{\theta}{2}\, 0.5 \cos\dfrac{\theta}{2} = (0.5)^2 \sin\dfrac{\theta}{2}\cos\dfrac{\theta}{2} = 0.25 \sin\dfrac{\theta}{2}\cos\dfrac{\theta}{2}$ square meters

$V = (0.25)(4) \sin\dfrac{\theta}{2}\cos\dfrac{\theta}{2}$ cubic meters $= \sin\dfrac{\theta}{2}\cos\dfrac{\theta}{2}$ cubic meters

119. $6 \sin\dfrac{\pi}{4} \cos\dfrac{\pi}{4} = 6\left[\dfrac{1}{2}\sin\left(\dfrac{\pi}{4} + \dfrac{\pi}{4}\right) + \sin\left(\dfrac{\pi}{4} - \dfrac{\pi}{4}\right)\right] = 3\left(\sin\dfrac{\pi}{2} + \sin 0\right) = 3$

121. $\sin 5\alpha \sin 4\alpha = \dfrac{1}{2}[\cos(5\alpha - 4\alpha) - \cos(5\alpha + 4\alpha)] = \dfrac{1}{2}[\cos \alpha - \cos 9\alpha]$

123. $\cos 5\theta + \cos 4\theta = 2 \cos\left(\dfrac{9\theta}{2}\right)\cos\left(\dfrac{\theta}{2}\right)$

125. $\sin\left(x + \dfrac{\pi}{4}\right) - \sin\left(x - \dfrac{\pi}{4}\right) = 2 \cos x \sin\dfrac{\pi}{4}$

$= \sqrt{2} \cos x$

127. $y = 1.5 \sin 8t - 0.5 \cos 8t$

$a = \dfrac{3}{2}, \; b = -\dfrac{1}{2}, \; B = 8, \; C = \arctan\!\left(-\dfrac{1/2}{3/2}\right)$

$y = \sqrt{\left(\dfrac{3}{2}\right)^2 + \left(\dfrac{1}{2}\right)^2}\, \sin\!\left(8t + \arctan\!\left(-\dfrac{1}{3}\right)\right)$

$y = \dfrac{1}{2}\sqrt{10}\, \sin\!\left(8t - \arctan\dfrac{1}{3}\right)$

129. The amplitude is $\dfrac{\sqrt{10}}{2}$.

131. If $\dfrac{\pi}{2} < \theta < \pi$, then $\cos\dfrac{\theta}{2} < 0$. False, if

$\dfrac{\pi}{2} < \theta < \pi \implies \dfrac{\pi}{4} < \dfrac{\theta}{2} < \dfrac{\pi}{2}$,

which is in Quadrant I $\implies \cos\!\left(\dfrac{\theta}{2}\right) > 0$.

133. $4\sin(-x)\cos(-x) = -2\sin 2x$. True.

$\begin{aligned}
4\sin(-x)\cos(-x) &= 4(-\sin x)(\cos x) \\
&= -4\sin x \cos x \\
&= -2(2\sin x \cos x) \\
&= -2\sin 2x
\end{aligned}$

135. Answers will vary. See page 352.

137. $y_1 = \sec^2\!\left(\dfrac{\pi}{2} - x\right) = \csc^2 x$

$y_2 = \cot^2 x$

$\csc^2 x = \cot^2 x + 1$

Let $y_3 = y_2 + 1 = \cot^2 x + 1 = y_1$.

Chapter 5   Practice Test

1. Find the value of the other five trigonometric functions, given $\tan x = \frac{4}{11}$, $\sec x < 0$.

2. Simplify $\dfrac{\sec^2 x + \csc^2 x}{\csc^2 x(1 + \tan^2 x)}$.

3. Rewrite as a single logarithm and simplify $\ln|\tan \theta| - \ln|\cot \theta|$.

4. True or false: $\cos\left(\dfrac{\pi}{2} - x\right) = \dfrac{1}{\csc x}$

5. Factor and simplify: $\sin^4 x + (\sin^2 x) \cos^2 x$

6. Multiply and simplify: $(\csc x + 1)(\csc x - 1)$

7. Rationalize the denominator and simplify:

   $\dfrac{\cos^2 x}{1 - \sin x}$

8. Verify:

   $\dfrac{1 + \cos \theta}{\sin \theta} + \dfrac{\sin \theta}{1 + \cos \theta} = 2 \csc \theta$

9. Verify:

   $\tan^4 x + 2 \tan^2 x + 1 = \sec^4 x$

10. Use the sum or difference formulas to determine:

    (a) $\sin 105°$ (b) $\tan 15°$

11. Simplify: $(\sin 42°) \cos 38° - (\cos 42°) \sin 38°$

12. Verify: $\tan\left(\theta + \dfrac{\pi}{4}\right) = \dfrac{1 + \tan \theta}{1 - \tan \theta}$

13. Write $\sin(\arcsin x - \arccos x)$ as an algebraic expression in x.

14. Use the double-angle formulas to determine:

    (a) $\cos 120°$ (b) $\tan 300°$

15. Use the half-angle formulas to determine:

    (a) $\sin 22.5°$ (b) $\tan \dfrac{\pi}{12}$

16. Given $\sin \theta = 4/5$, θ lies in Quadrant II, find $\cos \theta/2$.

17. Use the power-reducing identities to write $(\sin^2 x) \cos^2 x$ in terms of the first power of cosine.

18. Rewrite as a sum: $6(\sin 5\theta) \cos 2\theta$

19. Rewrite as a product: $\sin(x + \pi) + \sin(x - \pi)$

20. Verify: $\dfrac{\sin 9x + \sin 5x}{\cos 9x - \cos 5x} = -\cot 2x$

21. Verify: $(\cos u) \sin v = \frac{1}{2}[\sin(u + v) - \sin(u - v)]$

22. Find all solutions in the interval $[0, 2\pi)$:

    $4 \sin^2 x = 1$

23. Find all solutions in the interval $[0, 2\pi)$:

    $\tan^2 \theta + \left(\sqrt{3} - 1\right) \tan \theta - \sqrt{3} = 0$

24. Find all solutions in the interval $[0, 2\pi)$:

    $\sin 2x = \cos x$

25. Use the Quadratic Formula to find all solutions in the interval $[0, 2\pi)$:

    $\tan^2 x - 6 \tan x + 4 = 0$

C H A P T E R 6
Additional Topics in Trigonometry

Section 6.1   Law of Sines 253

Section 6.2   Law of Cosines 257

Section 6.3   Vectors in the Plane 262

Section 6.4   Vectors and Dot Products 269

Section 6.5   Trigonometric Form of a Complex Number 274

Review Exercises 286

Practice Test 296

CHAPTER 6
Additional Topics in Trigonometry

Section 6.1   Law of Sines

- If ABC is any oblique triangle with sides a, b, and c, then the Law of Sines says

$$\frac{a}{\sin A} = \frac{b}{\sin B} = \frac{c}{\sin C}.$$

- You should be able to use the Law of Sines to solve an oblique triangle for the remaining three parts, given:
  - (a) Two angles and any side (AAS or ASA)
  - (b) Two sides and an angle opposite one of them (SSA)
    1. If A is acute and $h = b \sin A$:
       - (a) $a < h$, no triangle is possible.
       - (b) $a = h$ or $a \geq b$, one triangle is possible.
       - (c) $h < a < b$, two triangles are possible.
    2. If A is obtuse and $h = b \sin A$:
       - (a) $a \leq b$, no triangle is possible.
       - (b) $a > b$, one triangle is possible.

- The area of any triangle equals one-half the product of the lengths of two sides times the sine of their included angle.

$$A = \tfrac{1}{2}ab \sin C = \tfrac{1}{2}ac \sin B = \tfrac{1}{2}bc \sin A$$

Vocabulary Check

1. oblique

2. $\dfrac{b}{\sin B}$

3. (a) Two; any; AAS; ASA

   (b) Two; an opposite; SSA

4. $\dfrac{1}{2}bc \sin A$; $\dfrac{1}{2}ab \sin C$; $\dfrac{1}{2}ac \sin B$

---

1. Given: $A = 25°$, $B = 60°$, $a = 12$

   $C = 180° - 25° - 60° = 95°$

   $b = \dfrac{a}{\sin A}(\sin B) = \dfrac{12}{\sin 25°}(\sin 60°) \approx 24.59$ in.

   $c = \dfrac{a}{\sin A}(\sin C) = \dfrac{12}{\sin 25°}(\sin 95°) \approx 28.29$ in.

3. Given: $B = 15°$, $C = 125°$, $c = 20$

   $A = 180° - 15° - 125° = 40°$

   $a = \dfrac{c}{\sin C}(\sin A) = \dfrac{20}{\sin 125°}(\sin 40°) \approx 15.69$ cm

   $b = \dfrac{c}{\sin C}(\sin B) = \dfrac{20}{\sin 125°}(\sin 15°) \approx 6.32$ cm

5. Given: $A = 80° 15'$, $B = 25° 30'$, $b = 2.8$

$C = 180° - 80° 15' - 25° 30' = 74° 15'$

$a = \dfrac{b}{\sin B}(\sin A) = \dfrac{2.8}{\sin 25° 30'}(\sin 80° 15') \approx 6.41 \text{ km}$

$c = \dfrac{b}{\sin B}(\sin C) = \dfrac{2.8}{\sin 25° 30'}(\sin 74° 15') \approx 6.26 \text{ km}$

7. Given: $A = 36°$, $a = 8$, $b = 5$

$\sin B = \dfrac{b \sin A}{a} = \dfrac{5 \sin(36°)}{8} \approx 0.3674 \implies B \approx 21.6°$

$C = 180° - A - B \approx 180° - 36° - 21.6° = 122.4°$

$c = \dfrac{a}{\sin A}(\sin C) = \dfrac{8}{\sin(36°)}\sin(122.4°) \approx 11.49$

9. Given: $A = 102.4°$, $C = 16.7°$, $a = 21.6$

$B = 180° - A - C = 180° - 102.4° - 16.7° = 60.9°$

$b = \dfrac{a}{\sin A}(\sin B) = \dfrac{21.6}{\sin 102.4°}(\sin 60.9°) \approx 19.32$

$c = \dfrac{a}{\sin A}(\sin C) = \dfrac{21.6}{\sin 102.4°}(\sin 16.7°) \approx 6.36$

11. Given: $A = 110° 15'$, $a = 48$, $b = 16$

$\sin B = \dfrac{b \sin A}{a} = \dfrac{16 \sin 110° 15'}{48} \approx 0.31273 \implies B \approx 18° 13'$

$C = 180° - A - B \approx 180° - 110° 15' - 18° 13' = 51° 32'$

$c = \dfrac{a}{\sin A}(\sin C) = \dfrac{48}{\sin 110° 15'}(\sin 51° 32') \approx 40.06$

13. Given: $A = 110°$, $a = 125$, $b = 100$

$\sin B = \dfrac{b \sin A}{a} = \dfrac{100 \sin 110°}{125} \approx 0.75175 \implies B \approx 48.74°$

$C = 180° - A - B \approx 21.26°$

$c = \dfrac{a}{\sin A}(\sin C) = \dfrac{125 \sin 21.26°}{\sin 110°} \approx 48.23$

15. Given: $A = 76°$, $a = 18$, $b = 20$

$\sin B = \dfrac{b \sin A}{a} = \dfrac{20 \sin 76°}{18} \approx 1.078$

No solution

17. Given: $A = 58°, a = 11.4, b = 12.8$

$$\sin B = \frac{b \sin A}{a} = \frac{12.8 \sin 58°}{11.4} \approx 0.9522 \Rightarrow B \approx 72.21° \text{ or } 107.79°$$

Case 1

$B \approx 72.21°$

$C = 180° - 58° - 72.21° = 49.79°$

$c = \dfrac{a}{\sin A}(\sin C) = \dfrac{11.4}{\sin 58°}(\sin 49.79°) \approx 10.27$

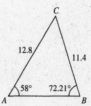

Case 2

$B \approx 107.79°$

$C \approx 180° - 58° - 107.79° = 14.21°$

$c = \dfrac{a}{\sin A}(\sin C) = \dfrac{11.4}{\sin 58°}(\sin 14.21°) \approx 3.30$

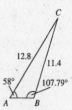

19. Area $= \frac{1}{2}ab \sin C$

$ = \frac{1}{2}(6)(10) \sin(110°)$

$ \approx 28.2$ square units

21. Area $= \frac{1}{2}bc \sin A$

$ = \frac{1}{2}(67)(85) \sin(38° 45')$

$ \approx 1782.3$ square units

23. Area $= \frac{1}{2}ac \sin B$

$ = \frac{1}{2}(103)(58) \sin 75° 15'$

$ \approx 2888.6$ square units

25. Angle $CAB = 70°$

Angle $B = 20° + 14° = 34°$

(a)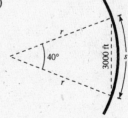

(b) $\dfrac{16}{\sin 70°} = \dfrac{h}{\sin 34°}$

(c) $h = \dfrac{16 \sin 34°}{\sin 70°} \approx 9.52$ meters

27. $\sin A = \dfrac{a \sin B}{b} = \dfrac{500 \sin(46°)}{720} \approx 0.4995$

$A \approx 29.97°$

$\angle ACD = 90° - 29.97° \approx 60°$

Bearing: S 60° W or (240° in plane navigation)

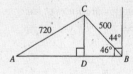

29. (a)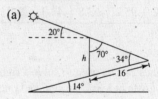

(b) $r = \dfrac{3000 \sin[1/2(180° - 40°)]}{\sin 40°} \approx 4385.71$ feet

(c) $s \approx 40°\left(\dfrac{\pi}{180°}\right)4385.71 \approx 3061.80$ feet

31. $\angle ACD = 65°$

$\angle ADC = 180° - 65° - 15° = 100°$

$\angle CDB = 180° - 100° = 80°$

$\angle B = 180° - 80° - 70° = 30°$

$a = \dfrac{b}{\sin B}(\sin A) = \dfrac{30}{\sin 30°}(\sin 15°) \approx 15.53 \text{ km}$

$c = \dfrac{b}{\sin B}(\sin C) = \dfrac{30}{\sin 30°}(\sin 135°) \approx 42.43 \text{ km}$

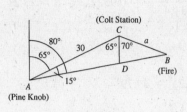

(Colt Station)
(Fire)
(Pine Knob)

33. $\dfrac{\sin(42° - \theta)}{10} = \dfrac{\sin 48°}{17}$

$\sin(42° - \theta) = \dfrac{10}{17}\sin 48° \approx 0.4371$

$42° - \theta \approx 25.919$

$\theta \approx 16.1°$

35. (a) $\sin \alpha = \dfrac{5.45}{58.36} \approx 0.0934$

$\alpha \approx 5.36°$

(b) $\dfrac{d}{\sin \beta} = \dfrac{58.36}{\sin \theta} \implies \sin \beta = \dfrac{d \sin \theta}{58.36}$

$\beta = \sin^{-1}\left[\dfrac{d \sin \theta}{58.36}\right]$

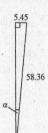

5.45

58.36

(c) $\theta + \beta + 90° + 5.36° = 180° \implies \beta = 84.64° - \theta$

$d = \sin \beta \left(\dfrac{58.36}{\sin \theta}\right) = \sin(84.64° - \theta)\dfrac{58.36}{\sin \theta}$

(d)

θ	10°	20°	30°	40°	50°	60°
d	324.1	154.2	95.2	63.8	43.3	28.1

37. False. If just the three angles are known, the triangle cannot be solved.

39. Yes, the Law of Sines can be used to solve a right triangle if you are given at least one side and one angle, or two sides.

41. Given: $A = 45°, B = 52°, a = 16$

$C = 180° - 45° - 52° = 83°$

$b = \dfrac{a}{\sin A}(\sin B) = \dfrac{16}{\sin 45°}(\sin 52°) \approx 17.83$

$c = \dfrac{a}{\sin A}(\sin C) = \dfrac{16}{\sin 45°}(\sin 83°) \approx 22.46$

$(a + b)\sin\left(\dfrac{C}{2}\right) = c\cos\left(\dfrac{A - B}{2}\right)$

$(16 + 17.83)\sin\left(\dfrac{83°}{2}\right) = 22.46\cos\left(\dfrac{45° - 52°}{2}\right)$

$22.42 = 22.42$

43. $\tan \theta = \dfrac{\sin \theta}{\cos \theta} = -\dfrac{12}{5}$

$\sec \theta = \dfrac{13}{5}$

$\cot \theta = -\dfrac{5}{12}$

$\csc \theta = -\dfrac{13}{12}$

45. $6 \sin 8\theta \cos 3\theta = 6\left(\dfrac{1}{2}\right)[\sin(8\theta + 3\theta) + \sin(8\theta - 3\theta)] = 3(\sin 11\theta + \sin 5\theta)$

47. $3 \cos \dfrac{\pi}{6} \sin \dfrac{5\pi}{3} = 3\left(\dfrac{1}{2}\right)\left[\sin\left(\dfrac{\pi}{6} + \dfrac{5\pi}{3}\right) - \sin\left(\dfrac{\pi}{6} - \dfrac{5\pi}{3}\right)\right]$

$\qquad\qquad\qquad = \dfrac{3}{2}\left[\sin\left(\dfrac{11\pi}{6}\right) - \sin\left(-\dfrac{3\pi}{2}\right)\right]$

$\qquad\qquad\qquad = \dfrac{3}{2}\left[-\dfrac{1}{2} - 1\right] = -\dfrac{9}{4}$

Section 6.2   Law of Cosines

■ If ABC is any oblique triangle with sides a, b, and c, then the Law of Cosines says:

(a) $a^2 = b^2 + c^2 - 2bc \cos A \quad$ or $\quad \cos A = \dfrac{b^2 + c^2 - a^2}{2bc}$

(b) $b^2 = a^2 + c^2 - 2ac \cos B \quad$ or $\quad \cos B = \dfrac{a^2 + c^2 - b^2}{2ac}$

(c) $c^2 = a^2 + b^2 - 2ab \cos C \quad$ or $\quad \cos C = \dfrac{a^2 + b^2 - c^2}{2ab}$

■ You should be able to use the Law of Cosines to solve an oblique triangle for the remaining three parts, given:

(a) Three sides (SSS)

(b) Two sides and their included angle (SAS)

■ Given any triangle with sides of lengths a, b, and c, then the area of the triangle is

$\qquad$ Area $= \sqrt{s(s - a)(s - b)(s - c)}$, where $s = \dfrac{a + b + c}{2}$.   (Heron's Formula)

Vocabulary Check

1. $c^2 = a^2 + b^2 - 2ab \cos C$

2. Heron's Area

3. $\frac{1}{2}bh$, $\sqrt{s(s - a)(s - b)(s - c)}$

1. Given: $a = 12$, $b = 16$, $c = 18$

$\cos A = \dfrac{b^2 + c^2 - a^2}{2bc} = \dfrac{16^2 + 18^2 - 12^2}{2(16)(18)} \approx 0.75694 \implies A \approx 40.80°$

$\sin B = \dfrac{b \sin A}{a} \approx 0.8712 \implies B \approx 60.61°$

$C \approx 180° - 60.61° - 40.80° = 78.59°$

3. Given: $a = 8.5$, $b = 9.2$, $c = 10.8$

$\cos A = \dfrac{b^2 + c^2 - a^2}{2bc} = \dfrac{9.2^2 + 10.8^2 - 8.5^2}{2(9.2)(10.8)} \approx 0.6493 \implies A \approx 49.51°$

$\sin B = \dfrac{b \sin A}{a} \approx \dfrac{9.2 \sin 49.51°}{8.5} \approx 0.82315 \implies B \approx 55.40°$

$C \approx 180° - 55.40° - 49.51° = 75.09°$

5. Given: $a = 10$, $c = 15$, $B = 20°$

$b^2 = a^2 + c^2 - 2ac \cos B = 100 + 225 - 2(10)(15) \cos 20° \approx 43.0922 \implies b \approx 6.56 \text{ mm}$

$\cos A = \dfrac{b^2 + c^2 - a^2}{2bc} \approx \dfrac{43.0922 + 225 - 100}{2(6.56)(15)} \approx 0.8541 \implies A \approx 31.40°$

$C \approx 180° - 20° - 31.40° = 128.60°$

7. Given: $a = 10.4$, $c = 12.5$, $B = 50° \, 30' = 50.5°$

$b^2 = a^2 + c^2 - 2ac \cos B = 10.4^2 + 12.5^2 - 2(10.4)(12.5) \cos 50.5° \approx 99.0297 \implies b \approx 9.95 \text{ ft}$

$\cos A = \dfrac{b^2 + c^2 - a^2}{2bc} \approx \dfrac{99.0297 + 12.5^2 - 10.4^2}{2(9.95)(12.5)} \approx 0.5914 \implies A \approx 53.75° = 53° \, 45'$

$C \approx 180° - 50.5° - 53.75° = 75.75° = 75° \, 45'$

9. Given: $a = 6$, $b = 8$, $c = 12$

$\cos A = \dfrac{b^2 + c^2 - a^2}{2bc} = \dfrac{64 + 144 - 36}{2(8)(12)} \approx 0.8958 \implies A \approx 26.4°$

$\sin B = \dfrac{b \sin A}{a} \approx \dfrac{8 \sin 26.4°}{6} \approx 0.5928 \implies B \approx 36.3°$

$C \approx 180° - 26.4° - 36.3° = 117.3°$

11. Given: $A = 50°$, $b = 15$, $c = 30$

$a^2 = b^2 + c^2 - 2bc \cos A = 225 + 900 - 2(15)(30) \cos 50° \approx 546.49 \implies a \approx 23.38$

$\cos B = \dfrac{a^2 + c^2 - b^2}{2ac} \approx \dfrac{546.49 + 900 - 225}{2(23.4)(30)} \approx 0.8708 \implies B \approx 29.4°$

$C = 180° - A - B \approx 180° - 50° - 29.5° = 100.6°$

13. Given: $a = 9$, $b = 12$, $c = 15$

$\cos C = \dfrac{a^2 + b^2 - c^2}{2ab} = \dfrac{81 + 144 - 225}{2(9)(12)} = 0 \implies C = 90°$

$\sin A = \dfrac{9}{15} = \dfrac{3}{5} \implies A \approx 36.9°$

$B \approx 180° - 90° - 36.9° = 53.1°$

15. Given: $a = 75.4$, $b = 48$, $c = 48$

$\cos A = \dfrac{b^2 + c^2 - a^2}{2bc} = \dfrac{48^2 + 48^2 - 75.4^2}{2(48)(48)} \approx -0.2338 \implies A \approx 103.5°$

$\sin B = \dfrac{b \sin A}{a} \approx \dfrac{48 \sin (103.5°)}{75.4} \approx 0.6190 \implies B \approx 38.2°$

$C = B \approx 38.2°$ (Because of roundoff error, $A + B + C \ne 180°$.)

17. Given: $B = 8° 15' = 8.25°$, $a = 26$, $c = 18$

$b^2 = a^2 + c^2 - 2ac \cos B = 26^2 + 18^2 - 2(26)(18) \cos(8.25°) \approx 73.6863 \implies b \approx 8.58$

$\sin C = \dfrac{c \sin B}{b} \approx \dfrac{18 \sin(8.25°)}{8.58} \approx 0.3 \implies C \approx 17.51° \approx 17° 31'$

$A = 180° - B - C \approx 180° - 8.25° - 17.51° = 154.24° \approx 154° 14'$

19. Given: $B = 75° 20' \approx 75.33°$, $a = 6.2$, $c = 9.5$

$b^2 = a^2 + c^2 - 2ac \cos B$

$\quad = 6.2^2 + 9.5^2 - 2(6.2)(9.5) \cos(75.33°)$

$\quad = 98.86$

$b \approx 9.94$

$\sin C = \dfrac{c \sin B}{b} = \dfrac{9.5 \sin(75.33°)}{9.94}$

$\quad \approx 0.9246 \implies C \approx 67.6°$

$A = 180° - B - C \approx 37.1°$

21. $d^2 = 4^2 + 8^2 - 2(4)(8) \cos 30°$

$\quad \approx 24.57 \implies d \approx 4.96$

$2\phi = 360° - 2\theta \implies \phi = 150°$

$c^2 = 4^2 + 8^2 - 2(4)(8) \cos 150° \approx 135.43$

$c \approx 11.64$

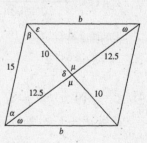

23. $\cos \phi = \dfrac{10^2 + 14^2 - 20^2}{2(10)(14)}$

$\quad \phi \approx 111.8°$

$2\theta \approx 360° - 2(111.80°)$

$\quad \theta = 68.2°$

$d^2 = 10^2 + 14^2 - 2(10)(14) \cos 68.2°$

$d \approx 13.86$

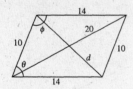

25. $\cos \alpha = \dfrac{15^2 + 12.5^2 - 10^2}{2(15)(12.5)} = 0.75 \implies \alpha \approx 41.41°$

$\cos \beta = \dfrac{15^2 + 10^2 - 12.5^2}{2(15)(10)} = 0.5625 \implies \beta \approx 55.77°$

$\delta = 180° - 41.41° - 55.77° \approx 82.82°$

$\mu = 180° - \delta \approx 97.18°$

$b^2 = 12.5^2 + 10^2 - 2(12.5)(10) \cos(97.18°) \approx 287.50$

$b \approx 16.96$

$\sin \omega = \dfrac{10}{16.96} \sin \mu \approx 0.585 \implies \omega \approx 35.8°$

$\sin \epsilon = \dfrac{12.5}{16.96} \sin \mu \approx 0.731 \implies \epsilon \approx 47°$

$\theta = \alpha + \omega \approx 77.2°$

$\phi = \beta + \epsilon \approx 102.8°$

27. Given: $a = 12$, $b = 24$, $c = 18$

$s = \dfrac{a + b + c}{2} = 27$

Area $= \sqrt{s(s - a)(s - b)(s - c)}$

$\quad = \sqrt{27(15)(3)(9)}$

$\quad \approx 104.57$ square inches

29. Given: $a = 5$, $b = 8$, $c = 10$

$s = \dfrac{a + b + c}{2} = \dfrac{23}{2} = 11.5$

Area $= \sqrt{s(s - a)(s - b)(s - c)}$

$\quad = \sqrt{11.5(6.5)(3.5)(1.5)}$

$\quad \approx 19.81$ square units

31. Given: $a = 1.24$, $b = 2.45$, $c = 1.25$

$$s = \frac{a + b + c}{2} = 2.47$$

$$\text{Area} = \sqrt{s(s - a)(s - b)(s - c)}$$

$$= \sqrt{2.47(1.23)(0.02)(1.22)}$$

$$\approx 0.27 \text{ square feet}$$

33. Given: $a = 3.5$, $b = 10.2$, $c = 9$

$$s = \frac{a + b + c}{2} = 11.35$$

$$\text{Area} = \sqrt{s(s - a)(s - b)(s - c)}$$

$$= \sqrt{11.35(7.85)(1.15)(2.35)}$$

$$\approx 15.52 \text{ square units}$$

35. Given: $a = 10.59$, $b = 6.65$, $c = 12.31$

$$s = \frac{a + b + c}{2} = 14.775$$

$$\text{Area} = \sqrt{s(s - a)(s - b)(s - c)}$$

$$= \sqrt{14.775(4.185)(8.125)(2.465)}$$

$$\approx 35.19 \text{ square units}$$

37. $B = 105° + 32° = 137°$

$$b^2 = a^2 + c^2 - 2ac \cdot \cos B$$

$$= 648^2 + 810^2 - 2(648)(810)\cos(137°)$$

$$= 1{,}843{,}749.862$$

$$b = 1357.8 \text{ miles}$$

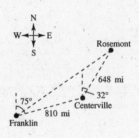

From the Law of Sines:

$$\frac{a}{\sin A} = \frac{b}{\sin B} \implies \sin A = \frac{a}{b}\sin B = \frac{648}{1357.8}\sin(137°) \approx 0.32548$$

$$\implies A \approx 19° \implies \text{Bearing S 56° W (or 236° for airplane navigation)}$$

39. $\cos B = \dfrac{1500^2 + 3600^2 - 2800^2}{2(1500)(3600)} \approx 0.6824 \implies B \approx 46.97°$

Bearing at B: $90° - 46.97°$: N 43.03° E

$$\cos C = \frac{1500^2 + 2800^2 - 3600^2}{2(1500)(2800)} \approx -0.3417 \implies C \approx 109.98°$$

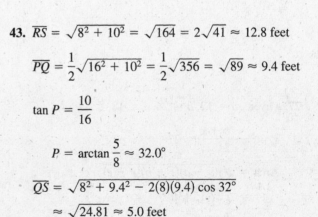

Bearing at C: $C - (90° - 46.97°) = 66.95°$

S 66.95° E

41. $C = 180° - 53° - 67° = 60°$

$$c^2 = a^2 + b^2 - 2ab\cos C$$

$$= 36^2 + 48^2 - 2(36)(48)(0.5) = 1872$$

$$c \approx 43.3 \text{ mi}$$

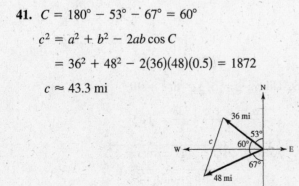

43. $\overline{RS} = \sqrt{8^2 + 10^2} = \sqrt{164} = 2\sqrt{41} \approx 12.8$ feet

$$\overline{PQ} = \frac{1}{2}\sqrt{16^2 + 10^2} = \frac{1}{2}\sqrt{356} = \sqrt{89} \approx 9.4 \text{ feet}$$

$$\tan P = \frac{10}{16}$$

$$P = \arctan\frac{5}{8} \approx 32.0°$$

$$\overline{QS} = \sqrt{8^2 + 9.4^2 - 2(8)(9.4)\cos 32°}$$

$$\approx \sqrt{24.81} \approx 5.0 \text{ feet}$$

45. $s = \dfrac{a + b + c}{2} = \dfrac{145 + 257 + 290}{2} = 346$

Area $= \sqrt{s(s - a)(s - b)(s - c)} = \sqrt{346(201)(89)(56)} \approx 18{,}617.7$ square feet

47. (a) $7^2 = 1.5^2 + x^2 - 2(1.5)(x) \cos \theta$

$49 = 2.25 + x^2 - 3x \cos \theta$

(b) $\qquad x^2 - 3x \cos \theta = 46.75$

$x^2 - 3x \cos \theta + \left(\dfrac{3 \cos \theta}{2}\right)^2 = 46.75 + \left(\dfrac{3 \cos \theta}{2}\right)^2$

$\left[x - \dfrac{3 \cos \theta}{2}\right]^2 = \dfrac{187}{4} + \dfrac{9 \cos^2 \theta}{4}$

$x - \dfrac{3 \cos \theta}{2} = \pm \sqrt{\dfrac{187 + 9 \cos^2 \theta}{4}}$

(c)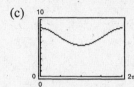

(d) Note that $x = 8.5$ when $\theta = 0$ and $\theta = 2\pi$, and $x = 5.5$ when $\theta = \pi$. Thus, the distance is

$2(8.5 - 5.5) = 2(3) = 6$ inches.

Choosing the positive values of x, we have

$x = \dfrac{1}{2}\left(3 \cos \theta + \sqrt{9 \cos^2 \theta + 187}\right).$

49. False. This is not a triangle! $5 + 10 < 16$

51. False. $s = \dfrac{a + b + c}{2}$, not $\dfrac{a + b + c}{3}$.

53. $\dfrac{1}{2}bc(1 - \cos A) = \dfrac{1}{2}bc\left[1 - \dfrac{b^2 + c^2 - a^2}{2bc}\right]$

$= \dfrac{1}{2}bc\left[\dfrac{2bc - b^2 - c^2 + a^2}{2bc}\right]$

$= \dfrac{1}{4}[a^2 - (b - c)^2]$

$= \dfrac{1}{4}[a - b + c)(a + b - c)]$

$= \left(\dfrac{a - b + c}{2}\right)\left(\dfrac{a + b - c}{2}\right)$

55. Since $0 < C < 180°$, $\cos\left(\dfrac{C}{2}\right) = \sqrt{\dfrac{1 + \cos C}{2}}.$

Hence, $\cos\left(\dfrac{C}{2}\right) = \sqrt{\dfrac{1 + (a^2 + b^2 - c^2)/(2ab)}{2}} = \sqrt{\dfrac{2ab + a^2 + b^2 - c^2}{4ab}}.$

On the other hand,

$s(s - c) = \dfrac{1}{2}(a + b + c)\left(\dfrac{1}{2}(a + b + c) - c\right)$

$= \dfrac{1}{2}(a + b + c)\dfrac{1}{2}(a + b - c)$

$= \dfrac{1}{4}((a + b)^2 - c^2)$

$= \dfrac{1}{4}(a^2 + b^2 + 2ab - c^2).$

Thus, $\sqrt{\dfrac{s(s - c)}{ab}} = \sqrt{\dfrac{a^2 + b^2 + 2ab - c^2}{4ab}}$ and we have verified that $\cos\left(\dfrac{C}{2}\right) = \sqrt{\dfrac{s(s - c)}{ab}}.$

57. Given: $a = 12$, $b = 30$, $A = 20°$

$a^2 = b^2 + c^2 - 2bc \cos A$

$12^2 = 30^2 + c^2 - 2(30)(c) \cos 20°$

$c^2 - (60 \cos 20°)c + 756 = 0$

Solving this quadratic equation, $c \approx 21.97, 34.41$.

For $c = 21.97$,

$$\cos B = \frac{a^2 + c^2 - b^2}{2ac} \approx \frac{12^2 + 21.97^2 - 30^2}{2(12)(21.97)} \approx -0.5184 \implies B \approx 121.2°$$

$C \approx 180° - 121.2° - 20° = 38.8°$

For $c = 34.41$,

$$\cos B = \frac{a^2 + c^2 - b^2}{2ac} \approx \frac{12^2 + 34.41^2 - 30^2}{2(12)(34.41)} \approx 0.5183 \implies B \approx 58.8°$$

$C \approx 180° - 58.8° - 20° = 101.2°$.

Using the Law of Sines, $\sin B = \dfrac{b \sin A}{a} = \dfrac{30 \sin 20°}{12} \approx 0.8551 \implies B \approx 58.8°$ or $121.2°$.

For $B = 58.8°$, $C = 180° - 58.8° - 20° = 101.2°$ and $c = \dfrac{a \sin C}{\sin A} \approx 34.42$.

For $B = 121.2°$, $C = 180° - 121.2° - 20° = 38.8°$ and $c = \dfrac{a \sin C}{\sin A} \approx 21.98$.

59. $\arcsin(-1) = -\dfrac{\pi}{2}$ because $\sin\left(-\dfrac{\pi}{2}\right) = -1$.

61. $\tan^{-1}\left(\sqrt{3}\right) = \dfrac{\pi}{3}$ because $\tan\left(\dfrac{\pi}{3}\right) = \sqrt{3}$.

Section 6.3 Vectors in the Plane

■ A vector **v** is the collection of all directed line segments that are equivalent to a given directed line segment $\overrightarrow{PQ}$.

■ You should be able to *geometrically* perform the operations of vector addition and scalar multiplication.

■ The component form of the vector with initial point $P = (p_1, p_2)$ and terminal point $Q = (q_1, q_2)$ is

$$\overrightarrow{PQ} = \langle q_1 - p_1, q_2 - p_2 \rangle = \langle v_1, v_2 \rangle = \mathbf{v}.$$

■ The magnitude of $\mathbf{v} = \langle v_1, v_2 \rangle$ is given by $\|\mathbf{v}\| = \sqrt{v_1{}^2 + v_2{}^2}$.

■ You should be able to perform the operations of scalar multiplication and vector addition in component form.

■ You should know the following properties of vector addition and scalar multiplication.

   (a) $\mathbf{u} + \mathbf{v} = \mathbf{v} + \mathbf{u}$ (b) $(\mathbf{u} + \mathbf{v}) + \mathbf{w} = \mathbf{u} + (\mathbf{v} + \mathbf{w})$

   (c) $\mathbf{u} + \mathbf{0} = \mathbf{u}$ (d) $\mathbf{u} + (-\mathbf{u}) = \mathbf{0}$

   (e) $c(d\mathbf{u}) = (cd)\mathbf{u}$ (f) $(c + d)\mathbf{u} = c\mathbf{u} + d\mathbf{u}$

   (g) $c(\mathbf{u} + \mathbf{v}) = c\mathbf{u} + c\mathbf{v}$ (h) $1(\mathbf{u}) = \mathbf{u}, 0\mathbf{u} = \mathbf{0}$

   (i) $\|c\mathbf{v}\| = |c| \, \|\mathbf{v}\|$

—CONTINUED—

Section 6.3 —CONTINUED—

■ A unit vector in the direction of $\mathbf{v}$ is given by $\mathbf{u} = \dfrac{\mathbf{v}}{\|\mathbf{v}\|}$.

■ The standard unit vectors are $\mathbf{i} = \langle 1, 0 \rangle$ and $\mathbf{j} = \langle 0, 1 \rangle$. $\mathbf{v} = \langle v_1, v_2 \rangle$ can be written as $\mathbf{v} = v_1 \mathbf{i} + v_2 \mathbf{j}$.

■ A vector $\mathbf{v}$ with magnitude $\|\mathbf{v}\|$ and direction θ can be written as $\mathbf{v} = a\mathbf{i} + b\mathbf{j} = \|\mathbf{v}\|(\cos\theta)\mathbf{i} + \|\mathbf{v}\|(\sin\theta)\mathbf{j}$ where $\tan\theta = b/a$.

Vocabulary Check

1. directed line segment

2. initial, terminal

3. magnitude

4. vector

5. standard position

6. unit vector

7. multiplication, addition

8. resultant

9. linear combination, horizontal, vertical

---

1. $\mathbf{u} = \langle 6 - 2, 5 - 4 \rangle = \langle 4, 1 \rangle = \mathbf{v}$

3. Initial point: $(0, 0)$

Terminal point: $(4, 3)$

$$\mathbf{v} = \langle 4 - 0, 3 - 0 \rangle = \langle 4, 3 \rangle$$

$$\|\mathbf{v}\| = \sqrt{4^2 + 3^2} = 5$$

5. Initial point: $(2, 2)$

Terminal point: $(-1, 4)$

$$\mathbf{v} = \langle -1 - 2, 4 - 2 \rangle = \langle -3, 2 \rangle$$

$$\|\mathbf{v}\| = \sqrt{(-3)^2 + 2^2} = \sqrt{13} \approx 3.61$$

7. Initial point: $(3, -2)$

Terminal point: $(3, 3)$

$$\mathbf{v} = \langle 3 - 3, 3 - (-2) \rangle = \langle 0, 5 \rangle$$

$$\|\mathbf{v}\| = 5$$

9. Initial point: $\left(\dfrac{2}{5}, 1 \right)$

Terminal point: $\left(1, \dfrac{2}{5} \right)$

$$\mathbf{v} = \left\langle 1 - \dfrac{2}{5}, \dfrac{2}{5} - 1 \right\rangle = \left\langle \dfrac{3}{5}, -\dfrac{3}{5} \right\rangle$$

$$\|\mathbf{v}\| = \sqrt{ \left(\dfrac{3}{5} \right)^2 + \left(-\dfrac{3}{5} \right)^2 } = \sqrt{ \dfrac{18}{25} } = \dfrac{3}{5}\sqrt{2}$$

11. Initial point: $\left(\dfrac{-2}{3}, -1 \right)$

Terminal point: $\left(\dfrac{1}{2}, \dfrac{4}{5} \right)$

$$\mathbf{v} = \left\langle \dfrac{1}{2} - \left(-\dfrac{2}{3} \right), \dfrac{4}{5} - (-1) \right\rangle = \left\langle \dfrac{7}{6}, \dfrac{9}{5} \right\rangle$$

$$\|\mathbf{v}\| = \sqrt{ \left(\dfrac{7}{6} \right)^2 + \left(\dfrac{9}{5} \right)^2 } = \dfrac{\sqrt{4141}}{30} \approx 2.1450$$

13. $-\mathbf{v}$

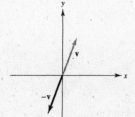

15. $\mathbf{u} + \mathbf{v}$

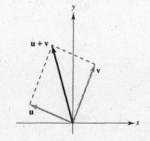

17. $\mathbf{u} + 2\mathbf{v}$

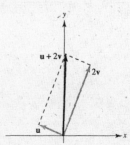

19.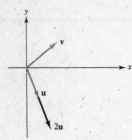

21.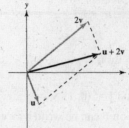

23.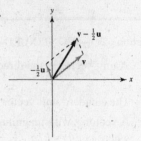

25. $\mathbf{u} = \langle 4, 2 \rangle$, $\mathbf{v} = \langle 7, 1 \rangle$

(a) $\mathbf{u} + \mathbf{v} = \langle 11, 3 \rangle$

(b) $\mathbf{u} - \mathbf{v} = \langle -3, 1 \rangle$

(c) $2\mathbf{u} - 3\mathbf{v} = \langle 8, 4 \rangle - \langle 21, 3 \rangle = \langle -13, 1 \rangle$

(d) $\mathbf{v} + 4\mathbf{u} = \langle 7, 1 \rangle + \langle 16, 8 \rangle = \langle 23, 9 \rangle$

27. $\mathbf{u} = \langle -6, -8 \rangle$, $\mathbf{v} = \langle 2, 4 \rangle$

(a) $\mathbf{u} + \mathbf{v} = \langle -4, -4 \rangle$

(b) $\mathbf{u} - \mathbf{v} = \langle -8, -12 \rangle$

(c) $2\mathbf{u} - 3\mathbf{v} = \langle -12, -16 \rangle - \langle 6, 12 \rangle = \langle -18, -28 \rangle$

(d) $\mathbf{v} + 4\mathbf{u} = \langle 2, 4 \rangle + 4\langle -6, -8 \rangle = \langle -22, -28 \rangle$

29. $\mathbf{u} = \mathbf{i} + \mathbf{j}$, $\mathbf{v} = 2\mathbf{i} - 3\mathbf{j}$

(a) $\mathbf{u} + \mathbf{v} = 3\mathbf{i} - 2\mathbf{j}$

(b) $\mathbf{u} - \mathbf{v} = -\mathbf{i} + 4\mathbf{j}$

(c) $2\mathbf{u} - 3\mathbf{v} = (2\mathbf{i} + 2\mathbf{j}) - (6\mathbf{i} - 9\mathbf{j}) = -4\mathbf{i} + 11\mathbf{j}$

(d) $\mathbf{v} + 4\mathbf{u} = (2\mathbf{i} - 3\mathbf{j}) + (4\mathbf{i} + 4\mathbf{j}) = 6\mathbf{i} + \mathbf{j}$

31. $\mathbf{w} = \mathbf{u} + \mathbf{v}$

33. $\mathbf{u} = \mathbf{w} - \mathbf{v}$

35. $\|\langle 6, 0 \rangle\| = 6$

Unit vector: $\frac{1}{6}\langle 6, 0 \rangle = \langle 1, 0 \rangle$

37. $\mathbf{v} = \langle -1, 1 \rangle$

$\|\mathbf{v}\| = \sqrt{2}$

Unit vector $= \frac{1}{\|\mathbf{v}\|}\mathbf{v} = \left\langle -\frac{1}{\sqrt{2}}, \frac{1}{\sqrt{2}} \right\rangle = \left\langle -\frac{\sqrt{2}}{2}, \frac{\sqrt{2}}{2} \right\rangle$

39. $\|\mathbf{v}\| = \|\langle -24, -7 \rangle\| = \sqrt{(-24)^2 + (-7)^2} = 25$

Unit vector: $\frac{1}{25}\langle -24, -7 \rangle = \left\langle -\frac{24}{25}, -\frac{7}{25} \right\rangle$

41. $\mathbf{u} = \frac{1}{\|\mathbf{v}\|}\mathbf{v} = \frac{1}{\sqrt{16 + 9}}(4\mathbf{i} - 3\mathbf{j}) = \frac{1}{5}(4\mathbf{i} - 3\mathbf{j}) = \frac{4}{5}\mathbf{i} - \frac{3}{5}\mathbf{j}$

43. $\mathbf{u} = \frac{1}{2}(2\mathbf{j}) = \mathbf{j}$

45. $8\left(\frac{1}{\|\mathbf{u}\|}\mathbf{u}\right) = 8\left(\frac{1}{\sqrt{5^2 + 6^2}}\langle 5, 6 \rangle\right) = \frac{8}{\sqrt{61}}\langle 5, 6 \rangle = \left\langle \frac{40\sqrt{61}}{61}, \frac{48\sqrt{61}}{61} \right\rangle$

47. $7\left(\dfrac{1}{\|\mathbf{u}\|}\mathbf{u}\right) = 7\left(\dfrac{1}{\sqrt{3^2 + 4^2}}\langle 3, 4\rangle\right)$

$\qquad\qquad = \dfrac{7}{5}\langle 3, 4\rangle$

$\qquad\qquad = \left\langle \dfrac{21}{5}, \dfrac{28}{5}\right\rangle$

$\qquad\qquad = \dfrac{21}{5}\mathbf{i} + \dfrac{28}{5}\mathbf{j}$

49. $8\left(\dfrac{1}{\|\mathbf{u}\|}\mathbf{u}\right) = 8\left(\dfrac{1}{2}\langle -2, 0\rangle\right)$

$\qquad\qquad = 4\langle -2, 0\rangle$

$\qquad\qquad = \langle -8, 0\rangle$

$\qquad\qquad = -8\mathbf{i}$

51. $\mathbf{v} = \langle 4 - (-3), 5 - 1\rangle = \langle 7, 4\rangle = 7\mathbf{i} + 4\mathbf{j}$

53. $\mathbf{v} = \langle 2 - (-1), 3 - (-5)\rangle = \langle 3, 8\rangle = 3\mathbf{i} + 8\mathbf{j}$

55. $\mathbf{v} = \dfrac{3}{2}\mathbf{u}$

$\qquad = \dfrac{3}{2}(2\mathbf{i} - \mathbf{j}) = 3\mathbf{i} - \dfrac{3}{2}\mathbf{j}$

$\qquad = \left\langle 3, -\dfrac{3}{2}\right\rangle$

57. $\mathbf{v} = \mathbf{u} + 2\mathbf{w}$

$\qquad = (2\mathbf{i} - \mathbf{j}) + 2(\mathbf{i} + 2\mathbf{j})$

$\qquad = 4\mathbf{i} + 3\mathbf{j} = \langle 4, 3\rangle$

59. $\mathbf{v} = \dfrac{1}{2}(3\mathbf{u} + \mathbf{w})$

$\qquad = \dfrac{1}{2}(3\langle 2, -1\rangle + \langle 1, 2\rangle)$

$\qquad = \left\langle \dfrac{7}{2}, -\dfrac{1}{2}\right\rangle$

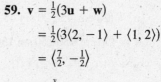

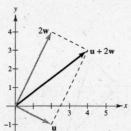

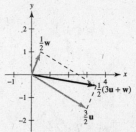

61. $\mathbf{v} = 5(\cos 30°\mathbf{i} + \sin 30°\mathbf{j})$

$\qquad \|\mathbf{v}\| = 5, \ \theta = 30°$

63. $\mathbf{v} = 6\mathbf{i} - 6\mathbf{j}$

$\qquad \|\mathbf{v}\| = \sqrt{6^2 + (-6)^2} = \sqrt{72} = 6\sqrt{2}$

$\qquad \tan\theta = -\dfrac{6}{6} = -1$

Since $\mathbf{v}$ lies in Quadrant IV, $\theta = 315°$.

65. $\mathbf{v} = -2\mathbf{i} + 5\mathbf{j}$

$\qquad \|\mathbf{v}\| = \sqrt{(-2)^2 + 5^2} = \sqrt{29}$

$\qquad \tan\theta = -\dfrac{5}{2}$

Since $\mathbf{v}$ lies in Quadrant II, $\theta \approx 111.8°$.

67. $\mathbf{v} = \langle 3\cos 0°, 3\sin 0°\rangle$

$\qquad = \langle 3, 0\rangle$

69. $\mathbf{v} = \left\langle 3\sqrt{2}\cos 150°, 3\sqrt{2}\sin 150°\right\rangle$

$\qquad = \left\langle -\dfrac{3\sqrt{6}}{2}, \dfrac{3\sqrt{2}}{2}\right\rangle$

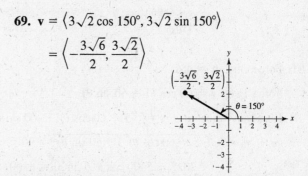

71. $\mathbf{v} = 2\left(\dfrac{1}{\sqrt{3^2 + 1^2}}\right)(\mathbf{i} + 3\mathbf{j})$

$$= \dfrac{2}{\sqrt{10}}(\mathbf{i} + 3\mathbf{j})$$

$$= \dfrac{\sqrt{10}}{5}\mathbf{i} + \dfrac{3\sqrt{10}}{5}\mathbf{j}$$

$$= \left\langle \dfrac{\sqrt{10}}{5}, \dfrac{3\sqrt{10}}{5} \right\rangle$$

$\tan \theta = \dfrac{3}{1} \implies \theta \approx 71.57°$

73. $\mathbf{u} = \langle 5 \cos 60°, 5 \sin 60° \rangle = \left\langle \dfrac{5}{2}, \dfrac{5\sqrt{3}}{2} \right\rangle$

$\mathbf{v} = \langle 5 \cos 90°, 5 \sin 90° \rangle = \langle 0, 5 \rangle$

$\mathbf{u} + \mathbf{v} = \left\langle \dfrac{5}{2}, \dfrac{5\sqrt{3}}{2} \right\rangle + \langle 0, 5 \rangle = \left\langle \dfrac{5}{2}, 5 + \dfrac{5}{2}\sqrt{3} \right\rangle$

75. $\mathbf{u} = \langle 20 \cos 45°, 20 \sin 45° \rangle = \left\langle 10\sqrt{2}, 10\sqrt{2} \right\rangle$

$\mathbf{v} = \langle 50 \cos 150°, 50 \sin 150° \rangle = \left\langle -25\sqrt{3}, 25 \right\rangle$

$\mathbf{u} + \mathbf{v} = \left\langle 10\sqrt{2} - 25\sqrt{3}, 10\sqrt{2} + 25 \right\rangle$

77. $\mathbf{v} = \mathbf{i} + \mathbf{j}$

$\mathbf{w} = 2(\mathbf{i} - \mathbf{j})$

$\mathbf{u} = \mathbf{v} - \mathbf{w} = -\mathbf{i} + 3\mathbf{j}$

$\|\mathbf{v}\| = \sqrt{2}$

$\|\mathbf{w}\| = 2\sqrt{2}$

$\|\mathbf{v} - \mathbf{w}\| = \sqrt{10}$

$\cos \alpha = \dfrac{\|\mathbf{v}\|^2 + \|\mathbf{w}\|^2 - \|\mathbf{v} - \mathbf{w}\|^2}{2\|\mathbf{v}\|\,\|\mathbf{w}\|} = \dfrac{2 + 8 - 10}{2\sqrt{2} \cdot 2\sqrt{2}} = 0$

$\alpha = 90°$

79. $\mathbf{u} = 400 \cos 25°\mathbf{i} + 400 \sin 25°\mathbf{j}$

$\mathbf{v} = 300 \cos 70°\mathbf{i} + 300 \sin 70°\mathbf{j}$

$\mathbf{u} + \mathbf{v} \approx 465.13\mathbf{i} + 450.96\mathbf{j}$

$\|\mathbf{u} + \mathbf{v}\| \approx \sqrt{(465.13)^2 + (450.96)^2} \approx 647.85$

$\alpha = \arctan\left(\dfrac{450.96}{465.13}\right) \approx 44.11°$

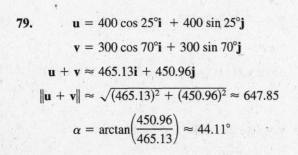

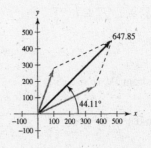

81. Force One: $\mathbf{u} = 45\mathbf{i}$

Force Two: $\mathbf{v} = 60 \cos \theta\mathbf{i} + 60 \sin \theta\mathbf{j}$

Resultant Force: $\mathbf{u} + \mathbf{v} = (45 + 60 \cos \theta)\mathbf{i} + 60 \sin \theta\mathbf{j}$

$\|\mathbf{u} + \mathbf{v}\| = \sqrt{(45 + 60 \cos \theta)^2 + (60 \sin \theta)^2} = 90$

$2025 + 5400 \cos \theta + 3600 = 8100$

$5400 \cos \theta = 2475$

$\cos \theta = \dfrac{2475}{5400} \approx 0.4583$

$\theta \approx 62.7°$

83. Horizontal component of velocity: $70 \cos 40° \approx 53.62$ ft/sec

Vertical component of velocity: $70 \sin 40° \approx 45.0$ ft/sec

85. Rope $\overrightarrow{AC}$: $\mathbf{u} = 10\mathbf{i} - 24\mathbf{j}$

The vector lies in Quadrant IV and its reference angle is $\arctan\left(\frac{12}{5}\right)$.

$\mathbf{u} = \|\mathbf{u}\|\left[\cos\left(\arctan\frac{12}{5}\right)\mathbf{i} - \sin\left(\arctan\frac{12}{5}\right)\mathbf{j}\right]$

Rope $\overrightarrow{BC}$: $\mathbf{v} = -20\mathbf{i} - 24\mathbf{j}$

The vector lies in Quadrant III and its reference angle is $\arctan\left(\frac{6}{5}\right)$.

$\mathbf{v} = \|\mathbf{v}\|\left[-\cos\left(\arctan\frac{6}{5}\right)\mathbf{i} - \sin\left(\arctan\frac{6}{5}\right)\mathbf{j}\right]$

Resultant: $\mathbf{u} + \mathbf{v} = -5000\mathbf{j}$

$\|\mathbf{u}\| \cos\left(\arctan\frac{12}{5}\right) - \|\mathbf{v}\| \cos\left(\arctan\frac{6}{5}\right) = 0$

$-\|\mathbf{u}\| \sin\left(\arctan\frac{12}{5}\right) - \|\mathbf{v}\| \sin\left(\arctan\frac{6}{5}\right) = -5000$

Solving this system of equations yields:

$T_{AC} = \|\mathbf{u}\| \approx 3611.1$ pounds

$T_{BC} = \|\mathbf{v}\| \approx 2169.5$ pounds

87. (a) Tow line 1: $\mathbf{u} = \|\mathbf{u}\|(\cos\theta\mathbf{i} + \sin\theta\mathbf{j})$

Tow line 2: $\mathbf{v} = \|\mathbf{u}\|(\cos(-\theta)\mathbf{i} + \sin(-\theta)\mathbf{j})$

Resultant: $\mathbf{u} + \mathbf{v} = 6000\mathbf{i} = [\|\mathbf{u}\|\cos\theta + \|\mathbf{u}\|\cos(-\theta)]\mathbf{i} \implies$

$6000 = 2\|\mathbf{u}\|\cos\theta \implies \|\mathbf{u}\| \approx 3000\sec\theta$

$T = \|\mathbf{u}\| = 3000\sec\theta$

Domain: $0° \le \theta < 90°$

(b)

θ	10°	20°	30°	40°	50°	60°
T	3046.3	3192.5	3464.1	3916.2	4667.2	6000.0

(c)

(d) The tension increases because the component in the direction of the motion of the barge decreases.

89. Airspeed: $\mathbf{v} = 860(\cos 302°\mathbf{i} + \sin 302°\mathbf{j})$

Groundspeed: $\mathbf{u} = 800(\cos 310°\mathbf{i} + \sin 310°\mathbf{j})$

$\mathbf{w} + \mathbf{v} = \mathbf{u}$

$\mathbf{w} = \mathbf{u} - \mathbf{v} = 800(\cos 310°\mathbf{i} + \sin 310°\mathbf{j}) - 860(\cos 302°\mathbf{i} + \sin 302°\mathbf{j}) \approx 58.50\mathbf{i} + 116.49\mathbf{j}$

$\|\mathbf{w}\| = \sqrt{58.50^2 + 116.49^2} \approx 130.35$ km/hr

$\theta = \arctan\left(\frac{116.49}{58.50}\right) \approx 63.3°$

Direction: N 26.7° E

91. (a) $\mathbf{u} = 220\mathbf{i}$, $\mathbf{v} = 150\cos 30°\mathbf{i} + 150\sin 30°\mathbf{j}$

$\mathbf{u} + \mathbf{v} = \left(220 + 75\sqrt{3}\right)\mathbf{i} + 75\mathbf{j}$

$\|\mathbf{u} + \mathbf{v}\| = \sqrt{\left(220 + 75\sqrt{3}\right)^2 + 75^2} \approx 357.85$ newtons

$\tan\theta = \dfrac{75}{220 + 75\sqrt{3}} \implies \theta \approx 12.1°$

(b) $\mathbf{u} + \mathbf{v} = 220\mathbf{i} + (150\cos\theta\mathbf{i} + 150\sin\theta\mathbf{j})$

$M = \|\mathbf{u} + \mathbf{v}\| = \sqrt{220^2 + 150^2(\cos^2\theta + \sin^2\theta) + 2(220)(150)\cos\theta}$

$= \sqrt{70{,}900 + 66{,}000\cos\theta} = 10\sqrt{709 + 660\cos\theta}$

$\alpha = \arctan\left(\dfrac{15\sin\theta}{22 + 15\cos\theta}\right)$

(c)

θ	0°	30°	60°	90°	120°	150°	180°
M	370.0	357.9	322.3	266.3	194.7	117.2	70.0
α	0°	12.1°	23.8°	34.3°	41.9°	39.8°	0°

(d)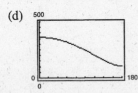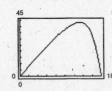

(e) For increasing θ, the two vectors tend to work against each other resulting in a decrease in the magnitude of the resultant.

93. True. See page 424.

95. True. In fact, $a = b = 0$.

97. True. $\mathbf{a}$ and $\mathbf{d}$ are parallel, and pointing in opposite directions.

99. True

101. True. $\mathbf{a} + \mathbf{w} = 2\mathbf{a} = -2\mathbf{d}$

103. False. $\mathbf{u} - \mathbf{v} = 2\mathbf{u}$ and

$-2(\mathbf{b} + \mathbf{t}) = -2(-2\mathbf{u}) = 4\mathbf{u}$

105. (a) The angle between them is $0°$.

(b) The angle between them is $180°$.

(c) No. At most it can be equal to the sum when the angle between them is $0°$.

107. Let $\mathbf{v} = (\cos\theta)\mathbf{i} + (\sin\theta)\mathbf{j}$.

$\|\mathbf{v}\| = \sqrt{\cos^2\theta + \sin^2\theta} = \sqrt{1} = 1$

Therefore, $\mathbf{v}$ is a unit vector for any value of θ.

109. $\mathbf{u} = \langle 5 - 1, 2 - 6\rangle = \langle 4, -4\rangle$

$\mathbf{v} = \langle 9 - 4, 4 - 5\rangle = \langle 5, -1\rangle$

$\mathbf{u} - \mathbf{v} = \langle -1, -3\rangle$

$\mathbf{v} - \mathbf{u} = \langle 1, 3\rangle$

111. $\left(\dfrac{6x^4}{7y^{-2}}\right)(14x^{-1}y^5) = \dfrac{12x^4y^5y^2}{x}$

$= 12x^3y^7,\ x \neq 0, y \neq 0$

113. $(18x)^0(4xy)^2(3x^{-1}) = \dfrac{16x^2y^2(3)}{x}$

$= 48xy^2,\ x \neq 0$

115. $(2.1 \times 10^9)(3.4 \times 10^{-4}) = 7.14 \times 10^5$

117. $\sin \theta = \dfrac{x}{7} \implies \sqrt{49 - x^2} = 7 \cos \theta$

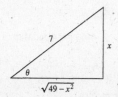

119. $\cot \theta = \dfrac{x}{10} \implies \sqrt{x^2 + 100} = 10 \cdot \csc \theta$

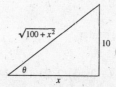

121. $\cos x(\cos x + 1) = 0$

$\cos x = 0 \implies x = \dfrac{\pi}{2} + n\pi$

$\cos x = -1 \implies x = \pi + 2n\pi$

123. $3 \sec x + 4 = 10$

$\sec x = 2$

$\cos x = \dfrac{1}{2}$

$x = \dfrac{\pi}{3} + 2n\pi, \dfrac{5\pi}{3} + 2n\pi$

Section 6.4   Vectors and Dot Products

■ Know the definition of the dot product of $\mathbf{u} = \langle u_1, u_2 \rangle$ and $\mathbf{v} = \langle v_1, v_2 \rangle$.

$\mathbf{u} \cdot \mathbf{v} = u_1 v_1 + u_2 v_2$

■ Know the following properties of the dot product:

1. $\mathbf{u} \cdot \mathbf{v} = \mathbf{v} \cdot \mathbf{u}$
2. $\mathbf{0} \cdot \mathbf{v} = 0$
3. $\mathbf{u} \cdot (\mathbf{v} + \mathbf{w}) = \mathbf{u} \cdot \mathbf{v} + \mathbf{u} \cdot \mathbf{w}$
4. $\mathbf{v} \cdot \mathbf{v} = \|\mathbf{v}\|^2$
5. $c(\mathbf{u} \cdot \mathbf{v}) = c\mathbf{u} \cdot \mathbf{v} = \mathbf{u} \cdot c\mathbf{v}$

■ If θ is the angle between two nonzero vectors $\mathbf{u}$ and $\mathbf{v}$, then

$\cos \theta = \dfrac{\mathbf{u} \cdot \mathbf{v}}{\|\mathbf{u}\| \, \|\mathbf{v}\|}$.

■ The vectors $\mathbf{u}$ and $\mathbf{v}$ are orthogonal if $\mathbf{u} \cdot \mathbf{v} = 0$.

■ Know the definition of vector components. $\mathbf{u} = \mathbf{w}_1 + \mathbf{w}_2$ where $\mathbf{w}_1$ and $\mathbf{w}_2$ are orthogonal, and $\mathbf{w}_1$ is parallel to $\mathbf{v}$. $\mathbf{w}_1$ is called the projection of $\mathbf{u}$ onto $\mathbf{v}$ and is denoted by

$\mathbf{w}_1 = \operatorname{proj}_{\mathbf{v}} \mathbf{u} = \left(\dfrac{\mathbf{u} \cdot \mathbf{v}}{\|\mathbf{v}\|^2} \right) \mathbf{v}$.

Then we have $\mathbf{w}_2 = \mathbf{u} - \mathbf{w}_1$.

■ Know the definition of work.

1. Projection form: $W = \|\operatorname{proj}_{\overrightarrow{PQ}} \mathbf{F}\| \, \|\overrightarrow{PQ}\|$
2. Dot product form: $W = \mathbf{F} \cdot \overrightarrow{PQ}$

Vocabulary Check

1. dot product

2. $\dfrac{\mathbf{u} \cdot \mathbf{v}}{\|\mathbf{u}\| \, \|\mathbf{v}\|}$

3. orthogonal

4. $\left(\dfrac{\mathbf{u} \cdot \mathbf{v}}{\|\mathbf{v}\|^2} \right) \mathbf{v}$

5. $\|\operatorname{proj}_{\overrightarrow{PQ}} \mathbf{F}\| \, \|\overrightarrow{PQ}\|, \, \mathbf{F} \cdot \overrightarrow{PQ}$

1. $\mathbf{u} \cdot \mathbf{v} = \langle 6, 3 \rangle \cdot \langle 2, -4 \rangle = 6(2) + 3(-4) = 0$

3. $\mathbf{u} \cdot \mathbf{v} = \langle 5, 1 \rangle \cdot \langle 3, -1 \rangle = 5(3) + 1(-1) = 14$

5. $\mathbf{u} = \langle 2, 2 \rangle$

   $\mathbf{u} \cdot \mathbf{u} = 2(2) + 2(2) = 8,$ scalar

7. $\mathbf{u} = \langle 2, 2 \rangle, \mathbf{v} = \langle -3, 4 \rangle$

   $\mathbf{u} \cdot 2\mathbf{v} = 2\mathbf{u} \cdot \mathbf{v} = 4(-3) + 4(4) = 4,$ scalar

9. $(3\mathbf{w} \cdot \mathbf{v})\mathbf{u} = (3\langle 1, -4 \rangle \cdot \langle -3, 4 \rangle)\langle 2, 2 \rangle$

   $= (3(-3) + (-12)(4))\langle 2, 2 \rangle$

   $= -57\langle 2, 2 \rangle$

   $= \langle -114, -114 \rangle,$ vector

11. $\mathbf{u} = \langle -5, 12 \rangle$

   $\|\mathbf{u}\| = \sqrt{\mathbf{u} \cdot \mathbf{u}} = \sqrt{(-5)^2 + 12^2} = 13$

13. $\mathbf{u} = 20\mathbf{i} + 25\mathbf{j}$

   $\|\mathbf{u}\| = \sqrt{\mathbf{u} \cdot \mathbf{u}} = \sqrt{(20)^2 + (25)^2} = \sqrt{1025} = 5\sqrt{41}$

15. $\mathbf{u} = -4\mathbf{j}$

   $\|\mathbf{u}\| = \sqrt{\mathbf{u} \cdot \mathbf{u}} = \sqrt{(-4)(-4)} = 4$

17. $\mathbf{u} = \langle -1, 0 \rangle, \mathbf{v} = \langle 0, 2 \rangle$

   $\cos \theta = \dfrac{\mathbf{u} \cdot \mathbf{v}}{\|\mathbf{u}\| \, \|\mathbf{v}\|} = \dfrac{0}{(1)(2)} = 0 \implies \theta = 90°$

19. $\mathbf{u} = 3\mathbf{i} + 4\mathbf{j}, \mathbf{v} = -2\mathbf{i} + 3\mathbf{j}$

   $\cos \theta = \dfrac{\mathbf{u} \cdot \mathbf{v}}{\|\mathbf{u}\| \, \|\mathbf{v}\|} = \dfrac{-6 + 12}{(5)(\sqrt{13})} = \dfrac{6}{5\sqrt{13}}$

   $\theta = \arccos\left(\dfrac{6}{5\sqrt{13}}\right) \approx 70.56°$

21. $\mathbf{u} = 2\mathbf{i}, \mathbf{v} = -3\mathbf{j}$

   $\cos \theta = \dfrac{\mathbf{u} \cdot \mathbf{v}}{\|\mathbf{u}\| \, \|\mathbf{v}\|} = \dfrac{0}{(2)(3)} = 0 \implies \theta = 90°$

23. $\mathbf{u} = \left(\cos \dfrac{\pi}{3}\right)\mathbf{i} + \left(\sin \dfrac{\pi}{3}\right)\mathbf{j} = \dfrac{1}{2}\mathbf{i} + \dfrac{\sqrt{3}}{2}\mathbf{j}$

   $\mathbf{v} = \left(\cos \dfrac{3\pi}{4}\right)\mathbf{i} + \left(\sin \dfrac{3\pi}{4}\right)\mathbf{j} = -\dfrac{\sqrt{2}}{2}\mathbf{i} + \dfrac{\sqrt{2}}{2}\mathbf{j}$

   $\|\mathbf{u}\| = \|\mathbf{v}\| = 1$

   $\cos \theta = \dfrac{\mathbf{u} \cdot \mathbf{v}}{\|\mathbf{u}\| \, \|\mathbf{v}\|} = \mathbf{u} \cdot \mathbf{v} = \left(\dfrac{1}{2}\right)\left(-\dfrac{\sqrt{2}}{2}\right) + \left(\dfrac{\sqrt{3}}{2}\right)\left(\dfrac{\sqrt{2}}{2}\right) = \dfrac{-\sqrt{2} + \sqrt{6}}{4}$

   $\theta = \arccos\left(\dfrac{-\sqrt{2} + \sqrt{6}}{4}\right) = 75° = \dfrac{5\pi}{12}$

25. $\mathbf{u} = 2\mathbf{i} - 4\mathbf{j}, \mathbf{v} = 3\mathbf{i} - 5\mathbf{j}$

   $\cos \theta = \dfrac{\mathbf{u} \cdot \mathbf{v}}{\|\mathbf{u}\| \, \|\mathbf{v}\|} = \dfrac{6 + 20}{\sqrt{20}\sqrt{34}}$

   $= \dfrac{13\sqrt{170}}{170} \implies \theta \approx 4.40°$

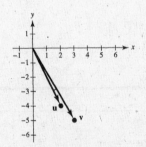

27. $\mathbf{u} = 6\mathbf{i} - 2\mathbf{j},\ \mathbf{v} = 8\mathbf{i} - 5\mathbf{j}$

$$\cos\theta = \frac{\mathbf{u}\cdot\mathbf{v}}{\|\mathbf{u}\|\,\|\mathbf{v}\|} = \frac{48+10}{\sqrt{40}\sqrt{89}}$$

$$= \frac{29\sqrt{890}}{890} \implies \theta \approx 13.57°$$

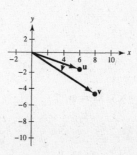

29. $P = (1,2),\ Q = (3,4),\ R = (2,5)$

$\overrightarrow{PQ} = \langle 2,2\rangle,\ \overrightarrow{PR} = \langle 1,3\rangle,\ \overrightarrow{QR} = \langle -1,1\rangle$

$$\cos\alpha = \frac{\overrightarrow{PQ}\cdot\overrightarrow{PR}}{\|\overrightarrow{PQ}\|\,\|\overrightarrow{PR}\|} = \frac{8}{(2\sqrt{2})(\sqrt{10})} \implies \alpha = \arccos\frac{2}{\sqrt{5}} \approx 26.6°$$

$$\cos\beta = \frac{\overrightarrow{PQ}\cdot\overrightarrow{QR}}{\|\overrightarrow{PQ}\|\,\|\overrightarrow{QR}\|} = 0 \implies \beta = 90°$$

Thus, $\gamma \approx 180° - 26.6° - 90° = 63.4°$.

31. $\mathbf{u}\cdot\mathbf{v} = \|\mathbf{u}\|\,\|\mathbf{v}\|\cos\theta$

$$= (9)(36)\cos\frac{3\pi}{4}$$

$$= 324\left(-\frac{\sqrt{2}}{2}\right)$$

$$= -162\sqrt{2}$$

33. $\mathbf{u} = \langle -12,30\rangle,\ \mathbf{v} = \left\langle \dfrac{1}{2}, -\dfrac{5}{4}\right\rangle$

$\mathbf{u} = -24\mathbf{v} \implies \mathbf{u}$ and $\mathbf{v}$ are parallel.

35. $\mathbf{u} = \frac{1}{4}(3\mathbf{i} - \mathbf{j}),\ \mathbf{v} = 5\mathbf{i} + 6\mathbf{j}$

$\mathbf{u} \neq k\mathbf{v} \implies$ Not parallel

$\mathbf{u}\cdot\mathbf{v} \neq 0 \implies$ Not orthogonal

Neither

37. $\mathbf{u} = 2\mathbf{i} - 2\mathbf{j},\ \mathbf{v} = -\mathbf{i} - \mathbf{j}$

$\mathbf{u}\cdot\mathbf{v} = 0 \implies \mathbf{u}$ and $\mathbf{v}$ are orthogonal.

39. $\mathbf{u}\cdot\mathbf{v} = \langle 2, -k\rangle \cdot \langle 3,2\rangle = 6 - 2k = 0 \implies k = 3$

41. $\mathbf{u}\cdot\mathbf{v} = \langle 1,4\rangle \cdot \langle 2k, -5\rangle = 2k - 20 = 0 \implies k = 10$

43. $\mathbf{u}\cdot\mathbf{v} = \langle -3k, 2\rangle \cdot \langle -6,0\rangle = 18k = 0 \implies k = 0$

45. $\mathbf{u} = \langle 3,4\rangle,\ \mathbf{v} = \langle 8,2\rangle$

$$\mathbf{w}_1 = \text{proj}_{\mathbf{v}}\mathbf{u} = \left(\frac{\mathbf{u}\cdot\mathbf{v}}{\|\mathbf{v}\|^2}\right)\mathbf{v}$$

$$= \left(\frac{32}{68}\right)\mathbf{v} = \frac{8}{17}\langle 8,2\rangle = \frac{16}{17}\langle 4,1\rangle$$

$$\mathbf{w}_2 = \mathbf{u} - \mathbf{w}_1 = \langle 3,4\rangle - \frac{16}{17}\langle 4,1\rangle = \frac{13}{17}\langle -1,4\rangle$$

$$\mathbf{u} = \mathbf{w}_1 + \mathbf{w}_2 = \frac{16}{17}\langle 4,1\rangle + \frac{13}{17}\langle -1,4\rangle$$

47. $\mathbf{u} = \langle 0,3\rangle,\ \mathbf{v} = \langle 2,15\rangle$

$$\mathbf{w}_1 = \text{proj}_{\mathbf{v}}\mathbf{u} = \left(\frac{\mathbf{u}\cdot\mathbf{v}}{\|\mathbf{v}\|^2}\right)\mathbf{v} = \frac{45}{229}\langle 2,15\rangle$$

$$\mathbf{w}_2 = \mathbf{u} - \mathbf{w}_1 = \langle 0,3\rangle - \frac{45}{229}\langle 2,15\rangle$$

$$= \left\langle -\frac{90}{229}, \frac{12}{229}\right\rangle = \frac{6}{229}\langle -15,2\rangle$$

$$\mathbf{u} = \mathbf{w}_1 + \mathbf{w}_2 = \frac{45}{229}\langle 2,15\rangle + \frac{6}{229}\langle -15,2\rangle$$

49. proj$_\mathbf{v}\mathbf{u} = \mathbf{u}$ since they are parallel.

$$\text{proj}_\mathbf{v}\mathbf{u} = \frac{\mathbf{u} \cdot \mathbf{v}}{\|\mathbf{v}\|^2}\mathbf{v}$$

$$= \frac{18 + 8}{36 + 16}\mathbf{v} = \frac{26}{52}\langle 6, 4 \rangle = \langle 3, 2 \rangle = \mathbf{u}$$

51. proj$_\mathbf{v}\mathbf{u} = \mathbf{0}$ since they are perpendicular.

$$\text{proj}_\mathbf{v}\mathbf{u} = \frac{\mathbf{u} \cdot \mathbf{v}}{\|\mathbf{v}\|^2}\mathbf{v} = \mathbf{0}, \text{ since } \mathbf{u} \cdot \mathbf{v} = 0.$$

53. $\mathbf{u} = \langle 2, 6 \rangle$

For $\mathbf{v}$ to be orthogonal to $\mathbf{u}$, $\mathbf{u} \cdot \mathbf{v}$ must equal 0.

Two possibilities: $\langle 6, -2 \rangle$ and $\langle -6, 2 \rangle$

55. $\mathbf{u} = \frac{1}{2}\mathbf{i} - \frac{3}{4}\mathbf{j}$

For $\mathbf{v}$ to be orthogonal to $\mathbf{u}$, $\mathbf{u} \cdot \mathbf{v}$ must equal 0.

Two possibilities: $\left\langle \frac{3}{4}, \frac{1}{2} \right\rangle$ and $\left\langle -\frac{3}{4}, -\frac{1}{2} \right\rangle$

57. $W = \|\text{proj}_{\overrightarrow{PQ}}\mathbf{v}\| \, \|\overrightarrow{PQ}\|$ where $\overrightarrow{PQ} = \langle 4, 7 \rangle$ and $\mathbf{v} = \langle 1, 4 \rangle$

$$\text{proj}_{\overrightarrow{PQ}}\mathbf{v} = \left(\frac{\mathbf{v} \cdot \overrightarrow{PQ}}{\|\overrightarrow{PQ}\|^2}\right)\overrightarrow{PQ} = \left(\frac{32}{65}\right)\langle 4, 7 \rangle$$

$$W = \|\text{proj}_{\overrightarrow{PQ}}\mathbf{v}\| \, \|\overrightarrow{PQ}\| = \left(\frac{32\sqrt{65}}{65}\right)\left(\sqrt{65}\right) = 32$$

59. (a) $\mathbf{u} \cdot \mathbf{v} = \langle 1245, 2600 \rangle \cdot \langle 12.20, 8.50 \rangle$

$$= 1245(12.20) + 2600(8.50) = 37,289$$

This is the total dollar value of the picture frames produced.

(b) Multiply $\mathbf{v}$ by 1.02.

61. (a) $\mathbf{F} = -30,000\vec{\mathbf{j}}$, Gravitational force

$$\mathbf{v} = \langle \cos(d°), \sin(d°) \rangle$$

$$\mathbf{w}_1 = \text{proj}_\mathbf{v}\mathbf{F} = \left(\frac{\mathbf{F} \cdot \mathbf{v}}{\|\mathbf{v}\|^2}\right)\mathbf{v} = (\mathbf{F} \cdot \mathbf{v})\mathbf{v} = -30,000\sin(d°)\langle \cos d°, \sin d° \rangle$$

$$= \langle -30,000\sin d° \cos d°, -30,000\sin^2 d° \rangle$$

Force needed: $30,000\sin(d°)$

(b)

d	0°	1°	2°	3°	4°	5°	6°	7°	8°	9°	10°
Force	0	523.6	1047.0	1570.1	2092.7	2614.7	3135.9	3656.1	4175.2	4693.0	5209.4

(c) $\mathbf{w}_2 = \mathbf{F} - \mathbf{w}_1 = -30,000\mathbf{j} + 2614.7(\cos(5°)\mathbf{i} + \sin(5°)\mathbf{j}) \approx 2604.75\mathbf{i} - 29,772.11\mathbf{j}$

$\|\mathbf{w}_2\| \approx 29,885.8$ pounds

63. (a) $\mathbf{F} = 15,691\langle \cos 30°, \sin 30° \rangle$

$$\overrightarrow{PQ} = d\langle 1, 0 \rangle$$

$$W = \mathbf{F} \cdot \overrightarrow{PQ} = 15,691\frac{\sqrt{3}}{2}d \approx 13,588.8d$$

(b)

d	0	200	400	800
Work	0	2,717,761	5,435,522	10,871,044

65. $\|\mathbf{F}\| = 250$, $\|\overrightarrow{PQ}\| = 100$, $\theta = 30°$

$$W = \|\mathbf{F}\| \, \|\overrightarrow{PQ}\| \cos \theta$$

$$= (250)(100)\cos 30°$$

$$= 25,000\frac{\sqrt{3}}{2}$$

$$= 12,500\sqrt{3} \text{ foot-pounds}$$

$$\approx 21,650.64 \text{ foot-pounds}$$

67. $W = (\cos 25°)(20)(40) \approx 725.05$ foot-pounds

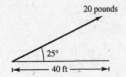

20 pounds

25°

40 ft

69. True. $\mathbf{u} \cdot \mathbf{v} = 0$

71. $\mathbf{u} \cdot \mathbf{v} = 0 \implies$ they are orthogonal (unit vectors).

73. (a) $\text{proj}_{\mathbf{v}}\mathbf{u} = \mathbf{u} \implies \mathbf{u}$ and $\mathbf{v}$ are parallel.

(b) $\text{proj}_{\mathbf{v}}\mathbf{u} = \mathbf{0} \implies \mathbf{u}$ and $\mathbf{v}$ are orthogonal.

75. Use the Law of Cosines on the triangle:

$$\|\mathbf{u} - \mathbf{v}\|^2 = \|\mathbf{u}\|^2 + \|\mathbf{v}\|^2 - 2\|\mathbf{u}\|\,\|\mathbf{v}\| \cos \theta$$

$$= \|\mathbf{u}\|^2 + \|\mathbf{v}\|^2 - 2\mathbf{u} \cdot \mathbf{v}$$

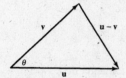

77. From trigonometry, $x = \cos \theta$ and $y = \sin \theta$. Thus, $\mathbf{u} = \langle x, y \rangle = \cos \theta \mathbf{i} + \sin \theta \mathbf{j}$.

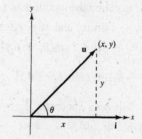

79. $g(x) = f(x - 4)$ is a horizontal shift of f four units to the right.

81. $g(x) = f(x) + 6$ is a vertical shift of f six units upward.

83. $\sqrt{-4} - 1 = 2i - 1 = -1 + 2i$

85. $3i(4 - 5i) = 12i + 15 = 15 + 12i$

87. $(1 + 3i)(1 - 3i) = 1 - (3i)^2 = 1 + 9 = 10$

89. $\dfrac{3}{1 + i} + \dfrac{2}{2 - 3i} = \dfrac{3}{1 + i} \cdot \dfrac{1 - i}{1 - i} + \dfrac{2}{2 - 3i} \cdot \dfrac{2 + 3i}{2 + 3i}$

$$= \dfrac{3 - 3i}{2} + \dfrac{4 + 6i}{13}$$

$$= \dfrac{39 - 39i + 8 + 12i}{26}$$

$$= \dfrac{47}{26} - \dfrac{27}{26}i$$

91. $-2i$

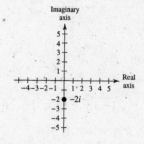

93. $1 + 8i$

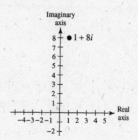

Section 6.5   Trigonometric Form of a Complex Number

■ You should be able to graphically represent complex numbers.

■ The absolute value of the complex numbers $z = a + bi$ is $|z| = \sqrt{a^2 + b^2}$.

■ The trigonometric form of the complex number $z = a + bi$ is $z = r(\cos \theta + i \sin \theta)$ where

    (a) $a = r \cos \theta$ (b) $b = r \sin \theta$

    (c) $r = \sqrt{a^2 + b^2}$; r is called the modulus of z.   (d) $\tan \theta = b/a$; θ is called the argument of z.

■ Given $z_1 = r_1(\cos \theta_1 + i \sin \theta_1)$ and $z_2 = r_2(\cos \theta_2 + i \sin \theta_2)$:

    (a) $z_1 z_2 = r_1 r_2 [\cos(\theta_1 + \theta_2) + i \sin(\theta_1 + \theta_2)]$

    (b) $\dfrac{z_1}{z_2} = \dfrac{r_1}{r_2}[\cos(\theta_1 - \theta_2) + i \sin(\theta_1 - \theta_2)]$, $z_2 \neq 0$

■ You should know DeMoivre's Theorem: If $z = r(\cos \theta + i \sin \theta)$, then for any positive integer n,

    $z^n = r^n(\cos n\theta + i \sin n\theta)$.

■ You should know that for any positive integer n, $z = r(\cos \theta + i \sin \theta)$ has n distinct nth roots given by

$$\sqrt[n]{r}\left[\cos\left(\frac{\theta + 2\pi k}{n}\right) + i \sin\left(\frac{\theta + 2\pi k}{n}\right)\right]$$

where $k = 0, 1, 2, \ldots, n - 1$.

Vocabulary Check

1. absolute value   **2.** trigonometric form, modulus, argument

3. DeMoivre's   **4.** nth root

1. $|6i| = 6$

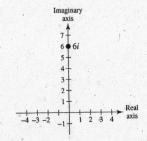

3. $|-4| = \sqrt{(-4)^2 + 0^2}$

$= \sqrt{16} = 4$

5. $|-4 + 4i| = \sqrt{(-4)^2 + (4)^2}$

$= \sqrt{32} = 4\sqrt{2}$

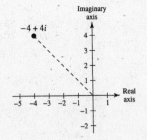

7. $|3 + 6i| = \sqrt{9 + 36}$

$= \sqrt{45} = 3\sqrt{5}$

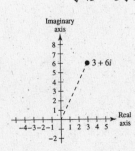

9. $z = 3i$

$r = \sqrt{0^2 + 3^2} = \sqrt{9} = 3$

$\tan \theta = \dfrac{3}{0}$, undefined $\Rightarrow \theta = \dfrac{\pi}{2}$

$z = 3\left(\cos \dfrac{\pi}{2} + i \sin \dfrac{\pi}{2}\right)$

11. $z = -2$

$r = \sqrt{(-2)^2 + 0^2} = 2$

$\tan \theta = \pi \Rightarrow \theta = \pi$

$z = 2(\cos \pi + i \sin \pi)$

13. $z = -2 - 2i$

$r = \sqrt{(-2)^2 + (-2)^2} = \sqrt{8} = 2\sqrt{2}$

$\tan \theta = \dfrac{-2}{-2} = 1$, θ is in Quadrant III.

$\theta = \dfrac{5\pi}{4}$

$z = 2\sqrt{2}\left(\cos \dfrac{5\pi}{4} + i \sin \dfrac{5\pi}{4}\right)$

15. $z = \sqrt{3} - i$

$r = \sqrt{\left(\sqrt{3}\right)^2 + (-1)^2} = 2$

$\tan \theta = \dfrac{-1}{\sqrt{3}} \implies \theta = \dfrac{11\pi}{6}$

$z = 2\left(\cos \dfrac{11\pi}{6} + i \sin \dfrac{11\pi}{6}\right)$

17. $z = 5 - 5i$

$r = \sqrt{5^2 + (-5)^2} = \sqrt{50} = 5\sqrt{2}$

$\tan \theta = -\dfrac{5}{5} = -1 \implies \theta = \dfrac{7\pi}{4}$

$z = 5\sqrt{2}\left(\cos \dfrac{7\pi}{4} + i \sin \dfrac{7\pi}{4}\right)$

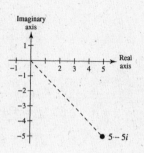

19. $z = \sqrt{3} + i$

$r = \sqrt{\left(\sqrt{3}\right)^2 + 1^2} = \sqrt{4} = 2$

$\tan \theta = \dfrac{1}{\sqrt{3}} = \dfrac{\sqrt{3}}{3} \implies \theta = \dfrac{\pi}{6}$

$z = 2\left(\cos \dfrac{\pi}{6} + i \sin \dfrac{\pi}{6}\right)$

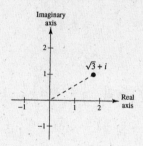

21. $z = -2(1 + \sqrt{3}i)$

$r = \sqrt{(-2)^2 + \left(-2\sqrt{3}\right)^2} = \sqrt{16} = 4$

$\tan \theta = \dfrac{\sqrt{3}}{1} = \sqrt{3} \implies \theta = \dfrac{4\pi}{3}$

$z = 4\left(\cos \dfrac{4\pi}{3} + i \sin \dfrac{4\pi}{3}\right)$

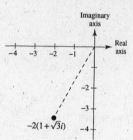

23. $z = -8i$

$r = \sqrt{0 + (-8)^2} = \sqrt{64} = 8$

$\tan \theta = -\dfrac{8}{0}$, undefined $\implies \theta = \dfrac{3\pi}{2}$

$z = 8\left(\cos \dfrac{3\pi}{2} + i \sin \dfrac{3\pi}{2}\right)$

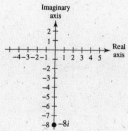

25. $z = -7 + 4i$

$r = \sqrt{49 + 16} = \sqrt{65}$

$\tan \theta = \dfrac{4}{-7} \implies \theta \approx 2.62$ radians or $150.26°$

$z = \sqrt{65}(\cos 150.26° + i \sin 150.26°)$

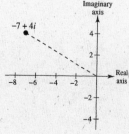

27. $z = 3$

$r = \sqrt{3^2 + 0^2} = 3$

$\tan \theta = \dfrac{0}{3} = 0 \implies \theta = 0°$

$z = 3(\cos 0° + i \sin 0°)$

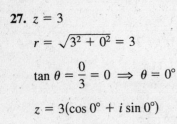

29. $z = 3 + \sqrt{3}\,i$

$r = \sqrt{9 + 3} = \sqrt{12} = 2\sqrt{3}$

$\tan \theta = \dfrac{\sqrt{3}}{3} \implies \theta = \dfrac{\pi}{6}$ or $30°$

$z = 2\sqrt{3}\left(\cos \dfrac{\pi}{6} + i \sin \dfrac{\pi}{6}\right)$

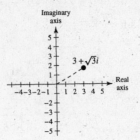

31. $z = -1 - 2i$

$r = \sqrt{1^2 + 2^2} = \sqrt{5}$

$\tan \theta = \dfrac{-2}{-1} = 2 \implies \theta \approx 243.4°$

$z = \sqrt{5}(\cos 243.4° + i \sin 243.4°)$

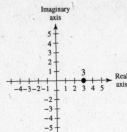

33. $z = 5 + 2i$

$r = \sqrt{25 + 4} = \sqrt{29} \approx 5.385$

$\tan \theta = \dfrac{2}{5} \implies \theta \approx 21.80°$

$z = \sqrt{29}(\cos 21.80° + i \sin 21.80°)$

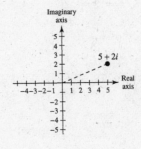

35. $z = 3\sqrt{2} - 7i$

$r = \sqrt{18 + 49} = \sqrt{67} \approx 8.185$

$\tan \theta = \dfrac{-7}{3\sqrt{2}} \approx -1.6499 \implies \theta \approx 301.22°$

$z = \sqrt{67}(\cos 301.22° + i \sin 301.22°)$

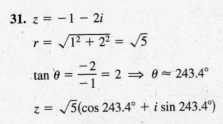

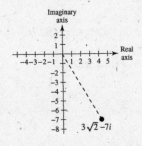

37. $2(\cos 120° + i \sin 120°) = 2\left(-\dfrac{1}{2} + \dfrac{\sqrt{3}}{2}i\right)$

$= -1 + \sqrt{3}\,i$

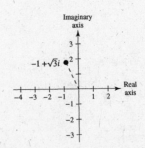

39. $\dfrac{3}{2}(\cos 330° + i \sin 330°) = \dfrac{3}{2}\left(\dfrac{\sqrt{3}}{2} - \dfrac{1}{2}i\right)$

$$= \dfrac{3\sqrt{3}}{4} - \dfrac{3}{4}i$$

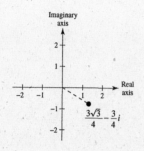

41. $3.75\left(\cos \dfrac{3\pi}{4} + i \sin \dfrac{3\pi}{4}\right) = -\dfrac{15\sqrt{2}}{8} + \dfrac{15\sqrt{2}}{8}i$

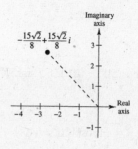

43. $6\left(\cos \dfrac{\pi}{3} + i \sin \dfrac{\pi}{3}\right) = 6\left(\dfrac{1}{2} + \dfrac{\sqrt{3}}{2}i\right) = 3 + 3\sqrt{3}i$

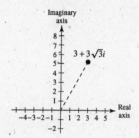

45. $4\left(\cos \dfrac{3\pi}{2} + i \sin \dfrac{3\pi}{2}\right) = 4(0 - i) = -4i$

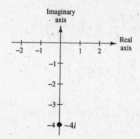

47. $3[\cos(18° \, 45') + i \sin(18° \, 45')] \approx 2.8408 + 0.9643i$

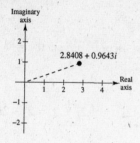

49. $5\left(\cos \dfrac{\pi}{9} + i \sin \dfrac{\pi}{9}\right) \approx 4.6985 + 1.7101i$

51. $9(\cos 58° + i \sin 58°) \approx 4.7693 + 7.6324i$

53.

The absolute value of each power is 1.

55. $\left[3\left(\cos \dfrac{\pi}{3} + i \sin \dfrac{\pi}{3}\right)\right]\left[4\left(\cos \dfrac{\pi}{6} + i \sin \dfrac{\pi}{6}\right)\right] = (3)(4)\left[\cos\left(\dfrac{\pi}{3} + \dfrac{\pi}{6}\right) + i \sin\left(\dfrac{\pi}{6} + \dfrac{\pi}{3}\right)\right] = 12\left(\cos \dfrac{\pi}{2} + i \sin \dfrac{\pi}{2}\right)$

57. $\left[\dfrac{5}{3}(\cos 140° + i \sin 140°)\right]\left[\dfrac{2}{3}(\cos 60° + i \sin 60°)\right] = \left(\dfrac{5}{3}\right)\left(\dfrac{2}{3}\right)[\cos(140° + 60°) + i \sin(140° + 60°)]$

$$= \dfrac{10}{9}(\cos 200° + i \sin 200°)$$

59. $\left[\frac{11}{20}(\cos 290° + i \sin 290°)\right]\left[\frac{2}{5}(\cos 200° + i \sin 200°)\right] = \left(\frac{11}{20}\right)\left(\frac{2}{5}\right)[\cos(290° + 200°) + i \sin(290° + 200°)]$

$$= \frac{11}{50}(\cos 490° + i \sin 490°)$$

$$= \frac{11}{50}(\cos 130° + i \sin 130°)$$

61. $\dfrac{\cos 50° + i \sin 50°}{\cos 20° + i \sin 20°} = \cos(50° - 20°) + i \sin(50° - 20°) = \cos 30° + i \sin 30°$

63. $\dfrac{2(\cos 120° + i \sin 120°)}{4(\cos 40° + i \sin 40°)} = \dfrac{1}{2}[\cos(120° - 40°) + i \sin(120° - 40°)] = \dfrac{1}{2}(\cos 80° + i \sin 80°)$

65. $\dfrac{18(\cos 54° + i \sin 54°)}{3(\cos 102° + i \sin 102°)} = 6(\cos(54° - 102°) + i \sin(54° - 102°))$

$$= 6(\cos(-48°) + i \sin(-48°)) = 6(\cos 312° + i \sin 312°)$$

67. (a) $2 - 2i = 2\sqrt{2}\left(\cos\dfrac{7\pi}{4} + i \sin\dfrac{7\pi}{4}\right)$

$1 + i = \sqrt{2}\left(\cos\dfrac{\pi}{4} + i \sin\dfrac{\pi}{4}\right)$

(b) $(2 - 2i)(1 + i) = 2\sqrt{2}\left(\cos\dfrac{7\pi}{4} + i \sin\dfrac{7\pi}{4}\right)\sqrt{2}\left(\cos\dfrac{\pi}{4} + i \sin\dfrac{\pi}{4}\right) = 4(\cos 2\pi + i \sin 2\pi) = 4$

(c) $(2 - 2i)(1 + i) = 2 + 2 = 4$

69. (a) $2 + 2i = 2\sqrt{2}(\cos 45° + i \sin 45°)$

$1 - i = \sqrt{2}[\cos(-45°) + i \sin(-45°)]$

(b) $(2 + 2i)(1 - i) = [2\sqrt{2}(\cos 45° + i \sin 45°)][\sqrt{2}(\cos(-45°) + i \sin(-45°))] = 4(\cos 0° + i \sin 0°) = 4$

(c) $(2 + 2i)(1 - i) = 2 - 2i + 2i - 2i^2 = 2 + 2 = 4$

71. (a) $-2i = 2[\cos(-90°) + i \sin(-90°)]$
                                         (c) $-2i(1 + i) = -2i - 2i^2 = -2i + 2 = 2 - 2i$

$1 + i = \sqrt{2}(\cos 45° + i \sin 45°)$

(b) $-2i(1 + i) = 2[\cos(-90°) + i \sin(-90°)][\sqrt{2}(\cos 45° + i \sin 45°)]$

$$= 2\sqrt{2}[\cos(-45°) + i \sin(-45°)]$$

$$= 2\sqrt{2}\left[\dfrac{1}{\sqrt{2}} - \dfrac{1}{\sqrt{2}}i\right] = 2 - 2i$$

73. (a) $-2i = 2\left(\cos\dfrac{3\pi}{2} + i \sin\dfrac{3\pi}{2}\right)$
                              (c) $-2i(\sqrt{3} - i) = -2\sqrt{3}i - 2$

$\sqrt{3} - i = 2\left(\cos\dfrac{11\pi}{6} + i \sin\dfrac{11\pi}{6}\right)$

(b) $-2i(\sqrt{3} - i) = 2\left(\cos\dfrac{3\pi}{2} + i \sin\dfrac{3\pi}{2}\right)2\left(\cos\dfrac{11\pi}{6} + i \sin\dfrac{11\pi}{6}\right)$

$$= 4\left(\cos\dfrac{20\pi}{6} + i \sin\dfrac{20\pi}{6}\right) = 4\left(-\dfrac{1}{2} - \dfrac{\sqrt{3}}{2}i\right) = -2 - 2\sqrt{3}i$$

75. (a) $2 = 2(\cos 0 + i \sin 0)$

$$1 - i = \sqrt{2}\left(\cos \frac{7\pi}{4} + i \sin \frac{7\pi}{4}\right)$$

(b) $2(1 - i) = 2\sqrt{2}\left(\cos \frac{7\pi}{4} + i \sin \frac{7\pi}{4}\right)$

$$= 2\sqrt{2}\left(\frac{\sqrt{2}}{2} - \frac{\sqrt{2}}{2}i\right)$$

$$= 2(1 - i) = 2 - 2i$$

(c) $2(1 - i) = 2 - 2i$

77. (a) $3 + 3i = 3\sqrt{2}\left(\cos \frac{\pi}{4} + i \sin \frac{\pi}{4}\right)$

$$1 - \sqrt{3}i = 2\left(\cos \frac{5\pi}{3} + i \sin \frac{5\pi}{3}\right)$$

(b) $\dfrac{3 + 3i}{1 - \sqrt{3}i} = \dfrac{3\sqrt{2}}{2}\left(\cos\left(\frac{\pi}{4} - \frac{5\pi}{3}\right) + i \sin\left(\frac{\pi}{4} - \frac{5\pi}{3}\right)\right)$

$$= \frac{3\sqrt{2}}{2}\left(\cos\left(-\frac{17\pi}{12}\right) + i \sin\left(-\frac{17\pi}{12}\right)\right)$$

$$\approx -0.549 + 2.049i$$

(c) $\dfrac{3 + 3i}{1 - \sqrt{3}i} \cdot \dfrac{1 + \sqrt{3}i}{1 + \sqrt{3}i} = \dfrac{3 - 3\sqrt{3} + \left(3 + 3\sqrt{3}\right)i}{4}$

$$\approx -0.549 + 2.049i$$

79. (a) $5 = 5(\cos 0 + i \sin 0)$

$$2 + 2i = 2\sqrt{2}\left(\cos \frac{\pi}{4} + i \sin \frac{\pi}{4}\right)$$

(b) $\dfrac{5}{2 + 2i} = \dfrac{5(\cos 0 + i \sin 0)}{2\sqrt{2}\left(\cos \dfrac{\pi}{4} + i \sin \dfrac{\pi}{4}\right)}$

$$= \frac{5}{2\sqrt{2}}\left(\cos\left(-\frac{\pi}{4}\right) + i \sin\left(-\frac{\pi}{4}\right)\right)$$

$$= \frac{5}{2\sqrt{2}}\left(\frac{\sqrt{2}}{2} - \frac{\sqrt{2}}{2}i\right) = \frac{5}{4} - \frac{5}{4}i$$

(c) $\dfrac{5}{2 + 2i} \cdot \dfrac{2 - 2i}{2 - 2i} = \dfrac{5(2 - 2i)}{4 + 4} = \dfrac{5}{4} - \dfrac{5}{4}i$

81. (a) $4i = 4\left(\cos \frac{\pi}{2} + i \sin \frac{\pi}{2}\right)$

$$-1 + i = \sqrt{2}\left(\cos \frac{3\pi}{4} + i \sin \frac{3\pi}{4}\right)$$

(b) $\dfrac{4i}{-1 + i} = \dfrac{4}{\sqrt{2}}\left(\cos\left(\frac{\pi}{2} - \frac{3\pi}{4}\right) + i \sin\left(\frac{\pi}{2} - \frac{3\pi}{4}\right)\right)$

$$= \frac{4}{\sqrt{2}}\left(\cos\left(-\frac{\pi}{4}\right) + i \sin\left(-\frac{\pi}{4}\right)\right)$$

$$= \frac{4}{\sqrt{2}}\left(\frac{\sqrt{2}}{2} - \frac{\sqrt{2}}{2}i\right) = 2 - 2i$$

(c) $\dfrac{4i}{-1 + i} \cdot \dfrac{-1 - i}{-1 - i} = \dfrac{-4i + 4}{2} = 2 - 2i$

83. Let $z = x + iy$ such that:

$$|z| = 2 \implies 2 = \sqrt{x^2 + y^2} \implies 4 = x^2 + y^2$$

Circle with radius of 2

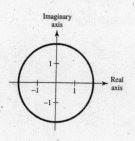

85. Let $z = x + iy$.

$$|z| = 4 \implies 4 = \sqrt{x^2 + y^2} \implies x^2 + y^2 = 16$$

Circle of radius 4

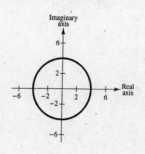

87. $\theta = \dfrac{\pi}{6}$

Let $z = x + iy$ such that:

$\tan \dfrac{\pi}{6} = \dfrac{y}{x} \Longrightarrow$

$\dfrac{y}{x} = \dfrac{1}{\sqrt{3}} \Longrightarrow y = \dfrac{1}{\sqrt{3}}x$

Line

89. $\theta = \dfrac{5\pi}{6}$

Let $z = x + iy$ such that:

$\tan \dfrac{5\pi}{6} = \dfrac{y}{x} \Longrightarrow$

$\dfrac{y}{x} = -\dfrac{\sqrt{3}}{3} \Longrightarrow y = -\dfrac{\sqrt{3}}{3}x$

Line

91. $(1 + i)^3 = \left[\sqrt{2}\left(\cos\dfrac{\pi}{4} + i\sin\dfrac{\pi}{4}\right) \right]^3$

$\qquad = \left(\sqrt{2}\right)^3\left(\cos\dfrac{3\pi}{4} + i\sin\dfrac{3\pi}{4}\right)$

$\qquad = 2\sqrt{2}\left(-\dfrac{\sqrt{2}}{2} + \dfrac{\sqrt{2}}{2}i\right)$

$\qquad = -2 + 2i$

93. $(-1 + i)^{10} = \left[\sqrt{2}\left(\cos\dfrac{3\pi}{4} + i\sin\dfrac{3\pi}{4}\right) \right]^{10}$

$\qquad = \left(\sqrt{2}\right)^{10}\left(\cos\dfrac{30\pi}{4} + i\sin\dfrac{30\pi}{4}\right)$

$\qquad = 32\left[\cos\left(\dfrac{3\pi}{2} + 6\pi\right) + i\sin\left(\dfrac{3\pi}{2} + 6\pi\right)\right]$

$\qquad = 32\left(\cos\dfrac{3\pi}{2} + i\sin\dfrac{3\pi}{2}\right)$

$\qquad = 32[0 + i(-1)] = -32i$

95. $2\left(\sqrt{3} + i\right)^5 = 2\left[2\left(\cos\dfrac{\pi}{6} + i\sin\dfrac{\pi}{6}\right) \right]^5$

$\qquad = 2\left[2^5\left(\cos\dfrac{5\pi}{6} + i\sin\dfrac{5\pi}{6}\right) \right]$

$\qquad = 64\left(-\dfrac{\sqrt{3}}{2} + \dfrac{1}{2}i\right)$

$\qquad = -32\sqrt{3} + 32i$

97. $[5(\cos 20° + i\sin 20°)]^3 = 5^3(\cos 60° + i\sin 60°)$

$\qquad = \dfrac{125}{2} + \dfrac{125\sqrt{3}}{2}i$

99. $\left(\cos\dfrac{5\pi}{4} + i\sin\dfrac{5\pi}{4}\right)^{10} = \cos\dfrac{25\pi}{2} + i\sin\dfrac{25\pi}{2}$

$\qquad = \cos\left(12\pi + \dfrac{\pi}{2}\right) + i\sin\left(12\pi + \dfrac{\pi}{2}\right) = \cos\dfrac{\pi}{2} + i\sin\dfrac{\pi}{2} = i$

101. $[2(\cos 1.25 + i\sin 1.25)]^4 = 2^4(\cos 5 + i\sin 5)$

$\qquad \approx 4.5386 - 15.3428i$

103. $[2(\cos\pi + i\sin\pi)]^8 = 2^8(\cos 8\pi + i\sin 8\pi)$

$\qquad = 256(1) = 256$

105. $(3 - 2i)^5 = -597 - 122i$

107. $[4(\cos 10° + i\sin 10°)]^6 = 4^6(\cos 60° + i\sin 60°)$

$\qquad = 4096\left(\dfrac{1}{2} + \dfrac{\sqrt{3}}{2}i\right)$

$\qquad = 2048 + 2048\sqrt{3}i$

109. $\left[3\left(\cos\dfrac{\pi}{8}+i\sin\dfrac{\pi}{8}\right)\right]^2 = 3^2\left(\cos\dfrac{\pi}{4}+i\sin\dfrac{\pi}{4}\right)$

$\qquad\qquad = 9\left(\dfrac{\sqrt{2}}{2}+i\dfrac{\sqrt{2}}{2}\right)$

$\qquad\qquad = \dfrac{9}{2}\sqrt{2}+\dfrac{9}{2}\sqrt{2}i$

111. $\left[-\dfrac{1}{2}\left(1+\sqrt{3}i\right)\right]^6 = \left[\cos\dfrac{4\pi}{3}+i\sin\dfrac{4\pi}{3}\right]^6$

$\qquad\qquad = \cos 8\pi + i\sin 8\pi$

$\qquad\qquad = 1$

113. (a) In trigonometric form we have:

$\qquad$ $2(\cos 30° + i\sin 30°)$

$\qquad$ $2(\cos 150° + i\sin 150°)$

$\qquad$ $2(\cos 270° + i\sin 270°)$

$\quad$ (b) There are three roots evenly spaced around a circle of radius 2. Therefore, they represent the cube roots of some number of modulus 8. Cubing them shows that they are all cube roots of $8i$.

$\quad$ (c) $[2(\cos 30° + i\sin 30°)]^3 = 8i$

$\qquad$ $[2(\cos 150° + i\sin 150°)]^3 = 8i$

$\qquad$ $[2(\cos 270° + i\sin 270°)]^3 = 8i$

115. (a) In trigonometric form we have:

$\qquad$ $\cos 120° + i\sin 120°$

$\qquad$ $\cos 240° + i\sin 240°$

$\qquad$ $\cos 0° + i\sin 0°$

$\quad$ (b) These are the three cube roots of 1.

$\quad$ (c) $(\cos 120° + i\sin 120°)^3 = 1$

$\qquad$ $(\cos 240° + i\sin 240°)^3 = 1$

$\qquad$ $(\cos 0° + i\sin 0°)^3 = 1$

117. $2i = 2\left(\cos\dfrac{\pi}{2}+i\sin\dfrac{\pi}{2}\right)$

$\quad$ Square roots:

$\quad$ $\sqrt{2}\left(\cos\dfrac{\pi}{4}+i\sin\dfrac{\pi}{4}\right) = 1+i$

$\quad$ $\sqrt{2}\left(\cos\dfrac{5\pi}{4}+i\sin\dfrac{5\pi}{4}\right) = -1-i$

119. $-3i = 3\left(\cos\dfrac{3\pi}{2}+i\sin\dfrac{3\pi}{2}\right)$

$\quad$ Square roots:

$\quad$ $\sqrt{3}\left(\cos\dfrac{3\pi}{4}+i\sin\dfrac{3\pi}{4}\right) = -\dfrac{\sqrt{6}}{2}+\dfrac{\sqrt{6}}{2}i$

$\quad$ $\sqrt{3}\left(\cos\dfrac{7\pi}{4}+i\sin\dfrac{7\pi}{4}\right) = \dfrac{\sqrt{6}}{2}-\dfrac{\sqrt{6}}{2}i$

121. $2-2i = 2\sqrt{2}\left(\cos\dfrac{7\pi}{4}+i\sin\dfrac{7\pi}{4}\right)$

$\quad$ Square roots:

$\quad$ $8^{1/4}\left(\cos\dfrac{7\pi}{8}+i\sin\dfrac{7\pi}{8}\right) \approx -1.554 + 0.644i$

$\quad$ $8^{1/4}\left(\cos\dfrac{15\pi}{8}+i\sin\dfrac{15\pi}{8}\right) \approx 1.554 - 0.644i$

123. $1+\sqrt{3}i = 2\left(\cos\dfrac{\pi}{3}+i\sin\dfrac{\pi}{3}\right)$

$\quad$ Square roots:

$\quad$ $\sqrt{2}\left(\cos\dfrac{\pi}{6}+i\sin\dfrac{\pi}{6}\right) = \dfrac{\sqrt{6}}{2}+\dfrac{\sqrt{2}}{2}i$

$\quad$ $\sqrt{2}\left(\cos\dfrac{7\pi}{6}+i\sin\dfrac{7\pi}{6}\right) = -\dfrac{\sqrt{6}}{2}-\dfrac{\sqrt{2}}{2}i$

125. (a) Square roots of $5(\cos 120° + i\sin 120°)$:

$\qquad$ $\sqrt{5}\left[\cos\left(\dfrac{120°+360°k}{2}\right)+i\sin\left(\dfrac{120°+360°k}{2}\right)\right],\ k = 0, 1$

$\qquad$ $\sqrt{5}(\cos 60° + i\sin 60°)$

$\qquad$ $\sqrt{5}(\cos 240° + i\sin 240°)$

$\quad$ (c) $\dfrac{\sqrt{5}}{2}+\dfrac{\sqrt{15}}{2}i,\ -\dfrac{\sqrt{5}}{2}-\dfrac{\sqrt{15}}{2}i$

(b)

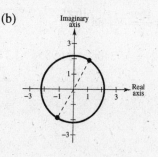

127. (a) Fourth roots of $16\left(\cos\dfrac{4\pi}{3} + i\sin\dfrac{4\pi}{3}\right)$:

$$\sqrt[4]{16}\left[\cos\left(\frac{(4\pi/3) + 2k\pi}{4}\right) + i\sin\left(\frac{(4\pi/3) + 2k\pi}{4}\right)\right],\ k = 0, 1, 2, 3$$

$$2\left(\cos\frac{\pi}{3} + i\sin\frac{\pi}{3}\right)$$

$$2\left(\cos\frac{5\pi}{6} + i\sin\frac{5\pi}{6}\right)$$

$$2\left(\cos\frac{4\pi}{3} + i\sin\frac{4\pi}{3}\right)$$

$$2\left(\cos\frac{11\pi}{6} + i\sin\frac{11\pi}{6}\right)$$

(c) $1 + \sqrt{3}i,\ -\sqrt{3} + i,\ -1 - \sqrt{3}i,\ \sqrt{3} - i$

(b)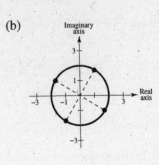

129. (a) Cube roots of $-27i = 27\left(\cos\dfrac{3\pi}{2} + i\sin\dfrac{3\pi}{2}\right)$:

$$(27)^{1/3}\left[\cos\left(\frac{(3\pi/2) + 2k\pi}{3}\right) + i\sin\left(\frac{(3\pi/2) + 2k\pi}{3}\right)\right],\quad k = 0, 1, 2$$

$$3\left(\cos\frac{\pi}{2} + i\sin\frac{\pi}{2}\right)$$

$$3\left(\cos\frac{7\pi}{6} + i\sin\frac{7\pi}{6}\right)$$

$$3\left(\cos\frac{11\pi}{6} + i\sin\frac{11\pi}{6}\right)$$

(c) $3i,\ -\dfrac{3\sqrt{3}}{2} - \dfrac{3}{2}i,\ \dfrac{3\sqrt{3}}{2} - \dfrac{3}{2}i$

(b)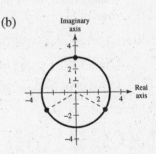

131. (a) Cube roots of $-\dfrac{125}{2}(1 + \sqrt{3}i) = 125\left(\cos\dfrac{4\pi}{3} + i\sin\dfrac{4\pi}{3}\right)$:

$$\sqrt[3]{125}\left[\cos\left(\frac{(4\pi/3) + 2k\pi}{3}\right) + i\sin\left(\frac{(4\pi/3) + 2k\pi}{3}\right)\right],\quad k = 0, 1, 2$$

$$5\left(\cos\frac{4\pi}{9} + i\sin\frac{4\pi}{9}\right)$$

$$5\left(\cos\frac{10\pi}{9} + i\sin\frac{10\pi}{9}\right)$$

$$5\left(\cos\frac{16\pi}{9} + i\sin\frac{16\pi}{9}\right)$$

(c) $0.8682 + 4.9240i,\ -4.6985 - 1.7101i,\ 3.8302 - 3.2139i$

(b)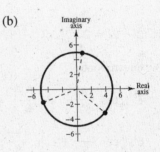

133. (a) Cube roots of $64i = 64\left(\cos\dfrac{\pi}{2} + i\sin\dfrac{\pi}{2}\right)$:

(b)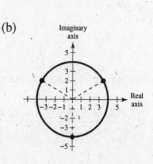

$$(64)^{1/3}\left[\cos\left(\dfrac{(\pi/2) + 2k\pi}{3}\right) + i\sin\left(\dfrac{(\pi/2) + 2k\pi}{3}\right)\right], \quad k = 0, 1, 2$$

$$4\left(\cos\dfrac{\pi}{6} + i\sin\dfrac{\pi}{6}\right)$$

$$4\left(\cos\dfrac{5\pi}{6} + i\sin\dfrac{5\pi}{6}\right)$$

$$4\left(\cos\dfrac{9\pi}{6} + i\sin\dfrac{9\pi}{6}\right) = 4\left(\cos\dfrac{3\pi}{2} + i\sin\dfrac{3\pi}{2}\right)$$

(c) $2\sqrt{3} + 2i, -2\sqrt{3} + 2i, -4i$

135. (a) Fifth roots of $1 = \cos 0 + i\sin 0$:

(b)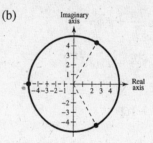

$$\cos\dfrac{2k\pi}{5} + i\sin\dfrac{2k\pi}{5}, \quad k = 0, 1, 2, 3, 4$$

$$\cos 0 + i\sin 0$$

$$\cos\dfrac{2\pi}{5} + i\sin\dfrac{2\pi}{5}$$

$$\cos\dfrac{4\pi}{5} + i\sin\dfrac{4\pi}{5}$$

$$\cos\dfrac{6\pi}{5} + i\sin\dfrac{6\pi}{5}$$

$$\cos\dfrac{8\pi}{5} + i\sin\dfrac{8\pi}{5}$$

(c) $1, 0.3090 + 0.9511i, -0.8090 + 0.5878i, -0.8090 - 0.5878i, 0.3090 - 0.9511i$

137. (a) Cube roots of $-125 = 125(\cos 180° + i\sin 180°)$ are:

(b)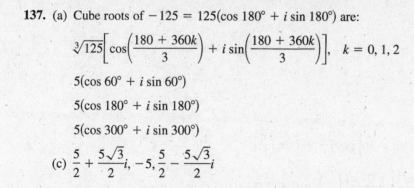

$$\sqrt[3]{125}\left[\cos\left(\dfrac{180 + 360k}{3}\right) + i\sin\left(\dfrac{180 + 360k}{3}\right)\right], \quad k = 0, 1, 2$$

$$5(\cos 60° + i\sin 60°)$$

$$5(\cos 180° + i\sin 180°)$$

$$5(\cos 300° + i\sin 300°)$$

(c) $\dfrac{5}{2} + \dfrac{5\sqrt{3}}{2}i, -5, \dfrac{5}{2} - \dfrac{5\sqrt{3}}{2}i$

139. (a) Fifth roots of $128(-1 + i) = 128\sqrt{2}(\cos 135° + i\sin 135°)$ are:

(b)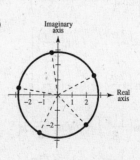

$$2\sqrt{2}(\cos 27° + i\sin 27°) = 2\sqrt{2}\left(\cos\dfrac{3\pi}{20} + i\sin\dfrac{3\pi}{20}\right)$$

$$2\sqrt{2}(\cos 99° + i\sin 99°)$$

$$2\sqrt{2}(\cos 171° + i\sin 171°)$$

$$2\sqrt{2}(\cos 243° + i\sin 243°)$$

$$2\sqrt{2}(\cos 315° + i\sin 315°)$$

(c) $2.52 + 1.28i, -0.44 + 2.79i, -2.79 + 0.44i, -1.28 - 2.52i, 2 - 2i$

141. $x^4 - i = 0$

$$x^4 = i$$

The solutions are the fourth roots of $i = \cos\dfrac{\pi}{2} + i\sin\dfrac{\pi}{2}$:

$$\sqrt[4]{1}\left[\cos\left(\frac{(\pi/2) + 2k\pi}{4}\right) + i\sin\left(\frac{(\pi/2) + 2k\pi}{4}\right)\right], \quad k = 0, 1, 2, 3$$

$$\cos\frac{\pi}{8} + i\sin\frac{\pi}{8}$$

$$\cos\frac{5\pi}{8} + i\sin\frac{5\pi}{8}$$

$$\cos\frac{9\pi}{8} + i\sin\frac{9\pi}{8}$$

$$\cos\frac{13\pi}{8} + i\sin\frac{13\pi}{8}$$

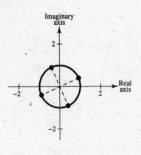

143. $x^5 = -243$

The solutions are the fifth roots of $-243 = 243[\cos\pi + i\sin\pi]$:

$$\sqrt[5]{243}\left[\cos\left(\frac{\pi + 2k\pi}{5}\right) + i\sin\left(\frac{\pi + 2k\pi}{5}\right)\right], \quad k = 0, 1, 2, 3, 4$$

$$3\left(\cos\frac{\pi}{5} + i\sin\frac{\pi}{5}\right)$$

$$3\left(\cos\frac{3\pi}{5} + i\sin\frac{3\pi}{5}\right)$$

$$3\left(\cos\frac{5\pi}{5} + i\sin\frac{5\pi}{5}\right) = -3$$

$$3\left(\cos\frac{7\pi}{5} + i\sin\frac{7\pi}{5}\right)$$

$$3\left(\cos\frac{9\pi}{5} + i\sin\frac{9\pi}{5}\right)$$

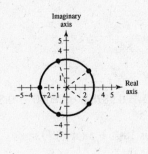

145. $x^4 = -16i$

The solutions are the fourth roots of $-16i = 16\left[\cos\dfrac{3\pi}{2} + i\sin\dfrac{3\pi}{2}\right]$:

$$\sqrt[4]{16}\left[\cos\left(\frac{(3\pi/2) + 2k\pi}{4}\right) + i\sin\left(\frac{(3\pi/2) + 2k\pi}{4}\right)\right], \quad k = 0, 1, 2, 3$$

$$2\left[\cos\left(\frac{3\pi}{8}\right) + i\sin\left(\frac{3\pi}{8}\right)\right]$$

$$2\left[\cos\left(\frac{7\pi}{8}\right) + i\sin\left(\frac{7\pi}{8}\right)\right]$$

$$2\left[\cos\left(\frac{11\pi}{8}\right) + i\sin\left(\frac{11\pi}{8}\right)\right]$$

$$2\left[\cos\left(\frac{15\pi}{8}\right) + i\sin\left(\frac{15\pi}{8}\right)\right]$$

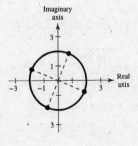

147. $x^3 - (1 - i) = 0$

$$x^3 = 1 - i = \sqrt{2}(\cos 315° + i \sin 315°)$$

The solutions are the cube roots of $1 - i$:

$$\sqrt[3]{\sqrt{2}}\left[\cos\left(\frac{315° + 360°k}{3}\right) + i \sin\left(\frac{315° + 360°k}{3}\right)\right], \quad k = 0, 1, 2$$

$\sqrt[6]{2}(\cos 105° + i \sin 105°)$

$\sqrt[6]{2}(\cos 225° + i \sin 225°)$

$\sqrt[6]{2}(\cos 345° + i \sin 345°)$

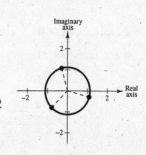

149. $E = I \cdot Z$

$= (10 + 2i)(4 + 3i)$

$= (40 - 6) + (30 + 8)i$

$= 34 + 38i$

151. $Z = \dfrac{E}{I}$

$= \dfrac{5 + 5i}{2 + 4i} \cdot \dfrac{2 - 4i}{2 - 4i}$

$= \dfrac{(10 + 20) + (10 - 20)i}{4 + 16}$

$= \dfrac{30 - 10i}{20}$

$= \dfrac{3}{2} - \dfrac{1}{2}i$

153. $I = \dfrac{E}{Z}$

$= \dfrac{12 + 24i}{12 + 20i} \cdot \dfrac{12 - 20i}{12 - 20i}$

$= \dfrac{(144 + 480) + (288 - 240)i}{144 + 400}$

$= \dfrac{624 + 48i}{544}$

$= \dfrac{39}{34} + \dfrac{3}{34}i$

155. True. $\left[\dfrac{1}{2}\left(1 - \sqrt{3}i\right)\right]^9 = \left[\dfrac{1}{2} - \dfrac{\sqrt{3}}{2}i\right]^9 = -1$

157. True

159. $z = r(\cos \theta + i \sin \theta)$

$\bar{z} = r(\cos \theta - i \sin \theta)$

$= r(\cos(-\theta) + i \sin(-\theta))$

161. $z = r(\cos \theta + i \sin \theta)$

$-z = -r(\cos \theta + i \sin \theta)$

$= r(-\cos \theta - i \sin \theta)$

$= r(\cos(\theta + \pi) + i \sin(\theta + \pi))$

163. $d = 16 \cos\left(\dfrac{\pi}{4}t\right)$

Maximum displacement: 16

Lowest possible t-value: $\dfrac{\pi}{4}t = \dfrac{\pi}{2} \Rightarrow t = 2$

165. $d = \dfrac{1}{8}\cos(12\pi t)$

Maximum displacement: $\dfrac{1}{8}$

Lowest possible t-value: $12\pi t = \dfrac{\pi}{2} \Rightarrow t = \dfrac{1}{24}$

167. $2\cos(x + \pi) + 2\cos(x - \pi) = 0$

$$4\cos x \cos \pi = 0$$

$$\cos x = 0$$

$$x = \frac{\pi}{2}, \frac{3\pi}{2}$$

169. $\sin\left(x - \frac{\pi}{3}\right) - \sin\left(x + \frac{\pi}{3}\right) = \frac{3}{2}$

$$\frac{1}{2}\left[\sin\left(x + \frac{\pi}{3}\right) - \sin\left(x - \frac{\pi}{3}\right)\right] = \left(-\frac{3}{2}\right)\frac{1}{2}$$

$$\cos x \sin \frac{\pi}{3} = -\frac{3}{4}$$

$$\cos x\left(\frac{\sqrt{3}}{2}\right) = -\frac{3}{4}$$

$$\cos x = -\frac{\sqrt{3}}{2}$$

$$x = \frac{5\pi}{6}, \frac{7\pi}{6}$$

Review Exercises for Chapter 6

1. Given: $A = 32°, B = 50°, a = 16$

$C = 180° - 32° - 50° = 98°$

$b = \dfrac{a \sin B}{\sin A} = \dfrac{16 \sin 50°}{\sin 32°} \approx 23.13$

$c = \dfrac{a \sin C}{\sin A} = \dfrac{16 \sin 98°}{\sin 32°} \approx 29.90$

3. Given: $B = 25°, C = 105°, c = 25$

$A = 180° - 25° - 105° = 50°$

$b = \dfrac{c \sin B}{\sin C} = \dfrac{25 \sin 25°}{\sin 105°} \approx 10.94$

$a = \dfrac{c \sin A}{\sin C} = \dfrac{25 \sin 50°}{\sin 105°} \approx 19.83$

5. Given: $A = 60° \ 15' = 60.25°, \quad B = 45° \ 30' = 45.5°, \ b = 4.8$

$C = 180° - 60.25° - 45.5° = 74.25° = 74° \ 15'$

$a = \dfrac{b \sin A}{\sin B} = \dfrac{4.8 \sin 60.25°}{\sin 45.5°} \approx 5.84$

$c = \dfrac{b \sin C}{\sin B} = \dfrac{4.8 \sin 74.25°}{\sin 45.5°} \approx 6.48$

7. Given: $A = 75°, \ a = 2.5, \ b = 16.5$

$\sin B = \dfrac{b \sin A}{a} = \dfrac{16.5 \sin 75°}{2.5} \approx 6.375 \Rightarrow$ no triangle formed

No solution

9. Given: $B = 115°, a = 9, b = 14.5$

$\sin A = \dfrac{a \sin B}{b} = \dfrac{9 \sin 115°}{14.5} \approx 0.5625 \Rightarrow A \approx 34.2°$

$C \approx 180° - 115° - 34.2° = 30.8°$

$c = \dfrac{b}{\sin B}(\sin C) \approx \dfrac{14.5}{\sin 115°}(\sin 30.8°) \approx 8.18$

11. Given: $C = 50°$, $a = 25$, $c = 22$

$$\sin A = \frac{a \sin C}{c} = \frac{25 \sin 50°}{22} \approx \frac{25(0.7660)}{22} \approx 0.8705 \implies A \approx 60.5° \text{ or } 119.5°$$

Case 1:

$A \approx 60.5°$

$B \approx 180° - 50° - 60.5° = 69.5°$

$b = \frac{c \sin B}{\sin C} \approx \frac{22(0.9367)}{0.7660} \approx 26.90$

Case 2:

$A \approx 119.5°$

$B \approx 180° - 50° - 119.5° = 10.5°$

$b = \frac{c \sin B}{\sin C} \approx 5.24$

13. $A = 27°$, $b = 5$, $c = 8$

Area $= \frac{1}{2}bc \sin A$

$= \frac{1}{2}(5)(8)(\sin 27°)$

≈ 9.08 square units

15. $C = 122°$, $b = 18$, $a = 29$

Area $= \frac{1}{2}ab \sin C$

$= \frac{1}{2}(29)(18) \sin 122°$

≈ 221.34 square units

17. $h = 50 \tan 17° \approx 15.3$ meters

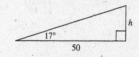

19. $\sin 28° = \dfrac{h}{75}$

$h = 75 \sin 28° \approx 35.21$ feet

$\cos 28° = \dfrac{x}{75}$

$x = 75 \cos 28° \approx 66.22$ feet

$\tan 45° = \dfrac{H}{x}$

$H = x \tan 45° \approx 66.22$ feet

Height of tree: $H - h \approx 31$ feet

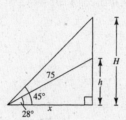

21. Given: $a = 18$, $b = 12$, $c = 15$

$\cos A = \dfrac{b^2 + c^2 - a^2}{2bc} = \dfrac{12^2 + 15^2 - 18^2}{2(12)(15)} = 0.125 \implies A \approx 82.82°$

$\cos B = \dfrac{a^2 + c^2 - b^2}{2ac} = \dfrac{18^2 + 15^2 - 12^2}{2(18)(15)} = 0.75 \implies B \approx 41.41°$

$C = 180° - A - B \approx 55.77°$

23. Given: $a = 9$, $b = 12$, $c = 20$

$\cos C = \dfrac{a^2 + b^2 - c^2}{2ab}$

$= \dfrac{81 + 144 - 400}{2(9)(12)}$

$\approx -0.8102 \implies C \approx 144.1°$

$\sin A = \dfrac{a \sin C}{c}$

$= \dfrac{9 \sin(144.1°)}{20}$

$\approx 0.264 \implies A \approx 15.3$

$B = 180° - 144.1° - 15.3° = 20.6°$

25. Given: $a = 6.5, b = 10.2, c = 16$

$$\cos A = \frac{b^2 + c^2 - a^2}{2bc} = \frac{10.2^2 + 16^2 - 6.5^2}{2(10.2)(16)} \approx 0.97 \implies A \approx 13.19°$$

$$\cos B = \frac{a^2 + c^2 - b^2}{2ac} = \frac{6.5^2 + 16^2 - 10.2^2}{2(6.5)(16)} \approx 0.93 \implies B \approx 20.98°$$

$$C = 180° - A - B \approx 145.83°$$

27. Given: $C = 65°, a = 25, b = 12$

$$c^2 = a^2 + b^2 - 2ab \cos C = 25^2 + 12^2 - 2(25)(12) \cos 65° \approx 515.4290 \implies c \approx 22.70$$

$$\sin A = \frac{a \sin C}{c} = \frac{25 \sin 65°}{22.70} \approx 0.998 \implies A \approx 86.38°$$

$$B = 180° - A - C \approx 28.62°$$

29. Given: $B = 110°, a = 4, c = 4$

$$b^2 = a^2 + c^2 - 2ac \cos B = 16 + 16 - 2(4)(4)(\cos 110) \approx 42.94 \implies b \approx 6.55$$

$$\sin A = \frac{a \sin B}{b} \approx \frac{4 \sin 110°}{6.55} \approx 0.5739 \implies A \approx 35°$$

$$c = a \implies C = A \approx 35°$$

31. Given: $B = 55° \, 30' = 55.5°, \, a = 12.4, \, c = 18.5$

$$b^2 = a^2 + c^2 - 2ac \cos B = 12.4^2 + 18.5^2 - 2(12.4)(18.5) \cos 55.5° \approx 236.1428 \implies b \approx 15.37$$

$$\sin A = \frac{a \sin B}{b} \approx \frac{12.4 \sin 55.5°}{15.37} \approx 0.665 \implies A \approx 41.68° \approx 41° \, 41'$$

$$C = 180° - A - B = 82.82° \approx 82° \, 49'$$

33. $a^2 = 5^2 + 8^2 - 2(5)(8) \cos 152° \approx 159.6 \implies a \approx 12.63$ ft

    $b^2 = 5^2 + 8^2 - 2(5)(8) \cos 28° \approx 18.36 \implies b \approx 4.285$ ft

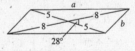

35. Angle between planes is $5° + 67° = 72°$.

In two hours, distances from airport are 850 miles and 1060 miles.

By the Law of Cosines,

$$d^2 = 850^2 + 1060^2 - 2(850)(1060) \cos(72°) \approx 1,289,251.376$$

$$d \approx 1135.5 \text{ miles.}$$

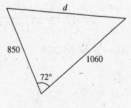

37. $a = 4, \, b = 5, \, c = 7$

$$s = \frac{a + b + c}{2} = \frac{4 + 5 + 7}{2} = 8$$

$$\text{Area} = \sqrt{s(s - a)(s - b)(s - c)}$$

$$= \sqrt{8(4)(3)(1)}$$

$$\approx 9.798 \text{ square units}$$

39. $a = 64.8, \, b = 49.2, \, c = 24.1$

$$s = \frac{a + b + c}{2} = \frac{64.8 + 49.2 + 24.1}{2} = 69.05$$

$$\text{Area} = \sqrt{s(s - a)(s - b)(s - c)}$$

$$= \sqrt{69.05(4.25)(19.85)(44.95)}$$

$$\approx 511.7 \text{ square units}$$

41. Initial point: $(-5, 4)$

Terminal point: $(2, -1)$

$\mathbf{v} = \langle 2 - (-5), -1 - 4 \rangle = \langle 7, -5 \rangle$

43. Initial point: $(0, 10)$

Terminal point: $(7, 3)$

$\mathbf{v} = \langle 7 - 0, 3 - 10 \rangle = \langle 7, -7 \rangle$

45. $8 \cos 120°\mathbf{i} + 8 \sin 120°\mathbf{j} = \langle -4, 4\sqrt{3} \rangle$

47. $2\mathbf{u}$

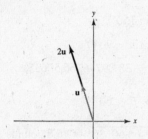

49. $2\mathbf{u} + \mathbf{v}$

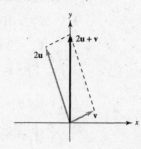

51. $\mathbf{u} - 2\mathbf{v}$

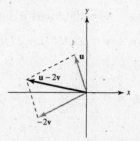

53. (a) $\mathbf{u} + \mathbf{v} = \langle -1, -3 \rangle + \langle -3, 6 \rangle = \langle -4, 3 \rangle$

(b) $\mathbf{u} - \mathbf{v} = \langle 2, -9 \rangle$

(c) $3\mathbf{u} = \langle -3, -9 \rangle$

(d) $2\mathbf{v} + 5\mathbf{u} = \langle -6, 12 \rangle + \langle -5, -15 \rangle = \langle -11, -3 \rangle$

55. (a) $\mathbf{u} + \mathbf{v} = \langle -5, 2 \rangle + \langle 4, 4 \rangle = \langle -1, 6 \rangle$

(b) $\mathbf{u} - \mathbf{v} = \langle -9, -2 \rangle$

(c) $3\mathbf{u} = \langle -15, 6 \rangle$

(d) $2\mathbf{v} + 5\mathbf{u} = \langle 8, 8 \rangle + \langle -25, 10 \rangle = \langle -17, 18 \rangle$

57. (a) $\mathbf{u} + \mathbf{v} = \langle 2, -1 \rangle + \langle 5, 3 \rangle = \langle 7, 2 \rangle$

(b) $\mathbf{u} - \mathbf{v} = \langle -3, -4 \rangle$

(c) $3\mathbf{u} = \langle 6, -3 \rangle$

(d) $2\mathbf{v} + 5\mathbf{u} = \langle 10, 6 \rangle + \langle 10, -5 \rangle = \langle 20, 1 \rangle$

59. (a) $\mathbf{u} + \mathbf{v} = \langle 4, 0 \rangle + \langle -1, 6 \rangle = \langle 3, 6 \rangle$

(b) $\mathbf{u} - \mathbf{v} = \langle 5, -6 \rangle$

(c) $3\mathbf{u} = \langle 12, 0 \rangle$

(d) $2\mathbf{v} + 5\mathbf{u} = \langle -2, 12 \rangle + \langle 20, 0 \rangle = \langle 18, 12 \rangle$

61. $3\mathbf{v} = 3(10\mathbf{i} + 3\mathbf{j}) = 30\mathbf{i} + 9\mathbf{j} = \langle 30, 9 \rangle$

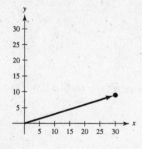

63. $\mathbf{w} = 4\mathbf{u} + 5\mathbf{v} = 4(6\mathbf{i} - 5\mathbf{j}) + 5(10\mathbf{i} + 3\mathbf{j})$

$= 74\mathbf{i} - 5\mathbf{j}$

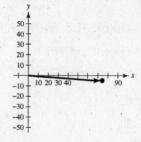

65. $\|\mathbf{u}\| = 6$

Unit vector: $\frac{1}{6}\langle 0, -6 \rangle = \langle 0, -1 \rangle$

67. $\|\mathbf{v}\| = \sqrt{5^2 + (-2)^2} = \sqrt{29}$

Unit vector: $\frac{1}{\sqrt{29}}\langle 5, -2 \rangle = \left\langle \frac{5}{\sqrt{29}}, \frac{-2}{\sqrt{29}} \right\rangle$

69. $\mathbf{u} = \langle 1 - (-8), -5 - 3 \rangle = \langle 9, -8 \rangle = 9\mathbf{i} - 8\mathbf{j}$

71. $\mathbf{v} = -10\mathbf{i} + 10\mathbf{j}$

$\|\mathbf{v}\| = \sqrt{(-10)^2 + (10)^2} = \sqrt{200} = 10\sqrt{2}$

$\tan \theta = \dfrac{10}{-10} = -1 \implies \theta = 135°$ since $\mathbf{v}$ is in Quadrant II.

$\mathbf{v} = 10\sqrt{2}(\cos 135°\mathbf{i} + \sin 135°\mathbf{j})$

73. $\mathbf{u} = 15[(\cos 20°)\mathbf{i} + (\sin 20°)\mathbf{j}]$

  $\mathbf{v} = 20[(\cos 63°)\mathbf{i} + (\sin 63°)\mathbf{j}]$

$\mathbf{u} + \mathbf{v} \approx 23.1752\mathbf{i} + 22.9504\mathbf{j}$

$\|\mathbf{u} + \mathbf{v}\| \approx 32.62$

$\tan \theta = \dfrac{22.9504}{23.1752} \implies \theta \approx 44.72°$

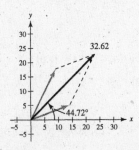

75. $\mathbf{F}_1 = 250\langle \cos 60°, \sin 60°\rangle$

  $\mathbf{F}_2 = 100\langle \cos 150°, \sin 150°\rangle$

  $\mathbf{F}_3 = 200\langle \cos(-90°), \sin(-90°)\rangle$

  $\mathbf{F} = \mathbf{F}_1 + \mathbf{F}_2 + \mathbf{F}_3 \approx \langle 38.39746, 66.50635\rangle$

$\tan \theta \approx \dfrac{66.50635}{38.39746} \implies \theta \approx 60°$

$\|\mathbf{F}\| = \sqrt{38.39746^2 + 66.50635^2} \approx 76.8 \text{ pounds}$

77. Rope One: $\mathbf{u} = \|\mathbf{u}\|(\cos 30°\mathbf{i} - \sin 30°\mathbf{j}) = \|\mathbf{u}\|\left(\dfrac{\sqrt{3}}{2}\mathbf{i} - \dfrac{1}{2}\mathbf{j}\right)$

Rope Two: $\mathbf{v} = \|\mathbf{u}\|(-\cos 30°\mathbf{i} - \sin 30°\mathbf{j}) = \|\mathbf{u}\|\left(-\dfrac{\sqrt{3}}{2}\mathbf{i} - \dfrac{1}{2}\mathbf{j}\right)$

Resultant: $\mathbf{u} + \mathbf{v} = -\|\mathbf{u}\|\mathbf{j} = -180\mathbf{j}$

  $\|\mathbf{u}\| = 180$

Therefore, the tension on each rope is $\|\mathbf{u}\| = 180$ pounds.

79. Airplane velocity: $\mathbf{u} = 430\langle \cos 315°, \sin 315°\rangle$

Wind velocity: $\mathbf{w} = 35\langle \cos 60°, \sin 60°\rangle$

  $\mathbf{u} + \mathbf{w} \approx \langle 321.5559, -273.7450\rangle$

$\|\mathbf{u} + \mathbf{w}\| \approx 422.3 \text{ mph}$

$\tan(\mathbf{u} + \mathbf{w}) = \dfrac{-273.7450}{321.5559}$

  $\approx -0.8513 \implies \theta \approx -40.4°$

The bearing from the north is $90° + 40.4° = 130.4°$.

81. $\mathbf{u} \cdot \mathbf{v} = \langle 0, -2\rangle \cdot \langle 1, 10\rangle = 0 - 20 = -20$

83. $\mathbf{u} \cdot \mathbf{v} = \langle 6, -1\rangle \cdot \langle 2, 5\rangle = 6(2) + (-1)(5) = 7$

85. $\mathbf{u} \cdot \mathbf{u} = \langle -3, -4\rangle \cdot \langle -3, -4\rangle$

  $= 9 + 16 = 25 = \|\mathbf{u}\|^2$

87. $4\mathbf{u} \cdot \mathbf{v} = 4\langle -3, -4\rangle \cdot \langle 2, 1\rangle = 4(-6 - 4) = -40$

89. $\mathbf{u} = \langle 2\sqrt{2}, -4 \rangle, \mathbf{v} = \langle -\sqrt{2}, 1 \rangle$

$\cos \theta = \dfrac{\mathbf{u} \cdot \mathbf{v}}{\|\mathbf{u}\| \, \|\mathbf{v}\|} = \dfrac{-8}{(\sqrt{24})(\sqrt{3})} \implies \theta \approx 160.5°$

91. $\mathbf{u} = \cos\dfrac{7\pi}{4}\mathbf{i} + \sin\dfrac{7\pi}{4}\mathbf{j} = \left\langle \dfrac{1}{\sqrt{2}}, -\dfrac{1}{\sqrt{2}} \right\rangle$

$\mathbf{v} = \cos\dfrac{5\pi}{6}\mathbf{i} + \sin\dfrac{5\pi}{6}\mathbf{j} = \left\langle -\dfrac{\sqrt{3}}{2}, \dfrac{1}{2} \right\rangle$

$\cos \theta = \dfrac{\mathbf{u} \cdot \mathbf{v}}{\|\mathbf{u}\| \, \|\mathbf{v}\|} = \dfrac{\left(-\sqrt{3}/(2\sqrt{2})\right) - \left(1/(2\sqrt{2})\right)}{(1)(1)}$

$\approx -0.966 \implies \theta \approx 165°$ or $\dfrac{11\pi}{12}$

93. $\cos \theta = \dfrac{\mathbf{u} \cdot \mathbf{v}}{\|\mathbf{u}\| \, \|\mathbf{v}\|}$

$= 0 \implies \theta = 90°$

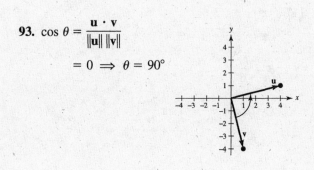

95. $\cos \theta = \dfrac{\mathbf{u} \cdot \mathbf{v}}{\|\mathbf{u}\| \, \|\mathbf{v}\|} = \dfrac{70 - 15}{\sqrt{74}\sqrt{109}}$

≈ 0.612

$\implies \theta \approx 52.2°$

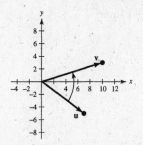

97. $\mathbf{u} = \langle 39, -12 \rangle, \mathbf{v} = \langle -26, 8 \rangle$

$\mathbf{u} \cdot \mathbf{v} = 39(-26) + (-12)(8)$

$= -1110 \neq 0 \implies \mathbf{u}$ and $\mathbf{v}$ are not orthogonal.

$\mathbf{v} = -\tfrac{2}{3}\mathbf{u} \implies \mathbf{u}$ and $\mathbf{v}$ are parallel.

99. $\mathbf{u} = \langle 8, 5 \rangle, \mathbf{v} = \langle -2, 4 \rangle$

$\mathbf{u} \cdot \mathbf{v} = 8(-2) + (5)(4)$

$= 4 \neq 0 \implies \mathbf{u}$ and $\mathbf{v}$ are not orthogonal.

$\mathbf{u} \neq k\mathbf{v} \implies \mathbf{u}$ and $\mathbf{v}$ are not parallel.

Neither

101. $\langle 1, -k \rangle \cdot \langle 1, 2 \rangle = 1 - 2k = 0 \implies k = \tfrac{1}{2}$

103. $\langle k, -1 \rangle \cdot \langle 2, -2 \rangle = 2k + 2 = 0 \implies k = -1$

105. $\mathbf{u} = \langle -4, 3 \rangle, \ \mathbf{v} = \langle -8, -2 \rangle$

$\text{proj}_{\mathbf{v}}\mathbf{u} = \left(\dfrac{\mathbf{u} \cdot \mathbf{v}}{\|\mathbf{v}\|^2}\right)\mathbf{v} = \left(\dfrac{26}{68}\right)\langle -8, -2 \rangle = -\dfrac{13}{17}\langle 4, 1 \rangle$

$\mathbf{u} - \text{proj}_{\mathbf{v}}\mathbf{u} = \langle -4, 3 \rangle - \left\langle \dfrac{-52}{17}, \dfrac{-13}{17} \right\rangle$

$= \left\langle -\dfrac{16}{17}, \dfrac{64}{17} \right\rangle$

$\mathbf{u} = \left\langle -\dfrac{52}{17}, -\dfrac{13}{17} \right\rangle + \left\langle -\dfrac{16}{17}, \dfrac{64}{17} \right\rangle$

107. $\mathbf{u} = \langle 2, 7 \rangle, \mathbf{v} = \langle 1, -1 \rangle$

$\text{proj}_{\mathbf{v}}\mathbf{u} = \left(\dfrac{\mathbf{u} \cdot \mathbf{v}}{\|\mathbf{v}\|^2}\right)\mathbf{v} = \dfrac{-5}{2}\langle 1, -1 \rangle = \left\langle -\dfrac{5}{2}, \dfrac{5}{2} \right\rangle$

$\mathbf{u} - \text{proj}_{\mathbf{v}}\mathbf{u} = \langle 2, 7 \rangle - \left\langle -\dfrac{5}{2}, \dfrac{5}{2} \right\rangle = \left\langle \dfrac{9}{2}, \dfrac{9}{2} \right\rangle$

$\mathbf{u} = \left\langle -\dfrac{5}{2}, \dfrac{5}{2} \right\rangle + \left\langle \dfrac{9}{2}, \dfrac{9}{2} \right\rangle$

109. 48 inches = 4 feet

Work = $18{,}000(4) = 72{,}000$ ft $\cdot$ lb

111. $|-i| = 1$

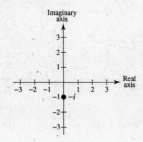

113. $|7 - 5i| = \sqrt{7^2 + (-5)^2} = \sqrt{74}$

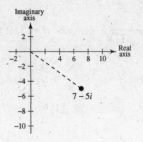

115. $z = 2 - 2i$

$|z| = \sqrt{4 + 4} = 2\sqrt{2}$

$\theta = \dfrac{7\pi}{4}$

$z = 2\sqrt{2}\left(\cos\dfrac{7\pi}{4} + i\sin\dfrac{7\pi}{4}\right)$

117. $z = -\sqrt{3} - i$

$|z| = \sqrt{3 + 1} = 2$

$\theta = \dfrac{7\pi}{6}$

$z = 2\left(\cos\dfrac{7\pi}{6} + i\sin\dfrac{7\pi}{6}\right)$

119. $z = -2i$

$|z| = 2$

$\theta = \dfrac{3\pi}{2}$

$z = 2\left(\cos\dfrac{3\pi}{2} + i\sin\dfrac{3\pi}{2}\right)$

121. $\left[\dfrac{5}{2}\left(\cos\dfrac{\pi}{2} + i\sin\dfrac{\pi}{2}\right)\right]\left[4\left(\cos\dfrac{\pi}{4} + i\sin\dfrac{\pi}{4}\right)\right] = 10\left[\cos\dfrac{3\pi}{4} + i\sin\dfrac{3\pi}{4}\right]$

123. $\dfrac{20(\cos 320° + i\sin 320°)}{5(\cos 80° + i\sin 80°)} = 4[\cos 240° + i\sin 240°]$

125. (a) $2 - 2i = 2\sqrt{2}\left(\cos\dfrac{7\pi}{4} + i\sin\dfrac{7\pi}{4}\right)$

$ 3 + 3i = 3\sqrt{2}\left(\cos\dfrac{\pi}{4} + i\sin\dfrac{\pi}{4}\right)$

(c) $(2 - 2i)(3 + 3i) = 6 + 6 = 12$

(b) $2\sqrt{2}\left(\cos\dfrac{7\pi}{4} + i\sin\dfrac{7\pi}{4}\right)3\sqrt{2}\left(\cos\dfrac{\pi}{4} + i\sin\dfrac{\pi}{4}\right) = 12(\cos 2\pi + i\sin 2\pi) = 12$

127. (a) $-i = \cos\dfrac{3\pi}{2} + i\sin\dfrac{3\pi}{2}$

$ 2 + 2i = 2\sqrt{2}\left(\cos\dfrac{\pi}{4} + i\sin\dfrac{\pi}{4}\right)$

(c) $-i(2 + 2i) = -2i + 2 = 2 - 2i$

(b) $\left(\cos\dfrac{3\pi}{2} + i\sin\dfrac{3\pi}{2}\right)2\sqrt{2}\left(\cos\dfrac{\pi}{4} + i\sin\dfrac{\pi}{4}\right) = 2\sqrt{2}\left(\cos\dfrac{7\pi}{4} + i\sin\dfrac{7\pi}{4}\right)$

$ = 2\sqrt{2}\left(\dfrac{\sqrt{2}}{2} + i\left(-\dfrac{\sqrt{2}}{2}\right)\right)$

$ = 2 - 2i$

129. (a) $3 - 3i = 3\sqrt{2}\left(\cos\dfrac{7\pi}{4} + i\sin\dfrac{7\pi}{4}\right)$

$2 + 2i = 2\sqrt{2}\left(\cos\dfrac{\pi}{4} + i\sin\dfrac{\pi}{4}\right)$

(b) $\dfrac{3\sqrt{2}\left(\cos\dfrac{7\pi}{4} + i\sin\dfrac{7\pi}{4}\right)}{2\sqrt{2}\left(\cos\dfrac{\pi}{4} + i\sin\dfrac{\pi}{4}\right)} = \dfrac{3}{2}\left(\cos\dfrac{3\pi}{2} + i\sin\dfrac{3\pi}{2}\right)$

$$= \dfrac{3}{2}(-i) = -\dfrac{3}{2}i$$

(c) $\dfrac{3 - 3i}{2 + 2i} \cdot \dfrac{(2 - 2i)}{(2 - 2i)} = \dfrac{6 - 12i - 6}{8} = -\dfrac{12i}{8} = -\dfrac{3}{2}i$

131. $\left[5\left(\cos\dfrac{\pi}{12} + i\sin\dfrac{\pi}{12}\right)\right]^4 = 5^4\left(\dfrac{4\pi}{12} + i\sin\dfrac{4\pi}{12}\right)$

$$= 625\left(\cos\dfrac{\pi}{3} + i\sin\dfrac{\pi}{3}\right)$$

$$= 625\left(\dfrac{1}{2} + \dfrac{\sqrt{3}}{2}i\right)$$

$$= \dfrac{625}{2} + \dfrac{625\sqrt{3}}{2}i$$

133. $(2 + 3i)^6 \approx \left[\sqrt{13}(\cos 56.3° + i\sin 56.3°)\right]^6$

$$= 13^3(\cos 337.9° + i\sin 337.9°)$$

$$\approx 13^3(0.9263 - 0.3769i)$$

$$\approx 2035 - 828i$$

135. $-\sqrt{3} + i = 2\left(\cos\dfrac{5\pi}{6} + i\sin\dfrac{5\pi}{6}\right)$

Square roots:

$\sqrt{2}\left(\cos\dfrac{5\pi}{12} + i\sin\dfrac{5\pi}{12}\right) \approx 0.3660 + 1.3660i$

$\sqrt{2}\left(\cos\dfrac{17\pi}{12} + i\sin\dfrac{17\pi}{12}\right) \approx -0.3660 - 1.3660i$

137. $-2i = 2\left(\cos\dfrac{3\pi}{2} + i\sin\dfrac{3\pi}{2}\right)$

Square roots:

$\sqrt{2}\left(\cos\dfrac{3\pi}{4} + i\sin\dfrac{3\pi}{4}\right) = \sqrt{2}\left(-\dfrac{\sqrt{2}}{2} + i\dfrac{\sqrt{2}}{2}\right) = -1 + i$

$\sqrt{2}\left(\cos\dfrac{7\pi}{4} + i\sin\dfrac{7\pi}{4}\right) = 1 - i$

139. $-2 - 2i = 2\sqrt{2}\left(\cos\dfrac{5\pi}{4} + i\sin\dfrac{5\pi}{4}\right)$

Square roots:

$2^{3/4}\left(\cos\dfrac{5\pi}{8} + i\sin\dfrac{5\pi}{8}\right) \approx -0.6436 + 1.5538i$

$2^{3/4}\left(\cos\dfrac{13\pi}{8} + i\sin\dfrac{13\pi}{8}\right) \approx 0.6436 - 1.5538i$

141. (a) Sixth roots of $-729i = 729\left(\cos\dfrac{3\pi}{2} + i\sin\dfrac{3\pi}{2}\right)$:

(b)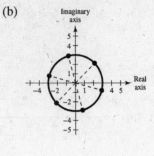

$$\sqrt[6]{729}\left(\cos\dfrac{(3\pi/2) + 2k\pi}{6} + i\sin\dfrac{(3\pi/2) + 2k\pi}{6}\right),\ k = 0, 1, 2, 3, 4, 5$$

$$3\left(\cos\dfrac{\pi}{4} + i\sin\dfrac{\pi}{4}\right)$$

$$3\left(\cos\dfrac{7\pi}{12} + i\sin\dfrac{7\pi}{12}\right)$$

$$3\left(\cos\dfrac{11\pi}{12} + i\sin\dfrac{11\pi}{12}\right)$$

$$3\left(\cos\dfrac{5\pi}{4} + i\sin\dfrac{5\pi}{4}\right)$$

$$3\left(\cos\dfrac{19\pi}{12} + i\sin\dfrac{19\pi}{12}\right)$$

$$3\left(\cos\dfrac{23\pi}{12} + i\sin\dfrac{23\pi}{12}\right)$$

(c) $\dfrac{3\sqrt{2}}{2} + \dfrac{3\sqrt{2}}{2}i,\ -0.7765 + 2.898i,\ -2.898 + 0.7765i,\ -\dfrac{3\sqrt{2}}{2} - \dfrac{3\sqrt{2}}{2}i,\ 0.7765 - 2.898i,\ 2.898 - 0.7765i$

143. (a) Cube roots of $8 = 8(\cos 0 + i\sin 0)$:

(b)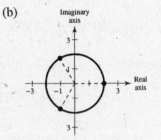

$$\sqrt[3]{8}\left(\cos\left(\dfrac{2k\pi}{3}\right) + i\sin\left(\dfrac{2k\pi}{3}\right)\right),\ k = 0, 1, 2$$

$$2(\cos 0 + i\sin 0)$$

$$2\left(\cos\dfrac{2\pi}{3} + i\sin\dfrac{2\pi}{3}\right)$$

$$2\left(\cos\dfrac{4\pi}{3} + i\sin\dfrac{4\pi}{3}\right)$$

(c) $2,\ -1 + \sqrt{3}i,\ -1 - \sqrt{3}i$

145. $x^4 + 256 = 0$

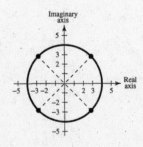

$$x^4 = -256 = 256(\cos\pi + i\sin\pi)$$

$$\sqrt[4]{-256} = 4\left[\cos\left(\dfrac{\pi + 2\pi k}{4}\right) + i\sin\left(\dfrac{\pi + 2\pi k}{4}\right)\right],\ k = 0, 1, 2, 3$$

$$4\left(\cos\dfrac{\pi}{4} + i\sin\dfrac{\pi}{4}\right) = \dfrac{4\sqrt{2}}{2} + \dfrac{4\sqrt{2}}{2}i = 2\sqrt{2} + 2\sqrt{2}i$$

$$4\left(\cos\dfrac{3\pi}{4} + i\sin\dfrac{3\pi}{4}\right) = -\dfrac{4\sqrt{2}}{2} + \dfrac{4\sqrt{2}}{2}i = -2\sqrt{2} + 2\sqrt{2}i$$

$$4\left(\cos\dfrac{5\pi}{4} + i\sin\dfrac{5\pi}{4}\right) = -\dfrac{4\sqrt{2}}{2} - \dfrac{4\sqrt{2}}{2}i = -2\sqrt{2} - 2\sqrt{2}i$$

$$4\left(\cos\dfrac{7\pi}{4} + i\sin\dfrac{7\pi}{4}\right) = \dfrac{4\sqrt{2}}{2} - \dfrac{4\sqrt{2}}{2}i = 2\sqrt{2} - 2\sqrt{2}i$$

147. $x^3 + 8i = 0$

$$x^3 = -8i$$

$$-8i = 8\left(\cos\frac{3\pi}{2} + i \sin\frac{3\pi}{2}\right)$$

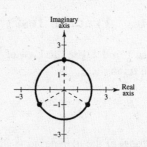

$$\sqrt[3]{-8i} = \sqrt[3]{8}\left[\cos\frac{(3\pi/2) + 2\pi k}{3} + i \sin\frac{(3\pi/2) + 2\pi k}{3}\right], \ k = 0, 1, 2$$

$$2\left(\cos\frac{\pi}{2} + i \sin\frac{\pi}{2}\right) = 2i$$

$$2\left(\cos\frac{7\pi}{6} + i \sin\frac{7\pi}{6}\right) = -\sqrt{3} - i$$

$$2\left(\cos\frac{11\pi}{6} + i \sin\frac{11\pi}{6}\right) = \sqrt{3} - i$$

149. True

151. Length and direction characterize vectors in plane.

153. $z_1 z_2 = 2(\cos\theta + i \sin\theta)2(\cos(\pi - \theta) + i \sin(\pi - \theta))$

$$= 4(\cos\theta + i \sin\theta)(-\cos\theta + i \sin\theta)$$

$$= 4(-\cos^2\theta - \sin^2\theta) = -4$$

$$\frac{z_1}{z_2} = \frac{2(\cos\theta + i \sin\theta)}{2(\cos(\pi - \theta) + i \sin(\pi - \theta))}$$

$$= \frac{\cos\theta + i \sin\theta}{-\cos\theta + i \sin\theta} \cdot \frac{-\cos\theta - i \sin\theta}{-\cos\theta - i \sin\theta}$$

$$= -\cos^2\theta + \sin^2\theta - 2\sin\theta\cos\theta i$$

$$= -(\cos\theta + i \sin\theta)^2$$

$$= -\left(\frac{z_1}{2}\right)^2$$

$$= -\frac{z_1{}^2}{4}$$

Chapter 6   Practice Test

For Exercises 1 and 2, use the Law of Sines to find the remaining sides and angles of the triangle.

1. $A = 40°$, $B = 12°$, $b = 100$ 2. $C = 150°$, $a = 5$, $c = 20$

3. Find the area of the triangle: $a = 3$, $b = 6$, $C = 130°$

4. Determine the number of solutions to the triangle: $a = 10$, $b = 35$, $A = 22.5°$

For Exercises 5 and 6, use the Law of Cosines to find the remaining sides and angles of the triangle.

5. $a = 49$, $b = 53$, $c = 38$ 6. $C = 29°$, $a = 100$, $b = 300$

7. Use Heron's Formula to find the area of the triangle: $a = 4.1$, $b = 6.8$, $c = 5.5$.

8. A ship travels 40 miles due east, then adjusts its course 12° southward. After traveling 70 miles in that direction, how far is the ship from its point of departure?

9. $\mathbf{w} = 4\mathbf{u} - 7\mathbf{v}$ where $\mathbf{u} = 3\mathbf{i} + \mathbf{j}$ and $\mathbf{v} = -\mathbf{i} + 2\mathbf{j}$. Find $\mathbf{w}$.

10. Find a unit vector in the direction of $\mathbf{v} = 5\mathbf{i} - 3\mathbf{j}$.

11. Find the dot product and the angle between $\mathbf{u} = 6\mathbf{i} + 5\mathbf{j}$ and $\mathbf{v} = 2\mathbf{i} - 3\mathbf{j}$.

12. $\mathbf{v}$ is a vector of magnitude 4 making an angle of 30° with the positive x-axis. Find $\mathbf{v}$ in component form.

13. Find the projection of $\mathbf{u}$ onto $\mathbf{v}$ given $\mathbf{u} = \langle 3, -1 \rangle$ and $\mathbf{v} = \langle -2, 4 \rangle$.

14. Give the trigonometric form of $z = 5 - 5i$.

15. Give the standard form of $z = 6(\cos 225° + i \sin 225°)$.

16. Multiply $[7(\cos 23° + i \sin 23°)][4(\cos 7° + i \sin 7°)]$.

17. Divide $\dfrac{9\left(\cos \dfrac{5\pi}{4} + i \sin \dfrac{5\pi}{4}\right)}{3(\cos \pi + i \sin \pi)}$.   18. Find $(2 + 2i)^8$.

19. Find the cube roots of $8\left(\cos \dfrac{\pi}{3} + i \sin \dfrac{\pi}{3}\right)$.   20. Find all the solutions to $x^4 + i = 0$.

C H A P T E R 7
Linear Systems and Matrices

Section 7.1   Solving Systems of Equations **298**

Section 7.2   Systems of Linear Equations in Two Variables **306**

Section 7.3   Multivariable Linear Systems **313**

Section 7.4   Matrices and Systems of Equations **326**

Section 7.5   Operations with Matrices **335**

Section 7.6   The Inverse of a Square Matrix **342**

Section 7.7   The Determinant of a Square Matrix **348**

Section 7.8   Applications of Matrices and Determinants **353**

Review Exercises **358**

Practice Test **371**

C H A P T E R 7
Linear Systems and Matrices

Section 7.1   Solving Systems of Equations

- You should be able to solve systems of equations by the method of substitution.
  1. Solve one of the equations for one of the variables.
  2. Substitute this expression into the other equation and solve.
  3. Back-substitute into the first equation to find the value of the other variable.
  4. Check your answer in each of the original equations.
- You should be able to find solutions graphically. (See Example 5 in textbook.)

Vocabulary Check

1. system, equations

2. solution

3. method, substitution

4. point, intersection

5. break-even point

---

1. (a) $4(0) - (-3) \overset{?}{=} 1$

   $6(0) + (-3) \overset{?}{=} -6$

   $3 \neq 1$

   $-3 \neq -6$

   No, $(0, -3)$ is not a solution.

   (b) $4(-1) - (-5) \overset{?}{=} 1$

   $6(-1) + (-5) \overset{?}{=} -6$

   $1 = 1$

   $-11 \neq -6$

   No, $(-1, -5)$ is not a solution.

   (c) $4\left(-\frac{3}{2}\right) - (3) \overset{?}{=} 1$

   $6\left(-\frac{3}{2}\right) + (3) \overset{?}{=} -6$

   $-9 \neq 1$

   $-6 = -6$

   No, $\left(-\frac{3}{2}, 3\right)$ is not a solution.

   (d) $4\left(-\frac{1}{2}\right) - (-3) \overset{?}{=} 1$

   $6\left(-\frac{1}{2}\right) + (-3) \overset{?}{=} -6$

   $1 = 1$

   $-6 = -6$

   Yes, $\left(-\frac{1}{2}, -3\right)$ is a solution.

3. (a) $\qquad 0 \overset{?}{=} -2e^{-2}$

   $3(-2) - 0 \overset{?}{=} 2$

   $0 \neq -2e^{-2}$

   $-6 \neq 2$

   No, $(-2, 0)$ is not a solution.

   (b) $\qquad -2 \overset{?}{=} -2e^{0}$

   $3(0) - (-2) \overset{?}{=} 2$

   $-2 = -2$

   $2 = 2$

   Yes, $(0, -2)$ is a solution.

—CONTINUED—

3. —CONTINUED—

(c) $\qquad -3 \overset{?}{=} -2e^0$

$\qquad 3(0) - (-3) \overset{?}{=} 2$

$\qquad\qquad -3 \neq -2$

$\qquad\qquad 3 \neq 2$

No, $(0, -3)$ is not a solution.

(d) $\qquad -5 \overset{?}{=} -2e^{-1}$

$\qquad 3(-1) - (-5) \overset{?}{=} 2$

$\qquad\qquad -5 \neq -2e^{-1}$

$\qquad\qquad 2 = 2$

No, $(-1, -5)$ is not a solution.

5. $\begin{cases} 2x + y = 6 & \text{Equation 1} \\ -x + y = 0 & \text{Equation 2} \end{cases}$

Solve for y in Equation 1: $y = 6 - 2x$

Substitute for y in Equation 2: $-x + (6 - 2x) = 0$

Solve for x: $-3x + 6 = 0 \implies x = 2$

Back-substitute $x = 2$: $y = 6 - 2(2) = 2$

Answer: $(2, 2)$

7. $\begin{cases} x - y = -4 & \text{Equation 1} \\ x^2 - y = -2 & \text{Equation 2} \end{cases}$

Solve for y in Equation 1: $y = x + 4$

Substitute for y in Equation 2: $x^2 - (x + 4) = -2$

Solve for x: $x^2 - x - 2 = 0$

$\qquad\qquad\qquad \implies (x + 1)(x - 2) = 0$

$\qquad\qquad\qquad \implies x = -1, 2$

Back-substitute $x = -1$: $y = -1 + 4 = 3$

Back-substitute $x = 2$: $y = 2 + 4 = 6$

Answer: $(-1, 3), (2, 6)$

9. $\begin{cases} 3x + y = 2 & \text{Equation 1} \\ x^3 - 2 + y = 0 & \text{Equation 2} \end{cases}$

Solve for y in Equation 1: $y = 2 - 3x$

Substitute for y in Equation 2: $x^3 - 2 + (2 - 3x) = 0$

Solve for x: $x^3 - 3x = 0 \implies x(x^2 - 3) = 0 \implies x = 0, \pm\sqrt{3}$

Back-substitute: $x = 0$: $y = 2$

$\qquad\qquad x = \sqrt{3}$: $y = 2 - 3\sqrt{3}$

$\qquad\qquad x = -\sqrt{3}$: $y = 2 + 3\sqrt{3}$

Answer: $(0, 2), \left(\sqrt{3}, 2 - 3\sqrt{3}\right), \left(-\sqrt{3}, 2 + 3\sqrt{3}\right)$

11. $\begin{cases} -\frac{7}{2}x - y = -18 & \text{Equation 1} \\ 8x^2 - 2y^3 = 0 & \text{Equation 2} \end{cases}$

Solve for x in Equation 1: $-\frac{7}{2}x = y - 18 \implies x = -\frac{2}{7}y + \frac{36}{7}$

Substitute for x in Equation 2: $8\left(-\frac{2}{7}y + \frac{36}{7}\right)^2 - 2y^3 = 0$

Solve for x: $\quad -2y^3 + 8\left(\frac{4}{49}y^2 - \frac{144}{49}y + \frac{36^2}{49}\right) = 0$

$\qquad\qquad 49y^3 - 16y^2 + 576y - 5184 = 0$

$\qquad\qquad (y - 4)(49y^2 + 180y + 1296) = 0$

Hence, $y = 4$ and $x = -\frac{2}{7}(4) + \frac{36}{7} = 4$.

Answer: $(4, 4)$

13. $\begin{cases} x - y = 0 & \text{Equation 1} \\ 5x - 3y = 10 & \text{Equation 2} \end{cases}$

Solve for y in Equation 1: $y = x$

Substitute for y in Equation 2: $5x - 3x = 10$

Solve for x: $2x = 10 \implies x = 5$

Back-substitute in Equation 1: $y = x = 5$

Answer: $(5, 5)$

15. $\begin{cases} 2x - y + 2 = 0 & \text{Equation 1} \\ 4x + y - 5 = 0 & \text{Equation 2} \end{cases}$

Solve for y in Equation 1: $y = 2x + 2$

Substitute for y in Equation 2:
$4x + (2x + 2) - 5 = 0$

Solve for x:
$4x + (2x + 2) - 5 = 0 \implies 6x - 3 = 0 \implies x = \frac{1}{2}$

Back-substitute $x = \frac{1}{2}$: $y = 2x + 2 = 2\left(\frac{1}{2}\right) + 2 = 3$

Answer: $\left(\frac{1}{2}, 3\right)$

17. $\begin{cases} 1.5x + 0.8y = 2.3 \implies 15x + 8y = 23 \\ 0.3x - 0.2y = 0.1 \implies 3x - 2y = 1 \end{cases}$

Solve for y in Equation 2: $-2y = 1 - 3x$

$$y = \frac{3x - 1}{2}$$

Substitute for y in Equation 1: $15x + 8\left(\frac{3x - 1}{2}\right) = 23$

$$15x + 12x - 4 = 23$$
$$27x = 27$$
$$x = 1$$

Then, $y = \dfrac{3x - 1}{2} = \dfrac{3(1) - 1}{2} = 1.$ *Answer:* $(1, 1)$

19. $\begin{cases} \frac{1}{5}x + \frac{1}{2}y = 8 & \text{Equation 1} \\ x + y = 20 & \text{Equation 2} \end{cases}$

Solve for x in Equation 2: $x = 20 - y$

Substitute for x in Equation 1: $\frac{1}{5}(20 - y) + \frac{1}{2}y = 8$

Solve for y: $4 + \frac{3}{10}y = 8 \implies y = \frac{40}{3}$

Back-substitute $y = \frac{40}{3}$: $x = 20 - y$

$$= 20 - \frac{40}{3} = \frac{20}{3}$$

Answer: $\left(\frac{20}{3}, \frac{40}{3}\right)$

21. $\begin{cases} -\frac{5}{3}x + y = 5 & \text{Equation 1} \\ -5x + 3y = 6 & \text{Equation 2} \end{cases}$

Solve for y in Equation 1: $y = 5 + \frac{5}{3}x$

Substitute for y in Equation 2: $-5x + 3\left(5 + \frac{5}{3}x\right) = 6$

Solve for x: $-5x + 15 + 5x = 6$

$$15 \neq 6 \quad \text{Inconsistent}$$

No solution

23. $\begin{cases} x^2 - 2x + y = 8 & \text{Equation 1} \\ x - y = -2 & \text{Equation 2} \end{cases}$

Solve for y in Equation 2: $y = x + 2$

Substitute for y in Equation 1:

$$x^2 - 2x + (x + 2) = 8$$
$$x^2 - x - 6 = 0$$
$$(x - 3)(x + 2) = 0$$
$$x = 3, -2$$

$x = 3 \implies y = 5$

$x = -2 \implies y = 0$

Answer: $(3, 5), (-2, 0)$

25. $\begin{cases} 2x^2 - y = 1 & \text{Equation 1} \\ x - y = 2 & \text{Equation 2} \end{cases}$

Solve for y in Equation 2: $y = x - 2$

Substitute for y in Equation 1: $2x^2 - (x - 2) = 1$

$$2x^2 - x + 1 = 0$$

No real solution

27. $\begin{cases} x^3 - y = 0 & \text{Equation 1} \\ x - y = 0 & \text{Equation 2} \end{cases}$

Solve for y in Equation 2: $y = x$

Substitute for y in Equation 1: $x^3 - x = 0$

Solve for x:

$x(x-1)(x+1) = 0 \implies x = 0, 1, -1$

Back-substitute: $x = 0 \implies y = 0$

$\qquad\qquad\qquad x = 1 \implies y = 1$

$\qquad\qquad\qquad x = -1 \implies y = -1$

Answer: $(0, 0), (1, 1), (-1, -1)$

29. $\begin{cases} -x + 2y = 2 \\ 3x + y = 15 \end{cases}$

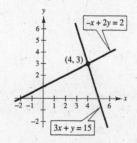

Point of intersection: $(4, 3)$

31. $\begin{cases} x - 3y = -2 \\ 5x + 3y = 17 \end{cases}$

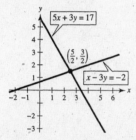

Point of intersection: $\left(\frac{5}{2}, \frac{3}{2} \right)$

33. $\begin{cases} x^2 + y = 1 \\ x + y = 2 \end{cases}$

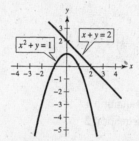

No solution

35. $\begin{cases} -x + y = 3 \implies y_1 = x + 3 \\ x^2 - 6x - 27 + y^2 = 0 \implies y_2 = \sqrt{6x - x^2 + 27} \\ \qquad\qquad\qquad\qquad\qquad y_3 = -\sqrt{6x - x^2 + 27} \end{cases}$

Points of intersection: $(-3, 0), (3, 6)$

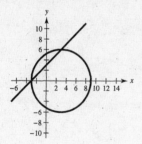

37. $\begin{cases} 7x + 8y = 24 \implies y_1 = -\frac{7}{8}x + 3 \\ x - 8y = 8 \implies y_2 = \frac{1}{8}x - 1 \end{cases}$

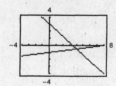

Point of intersection: $\left(4, -\frac{1}{2} \right)$

39. $x - y^2 = -1 \implies y^2 = x + 1 \implies y_1 = \sqrt{x+1}$

$\qquad\qquad\qquad\qquad\qquad\qquad\qquad y_2 = -\sqrt{x+1}$

$x - y = 5 \implies y_3 = x - 5$

Points of intersection: $(8, 3), (3, -2)$

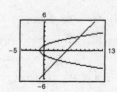

41. $x^2 + y^2 = 8 \implies y_1 = \sqrt{8 - x^2}, y_2 = -\sqrt{8 - x^2}$

$\qquad y = x^2 \implies y_3 = x^2$

Points of intersection: $(1.540, 2.372), (-1.540, 2.372)$

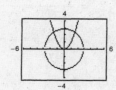

43. $\begin{cases} y = e^x \\ x - y + 1 = 0 \implies y = x + 1 \end{cases}$

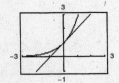

Point of intersection: $(0, 1)$

45. $\begin{cases} x + 2y = 8 & \implies y_1 = 4 - x/2 \\ y = 2 + \ln x & \implies y_2 = 2 + \ln x \end{cases}$

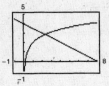

Point of intersection: $\approx (2.318, 2.841)$

47. $\begin{cases} y = \sqrt{x} + 4 \\ y = 2x + 1 \end{cases}$

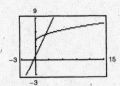

Point of intersection: $\left(\frac{9}{4}, \frac{11}{2}\right)$

49. $\begin{cases} x^2 + y^2 = 169 \implies y_1 = \sqrt{169 - x^2} \text{ and} \\ \qquad\qquad\qquad y_2 = -\sqrt{169 - x^2} \\ x^2 - 8y = 104 \implies y_3 = \frac{1}{8}x^2 - 13 \end{cases}$

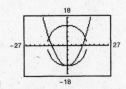

Points of intersection: $(0, -13), (\pm 12, 5)$

51. $\begin{cases} y = 2x & \text{Equation 1} \\ y = x^2 + 1 & \text{Equation 2} \end{cases}$

Substitute for y in Equation 2: $2x = x^2 + 1$

Solve for x:
$x^2 - 2x + 1 = (x - 1)^2 = 0 \implies x = 1$

Back-substitute $x = 1$ in Equation 1: $y = 2x = 2$

Answer: $(1, 2)$

53. $\begin{cases} 3x - 7y + 6 = 0 & \text{Equation 1} \\ \quad x^2 - y^2 = 4 & \text{Equation 2} \end{cases}$

Solve for y in Equation 1: $y = \dfrac{3x + 6}{7}$

Substitute for y in Equation 2: $x^2 - \left(\dfrac{3x + 6}{7}\right)^2 = 4$

Solve for x: $\quad x^2 - \left(\dfrac{9x^2 + 36x + 36}{49}\right) = 4$

$$49x^2 - (9x^2 + 36x + 36) = 196$$

$$40x^2 - 36x - 232 = 0$$

$$10x^2 - 9x - 58 = 0 \implies x = \frac{9 \pm \sqrt{81 + 40(58)}}{20} \implies x = \frac{29}{10}, -2$$

Back-substitute $x = \dfrac{29}{10}$: $y = \dfrac{3x + 6}{7} = \dfrac{3(29/10) + 6}{7} = \dfrac{21}{10}$

Back-substitute $x = -2$: $y = \dfrac{3x + 6}{7} = 0$

Answers: $\left(\frac{29}{10}, \frac{21}{10}\right), (-2, 0)$

55. $x^2 + y^2 = 1$

$x + y = 4$

Graphing $y_1 = \sqrt{1 - x^2}$, $y_2 = -\sqrt{1 - x^2}$ and $y_3 = 4 - x$, you see that there are no points of intersection.

No solution

57. $\begin{cases} y = 2x + 1 \\ y = \sqrt{x + 2} \end{cases}$

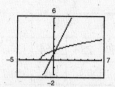

Point of intersection: $\left(\frac{1}{4}, \frac{3}{2}\right)$

59. $\begin{cases} y - e^{-x} = 1 \implies y = e^{-x} + 1 \\ y - \ln x = 3 \implies y = \ln x + 3 \end{cases}$

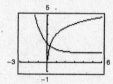

Point of intersection: Approximately $(0.287, 1.751)$

61. $\begin{cases} y = x^3 - 2x^2 + 1 \qquad \text{Equation 1} \\ y = 1 - x^2 \qquad\qquad\;\; \text{Equation 2} \end{cases}$

Substitute for y in Equation 2:

$x^3 - 2x^2 + 1 = 1 - x^2$

Solve for x: $x^3 - x^2 = 0$

$x^2(x - 1) = 0 \implies x = 0, 1$

Back-substitute: $x = 0 \implies y = 1$

$x = 1 \implies y = 0$

Answer: $(0, 1), (1, 0)$

63. $\begin{cases} xy - 1 = 0 \qquad\quad \text{Equation 1} \\ 2x - 4y + 7 = 0 \quad\; \text{Equation 2} \end{cases}$

Solve for y in Equation 1: $y = \dfrac{1}{x}$

Substitute for y in Equation 2: $2x - 4\left(\dfrac{1}{x}\right) + 7 = 0$

Solve for x: $2x^2 - 4 + 7x = 0$

$\implies (2x - 1)(x + 4) = 0 \implies x = \dfrac{1}{2}, -4$

Back-substitute $x = \dfrac{1}{2}$: $y = \dfrac{1}{1/2} = 2$

Back-substitute $x = -4$: $y = \dfrac{1}{-4} = -\dfrac{1}{4}$

Two points of intersection: $\left(\dfrac{1}{2}, 2\right), \left(-4, -\dfrac{1}{4}\right)$

65. $C = 8650x + 250,000, \; R = 9950x$

$R = C$

$9950x = 8650x + 250,000$

$1300x = 250,000$

$x \approx 192$ units

$R \approx \$1,910,400$ (Answers will vary.)

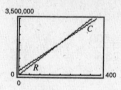

67. $C = 5.5\sqrt{x} + 10,000, \; R = 3.29x$

$R = C$

$3.29x = 5.5\sqrt{x} + 10,000$

$3.29x - 10,000 = 5.5\sqrt{x}$

$10.8241x^2 - 65,800x + 100,000,000 = 30.25x$

$10.8241x^2 - 65,830.25x + 100,000,000 = 0$

$x \approx 3133$ units

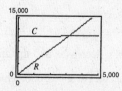

In order for the revenue to break even with the cost, 3133 units must be sold, $R \approx \$10,308$.

69. $N = 360 - 24x$ Animated film

     $N = 24 + 18x$ Horror film

(a)

Week x	1	2	3	4	5	6	7	8	9	10	11	12
Animated	336	312	288	264	240	216	192	168	144	120	96	72
Horror	42	60	78	96	114	132	150	168	186	204	222	240

(b) For $x = 8, N = 168$

(c) $360 - 24x = 24 + 18x$

         $336 = 42x$

           $x = 8$

      $N = 24 + 18(8) = 168$

(d) The answers are the same.

(e) During week 8 the same number (168) were rented.

71. (a) $C = 35.45x + 16,000$

     $R = 55.95x$

(b)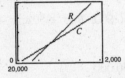

    $C = R$ for $x \approx 780$ units

    Algebraically,

    $35.45x + 16,000 = 55.95x$

          $16,000 = 20.5x$

              $x = \dfrac{16,000}{20.5} \approx 780$ units

73. $0.06x = 0.03x + 350$

    $0.03x = 350$

         $x \approx \$11,666.67$

To make the straight commission offer better, you would have to sell more than \$11,666.67 per week.

75. (a) $\begin{cases} x + y = 20,000 \\ 0.065x + 0.085y = 1600 \end{cases}$

(b)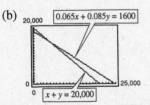

As x increases, y decreases and the amount of interest decreases.

(c) The curves intersect at $x = 5000$. Thus, \$5000 should be invested at 6.5%.

77. $M = 47.4t + 5104$

$P = 76.5t + 4875$

(a)

Year	1990	1994	1998	2002	2006	2010
Missouri	5104	5294	5483	5673	5862	6052
Tennessee	4875	5181	5487	5793	6099	6405

(b) From 1998 to 2010

(c)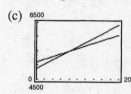

Point of intersection: (7.87, 5477.0)

(d) $47.4t + 5104 = 76.5t + 4875$

$$229 = 29.1t$$

$$t = 229/29.1 \approx 7.87$$

(e) The population of Tennessee surpassed that of Missouri during 1997.

79. $2l + 2w = 30 \Rightarrow l + w = 15$

$l = w + 3 \Rightarrow (w + 3) + w = 15$

$$2w = 12$$

$$w = 6$$

$l = w + 3 = 9$

Dimensions: 6 meters × 9 meters

81. $2l + 2w = 40 \Rightarrow l + w = 20 \Rightarrow w = 20 - l$

$lw = 96 \Rightarrow l(20 - l) = 96$

$$20l - l^2 = 96$$

$$0 = l^2 - 20l + 96$$

$$0 = (l - 8)(l - 12)$$

$$l = 8 \text{ or } l = 12$$

$l = 12, w = 8$

If the length is supposed to be greater than the width, we have $l = 12$ miles and $w = 8$ miles.

83. False. You could solve for x first.

85. The system has no solution if you arrive at a false statement, ie. $4 = 8$, or you have a quadratic equation with a negative discriminant, which would yield imaginary roots.

87. (a) The line $y = 2x$ intersects the parabola $y = x^2$ at two points, $(0, 0)$ and $(2, 4)$.

(b) The line $y = 0$ intersects $y = x^2$ at $(0, 0)$ only.

(c) The line $y = x - 2$ does not intersect $y = x^2$.

(Other answers possible.)

89. Answers will vary. For example,

$y = x - 3$

$2y = x - 4$

91. $(-2, 7), (5, 5)$

$$m = \frac{5 - 7}{5 - (-2)} = -\frac{2}{7}$$

$$y - 7 = -\frac{2}{7}(x - (-2))$$

$$7y - 49 = -2x - 4$$

$$2x + 7y - 45 = 0$$

93. $(6, 3), (10, 3)$

$$m = \frac{3 - 3}{10 - 6} = 0$$

The line is horizontal.

$$y = 3 \implies y - 3 = 0$$

95. $\left(\frac{3}{5}, 0\right), (4, 6)$

$$m = \frac{6 - 0}{4 - \frac{3}{5}} = \frac{6}{\frac{17}{5}} = \frac{30}{17}$$

$$y - 6 = \frac{30}{17}(x - 4)$$

$$17y - 102 = 30x - 120$$

$$0 = 30x - 17y - 18$$

97. Domain: all $x \neq 6$

Vertical asymptotes: $x = 6$

Horizontal asymptote: $y = 0$

99. Domain: all $x \neq \pm 4$

Vertical asymptotes: $x = \pm 4$

Horizontal asymptote: $y = 1$

101. Domain: all real numbers x

Horizontal asymptote: $y = 0$

Section 7.2   Systems of Linear Equations in Two Variables

■ You should be able to solve a linear system by the method of elimination.

1. Obtain coefficients for either x or y that differ only in sign. This is done by multiplying all the terms of one or both equations by appropriate constants.

2. Add the equations to eliminate one of the variables and then solve for the remaining variable.

3. Use back-substitution into either original equation and solve for the other variable.

4. Check your answer.

■ You should know that for a system of two linear equations, one of the following is true.

(a) There are infinitely many solutions; the lines are identical. The system is consistent.

(b) There is no solution; the lines are parallel. The system is inconsistent.

(c) There is one solution; the lines intersect at one point. The system is consistent.

Vocabulary Check

1. method, elimination   **2.** equivalent   **3.** consistent, inconsistent

1. $\begin{cases} 2x + y = 5 & \text{Equation 1} \\ x - y = 1 & \text{Equation 2} \end{cases}$

Add to eliminate y: $3x = 6 \implies x = 2$

Substitute $x = 2$ in Equation 2: $2 - y = 1 \implies y = 1$

Answer: $(2, 1)$

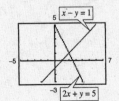

3. $\begin{cases} x + y = 0 & \text{Equation 1} \\ 3x + 2y = 1 & \text{Equation 2} \end{cases}$

Multiply Equation 1 by -2: $-2x - 2y = 0$

Add this to Equation 2 to eliminate y: $x = 1$

Substitute $x = 1$ in Equation 1: $1 + y = 0 \Rightarrow y = -1$

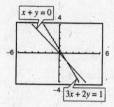

Answer: $(1, -1)$

5. $\begin{cases} x - y = 2 & \text{Equation 1} \\ -2x + 2y = 5 & \text{Equation 2} \end{cases}$

Multiply Equation 1 by 2: $2x - 2y = 4$

Add this to Equation 2: $0 = 9$

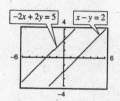

There are no solutions.

7. $\begin{cases} x + 2y = 4 & \text{Equation 1} \\ x - 2y = 1 & \text{Equation 2} \end{cases}$

Add to eliminate y:

$$2x = 5$$

$$x = \frac{5}{2}$$

Substitute $x = \frac{5}{2}$ in Equation 1:

$$\frac{5}{2} + 2y = 4 \Rightarrow y = \frac{3}{4}$$

Answer: $\left(\frac{5}{2}, \frac{3}{4}\right)$

9. $\begin{cases} 2x + 3y = 18 & \text{Equation 1} \\ 5x - y = 11 & \text{Equation 2} \end{cases}$

Multiply Equation 2 by 3: $15x - 3y = 33$

Add this to Equation 1 to eliminate y:

$$17x = 51 \Rightarrow x = 3$$

Substitute $x = 3$ in Equation 1:

$$6 + 3y = 18 \Rightarrow y = 4$$

Answer: $(3, 4)$

11. $\begin{cases} 3r + 2s = 10 & \text{Equation 1} \\ 2r + 5s = 3 & \text{Equation 2} \end{cases}$

Multiply Equation 1 by 2 and Equation 2 by -3: $6r + 4s = 20$

$$-6r - 15s = -9$$

Add to eliminate r: $-11s = 11 \Rightarrow s = -1$

Substitute $s = -1$ in Equation 1: $3r + 2(-1) = 10 \Rightarrow r = 4$

Answer: $(4, -1)$

13. $\begin{cases} 5u + 6v = 24 & \text{Equation 1} \\ 3u + 5v = 18 & \text{Equation 2} \end{cases}$

Multiply Equation 1 by 3 and Equation 2 by (-5): $15u + 18v = 72$

$$-15u - 25v = -90$$

Add to eliminate u: $-7v = -18 \Rightarrow v = \frac{18}{7}$

Substitute $v = \frac{18}{7}$ in Equation 2: $3u + 5\left(\frac{18}{7}\right) = 18 \Rightarrow u = \frac{12}{7}$

Answer: $\left(\frac{12}{7}, \frac{18}{7}\right)$

15. $\begin{cases} 1.8x + 1.2y = 4 & \text{Equation 1} \\ 9x + 6y = 3 & \text{Equation 2} \end{cases}$

Multiply Equation 1 by (-5): $-9x - 6y = -20$

Add this to Equation 2: $0 = -17$

Inconsistent; no solution

17. $2x - 5y = 0$

$\quad x - y = 3 \implies y = x - 3$

$2x - 5(x - 3) = 0$

$\quad\quad -3x = -15$

$\quad\quad\quad x = 5, y = 2$

Matches (b)

One solution; consistent

19. $\left.\begin{array}{l} 2x - 5y = 0 \\ 2x - 3y = -4 \end{array}\right\} \quad \begin{array}{l} -2y = 4 \\ y = -2, x = -5 \end{array}$

One solution; consistent

Matches (c)

21. $\begin{cases} 4x + 3y = 3 & \text{Equation 1} \\ 3x + 11y = 13 & \text{Equation 2} \end{cases}$

Multiply Equation 1 by 3 and Equation 2 by -4:

$\begin{cases} 12x + 9y = 9 \\ -12x - 44y = -52 \end{cases}$

Add to eliminate x: $-35y = -43 \implies y = \frac{43}{35}$

Substitute $y = \frac{43}{35}$ into Equation 1:

$4x + 3\left(\frac{43}{35}\right) = 3 \implies x = -\frac{6}{35}$

Answer: $\left(-\frac{6}{35}, \frac{43}{35}\right)$

23. $\begin{cases} \frac{2}{5}x - \frac{3}{2}y = 4 & \text{Equation 1} \\ \frac{1}{5}x - \frac{3}{4}y = -2 & \text{Equation 2} \end{cases}$

Multiply Equation 2 by -2 and add to Equation 1:

$0 = 8$

Inconsistent; no solution

25. $\begin{cases} \frac{3}{4}x + y = \frac{1}{8} & \text{Equation 1} \\ \frac{9}{4}x + 3y = \frac{3}{8} & \text{Equation 2} \end{cases}$

Multiply Equation 1 by -3: $-\frac{9}{4}x - 3y = -\frac{3}{8}$

Add this to Equation 2: $0 = 0$

There are an infinite number of solutions.

The solutions consist of all (x, y) satisfying $\frac{3}{4}x + y = \frac{1}{8}$, or $6x + 8y = 1$.

27. $\begin{cases} \dfrac{x + 3}{4} + \dfrac{y - 1}{3} = 1 & \text{Equation 1} \\ \quad\quad 2x - y = 12 & \text{Equation 2} \end{cases}$

Multiply Equation 1 by 12 and Equation 2 by 4:

$\begin{cases} 3x + 4y = 7 \\ 8x - 4y = 48 \end{cases}$

Add to eliminate y: $11x = 55 \implies x = 5$

Substitute $x = 5$ into Equation 2:

$2(5) - y = 12 \implies y = -2$

Answer: $(5, -2)$

29. $\begin{cases} \dfrac{x - 1}{2} + \dfrac{y + 2}{3} = 4 & \text{Equation 1} \\ \quad\quad x - 2y = 5 & \text{Equation 2} \end{cases}$

Multiply Equation 1 by 6:

$3(x - 1) + 2(y + 2) = 24 \implies 3x + 2y = 23$

Add this to Equation 2 to eliminate y:

$4x = 28 \implies x = 7$

Substitute $x = 7$ in Equation 2:

$7 - 2y = 5 \implies y = 1$

Answer: $(7, 1)$

31. $\begin{cases} 2.5x - 3y = 1.5 & \text{Equation 1} \\ 10x - 12y = 6 & \text{Equation 2 multiplied by 5} \end{cases}$

Multiply Equation 1 by (-4):

$$-10x + 12y = -6$$

Add this to Equation 2 to eliminate x:

$$0 = 0$$

The solution set consists of all points lying on the line $10x - 12y = 6$.

All points on the line $5x - 6y = 3$.

Let $x = a$, then $y = \frac{5}{6}a - \frac{1}{2}$.

Answer: $\left(a, \frac{5}{6}a - \frac{1}{2}\right)$, where a is any real number.

35. $\begin{cases} 0.05x - 0.03y = 0.21 & \text{Equation 1} \\ 0.07x + 0.02y = 0.16 & \text{Equation 2} \end{cases}$

Multiply Equation 1 by 200 and Equation 2 by 300:

$$\begin{cases} 10x - 6y = 42 \\ 21x + 6y = 48 \end{cases}$$

Add to eliminate y: $31x = 90$

$$x = \frac{90}{31}$$

Substitute $x = \frac{90}{31}$ in Equation 2:

$$0.07\left(\frac{90}{31}\right) + 0.02y = 0.16$$
$$y = -\frac{67}{31}$$

Answer: $\left(\frac{90}{31}, -\frac{67}{31}\right)$

39. Let $X = \dfrac{1}{x}$ and $Y = \dfrac{1}{y}$.

$\begin{cases} X + 2Y = 5 & \text{Equation 1} \\ 3X - 4Y = -5 & \text{Equation 2} \end{cases}$

Multiply Equation 1 by 2:

$\begin{cases} 2X + 4Y = 10 \\ 3X - 4Y = -5 \end{cases}$

Adding the equations eliminates Y:

$$5X = 5 \implies X = 1.$$

Hence,

$$2Y = 5 - X = 4 \implies Y = 2.$$

$$x = \frac{1}{X} = 1, \, y = \frac{1}{Y} = \frac{1}{2}$$

Answer: $\left(1, \frac{1}{2}\right)$

33. $\begin{cases} 0.2x - 0.5y = -27.8 & \text{Equation 1} \\ 0.3x + 0.4y = 68.7 & \text{Equation 2} \end{cases}$

Multiply Equation 1 by 40 and Equation 2 by 50:

$\begin{cases} 8x - 20y = -1112 \\ 15x + 20y = 3435 \end{cases}$

Adding the equation eliminates y:

$$23x = 2323 \implies x = 101$$

Substitute $x = 101$ into Equation 1:

$$8(101) - 20y = -1112 \implies y = 96$$

Answer: $(101, 96)$

37. Let $X = \dfrac{1}{x}$ and $Y = \dfrac{1}{y}$.

$\begin{cases} X + 3Y = 2 & \text{Equation 1} \\ 4X - Y = -5 & \text{Equation 2} \end{cases}$

Multiply Equation 1 by 4:

$\begin{cases} 4X + 12Y = 8 \\ 4X - Y = -5 \end{cases}$

Subtract to eliminates X:

$$13Y = 13 \implies Y = 1.$$

Hence,

$$X = 2 - 3Y = 2 - 3(1) = -1$$

$$x = \frac{1}{X} = -1, \, y = \frac{1}{Y} = 1$$

Answer: $(-1, 1)$

41. $\begin{cases} 2x - 5y = 0 \implies y = \frac{2}{5}x \\ x - y = 3 \implies y = x - 3 \end{cases}$

The system is consistent.
There is one solution, $(5, 2)$.

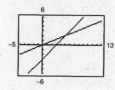

43. $\begin{cases} \frac{3}{5}x - y = 3 \implies y = \frac{3}{5}x - 3 \\ -3x + 5y = 9 \implies y = \frac{1}{5}(3x + 9) = \frac{3}{5}x + \frac{9}{5} \end{cases}$

The lines are parallel.
The system is inconsistent.

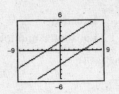

45. $\begin{cases} 8x - 14y = 5 \implies y = (8x - 5)/14 = \frac{4}{7}x - \frac{5}{14} \\ 2x - 3.5y = 1.25 \implies y = (2x - 1.25)/3.5 = \frac{4}{7}x - \frac{5}{14} \end{cases}$

The system is consistent. The solution set consists of all points on the line $y = \frac{4}{7}x - \frac{5}{14}$, or $8x - 14y = 5$.

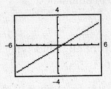

47. $\begin{cases} 6y = 42 \implies y = 7 \\ 6x - y = 16 \implies y = 6x - 16 \end{cases}$

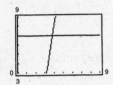

Answer: $\left(\frac{23}{6}, 7\right) \approx (3.833, 7)$

49. $\begin{cases} \frac{3}{2}x - \frac{1}{5}y = 8 \implies y = 5\left(\frac{3}{2}x - 8\right) \\ -2x + 3y = 3 \implies y = \frac{1}{3}(3 + 2x) \end{cases}$

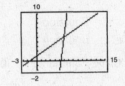

Answer: $(6, 5)$

51. $\frac{1}{3}x + y = -\frac{1}{3} \implies y = -\frac{1}{3} - \frac{1}{3}x$

$5x - 3y = 7 \implies y = \frac{1}{3}(5x - 7)$

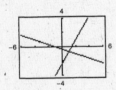

Answer: $(1, -0.667)$

53. $\begin{cases} 0.5x + 2.2y = 9 \implies y = 1/2.2(9 - 0.5x) \\ 6x + 0.4y = -22 \implies y = 1/0.4(-22 - 6x) \end{cases}$

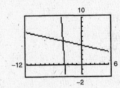

Answer: $(-4, 5)$

55. $\begin{cases} 3x - 5y = 7 & \text{Equation 1} \\ 2x + y = 9 & \text{Equation 2} \end{cases}$

Multiply Equation 2 by 5:

$10x + 5y = 45$

Add this to Equation 1:

$13x = 52 \implies x = 4$

Back-substitute $x = 4$ into Equation 2:

$2(4) + y = 9 \implies y = 1$

Answer: $(4, 1)$

57. $\begin{cases} y = 4x + 3 \\ y = -5x - 12 \end{cases}$

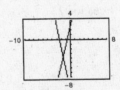

The lines intersect at
$\left(-\frac{5}{3}, -\frac{11}{3}\right) \approx (-1.667, -3.667)$

59. $\begin{cases} x - 5y = 21 \\ 6x + 5y = 21 \end{cases}$

Adding the equations, $7x = 42 \implies x = 6$.

Back-substituting, $x - 5y = 6 - 5y = 21 \implies$

$-5y = 15 \implies y = -3$

Answer: $(6, -3)$

61. $\begin{cases} -2x + 8y = 19 \qquad \text{Equation 1} \\ y = x - 3 \qquad\qquad \text{Equation 2} \end{cases}$

Substitute into Equation 1,

$-2x + 8(x - 3) = 19 \implies 6x = 43$

$\implies x = \frac{43}{6}$

Back-substituting, $y = x - 3 = \frac{43}{6} - 3 = \frac{25}{6}$

Answer: $\left(\frac{43}{6}, \frac{25}{6}\right)$

63. There are infinitely many systems that have the solution $(0, 8)$. One possible system is

$\begin{cases} x + y = 8 \\ -x + y = 8 \end{cases}$

65. There are infinitely many systems that have the solution $\left(3, \frac{5}{2}\right)$. One possible system:

$2(3) + 2\left(\frac{5}{2}\right) = 11 \implies 2x + 2y = 11$

$3 - 4\left(\frac{5}{2}\right) = -7 \implies x - 4y = -7$

67. Demand = Supply

$50 - 0.5x = 0.125x$

$50 = 0.625x$

$x = 80 \text{ units}$

$p = \$10$

Answer: $(80, 10)$

69. Demand = Supply

$140 - 0.00002x = 80 + 0.00001x$

$60 = 0.00003x$

$x = 2{,}000{,}000 \text{ units}$

$p = \$100.00$

Answer: $(2{,}000{,}000, 100)$

71. Let $x = $ the ground speed and $y = $ the wind speed.

$\begin{cases} 3.6(x - y) = 1800 \quad \text{Equation 1} \\ 3(x + y) = 1800 \quad \text{Equation 2} \end{cases}$

$\begin{aligned} x - y &= 500 \\ x + y &= 600 \\ \hline 2x \quad\quad &= 1100 \\ x \quad\quad &= 550 \end{aligned}$

Substituting $x = 550$ in Equation 2:

$550 + y = 600$

$y = 50$

Answer: $x = 550$ mph, $y = 50$ mph

73. (a) $\begin{cases} A + C = 1175 \qquad\qquad \text{Equation 1} \\ 5A + 3.5C = 5087.5 \quad\;\; \text{Equation 2} \end{cases}$

(b) Multiply Equation 1 by 5:

$5A + 5C = 5875$

$5A + 3.5C = 5087.5$

Subtracting eliminates A:

$1.5C = 787.5$

$C = 525$

Hence, $A = 1175 - 525 = 650$.

650 adult tickets and 525 child tickets

(c)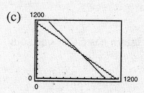

Let $C = y_1 = 1175 - x$

$C = y_2 = \dfrac{1}{3.5}(5087.5 - 5x)$

Point of intersection: $(A, C) = (650, 525)$

75. Let M = number of oranges and
let R = number of grapefruit.

$$\begin{cases} M + R = 16 & \text{Equation 1} \\ 0.95M + 1.05R = 15.90 & \text{Equation 2} \end{cases}$$

Solving for R in Equation 1: $R = 16 - M$.

Substituting into Equation 2:

$0.95M + 1.05(16 - M) = 15.9$

$0.95M + 16.8 - 1.05M = 15.9$

$0.9 = 0.1M$

$M = 9$

Hence, $R = 16 - 9 = 7$.

9 oranges and 7 grapefruit

77. Let m = number of movies and
let v = number of videos.

$$\begin{cases} m + v = 310 & \text{Equation 1} \\ 3m + 2.5v = 867.5 & \text{Equation 2} \end{cases}$$

Solve for m in Equation 1: $m = 310 - v$.

Substitute for m Equation 2:

$3(310 - v) + 2.5v = 867.5$

$62.5 = 0.5v$

$v = 125$

$m = 310 - 125 = 185$

185 movies and 125 videos

79. $\begin{cases} 5b + 10a = 20.2 \implies -10b - 20a = -40.4 \\ 10b + 30a = 50.1 \implies \underline{10b + 30a = 50.1} \end{cases}$

$10a = 9.7$

$a = 0.97$

$b = 2.10$

Least squares regression line: $y = 0.97x + 2.10$

81. $\begin{cases} 5b + 10a = 2.7 & \text{Equation 1} \\ 10b + 30a = -19.6 & \text{Equation 2} \end{cases}$

Multiply Equation 1 by -2:

$-10b - 20a = -5.4$

$10b + 30a = -19.6$

Adding,

$10a = -25.0$

$a = -2.5$

$b = (2.7 + 10(2.5))/5 = 5.54$

$y = -2.5x + 5.54$

83. (a) $\begin{cases} 4b + 7a = 174 \implies 28b + 49a = 1218 \\ 7b + 13.5a = 322 \implies -28b - 54a = -1288 \end{cases}$

Adding, $-5a = -70 \implies a = 14$, $b = 19$.

Thus, $y = 14x + 19$

(b) Using a graphing utility, you obtain $y = 14x + 19$.

(c)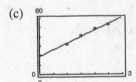

(d) If $x = 1.6$, (160 pounds/acre),
$y = 14(1.6) + 19 = 41.4$ bushels per acre.

85. True. A consistent linear system has either one solution or an infinite number of solutions.

87. $\begin{cases} 100y - x = 200 & \text{Equation 1} \\ 99y - x = -198 & \text{Equation 2} \end{cases}$

Subtract Equation 2 from Equation 1 to eliminate x: $y = 398$

Substitute $y = 398$ into Equation 1: $100(398) - x = 200 \implies x = 39,600$

Answer: (39,600, 398)

The lines are not parallel. The scale on the axes must be changed to see the point of intersection.

89. No, it is not possible for a consistent system of linear equations to have exactly two solutions. Either the lines will intersect once or they will coincide and then the system would have infinite solutions.

91. $\begin{cases} 4x - 8y = -3 & \text{Equation 1} \\ 2x + ky = 16 & \text{Equation 2} \end{cases}$

Multiply Equation 2 by -2: $-4x - 2ky = -32$

Add this to Equation 1: $-8y - 2ky = -35$

The system is inconsistent if $-8y - 2ky = 0$.

This occurs when $k = -4$. Note that for $k = -4$, the two original equations represent parallel lines.

93. $u \sin x + v \cos x = 0$

$u \cos x - v \sin x = \sec x$

Multiply the first equation by $\sin x$, the second by $\cos x$, and add the equations:

$u \sin^2 x + u \cos^2 x = \sec x \cdot \cos x$

$u = 1$

Hence, $v \cos x = -u \sin x = -\sin x$

$v = -\tan x.$

95. $-11 - 6x \geq 33$

$-6x \geq 44$

$x \leq -\dfrac{44}{6} = -\dfrac{22}{3}$

97. $|x - 8| < 10$

$-10 < x - 8 < 10$

$-2 < x < 18$

99. $2x^2 + 3x - 35 < 0$

$(2x - 7)(x + 5) < 0$

Critical numbers: $\dfrac{7}{2}, -5$. Testing the three intervals,

$-5 < x < \dfrac{7}{2}.$

101. $\ln x + \ln 6 = \ln 6x$

103. $\log_9 12 - \log_9 x = \log_9 \dfrac{12}{x}$

105. $2 \ln x - \ln(x + 2) = \ln x^2 - \ln(x + 2)$

$= \ln \left[\dfrac{x^2}{x + 2} \right]$

107. Answers will vary.

Section 7.3   Multivariable Linear Systems

■ You should know the operations that lead to equivalent systems of linear equations:

(a) Interchange any two equations.

(b) Multiply all terms of an equation by a nonzero constant.

(c) Replace an equation by the sum of itself and a constant multiple of any other equation in the system.

■ You should be able to use the method of elimination.

Vocabulary Check

1. row-echelon

2. ordered triple

3. Gaussian

4. independent, dependent

5. nonsquare

6. three-dimensional

7. partial fraction decomposition

1. (a) $3(3) - 5 + (-3) \overset{?}{=} 1$ Yes

$2(3) - 3(-3) \overset{?}{=} -14$ No

$5(5) + 2(-3) \overset{?}{=} 8$ No

No, $(3, 5, -3)$ is not a solution.

(c) $3(-4) - 1 + 2 \overset{?}{=} 1$ No

$2(-4) - 3(2) \overset{?}{=} -14$ Yes

$5(1) + 2(2) \overset{?}{=} 8$ No

No, $(-4, 1, 2)$ is not a solution.

(b) $3(-1) - (0) + 4 \overset{?}{=} 1$ Yes

$2(-1) - 3(4) \overset{?}{=} -14$ Yes

$5(0) + 2(4) \overset{?}{=} 8$ Yes

Yes, $(-1, 0, 4)$ is a solution.

(d) $3(1) - 0 + 4 \overset{?}{=} 1$ No

$2(1) - 3(4) \overset{?}{=} -14$ No

$5(0) + 2(4) \overset{?}{=} 8$ Yes

No, $(1, 0, 4)$ is not a solution.

3. (a) $4(0) + 1 - 1 \overset{?}{=} 0$ Yes

$-8(0) - 6(1) + 1 \overset{?}{=} -\frac{7}{4}$ No

$3(0) - 1 \overset{?}{=} -\frac{9}{4}$ No

No, $(0, 1, 1)$ is not a solution.

(c) $4\left(-\frac{1}{2}\right) + \frac{3}{4} - \left(-\frac{5}{4}\right) \overset{?}{=} 0$ Yes

$-8\left(-\frac{1}{2}\right) - 6\left(\frac{3}{4}\right) - \frac{5}{4} \overset{?}{=} -\frac{7}{4}$ Yes

$3\left(-\frac{1}{2}\right) - \frac{3}{4} \overset{?}{=} -\frac{9}{4}$ Yes

Yes, $\left(-\frac{1}{2}, \frac{3}{4}, -\frac{5}{4}\right)$ is a solution.

(b) $4\left(-\frac{3}{2}\right) + \frac{5}{4} - \left(-\frac{5}{4}\right) \overset{?}{=} 0$ No

$-8\left(-\frac{3}{2}\right) - 6\left(\frac{5}{4}\right) + \left(-\frac{5}{4}\right) \overset{?}{=} -\frac{7}{4}$ No

$3\left(-\frac{3}{2}\right) - \left(\frac{5}{4}\right) \overset{?}{=} -\frac{9}{4}$ No

No, $\left(-\frac{3}{2}, \frac{5}{4}, -\frac{5}{4}\right)$ is not a solution.

(d) $4\left(-\frac{1}{2}\right) + 2 - 0 \overset{?}{=} 0$ Yes

$-8\left(-\frac{1}{2}\right) - 6(2) + 0 \overset{?}{=} -\frac{7}{2}$ No

$3\left(-\frac{1}{2}\right) - 2 \overset{?}{=} -\frac{9}{4}$ No

No, $\left(-\frac{1}{2}, 2, 0\right)$ is not a solution.

5. $\begin{cases} 2x - y + 5z = 16 & \text{Equation 1} \\ y + 2z = 2 & \text{Equation 2} \\ z = 2 & \text{Equation 3} \end{cases}$

Back-substitute $z = 2$ into Equation 2:

$y + 2(2) = 2$

$y = -2$

Back-substitute $z = 2$ and $y = -2$ into Equation 1:

$2x - (-2) + 5(2) = 16$

$2x = 4$

$x = 2$

Answer: $(2, -2, 2)$

7. $\begin{cases} 2x + y - 3z = 10 & \text{Equation 1} \\ y + z = 12 & \text{Equation 2} \\ z = 2 & \text{Equation 3} \end{cases}$

Back-substitute $z = 2$ into Equation 2:

$y + 2 = 12$

$y = 10$

Back-substitute $y = 10$ and $z = 2$ into Equation 1:

$2x + 10 - 3(2) = 10$

$2x = 6$

$x = 3$

Answer: $(3, 10, 2)$

9. $\begin{cases} 4x - 2y + z = 8 & \text{Equation 1} \\ \quad\ - y + z = 4 & \text{Equation 2} \\ \qquad\qquad z = 2 & \text{Equation 3} \end{cases}$

Back-substitute $z = 2$ into Equation 2:

$$-y + 2 = 4$$

$$y = -2$$

Back-substitute $y = -2$ and $z = 2$ into Equation 1:

$$4x - 2(-2) + 2 = 8$$

$$4x = 2$$

$$x = \tfrac{1}{2}$$

Answer: $\left(\tfrac{1}{2}, -2, 2\right)$

13. $\begin{cases} x - 2y + \ z = 1 & \text{Equation 1} \\ 2x - \ y + 3z = 0 & \text{Equation 2} \\ 3x - \ y - 4z = 1 & \text{Equation 3} \end{cases}$

Add -2 times Equation 1 to Equation 2.

$$3y + z = -2 \quad \text{New Equation 2}$$

This is the first step in putting the system in row-echelon form.

17. $\begin{cases} 2x \qquad + 2z = 2 & \text{Equation 1} \\ 5x + 3y \qquad = 4 & \text{Equation 2} \\ \qquad 3y - 4z = 4 & \text{Equation 3} \end{cases}$

$\begin{cases} x \qquad + \ z = 1 & \left(\tfrac{1}{2}\right) \text{Eq. 1} \\ 5x + 3y \qquad = 4 \\ \qquad 3y - 4z = 4 \end{cases}$

$\begin{cases} x \qquad + \ z = \ \ 1 \\ \quad 3y - 5z = -1 & (-5) \text{Eq. 1} + \text{Eq. 2} \\ \quad 3y - 4z = \ \ 4 \end{cases}$

$\begin{cases} x \qquad + \ z = \ \ 1 \\ \quad 3y - 5z = -1 \\ \qquad\quad z = \ \ 5 & (-1) \text{Eq. 2} + \text{Eq. 3} \end{cases}$

$$3y - 5(5) = -1 \implies y = 8$$

$$x + 5 = 1 \implies x = -4$$

Answer: $(-4, 8, 5)$

11. $\begin{cases} \ \ x - 2y + 3z = 5 & \text{Equation 1} \\ -x + 3y - 5z = 4 & \text{Equation 2} \\ \ 2x \qquad - 3z = 0 & \text{Equation 3} \end{cases}$

Add Equation 1 to Equation 2.

$$y - 2z = 9 \quad \text{New Equation 2}$$

This is the first step in putting the system in row-echelon form.

$\begin{cases} \ \ x - 2y + 3z = 5 \\ \qquad\ \ y - 2z = 9 \\ \ 2x \qquad - 3z = 0 \end{cases}$

15. $\begin{cases} x + y + z = 6 & \text{Equation 1} \\ 2x - y + z = 3 & \text{Equation 2} \\ 3x \qquad - z = 0 & \text{Equation 3} \end{cases}$

$\begin{cases} x + \ y + \ z = \quad 6 \\ \quad -3y - \ z = \ -9 & (-2) \text{Eq. 1} + \text{Eq. 2} \\ \quad -3y - 4z = -18 & (-3) \text{Eq. 1} + \text{Eq. 3} \end{cases}$

$\begin{cases} x + \ y + z = \quad 6 \\ \quad -3y - z = -9 \\ \qquad\quad -3z = -9 & (-1) \text{Eq. 2} + \text{Eq. 3} \end{cases}$

$$-3z = -9 \implies z = 3$$

$$-3y - 3 = -9 \implies y = 2$$

$$x + 2 + 3 = 6 \implies x = 1$$

Answer: $(1, 2, 3)$

19. $\begin{cases} 4x + \ y - 3z = \ 11 & \text{Equation 1} \\ 2x - 3y + 2z = \ \ 9 & \text{Equation 2} \\ \ x + \ y + \ z = -3 & \text{Equation 3} \end{cases}$

$\begin{cases} \ x + \ y + \ z = -3 & \text{Interchange equations} \\ 2x - 3y + 2z = \ \ 9 & \text{1 and 3} \\ 4x + \ y - 3z = \ 11 \end{cases}$

$\begin{cases} x + \quad y + \ z = -3 \\ \qquad -5y \qquad = 15 & (-2) \text{Eq. 1} + \text{Eq. 2} \\ \qquad -3y - 7z = 23 & (-4) \text{Eq. 1} + \text{Eq. 3} \end{cases}$

$$y = -3 \implies -3(-3) - 7z = 23$$

$$\implies -7z = 14$$

$$\implies z = -2$$

$$x + (-3) + (-2) = -3 \implies x = 2$$

Answer: $(2, -3, -2)$

21. $\begin{cases} x + y - 2z = 3 \\ 3x - 2y + 4z = 1 \\ 2x - 3y + 6z = 8 \end{cases}$ Interchange the equations.

$\begin{cases} x + y - 2z = 3 \\ -5y + 10z = -8 \\ -5y + 10z = 2 \end{cases}$ -3 Eq. 1 + Eq. 2
-2 Eq. 1 + Eq. 3

$\begin{cases} x + y - 2z = 3 \\ -5y + 10z = -8 \\ 0 = 10 \end{cases}$ $-$Eq. 2 + Eq. 3

Inconsistent; no solution.

25. $\begin{cases} x + 2y - 7z = -4 \\ 2x + y + z = 13 \\ 3x + 9y - 36z = -33 \end{cases}$ Equation 1
Equation 2
Equation 3

$\begin{cases} x + 2y - 7z = -4 \\ -3y + 15z = 21 \\ 3y - 15z = -21 \end{cases}$ -2 Eq. 1 + Eq. 2
-3 Eq. 1 + Eq. 3

$\begin{cases} x + 2y - 7z = -4 \\ -3y + 15z = 21 \\ 0 = 0 \end{cases}$ Eq. 2 + Eq. 3

$\begin{cases} x + 2y - 7z = -4 \\ y - 5z = -7 \end{cases}$ $-\frac{1}{3}$ Eq. 2

$\begin{cases} x + 3z = 10 \\ y - 5z = -7 \end{cases}$ -2 Eq. 2 + Eq. 1

Let $z = a$, then:

$y = 5a - 7$

$x = -3a + 10$

Answer: $(-3a + 10, 5a - 7, a)$

23. $\begin{cases} 3x + 3y + 5z = 1 \\ 3x + 5y + 9z = 0 \\ 5x + 9y + 17z = 0 \end{cases}$

$\begin{cases} 6x + 6y + 10z = 2 \\ 3x + 5y + 9z = 0 \\ 5x + 9y + 17z = 0 \end{cases}$ 2 Eq. 1

$\begin{cases} x - 3y - 7z = 2 \\ 3x + 5y + 9z = 0 \\ 5x + 9y + 17z = 0 \end{cases}$ $-$Eq. 3 + Eq. 1

$\begin{cases} x - 3y - 7z = 2 \\ 14y + 30z = -6 \\ 24y + 52z = -10 \end{cases}$ -3 Eq. 1 + Eq. 2
-5 Eq. 1 + Eq. 3

$\begin{cases} x - 3y - 7z = 2 \\ 84y + 180z = -36 \\ 84y + 182z = -35 \end{cases}$ 6 Eq. 2
3.5 Eq. 3

$\begin{cases} x - 3y - 7z = 2 \\ 84y + 180z = -36 \\ 2z = 1 \end{cases}$ $-$Eq. 2 + Eq. 3

$2z = 1 \implies z = \frac{1}{2}$

$84y + 180\left(\frac{1}{2}\right) = -36 \implies y = -\frac{3}{2}$

$x - 3\left(-\frac{3}{2}\right) - 7\left(\frac{1}{2}\right) = 2 \implies x = 1$

Answer: $\left(1, -\frac{3}{2}, \frac{1}{2}\right)$

27. $\begin{cases} x - y + 2z = 6 \\ 2x + y + z = 3 \\ x + y + z = 2 \end{cases}$ Equation 1
Equation 2
Equation 3

$\begin{cases} x - y + 2z = 6 \\ 3y - 3z = -9 \\ 2y - z = -4 \end{cases}$ (-2) Eq. 1 + Eq. 2
(-1) Eq. 1 + Eq. 3

$\begin{cases} x - y + 2z = 6 \\ y - z = -3 \\ 2y - z = -4 \end{cases}$ $\left(\frac{1}{3}\right)$ Eq. 2

$\begin{cases} x - y + 2z = 6 \\ y - z = -3 \\ z = 2 \end{cases}$ (-2) Eq. 2 + Eq. 3

$y - z = -3 \implies y = 2 - 3 = -1$

$x - y + 2z = 6 \implies x = 6 + (-1) - 2(2) = 1$

Answer: $(1, -1, 2)$

29. $\begin{cases} 3x - 3y + 6z = 6 \\ x + 2y - z = 5 \\ 5x - 8y + 13z = 7 \end{cases}$ Equation 1
Equation 2
Equation 3

$\begin{cases} x - y + 2z = 2 \\ x + 2y - z = 5 \\ 5x - 8y + 13z = 7 \end{cases}$ $\left(\frac{1}{3}\right)$ Equation 1

$\begin{cases} x - y + 2z = 2 \\ 3y - 3z = 3 \\ -3y + 3z = -3 \end{cases}$ (-1) Eq. 1 + Eq. 2
(-5) Eq. 1 + Eq. 3

$\begin{cases} x - y + 2z = 2 \\ y - z = 1 \\ 0 = 0 \end{cases}$

$\begin{cases} x + z = 3 \\ y - z = 1 \end{cases}$

Let $z = a$, then:

$y = a + 1$

$x = -a + 3$

Answer: $(-a + 3, a + 1, a)$

33. $\begin{cases} x + 4z = 1 \\ x + y + 10z = 10 \\ 2x - y + 2z = -5 \end{cases}$

$\begin{cases} x + 4z = 1 \\ y + 6z = 9 \\ -y - 6z = -7 \end{cases}$ $-$Eq. 1 + Eq. 2
-2 Eq. + Eq. 3

$\begin{cases} x + 4z = 1 \\ y + 6z = 9 \\ 0 = 2 \end{cases}$ Eq. 2 + Eq. 3

No solution; inconsistent.

37. $\begin{cases} x - 2y + 5z = 2 \\ 4x - z = 0 \end{cases}$

$\begin{cases} x - 2y + 5z = 2 \\ 8y - 21z = -8 \end{cases}$ -4 Eq. 1 + Eq. 2

$\begin{cases} x - 2y + 5z = 2 \\ y - \frac{21}{8}z = -1 \end{cases}$ $\frac{1}{8}$ Eq. 2

$\begin{cases} x - \frac{1}{4}z = 0 \\ y - \frac{21}{8}z = -1 \end{cases}$ 2 Eq. 2 + Eq. 1

Let $z = a$. Then $y = \frac{21}{8}a - 1$ and $x = \frac{1}{4}a$

Answer: $\left(\frac{1}{4}a, \frac{21}{8}a - 1, a\right)$

31. $\begin{cases} x - 2y + 3z = 4 \\ 3x - y + 2z = 0 \\ x + 3y - 4z = -2 \end{cases}$ Equation 1
Equation 2
Equation 3

$\begin{cases} x - 2y + 3z = 4 \\ 5y - 7z = -12 \\ 5y - 7z = -6 \end{cases}$ -3 Eq. 1 + Eq. 2
-1 Eq. 1 + Eq. 3

$\begin{cases} x - 2y + 3z = 4 \\ 5y - 7z = -12 \\ 0 = 6 \end{cases}$ $-$Eq. 2 + Eq. 3

No solution; inconsistent.

35. $\begin{cases} x + 2y + z = 1 \\ x - 2y + 3z = -3 \\ 2x + y + z = -1 \end{cases}$ Equation 1
Equation 2
Equation 3

$\begin{cases} x + 2y + z = 1 \\ -4y + 2z = -4 \\ -3y - z = -3 \end{cases}$ (-1) Eq. 1 + Eq. 2
(-2) Eq. 1 + Eq. 3

$\begin{cases} x + 2y + z = 1 \\ y - \frac{1}{2}z = 1 \\ 3y + z = 3 \end{cases}$ $\left(-\frac{1}{4}\right)$ Eq. 2
(-1) Eq. 3

$\begin{cases} x + 2y + z = 1 \\ y - \frac{1}{2}z = 1 \\ \frac{5}{2}z = 0 \end{cases}$ (-3) Eq. 2 + Eq. 3

$z = 0$

$y = 1 - 0 = 1$

$x + 2y + z = 1 \implies x = 1 - 2 = -1$

Answer: $(-1, 1, 0)$

39. $\begin{cases} 2x - 3y + z = -2 \\ -4x + 9y = 7 \end{cases}$

$\begin{cases} 2x - 3y + z = -2 \\ 3y + 2z = 3 \end{cases}$ 2 Eq. 1 + Eq. 2

$\begin{cases} 2x + 3z = 1 \\ 3y + 2z = 3 \end{cases}$ Eq. 2 + Eq. 1

Let $z = a$, then:

$$y = -\frac{2}{3}a + 1$$

$$x = -\frac{3}{2}a + \frac{1}{2}$$

Answer: $\left(-\frac{3}{2}a + \frac{1}{2}, -\frac{2}{3}a + 1, a\right)$

41. $\begin{cases} x - 3y + 2z = 18 & \text{Equation 1} \\ 5x - 13y + 12z = 80 & \text{Equation 2} \end{cases}$

$\begin{cases} x - 3y + 2z = 18 \\ 2y + 2z = -10 \end{cases}$ -5 Eq. 1 + Eq. 2

$\begin{cases} x - 3y + 2z = 18 \\ y + z = -5 \end{cases}$ $\frac{1}{2}$ Eq. 2

$\begin{cases} x + 5z = 3 \\ y + z = -5 \end{cases}$ 3 Eq. 2 + Eq. 1

Let $z = a$, then $y = -a - 5$, and $x = -5a + 3$.

Answer: $(-5a + 3, -a - 5, a)$

43. $\begin{cases} x - y + 2z - w = 0 & \text{Equation 1} \\ 2x + y + z - w = 0 & \text{Equation 2} \\ x + y - w = -1 & \text{Equation 3} \\ x + y - 2z + w = 1 & \text{Equation 4} \end{cases}$

$\begin{cases} x - y + 2z - w = 0 \\ 3y - 3z + w = 0 \\ 2y - 2z = -1 \\ 2y - 4z + 2w = 1 \end{cases}$ $\begin{matrix}(-2) \text{ Eq. 1 + Eq. 2} \\ (-1) \text{ Eq. 1 + Eq. 3} \\ (-1) \text{ Eq. 1 + Eq. 4}\end{matrix}$

$\begin{cases} x - y + 2z - w = 0 \\ 2y - 2z = -1 \\ 3y - 3z + w = 0 \\ 2y - 4z + 2w = 1 \end{cases}$ $\begin{matrix}\text{Interchange} \\ \text{Eq. 2 and 3}\end{matrix}$

$\begin{cases} x - y + 2z - w = 0 \\ 2y - 2z = -1 \\ w = \frac{3}{2} \\ -2z + 2w = 2 \end{cases}$ $\begin{matrix}\left(-\frac{3}{2}\right) \text{ Eq. 2 + Eq. 3} \\ (-1) \text{ Eq. 2 + Eq. 4}\end{matrix}$

$w = \frac{3}{2}$

$-2z + 2w = 2 \implies -2z = 2 - 2\left(\frac{3}{2}\right) = -1 \implies z = \frac{1}{2}$

$2y - 2z = -1 \implies 2y = -1 + 2\left(\frac{1}{2}\right) = 0 \implies y = 0$

$x - y + 2z - w = 0 \implies x = -2\left(\frac{1}{2}\right) + \frac{3}{2} = \frac{1}{2}$

Answer: $\left(\frac{1}{2}, 0, \frac{1}{2}, \frac{3}{2}\right)$

45. Let $X = \frac{1}{x}, Y = \frac{1}{y}, Z = \frac{1}{z}$.

$\begin{cases} X + 2Y - 3Z = 3 & \text{Equation 1} \\ X - 2Y + Z = 1 & \text{Equation 2} \\ 2X + 2Y - 3Z = 4 & \text{Equation 3} \end{cases}$

$\begin{cases} X + 2Y - 3Z = 3 \\ -4Y + 4Z = -2 \\ -2Y + 3Z = -2 \end{cases}$ $\begin{matrix}(-1) \text{ Eq. 1 + Eq. 2} \\ (-2) \text{ Eq. 1 + Eq. 3}\end{matrix}$

$\begin{cases} X + 2Y - 3Z = 3 \\ 2Y - 2Z = 1 \\ -2Y + 3Z = -2 \end{cases}$ $\left(-\frac{1}{2}\right)$ Eq. 2

$\begin{cases} X + 2Y - 3Z = 3 \\ 2Y - 2Z = 1 \\ Z = -1 \end{cases}$ Eq. 2 + Eq. 3

$2Y - 2Z = 1 \implies 2Y = 1 + 2Z = -1 \implies Y = -\frac{1}{2}$

$X + 2Y - 3Z = 3 \implies X = 3 - 2\left(-\frac{1}{2}\right) + 3(-1) = 1$

$x = \frac{1}{X} = 1, y = \frac{1}{Y} = -2, z = \frac{1}{Z} = -1$

Answer: $(1, -2, -1)$

47. Let $X = \dfrac{1}{x}, Y = \dfrac{1}{y}, Z = \dfrac{1}{z}.$

$$\begin{cases} 2X - Y + 2Z = 4 & \text{Equation 1} \\ X + 2Y - 2Z = -2 & \text{Equation 2} \\ 3X + 3Y + 4Z = 2 & \text{Equation 3} \end{cases}$$

$$\begin{cases} X + 2Y - 2Z = -2 & \text{Interchange} \\ 2X - Y + 2Z = 4 & \text{Eq. 1 and 2} \\ 3X + 3Y + 4Z = 2 \end{cases}$$

$$\begin{cases} X + 2Y - 2Z = -2 \\ \quad\;\; -5Y + 6Z = 8 & (-2)\text{ Eq. } 1 + \text{Eq. } 2 \\ \quad\;\; -3Y + 10Z = 8 & (-3)\text{ Eq. } 1 + \text{Eq. } 3 \end{cases}$$

$$\begin{cases} X + 2Y - 2Z = -2 \\ \quad\;\; -5Y + 6Z = 8 \\ \quad\quad\quad\; \frac{32}{5}Z = \frac{16}{5} & (-\frac{3}{5})\text{Eq. } 2 + \text{Eq. } 3 \end{cases}$$

$$Z = \frac{1}{2}$$

$$-5Y + 6Z = 8 \implies -5Y = 8 - 6\left(\frac{1}{2}\right) = 5 \implies Y = -1$$

$$X + 2Y - 2Z = -2 \implies X = -2 - 2(-1) + 2\left(\frac{1}{2}\right) = 1$$

$$x = \frac{1}{X} = 1, y = \frac{1}{Y} = -1, z = \frac{1}{Z} = 2$$

Answer: $(1, -1, 2)$

49. There are an infinite number of linear systems that have $(4, -1, 2)$ as their solution. One such system is as follows:

$$3(4) + (-1) - (2) = 9 \implies 3x + y - z = 9$$
$$(4) + 2(-1) - (2) = 0 \implies x + 2y - z = 0$$
$$-(4) + (-1) + 3(2) = 1 \implies -x + y + 3z = 1$$

51. There are an infinite numbers of linear systems that have $\left(3, -\frac{1}{2}, \frac{7}{4}\right)$ as their solution. One such system is:

$$1(3) + 2\left(-\frac{1}{2}\right) + 4\left(\frac{7}{4}\right) = 9 \implies x + 2y + 4z = 9$$
$$4\left(-\frac{1}{2}\right) + 8\left(\frac{7}{4}\right) = 12 \implies \quad\;\; 4y + 8z = 12$$
$$4\left(\frac{7}{4}\right) = 7 \implies \quad\quad\quad 4z = 7$$

53. $2x + 3y + 4z = 12$

$(6, 0, 0), \; (0, 4, 0), \; (0, 0, 3), \; (4, 0, 1)$

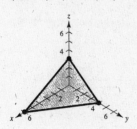

55. $2x + y + z = 4$

$(2, 0, 0), \; (0, 4, 0), \; (0, 0, 4), \; (0, 2, 2)$

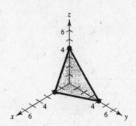

57. $\dfrac{7}{x^2 - 14x} = \dfrac{7}{x(x - 14)} = \dfrac{A}{x} + \dfrac{B}{x - 14}$

59. $\dfrac{12}{x^3 - 10x^2} = \dfrac{12}{x^2(x - 10)} = \dfrac{A}{x} + \dfrac{B}{x^2} + \dfrac{C}{x - 10}$

61. $\dfrac{4x^2 + 3}{(x - 5)^3} = \dfrac{A}{(x - 5)} + \dfrac{B}{(x - 5)^2} + \dfrac{C}{(x - 5)^3}$

63. $\dfrac{1}{x^2 - 1} = \dfrac{A}{x + 1} + \dfrac{B}{x - 1}$

$1 = A(x - 1) + B(x + 1) = (A + B)x + (B - A)$

$\begin{cases} A + B = 0 \\ -A + B = 1 \end{cases}$

$2B = 1 \implies B = \frac{1}{2} \implies A = -\frac{1}{2}$

$\dfrac{1}{x^2 - 1} = \dfrac{-1/2}{x + 1} + \dfrac{1/2}{x - 1} = \dfrac{1}{2}\left[\dfrac{1}{x - 1} - \dfrac{1}{x + 1}\right]$

65. $\dfrac{1}{x^2 + x} = \dfrac{1}{x(x + 1)} = \dfrac{A}{x} + \dfrac{B}{x + 1}$

$1 = A(x + 1) + Bx = (A + B)x + A$

$\begin{cases} A + B = 0 \\ A = 1 \implies B = -1 \end{cases}$

$\dfrac{1}{x^2 + x} = \dfrac{1}{x} + \dfrac{-1}{x + 1} = \dfrac{1}{x} - \dfrac{1}{x + 1}$

67. $\dfrac{1}{2x^2 + x} = \dfrac{1}{x(2x + 1)} = \dfrac{A}{2x + 1} + \dfrac{B}{x}$

$1 = Ax + B(2x + 1) = (A + 2B)x + B$

$\begin{cases} A + 2B = 0 \\ B = 1 \implies A = -2 \end{cases}$

$\dfrac{1}{2x^2 + x} = \dfrac{-2}{2x + 1} + \dfrac{1}{x} = \dfrac{1}{x} - \dfrac{2}{2x + 1}$

69. $\dfrac{5 - x}{2x^2 + x - 1} = \dfrac{5 - x}{(2x - 1)(x + 1)}$

$\qquad\qquad = \dfrac{A}{2x - 1} + \dfrac{B}{x + 1}$

$5 - x = A(x + 1) + B(2x - 1)$

$\qquad = (A + 2B)x + (A - B)$

$\begin{cases} A + 2B = -1 \implies A = -1 - 2B \\ A - B = 5 \end{cases}$

$(-1 - 2B) - B = 5 \implies B = -2$ and $A = 3$

$\dfrac{5 - x}{2x^2 + x - 1} = \dfrac{3}{2x - 1} + \dfrac{-2}{x + 1}$

71. $\dfrac{x^2 + 12x + 12}{x^3 - 4x} = \dfrac{x^2 + 12x + 12}{x(x - 2)(x + 2)} = \dfrac{A}{x} + \dfrac{B}{x + 2} + \dfrac{C}{x - 2}$

$x^2 + 12x + 12 = A(x + 2)(x - 2) + Bx(x - 2) + Cx(x + 2)$

$\qquad\qquad\qquad = (A + B + C)x^2 + (-2B + 2C)x + (-4A)$

$\begin{cases} A + B + C = 1 \\ -2B + 2C = 12 \\ -4A = 12 \implies A = -3 \end{cases}$

$\begin{cases} B + C = 4 \\ -B + C = 6 \end{cases}$

$2C = 10 \implies C = 5 \implies B = -1$

$\dfrac{x^2 + 12x + 12}{x^3 - 4x} = \dfrac{-3}{x} + \dfrac{-1}{x + 2} + \dfrac{5}{x - 2}$

73. $\dfrac{4x^2 + 2x - 1}{x^2(x + 1)} = \dfrac{A}{x} + \dfrac{B}{x^2} + \dfrac{C}{x + 1}$

$4x^2 + 2x - 1 = Ax(x + 1) + B(x + 1) + Cx^2$

$\qquad\qquad = (A + C)x^2 + (A + B)x + B$

$\begin{cases} A + C = 4 \\ A + B = 2 \\ B = -1 \end{cases}$

$B = -1 \implies A = 3 \implies C = 1$

$\dfrac{4x^2 + 2x - 1}{x^2(x + 1)} = \dfrac{3}{x} + \dfrac{-1}{x^2} + \dfrac{1}{x + 1}$

75. $\dfrac{27 - 7x}{x(x - 3)^2} = \dfrac{A}{x} + \dfrac{B}{x - 3} + \dfrac{C}{(x - 3)^2}$

$27 - 7x = A(x - 3)^2 + Bx(x - 3) + Cx$

$\qquad = (A + B)x^2 + (-6A - 3B + C)x + 9A$

$\begin{cases} A + B = 0 \\ -6A - 3B + C = -7 \\ 9A = 27 \end{cases}$

$A = 3 \implies B = -3 \implies C = -7 + 18 - 9 = 2$

$\dfrac{27 - 7x}{x(x - 3)^2} = \dfrac{3}{x} + \dfrac{-3}{x - 3} + \dfrac{2}{(x - 3)^2}$

77. $\dfrac{2x^3 - x^2 + x + 5}{x^2 + 3x + 2} = 2x - 7 + \dfrac{18x + 19}{(x + 1)(x + 2)}$

$\dfrac{18x + 19}{(x + 1)(x + 2)} = \dfrac{A}{x + 1} + \dfrac{B}{x + 2}$

$18x + 19 = A(x + 2) + B(x + 1) = (A + B)x + (2A + B)$

$\begin{cases} A + B = 18 \\ 2A + B = 19 \end{cases}$

$A = 1 \implies B = 17$

$\dfrac{2x^3 - x^2 + x + 5}{x^2 + 3x + 2} = 2x - 7 + \dfrac{1}{x + 1} + \dfrac{17}{x + 2}$

79. $\dfrac{x^4}{(x - 1)^3} = x + 3 + \dfrac{6x^2 - 8x + 3}{(x - 1)^3}$

$\dfrac{6x^2 - 8x + 3}{(x - 1)^3} = \dfrac{A}{x - 1} + \dfrac{B}{(x - 1)^2} + \dfrac{C}{(x - 1)^3}$

$6x^2 - 8x + 3 = A(x - 1)^2 + B(x - 1) + C = Ax^2 + (-2A + B)x + (A - B + C)$

$\begin{cases} A \qquad\qquad = 6 \\ -2A + B \qquad = -8 \\ A - B + C = 3 \end{cases}$

$A = 6 \implies B = -8 + 2(6) = 4 \implies C = 3 - 6 + 4 = 1$

$\dfrac{x^4}{(x - 1)^3} = \dfrac{6}{x - 1} + \dfrac{4}{(x - 1)^2} + \dfrac{1}{(x - 1)^3} + x + 3$

81. $\dfrac{x - 12}{x(x - 4)} = \dfrac{A}{x} + \dfrac{B}{x - 4}$

$x - 12 = A(x - 4) + Bx$

$\begin{cases} A + B = 1 \\ -4A \qquad = -12 \end{cases} \implies A = 3, B = -2$

$\dfrac{x - 12}{x(x - 4)} = \dfrac{3}{x} - \dfrac{2}{x - 4}$

$y = \dfrac{x - 12}{x(x - 4)}$ $y = \dfrac{3}{x}, y = -\dfrac{2}{x - 4}$

Vertical asymptotes: $x = 0$ and $x = 4$ Vertical asymptotes: $x = 0$ and $x = 4$

The combination of the vertical asymptotes of the terms of the decompositions are the same as the vertical asymptotes of the rational function.

83. $s = \frac{1}{2}at^2 + v_0t + s_0$

(1, 128), (2, 80), (3, 0)

$$\begin{cases} 128 = \frac{1}{2}a + v_0 + s_0 \Rightarrow a + 2v_0 + 2s_0 = 256 \\ 80 = 2a + 2v_0 + s_0 \Rightarrow 2a + 2v_0 + s_0 = 80 \\ 0 = \frac{9}{2}a + 3v_0 + s_0 \Rightarrow 9a + 6v_0 + 2s_0 = 0 \end{cases}$$

Solving the system, $a = -32, v_0 = 0, s_0 = 144$.

Thus, $s = \frac{1}{2}(-32)t^2 + (0)t + 144$

$\qquad = -16t^2 + 144$.

85. $s = \frac{1}{2}at^2 + v_0t + a_0$

(1, 452), (2, 372), (3, 260)

$$\begin{cases} 452 = \frac{1}{2}a + v_0 + s_0 \Rightarrow a + 2v_0 + 2s_0 = 904 \\ 372 = 2a + 2v_0 + s_0 \Rightarrow 2a + 2v_0 + s_0 = 372 \\ 260 = \frac{9}{2}a + 3v_0 + s_0 \Rightarrow 9a + 6v_0 + 2s_0 = 520 \end{cases}$$

Solving the system, $a = -32, v_0 = -32, s_0 = 500$

Thus, $s = \frac{1}{2}(-32)t^2 - 32t + 500$

$\qquad = -16t^2 - 32t + 500$

87. $y = ax^2 + bx + c$ passing through (0, 0), (2, −2), (4, 0)

$$\begin{cases} (0, 0): \quad 0 = 4a + 2b + c \Rightarrow \quad c = -4a - 2b \\ (2, -2): -2 = 4a + 2b + c \Rightarrow -1 = \quad 2a + b \\ (4, 0): \quad 0 = 16a + 4b + c \Rightarrow \quad 0 = \quad 4a + b \end{cases}$$

Answer: $a = \frac{1}{2}, b = -2, c = 0$

The equation of the parabola is $y = \frac{1}{2}x^2 - 2x$.

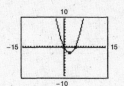

89. $y = ax^2 + bx + c$ passing through (2, 0), (3, −1), (4, 0)

$$\begin{cases} (2, 0): \quad 0 = 4a + 2b + c \quad \Rightarrow \quad c = -4a - 2b \\ (3, -1): -1 = 9a + 3b + c \quad \Rightarrow -1 = \quad 5a + b \\ (4, 0): \quad 0 = 16a + 4b + c \quad \Rightarrow \quad 0 = 12a + 2b \end{cases}$$

Answer: $a = 1, b = -6, c = 8$

The equation of the parabola is $y = x^2 - 6x + 8$.

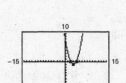

91. $x^2 + y^2 + Dx + Ey + F = 0$ passing through (0, 0), (2, 2), (4, 0)

(0, 0): $\qquad\qquad\qquad F = 0$

(2, 2): $8 + 2D + 2E + F = 0 \Rightarrow D + E = -4$

(4, 0): $16 + 4D \qquad + F = 0 \Rightarrow D = -4$ and $E = 0$

The equation of the circle is $x^2 + y^2 - 4x = 0$.

To graph, let $y_1 = \sqrt{4x - x^2}$ and $y_2 = -\sqrt{4x - x^2}$.

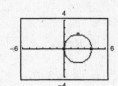

93. $x^2 + y^2 + Dx + Ey + F = 0$ passes through (−3, −1), (2, 4), (−6, 8).

(−3, −1): $10 - 3D - E + F = 0 \Rightarrow \quad 10 = \quad 3D + E - F$

(2, 4): $20 + 2D + 4E + F = 0 \Rightarrow \quad 20 = -2D - 4E - F$

(−6, 8): $100 - 6D + 8E + F = 0 \Rightarrow 100 = \quad 6D - 8E - F$

Answer: $D = 6, E = -8, F = 0$

The equation of the circle is $x^2 + y^2 + 6x - 8y = 0$. To graph, complete the squares first, then solve for y.

$(x^2 + 6x + 9) + (y^2 - 8y + 16) = 0 + 9 + 16$

$\qquad\qquad (x + 3)^2 + (y - 4)^2 = 25$

$\qquad\qquad\qquad\qquad (y - 4)^2 = 25 - (x + 3)^2$

$\qquad\qquad\qquad\qquad\quad y - 4 = \pm\sqrt{25 - (x + 3)^2}$

$\qquad\qquad\qquad\qquad\qquad y = 4 \pm \sqrt{25 - (x + 3)^2}$

Let $y_1 = 4 + \sqrt{25 - (x + 3)^2}$ and $y_2 = 4 - \sqrt{25 - (x + 3)^2}$.

95. Let x = amount at 8%.

Let y = amount at 9%.

Let z = amount at 10%.

$$\begin{cases} x + y + z = 775{,}000 \\ 0.08x + 0.09y + 0.10z = 67{,}000 \\ x = 4z \end{cases}$$

$$\begin{cases} x + y + z = 775{,}000 \\ 8x + 9y + 10z = 6{,}700{,}000 \\ x - 4z = 0 \end{cases}$$

Solving the system, $x =$ \$366,666.67 at 8%,
$y =$ \$316,666.67 at 9%, $z =$ \$91,666.67 at 10%.

97. Let C = amount in certificates of deposit.

Let M = amount in municipal bonds.

Let B = amount in blue chip stocks.

Let G = amount in growth or speculative stocks.

$$\begin{cases} C + M + B + G = 500{,}000 \\ 0.08C + 0.09M + 0.12B + 0.15G = 0.10(500{,}000) \\ M = \frac{1}{4}(500{,}000) \end{cases}$$

Solving the system

$C = 156{,}250 + 0.75s$

$M = 125{,}000$

$B = 218{,}750 - 1.75s$

$G = s$

99. Let x = number of 1-point free throws.

Let y = number of 2-point field goals.

Let z = number of 3-point basket.

$$\begin{cases} x + 2y + 3z = 84 \\ -x + y = 6 \\ y - 4z = 0 \end{cases}$$

Solving the system, $x = 18$, $y = 24$, $z = 6$.

18 free throws, 24 2-point field goals,
6 3-point baskets.

101. Let x = number of touchdowns.

Let y = number of extra-point kicks.

Let z = number of field goals.

$$\begin{cases} x + y + z = 9 \\ 6x + y + 3z = 31 \\ x - 4z = 0 \\ x - y = 0 \end{cases}$$

Solving the system, $x = 4$, $y = 4$, $z = 1$.

4 touchdowns, 4 extra-points and 1 field goal

103.
$$\begin{cases} I_1 - I_2 + I_3 = 0 & \text{Equation 1} \\ 3I_1 + 2I_2 = 7 & \text{Equation 2} \\ 2I_2 + 4I_3 = 8 & \text{Equation 3} \end{cases}$$

$$\begin{cases} I_1 - I_2 + I_3 = 0 \\ 5I_2 - 3I_3 = 7 & -3 \text{ Eq. 1} + \text{Eq. 2} \\ 2I_2 + 4I_3 = 8 \end{cases}$$

$$\begin{cases} I_1 - I_2 + I_3 = 0 \\ 10I_2 - 6I_3 = 14 & 2 \text{ Eq. 2} \\ 10I_2 + 20I_3 = 40 & 5 \text{ Eq. 3} \end{cases}$$

$$\begin{cases} I_1 - I_2 + I_3 = 0 \\ 10I_2 - 6I_3 = 14 \\ 26I_3 = 26 & -\text{Eq. 2} + \text{Eq. 3} \end{cases}$$

$26I_3 = 26 \Rightarrow I_3 = 1$

$10I_2 - 6(1) = 14 \Rightarrow I_2 = 2$

$I_1 - 2 + 1 = 0 \Rightarrow I_1 = 1$

Answer: $I_1 = 1$ ampere, $I_2 = 2$ amperes,
$I_3 = 1$ ampere

105. Let x = number of par-3 holes.

Let y = number of par-4 holes.

Let z = number of par-5 holes.

$$\begin{cases} 3x + 4y + 5z = 72 \\ y - 2z = 2 \\ x - z = 0 \end{cases}$$

Solving the system, $x = 4$, $y = 10$, $z = 4$.

4 par-3 holes, 10 par-4 holes, and 4 par-5 holes

107. Least squares regression parabola through
$(-4, 5), (-2, 6), (2, 6), (4, 2)$

$$\begin{cases} 4c & + 40a = 19 \\ 40b & = -12 \\ 40c & + 544a = 160 \end{cases}$$

Solving the system, $a = -\frac{5}{24}$, $b = -\frac{3}{10}$, and $c = \frac{41}{6}$.

Thus, $y = -\frac{5}{24}x^2 - \frac{3}{10}x + \frac{41}{6}$.

109. Least squares regression parabola through
$(0, 0), (2, 2), (3, 6), (4, 12)$

$$\begin{cases} 4c + 9b + 29a = 20 \\ 9c + 29b + 99a = 70 \\ 29c + 99b + 353a = 254 \end{cases}$$

Solving the system, $a = 1$, $b = -1$, and $c = 0$.
Thus, $y = x^2 - x$.

111. (a) $\begin{cases} a(30)^2 + b(30) + c = 55 \\ a(40)^2 + b(40) + c = 105 \\ a(50)^2 + b(50) + c = 188 \end{cases}$

Solving the system, $a = 0.165$, $b = -6.55$
and $c = 103$.

$y = 0.165x^2 - 6.55x + 103$

(b)

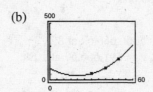

(c) For $x = 70$, $y = 453$ feet.

113. (a) $\dfrac{2000(4 - 3x)}{(11 - 7x)(7 - 4x)} = \dfrac{A}{11 - 7x} + \dfrac{B}{7 - 4x}$, $0 \le x \le 1$

$2000(4 - 3x) = A(7 - 4x) + B(11 - 7x)$

$\begin{cases} -6000 = -4A - 7B \\ 8000 = 7A + 11B \end{cases} \Rightarrow \begin{matrix} A = -2000 \\ B = 2000 \end{matrix}$

$\dfrac{2000(4 - 3x)}{(11 - 7x)(7 - 4x)} = \dfrac{-2000}{11 - 7x} + \dfrac{2000}{7 - 4x}$

$\qquad\qquad = \dfrac{2000}{7 - 4x} - \dfrac{2000}{11 - 7x}$

(b) $y_1 = \dfrac{2000}{7 - 4x}$

$\quad y_2 = \dfrac{2000}{11 - 7x}$

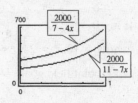

115. False. The coefficient of y in the second equation is not 1.

117. False. The correct form is

$$\frac{A}{x + 10} + \frac{B}{x - 10} + \frac{C}{(x - 10)^2}.$$

119. $\dfrac{1}{a^2 - x^2} = \dfrac{1}{(a + x)(a - x)} = \dfrac{A}{a + x} + \dfrac{B}{a - x}$

$\qquad 1 = A(a - x) + B(a + x) = (-A + B)x + (Aa + Ba)$

$\begin{cases} -A + B = 0 \\ Aa + Ba = 1 \end{cases} \Rightarrow A = \dfrac{1}{2a}, B = \dfrac{1}{2a}$

$\dfrac{1}{a^2 - x^2} = \dfrac{1/2a}{a + x} + \dfrac{1/2a}{a - x} = \dfrac{1}{2a}\left[\dfrac{1}{a + x} + \dfrac{1}{a - x}\right]$

121. $\dfrac{1}{y(a - y)} = \dfrac{A}{y} + \dfrac{B}{a - y}$

$\qquad 1 = A(a - y) + By = (-A + B)y + aA$

$A = \dfrac{1}{a}, B = \dfrac{1}{a}$

$\dfrac{1}{y(a - y)} = \dfrac{1}{a}\left(\dfrac{1}{y} + \dfrac{1}{a - y}\right)$

123. No, they are not equivalent. In the second system, the constant in the second equation should be -11 and the coefficient of z in the third equation should be 2.

125. $\begin{cases} y + \quad \lambda = \\ \quad x + \lambda = \end{cases} \Rightarrow x = y = -\lambda$
$\begin{cases} x + y - 10 = \end{cases} \Rightarrow 2x - 10 = 0$

$$x = 5$$
$$y = 5$$
$$\lambda = -5$$

127. $\begin{cases} 2x - 2x\lambda = 0 \Rightarrow x = x\lambda \\ -2y + \lambda = 0 \Rightarrow 2y = \lambda \\ y - x^2 = 0 \Rightarrow y = x^2 \end{cases}$

From the first equation, $x = 0$ or $\lambda = 1$.

If $x = 0$, then $y = 0^2 = 0$ and $\lambda = 0$.

If $x \ne 0$, then $\lambda = 1 \Rightarrow y = \frac{1}{2}$ and $x = \pm\sqrt{\frac{1}{2}}$.

Thus, the solutions are:

(1) $x = y = \lambda = 0$

(2) $x = \dfrac{\sqrt{2}}{2}$, $y = \dfrac{1}{2}$, $\lambda = 1$

(3) $x = -\dfrac{\sqrt{2}}{2}$, $y = \dfrac{1}{2}$, $\lambda = 1$

129. $y = -3x + 7$

131. $y = -2x^2$

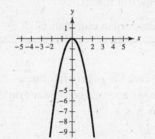

133. $y = -x^2(x - 3)$

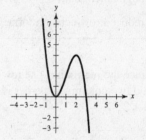

135. (a) $f(x) = x^3 + x^2 - 12x$

$= x(x^2 + x - 12) = x(x + 4)(x - 3)$

$\Rightarrow x = 0, -4, 3$

(b)

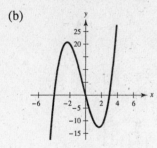

137. (a) $f(x) = 2x^3 + 5x^2 - 21x - 36$

$= (2x + 3)(x + 4)(x - 3)$

$\Rightarrow x = -\frac{3}{2}, -4, 3$

(b)

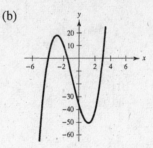

139.

x	-2	-1	0	1	2
y	16	8	4	2	1

$y = \left(\frac{1}{2}\right)^{x-2}$

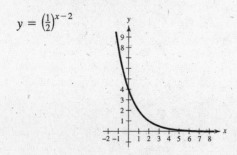

141.

x	-2	-1	0	1	2
y	$-\frac{1}{2}$	0	1	3	7

$y = 2^{x+1} - 1$

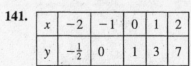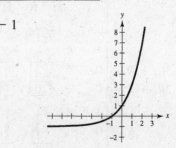

143. Answers will vary.

Section 7.4 Matrices and Systems of Equations

- You should be able to use elementary row operations to produce a row-echelon form (or reduced row-echelon form) of a matrix.

  1. Interchange two rows

  2. Multiply a row by a nonzero constant

  3. Add a multiple of one row to another row

- You should be able to use either Gaussian elimination with back-substitution or Gauss-Jordan elimination to solve a system of linear equations.

Vocabulary Check

1. matrix

2. square

3. row matrix, column matrix

4. augmented matrix

5. coefficient matrix

6. row-equivalent

7. reduced row-echelon form

8. Gauss-Jordan elimination

---

1. Since the matrix has one row and two columns, its order is 1×2.

3. Since the matrix has three rows and one column, its order is 3×1.

5. Since the matrix has two rows and two columns, its order is 2×2.

7. $\begin{aligned} 6x - 7y &= 11 \\ -2x + 5y &= -1 \end{aligned}$

$\begin{bmatrix} 6 & -7 & \vdots & 11 \\ -2 & 5 & \vdots & -1 \end{bmatrix}$

9. $\begin{cases} x + 10y - 2z = 2 \\ 5x - 3y + 4z = 0 \\ 2x + y = 6 \end{cases}$

$\begin{bmatrix} 1 & 10 & -2 & \vdots & 2 \\ 5 & -3 & 4 & \vdots & 0 \\ 2 & 1 & 0 & \vdots & 6 \end{bmatrix}$

11. $\begin{aligned} 3x + 4y &= 9 \\ x - y &= -3 \end{aligned}$

$\begin{bmatrix} 3 & 4 & \vdots & 9 \\ 1 & -1 & \vdots & -3 \end{bmatrix}$

13. $\begin{bmatrix} 9 & 12 & 3 & \vdots & 0 \\ -2 & 18 & 5 & \vdots & 10 \\ 1 & 7 & -8 & \vdots & -4 \end{bmatrix}$

$\begin{cases} 9x + 12y + 3z = 0 \\ -2x + 18y + 5z = 10 \\ x + 7y - 8z = -4 \end{cases}$

15. $\begin{bmatrix} 1 & 4 & 3 \\ 2 & 10 & 5 \end{bmatrix}$

$-2R_1 + R_2 \rightarrow \begin{bmatrix} 1 & 4 & 3 \\ 0 & \boxed{2} & -1 \end{bmatrix}$

17. $\begin{bmatrix} 1 & 1 & 4 & -1 \\ 3 & 8 & 10 & 3 \\ -2 & 1 & 12 & 6 \end{bmatrix}$

$\begin{aligned} -3R_1 + R_2 &\rightarrow \\ 2R_1 + R_3 &\rightarrow \end{aligned} \begin{bmatrix} 1 & 1 & 4 & -1 \\ 0 & 5 & \boxed{-2} & \boxed{6} \\ 0 & 3 & \boxed{20} & \boxed{4} \end{bmatrix}$

$\tfrac{1}{5}R_2 \rightarrow \begin{bmatrix} 1 & 1 & 4 & -1 \\ 0 & 1 & -\tfrac{2}{5} & \tfrac{6}{5} \\ 0 & 3 & \boxed{20} & \boxed{4} \end{bmatrix}$

19. Add -3 times Row 2 to Row 1.

21. Interchange Rows 1 and 2.

23. $\begin{bmatrix} 1 & 0 & 0 & 0 \\ 0 & 1 & 1 & 5 \\ 0 & 0 & 0 & 0 \end{bmatrix}$

This matrix is in reduced row-echelon form.

25. $\begin{bmatrix} 3 & 0 & 3 & 7 \\ 0 & -2 & 0 & 4 \\ 0 & 0 & 1 & 5 \end{bmatrix}$

The first nonzero entries in rows one and two are not one. The matrix is not in row-echelon form.

27. $\begin{bmatrix} 1 & 0 & 0 & 1 \\ 0 & 1 & 0 & -1 \\ 0 & 0 & 0 & 2 \end{bmatrix}$

The first nonzero entry in row three is two, not one. The matrix is not in row-echelon form.

29. $\begin{bmatrix} 1 & 2 & 3 \\ 2 & -1 & -4 \\ 3 & 1 & -1 \end{bmatrix}$

(a) $\begin{bmatrix} 1 & 2 & 3 \\ 0 & -5 & -10 \\ 3 & 1 & -1 \end{bmatrix}$ (b) $\begin{bmatrix} 1 & 2 & 3 \\ 0 & -5 & -10 \\ 0 & -5 & -10 \end{bmatrix}$ (c) $\begin{bmatrix} 1 & 2 & 3 \\ 0 & -5 & -10 \\ 0 & 0 & 0 \end{bmatrix}$

(d) $\begin{bmatrix} 1 & 2 & 3 \\ 0 & 1 & 2 \\ 0 & 0 & 0 \end{bmatrix}$ (e) $\begin{bmatrix} 1 & 0 & -1 \\ 0 & 1 & 2 \\ 0 & 0 & 0 \end{bmatrix}$ This matrix is in reduced row-echelon form.

31. (See Exercise 29.) (Answer is a series of screens.)

(a)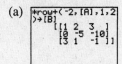
```
*row+(-2,[A],1,2
)→[B]
    [[1 2 3  ]
     [0 -5 -10]
     [3 1 -1 ]]
```

(b)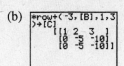
```
*row+(-3,[B],1,3
)→[C]
    [[1 2 3  ]
     [0 -5 -10]
     [0 -5 -10]]
```

(c)
```
*row+(-1,[C],2,3
)→[D]
    [[1 2 3  ]
     [0 -5 -10]
     [0 0 0 ]]
```

(d)
```
*row(-1/5,[D],2)
→[E]
    [[1 2 3]
     [0 1 2]
     [0 0 0]]
```

(e)
```
*row+(-2,[E],2,1
)
    [[1 0 -1]
     [0 1 2]
     [0 0 0]]
```

33. $\begin{bmatrix} 1 & 2 & 3 & 0 \\ -1 & 4 & 0 & -5 \\ 2 & 6 & 3 & 10 \end{bmatrix}$

$\begin{matrix} R_1 + R_2 \to \\ -2R_1 + R_3 \to \end{matrix} \begin{bmatrix} 1 & 2 & 3 & 0 \\ 0 & 6 & 3 & -5 \\ 0 & 2 & -3 & 10 \end{bmatrix}$

$\frac{1}{6}R_2 \to \begin{bmatrix} 1 & 2 & 3 & 0 \\ 0 & 1 & \frac{1}{2} & -\frac{5}{6} \\ 0 & 2 & -3 & 10 \end{bmatrix}$

$-2R_2 + R_3 \to \begin{bmatrix} 1 & 2 & 3 & 0 \\ 0 & 1 & \frac{1}{2} & -\frac{5}{6} \\ 0 & 0 & -4 & \frac{35}{3} \end{bmatrix}$

$-\frac{1}{4}R_3 \begin{bmatrix} 1 & 2 & 3 & 0 \\ 0 & 1 & \frac{1}{2} & -\frac{5}{6} \\ 0 & 0 & 1 & -\frac{35}{12} \end{bmatrix}$

(Answers may vary.)

35.
$$\begin{bmatrix} 1 & -1 & -1 & 1 \\ 5 & -4 & 1 & 8 \\ -6 & 8 & 18 & 0 \end{bmatrix}$$

$$\begin{matrix} -5R_1 + R_2 \to \\ 6R_1 + R_3 \to \end{matrix} \begin{bmatrix} 1 & -1 & -1 & 1 \\ 0 & 1 & 6 & 3 \\ 0 & 2 & 12 & 6 \end{bmatrix}$$

$$-2R_2 + R_3 \to \begin{bmatrix} 1 & -1 & -1 & 1 \\ 0 & 1 & 6 & 3 \\ 0 & 0 & 0 & 0 \end{bmatrix}$$

37.
$$\begin{bmatrix} 3 & 3 & 3 \\ -1 & 0 & -4 \\ 2 & 4 & -2 \end{bmatrix}$$

$$\tfrac{1}{3}R_1 \to \begin{bmatrix} 1 & 1 & 1 \\ -1 & 0 & -4 \\ 2 & 4 & -2 \end{bmatrix}$$

$$\begin{matrix} R_1 + R_2 \to \\ -2R_1 + R_3 \to \end{matrix} \begin{bmatrix} 1 & 1 & 1 \\ 0 & 1 & -3 \\ 0 & 2 & -4 \end{bmatrix}$$

$$\begin{matrix} -R_2 + R_1 \to \\ -2R_2 + R_3 \to \end{matrix} \begin{bmatrix} 1 & 0 & 4 \\ 0 & 1 & -3 \\ 0 & 0 & 2 \end{bmatrix}$$

$$\tfrac{1}{2}R_3 \to \begin{bmatrix} 1 & 0 & 4 \\ 0 & 1 & -3 \\ 0 & 0 & 1 \end{bmatrix}$$

$$\begin{matrix} -4R_3 + R_1 \to \\ 3R_3 + R_2 \to \end{matrix} \begin{bmatrix} 1 & 0 & 0 \\ 0 & 1 & 0 \\ 0 & 0 & 1 \end{bmatrix}$$

39.
$$\begin{bmatrix} -4 & 1 & 0 & 6 \\ 1 & -2 & 3 & -4 \end{bmatrix}$$

$$\begin{matrix} R_1 \to \\ R_2 \to \end{matrix} \begin{bmatrix} 1 & -2 & 3 & -4 \\ -4 & 1 & 0 & 6 \end{bmatrix}$$

$$4R_1 + R_2 \to \begin{bmatrix} 1 & -2 & 3 & -4 \\ 0 & -7 & 12 & -10 \end{bmatrix}$$

$$-\tfrac{1}{7}R_2 \to \begin{bmatrix} 1 & -2 & 3 & -4 \\ 0 & 1 & -\tfrac{12}{7} & \tfrac{10}{7} \end{bmatrix}$$

$$2R_2 + R_1 \to \begin{bmatrix} 1 & 0 & -\tfrac{3}{7} & -\tfrac{8}{7} \\ 0 & 1 & -\tfrac{12}{7} & \tfrac{10}{7} \end{bmatrix}$$

41. $x - 2y = 4$

$\qquad y = -3$

$x = 2y + 4 = 2(-3) + 4 = -2$

Answer: $(x, y) = (-2, -3)$

43. $\begin{cases} x - y + 2z = 4 \\ \qquad y - z = 2 \\ \qquad\qquad z = -2 \end{cases}$

$y - (-2) = 2$

$\qquad y = 0$

$x - 0 + 2(-2) = 4$

$\qquad x = 8$

Answer: $(8, 0, -2)$

45. $\begin{bmatrix} 1 & 0 & \vdots & 7 \\ 0 & 1 & \vdots & -5 \end{bmatrix}$

$x = 7$

$y = -5$

Answer: $(7, -5)$

47. $\begin{bmatrix} 1 & 0 & 0 & \vdots & -4 \\ 0 & 1 & 0 & \vdots & -8 \\ 0 & 0 & 1 & \vdots & 2 \end{bmatrix}$

$x = -4$

$y = -8$

$z = 2$

Answer: $(-4, -8, 2)$

49. $\begin{cases} x + 2y = 7 \\ 2x + y = 8 \end{cases}$

$$\begin{bmatrix} 1 & 2 & \vdots & 7 \\ 2 & 1 & \vdots & 8 \end{bmatrix}$$

$-2R_1 + R_2 \rightarrow \begin{bmatrix} 1 & 2 & \vdots & 7 \\ 0 & -3 & \vdots & -6 \end{bmatrix}$

$-\frac{1}{3}R_2 \rightarrow \begin{bmatrix} 1 & 2 & \vdots & 7 \\ 0 & 1 & \vdots & 2 \end{bmatrix}$

$y = 2$

$x + 2(2) = 7 \implies x = 3$

Answer: $(3, 2)$

51. $\begin{cases} -x + y = -22 \\ 3x + 4y = 4 \\ 4x - 8y = 32 \end{cases}$

$$\begin{bmatrix} -1 & 1 & \vdots & -22 \\ 3 & 4 & \vdots & 4 \\ 4 & -8 & \vdots & 32 \end{bmatrix}$$

$\begin{matrix} 3R_1 + R_2 \rightarrow \\ 4R_1 + R_3 \rightarrow \end{matrix} \begin{bmatrix} -1 & 1 & \vdots & -22 \\ 0 & 7 & \vdots & -62 \\ 0 & -4 & \vdots & -56 \end{bmatrix}$

$\begin{matrix} R_2 \rightarrow \\ R_3 \rightarrow \end{matrix} \begin{bmatrix} -1 & 1 & \vdots & -22 \\ 0 & -4 & \vdots & -56 \\ 0 & 7 & \vdots & -62 \end{bmatrix}$

$\begin{matrix} -\frac{1}{4}R_2 \rightarrow \\ -7R_2 + R_3 \rightarrow \end{matrix} \begin{bmatrix} -1 & 1 & \vdots & -22 \\ 0 & 1 & \vdots & 14 \\ 0 & 0 & \vdots & -160 \end{bmatrix}$

No solution, inconsistent

53. $\begin{bmatrix} 3 & 2 & -1 & 1 & \vdots & 0 \\ 1 & -1 & 4 & 2 & \vdots & 25 \\ -2 & 1 & 2 & -1 & \vdots & 2 \\ 1 & 1 & 1 & 1 & \vdots & 6 \end{bmatrix}$

$$\begin{bmatrix} 1 & -1 & 4 & 2 & \vdots & 25 \\ 0 & 5 & -13 & -5 & \vdots & -75 \\ 0 & -1 & 10 & 3 & \vdots & 52 \\ 0 & 2 & -3 & -1 & \vdots & -19 \end{bmatrix}$$

$$\begin{bmatrix} 1 & -1 & 4 & 2 & \vdots & 25 \\ 0 & 1 & -10 & -3 & \vdots & -52 \\ 0 & 0 & 37 & 10 & \vdots & 185 \\ 0 & 0 & 17 & 5 & \vdots & 85 \end{bmatrix}$$

$$\begin{bmatrix} 1 & -1 & 4 & 2 & \vdots & 25 \\ 0 & 1 & -10 & -3 & \vdots & -52 \\ 0 & 0 & 37 & 10 & \vdots & 185 \\ 0 & 0 & 0 & \frac{15}{37} & \vdots & 0 \end{bmatrix}$$

$w = 0, z = \frac{185}{37} = 5, y = -52 + 10(5) = -2$

$x = 25 + (-2) - 4(5) = 3$

Answer: $(3, -2, 5, 0)$

55. $\begin{cases} x \quad\;\;\; - 3z = -2 \\ 3x + y - 2z = \;\;\;5 \\ 2x + 2y + z = \;\;\;4 \end{cases}$

$$\begin{bmatrix} 1 & 0 & -3 & \vdots & -2 \\ 3 & 1 & -2 & \vdots & 5 \\ 2 & 2 & 1 & \vdots & 4 \end{bmatrix}$$

$\begin{matrix} \\ -3R_1 + R_2 \rightarrow \\ -2R_1 + R_3 \rightarrow \end{matrix}\begin{bmatrix} 1 & 0 & -3 & \vdots & -2 \\ 0 & 1 & 7 & \vdots & 11 \\ 0 & 2 & 7 & \vdots & 8 \end{bmatrix}$

$\begin{matrix} \\ \\ -2R_2 + R_3 \rightarrow \end{matrix}\begin{bmatrix} 1 & 0 & -3 & \vdots & -2 \\ 0 & 1 & 7 & \vdots & 11 \\ 0 & 0 & -7 & \vdots & -14 \end{bmatrix}$

$\begin{matrix} \\ \\ -\frac{1}{7}R_3 \rightarrow \end{matrix}\begin{bmatrix} 1 & 0 & -3 & \vdots & -2 \\ 0 & 1 & 7 & \vdots & 11 \\ 0 & 0 & 1 & \vdots & 2 \end{bmatrix}$

$\begin{matrix} 3R_3 + R_1 \rightarrow \\ -7R_3 + R_2 \rightarrow \\ \end{matrix}\begin{bmatrix} 1 & 0 & 0 & \vdots & 4 \\ 0 & 1 & 0 & \vdots & -3 \\ 0 & 0 & 1 & \vdots & 2 \end{bmatrix}$

Answer: $(4, -3, 2)$

57. $\begin{cases} x + y - 5z = 3 \\ x \quad\;\;\; - 2z = 1 \\ 2x - y - z = 0 \end{cases}$

$$\begin{bmatrix} 1 & 1 & -5 & \vdots & 3 \\ 1 & 0 & -2 & \vdots & 1 \\ 2 & -1 & -1 & \vdots & 0 \end{bmatrix}$$

$\begin{matrix} \\ -R_1 + R_2 \rightarrow \\ -2R_1 + R_3 \rightarrow \end{matrix}\begin{bmatrix} 1 & 1 & -5 & \vdots & 3 \\ 0 & -1 & 3 & \vdots & -2 \\ 0 & -3 & 9 & \vdots & -6 \end{bmatrix}$

$\begin{matrix} \\ \\ -3R_2 + R_3 \rightarrow \end{matrix}\begin{bmatrix} 1 & 1 & -5 & \vdots & 3 \\ 0 & -1 & 3 & \vdots & -2 \\ 0 & 0 & 0 & \vdots & 0 \end{bmatrix}$

$\begin{matrix} R_2 + R_1 \rightarrow \\ -R_2 \rightarrow \\ \end{matrix}\begin{bmatrix} 1 & 0 & -2 & \vdots & 1 \\ 0 & 1 & -3 & \vdots & 2 \\ 0 & 0 & 0 & \vdots & 0 \end{bmatrix}$

Let $z = a$, any real number

$y - 3a = 2 \implies y = 3a + 2$

$x - 2a = 1 \implies x = 2a + 1$

Answer: $(2a + 1, 3a + 2, a)$

59. $\begin{cases} -x + y - z = -14 \\ 2x - y + z = \;\;\;21 \\ 3x + 2y + z = \;\;\;19 \end{cases}$

$$\begin{bmatrix} -1 & 1 & -1 & \vdots & -14 \\ 2 & -1 & 1 & \vdots & 21 \\ 3 & 2 & 1 & \vdots & 19 \end{bmatrix}$$

$\begin{matrix} \\ 2R_1 + R_2 \rightarrow \\ 3R_1 + R_3 \rightarrow \end{matrix}\begin{bmatrix} -1 & 1 & -1 & \vdots & -14 \\ 0 & 1 & -1 & \vdots & -7 \\ 0 & 5 & -2 & \vdots & -23 \end{bmatrix}$

$\begin{matrix} -R_1 \rightarrow \\ \\ -5R_2 + R_3 \rightarrow \end{matrix}\begin{bmatrix} 1 & -1 & 1 & \vdots & 14 \\ 0 & 1 & -1 & \vdots & -7 \\ 0 & 0 & 3 & \vdots & 12 \end{bmatrix}$

$\begin{matrix} \\ \\ \frac{1}{3}R_3 \rightarrow \end{matrix}\begin{bmatrix} 1 & -1 & 1 & \vdots & 14 \\ 0 & 1 & -1 & \vdots & -7 \\ 0 & 0 & 1 & \vdots & 4 \end{bmatrix}$

$\begin{matrix} -R_3 + R_1 \rightarrow \\ R_3 + R_2 \rightarrow \\ \end{matrix}\begin{bmatrix} 1 & -1 & 0 & \vdots & 10 \\ 0 & 1 & 0 & \vdots & -3 \\ 0 & 0 & 1 & \vdots & 4 \end{bmatrix}$

$\begin{matrix} R_2 + R_1 \rightarrow \\ \\ \end{matrix}\begin{bmatrix} 1 & 0 & 0 & \vdots & 7 \\ 0 & 1 & 0 & \vdots & -3 \\ 0 & 0 & 1 & \vdots & 4 \end{bmatrix}$

Answer: $(7, -3, 4)$

61. $\begin{cases} 3x + 3y + 12z = \;\;6 \\ x + y + 4z = \;\;2 \\ 2x + 5y + 20z = 10 \\ -x + 2y + 8z = \;\;4 \end{cases}$

$$\begin{bmatrix} 3 & 3 & 12 & \vdots & 6 \\ 1 & 1 & 4 & \vdots & 2 \\ 2 & 5 & 20 & \vdots & 10 \\ -1 & 2 & 8 & \vdots & 4 \end{bmatrix} \implies \begin{bmatrix} 1 & 0 & 0 & \vdots & 0 \\ 0 & 1 & 4 & \vdots & 2 \\ 0 & 0 & 0 & \vdots & 0 \\ 0 & 0 & 0 & \vdots & 0 \end{bmatrix}$$

Let $z = a$, any real number

$y = -4a + 2$

$x = 0$

Answer: $(0, -4a + 2, a)$

63. $\begin{cases} 2x + 10y + 2z = 6 \\ x + 5y + 2z = 6 \\ x + 5y + z = 3 \\ -3x - 15y - 3z = -9 \end{cases}$ $\begin{bmatrix} 2 & 10 & 2 & \vdots & 6 \\ 1 & 5 & 2 & \vdots & 6 \\ 1 & 5 & 1 & \vdots & 3 \\ -3 & -15 & -3 & \vdots & -9 \end{bmatrix} \Rightarrow \begin{bmatrix} 1 & 5 & 0 & \vdots & 0 \\ 0 & 0 & 1 & \vdots & 3 \\ 0 & 0 & 0 & \vdots & 0 \\ 0 & 0 & 0 & \vdots & 0 \end{bmatrix}$

$z = 3, y = a, x = -5a$

Answer: $(-5a, a, 3)$

65. Yes, the systems yield the same solutions.

(a) $z = -3; y = 5(-3) + 16 = 1;$
$x = 2(1) - (-3) - 6 = -1$

Answer: $(-1, 1, -3)$

(b) $z = -3, y = -3(-3) - 8 = 1,$
$x = -1 + 2(-3) + 6 = -1$

Answer: $(-1, 1, -3)$

67. No, solutions are different.

(a) $z = 8, y = 7(8) - 54 = 2,$
$x = 4(2) - 5(8) + 27 = -5$

Answer: $(-5, 2, 8)$

(b) $z = 8, y = -5(8) + 42 = 2,$
$x = 6(2) - 8 + 15 = 19$

Answer: $(19, 2, 8)$

69. $f(x) = ax^2 + bx + c$

$\begin{cases} f(1) = a + b + c = 8 \\ f(2) = 4a + 2b + c = 13 \\ f(3) = 9a + 3b + c = 20 \end{cases}$

$\begin{bmatrix} 1 & 1 & 1 & \vdots & 8 \\ 4 & 2 & 1 & \vdots & 13 \\ 9 & 3 & 1 & \vdots & 20 \end{bmatrix}$

$\begin{matrix} \\ -4R_1 + R_2 \rightarrow \\ -9R_1 + R_3 \rightarrow \end{matrix} \begin{bmatrix} 1 & 1 & 1 & \vdots & 8 \\ 0 & -2 & -3 & \vdots & -19 \\ 0 & -6 & -8 & \vdots & -52 \end{bmatrix}$

$\begin{matrix} \\ -\frac{1}{2}R_2 \rightarrow \\ -3R_2 + R_3 \rightarrow \end{matrix} \begin{bmatrix} 1 & 1 & 1 & \vdots & 8 \\ 0 & 1 & \frac{3}{2} & \vdots & \frac{19}{2} \\ 0 & 0 & 1 & \vdots & 5 \end{bmatrix}$

$c = 5$

$b + \frac{3}{2}(5) = \frac{19}{2} \implies b = 2$

$a + 2 + 5 = 8 \implies a = 1$

Answer: $y = x^2 + 2x + 5$

71. $f(x) = ax^2 + bx + c$

$\begin{cases} f(1) = a + b + c = 2 \\ f(-2) = 4a - 2b + c = 11 \\ f(3) = 9a + 3b + c = 16 \end{cases}$

$\begin{bmatrix} 1 & 1 & 1 & \vdots & 2 \\ 4 & -2 & 1 & \vdots & 11 \\ 9 & 3 & 1 & \vdots & 16 \end{bmatrix}$

$\begin{matrix} \\ -4R_1 + R_2 \rightarrow \\ -9R_1 + R_3 \rightarrow \end{matrix} \begin{bmatrix} 1 & 1 & 1 & \vdots & 2 \\ 0 & -6 & -3 & \vdots & 3 \\ 0 & -6 & -8 & \vdots & -2 \end{bmatrix}$

$\begin{matrix} \\ \\ (-1)R_2 + R_3 \rightarrow \end{matrix} \begin{bmatrix} 1 & 1 & 1 & \vdots & 2 \\ 0 & -6 & -3 & \vdots & 3 \\ 0 & 0 & -5 & \vdots & -5 \end{bmatrix}$

$-5c = -5 \implies c = 1$

$-6b - 3c = 3 \implies -6b = 3 + 3 = 6 \implies b = -1$

$a + b + c = 2 \implies a = 2 + 1 - 1 = 2$

Answer: $y = 2x^2 - x + 1$

73. $f(x) = ax^2 + bx + c$

$f(-2) = 4a - 2b + c = -15$
$f(-1) = a - b + c = 7$
$f(1) = a + b + c = -3$

Solving the system, $a = -9, b = -5, c = 11.$

$f(x) = -9x^2 - 5x + 11$

75. $f(x) = ax^3 + bx^2 + cx + d$

$f(-2) = -8a + 4b - 2c + d = -7$
$f(-1) = -a + b - c + d = 2$
$f(1) = a + b + c + d = -4$
$f(2) = 8a + 4b + 2c + d = -7$

Solving the system,
$a = 1, b = -2, c = -4, d = 1.$

$f(x) = x^3 - 2x^2 - 4x + 1$

77. $x =$ amount at 7%, $y =$ amount at 8%, $z =$ amount at 10%

$$\begin{cases} x + \quad y + \quad z = 1{,}500{,}000 \\ 0.07x + 0.08y + 0.1z = \quad 130{,}500 \\ 4x \quad\quad - \quad z = \quad\quad\quad 0 \end{cases}$$

$$\begin{bmatrix} 1 & 1 & 1 & \vdots & 1{,}500{,}000 \\ 0.07 & 0.08 & 0.1 & \vdots & 130{,}500 \\ 4 & 0 & -1 & \vdots & 0 \end{bmatrix}$$

$$\begin{matrix} \\ -0.07R_1 + R_2 \rightarrow \\ -4R_1 + R_3 \rightarrow \end{matrix} \begin{bmatrix} 1 & 1 & 1 & \vdots & 1{,}500{,}000 \\ 0 & 0.01 & 0.03 & \vdots & 25{,}500 \\ 0 & -4 & -5 & \vdots & -6{,}000{,}000 \end{bmatrix}$$

$$\begin{matrix} \\ 100R_2 \rightarrow \\ 4R_2 + R_3 \rightarrow \end{matrix} \begin{bmatrix} 1 & 1 & 1 & \vdots & 1{,}500{,}000 \\ 0 & 1 & 3 & \vdots & 2{,}550{,}000 \\ 0 & 0 & 7 & \vdots & 4{,}200{,}000 \end{bmatrix}$$

$7z = 4{,}200{,}000 \implies z = 600{,}00$

$y + 3(600{,}000) = 2{,}550{,}000 \implies y = 750{,}000$

$x + 750{,}000 + 600{,}000 = 1{,}500{,}000 \implies x = 150{,}000$

Answers: \$150,000 at 7%, \$750,000 at 8%, \$600,000 at 10%

79. $\begin{cases} I_1 - I_2 + I_3 = 0 \\ 2I_1 + 2I_2 \quad\quad = 7 \\ \quad\quad 2I_2 + 4I_3 = 8 \end{cases}$

$$\begin{bmatrix} 1 & -1 & 1 & \vdots & 0 \\ 2 & 2 & 0 & \vdots & 7 \\ 0 & 2 & 4 & \vdots & 8 \end{bmatrix}$$

$$\begin{matrix} \\ -2R_1 + R_2 \rightarrow \\ \\ \end{matrix} \begin{bmatrix} 1 & -1 & 1 & \vdots & 0 \\ 0 & 4 & -2 & \vdots & 7 \\ 0 & 2 & 4 & \vdots & 8 \end{bmatrix}$$

$$\begin{matrix} \\ R_3 \rightarrow \\ R_2 \rightarrow \end{matrix} \begin{bmatrix} 1 & -1 & 1 & \vdots & 0 \\ 0 & 2 & 4 & \vdots & 8 \\ 0 & 4 & -2 & \vdots & 7 \end{bmatrix}$$

$$\begin{matrix} \\ \tfrac{1}{2}R_2 \rightarrow \\ \\ \end{matrix} \begin{bmatrix} 1 & -1 & 1 & \vdots & 0 \\ 0 & 1 & 2 & \vdots & 4 \\ 0 & 4 & -2 & \vdots & 7 \end{bmatrix}$$

$$\begin{matrix} \\ \\ -4R_2 + R_3 \rightarrow \end{matrix} \begin{bmatrix} 1 & -1 & 1 & \vdots & 0 \\ 0 & 1 & 2 & \vdots & 4 \\ 0 & 0 & -10 & \vdots & -9 \end{bmatrix}$$

$$\begin{matrix} \\ \\ -\tfrac{1}{10}R_3 \rightarrow \end{matrix} \begin{bmatrix} 1 & -1 & 1 & \vdots & 0 \\ 0 & 1 & 2 & \vdots & 4 \\ 0 & 0 & 1 & \vdots & \tfrac{9}{10} \end{bmatrix}$$

$I_3 = \tfrac{9}{10}$ amperes; $I_2 + 2\left(\tfrac{9}{10}\right) = 4 \implies I_2 = \tfrac{11}{5}$ amperes; $I_1 - \tfrac{11}{5} + \tfrac{9}{10} = 0 \implies I_1 = \tfrac{13}{10}$ amperes

81. (a) (2, 55.37) $\begin{cases} 4a + 2b + c = 55.37 \\ 9a + 3b + c = 59.52 \\ 16a + 4b + c = 63.59 \end{cases}$
(3, 59.52)
(4, 63.59)

Solving the system using matrices,

$$\begin{bmatrix} 4 & 2 & 1 & \vdots & 55.37 \\ 9 & 3 & 1 & \vdots & 59.52 \\ 16 & 4 & 1 & \vdots & 63.59 \end{bmatrix} \Rightarrow \begin{bmatrix} 1 & 0 & 0 & \vdots & -0.04 \\ 0 & 1 & 0 & \vdots & 4.35 \\ 0 & 0 & 1 & \vdots & 46.83 \end{bmatrix}$$

you obtain $a = -0.04$, $b = 4.35$, $c = 46.83$.

$$y = -0.04t^2 + 4.35t + 46.83$$

(b)

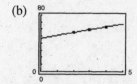

(c) For 2005, $t = 5$ and $y \approx 67.58$ dollars

For 2010, $t = 10$ and $y \approx 86.33$ dollars

For 2015, $t = 15$ and $y \approx 103.08$ dollars

83. Let x = number of pounds of glossy.

Let y = number of pounds of semi-glossy

Let z = number of pounds of matte

$$\begin{cases} x + y + z = 100 \\ 5.5x + 4.25y + 3.75z = 480 \\ y + z = 50 \end{cases}$$

Solving the system, $x = 50$, $y = 35$ and $z = 15$.

50 pounds of glossy, 35 pounds of semi-glossy and 15 pounds of matte

85. (a)
$$\begin{cases} x_1 + x_3 = 600 \\ x_1 = x_2 + x_4 \implies x_1 - x_2 - x_4 = 0 \\ x_2 + x_5 = 500 \\ x_3 + x_6 = 600 \\ x_4 + x_7 = x_6 \implies x_4 - x_6 + x_7 = 0 \\ x_5 + x_7 = 500 \end{cases}$$

$$\begin{bmatrix} 1 & 0 & 1 & 0 & 0 & 0 & 0 & \vdots & 600 \\ 1 & -1 & 0 & -1 & 0 & 0 & 0 & \vdots & 0 \\ 0 & 1 & 0 & 0 & 1 & 0 & 0 & \vdots & 500 \\ 0 & 0 & 1 & 0 & 0 & 1 & 0 & \vdots & 600 \\ 0 & 0 & 0 & 1 & 0 & -1 & 1 & \vdots & 0 \\ 0 & 0 & 0 & 0 & 1 & 0 & 1 & \vdots & 500 \end{bmatrix}$$

$$\begin{matrix} \\ -R_1 + R_2 \to \\ R_2 + R_3 \to \\ R_3 + R_4 \to \\ R_4 + R_5 \to \\ -R_5 + R_6 \to \end{matrix} \begin{bmatrix} 1 & 0 & 1 & 0 & 0 & 0 & 0 & \vdots & 600 \\ 0 & -1 & -1 & -1 & 0 & 0 & 0 & \vdots & -600 \\ 0 & 0 & -1 & -1 & 1 & 0 & 0 & \vdots & -100 \\ 0 & 0 & 0 & -1 & 1 & 1 & 0 & \vdots & 500 \\ 0 & 0 & 0 & 0 & 1 & 0 & 1 & \vdots & 500 \\ 0 & 0 & 0 & 0 & 0 & 0 & 0 & \vdots & 0 \end{bmatrix}$$

$$\begin{matrix} \\ -R_3 + R_2 \to \\ -R_4 + R_3 \to \\ -R_4 \to \\ \\ \end{matrix} \begin{bmatrix} 1 & 0 & 1 & 0 & 0 & 0 & 0 & \vdots & 600 \\ 0 & -1 & 0 & 0 & -1 & 0 & 0 & \vdots & -500 \\ 0 & 0 & -1 & 0 & 0 & -1 & 0 & \vdots & -600 \\ 0 & 0 & 0 & 1 & -1 & -1 & 0 & \vdots & -500 \\ 0 & 0 & 0 & 0 & 1 & 0 & 1 & \vdots & 500 \\ 0 & 0 & 0 & 0 & 0 & 0 & 0 & \vdots & 0 \end{bmatrix}$$

Let $x_7 = t$ and $x_6 = s$, then:

$x_5 = 500 - t$

$x_4 = -500 + s + (500 - t) = s - t$

$x_3 = 600 - s$

$x_2 = 500 - (500 - t) = t$

$x_1 = 600 - (600 - s) = s$

(b) If $x_6 = x_7 = 0$, then $s = t = 0$, and

$x_1 = 0$

$x_2 = 0$

$x_3 = 600$

$x_4 = 0$

$x_5 = 500$

$x_6 = x_7 = 0$

(c) If $x_5 = 1000$ and $x_6 = 0$, then $s = 0$ and $t = -500$. Thus:

$x_1 = 0$

$x_2 = -500$

$x_3 = 600$

$x_4 = 500$

$x_5 = 1000$

$x_6 = 0$

$x_7 = -500$

87. False. It is a 2×4 matrix.

89. $\begin{cases} x + 3z = -2 & \text{Equation 1} \\ y + 4z = 1 & \text{Equation 2} \end{cases}$

(Equation 1) + (Equation 2) → new Equation 1

(Equation 1) + 2(Equation 2) → new Equation 2

2(Equation 1) + (Equation 2) → new Equation 3

$\begin{cases} x + y + 7z = -1 \\ x + 2y + 11z = 0 \\ 2x + y + 10z = -3 \end{cases}$

91. The row operation $-2R_1 + R_2$ was not performed on the last column. Nor was $-R_2 + R_1$.

93. $f(x) = \dfrac{7}{-x - 1}$

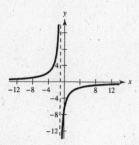

Asymptotes: $x = -1, y = 0$

95. $f(x) = \dfrac{x^2 - 2x - 3}{x - 4} = x + 2 + \dfrac{5}{x - 4}$

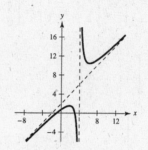

Asymptotes: $x = 4, y = x + 2$

Section 7.5   Operations with Matrices

- $A = B$ if and only if they have the same order and $a_{ij} = b_{ij}$.
- You should be able to perform the operations of matrix addition, scalar multiplication, and matrix multiplication.
- Some properties of matrix addition and scalar multiplication are:
  - (a) $A + B = B + A$
  - (b) $A + (B + C) = (A + B) + C$
  - (c) $(cd)A = c(dA)$
  - (d) $1A = A$
  - (e) $c(A + B) = cA + cB$
  - (f) $(c + d)A = cA + dA$
- Some properties of matrix multiplication are:
  - (a) $A(BC) = (AB)C$
  - (b) $A(B + C) = AB + AC$
  - (c) $(A + B)C = AC + BC$
  - (d) $c(AB) = (cA)B = A(cB)$
- You should remember that $AB \neq BA$ in general.

Vocabulary Check

1. equal

2. scalars

3. zero, *0*

4. identity

5. (a) iii   (b) i   (c) iv   (d) v   (e) ii

6. (a) ii   (b) iv   (c) i   (d) iii

1. $x = -4$, $y = 22$

3. $2x + 7 = 5 \implies x = -1$

$3y = 12 \implies y = 4$

$3z - 14 = 4 \implies z = 6$

5. (a) $A + B = \begin{bmatrix} 1 & -1 \\ 2 & -1 \end{bmatrix} + \begin{bmatrix} 2 & -1 \\ -1 & 8 \end{bmatrix} = \begin{bmatrix} 1+2 & -1-1 \\ 2-1 & -1+8 \end{bmatrix} = \begin{bmatrix} 3 & -2 \\ 1 & 7 \end{bmatrix}$

(b) $A - B = \begin{bmatrix} 1 & -1 \\ 2 & -1 \end{bmatrix} - \begin{bmatrix} 2 & -1 \\ -1 & 8 \end{bmatrix} = \begin{bmatrix} 1-2 & -1+1 \\ 2+1 & -1-8 \end{bmatrix} = \begin{bmatrix} -1 & 0 \\ 3 & -9 \end{bmatrix}$

(c) $3A = 3\begin{bmatrix} 1 & -1 \\ 2 & -1 \end{bmatrix} = \begin{bmatrix} 3(1) & 3(-1) \\ 3(2) & 3(-1) \end{bmatrix} = \begin{bmatrix} 3 & -3 \\ 6 & -3 \end{bmatrix}$

(d) $3A - 2B = \begin{bmatrix} 3 & -3 \\ 6 & -3 \end{bmatrix} - 2\begin{bmatrix} 2 & -1 \\ -1 & 8 \end{bmatrix} = \begin{bmatrix} 3 & -3 \\ 6 & -3 \end{bmatrix} + \begin{bmatrix} -4 & 2 \\ 2 & -16 \end{bmatrix} = \begin{bmatrix} -1 & -1 \\ 8 & -19 \end{bmatrix}$

7. $A = \begin{bmatrix} 8 & -1 \\ 2 & 3 \\ -4 & 5 \end{bmatrix}$, $B = \begin{bmatrix} 1 & 6 \\ -1 & -5 \\ 1 & 10 \end{bmatrix}$

(a) $A + B = \begin{bmatrix} 9 & 5 \\ 1 & -2 \\ -3 & 15 \end{bmatrix}$ (b) $A - B = \begin{bmatrix} 7 & -7 \\ 3 & 8 \\ -5 & -5 \end{bmatrix}$

(c) $3A = \begin{bmatrix} 24 & -3 \\ 6 & 9 \\ -12 & 15 \end{bmatrix}$ (d) $3A - 2B = \begin{bmatrix} 24 & -3 \\ 6 & 9 \\ -12 & 15 \end{bmatrix} - \begin{bmatrix} 2 & 12 \\ -2 & -10 \\ 2 & 20 \end{bmatrix} = \begin{bmatrix} 22 & -15 \\ 8 & 19 \\ -14 & -5 \end{bmatrix}$

9. $A = \begin{bmatrix} 4 & 5 & -1 & 3 & 4 \\ 1 & 2 & -2 & -1 & 0 \end{bmatrix}$, $B = \begin{bmatrix} 1 & 0 & -1 & 1 & 0 \\ -6 & 8 & 2 & -3 & -7 \end{bmatrix}$

(a) $A + B = \begin{bmatrix} 5 & 5 & -2 & 4 & 4 \\ -5 & 10 & 0 & -4 & -7 \end{bmatrix}$

(b) $A - B = \begin{bmatrix} 3 & 5 & 0 & 2 & 4 \\ 7 & -6 & -4 & 2 & 7 \end{bmatrix}$

(c) $3A = \begin{bmatrix} 12 & 15 & -3 & 9 & 12 \\ 3 & 6 & -6 & -3 & 0 \end{bmatrix}$

(d) $3A - 2B = \begin{bmatrix} 12 & 15 & -3 & 9 & 12 \\ 3 & 6 & -6 & -3 & 0 \end{bmatrix} - \begin{bmatrix} 2 & 0 & -2 & 2 & 0 \\ -12 & 16 & 4 & -6 & -14 \end{bmatrix}$

$= \begin{bmatrix} 10 & 15 & -1 & 7 & 12 \\ 15 & -10 & -10 & 3 & 14 \end{bmatrix}$

11. $A = \begin{bmatrix} 6 & 0 & 3 \\ -1 & -4 & 0 \end{bmatrix}$, $B = \begin{bmatrix} 8 & -1 \\ 4 & -3 \end{bmatrix}$

(a) $A + B$ is not possible.

(b) $A - B$ is not possible.

(c) $3A = \begin{bmatrix} 18 & 0 & 9 \\ -3 & -12 & 0 \end{bmatrix}$

(d) $3A - 2B$ is not possible.

13. $\begin{bmatrix} -5 & 0 \\ 3 & -6 \end{bmatrix} + \begin{bmatrix} 7 & 1 \\ -2 & -1 \end{bmatrix} + \begin{bmatrix} -10 & -8 \\ 14 & 6 \end{bmatrix} = \begin{bmatrix} -5 & 0 \\ 3 & -6 \end{bmatrix} + \begin{bmatrix} -3 & -7 \\ 12 & 5 \end{bmatrix} = \begin{bmatrix} -8 & -7 \\ 15 & -1 \end{bmatrix}$

15. $4\left(\begin{bmatrix} -4 & 0 & 1 \\ 0 & 2 & 3 \end{bmatrix} - \begin{bmatrix} 2 & 1 & -2 \\ 3 & -6 & 0 \end{bmatrix}\right) = 4\begin{bmatrix} -6 & -1 & 3 \\ -3 & 8 & 3 \end{bmatrix} = \begin{bmatrix} -24 & -4 & 12 \\ -12 & 32 & 12 \end{bmatrix}$

17. $\begin{bmatrix} 2 & 5 \\ -1 & -4 \end{bmatrix} + \begin{bmatrix} -3 & 0 \\ 2 & 2 \end{bmatrix} = \begin{bmatrix} -1 & 5 \\ 1 & -2 \end{bmatrix}$

19. $-\frac{1}{2}\begin{bmatrix} 3.211 & 6.829 \\ -1.004 & 4.914 \\ 0.055 & -3.889 \end{bmatrix} - 8\begin{bmatrix} 1.630 & -3.090 \\ 5.256 & 8.335 \\ -9.768 & 4.251 \end{bmatrix} = \begin{bmatrix} -14.645 & 21.305 \\ -41.546 & -69.137 \\ 78.117 & -32.064 \end{bmatrix}$

21. $X = 3\begin{bmatrix} -2 & -1 \\ 1 & 0 \\ 3 & -4 \end{bmatrix} - 2\begin{bmatrix} 0 & 3 \\ 2 & 0 \\ -4 & -1 \end{bmatrix} = \begin{bmatrix} -6 & -3 \\ 3 & 0 \\ 9 & -12 \end{bmatrix} - \begin{bmatrix} 0 & 6 \\ 4 & 0 \\ -8 & -2 \end{bmatrix} = \begin{bmatrix} -6 & -9 \\ -1 & 0 \\ 17 & -10 \end{bmatrix}$

23. $X = -\frac{3}{2}A + \frac{1}{2}B = -\frac{3}{2}\begin{bmatrix} -2 & -1 \\ 1 & 0 \\ 3 & -4 \end{bmatrix} + \frac{1}{2}\begin{bmatrix} 0 & 3 \\ 2 & 0 \\ -4 & -1 \end{bmatrix} = \begin{bmatrix} 3 & 3 \\ -\frac{1}{2} & 0 \\ -\frac{13}{2} & \frac{11}{2} \end{bmatrix}$

25. A is 3×2 and B is $3 \times 3 \implies AB$ is not defined.   **27.** $AB = \begin{bmatrix} -1 & 6 \\ -4 & 5 \\ 0 & 3 \end{bmatrix}\begin{bmatrix} 2 & 3 \\ 0 & 9 \end{bmatrix} = \begin{bmatrix} -2 & 51 \\ -8 & 33 \\ 0 & 27 \end{bmatrix}$

29. A is 3×3, B is $3 \times 3 \implies AB$ is 3×3.

$$AB = \begin{bmatrix} 5 & 0 & 0 \\ 0 & -8 & 0 \\ 0 & 0 & 7 \end{bmatrix}\begin{bmatrix} \frac{1}{5} & 0 & 0 \\ 0 & -\frac{1}{8} & 0 \\ 0 & 0 & \frac{1}{2} \end{bmatrix} = \begin{bmatrix} 1 & 0 & 0 \\ 0 & 1 & 0 \\ 0 & 0 & \frac{7}{2} \end{bmatrix}$$

31. $AB = \begin{bmatrix} 5 \\ 6 \end{bmatrix}\begin{bmatrix} -3 & -1 & -5 & -9 \end{bmatrix} = \begin{bmatrix} -15 & -5 & -25 & -45 \\ -18 & -6 & -30 & -54 \end{bmatrix}$

33. (a) $AB = \begin{bmatrix} 1 & 2 \\ 5 & 2 \end{bmatrix}\begin{bmatrix} 2 & -1 \\ -1 & 8 \end{bmatrix} = \begin{bmatrix} 2-2 & -1+16 \\ 10-2 & -5+16 \end{bmatrix} = \begin{bmatrix} 0 & 15 \\ 8 & 11 \end{bmatrix}$

     (b) $BA = \begin{bmatrix} 2 & -1 \\ -1 & 8 \end{bmatrix}\begin{bmatrix} 1 & 2 \\ 5 & 2 \end{bmatrix} = \begin{bmatrix} 2-5 & 4-2 \\ -1+40 & -2+16 \end{bmatrix} = \begin{bmatrix} -3 & 2 \\ 39 & 14 \end{bmatrix}$

     (c) $A^2 = \begin{bmatrix} 1 & 2 \\ 5 & 2 \end{bmatrix}\begin{bmatrix} 1 & 2 \\ 5 & 2 \end{bmatrix} = \begin{bmatrix} 1+10 & 2+4 \\ 5+10 & 10+4 \end{bmatrix} = \begin{bmatrix} 11 & 6 \\ 15 & 14 \end{bmatrix}$

35. (a) $AB = \begin{bmatrix} 3 & -1 \\ 1 & 3 \end{bmatrix}\begin{bmatrix} 1 & -3 \\ 3 & 1 \end{bmatrix} = \begin{bmatrix} 3-3 & -9-1 \\ 1+9 & -3+3 \end{bmatrix} = \begin{bmatrix} 0 & -10 \\ 10 & 0 \end{bmatrix}$

     (b) $BA = \begin{bmatrix} 1 & -3 \\ 3 & 1 \end{bmatrix}\begin{bmatrix} 3 & -1 \\ 1 & 3 \end{bmatrix} = \begin{bmatrix} 3-3 & -1-9 \\ 9+1 & -3+3 \end{bmatrix} = \begin{bmatrix} 0 & -10 \\ 10 & 0 \end{bmatrix}$

     (c) $A^2 = \begin{bmatrix} 3 & -1 \\ 1 & 3 \end{bmatrix}\begin{bmatrix} 3 & -1 \\ 1 & 3 \end{bmatrix} = \begin{bmatrix} 9-1 & -3-3 \\ 3+3 & -1+9 \end{bmatrix} = \begin{bmatrix} 8 & -6 \\ 6 & 8 \end{bmatrix}$

37. (a) $AB = \begin{bmatrix} 7 \\ 8 \\ -1 \end{bmatrix}\begin{bmatrix} 1 & 1 & 2 \end{bmatrix} = \begin{bmatrix} 7 & 7 & 14 \\ 8 & 8 & 16 \\ -1 & -1 & -2 \end{bmatrix}$

39. $AB = \begin{bmatrix} 70 & -17 & 73 \\ 32 & 11 & 6 \\ 16 & -38 & 70 \end{bmatrix}$

(b) $BA = \begin{bmatrix} 1 & 1 & 2 \end{bmatrix}\begin{bmatrix} 7 \\ 8 \\ -1 \end{bmatrix} = [7 + 8 - 2] = [13]$

(c) A^2 is not defined.

41. $\begin{bmatrix} -3 & 8 & -6 & 8 \\ -12 & 15 & 9 & 6 \\ 5 & -1 & 1 & 5 \end{bmatrix}\begin{bmatrix} 3 & 1 & 6 \\ 24 & 15 & 14 \\ 16 & 10 & 21 \\ 8 & -4 & 10 \end{bmatrix} = \begin{bmatrix} 151 & 25 & 48 \\ 516 & 279 & 387 \\ 47 & -20 & 87 \end{bmatrix}$

43. A is 2×4 and B is $2 \times 4 \implies AB$ is not defined.

45. $\left(\begin{bmatrix} 3 & 1 \\ 0 & -2 \end{bmatrix}\begin{bmatrix} 1 & 0 \\ -2 & 2 \end{bmatrix}\right)\begin{bmatrix} 1 & 0 \\ 2 & 4 \end{bmatrix} = \begin{bmatrix} 1 & 2 \\ 4 & -4 \end{bmatrix}\begin{bmatrix} 1 & 0 \\ 2 & 4 \end{bmatrix} = \begin{bmatrix} 5 & 8 \\ -4 & -16 \end{bmatrix}$

47. $\begin{bmatrix} 0 & 2 & -2 \\ 4 & 1 & 2 \end{bmatrix}\left(\begin{bmatrix} 4 & 0 \\ 0 & -1 \\ -1 & 2 \end{bmatrix} + \begin{bmatrix} -2 & 3 \\ -3 & 5 \\ 0 & -3 \end{bmatrix}\right) = \begin{bmatrix} 0 & 2 & -2 \\ 4 & 1 & 2 \end{bmatrix}\begin{bmatrix} 2 & 3 \\ -3 & 4 \\ -1 & -1 \end{bmatrix} = \begin{bmatrix} -4 & 10 \\ 3 & 14 \end{bmatrix}$

49. $\begin{bmatrix} 1 & 2 & \vdots & 4 \\ 3 & 2 & \vdots & 0 \end{bmatrix}$

(a) $\begin{bmatrix} 1 & 2 \\ 3 & 2 \end{bmatrix}\begin{bmatrix} 2 \\ 1 \end{bmatrix} = \begin{bmatrix} 4 \\ 8 \end{bmatrix} \implies \begin{bmatrix} 2 \\ 1 \end{bmatrix}$ is not a solution.

(b) $\begin{bmatrix} 1 & 2 \\ 3 & 2 \end{bmatrix}\begin{bmatrix} -2 \\ 3 \end{bmatrix} = \begin{bmatrix} 4 \\ 0 \end{bmatrix} \implies \begin{bmatrix} -2 \\ 3 \end{bmatrix}$ is a solution.

(c) $\begin{bmatrix} 1 & 2 \\ 3 & 2 \end{bmatrix}\begin{bmatrix} -4 \\ 4 \end{bmatrix} = \begin{bmatrix} 4 \\ -4 \end{bmatrix} \implies \begin{bmatrix} -4 \\ 4 \end{bmatrix}$ is not a solution.

(d) $\begin{bmatrix} 1 & 2 \\ 3 & 2 \end{bmatrix}\begin{bmatrix} 2 \\ -3 \end{bmatrix} = \begin{bmatrix} -4 \\ 0 \end{bmatrix} \implies \begin{bmatrix} 2 \\ -3 \end{bmatrix}$ is not a solution.

51. $\begin{bmatrix} -2 & -3 & \vdots & -6 \\ 4 & 2 & \vdots & 20 \end{bmatrix}$

(a) $\begin{bmatrix} -2 & -3 \\ 4 & 2 \end{bmatrix}\begin{bmatrix} 3 \\ 0 \end{bmatrix} = \begin{bmatrix} -6 \\ 12 \end{bmatrix} \implies \begin{bmatrix} 3 \\ 0 \end{bmatrix}$ is not a solution.

(c) $\begin{bmatrix} -2 & -3 \\ 4 & 2 \end{bmatrix}\begin{bmatrix} -6 \\ 6 \end{bmatrix} = \begin{bmatrix} -6 \\ -12 \end{bmatrix} \implies \begin{bmatrix} -6 \\ 6 \end{bmatrix}$ is not a solution.

(b) $\begin{bmatrix} -2 & -3 \\ 4 & 2 \end{bmatrix}\begin{bmatrix} 6 \\ -2 \end{bmatrix} = \begin{bmatrix} -6 \\ 20 \end{bmatrix} \implies \begin{bmatrix} 6 \\ -2 \end{bmatrix}$ is a solution.

(d) $\begin{bmatrix} -2 & -3 \\ 4 & 2 \end{bmatrix}\begin{bmatrix} 4 \\ 2 \end{bmatrix} = \begin{bmatrix} -14 \\ 20 \end{bmatrix} \implies \begin{bmatrix} 4 \\ 2 \end{bmatrix}$ is not a solution.

53. (a) $A = \begin{bmatrix} -1 & 1 \\ -2 & 1 \end{bmatrix}$, $X = \begin{bmatrix} x_1 \\ x_2 \end{bmatrix}$, $B = \begin{bmatrix} 4 \\ 0 \end{bmatrix}$

(b) By Gauss-Jordan elimination on

$$\begin{bmatrix} -1 & 1 & \vdots & 4 \\ -2 & 1 & \vdots & 0 \end{bmatrix}$$

$$\begin{matrix} -R_1 \to \\ 2R_1 + R_2 \to \end{matrix} \begin{bmatrix} 1 & -1 & \vdots & -4 \\ 0 & -1 & \vdots & -8 \end{bmatrix}$$

$$\begin{matrix} -R_2 + R_1 \to \\ -R_2 \to \end{matrix} \begin{bmatrix} 1 & 0 & \vdots & 4 \\ 0 & 1 & \vdots & 8 \end{bmatrix},$$

we have $x_1 = 4$ and $x_2 = 8$. Thus, $X = \begin{bmatrix} 4 \\ 8 \end{bmatrix}$.

55. (a) $A = \begin{bmatrix} -2 & -3 \\ 6 & 1 \end{bmatrix}$, $X = \begin{bmatrix} x_1 \\ x_2 \end{bmatrix}$, $B = \begin{bmatrix} -4 \\ -36 \end{bmatrix}$

(b)
$$\begin{bmatrix} -2 & -3 & \vdots & -4 \\ 6 & 1 & \vdots & -36 \end{bmatrix}$$

$$3R_1 + R_2 \to \begin{bmatrix} -2 & -3 & \vdots & -4 \\ 0 & -8 & \vdots & -48 \end{bmatrix}$$

$$\left(-\tfrac{1}{8}\right)R_2 \to \begin{bmatrix} -2 & -3 & \vdots & -4 \\ 0 & 1 & \vdots & 6 \end{bmatrix}$$

$$3R_2 + R_1 \to \begin{bmatrix} -2 & 0 & \vdots & 14 \\ 0 & 1 & \vdots & 6 \end{bmatrix}$$

$$-\tfrac{1}{2}R_1 \to \begin{bmatrix} 1 & 0 & \vdots & -7 \\ 0 & 1 & \vdots & 6 \end{bmatrix}$$

$x_1 = -7, x_2 = 6.$

Answer: $X = \begin{bmatrix} -7 \\ 6 \end{bmatrix}$

57. (a) $A = \begin{bmatrix} 1 & -2 & 3 \\ -1 & 3 & -1 \\ 2 & -5 & 5 \end{bmatrix}$, $X = \begin{bmatrix} x_1 \\ x_2 \\ x_3 \end{bmatrix}$, $B = \begin{bmatrix} 9 \\ -6 \\ 17 \end{bmatrix}$

(b)
$$\begin{bmatrix} 1 & -2 & 3 & \vdots & 9 \\ -1 & 3 & -1 & \vdots & -6 \\ 2 & -5 & 5 & \vdots & 17 \end{bmatrix}$$

$$\begin{matrix} R_1 + R_2 \to \\ -2R_1 + R_3 \to \end{matrix} \begin{bmatrix} 1 & -2 & 3 & \vdots & 9 \\ 0 & 1 & 2 & \vdots & 3 \\ 0 & -1 & -1 & \vdots & -1 \end{bmatrix}$$

$$\begin{matrix} 2R_2 + R_1 \to \\ \\ R_2 + R_3 \to \end{matrix} \begin{bmatrix} 1 & 0 & 7 & \vdots & 15 \\ 0 & 1 & 2 & \vdots & 3 \\ 0 & 0 & 1 & \vdots & 2 \end{bmatrix}$$

$$\begin{matrix} -7R_3 + R_1 \to \\ -2R_3 + R_2 \to \\ \\ \end{matrix} \begin{bmatrix} 1 & 0 & 0 & \vdots & 1 \\ 0 & 1 & 0 & \vdots & -1 \\ 0 & 0 & 1 & \vdots & 2 \end{bmatrix}$$

$x_1 = 1, x_2 = -1, x_3 = 2.$

Answer: $X = \begin{bmatrix} 1 \\ -1 \\ 2 \end{bmatrix}$.

59. (a) $A = \begin{bmatrix} 1 & -5 & 2 \\ -3 & 1 & -1 \\ 0 & -2 & 5 \end{bmatrix}$, $X = \begin{bmatrix} x_1 \\ x_2 \\ x_3 \end{bmatrix}$, $B = \begin{bmatrix} -20 \\ 8 \\ -16 \end{bmatrix}$

(b) $\begin{bmatrix} 1 & -5 & 2 & \vdots & -20 \\ -3 & 1 & -1 & \vdots & 8 \\ 0 & -2 & 5 & \vdots & -16 \end{bmatrix}$ $3R_1 + R_2 \rightarrow \begin{bmatrix} 1 & -5 & 2 & \vdots & -20 \\ 0 & -14 & 5 & \vdots & -52 \\ 0 & -2 & 5 & \vdots & -16 \end{bmatrix}$

$\begin{matrix} R_2 \\ R_3 \end{matrix} \begin{bmatrix} 1 & -5 & 2 & \vdots & -20 \\ 0 & -2 & 5 & \vdots & -16 \\ 0 & -14 & 5 & \vdots & -52 \end{bmatrix}$

$-7R_2 + R_3 \rightarrow \begin{bmatrix} 1 & -5 & 2 & \vdots & -20 \\ 0 & -2 & 5 & \vdots & -16 \\ 0 & 0 & -30 & \vdots & 60 \end{bmatrix}$

$-\frac{1}{30}R_3 \rightarrow \begin{bmatrix} 1 & -5 & 2 & \vdots & -20 \\ 0 & -2 & 5 & \vdots & -16 \\ 0 & 0 & 1 & \vdots & -2 \end{bmatrix}$

$\begin{matrix} -2R_3 + R_1 \rightarrow \\ -5R_3 + R_2 \rightarrow \end{matrix} \begin{bmatrix} 1 & -5 & 0 & \vdots & -16 \\ 0 & -2 & 0 & \vdots & -6 \\ 0 & 0 & 1 & \vdots & -2 \end{bmatrix}$

$\left(-\frac{1}{2}\right)R_2 \rightarrow \begin{bmatrix} 1 & -5 & 0 & \vdots & -16 \\ 0 & 1 & 0 & \vdots & 3 \\ 0 & 0 & 1 & \vdots & -2 \end{bmatrix}$

$5R_2 + R_1 \rightarrow \begin{bmatrix} 1 & 0 & 0 & \vdots & -1 \\ 0 & 1 & 0 & \vdots & 3 \\ 0 & 0 & 1 & \vdots & -2 \end{bmatrix}$

$x_1 = -1, x_2 = 3, x_3 = -2$ *Answer:* $X = \begin{bmatrix} -1 \\ 3 \\ -2 \end{bmatrix}$

61. (a) $A(B + C) = \begin{bmatrix} 7 & -2 & 5 \\ -6 & 13 & -8 \\ 16 & 11 & -3 \end{bmatrix}$

(b) $AB + AC = \begin{bmatrix} 7 & -2 & 5 \\ -6 & 13 & -8 \\ 16 & 11 & -3 \end{bmatrix}$

The answers are the same.

63. (a) $(A + B)^2 = \begin{bmatrix} 26 & 11 & 0 \\ 11 & 20 & -3 \\ 11 & 14 & 0 \end{bmatrix}$

(b) $A^2 + AB + BA + B^2 = \begin{bmatrix} 26 & 11 & 0 \\ 11 & 20 & -3 \\ 11 & 14 & 0 \end{bmatrix}$

The answers are the same.

65. (a) $A(BC) = \begin{bmatrix} 25 & -34 & 28 \\ -53 & 34 & -7 \\ -76 & 30 & 21 \end{bmatrix}$

(b) $(AB)C = \begin{bmatrix} 25 & -34 & 28 \\ -53 & 34 & -7 \\ -76 & 30 & 21 \end{bmatrix}$

The answers are the same.

67. (a) $A + cB = \begin{bmatrix} -1 & 10 & -4 \\ -5 & -1 & 0 \end{bmatrix}$

(b) $A + cB = \begin{bmatrix} 1 & 2 & -2 \\ -1 & 1 & 0 \end{bmatrix} + 2\begin{bmatrix} -1 & 4 & -1 \\ -2 & -1 & 0 \end{bmatrix} = \begin{bmatrix} -1 & 10 & -4 \\ -5 & -1 & 0 \end{bmatrix}$

69. (*a*), (*b*) $c(AB)$ Not possible

Number of columns of A (3) does not equal the number of rows of B (2).

71. (*a*), (*b*) $CA - BC$ Not possible

CA is 3×3, BC is 2×2.

73. (a) $cd\,A = \begin{bmatrix} -6 & -12 & 12 \\ 6 & -6 & 0 \end{bmatrix}$

(b) $cd\,A = 2(-3)\begin{bmatrix} 1 & 2 & -2 \\ -1 & 1 & 0 \end{bmatrix} = \begin{bmatrix} -6 & -12 & 12 \\ 6 & -6 & 0 \end{bmatrix}$

75. $A = \begin{bmatrix} 2 & 0 \\ 4 & 5 \end{bmatrix}$

$f(A) = A^2 - 5A + 2I = \begin{bmatrix} 2 & 0 \\ 4 & 5 \end{bmatrix}\begin{bmatrix} 2 & 0 \\ 4 & 5 \end{bmatrix} - 5\begin{bmatrix} 2 & 0 \\ 4 & 5 \end{bmatrix} + 2\begin{bmatrix} 1 & 0 \\ 0 & 1 \end{bmatrix} = \begin{bmatrix} -4 & 0 \\ 8 & 2 \end{bmatrix}$

77. $1.20\begin{bmatrix} 70 & 50 & 25 \\ 35 & 100 & 70 \end{bmatrix} = \begin{bmatrix} 84 & 60 & 30 \\ 42 & 120 & 84 \end{bmatrix}$

79. $BA = \begin{bmatrix} 3.50 & 6.00 \end{bmatrix}\begin{bmatrix} 125 & 100 & 75 \\ 100 & 175 & 125 \end{bmatrix} = \begin{bmatrix} 1037.50 & 1400 & 1012.50 \end{bmatrix}$

The entries in the last matrix BA represent the profit for both crops at each of the three outlets.

81. $ST = \begin{bmatrix} 3 & 2 & 2 & 3 & 0 \\ 0 & 2 & 3 & 4 & 3 \\ 4 & 2 & 1 & 3 & 2 \end{bmatrix}\begin{bmatrix} 840 & 1100 \\ 1200 & 1350 \\ 1450 & 1650 \\ 2650 & 3000 \\ 3050 & 3200 \end{bmatrix} = \begin{bmatrix} \$15{,}770 & \$18{,}300 \\ \$26{,}500 & \$29{,}250 \\ \$21{,}260 & \$24{,}150 \end{bmatrix}$

The entries represent the wholesale and retail prices of the inventory at each outlet.

83. $P^2 = \begin{bmatrix} 0.6 & 0.1 & 0.1 \\ 0.2 & 0.7 & 0.1 \\ 0.2 & 0.2 & 0.8 \end{bmatrix}\begin{bmatrix} 0.6 & 0.1 & 0.1 \\ 0.2 & 0.7 & 0.1 \\ 0.2 & 0.2 & 0.8 \end{bmatrix} = \begin{bmatrix} 0.40 & 0.15 & 0.15 \\ 0.28 & 0.53 & 0.17 \\ 0.32 & 0.32 & 0.68 \end{bmatrix}$

This product represents the changes in party affiliation after *two* elections.

85. True

For 87–93, A is of order 2×3, B is of order 2×3, C is of order 3×2 and D is of order 2×2.

87. $A + 2C$ is not possible. A and C are not of the same order.

89. AB is not possible. The number of columns of A does not equal the number of rows of B.

91. $BC - D$ is possible. The resulting order is 2×2.

93. $D(A - 3B)$ is possible. The resulting order is 2×3.

95. $(A + B)^2 = \begin{bmatrix} 1 & 0 \\ 2 & 1 \end{bmatrix}$

$A^2 + 2AB + B^2 = \begin{bmatrix} 0 & 0 \\ 3 & 2 \end{bmatrix}$

97. $(A + B)(A - B) = \begin{bmatrix} 3 & -2 \\ 4 & 3 \end{bmatrix}$

$A^2 - B^2 = \begin{bmatrix} 2 & -2 \\ 5 & 4 \end{bmatrix}$

99. $AC = \begin{bmatrix} 0 & 1 \\ 0 & 1 \end{bmatrix} \begin{bmatrix} 2 & 3 \\ 2 & 3 \end{bmatrix} = \begin{bmatrix} 2 & 3 \\ 2 & 3 \end{bmatrix}$

$BC = \begin{bmatrix} 1 & 0 \\ 1 & 0 \end{bmatrix} \begin{bmatrix} 2 & 3 \\ 2 & 3 \end{bmatrix} = \begin{bmatrix} 2 & 3 \\ 2 & 3 \end{bmatrix}$

$AC = BC$, but $A \neq B$.

101. (a) $A^2 = \begin{bmatrix} i & 0 \\ 0 & i \end{bmatrix} \begin{bmatrix} i & 0 \\ 0 & i \end{bmatrix} = \begin{bmatrix} -1 & 0 \\ 0 & -1 \end{bmatrix}$ and $i^2 = -1$

$A^3 = A^2A = \begin{bmatrix} -1 & 0 \\ 0 & -1 \end{bmatrix} \begin{bmatrix} i & 0 \\ 0 & i \end{bmatrix} = \begin{bmatrix} -i & 0 \\ 0 & -i \end{bmatrix}$ and $i^3 = -i$

$A^4 = A^3A = \begin{bmatrix} -i & 0 \\ 0 & -i \end{bmatrix} \begin{bmatrix} i & 0 \\ 0 & i \end{bmatrix} = \begin{bmatrix} 1 & 0 \\ 0 & 1 \end{bmatrix}$ and $i^4 = 1$

(b) $B^2 = \begin{bmatrix} 0 & -i \\ i & 0 \end{bmatrix} \begin{bmatrix} 0 & -i \\ i & 0 \end{bmatrix} = \begin{bmatrix} 1 & 0 \\ 0 & 1 \end{bmatrix}$,

The identity matrix

103. (a) $A = \begin{bmatrix} 0 & 2 \\ 0 & 0 \end{bmatrix}$, $B = \begin{bmatrix} 0 & 2 & 3 \\ 0 & 0 & 4 \\ 0 & 0 & 0 \end{bmatrix}$

(b) A^2 and B^3 are both zero matrices.

(c) If A is 4×4, then A^4 will be the zero matrix.

(d) If A is $n \times n$, then A^n is the zero matrix.

105. $3 \ln 4 - \frac{1}{3}\ln(x^2 + 3) = \ln 4^3 - \ln(x^2 + 3)^{1/3} = \ln\left[\dfrac{64}{(x^2 + 3)^{1/3}}\right]$

107. $\frac{1}{2}[2\ln(x + 5) + \ln x - \ln(x - 8)] = \ln(x + 5) + \ln x^{1/2} - \ln(x - 8)^{1/2}$

$$= \ln\left[\dfrac{(x + 5)\sqrt{x}}{\sqrt{x - 8}}\right]$$

Section 7.6   The Inverse of a Square Matrix

- ■ You should be able to find the inverse, if it exists, of a square matrix.
  - (a) Write the $n \times 2n$ matrix that consists of the given matrix A on the left and the $n \times n$ identity matrix I on the right to obtain $[A \; \vdots \; I]$. Note that we separate the matrices A and I by a dotted line. We call this process **adjoining** the matrices A and I.
  - (b) If possible, row reduce A to I using elementary row operations on the *entire* matrix $[A \; \vdots \; I]$. The result will be the matrix $[I \; \vdots \; A^{-1}]$. If this is not possible, then A is not invertible.
  - (c) Check your work by multiplying to see that $AA^{-1} = I = A^{-1}A$.
- ■ You should be able to use inverse matrices to solve systems of equation.
- ■ You should be able to find inverses using a graphing utility.

Vocabulary Check

1. square   **2.** inverse   **3.** nonsingular, singular

1. $AB = \begin{bmatrix} 2 & 1 \\ 5 & 3 \end{bmatrix} \begin{bmatrix} 3 & -1 \\ -5 & 2 \end{bmatrix} = \begin{bmatrix} 2(3) + 1(-5) & 2(-1) + 1(2) \\ 5(3) + 3(-5) & 5(-1) + 3(2) \end{bmatrix} = \begin{bmatrix} 1 & 0 \\ 0 & 1 \end{bmatrix}$

$BA = \begin{bmatrix} 3 & -1 \\ -5 & 2 \end{bmatrix} \begin{bmatrix} 2 & 1 \\ 5 & 3 \end{bmatrix} = \begin{bmatrix} 3(2) + (-1)(5) & 3(1) + (-1)(3) \\ -5(2) + 2(5) & -5(1) + 2(3) \end{bmatrix} = \begin{bmatrix} 1 & 0 \\ 0 & 1 \end{bmatrix}$

3. $AB = \begin{bmatrix} 1 & 2 \\ 3 & 4 \end{bmatrix} \begin{bmatrix} -2 & 1 \\ \frac{3}{2} & -\frac{1}{2} \end{bmatrix} = \begin{bmatrix} -2 + 3 & 1 - 1 \\ -6 + 6 & 3 - 2 \end{bmatrix} = \begin{bmatrix} 1 & 0 \\ 0 & 1 \end{bmatrix}$

$BA = \begin{bmatrix} -2 & 1 \\ \frac{3}{2} & -\frac{1}{2} \end{bmatrix} \begin{bmatrix} 1 & 2 \\ 3 & 4 \end{bmatrix} = \begin{bmatrix} -2 + 3 & -4 + 4 \\ \frac{3}{2} - \frac{3}{2} & 3 - 2 \end{bmatrix} = \begin{bmatrix} 1 & 0 \\ 0 & 1 \end{bmatrix}$

5. $AB = \begin{bmatrix} 2 & -17 & 11 \\ -1 & 11 & -7 \\ 0 & 3 & -2 \end{bmatrix} \begin{bmatrix} 1 & 1 & 2 \\ 2 & 4 & -3 \\ 3 & 6 & -5 \end{bmatrix}$

$= \begin{bmatrix} 2 - 34 + 33 & 2 - 68 + 66 & 4 + 51 - 55 \\ -1 + 22 - 21 & -1 + 44 - 42 & -2 - 33 + 35 \\ 6 - 6 & 12 - 12 & -9 + 10 \end{bmatrix} = \begin{bmatrix} 1 & 0 & 0 \\ 0 & 1 & 0 \\ 0 & 0 & 1 \end{bmatrix}$

$BA = \begin{bmatrix} 1 & 1 & 2 \\ 2 & 4 & -3 \\ 3 & 6 & -5 \end{bmatrix} \begin{bmatrix} 2 & -17 & 11 \\ -1 & 11 & -7 \\ 0 & 3 & -2 \end{bmatrix} = \begin{bmatrix} 2 - 1 & -17 + 11 + 6 & 11 - 7 - 4 \\ 4 - 4 & -34 + 44 - 9 & 22 - 28 + 6 \\ 6 - 6 & -51 + 66 - 15 & 33 - 42 + 10 \end{bmatrix} = \begin{bmatrix} 1 & 0 & 0 \\ 0 & 1 & 0 \\ 0 & 0 & 1 \end{bmatrix}$

7. $AB = \begin{bmatrix} -1 & -4 \\ 1 & 2 \end{bmatrix} \begin{bmatrix} 1 & 2 \\ -\frac{1}{2} & -\frac{1}{2} \end{bmatrix} = \begin{bmatrix} 1 & 0 \\ 0 & 1 \end{bmatrix}; BA = \begin{bmatrix} 1 & 0 \\ 0 & 1 \end{bmatrix}$

9. $AB = \begin{bmatrix} 1.6 & 2 \\ -3.5 & -4.5 \end{bmatrix} \begin{bmatrix} 22.5 & 10 \\ -17.5 & -8 \end{bmatrix} = \begin{bmatrix} 1 & 0 \\ 0 & 1 \end{bmatrix}; BA = \begin{bmatrix} 1 & 0 \\ 0 & 1 \end{bmatrix}$

11. $[A \;\vdots\; I] = \begin{bmatrix} 2 & 0 & \vdots & 1 & 0 \\ 0 & 3 & \vdots & 0 & 1 \end{bmatrix}$

$\begin{array}{c} \frac{1}{2}R_1 \rightarrow \\ \frac{1}{3}R_2 \rightarrow \end{array} \begin{bmatrix} 1 & 0 & \vdots & \frac{1}{2} & 0 \\ 0 & 1 & \vdots & 0 & \frac{1}{3} \end{bmatrix} = [I \;\vdots\; A^{-1}]$

$A^{-1} = \begin{bmatrix} \frac{1}{2} & 0 \\ 0 & \frac{1}{3} \end{bmatrix} = \frac{1}{6} \begin{bmatrix} 3 & 0 \\ 0 & 2 \end{bmatrix}$

13. $[A \;\vdots\; I] = \begin{bmatrix} 1 & -2 & \vdots & 1 & 0 \\ 2 & -3 & \vdots & 0 & 1 \end{bmatrix}$

$-2R_1 + R_2 \rightarrow \begin{bmatrix} 1 & -2 & \vdots & 1 & 0 \\ 0 & 1 & \vdots & -2 & 1 \end{bmatrix}$

$2R_2 + R_1 \rightarrow \begin{bmatrix} 1 & 0 & \vdots & -3 & 2 \\ 0 & 1 & \vdots & -2 & 1 \end{bmatrix}$

$A^{-1} = \begin{bmatrix} -3 & 2 \\ -2 & 1 \end{bmatrix}$

15. $A = \begin{bmatrix} 2 & 7 & 1 \\ -3 & -9 & 2 \end{bmatrix}$

A has no inverse because it is not square.

17. $[A \mathbin{\vdots} I] = \begin{bmatrix} 1 & 1 & 1 & \vdots & 1 & 0 & 0 \\ 3 & 5 & 4 & \vdots & 0 & 1 & 0 \\ 3 & 6 & 5 & \vdots & 0 & 0 & 1 \end{bmatrix}$

$\begin{matrix} \\ -3R_1 + R_2 \rightarrow \\ -3R_1 + R_3 \rightarrow \end{matrix} \begin{bmatrix} 1 & 1 & 1 & \vdots & 1 & 0 & 0 \\ 0 & 2 & 1 & \vdots & -3 & 1 & 0 \\ 0 & 3 & 2 & \vdots & -3 & 0 & 1 \end{bmatrix}$

$\begin{matrix} -R_2 + R_1 \rightarrow \\ \tfrac{1}{2}R_2 \rightarrow \\ -3R_2 + R_3 \rightarrow \end{matrix} \begin{bmatrix} 1 & 0 & \tfrac{1}{2} & \vdots & \tfrac{5}{2} & -\tfrac{1}{2} & 0 \\ 0 & 1 & \tfrac{1}{2} & \vdots & -\tfrac{3}{2} & \tfrac{1}{2} & 0 \\ 0 & 0 & \tfrac{1}{2} & \vdots & \tfrac{3}{2} & -\tfrac{3}{2} & 1 \end{bmatrix}$

$\begin{matrix} -R_3 + R_1 \rightarrow \\ -R_3 + R_2 \rightarrow \\ 2R_3 \rightarrow \end{matrix} \begin{bmatrix} 1 & 0 & 0 & \vdots & 1 & 1 & -1 \\ 0 & 1 & 0 & \vdots & -3 & 2 & -1 \\ 0 & 0 & 1 & \vdots & 3 & -3 & 2 \end{bmatrix}$

$= [I \mathbin{\vdots} A^{-1}]$

$A^{-1} = \begin{bmatrix} 1 & 1 & -1 \\ -3 & 2 & -1 \\ 3 & -3 & 2 \end{bmatrix}$

19. $[A \mathbin{\vdots} I] = \begin{bmatrix} -5 & 0 & 0 & \vdots & 1 & 0 & 0 \\ 2 & 0 & 0 & \vdots & 0 & 1 & 0 \\ -1 & 5 & 7 & \vdots & 0 & 0 & 1 \end{bmatrix}$

$\begin{matrix} (-\tfrac{1}{5})R_1 \\ (-2)R_1 + R_2 \end{matrix} \begin{bmatrix} 1 & 0 & 0 & \vdots & -\tfrac{1}{5} & 0 & 0 \\ 0 & 0 & 0 & \vdots & \tfrac{2}{5} & 1 & 0 \\ -1 & 5 & 7 & \vdots & 0 & 0 & 1 \end{bmatrix}$

Not invertible   (row of zeros)

A^{-1} does not exist.

21.   Not invertible.

A^{-1} does not exist.

23. $A = \begin{bmatrix} -\tfrac{1}{2} & \tfrac{3}{4} & \tfrac{1}{4} \\ 1 & 0 & -\tfrac{3}{2} \\ 0 & -1 & \tfrac{1}{2} \end{bmatrix}$

$A^{-1} = \begin{bmatrix} -12 & -5 & -9 \\ -4 & -2 & -4 \\ -8 & -4 & -6 \end{bmatrix}$

25. $A = \begin{bmatrix} 0.1 & 0.2 & 0.3 \\ -0.3 & 0.2 & 0.2 \\ 0.5 & 0.4 & 0.4 \end{bmatrix}$

$A^{-1} = \dfrac{5}{11}\begin{bmatrix} 0 & -4 & 2 \\ -22 & 11 & 11 \\ 22 & -6 & -8 \end{bmatrix}$

27. $A = \begin{bmatrix} -1 & 0 & 1 & 0 \\ 0 & 2 & 0 & -1 \\ 2 & 0 & -1 & 0 \\ 0 & -1 & 0 & 1 \end{bmatrix}$

$A^{-1} = \begin{bmatrix} 1 & 0 & 1 & 0 \\ 0 & 1 & 0 & 1 \\ 2 & 0 & 1 & 0 \\ 0 & 1 & 0 & 2 \end{bmatrix}$

29. $\begin{bmatrix} 5 & 1 \\ -2 & -2 \end{bmatrix}^{-1} = \dfrac{1}{5(-2)-(-2)(1)}\begin{bmatrix} -2 & -1 \\ 2 & 5 \end{bmatrix} = \dfrac{1}{-8}\begin{bmatrix} -2 & -1 \\ 2 & 5 \end{bmatrix} = \begin{bmatrix} \tfrac{1}{4} & \tfrac{1}{8} \\ -\tfrac{1}{4} & -\tfrac{5}{8} \end{bmatrix}$

31. $\begin{bmatrix} \tfrac{7}{2} & -\tfrac{3}{4} \\ \tfrac{1}{5} & \tfrac{4}{5} \end{bmatrix}^{-1} = \dfrac{1}{(\tfrac{7}{2})(\tfrac{4}{5})-(-\tfrac{3}{4})(\tfrac{1}{5})}\begin{bmatrix} \tfrac{4}{5} & \tfrac{3}{4} \\ -\tfrac{1}{5} & \tfrac{7}{2} \end{bmatrix} = \dfrac{20}{59}\begin{bmatrix} \tfrac{4}{5} & \tfrac{3}{4} \\ -\tfrac{1}{5} & \tfrac{7}{2} \end{bmatrix} = \dfrac{1}{59}\begin{bmatrix} 16 & 15 \\ -4 & 70 \end{bmatrix}$

33. $\begin{bmatrix} 2 & 3 \\ -1 & 5 \end{bmatrix}^{-1} = \dfrac{1}{2(5)-(3)(-1)}\begin{bmatrix} 5 & -3 \\ 1 & 2 \end{bmatrix} = \dfrac{1}{13}\begin{bmatrix} 5 & -3 \\ 1 & 2 \end{bmatrix} = \begin{bmatrix} \tfrac{5}{13} & -\tfrac{3}{13} \\ \tfrac{1}{13} & \tfrac{2}{13} \end{bmatrix}$

35. $\begin{bmatrix} -1 & 0 \\ 3 & -2 \end{bmatrix}^{-1} = \dfrac{1}{(-1)(-2)-0(3)}\begin{bmatrix} -2 & 0 \\ -3 & -1 \end{bmatrix} = \dfrac{1}{2}\begin{bmatrix} -2 & 0 \\ -3 & -1 \end{bmatrix} = \begin{bmatrix} -1 & 0 \\ -\tfrac{3}{2} & -\tfrac{1}{2} \end{bmatrix}$

37. $\begin{bmatrix} 1 & 2 \\ -2 & 0 \end{bmatrix}^{-1} = \begin{bmatrix} 0 & -\frac{1}{2} \\ \frac{1}{2} & \frac{1}{4} \end{bmatrix} \Rightarrow k = 0$

39. $\begin{bmatrix} -1 & 2 \\ -3 & 1 \end{bmatrix}^{-1} = \begin{bmatrix} \frac{1}{5} & -\frac{2}{5} \\ \frac{3}{5} & -\frac{1}{5} \end{bmatrix} \Rightarrow k = \frac{3}{5}$

41. $\begin{bmatrix} x \\ y \end{bmatrix} = \begin{bmatrix} -3 & 2 \\ -2 & 1 \end{bmatrix} \begin{bmatrix} 5 \\ 10 \end{bmatrix} = \begin{bmatrix} 5 \\ 0 \end{bmatrix}$

Answer: $(5, 0)$

43. $\begin{bmatrix} x \\ y \end{bmatrix} = \begin{bmatrix} -3 & 2 \\ -2 & 1 \end{bmatrix} \begin{bmatrix} 4 \\ 2 \end{bmatrix} \begin{bmatrix} -8 \\ -6 \end{bmatrix}$

Answer: $(-8, -6)$

45. $\begin{bmatrix} x \\ y \\ z \end{bmatrix} = \begin{bmatrix} 1 & 1 & -1 \\ -3 & 2 & -1 \\ 3 & -3 & 2 \end{bmatrix} \begin{bmatrix} 0 \\ 5 \\ 2 \end{bmatrix} = \begin{bmatrix} 3 \\ 8 \\ -11 \end{bmatrix}$

Answer: $(3, 8, -11)$

47. $\begin{bmatrix} x_1 \\ x_2 \\ x_3 \\ x_4 \end{bmatrix} = \begin{bmatrix} -24 & 7 & 1 & -2 \\ -10 & 3 & 0 & -1 \\ -29 & 7 & 3 & -2 \\ 12 & -3 & -1 & 1 \end{bmatrix} \begin{bmatrix} 0 \\ 1 \\ -1 \\ 2 \end{bmatrix} = \begin{bmatrix} 2 \\ 1 \\ 0 \\ 0 \end{bmatrix}$

Answer: $(2, 1, 0, 0)$

49. $A = \begin{bmatrix} 3 & 4 \\ 5 & 3 \end{bmatrix}$

$A^{-1} = \dfrac{1}{9 - 20} \begin{bmatrix} 3 & -4 \\ -5 & 3 \end{bmatrix}$

$\begin{bmatrix} x \\ y \end{bmatrix} = -\dfrac{1}{11} \begin{bmatrix} 3 & -4 \\ -5 & 3 \end{bmatrix} \begin{bmatrix} -2 \\ 4 \end{bmatrix} = -\dfrac{1}{11} \begin{bmatrix} -22 \\ 22 \end{bmatrix} = \begin{bmatrix} 2 \\ -2 \end{bmatrix}$

Answer: $(2, -2)$

51. $A = \begin{bmatrix} -0.4 & 0.8 \\ 2 & -4 \end{bmatrix}$

$A^{-1} = \dfrac{1}{1.6 - 1.6} \begin{bmatrix} -4 & -0.8 \\ -2 & -0.4 \end{bmatrix}$

A^{-1} does not exist.

[The system actually has no solution.]

53. $A = \begin{bmatrix} -\frac{1}{4} & \frac{3}{8} \\ \frac{3}{2} & \frac{3}{4} \end{bmatrix}$

$A^{-1} = \begin{bmatrix} -1 & \frac{1}{2} \\ 2 & \frac{1}{3} \end{bmatrix}$

$\begin{bmatrix} x \\ y \end{bmatrix} = A^{-1}b = \begin{bmatrix} -1 & \frac{1}{2} \\ 2 & \frac{1}{3} \end{bmatrix} \begin{bmatrix} -2 \\ -12 \end{bmatrix} = \begin{bmatrix} -4 \\ -8 \end{bmatrix}$

Answer: $(-4, -8)$

55. $A = \begin{bmatrix} 4 & -1 & 1 \\ 2 & 2 & 3 \\ 5 & -2 & 6 \end{bmatrix}$

$A^{-1} = \dfrac{1}{55} \begin{bmatrix} 18 & 4 & -5 \\ 3 & 19 & -10 \\ -14 & 3 & 10 \end{bmatrix}$

$\begin{bmatrix} x \\ y \\ z \end{bmatrix} = \dfrac{1}{55} \begin{bmatrix} 18 & 4 & -5 \\ 3 & 19 & -10 \\ -14 & 3 & 10 \end{bmatrix} \begin{bmatrix} -5 \\ 10 \\ 1 \end{bmatrix}$

$= \dfrac{1}{55} \begin{bmatrix} -55 \\ 165 \\ 110 \end{bmatrix} = \begin{bmatrix} -1 \\ 3 \\ 2 \end{bmatrix}$

Answer: $(-1, 3, 2)$

57. $A = \begin{bmatrix} 5 & -3 & 2 \\ 2 & 2 & -3 \\ -1 & 7 & -8 \end{bmatrix}$

A^{-1} does not exist.

The system actually has an infinite number of solutions of the form

$x = 0.3125t + 0.8125$

$y = 1.1875t + 0.6875$

$z = t$

where t is any real number.

59. $\begin{bmatrix} 7 & -3 & 0 & 2 & \vdots & 41 \\ -2 & 1 & 0 & -1 & \vdots & -13 \\ 4 & 0 & 1 & -2 & \vdots & 12 \\ -1 & 1 & 0 & -1 & \vdots & -8 \end{bmatrix}$ row reduces to $\begin{bmatrix} 1 & 0 & 0 & 0 & \vdots & 5 \\ 0 & 1 & 0 & 0 & \vdots & 0 \\ 0 & 0 & 1 & 0 & \vdots & -2 \\ 0 & 0 & 0 & 1 & \vdots & 3 \end{bmatrix}$

Answer: $(5, 0, -2, 3)$

61. (a) $(x, y) = (-3, 2)$ (b) $(x + h, y + k) = (-3 + 2, 2 + (-1)) = (-1, 1)$

(c) $B = AX = \begin{bmatrix} -1 \\ 1 \\ 1 \end{bmatrix}$ (d) $A^{-1} = \begin{bmatrix} 1 & 0 & -2 \\ 0 & 1 & 1 \\ 0 & 0 & 1 \end{bmatrix}$ (e) $A^{-1}B = \begin{bmatrix} -3 \\ 2 \\ 1 \end{bmatrix}$

The point $(-1, 1)$ has been translated back to $(-3, 2)$.

63. (a) $(x, y) = (2, -4)$ (b) $(x + h, y + k) = (2 - 3, -4 + 4) = (-1, 0)$

(c) $B = AX = \begin{bmatrix} -1 \\ 0 \\ 1 \end{bmatrix}$ (d) $A^{-1} = \begin{bmatrix} 1 & 0 & 3 \\ 0 & 1 & -4 \\ 0 & 0 & 1 \end{bmatrix}$ (e) $A^{-1}B = \begin{bmatrix} 2 \\ -4 \\ 1 \end{bmatrix}$

The point $(-1, 0)$ has been translated back to $(2, -4)$.

For 65 and 67 use $A = \begin{bmatrix} 1 & 1 & 1 \\ 0.065 & 0.07 & 0.09 \\ 0 & 2 & -1 \end{bmatrix}$. **Using the methods of this section, we have** $A^{-1} = \frac{1}{11}\begin{bmatrix} 50 & -600 & -4 \\ -13 & 200 & 5 \\ -26 & 400 & -1 \end{bmatrix}$.

65. $X = A^{-1}B = \frac{1}{11}\begin{bmatrix} 50 & -600 & -4 \\ -13 & 200 & 5 \\ -26 & 400 & -1 \end{bmatrix}\begin{bmatrix} 25{,}000 \\ 1900 \\ 0 \end{bmatrix} = \begin{bmatrix} 10{,}000 \\ 5000 \\ 10{,}000 \end{bmatrix}$

Answer: \$10,000 in AAA bonds, \$5000 in A-bonds and \$10,000 in B-bonds.

67. $X = A^{-1}B = \frac{1}{11}\begin{bmatrix} 50 & -600 & -4 \\ -13 & 200 & 5 \\ -26 & 400 & -1 \end{bmatrix}\begin{bmatrix} 65{,}000 \\ 5050 \\ 0 \end{bmatrix} = \begin{bmatrix} 20{,}000 \\ 15{,}000 \\ 30{,}000 \end{bmatrix}$

Answer: \$20,000 in AAA bonds, \$15,000 in A-bonds and \$30,000 in B-bonds.

69. $A = \begin{bmatrix} 2 & 0 & 4 \\ 0 & 1 & 4 \\ 1 & 1 & -1 \end{bmatrix}$

$A^{-1} = \frac{1}{14}\begin{bmatrix} 5 & -4 & 4 \\ -4 & 6 & 8 \\ 1 & 2 & -2 \end{bmatrix}$

$\begin{bmatrix} I_1 \\ I_2 \\ I_3 \end{bmatrix} = \frac{1}{14}\begin{bmatrix} 5 & -4 & 4 \\ -4 & 6 & 8 \\ 1 & 2 & -2 \end{bmatrix}\begin{bmatrix} 14 \\ 28 \\ 0 \end{bmatrix} = \begin{bmatrix} -3 \\ 8 \\ 5 \end{bmatrix}$

Answer: $I_1 = -3$ amps, $I_2 = 8$ amps, $I_3 = 5$ amps

For Exercises 71 and 73, let x = number of muffins, y = number of bones, z = number of cookies.

$$\begin{bmatrix} 2 & 1 & 2 \\ 3 & 1 & 1 \\ 2 & 1 & 1.5 \end{bmatrix} \begin{bmatrix} x \\ y \\ z \end{bmatrix} = \begin{bmatrix} \text{Beef} \\ \text{Chicken} \\ \text{Liver} \end{bmatrix}$$

$$\quad A \qquad\qquad X$$

$$A^{-1} = \begin{bmatrix} 1 & 1 & -2 \\ -5 & -2 & 8 \\ 2 & 0 & -2 \end{bmatrix}$$

71. $A^{-1} \begin{bmatrix} 700 \\ 500 \\ 600 \end{bmatrix} = \begin{bmatrix} 0 \\ 300 \\ 200 \end{bmatrix}$

300 units of bones

200 units of cookies

73. $A^{-1} \begin{bmatrix} 800 \\ 750 \\ 725 \end{bmatrix} = \begin{bmatrix} 100 \\ 300 \\ 150 \end{bmatrix}$

100 units of muffins

300 units of bones

150 units of cookies

75. (a) $\begin{aligned} f + \quad h + \quad s &= 10 \\ 2f + 2.5h + 3s &= 26 \\ h - \quad s &= 0 \end{aligned}$

(b) $\begin{bmatrix} 1 & 1 & 1 \\ 2 & 2.5 & 3 \\ 0 & 1 & -1 \end{bmatrix} \begin{bmatrix} f \\ h \\ s \end{bmatrix} = \begin{bmatrix} 10 \\ 26 \\ 0 \end{bmatrix}$

$$\quad A \qquad\qquad X \ = \ B$$

(c) $X = A^{-1}B = \begin{bmatrix} \frac{11}{3} & -\frac{4}{3} & -\frac{1}{3} \\ -\frac{4}{3} & \frac{2}{3} & \frac{2}{3} \\ -\frac{4}{3} & \frac{2}{3} & -\frac{1}{3} \end{bmatrix} \begin{bmatrix} 10 \\ 26 \\ 0 \end{bmatrix} = \begin{bmatrix} 2 \\ 4 \\ 4 \end{bmatrix}$

2 pounds French vanilla

4 pounds hazelnut

4 pounds Swiss chocolate

77. (a) $\begin{array}{l} (2, 5343) \\ (3, 5589) \\ (4, 6309) \end{array}$: $\begin{cases} 4a + 2b + c = 5343 \\ 9a + 3b + c = 5589 \\ 16a + 4b + c = 6309 \end{cases}$

(b) $A = \begin{bmatrix} 4 & 2 & 1 \\ 9 & 3 & 1 \\ 16 & 4 & 1 \end{bmatrix}, A^{-1} = \begin{bmatrix} \frac{1}{2} & -1 & \frac{1}{2} \\ -\frac{7}{2} & 6 & -\frac{5}{2} \\ 6 & -8 & 3 \end{bmatrix}$

$$A^{-1} \begin{bmatrix} 5343 \\ 5589 \\ 6309 \end{bmatrix} = \begin{bmatrix} a \\ b \\ c \end{bmatrix} = \begin{bmatrix} 237 \\ -939 \\ 6273 \end{bmatrix}$$

$$y = 237t^2 - 939t + 6273$$

(c)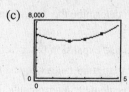

(d) For 2005, $t = 5$ and $y = 7503$ thousand.

For 2010, $t = 10$ and $y = 20{,}583$ thousand.

For 2015, $t = 15$ and $y = 45{,}513$ thousand.

(e) Answers will vary.

79. True. $AA^{-1} = A^{-1}A = I$

81. $AA^{-1} = \begin{bmatrix} a & b \\ c & d \end{bmatrix} \left(\dfrac{1}{ad - bc} \right) \begin{bmatrix} d & -b \\ -c & a \end{bmatrix} = \dfrac{1}{ad - bc} \begin{bmatrix} a & b \\ c & d \end{bmatrix} \begin{bmatrix} d & -b \\ -c & a \end{bmatrix}$

$$= \dfrac{1}{ad - bc} \begin{bmatrix} ad - bc & 0 \\ 0 & ad - bc \end{bmatrix} = \begin{bmatrix} 1 & 0 \\ 0 & 1 \end{bmatrix}$$

$$A^{-1}A = \dfrac{1}{ad - bc} \begin{bmatrix} d & -b \\ -c & a \end{bmatrix} \begin{bmatrix} a & b \\ c & d \end{bmatrix} = \dfrac{1}{ad - bc} \begin{bmatrix} ad - bc & 0 \\ 0 & ad - bc \end{bmatrix} = \begin{bmatrix} 1 & 0 \\ 0 & 1 \end{bmatrix}$$

83. $\dfrac{\left(\dfrac{9}{x}\right)}{\left(\dfrac{6}{x}+2\right)} = \dfrac{\left(\dfrac{9}{x}\right)}{\left(\dfrac{6+2x}{x}\right)} = \dfrac{9}{x} \cdot \dfrac{x}{6+2x} = \dfrac{9}{6+2x}, \quad x \neq 0$

85. $\dfrac{\dfrac{4}{x^2-9}+\dfrac{2}{x-2}}{\dfrac{1}{x+3}+\dfrac{1}{x-3}} \cdot \dfrac{(x^2-9)(x-2)}{(x^2-9)(x-2)}$

$= \dfrac{4(x-2)+2(x^2-9)}{(x-3)(x-2)+(x+3)(x-2)}$

$= \dfrac{2x^2+4x-26}{2x^2-4x}$

$= \dfrac{x^2+2x-13}{x(x-2)}, \quad x \neq \pm 3$

87. $e^{2x}+2e^x-15=(e^x+5)(e^x-3)=0 \implies e^x=3 \implies x=\ln 3 \approx 1.099$

89. $7\ln 3x = 12$

$\ln 3x = \dfrac{12}{7}$

$3x = e^{12/7}$

$x = \tfrac{1}{3}e^{12/7} \approx 1.851$

91. Answers will vary.

Section 7.7   The Determinant of a Square Matrix

■ You should be able to determine the determinant of a matrix of order 2×2 by using the products of the diagonals.

■ You should be able to use expansion by cofactors to find the determinant of a matrix of order 3 or greater.

■ The determinant of a triangular matrix equals the product of the entries on the main diagonal.

■ You should be able to calculate determinants using a graphing utility.

Vocabulary Check

1. determinant   **2.** minor   **3.** cofactor

4. expanding by cofactors   **5.** triangular   **6.** diagonal

---

1. $|4| = 4$

3. $\begin{vmatrix} 8 & 4 \\ 2 & 3 \end{vmatrix} = 8(3) - 4(2) = 24 - 8 = 16$

5. $\begin{vmatrix} 6 & 2 \\ -5 & 3 \end{vmatrix} = 6(3) - (2)(-5) = 18 + 10 = 28$

7. $\begin{vmatrix} -7 & 6 \\ \tfrac{1}{2} & 3 \end{vmatrix} = -7(3) - 6\left(\tfrac{1}{2}\right) = -21 - 3 = -24$

9. $\begin{vmatrix} 2 & -1 & 0 \\ 4 & 2 & 1 \\ 4 & 2 & 1 \end{vmatrix} = 2\begin{vmatrix} 2 & 1 \\ 2 & 1 \end{vmatrix} - 4\begin{vmatrix} -1 & 0 \\ 2 & 1 \end{vmatrix} + 4\begin{vmatrix} -1 & 0 \\ 2 & 1 \end{vmatrix} = 2(0) - 4(-1) + 4(-1) = 0$

11. $\begin{vmatrix} -1 & 2 & -5 \\ 0 & 3 & 4 \\ 0 & 0 & 3 \end{vmatrix} = (-1)(3)(3) = -9$ (Upper Triangular)

13. $\begin{vmatrix} 0.3 & 0.2 & 0.2 \\ 0.2 & 0.2 & 0.2 \\ -0.4 & 0.4 & 0.3 \end{vmatrix} = -0.002$

15. $\begin{bmatrix} 3 & 4 \\ 2 & -5 \end{bmatrix}$

(a) $M_{11} = -5$ (b) $C_{11} = M_{11} = -5$

$M_{12} = 2$ $C_{12} = -M_{12} = -2$

$M_{21} = 4$ $C_{21} = -M_{21} = -4$

$M_{22} = 3$ $C_{22} = M_{22} = 3$

17. $\begin{bmatrix} -4 & 6 & 3 \\ 7 & -2 & 8 \\ 1 & 0 & -5 \end{bmatrix}$

(a) $M_{11} = \begin{vmatrix} -2 & 8 \\ 0 & -5 \end{vmatrix} = 10$ $M_{23} = \begin{vmatrix} -4 & 6 \\ 1 & 0 \end{vmatrix} = -6$ (b) $C_{11} = 10$

$M_{12} = \begin{vmatrix} 7 & 8 \\ 1 & -5 \end{vmatrix} = -43$ $M_{31} = \begin{vmatrix} 6 & 3 \\ -2 & 8 \end{vmatrix} = 54$ $C_{12} = 43$

$\qquad\qquad\qquad\qquad\qquad\qquad\qquad\qquad\qquad C_{13} = 2$

$M_{13} = \begin{vmatrix} 7 & -2 \\ 1 & 0 \end{vmatrix} = 2$ $M_{32} = \begin{vmatrix} -4 & 3 \\ 7 & 8 \end{vmatrix} = -53$ $C_{21} = 30$

$\qquad\qquad\qquad\qquad\qquad\qquad\qquad\qquad\qquad C_{22} = 17$

$M_{21} = \begin{vmatrix} 6 & 3 \\ 0 & -5 \end{vmatrix} = -30$ $M_{33} = \begin{vmatrix} -4 & 6 \\ 7 & -2 \end{vmatrix} = -34$ $C_{23} = 6$

$\qquad\qquad\qquad\qquad\qquad\qquad\qquad\qquad\qquad C_{31} = 54$

$M_{22} = \begin{vmatrix} -4 & 3 \\ 1 & -5 \end{vmatrix} = 17$ $\qquad\qquad\qquad\qquad\qquad C_{32} = 53$

$\qquad\qquad\qquad\qquad\qquad\qquad\qquad\qquad\qquad C_{33} = -34$

19. (a) $\begin{vmatrix} -3 & 2 & 1 \\ 4 & 5 & 6 \\ 2 & -3 & 1 \end{vmatrix} = -3\begin{vmatrix} 5 & 6 \\ -3 & 1 \end{vmatrix} - 2\begin{vmatrix} 4 & 6 \\ 2 & 1 \end{vmatrix} + \begin{vmatrix} 4 & 5 \\ 2 & -3 \end{vmatrix} = -3(23) - 2(-8) - 22 = -75$

(b) $\begin{vmatrix} -3 & 2 & 1 \\ 4 & 5 & 6 \\ 2 & -3 & 1 \end{vmatrix} = -2\begin{vmatrix} 4 & 6 \\ 2 & 1 \end{vmatrix} + 5\begin{vmatrix} -3 & 1 \\ 2 & 1 \end{vmatrix} + 3\begin{vmatrix} -3 & 1 \\ 4 & 6 \end{vmatrix} = -2(-8) + 5(-5) + 3(-22) = -75$

21. (a) $\begin{vmatrix} 6 & 0 & -3 & 5 \\ 4 & 13 & 6 & -8 \\ -1 & 0 & 7 & 4 \\ 8 & 6 & 0 & 2 \end{vmatrix} = -4\begin{vmatrix} 0 & -3 & 5 \\ 0 & 7 & 4 \\ 6 & 0 & 2 \end{vmatrix} + 13\begin{vmatrix} 6 & -3 & 5 \\ -1 & 7 & 4 \\ 8 & 0 & 2 \end{vmatrix} - 6\begin{vmatrix} 6 & 0 & 5 \\ -1 & 0 & 4 \\ 8 & 6 & 2 \end{vmatrix} - 8\begin{vmatrix} 6 & 0 & -3 \\ -1 & 0 & 7 \\ 8 & 6 & 0 \end{vmatrix}$

$= -4(-282) + 13(-298) - 6(-174) - 8(-234) = 170$

—CONTINUED—

21. —CONTINUED—

(b)
$$\begin{vmatrix} 6 & 0 & -3 & 5 \\ 4 & 13 & 6 & -8 \\ -1 & 0 & 7 & 4 \\ 8 & 6 & 0 & 2 \end{vmatrix} = 0\begin{vmatrix} 4 & 6 & -8 \\ -1 & 7 & 4 \\ 8 & 0 & 2 \end{vmatrix} + 13\begin{vmatrix} 6 & -3 & 5 \\ -1 & 7 & 4 \\ 8 & 0 & 2 \end{vmatrix} + 0\begin{vmatrix} 6 & -3 & 5 \\ 4 & 6 & -8 \\ 8 & 0 & 2 \end{vmatrix} + 6\begin{vmatrix} 6 & -3 & 5 \\ 4 & 6 & -8 \\ -1 & 7 & 4 \end{vmatrix}$$

$$= 0 + 13(-298) + 0 + 6(674) = 170$$

23. Expand by Column 3.

$$\begin{vmatrix} 1 & 4 & -2 \\ 3 & 2 & 0 \\ -1 & 4 & 3 \end{vmatrix} = -2\begin{vmatrix} 3 & 2 \\ -1 & 4 \end{vmatrix} + 3\begin{vmatrix} 1 & 4 \\ 3 & 2 \end{vmatrix} = -2(14) + 3(-10) = -58$$

25. Expand by Column 3.

$$\begin{vmatrix} 2 & 6 & 6 & 2 \\ 2 & 7 & 3 & 6 \\ 1 & 5 & 0 & 1 \\ 3 & 7 & 0 & 7 \end{vmatrix} = 6\begin{vmatrix} 2 & 7 & 6 \\ 1 & 5 & 1 \\ 3 & 7 & 7 \end{vmatrix} - 3\begin{vmatrix} 2 & 6 & 2 \\ 1 & 5 & 1 \\ 3 & 7 & 7 \end{vmatrix} = 6(-20) - 3(16) = -168$$

27. Expand by Column 2.

$$\begin{vmatrix} 3 & 2 & 4 & -1 & 5 \\ -2 & 0 & 1 & 3 & 2 \\ 1 & 0 & 0 & 4 & 0 \\ 6 & 0 & 2 & -1 & 0 \\ 3 & 0 & 5 & 1 & 0 \end{vmatrix} = -2\begin{vmatrix} -2 & 1 & 3 & 2 \\ 1 & 0 & 4 & 0 \\ 6 & 2 & -1 & 0 \\ 3 & 5 & 1 & 0 \end{vmatrix} = (-2)(-2)\begin{vmatrix} 1 & 0 & 4 \\ 6 & 2 & -1 \\ 3 & 5 & 1 \end{vmatrix} = 4(103) = 412$$

29. $$\begin{vmatrix} 4 & 0 & 0 & 0 \\ 6 & -5 & 0 & 0 \\ 1 & 3 & 1 & 0 \\ 1 & -2 & 7 & 3 \end{vmatrix} = (4)(-5)(1)(3) = -60 \quad \text{(Lower Triangular)}$$

31. $\det(A) = (-6)(-1)(-7)(-2)(-2) = -168$

(Upper Triangular)

33. $$\begin{vmatrix} 1 & -1 & 8 & 4 \\ 2 & 6 & 0 & -4 \\ 2 & 0 & 2 & 6 \\ 0 & 2 & 8 & 0 \end{vmatrix} = -336$$

35. $$\begin{vmatrix} 3 & -2 & 4 & 3 & 1 \\ -1 & 0 & 2 & 1 & 0 \\ 5 & -1 & 0 & 3 & 2 \\ 4 & 7 & -8 & 0 & 0 \\ 1 & 2 & 3 & 0 & 2 \end{vmatrix} = 410$$

37. (a) $\begin{vmatrix} -1 & 0 \\ 0 & 3 \end{vmatrix} = -3$

(b) $\begin{vmatrix} 2 & 0 \\ 0 & -1 \end{vmatrix} = -2$

(c) $\begin{bmatrix} -1 & 0 \\ 0 & 3 \end{bmatrix}\begin{bmatrix} 2 & 0 \\ 0 & -1 \end{bmatrix} = \begin{bmatrix} -2 & 0 \\ 0 & -3 \end{bmatrix}$

(d) $\begin{vmatrix} -2 & 0 \\ 0 & -3 \end{vmatrix} = 6$ **[Note:** $|AB| = |A|\,|B|$**]**

39. (a) $\begin{vmatrix} -1 & 2 & 1 \\ 1 & 0 & 1 \\ 0 & 1 & 0 \end{vmatrix} = 2$

(b) $\begin{vmatrix} -1 & 0 & 0 \\ 0 & 2 & 0 \\ 0 & 0 & 3 \end{vmatrix} = -6$

(c) $\begin{bmatrix} -1 & 2 & 1 \\ 1 & 0 & 1 \\ 0 & 1 & 0 \end{bmatrix} \begin{bmatrix} -1 & 0 & 0 \\ 0 & 2 & 0 \\ 0 & 0 & 3 \end{bmatrix} = \begin{bmatrix} 1 & 4 & 3 \\ -1 & 0 & 3 \\ 0 & 2 & 0 \end{bmatrix}$

(d) $\begin{vmatrix} 1 & 4 & 3 \\ -1 & 0 & 3 \\ 0 & 2 & 0 \end{vmatrix} = -12$ [**Note:** $|AB| = |A|\,|B|$]

41. (a) $|A| = -25$

(b) $|B| = -220$

(c) $AB = \begin{bmatrix} -7 & -16 & -1 & -28 \\ -4 & -14 & -11 & 8 \\ 13 & 4 & 4 & -4 \\ -2 & 3 & 2 & 2 \end{bmatrix}$

(d) $|AB| = 5500$ [Note: $|AB| = |A|\,|B|$]

43. $\begin{vmatrix} w & x \\ y & z \end{vmatrix} = wz - xy$

$-\begin{vmatrix} y & z \\ w & x \end{vmatrix} = -(xy - wz) = wz - xy$

Thus, $\begin{vmatrix} w & x \\ y & z \end{vmatrix} = -\begin{vmatrix} y & z \\ w & x \end{vmatrix}$.

45. $\begin{vmatrix} w & x \\ y & z \end{vmatrix} = wz - xy$

$\begin{vmatrix} w & x + cw \\ y & z + cy \end{vmatrix} = w(z + cy) - y(x + cw) = wz - xy$

Thus, $\begin{vmatrix} w & x \\ y & z \end{vmatrix} = \begin{vmatrix} w & x + cw \\ y & z + cy \end{vmatrix}$.

47. $\begin{vmatrix} 1 & x & x^2 \\ 1 & y & y^2 \\ 1 & z & z^2 \end{vmatrix} = \begin{vmatrix} y & y^2 \\ z & z^2 \end{vmatrix} - \begin{vmatrix} x & x^2 \\ z & z^2 \end{vmatrix} + \begin{vmatrix} x & x^2 \\ y & y^2 \end{vmatrix}$

$= (yz^2 - y^2z) - (xz^2 - x^2z) + (xy^2 - x^2y)$

$= yz^2 - xz^2 - y^2z + x^2z + xy(y - x)$

$= z^2(y - x) - z(y^2 - x^2) + xy(y - x)$

$= z^2(y - x) - z(y - x)(y + x) + xy(y - x)$

$= (y - x)[z^2 - z(y + x) + xy]$

$= (y - x)[z^2 - zy - zx + xy]$

$= (y - x)[z^2 - zx - zy + xy]$

$= (y - x)[z(z - x) - y(z - x)]$

$= (y - x)(z - x)(z - y)$

49. $\begin{vmatrix} x & 2 \\ 1 & x \end{vmatrix} = 2$

$x^2 - 2 = 2$

$x^2 = 4$

$x = \pm 2$

51. $\begin{vmatrix} 2x & -3 \\ -2 & 2x \end{vmatrix} = 3$

$4x^2 - 6 = 3$

$4x^2 = 9$

$x^2 = \frac{9}{4}$

$x = \pm\frac{3}{2}$

53. $\begin{vmatrix} x & 1 \\ 2 & x - 2 \end{vmatrix} = -1$

$x^2 - 2x - 2 = -1$

$x^2 - 2x - 1 = 0$

$x = \dfrac{2 \pm \sqrt{4 + 4}}{2}$

$x = 1 \pm \sqrt{2}$

55. $\begin{vmatrix} x+3 & 2 \\ 1 & x+2 \end{vmatrix} = 0$

$(x+3)(x+2) - 2 = 0$

$x^2 + 5x + 4 = 0$

$(x+4)(x+1) = 0$

$x = -4, -1$

57. $\begin{vmatrix} 2x & 1 \\ -1 & x-1 \end{vmatrix} = x$

$2x^2 - 2x + 1 = x$

$2x^2 - 3x + 1 = 0$

$(x-1)(2x-1) = 0$

$x = 1, \frac{1}{2}$

59. $\begin{vmatrix} 1 & 2 & x \\ -1 & 3 & 2 \\ 3 & -2 & 1 \end{vmatrix} = 0$

$1\begin{vmatrix} 3 & 2 \\ -2 & 1 \end{vmatrix} - 2\begin{vmatrix} -1 & 2 \\ 3 & 1 \end{vmatrix} + x\begin{vmatrix} -1 & 3 \\ 3 & -2 \end{vmatrix} = 0$

$7 - 2(-7) + x(-7) = 0$

$21 = 7x$

$x = 3$

61. $\begin{vmatrix} 4u & -1 \\ -1 & 2v \end{vmatrix} = 8uv - 1$

63. $\begin{vmatrix} e^{2x} & e^{3x} \\ 2e^{2x} & 3e^{3x} \end{vmatrix} = 3e^{5x} - 2e^{5x} = e^{5x}$

65. $\begin{vmatrix} x & \ln x \\ 1 & \dfrac{1}{x} \end{vmatrix} = 1 - \ln x$

67. True. Expand along the row of zeros.

69. Let $A = \begin{bmatrix} 1 & 3 \\ -2 & 4 \end{bmatrix}$ and $B = \begin{bmatrix} -4 & 0 \\ 3 & 5 \end{bmatrix}$.

$|A| = \begin{vmatrix} 1 & 3 \\ -2 & 4 \end{vmatrix} = 10, \quad |B| = \begin{vmatrix} -4 & 0 \\ 3 & 5 \end{vmatrix} = -20$

$A + B = \begin{bmatrix} -3 & 3 \\ 1 & 9 \end{bmatrix}, \quad |A + B| = \begin{vmatrix} -3 & 3 \\ 1 & 9 \end{vmatrix} = -30$

Thus, $|A + B| \neq |A| + |B|$. Your answer may differ, depending on how you choose A and B.

71. (a) $|A| = 6$

(b) $A^{-1} = \begin{bmatrix} \frac{1}{3} & -\frac{1}{3} \\ \frac{1}{3} & \frac{1}{6} \end{bmatrix}$

(c) $\det(A^{-1}) = \dfrac{1}{6}$

(d) In general, $\det(A^{-1}) = \dfrac{1}{\det A}$.

73. (a) $|A| = 2$

(b) $A^{-1} = \begin{bmatrix} -4 & -5 & 1.5 \\ -1 & -1 & 0.5 \\ -1 & -1 & 0 \end{bmatrix}$

(c) $\det(A^{-1}) = \dfrac{1}{2}$

(d) In general, $\det(A^{-1}) = \dfrac{1}{\det A}$.

75. (a) Columns 2 and 3 are interchanged.

(b) Rows 1 and 3 are interchanged.

77. (a) 5 is factored out of the first row of A.

(b) 4 and 3 are factored out of columns 2 and 3.

79. $x^2 - 3x + 2 = (x - 2)(x - 1)$

81. $4y^2 - 12y + 9 = (2y - 3)^2$

83. $3x - 10y = 46$

$\qquad x + y = -2$

$\qquad\qquad y = -x - 2$

$\qquad 3x - 10(-x - 2) = 46$

$\qquad\qquad\qquad 13x = 26$

$\qquad\qquad\qquad\quad x = 2$

$\qquad\qquad\qquad\quad y = -2 - 2 = -4$

Answer: $(2, -4)$

Section 7.8   Applications of Matrices and Determinants

- You should be able to find the area of a triangle with vertices (x_1, y_1), (x_2, y_2), and (x_3, y_3).

$$\text{Area} = \pm\frac{1}{2}\begin{vmatrix} x_1 & y_1 & 1 \\ x_2 & y_2 & 1 \\ x_3 & y_3 & 1 \end{vmatrix}$$

  The $\pm$ symbol indicates that the appropriate sign should be chosen so that the area is positive.

- You should be able to test to see if three points, (x_1, y_1), (x_2, y_2), and (x_3, y_3), are collinear.

$$\begin{vmatrix} x_1 & y_1 & 1 \\ x_2 & y_2 & 1 \\ x_3 & y_3 & 1 \end{vmatrix} = 0, \text{ if and only if they are collinear.}$$

- You should be able to use Cramer's Rule to solve a system of linear equations.

- Now you should be able to solve a system of linear equations by substitution, elimination, elementary row operations on an augmented matrix, using the inverse matrix, or Cramer's Rule.

- You should be able to encode and decode messages by using an invertible $n \times n$ matrix.

Vocabulary Check

1. collinear   **2.** Cramer's Rule   **3.** cryptogram   **4.** uncoded, coded

1. Vertices: $(-2, 4), (2, 3), (-1, 5)$

$$\frac{1}{2}\begin{vmatrix} -2 & 4 & 1 \\ 2 & 3 & 1 \\ -1 & 5 & 1 \end{vmatrix} = \frac{1}{2}\left[-2\begin{vmatrix} 3 & 1 \\ 5 & 1 \end{vmatrix} - 4\begin{vmatrix} 2 & 1 \\ -1 & 1 \end{vmatrix} + \begin{vmatrix} 2 & 3 \\ -1 & 5 \end{vmatrix} \right]$$

$$= \frac{1}{2}\left[-2(-2) - 4(3) + 13 \right] = \frac{1}{2}(5) = \frac{5}{2}$$

Area $= \frac{5}{2}$ square units

3. Vertices: $\left(0, \frac{1}{2}\right), \left(\frac{5}{2}, 0\right), (4, 3)$

$$\frac{1}{2}\begin{vmatrix} 0 & \frac{1}{2} & 1 \\ \frac{5}{2} & 0 & 1 \\ 4 & 3 & 1 \end{vmatrix} = \frac{1}{2}\left[-\frac{1}{2}\left(-\frac{3}{2}\right) + \frac{15}{2}\right] = \frac{1}{2}\left[\frac{33}{4}\right] = \frac{33}{8}.$$

Area $= \frac{33}{8}$ square units

5. $\begin{vmatrix} -3 & 2 & 1 \\ 1 & 2 & 1 \\ -1 & -4 & 1 \end{vmatrix} = -3(6) - 2(2) + 1(-2) = -24$

Area rhombus $= |-24| = 24$ square units

7. $4 = \pm\frac{1}{2}\begin{vmatrix} -1 & 5 & 1 \\ -2 & 0 & 1 \\ x & 2 & 1 \end{vmatrix}$

$8 = \pm[(-1)(-2) - 5(-2 - x) + 1(-4)]$

$8 = \pm[5x + 8]$

$5x + 8 = 8$ or $5x + 8 = -8$

$x = 0$ or $x = -\frac{16}{5}$

$x = 0, -\frac{16}{5}$

9. Points: $(3, -1), (0, -3), (12, 5)$

$\begin{vmatrix} 3 & -1 & 1 \\ 0 & -3 & 1 \\ 12 & 5 & 1 \end{vmatrix} = 3(-8) + 12(2) = 0$

The points are collinear.

11. Points: $\left(2, -\frac{1}{2}\right), (-4, 4), (6, -3)$

$\begin{vmatrix} 2 & -\frac{1}{2} & 1 \\ -4 & 4 & 1 \\ 6 & -3 & 1 \end{vmatrix} = 2(7) + \frac{1}{2}(-10) + 1(-12)$

$= -3 \neq 0$

The points are not collinear.

13. $\begin{vmatrix} 1 & -2 & 1 \\ x & 2 & 1 \\ 5 & 6 & 1 \end{vmatrix} = 0$

$1(-4) + 2(x - 5) + 1(6x - 10) = 0$

$8x - 24 = 0$

$x = 3$

15. $\begin{cases} -7x + 11y = -1 \\ 3x - 9y = 9 \end{cases}$

$x = \dfrac{\begin{vmatrix} -1 & 11 \\ 9 & -9 \end{vmatrix}}{\begin{vmatrix} -7 & 11 \\ 3 & -9 \end{vmatrix}} = \dfrac{-90}{30} = -3$

$y = \dfrac{\begin{vmatrix} -7 & -1 \\ 3 & 9 \end{vmatrix}}{\begin{vmatrix} -7 & 11 \\ 3 & -9 \end{vmatrix}} = \dfrac{-60}{30} = -2$

Answer: $(-3, -2)$

17. $\begin{cases} 3x + 2y = -2 \\ 6x + 4y = 4 \end{cases}$

$\begin{vmatrix} 3 & 2 \\ 6 & 4 \end{vmatrix} = 12 - 17 = 0$

Cramer's rule cannot be used.

(In fact, the system is inconsistent.)

19. $\begin{cases} -0.4x + 0.8y = 1.6 \\ 0.2x + 0.3y = 2.2 \end{cases}$

$D = \begin{vmatrix} -0.4 & 0.8 \\ 0.2 & 0.3 \end{vmatrix} = -0.28$

$x = \dfrac{\begin{vmatrix} 1.6 & 0.8 \\ 2.2 & 0.3 \end{vmatrix}}{-0.28} = \dfrac{-1.28}{-0.28} = \dfrac{32}{7}$

$y = \dfrac{\begin{vmatrix} -0.4 & 1.6 \\ 0.2 & 2.2 \end{vmatrix}}{-0.28} = \dfrac{-1.20}{-0.28} = \dfrac{30}{7}$

Answer: $\left(\dfrac{32}{7}, \dfrac{30}{7} \right)$

21. $\begin{cases} 4x - y + z = -5 \\ 2x + 2y + 3z = 10 \\ 5x - 2y + 6z = 1 \end{cases}$ $D = \begin{vmatrix} 4 & -1 & 1 \\ 2 & 2 & 3 \\ 5 & -2 & 6 \end{vmatrix} = 55$

$x = \dfrac{\begin{vmatrix} -5 & -1 & 1 \\ 10 & 2 & 3 \\ 1 & -2 & 6 \end{vmatrix}}{55} = \dfrac{-55}{55} = -1, \quad y = \dfrac{\begin{vmatrix} 4 & -5 & 1 \\ 2 & 10 & 3 \\ 5 & 1 & 6 \end{vmatrix}}{55} = \dfrac{165}{55} = 3, \quad z = \dfrac{\begin{vmatrix} 4 & -1 & -5 \\ 2 & 2 & 10 \\ 5 & -2 & 1 \end{vmatrix}}{55} = \dfrac{110}{55} = 2$

Answer: $(-1, 3, 2)$

23. (a) $\begin{cases} 3x + 3y + 5z = 1 \\ 3x + 5y + 9z = 2 \\ 5x + 9y + 17z = 4 \end{cases}$

$\begin{cases} 3x + 3y + 5z = 1 \\ 2y + 4z = 1 \\ 4y + \frac{26}{3}z = \frac{7}{3} \end{cases}$

$\begin{cases} 3x + 3y + 5z = 1 \\ 2y + 4z = 1 \\ \frac{2}{3}z = \frac{1}{3} \end{cases}$

$z = \frac{1}{2}$

$2y + 4\left(\frac{1}{2}\right) = 1 \implies y = -\frac{1}{2}$

$3x + 3\left(-\frac{1}{2}\right) + 5\left(\frac{1}{2}\right) = 1 \implies x = 0$

Answer: $\left(0, -\frac{1}{2}, \frac{1}{2}\right)$

(b) $D = \begin{vmatrix} 3 & 3 & 5 \\ 3 & 5 & 9 \\ 5 & 9 & 17 \end{vmatrix} = 4$

$x = \dfrac{\begin{vmatrix} 1 & 3 & 5 \\ 2 & 5 & 9 \\ 4 & 9 & 17 \end{vmatrix}}{4} = 0$

$y = \dfrac{\begin{vmatrix} 3 & 1 & 5 \\ 3 & 2 & 9 \\ 5 & 4 & 17 \end{vmatrix}}{4} = -\dfrac{1}{2}$

$z = \dfrac{\begin{vmatrix} 3 & 3 & 1 \\ 3 & 5 & 2 \\ 5 & 9 & 4 \end{vmatrix}}{4} = \dfrac{1}{2}$

Answer: $\left(0, -\frac{1}{2}, \frac{1}{2}\right)$

25. (a) $\begin{cases} 2x - y + z = 5 \\ x - 2y - z = 1 \\ 3x + y + z = 4 \end{cases}$

$\begin{cases} x - 2y - z = 1 \\ \quad\;\; 3y + 3z = 3 \\ \quad\;\; 7y + 4z = 1 \end{cases}$

$\begin{cases} x - 2y - \quad z = 1 \\ \qquad y + \quad z = 1 \\ \qquad\qquad -3z = -6 \end{cases}$

$z = 2$

$y + 2 = 1 \implies y = -1$

$x - 2(-1) - 2 = 1 \implies x = 1$

Answer: $(1, -1, 2)$

(b) $D = \begin{vmatrix} 2 & -1 & 1 \\ 1 & -2 & -1 \\ 3 & 1 & 1 \end{vmatrix} = 9$

$x = \dfrac{\begin{vmatrix} 5 & -1 & 1 \\ 1 & -2 & -1 \\ 4 & 1 & 1 \end{vmatrix}}{9} = 1$

$y = \dfrac{\begin{vmatrix} 2 & 5 & 1 \\ 1 & 1 & -1 \\ 3 & 4 & 1 \end{vmatrix}}{9} = -1$

$z = \dfrac{\begin{vmatrix} 2 & -1 & 5 \\ 1 & -2 & 1 \\ 3 & 1 & 4 \end{vmatrix}}{9} = 2$

Answer: $(1, -1, 2)$

27. (a) $D = \begin{vmatrix} 5 & 10 & 30 \\ 10 & 30 & 100 \\ 30 & 100 & 354 \end{vmatrix} = 700$

$c = \dfrac{\begin{vmatrix} 5543 & 10 & 30 \\ 12{,}447 & 30 & 100 \\ 38{,}333 & 100 & 354 \end{vmatrix}}{700} = \dfrac{548{,}580}{700} \approx 784$

$b = \dfrac{\begin{vmatrix} 5 & 5543 & 30 \\ 10 & 12{,}447 & 100 \\ 30 & 38{,}333 & 354 \end{vmatrix}}{700} = \dfrac{169{,}070}{700} \approx 241.5$

$a = \dfrac{\begin{vmatrix} 5 & 10 & 5543 \\ 10 & 30 & 12{,}447 \\ 30 & 100 & 38{,}333 \end{vmatrix}}{700} = \dfrac{-18{,}450}{700} \approx -26.36$

$y = -26.36t^2 + 241.5t + 784$

(b)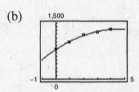

(c) No, because of the negative t^2 coefficient.

29. The uncoded row matrices are the rows of the 6×3 matrix on the left.

$\begin{array}{ccc} C & A & L \\ L & & M \\ E & & T \\ O & M & O \\ R & R & O \\ W & & \end{array} \begin{bmatrix} 3 & 1 & 12 \\ 12 & 0 & 13 \\ 5 & 0 & 20 \\ 15 & 13 & 15 \\ 18 & 18 & 15 \\ 23 & 0 & 0 \end{bmatrix} \begin{bmatrix} 1 & -1 & 0 \\ 1 & 0 & -1 \\ -6 & 2 & 3 \end{bmatrix} = \begin{bmatrix} -68 & 21 & 35 \\ -66 & 14 & 39 \\ -115 & 35 & 60 \\ -62 & 15 & 32 \\ -54 & 12 & 27 \\ 23 & -23 & 0 \end{bmatrix}$

Answer: $[-68, 21, 35], [-66, 14, 39], [-115, 35, 60]$

$[-62, 15, 32], [-54, 12, 27], [23, -23, 0]$

31.
$$\begin{matrix} G & O & N \\ E & & F \\ I & S & H \\ I & N & G \end{matrix}\begin{bmatrix} 7 & 15 & 14 \\ 5 & 0 & 6 \\ 9 & 19 & 8 \\ 9 & 14 & 7 \end{bmatrix}\begin{bmatrix} 1 & 2 & 2 \\ 3 & 7 & 9 \\ -1 & -4 & -7 \end{bmatrix} = \begin{bmatrix} 38 & 63 & 51 \\ -1 & -14 & -32 \\ 58 & 119 & 133 \\ 44 & 88 & 95 \end{bmatrix}$$

Cryptogram: 38   63   51   −1   −14   −32   58   119   133   44   88   95

33. $A^{-1} = \begin{bmatrix} 1 & 2 \\ 3 & 5 \end{bmatrix}^{-1} = \begin{bmatrix} -5 & 2 \\ 3 & -1 \end{bmatrix}$

$$\begin{bmatrix} 11 & 21 \\ 64 & 112 \\ 25 & 50 \\ 29 & 53 \\ 23 & 46 \\ 40 & 75 \\ 55 & 92 \end{bmatrix}\begin{bmatrix} -5 & 2 \\ 3 & -1 \end{bmatrix} = \begin{bmatrix} 8 & 1 \\ 16 & 16 \\ 25 & 0 \\ 14 & 5 \\ 23 & 0 \\ 25 & 5 \\ 1 & 18 \end{bmatrix}\begin{matrix} H & A \\ P & P \\ Y & \\ N & E \\ W & \\ Y & E \\ A & R \end{matrix}$$ Message: HAPPY NEW YEAR

35. $A^{-1} = \begin{bmatrix} 1 & 2 & 1 \\ -1 & 0 & 2 \\ 1 & -1 & -2 \end{bmatrix}^{-1} = \begin{bmatrix} \frac{2}{3} & 1 & \frac{4}{3} \\ 0 & -1 & -1 \\ \frac{1}{3} & 1 & \frac{2}{3} \end{bmatrix}$

$$\begin{bmatrix} 38 & 36 & -1 \\ 11 & 17 & 11 \\ 42 & 15 & -27 \\ -5 & 18 & 37 \\ 26 & 28 & 17 \\ 8 & 24 & 20 \\ 32 & 20 & -7 \\ 23 & -1 & -19 \end{bmatrix}\begin{bmatrix} \frac{2}{3} & 1 & \frac{4}{3} \\ 0 & -1 & -1 \\ \frac{1}{3} & 1 & \frac{2}{3} \end{bmatrix} = \begin{bmatrix} 25 & 1 & 14 \\ 11 & 5 & 5 \\ 19 & 0 & 23 \\ 9 & 14 & 0 \\ 23 & 15 & 18 \\ 12 & 4 & 0 \\ 19 & 5 & 18 \\ 9 & 5 & 19 \end{bmatrix}\begin{matrix} Y & A & N \\ K & E & E \\ S & - & W \\ I & N & - \\ W & O & R \\ L & D & - \\ S & E & R \\ I & E & S \end{matrix}$$

YANKEES WIN WORLD SERIES

37. True. Cramer's Rule requires that the determinant of the coefficient matrix be nonzero.

39. Answers will vary.

41.
$$y - 5 = \frac{5 - 3}{-1 - 7}(x + 1) = \frac{-1}{4}(x + 1)$$
$$4y - 20 = -x - 1$$
$$4y + x = 19$$
$$x + 4y - 19 = 0$$

43.
$$y + 3 = \frac{-3 + 1}{3 - 10}(x - 3) = \frac{2}{7}(x - 3)$$
$$7y + 21 = 2x - 6$$
$$7y - 2x = -27$$
$$2x - 7y - 27 = 0$$

45. $f(x) = \dfrac{2x^2}{x^2 + 4}$.

Horizontal asymptote: $y = 2$

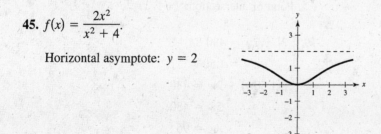

Review Exercises for Chapter 7

1. $\begin{cases} x + y = 2 \implies \qquad y = 2 - x \\ x - y = 0 \implies x - (2 - x) = 0 \end{cases}$

$$2x - 2 = 0$$
$$x = 1$$
$$y = 2 - 1 = 1$$

Answer: $(1, 1)$

3. $\begin{cases} x^2 - y^2 = 9 \\ x \; - y \; = 1 \implies \qquad x = y + 1 \end{cases}$

$$(y + 1)^2 - y^2 = 9$$
$$2y + 1 = 9$$
$$y = 4$$
$$x = 5$$

Answer: $(5, 4)$

5. $\begin{cases} y = 2x^2 \\ y = x^4 - 2x^2 \implies 2x^2 = x^4 - 2x^2 \end{cases}$

$$0 = x^4 - 4x^2$$
$$0 = x^2(x^2 - 4)$$
$$0 = x^2(x + 2)(x - 2)$$
$$x = 0, x = -2, x = 2$$
$$y = 0, y = 8, y = 8$$

Answer: $(0, 0), (-2, 8), (2, 8)$

7. $\begin{cases} 5x + 6y = 7 \implies y_1 = \dfrac{1}{6}(7 - 5x) \\ -x - 4y = 0 \implies y_2 = -\dfrac{x}{4} \end{cases}$

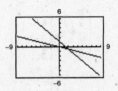

Answer: $\left(2, -\frac{1}{2}\right)$

9. $\begin{cases} y^2 - 4x = 0 \implies y^2 = 4x \implies y = \pm 2\sqrt{x} \\ x + y = 0 \implies y = -x \end{cases}$

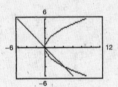

Points of intersection: $(0, 0), (4, -4)$

11. $\begin{cases} y = 3 - x^2 \\ y = 2x^2 + x + 1 \end{cases}$

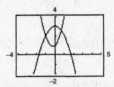

Points of intersection: $(0.67, 2.56), (-1, 2)$

13. $\begin{cases} y = 2(6 - x) \\ y = 2^{x-2} \end{cases}$

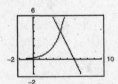

Point of intersection: $(4, 4)$

15. $\begin{cases} y = \ln(x + 2) + 1 \\ y = -x \end{cases}$

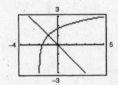

Point of intersection: $(-1, 1)$

17. Revenue $= 4.95x$

Cost $= 2.85x + 10,000$

Break even when Revenue $=$ Cost

$$4.95x = 2.85x + 10,000$$
$$2.10x = 10,000$$
$$x \approx 4762 \text{ units}$$

19. $\begin{cases} 2l + 2w = 480 \\ \qquad\quad l = 1.50w \end{cases}$

$$2(1.50w) + 2w = 480$$
$$5w = 480$$
$$w = 96$$
$$l = 144$$

The dimensions are 96×144 meters.

21. $\begin{cases} 2x - y = 2 \implies 16x - 8y = 16 \\ 6x + 8y = 39 \implies 6x + 8y = 39 \end{cases}$

$$22x \qquad = 55$$

$$x = \tfrac{55}{22} = \tfrac{5}{2}$$

$$y = 3$$

Answer: $\left(\tfrac{5}{2}, 3\right)$

23. $\begin{cases} 1/5x + 3/10y = 7/50 \implies 20x + 30y = 14 \implies 20x + 30y = 14 \\ 2/5x + 1/2y = 1/5 \implies 4x + 5y = 2 \implies -20x - 25y = -10 \end{cases}$

$$5y = 4$$

$$y = \tfrac{4}{5}$$

$$x = -\tfrac{1}{2}$$

Answer: $\left(-\tfrac{1}{2}, \tfrac{4}{5}\right)$ or $(-0.5, 0.8)$

25. $\begin{cases} 3x - 2y = 0 \implies 3x - 2y = 0 \\ 3x + 2(y + 5) = 10 \implies 3x + 2y = 0 \end{cases}$

$$6x \qquad = 0$$

$$x = 0$$

$$y = 0$$

Answer: $(0, 0)$

27. $\begin{cases} 1.25x - 2y = 3.5 \implies 5x - 8y = 14 \\ 5x - 8y = 14 \implies -5x + 8y = -14 \end{cases}$

$$0 = 0$$

Infinite number of solutions

Let $y = a$, then $5x - 8a = 14 \implies x = \tfrac{14}{5} + \tfrac{8}{5}a$.

Answer: $\left(\tfrac{14}{5} + \tfrac{8}{5}a, a\right)$

29. $\begin{cases} 3x + 2y = 0 \implies y = -\tfrac{3}{2}x \\ x - y = 4 \implies y = x - 4 \end{cases}$

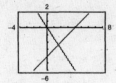

Consistent.

Answer: $(1.6, -2.4)$

31. $\begin{cases} \tfrac{1}{4}x - \tfrac{1}{5}y = 2 \implies y = \tfrac{5}{4}x - 10 \\ -5x + 4y = 8 \implies y = \tfrac{1}{4}(8 + 5x) = \tfrac{5}{4}x + 2 \end{cases}$

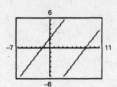

Inconsistent; lines are parallel.

33. $\begin{cases} 2x - 2y = 8 \implies y = x - 4 \\ 4x - 1.5y = -5.5 \implies y = \tfrac{8}{3}x + \tfrac{11}{3} \end{cases}$

Consistent

Answer: $(-4.6, -8.6)$

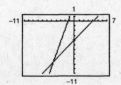

35. Demand = Supply

$$37 - 0.0002x = 22 + 0.00001x$$

$$15 = 0.00021x$$

$$x = \frac{500,000}{7}, \, p = \frac{159}{7}$$

Point of equilibrium: $\left(\dfrac{500,000}{7}, \dfrac{159}{7}\right)$

37. Let x = speed of the slower plane.

Let y = speed of the faster plane.

Then, distance of first plane + distance of second plane = 275 miles.

(rate of first plane)(time) + (rate of second plane)(time) = 275 miles

$$\begin{cases} x\left(\frac{40}{60}\right) + y\left(\frac{40}{60}\right) = 275 \\ y = x + 25 \end{cases}$$

$$\frac{2}{3}x + \frac{2}{3}(x + 25) = 275$$

$$4x + 50 = 825$$

$$4x = 775$$

$$x = 193.75 \text{ mph}$$

$$y = x + 25 = 218.75 \text{ mph}$$

39. $z = -2$

$-y + z = -5 \implies -y - 2 = -5 \implies y = 3$

$x - 4y + 3z = -14$

$\qquad\qquad \implies x = 4(3) - 3(-2) - 14 = 4$

Answer: $(4, 3, -2)$

41.
$$\begin{cases} x + 3y - z = 13 \\ 2x \qquad - 5z = 23 \\ 4x - y - 2z = 14 \end{cases}$$

$$\begin{cases} x + 3y - z = 13 \\ \quad - 6y - 3z = -3 \\ \quad - 13y + 2z = -38 \end{cases}$$

$$\begin{cases} x + 3y - z = 13 \\ \quad - 6y - 3z = -3 \\ \qquad\qquad \frac{17}{2}z = -\frac{63}{2} \end{cases}$$

$$\frac{17}{2}z = -\frac{63}{2} \implies z = -\frac{63}{17}$$

$$-6y - 3\left(-\frac{63}{17}\right) = -3 \implies y = \frac{40}{17}$$

$$x + 3\left(\frac{40}{17}\right) - \left(-\frac{63}{17}\right) = 13 \implies x = \frac{38}{17}$$

Answer: $\left(\frac{38}{17}, \frac{40}{17}, -\frac{63}{17}\right)$

43.
$$\begin{cases} x - 2y + z = -6 \\ 2x - 3y \qquad = -7 \\ -x + 3y - 3z = 11 \end{cases}$$

$$\begin{cases} x - 2y + z = -6 \\ \quad y - 2z = 5 \qquad -2 \text{ Eq.} 1 + \text{Eq.} 2 \\ \quad y - 2z = 5 \qquad \text{Eq.} 1 + \text{Eq.} 3 \end{cases}$$

$$\begin{cases} x - 2y + z = -6 \\ \quad y - 2z = 5 \\ \qquad\quad 0 = 0 \qquad -\text{Eq.} 2 + \text{Eq.} 3 \end{cases}$$

Let $z = a$, then $y = 2a + 5$.

$$x - 2(2a + 5) + a = -6$$

$$x - 3a - 10 = -6$$

$$x = 3a + 4$$

Answer: $(3a + 4, 2a + 5, a)$ where a is any real number.

45.
$$\begin{cases} x - 2y + 3z = -5 \\ 2x + 4y + 5z = 1 \\ x + 2y + z = 0 \end{cases}$$

$$\begin{cases} x - 2y + 3z = -5 \\ \quad 8y - z = 11 \\ \quad 4y - 2z = 5 \end{cases}$$

$$\begin{cases} x - 2y + 3z = -5 \\ \quad 4y - 2z = 5 \\ \qquad\quad 3z = 1 \end{cases}$$

$$z = \frac{1}{3}$$

$$4y = 5 + 2\left(\frac{1}{3}\right) = \frac{17}{3} \implies y = \frac{17}{12}$$

$$x = 2\left(\frac{17}{12}\right) - 3\left(\frac{1}{3}\right) - 5 = -\frac{19}{6}$$

Answer: $\left(-\frac{19}{6}, \frac{17}{12}, \frac{1}{3}\right)$

47. $\begin{cases} 5x - 12y + 7z = 16 & \text{Equation 1} \\ 3x - 7y + 4z = 9 & \text{Equation 2} \end{cases}$

3 times Eq. 1 and (-5) times Eq. 2:

$\begin{cases} 15x - 36y + 21z = 48 \\ -15x + 35y - 20z = -45 \end{cases}$

Adding, $-y + z = 3 \implies y = z - 3$.

$$5x - 12(z - 3) + 7z = 16$$
$$5x - 5z + 36 = 16$$
$$5x = 5z - 20$$
$$x = z - 4$$

Let $z = a$, then $x = a - 4$ and $y = a - 3$.

Answer: $(a - 4, a - 3, a)$ where a is any real number.

49.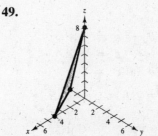

$(4, 0, 0), (0, -2, 0), (0, 0, 8), (1, 0, 6)$

51. $\dfrac{4 - x}{x^2 + 6x + 8} = \dfrac{A}{x + 2} + \dfrac{B}{x + 4}$

$4 - x = A(x + 4) + B(x + 2) = (A + B)x + (4A + 2B)$

$\begin{cases} A + B = -1 \\ 4A + 2B = 4 \end{cases} \implies A = 3, B = -4$

$\dfrac{4 - x}{x^2 + 6x + 8} = \dfrac{3}{x + 2} - \dfrac{4}{x + 4}$

53. $\dfrac{x^2 + 2x}{x^3 - x^2 + x - 1} = \dfrac{A}{x - 1} + \dfrac{Bx + C}{x^2 + 1}$

$x^2 + 2x = A(x^2 + 1) + (Bx + C)(x - 1) = (A + B)x^2 + (C - B)x + (A - C)$

$\begin{cases} A + B = 1 \\ -B + C = 2 \\ A - C = 0 \end{cases} \implies A = \frac{3}{2}, B = -\frac{1}{2}, C = \frac{3}{2}$

$\dfrac{x^2 + 2x}{x^3 - x^2 + x - 1} = \dfrac{3/2}{x - 1} + \dfrac{-(1/2)x + 3/2}{x^2 + 1} = \dfrac{1}{2}\left(\dfrac{3}{x - 1} - \dfrac{x - 3}{x^2 + 1}\right)$

55. $\dfrac{x^2 + 3x - 3}{x^3 + 2x^2 + x + 2} = \dfrac{x^2 + 3x - 3}{(x + 2)(x^2 + 1)} = \dfrac{A}{x + 2} + \dfrac{Bx + C}{x^2 + 1}$

$x^2 + 3x - 3 = A(x^2 + 1) + (Bx + C)(x + 2)$

$\begin{cases} A + B = 1 \\ 2B + C = 3 \\ A + 2C = -3 \end{cases}$

$\begin{cases} A + B = 1 \\ 2B + C = 3 \\ -B + 2C = -4 \end{cases}$

$\begin{cases} A + B = 1 \\ B - 2C = 4 \\ 5C = -5 \end{cases}$

$C = -1, B = 2(-1) + 4 = 2, A = 1 - 2 = -1$

$\dfrac{x^2 + 3x - 3}{x^3 + 2x^2 + x + 2} = \dfrac{-1}{x + 2} + \dfrac{2x - 1}{x^2 + 1}$

57. $y = ax^2 + bx + c$

$(-1, -4)$: $a - b + c = -4$
$(1, -2)$: $a + b + c = -2$
$(2, 5)$: $4a + 2b + c = 5$

Solving the system, $a = 2$, $b = 1$, $c = -5$.

$y = 2x^2 + x - 5$

59. Let x = gallons of spray X

Let y = gallons of spray Y

Let z = gallons of spray Z

$\begin{cases} \text{Chemical A: } \frac{1}{5}x + \frac{1}{3}z = 6 \\ \text{Chemical B: } \frac{2}{5}x + \frac{1}{3}z = 8 \\ \text{Cehmical C: } \frac{2}{5}x + y + \frac{1}{3}z = 13 \end{cases}$

Subtracting Eq. 2 − Eq. 1 gives $\frac{1}{5}x = 2 \implies x = 10$.

Then $z = 12$ and $y = 5$.

Answer: 10 gallons of spray X

5 gallons of spray Y

12 gallons of spray Z

61. Order 3×1

63. Order 1×1

65. $\begin{bmatrix} 3 & -10 & : & 15 \\ 5 & 4 & : & 22 \end{bmatrix}$

67. $\begin{bmatrix} 8 & -7 & 4 & : & 12 \\ 3 & -5 & 2 & : & 20 \\ 5 & 3 & -3 & : & 26 \end{bmatrix}$

69. $\begin{bmatrix} 5 & 1 & 7 & : & -9 \\ 4 & 2 & 0 & : & 10 \\ 9 & 4 & 2 & : & 3 \end{bmatrix}$

$\begin{cases} 5x + y + 7z = -9 \\ 4x + 2y = 10 \\ 9x + 4y + 2z = 3 \end{cases}$

71. $\begin{bmatrix} 0 & 1 & 1 \\ 1 & 2 & 3 \\ 2 & 2 & 2 \end{bmatrix}$

$\begin{matrix} R_1 + R_2 \to \\ -R_1 + R_2 \to \\ -2R_1 + R_3 \to \end{matrix} \begin{bmatrix} 1 & 3 & 4 \\ 0 & -1 & -1 \\ 0 & -4 & -6 \end{bmatrix}$

$\begin{matrix} 3R_2 + R_1 \to \\ -R_2 \to \\ -4R_2 + R_3 \to \end{matrix} \begin{bmatrix} 1 & 0 & 1 \\ 0 & 1 & 1 \\ 0 & 0 & -2 \end{bmatrix}$

73. $\begin{bmatrix} 3 & -2 & 1 & 0 \\ 4 & -3 & 0 & 1 \end{bmatrix} \implies \begin{bmatrix} 1 & 0 & 3 & -2 \\ 0 & 1 & 4 & -3 \end{bmatrix}$

75. $\begin{bmatrix} 1.5 & 3.6 & 4.2 \\ 0.2 & 1.4 & 1.8 \\ 2.0 & 4.4 & 6.4 \end{bmatrix} \implies \begin{bmatrix} 1 & 0 & 0 \\ 0 & 1 & 0 \\ 0 & 0 & 1 \end{bmatrix}$

77. $\begin{bmatrix} 5 & 4 & : & 2 \\ -1 & 1 & : & -22 \end{bmatrix}$

$\begin{matrix} 4R_2 + R_1 \to \\ R_1 + R_2 \to \end{matrix} \begin{bmatrix} 1 & 8 & : & -86 \\ 0 & 9 & : & -108 \end{bmatrix}$

$9y = -108$

$y = -12$

$x = -8(-12) - 86 = 10$

Answer: $(10, -12)$

79.
$$\begin{bmatrix} 2 & 1 & \vdots & 0.3 \\ 3 & -1 & \vdots & -1.3 \end{bmatrix}$$

$$-R_1 + R_2 \rightarrow \begin{bmatrix} 2 & 1 & \vdots & 0.3 \\ 1 & -2 & \vdots & -1.6 \end{bmatrix}$$

$$\begin{bmatrix} 1 & -2 & \vdots & -1.6 \\ 2 & 1 & \vdots & 0.3 \end{bmatrix}$$

$$-2R_1 + R_2 \rightarrow \begin{bmatrix} 1 & -2 & \vdots & -1.6 \\ 0 & 5 & \vdots & 3.5 \end{bmatrix}$$

$5y = 3.5 \implies y = 0.7$

$x = 2(0.7) - 1.6 = -0.2$

$x = -0.2, \; y = 0.7$

Answer: $(-0.2, 0.7)$

81.
$$\begin{bmatrix} 2 & 3 & 3 & \vdots & 3 \\ 6 & 6 & 12 & \vdots & 13 \\ 12 & 9 & -1 & \vdots & 2 \end{bmatrix}$$

$$\begin{matrix} -3R_1 + R_2 \rightarrow \\ -6R_1 + R_3 \rightarrow \end{matrix} \begin{bmatrix} 2 & 3 & 3 & \vdots & 3 \\ 0 & -3 & 3 & \vdots & 4 \\ 0 & -9 & -19 & \vdots & -16 \end{bmatrix}$$

$$\begin{matrix} R_2 + R_1 \rightarrow \\ \\ -3R_2 + R_3 \rightarrow \end{matrix} \begin{bmatrix} 2 & 0 & 6 & \vdots & 7 \\ 0 & -3 & 3 & \vdots & 4 \\ 0 & 0 & -28 & \vdots & -28 \end{bmatrix}$$

$$\begin{matrix} \frac{1}{2}R_1 \rightarrow \\ -\frac{1}{3}R_2 \rightarrow \\ -\frac{1}{28}R_3 \rightarrow \end{matrix} \begin{bmatrix} 1 & 0 & 3 & \vdots & \frac{7}{2} \\ 0 & 1 & -1 & \vdots & -\frac{4}{3} \\ 0 & 0 & 1 & \vdots & 1 \end{bmatrix}$$

$z = 1$

$y - 1 = -\frac{4}{3} \implies y = -\frac{1}{3}$

$x + 3(1) = \frac{7}{2} \implies x = \frac{1}{2}$

Answer: $\left(\frac{1}{2}, -\frac{1}{3}, 1\right)$

83. $\begin{cases} x + 2y - z = 1 \\ \qquad y + z = 0 \end{cases}$

$$\begin{bmatrix} 1 & 2 & -1 & \vdots & 1 \\ 0 & 1 & 1 & \vdots & 0 \end{bmatrix} \implies \begin{bmatrix} 1 & 0 & -3 & \vdots & 1 \\ 0 & 1 & 1 & \vdots & 0 \end{bmatrix}$$

$y = -z$

$x = 1 + 3z$

Answer: $(1 + 3a, -a, a)$, a is a real number.

85.
$$\begin{bmatrix} -1 & 1 & 2 & \vdots & 1 \\ 2 & 3 & 1 & \vdots & -2 \\ 5 & 4 & 2 & \vdots & 4 \end{bmatrix}$$

$$\begin{matrix} -R_1 \rightarrow \\ 2R_1 + R_2 \rightarrow \\ 5R_1 + R_3 \rightarrow \end{matrix} \begin{bmatrix} 1 & -1 & -2 & \vdots & -1 \\ 0 & 5 & 5 & \vdots & 0 \\ 0 & 9 & 12 & \vdots & 9 \end{bmatrix}$$

$$\frac{1}{5}R_2 \rightarrow \begin{bmatrix} 1 & -1 & -2 & \vdots & -1 \\ 0 & 1 & 1 & \vdots & 0 \\ 0 & 9 & 12 & \vdots & 9 \end{bmatrix}$$

$$\begin{matrix} R_2 + R_1 \rightarrow \\ \\ -9R_2 + R_3 \rightarrow \end{matrix} \begin{bmatrix} 1 & 0 & -1 & \vdots & -1 \\ 0 & 1 & 1 & \vdots & 0 \\ 0 & 0 & 3 & \vdots & 9 \end{bmatrix}$$

$$\frac{1}{3}R_3 \rightarrow \begin{bmatrix} 1 & 0 & -1 & \vdots & -1 \\ 0 & 1 & 1 & \vdots & 0 \\ 0 & 0 & 1 & \vdots & 3 \end{bmatrix}$$

$$\begin{matrix} R_3 + R_1 \rightarrow \\ -R_3 + R_2 \rightarrow \end{matrix} \begin{bmatrix} 1 & 0 & 0 & \vdots & 2 \\ 0 & 1 & 0 & \vdots & -3 \\ 0 & 0 & 1 & \vdots & 3 \end{bmatrix}$$

$x = 2, \; y = -3, \; z = 3$

Answer: $(2, -3, 3)$

87. $\begin{cases} x + y + 2z = 4 \\ x - y + 4z = 1 \\ 2x - y + 2z = 1 \end{cases}$
$$\begin{bmatrix} 1 & 1 & 2 & \vdots & 4 \\ 1 & -1 & 4 & \vdots & 1 \\ 2 & -1 & 2 & \vdots & 1 \end{bmatrix} \implies \begin{bmatrix} 1 & 0 & 0 & \vdots & 1 \\ 0 & 1 & 0 & \vdots & 2 \\ 0 & 0 & 1 & \vdots & \frac{1}{2} \end{bmatrix}$$

Answer: $\left(1, 2, \frac{1}{2}\right)$

89. $\begin{bmatrix} 1 & 2 & -1 & \vdots & 7 \\ 0 & -1 & -1 & \vdots & 4 \\ 4 & 0 & -1 & \vdots & 16 \end{bmatrix}$ reduces to $\begin{bmatrix} 1 & 0 & 0 & \vdots & 3 \\ 0 & 1 & 0 & \vdots & 0 \\ 0 & 0 & 1 & \vdots & -4 \end{bmatrix}$

Answer: $(3, 0, -4)$

91. $\begin{bmatrix} 3 & -1 & 5 & -2 & \vdots & -44 \\ 1 & 6 & 4 & -1 & \vdots & 1 \\ 5 & -1 & 1 & 3 & \vdots & -15 \\ 0 & 4 & -1 & -8 & \vdots & 58 \end{bmatrix}$ reduces to $\begin{bmatrix} 1 & 0 & 0 & 0 & \vdots & 2 \\ 0 & 1 & 0 & 0 & \vdots & 6 \\ 0 & 0 & 1 & 0 & \vdots & -10 \\ 0 & 0 & 0 & 1 & \vdots & -3 \end{bmatrix}$

Answer: $(2, 6, -10, -3)$

93. $x = 12$

$y = -7$

95. $x + 3 = 5x - 1 \Rightarrow x = 1$

$-4y = -44 \Rightarrow y = 11$

$y + 5 = 16 \Rightarrow y = 11$

$6x = 6 \Rightarrow x = 1$

Answer: $x = 1, y = 11$

97. (a) $A + B = \begin{bmatrix} 7 & 3 \\ -1 & 5 \end{bmatrix} + \begin{bmatrix} 10 & -20 \\ 14 & -3 \end{bmatrix} = \begin{bmatrix} 17 & -17 \\ 13 & 2 \end{bmatrix}$

(b) $A - B = \begin{bmatrix} -3 & 23 \\ -15 & 8 \end{bmatrix}$ (c) $4A = \begin{bmatrix} 28 & 12 \\ -4 & 20 \end{bmatrix}$

(d) $A + 3B = \begin{bmatrix} 7 & 3 \\ -1 & 5 \end{bmatrix} + \begin{bmatrix} 30 & -60 \\ 42 & -9 \end{bmatrix} = \begin{bmatrix} 37 & -57 \\ 41 & -4 \end{bmatrix}$

99. (a) $A + B = \begin{bmatrix} 6 & 0 & 7 \\ 5 & -1 & 2 \\ 3 & 2 & 3 \end{bmatrix} + \begin{bmatrix} 0 & 5 & 1 \\ -4 & 8 & 6 \\ 2 & -1 & 1 \end{bmatrix} = \begin{bmatrix} 6 & 5 & 8 \\ 1 & 7 & 8 \\ 5 & 1 & 4 \end{bmatrix}$

(b) $A - B = \begin{bmatrix} 6 & -5 & 6 \\ 9 & -9 & -4 \\ 1 & 3 & 2 \end{bmatrix}$

(c) $4A = \begin{bmatrix} 24 & 0 & 28 \\ 20 & -4 & 8 \\ 12 & 8 & 12 \end{bmatrix}$

(d) $A + 3B = \begin{bmatrix} 6 & 0 & 7 \\ 5 & -1 & 2 \\ 3 & 2 & 3 \end{bmatrix} + \begin{bmatrix} 0 & 15 & 3 \\ -12 & 24 & 18 \\ 6 & -3 & 3 \end{bmatrix} = \begin{bmatrix} 6 & 15 & 10 \\ -7 & 23 & 20 \\ 9 & -1 & 6 \end{bmatrix}$

101. $\begin{bmatrix} 2 & 1 & 0 \\ 0 & 5 & -4 \end{bmatrix} - 3 \begin{bmatrix} 5 & 3 & -6 \\ 0 & -2 & 5 \end{bmatrix} = \begin{bmatrix} 2 & 1 & 0 \\ 0 & 5 & -4 \end{bmatrix} - \begin{bmatrix} 15 & 9 & -18 \\ 0 & -6 & 15 \end{bmatrix}$

$= \begin{bmatrix} -13 & -8 & 18 \\ 0 & 11 & -19 \end{bmatrix}$

103. $-\begin{bmatrix} 8 & -1 \\ -2 & 4 \end{bmatrix} - 5 \begin{bmatrix} -2 & 0 \\ 3 & -1 \end{bmatrix} + \begin{bmatrix} 7 & -8 \\ 4 & 3 \end{bmatrix} = \begin{bmatrix} -8 & 1 \\ 2 & -4 \end{bmatrix} - \begin{bmatrix} -10 & 0 \\ 15 & -5 \end{bmatrix} + \begin{bmatrix} 7 & -8 \\ 4 & 3 \end{bmatrix}$

$= \begin{bmatrix} 9 & -7 \\ -9 & 4 \end{bmatrix}$

105. $\begin{bmatrix} 32 & -\frac{17}{2} & -\frac{3}{2} \\ 6 & 46 & 33 \end{bmatrix}$

107. $X = 3A - 2B = 3\begin{bmatrix} -4 & 0 \\ 1 & -5 \\ -3 & 2 \end{bmatrix} - 2\begin{bmatrix} 1 & 2 \\ -2 & 1 \\ 4 & 4 \end{bmatrix} = \begin{bmatrix} -14 & -4 \\ 7 & -17 \\ -17 & -2 \end{bmatrix}$

109. $X = \frac{1}{3}[B - 2A] = \frac{1}{3}\left(\begin{bmatrix} 1 & 2 \\ -2 & 1 \\ 4 & 4 \end{bmatrix} - 2\begin{bmatrix} -4 & 0 \\ 1 & -5 \\ -3 & 2 \end{bmatrix}\right) = \frac{1}{3}\begin{bmatrix} 9 & 2 \\ -4 & 11 \\ 10 & 0 \end{bmatrix}$

111. $\begin{bmatrix} 1 & 2 \\ 5 & -4 \\ 6 & 0 \end{bmatrix}\begin{bmatrix} 6 & -2 & 8 \\ 4 & 0 & 0 \end{bmatrix} = \begin{bmatrix} 1(6) + 2(4) & 1(-2) + 2(0) & 1(8) + 2(0) \\ 5(6) + (-4)(4) & 5(-2) + (-4)(0) & 5(8) + (-4)(0) \\ 6(6) + (0)(4) & 6(-2) + (0)(0) & 6(8) + (0)(0) \end{bmatrix} = \begin{bmatrix} 14 & -2 & 8 \\ 14 & -10 & 40 \\ 36 & -12 & 48 \end{bmatrix}$

113. $AB = \begin{bmatrix} 3 & -2 & 0 \\ 1 & 4 & 9 \end{bmatrix}\begin{bmatrix} 7 & 0 \\ 5 & 3 \\ -1 & 3 \end{bmatrix} = \begin{bmatrix} 11 & -6 \\ 18 & 39 \end{bmatrix}$ **115.** $\begin{bmatrix} 4 & 1 \\ 11 & -7 \\ 12 & 3 \end{bmatrix}\begin{bmatrix} 3 & -5 & 6 \\ 2 & -2 & -2 \end{bmatrix} = \begin{bmatrix} 14 & -22 & 22 \\ 19 & -41 & 80 \\ 42 & -66 & 66 \end{bmatrix}$

117. $\begin{bmatrix} 2 & 1 \\ 6 & 0 \end{bmatrix}\left(\begin{bmatrix} 4 & 2 \\ -3 & 1 \end{bmatrix} + \begin{bmatrix} -2 & 4 \\ 0 & 4 \end{bmatrix}\right) = \begin{bmatrix} 2 & 1 \\ 6 & 0 \end{bmatrix}\begin{bmatrix} 2 & 6 \\ -3 & 5 \end{bmatrix}$

$= \begin{bmatrix} 2(2) + 1(-3) & 2(6) + 1(5) \\ 6(2) + 0 & 6(6) + 0 \end{bmatrix}$

$= \begin{bmatrix} 1 & 17 \\ 12 & 36 \end{bmatrix}$

119. (a) $AB = \begin{bmatrix} 40 & 64 & 52 \\ 60 & 82 & 76 \\ 76 & 96 & 84 \end{bmatrix}\begin{bmatrix} 2.65 & 0.25 \\ 2.81 & 0.30 \\ 2.93 & 0.35 \end{bmatrix} = \begin{bmatrix} 438.2 & 47.4 \\ 612.1 & 66.2 \\ 717.28 & 77.2 \end{bmatrix}$

This is the sales and profit for milk on Friday, Saturday and Sunday.

(b) Profit $= 47.4 + 66.2 + 77.2 = \$190.80$

121. $AB = \begin{bmatrix} -4 & -1 \\ 7 & 2 \end{bmatrix}\begin{bmatrix} -2 & -1 \\ 7 & 4 \end{bmatrix} = \begin{bmatrix} 1 & 0 \\ 0 & 1 \end{bmatrix}; BA = I_2$

123. $\begin{bmatrix} -6 & 5 & \vdots & 1 & 0 \\ -5 & 4 & \vdots & 0 & 1 \end{bmatrix}$ row reduces to $\begin{bmatrix} 1 & 0 & \vdots & 4 & -5 \\ 0 & 1 & \vdots & 5 & -6 \end{bmatrix}$

$\begin{bmatrix} -6 & 5 \\ -5 & 4 \end{bmatrix}^{-1} = \begin{bmatrix} 4 & -5 \\ 5 & -6 \end{bmatrix}$

125. $\begin{bmatrix} -1 & -2 & -2 & \vdots & 1 & 0 & 0 \\ 3 & 7 & 9 & \vdots & 0 & 1 & 0 \\ 1 & 4 & 7 & \vdots & 0 & 0 & 1 \end{bmatrix}$ row reduces to $\begin{bmatrix} 1 & 0 & 0 & \vdots & 13 & 6 & -4 \\ 0 & 1 & 0 & \vdots & -12 & -5 & 3 \\ 0 & 0 & 1 & \vdots & 5 & 2 & -1 \end{bmatrix}$

$\begin{bmatrix} -1 & -2 & -2 \\ 3 & 7 & 9 \\ 1 & 4 & 7 \end{bmatrix}^{-1} = \begin{bmatrix} 13 & 6 & -4 \\ -12 & -5 & 3 \\ 5 & 2 & -1 \end{bmatrix}$

127. $\begin{bmatrix} 2 & 6 \\ 3 & -6 \end{bmatrix}^{-1} = \begin{bmatrix} \frac{1}{5} & \frac{1}{5} \\ \frac{1}{10} & -\frac{1}{15} \end{bmatrix}$

129. $\begin{bmatrix} 1 & 2 & 0 \\ -1 & 1 & 1 \\ 0 & -1 & 0 \end{bmatrix}^{-1} = \begin{bmatrix} 1 & 0 & 2 \\ 0 & 0 & -1 \\ 1 & 1 & 3 \end{bmatrix}$

131. $\begin{bmatrix} -7 & 2 \\ -8 & 2 \end{bmatrix}^{-1} = \dfrac{1}{(-7)(2) - (2)(-8)}\begin{bmatrix} 2 & -2 \\ 8 & -7 \end{bmatrix} = \begin{bmatrix} 1 & -1 \\ 4 & -\frac{7}{2} \end{bmatrix}$

133. $\begin{bmatrix} -1 & 10 \\ 2 & 20 \end{bmatrix}^{-1} = \dfrac{1}{(-1)(20) - 10(2)}\begin{bmatrix} 20 & -10 \\ -2 & -1 \end{bmatrix} = \dfrac{1}{-40}\begin{bmatrix} 20 & -10 \\ -2 & -1 \end{bmatrix} = \begin{bmatrix} -\frac{1}{2} & \frac{1}{4} \\ \frac{1}{20} & \frac{1}{40} \end{bmatrix}$

135. $\begin{cases} x + 5y = -1 \\ 3x - 5y = 5 \end{cases}$

$\begin{bmatrix} 1 & 5 \\ 3 & -5 \end{bmatrix}^{-1} = \begin{bmatrix} \frac{1}{4} & \frac{1}{4} \\ \frac{3}{20} & -\frac{1}{20} \end{bmatrix}$

$\begin{bmatrix} x \\ y \end{bmatrix} = \begin{bmatrix} \frac{1}{4} & \frac{1}{4} \\ \frac{3}{20} & -\frac{1}{20} \end{bmatrix}\begin{bmatrix} -1 \\ 5 \end{bmatrix} = \begin{bmatrix} 1 \\ -\frac{2}{5} \end{bmatrix}$

Answer: $\left(1, -\frac{2}{5}\right)$

137. $\begin{bmatrix} 3 & 2 & -1 \\ 1 & -1 & 2 \\ 5 & 1 & 1 \end{bmatrix}^{-1} = \begin{bmatrix} -1 & -1 & 1 \\ 3 & \frac{8}{3} & -\frac{7}{3} \\ 2 & \frac{7}{3} & -\frac{5}{3} \end{bmatrix}$

$\begin{bmatrix} x \\ y \\ z \end{bmatrix} = \begin{bmatrix} -1 & -1 & 1 \\ 3 & \frac{8}{3} & -\frac{7}{3} \\ 2 & \frac{7}{3} & -\frac{5}{3} \end{bmatrix}\begin{bmatrix} 6 \\ -1 \\ 7 \end{bmatrix} = \begin{bmatrix} 2 \\ -1 \\ -2 \end{bmatrix}$

Answer: $(2, -1, -2)$

139. $\begin{bmatrix} 1 & 2 & 1 & -1 \\ 2 & 1 & 1 & 1 \\ 1 & -1 & -3 & 0 \\ 0 & 0 & 1 & 1 \end{bmatrix}^{-1} = \begin{bmatrix} -1 & \frac{4}{3} & -\frac{2}{3} & -\frac{7}{3} \\ 2 & -\frac{5}{3} & \frac{4}{3} & \frac{11}{3} \\ -1 & 1 & -1 & -2 \\ 1 & -1 & 1 & 3 \end{bmatrix}^{-1}$

$\begin{bmatrix} x \\ y \\ z \\ w \end{bmatrix} = \begin{bmatrix} -1 & \frac{4}{3} & -\frac{2}{3} & -\frac{7}{3} \\ 2 & -\frac{5}{3} & \frac{4}{3} & \frac{11}{3} \\ -1 & 1 & -1 & -2 \\ 1 & -1 & 1 & 3 \end{bmatrix}\begin{bmatrix} -2 \\ 1 \\ 0 \\ 1 \end{bmatrix} = \begin{bmatrix} 1 \\ -2 \\ 1 \\ 0 \end{bmatrix}$

Answer: $(1, -2, 1, 0)$

141. $\begin{cases} x + 2y = -1 \\ 3x + 4y = -5 \end{cases}$

$\begin{bmatrix} 1 & 2 \\ 3 & 4 \end{bmatrix}^{-1} = \begin{bmatrix} -2 & 1 \\ \frac{3}{2} & -\frac{1}{2} \end{bmatrix} \Rightarrow \begin{bmatrix} x \\ y \end{bmatrix} = \begin{bmatrix} -2 & 1 \\ \frac{3}{2} & -\frac{1}{2} \end{bmatrix}\begin{bmatrix} -1 \\ -5 \end{bmatrix} = \begin{bmatrix} -3 \\ 1 \end{bmatrix}$

$x = -3, y = 1$

Answer: $(-3, 1)$

143. $\begin{cases} -3x - 3y - 4z = 2 \\ \qquad\quad y + z = -1 \\ 4x + 3y + 4z = -1 \end{cases}$

$$\begin{bmatrix} -3 & -3 & -4 \\ 0 & 1 & 1 \\ 4 & 3 & 4 \end{bmatrix}^{-1} = \begin{bmatrix} 1 & 0 & 1 \\ 4 & 4 & 3 \\ 4 & -3 & -3 \end{bmatrix} \Rightarrow \begin{bmatrix} x \\ y \\ z \end{bmatrix} = \begin{bmatrix} 1 & 0 & 1 \\ 4 & 4 & 3 \\ -4 & -3 & -3 \end{bmatrix} \begin{bmatrix} 2 \\ -1 \\ -1 \end{bmatrix} = \begin{bmatrix} 1 \\ 1 \\ -2 \end{bmatrix}$$

$x = 1, y = 1, z = -2$

Answer: $(1, 1, -2)$

145. $\begin{vmatrix} 8 & 5 \\ 2 & -4 \end{vmatrix} = 8(-4) - 2(5) = -42$

147. $\begin{vmatrix} 50 & -30 \\ 10 & 5 \end{vmatrix} = 50(5) - (-30)(10) = 550$

149. $A = \begin{bmatrix} 2 & -1 \\ 7 & 4 \end{bmatrix}$

Minors: $M_{11} = 4$ $M_{21} = -1$

            $M_{12} = 7$ $M_{22} = 2$

Cofactors: $C_{11} = 4$ $C_{21} = 1$

            $C_{12} = -7$ $C_{22} = 2$

151. $A = \begin{bmatrix} 3 & 2 & -1 \\ -2 & 5 & 0 \\ 1 & 8 & 6 \end{bmatrix}$

Minors: $M_{11} = \begin{vmatrix} 5 & 0 \\ 8 & 6 \end{vmatrix} = 30, M_{12} = \begin{vmatrix} -2 & 0 \\ 1 & 6 \end{vmatrix} = -12, M_{13} = \begin{vmatrix} -2 & 5 \\ 1 & 8 \end{vmatrix} = -21$

          $M_{21} = \begin{vmatrix} 2 & -1 \\ 8 & 6 \end{vmatrix} = 20, M_{22} = \begin{vmatrix} 3 & -1 \\ 1 & 6 \end{vmatrix} = 19, M_{23} = \begin{vmatrix} 3 & 2 \\ 1 & 8 \end{vmatrix} = 22$

          $M_{31} = \begin{vmatrix} 2 & -1 \\ 5 & 0 \end{vmatrix} = 5, M_{32} = \begin{vmatrix} 3 & -1 \\ -2 & 0 \end{vmatrix} = -2, M_{33} = \begin{vmatrix} 3 & 2 \\ -2 & 5 \end{vmatrix} = 19$

Cofactors: $C_{11} = 30, C_{12} = 12, C_{13} = -21$

            $C_{21} = -20, C_{22} = 19, C_{23} = -22$

            $C_{31} = 5, C_{32} = 2, C_{33} = 19$

153. $\begin{vmatrix} -2 & 4 & 1 \\ -6 & 0 & 2 \\ 5 & 3 & 4 \end{vmatrix} = 6 \begin{vmatrix} 4 & 1 \\ 3 & 4 \end{vmatrix} - 2 \begin{vmatrix} -2 & 4 \\ 5 & 3 \end{vmatrix} = 6(13) - 2(-26) = 130$

155. $\begin{vmatrix} 1 & 0 & -2 \\ 0 & 1 & 0 \\ -2 & 0 & 1 \end{vmatrix} = 1 \begin{vmatrix} 1 & -2 \\ -2 & 1 \end{vmatrix} = 1(1) - (-2)(-2) = 1 - 4 = -3$

157.
$$\begin{vmatrix} 3 & 0 & -4 & 0 \\ 0 & 8 & 1 & 2 \\ 6 & 1 & 8 & 2 \\ 0 & 3 & -4 & 1 \end{vmatrix} = 3\begin{vmatrix} 8 & 1 & 2 \\ 1 & 8 & 2 \\ 3 & -4 & 1 \end{vmatrix} + (-4)\begin{vmatrix} 0 & 8 & 2 \\ 6 & 1 & 2 \\ 0 & 3 & 1 \end{vmatrix}$$
(Expansion along Row 1)

$$= 3[8(8 - (-8)) - 1(1 - 6) + 2(-4 - 24)] - 4[0 - 6(8 - 6) + 0]$$

$$= 3[128 + 5 - 56] - 4[-12]$$

$$= 279$$

159. $\det(A) = 8(-1)(4)(3) = -96$ (Upper Triangular)

161. $(1, 0)$, $(5, 0)$, $(5, 8)$

$$\frac{1}{2}\begin{vmatrix} 1 & 0 & 1 \\ 5 & 0 & 1 \\ 5 & 8 & 1 \end{vmatrix} = \frac{1}{2}(32) = 16$$

Area $= 16$ square units

163. $\frac{1}{2}\begin{vmatrix} \frac{1}{2} & 1 & 1 \\ 2 & -\frac{5}{2} & 1 \\ \frac{3}{2} & 1 & 1 \end{vmatrix} = \frac{1}{2}(\frac{7}{2}) = \frac{7}{4}$

Area $= \frac{7}{4}$ square units

165. $\frac{1}{2}\begin{vmatrix} 2 & 4 & 1 \\ 5 & 6 & 1 \\ 4 & 1 & 1 \end{vmatrix} = \frac{1}{2}(-13)$

Area $= \frac{13}{2}$ square units

167. The figure is a rhombus.

$$\begin{vmatrix} -2 & -1 & 1 \\ 4 & 9 & 1 \\ -2 & -9 & 1 \end{vmatrix} = -48$$

Area $= 48$ square units

169. $\begin{vmatrix} -1 & 7 & 1 \\ 2 & 5 & 1 \\ 4 & 1 & 1 \end{vmatrix} = -8 \neq 0$

Not collinear

171. $x = \dfrac{\begin{vmatrix} 5 & 2 \\ 1 & 1 \end{vmatrix}}{\begin{vmatrix} 1 & 2 \\ -1 & 1 \end{vmatrix}} = \dfrac{3}{3} = 1$

$y = \dfrac{\begin{vmatrix} 1 & 5 \\ -1 & 1 \end{vmatrix}}{\begin{vmatrix} 1 & 2 \\ -1 & 1 \end{vmatrix}} = \dfrac{6}{3} = 2$

Answer: $(1, 2)$

173. $x = \dfrac{\begin{vmatrix} 6 & -2 \\ -23 & 3 \end{vmatrix}}{\begin{vmatrix} 5 & -2 \\ -11 & 3 \end{vmatrix}} = \dfrac{-28}{-7} = 4$

$y = \dfrac{\begin{vmatrix} 5 & 6 \\ -11 & -23 \end{vmatrix}}{\begin{vmatrix} 5 & -2 \\ -11 & 3 \end{vmatrix}} = \dfrac{-49}{-7} = 7$

Answer: $(4, 7)$

175. $x = \dfrac{\begin{vmatrix} -11 & 3 & -5 \\ -3 & -1 & 1 \\ 15 & -4 & 6 \end{vmatrix}}{\begin{vmatrix} -2 & 3 & -5 \\ 4 & -1 & 1 \\ -1 & -4 & 6 \end{vmatrix}} = \dfrac{-14}{14} = -1$

$y = \dfrac{\begin{vmatrix} -2 & -11 & -5 \\ 4 & -3 & 1 \\ -1 & 15 & 6 \end{vmatrix}}{14} = \dfrac{56}{14} = 4$

$z = \dfrac{\begin{vmatrix} -2 & 3 & -11 \\ 4 & -1 & -3 \\ -1 & -4 & 15 \end{vmatrix}}{14} = \dfrac{70}{14} = 5$

Answer: $(-1, 4, 5)$

177. $\begin{vmatrix} 1 & -3 & 2 \\ 2 & 2 & -3 \\ 1 & -7 & 8 \end{vmatrix} = 20$

$|A_1| = 0$

$|A_2| = -48$

$|A_3| = -52$

$x = 0, \; y = -\dfrac{48}{20} = -2.4, \; z = -\dfrac{52}{20} = -2.6$

Answer: $(0, -2.4, -2.6)$

179. (a) $\begin{cases} x - 3y + 2z = 5 \\ 2x + y - 4z = -1 \\ 2x + 4y + 2z = 3 \end{cases}$

$\begin{cases} x - 3y + 2z = 5 \\ 7y - 8z = -11 \\ 10y - 2z = -7 \end{cases}$

$\begin{cases} x - 3y + 2z = 5 \\ 7y - 8z = -11 \\ \frac{66}{7}z = \frac{61}{7} \end{cases}$

$z = \dfrac{61}{66}$

$7y - 8\left(\dfrac{61}{66}\right) = -11 \Rightarrow y = -\dfrac{17}{33}$

$x - 3\left(-\dfrac{17}{33}\right) + 2\left(\dfrac{61}{66}\right) = 5 \Rightarrow x = \dfrac{53}{33}$

Answer: $\left(\dfrac{53}{33}, -\dfrac{17}{33}, \dfrac{61}{66}\right)$

(b) $D = \begin{vmatrix} 1 & -3 & 2 \\ 2 & 1 & -4 \\ 2 & 4 & 2 \end{vmatrix} = 66$

$x = \dfrac{\begin{vmatrix} 5 & -3 & 2 \\ -1 & 1 & -4 \\ 3 & 4 & 2 \end{vmatrix}}{66} = \dfrac{106}{66} = \dfrac{53}{33}$

$y = \dfrac{\begin{vmatrix} 1 & 5 & 2 \\ 2 & -1 & -4 \\ 2 & 3 & 2 \end{vmatrix}}{66} = \dfrac{-34}{66} = \dfrac{-17}{33}$

$z = \dfrac{\begin{vmatrix} 1 & -3 & 5 \\ 2 & 1 & -1 \\ 2 & 4 & 3 \end{vmatrix}}{66} = \dfrac{61}{66}$

Answer: $\left(\dfrac{53}{33}, -\dfrac{17}{33}, \dfrac{61}{66}\right)$

181. I_HAVE_A_DREAM

[9 0 8] [1 22 5] [0 1 0] [4 18 5] [1 13 0]

$[9 \; 0 \; 8]A = [-30 \; -2 \; 24]$

$[1 \; 22 \; 5]A = [38 \; 8 \; -51]$

$[0 \; 1 \; 0]A = [3 \; 0 \; -3]$

$[4 \; 18 \; 5]A = [32 \; 2 \; -39]$

$[1 \; 13 \; 0]A = [41 \; -2 \; -39]$

Cryptogram: $-30 \; -2 \; 24 \; 38 \; 8 \; -51 \; 3 \; 0 \; -3$

$ 32 \; 2 \; -39 \; 41 \; -2 \; -39$

183. $A^{-1} = \begin{bmatrix} \frac{1}{2} & -\frac{1}{4} & \frac{1}{4} \\ -\frac{1}{4} & -\frac{3}{8} & -\frac{1}{8} \\ -\frac{1}{2} & -\frac{1}{4} & \frac{1}{4} \end{bmatrix}$

$\begin{bmatrix} 32 & -46 & 37 \\ 9 & -48 & 15 \\ 3 & -14 & 10 \\ -1 & -6 & 2 \\ -8 & -22 & -3 \end{bmatrix} \begin{bmatrix} \frac{1}{2} & -\frac{1}{4} & \frac{1}{4} \\ -\frac{1}{4} & -\frac{3}{8} & -\frac{1}{8} \\ -\frac{1}{2} & -\frac{1}{4} & \frac{1}{4} \end{bmatrix} = \begin{bmatrix} 9 & 0 & 23 \\ 9 & 12 & 12 \\ 0 & 2 & 5 \\ 0 & 2 & 1 \\ 3 & 11 & 0 \end{bmatrix} = \begin{matrix} I & - & W \\ I & L & L \\ - & B & E \\ - & B & A \\ C & K & - \end{matrix}$ Message: I WILL BE BACK

185. $A^{-1} = \begin{bmatrix} \frac{2}{3} & \frac{1}{3} & \frac{1}{3} \\ -\frac{1}{3} & \frac{1}{3} & \frac{1}{3} \\ \frac{1}{6} & \frac{1}{3} & -\frac{1}{6} \end{bmatrix}$

$\begin{bmatrix} 21 & -11 & 14 \\ 29 & -11 & -18 \\ 32 & -6 & -26 \\ 31 & -19 & -12 \\ 10 & 6 & 26 \\ 13 & -11 & -2 \\ 37 & 28 & -8 \\ 5 & 13 & 36 \end{bmatrix} \begin{bmatrix} \frac{2}{3} & \frac{1}{3} & \frac{1}{3} \\ -\frac{1}{3} & \frac{1}{3} & \frac{1}{3} \\ \frac{1}{6} & \frac{1}{3} & -\frac{1}{6} \end{bmatrix} = \begin{bmatrix} 20 & 8 & 1 \\ 20 & 0 & 9 \\ 19 & 0 & 13 \\ 25 & 0 & 6 \\ 9 & 14 & 1 \\ 12 & 0 & 1 \\ 14 & 19 & 23 \\ 5 & 18 & 0 \end{bmatrix} = \begin{matrix} T & H & A \\ T & - & I \\ S & - & M \\ Y & - & F \\ I & N & A \\ L & - & A \\ N & S & W \\ E & R & - \end{matrix}$ Message: THAT IS MY FINAL ANSWER

187. (a) $\begin{cases} 4b + 44a = 65 \\ 44b + 504a = 723.2 \end{cases}$

Using Cramer's Rule, you obtain,

$a = 0.41$ and $b = 11.74$

$y = 0.41t + 11.74$

(c) $y = 20$ when $t \approx 20.15$, or 2010

(b)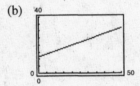

(d) $0.41t + 11.74 > 20$

$0.41t > 8.26$

$t > 20.15$

That is, year 2010.

189. True. Expansion by Row 3 gives

$\begin{vmatrix} a_{11} & a_{12} & a_{13} \\ a_{21} & a_{22} & a_{23} \\ a_{31} + c_1 & a_{32} + c_2 & a_{33} + c_3 \end{vmatrix} = (a_{31} + c_1)\begin{vmatrix} a_{12} & a_{13} \\ a_{22} & a_{23} \end{vmatrix} - (a_{32} + c_2)\begin{vmatrix} a_{11} & a_{13} \\ a_{21} & a_{23} \end{vmatrix} + (a_{33} + c_3)\begin{vmatrix} a_{11} & a_{12} \\ a_{21} & a_{22} \end{vmatrix}$

$= a_{31}\begin{vmatrix} a_{12} & a_{13} \\ a_{22} & a_{23} \end{vmatrix} - a_{32}\begin{vmatrix} a_{11} & a_{13} \\ a_{21} & a_{23} \end{vmatrix} + a_{33}\begin{vmatrix} a_{11} & a_{12} \\ a_{21} & a_{22} \end{vmatrix}$

$+ c_1\begin{vmatrix} a_{12} & a_{13} \\ a_{22} & a_{23} \end{vmatrix} - c_2\begin{vmatrix} a_{11} & a_{13} \\ a_{21} & a_{23} \end{vmatrix} + c_3\begin{vmatrix} a_{11} & a_{12} \\ a_{21} & a_{22} \end{vmatrix}$

$= \begin{vmatrix} a_{11} & a_{12} & a_{13} \\ a_{21} & a_{22} & a_{23} \\ a_{31} & a_{32} & a_{33} \end{vmatrix} + \begin{vmatrix} a_{11} & a_{12} & a_{13} \\ a_{21} & a_{22} & a_{23} \\ c_1 & c_2 & c_3 \end{vmatrix}$

Note: Expand each of these matrices by Row 3 to see the previous step.

191. A square $n \times n$ matrix has an inverse if $\det(A) \neq 0$.

Chapter 7   Practice Test

For Exercises 1–3, solve the given system by the method of substitution.

1. $x + y = 1$
$3x - y = 15$

2. $x - 3y = -3$
$x^2 + 6y = 5$

3. $x + y + z = 6$
$2x - y + 3z = 0$
$5x + 2y - z = -3$

4. Find the two numbers whose sum is 110 and product is 2800.

5. Find the dimensions of a rectangle if its perimeter is 170 feet and its area is 2800 square feet.

For Exercises 6–7, solve the linear system by elimination.

6. $2x + 15y = 4$
$x - 3y = 23$

7. $x + y = 2$
$38x - 19y = 7$

8. Use a graphing utility to graph the two equations. Use the graph to approximate the solution of the system. Verify your answer analytically.

$0.4x + 0.5y = 0.112$

$0.3x - 0.7y = -0.131$

9. Herbert invests \$17,000 in two funds that pay 11% and 13% simple interest, respectively. If he receives \$2080 in yearly interest, how much is invested in each fund?

10. Find the least squares regression line for the points $(4, 3)$, $(1, 1)$, $(-1, -2)$, and $(-2, -1)$.

For Exercises 11–13, solve the system of equations.

11. $x + y = -2$
$2x - y + z = 11$
$4y - 3z = -20$

12. $4x - y + 5z = 4$
$2x + y - z = 0$
$2x + 4y + 8z = 0$

13. $3x + 2y - z = 5$
$6x - y + 5z = 2$

14. Find the equation of the parabola $y = ax^2 + bx + c$ passing through the points $(0, -1)$, $(1, 4)$ and $(2, 13)$.

15. Find the position equation $s = \frac{1}{2}at^2 + v_0 t + s_0$ given that $s = 12$ feet after 1 second, $s = 5$ feet after 2 seconds, and $s = 4$ feet after 3 seconds.

16. Write the matrix in reduced row-echelon form.

$$\begin{bmatrix} 1 & -2 & 4 \\ 3 & -5 & 9 \end{bmatrix}$$

For Exercises 17–19, use matrices to solve the system of equations.

17. $3x + 5y = 3$
$2x - y = -11$

18. $2x + 3y = -3$
$3x + 2y = 8$
$x + y = 1$

19. $x + 3z = -5$
$2x + y = 0$
$3x + y - z = 3$

20. Multiply $\begin{bmatrix} 1 & 4 & 5 \\ 2 & 0 & -3 \end{bmatrix} \begin{bmatrix} 1 & 6 \\ 0 & -7 \\ -1 & 2 \end{bmatrix}$

21. Given $A = \begin{bmatrix} 9 & 1 \\ -4 & 8 \end{bmatrix}$ and $B = \begin{bmatrix} 6 & -2 \\ 3 & 5 \end{bmatrix}$, find $3A - 5B$.

22. Find $f(A)$:

$$f(x) = x^2 - 7x + 8, \quad A = \begin{bmatrix} 3 & 0 \\ 7 & 1 \end{bmatrix}$$

23. True or false:

$(A + B)(A + 3B) = A^2 + 4AB + 3B^2$ where A and B are matrices.

(Assume that A^2, AB, and B^2 exist.)

For Exercises 24 and 25, find the inverse of the matrix, if it exists.

24. $\begin{bmatrix} 1 & 2 \\ 3 & 5 \end{bmatrix}$

25. $\begin{bmatrix} 1 & 1 & 1 \\ 3 & 6 & 5 \\ 6 & 10 & 8 \end{bmatrix}$

26. Use an inverse matrix to solve the systems.

(a) $\begin{aligned} x + 2y &= 4 \\ 3x + 5y &= 1 \end{aligned}$ (b) $\begin{aligned} x + 2y &= 3 \\ 3x + 5y &= -2 \end{aligned}$

For Exercises 27 and 28, find the determinant of the matrix.

27. $\begin{bmatrix} 6 & -1 \\ 3 & 4 \end{bmatrix}$

28. $\begin{bmatrix} 1 & 3 & -1 \\ 5 & 9 & 0 \\ 6 & 2 & -5 \end{bmatrix}$

29. Use a graphing utility to find the determinant of the matrix.

$$\begin{bmatrix} 1 & 4 & 2 & 3 \\ 0 & 1 & -2 & 0 \\ 3 & 5 & -1 & 1 \\ 2 & 0 & 6 & 1 \end{bmatrix}$$

30. Evaluate $\begin{vmatrix} 6 & 4 & 3 & 0 & 6 \\ 0 & 5 & 1 & 4 & 8 \\ 0 & 0 & 2 & 7 & 3 \\ 0 & 0 & 0 & 9 & 2 \\ 0 & 0 & 0 & 0 & 1 \end{vmatrix}$.

31. Use a determinant to find the area of the triangle with vertices $(0, 7)$, $(5, 0)$, and $(3, 9)$.

32. Use a determinant to find the equation of the line through $(2, 7)$ and $(-1, 4)$.

For Exercises 33–35, use Cramer's Rule to find the indicated value.

33. Find x.

$\begin{aligned} 6x - 7y &= 4 \\ 2x + 5y &= 11 \end{aligned}$

34. Find z.

$\begin{aligned} 3x \quad\;\; + z &= 1 \\ y + 4z &= 3 \\ x - y \quad\;\; &= 2 \end{aligned}$

35. Find y.

$\begin{aligned} 721.4x - 29.1y &= 33.77 \\ 45.9x + 105.6y &= 19.85 \end{aligned}$

C H A P T E R 8
Sequences, Series, and Probability

Section 8.1   Sequences and Series 374

Section 8.2   Arithmetic Sequences and Partial Sums 382

Section 8.3   Geometric Sequences and Series 386

Section 8.4   Mathematical Induction 393

Section 8.5   The Binomial Theorem . . . . : . 400

Section 8.6   Counting Principles 406

Section 8.7   Probability 409

Review Exercises 412

Practice Test 419

CHAPTER 8
Sequences, Series, and Probability

Section 8.1   Sequences and Series

- ■ Given the general nth term in a sequence, you should be able to find, or list, some of the terms.
- ■ You should be able to find an expression for the nth term of a sequence.
- ■ You should be able to use and evaluate factorials.
- ■ You should be able to use sigma notation for a sum.

Vocabulary Check

1. infinite sequence
2. terms
3. finite
4. recursively
5. factorial
6. summation notation
7. index, upper limit, lower limit
8. series
9. nth partial sum

---

1. $a_n = 2n + 5$

$a_1 = 2(1) + 5 = 7$

$a_2 = 2(2) + 5 = 9$

$a_3 = 2(3) + 5 = 11$

$a_4 = 2(4) + 5 = 13$

$a_5 = 2(5) + 5 = 15$

3. $a_n = 2^n$

$a_1 = 2^1 = 2$

$a_2 = 2^2 = 4$

$a_3 = 2^3 = 8$

$a_4 = 2^4 = 16$

$a_5 = 2^5 = 32$

5. $a_n = \left(-\frac{1}{2}\right)^n$

$a_1 = \left(-\frac{1}{2}\right)^1 = -\frac{1}{2}$

$a_2 = \left(-\frac{1}{2}\right)^2 = \frac{1}{4}$

$a_3 = \left(-\frac{1}{2}\right)^3 = -\frac{1}{8}$

$a_4 = \left(-\frac{1}{2}\right)^4 = \frac{1}{16}$

$a_5 = \left(-\frac{1}{2}\right)^5 = -\frac{1}{32}$

7. $a_n = \dfrac{n + 1}{n}$

$a_1 = \dfrac{1 + 1}{1} = 2$

$a_2 = \dfrac{3}{2}$

$a_3 = \dfrac{4}{3}$

$a_4 = \dfrac{5}{4}$

$a_5 = \dfrac{6}{5}$

9. $a_n = \dfrac{n}{n^2 + 1}$

$a_1 = \dfrac{1}{1^2 + 1} = \dfrac{1}{2}$

$a_2 = \dfrac{2}{2^2 + 1} = \dfrac{2}{5}$

$a_3 = \dfrac{3}{3^2 + 1} = \dfrac{3}{10}$

$a_4 = \dfrac{4}{4^2 + 1} = \dfrac{4}{17}$

$a_5 = \dfrac{5}{5^2 + 1} = \dfrac{5}{26}$

11. $a_n = \dfrac{1 + (-1)^n}{n}$

$a_1 = 0$

$a_2 = \dfrac{2}{2} = 1$

$a_3 = 0$

$a_4 = \dfrac{2}{4} = \dfrac{1}{2}$

$a_5 = 0$

13. $a_n = 1 - \dfrac{1}{2^n}$

$a_1 = 1 - \dfrac{1}{2^1} = \dfrac{1}{2}$

$a_2 = 1 - \dfrac{1}{2^2} = 1 - \dfrac{1}{4} = \dfrac{3}{4}$

$a_3 = 1 - \dfrac{1}{2^3} = \dfrac{7}{8}$

$a_4 = 1 - \dfrac{1}{2^4} = \dfrac{15}{16}$

$a_5 = 1 - \dfrac{1}{2^5} = \dfrac{31}{32}$

15. $a_n = \dfrac{1}{n^{3/2}}$

$a_1 = \dfrac{1}{1} = 1$

$a_2 = \dfrac{1}{2^{3/2}}$

$a_3 = \dfrac{1}{3^{3/2}}$

$a_4 = \dfrac{1}{4^{3/2}} = \dfrac{1}{8}$

$a_5 = \dfrac{1}{5^{3/2}}$

17. $a_n = \dfrac{(-1)^n}{n^2}$

$a_1 = \dfrac{-1}{1} = -1$

$a_2 = \dfrac{1}{4}$

$a_3 = \dfrac{-1}{9}$

$a_4 = \dfrac{1}{16}$

$a_5 = \dfrac{-1}{25}$

19. $a_n = (2n - 1)(2n + 1)$

$a_1 = (1)(3) = 3$

$a_2 = (3)(5) = 15$

$a_3 = (5)(7) = 35$

$a_4 = (7)(9) = 63$

$a_5 = (9)(11) = 99$

21. $a_{25} = (-1)^{25}[3(25) - 2]$

$\qquad = -73$

23. $a_{10} = \dfrac{10^2}{10^2 + 1} = \dfrac{100}{101}$

25. $a_6 = \dfrac{2^6}{2^6 + 1} = \dfrac{64}{65}$

27. $a_n = \dfrac{2}{3}n$

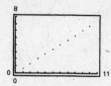

29. $a_n = 16(-0.5)^{n-1}$

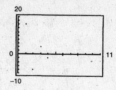

31. $a_n = \dfrac{2n}{n + 1}$

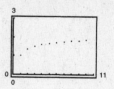

33. $a_n = 2(3n - 1) + 5$

n	1	2	3	4	5	6	7	8	9	10
a_n	9	15	21	27	33	39	45	51	57	63

35. $a_n = 1 + \dfrac{n + 1}{n}$

n	1	2	3	4	5	6	7	8	9	10
a_n	3	2.5	2.33	2.25	2.2	2.17	2.14	2.13	2.11	2.1

37. $a_n = (-1)^n + 1$

n	1	2	3	4	5	6	7	8	9	10
a_n	0	2	0	2	0	2	0	2	0	2

39. $a_n = \dfrac{8}{n+1}$

$a_n \to 0$ as $n \to \infty$

$a_1 = 4,\ a_{10} = \dfrac{8}{11}$

Matches graph (c).

41. $a_n = 4(0.5)^{n-1}$

$a_n \to 0$ as $n \to \infty$

$a_1 = 4,\ a_{10} \approx 0.008$

Matches graph (d).

43. $1, 4, 7, 10, 13, \ldots$

$a_n = 1 + (n-1)3 = 3n - 2$

45. $0, 3, 8, 15, 24, \ldots$

$a_n = n^2 - 1$

47. $\dfrac{2}{3}, \dfrac{3}{4}, \dfrac{4}{5}, \dfrac{5}{6}, \dfrac{6}{7}, \ldots$

$a_n = \dfrac{n+1}{n+2}$

49. $\dfrac{1}{2}, \dfrac{-1}{4}, \dfrac{1}{8}, \dfrac{-1}{16}, \ldots$

$a_n = \dfrac{(-1)^{n+1}}{2^n}$

51. $1 + \dfrac{1}{1}, 1 + \dfrac{1}{2}, 1 + \dfrac{1}{3}, 1 + \dfrac{1}{4}, 1 + \dfrac{1}{5}, \ldots$

$a_n = 1 + \dfrac{1}{n}$

53. $1, \dfrac{1}{2}, \dfrac{1}{6}, \dfrac{1}{24}, \dfrac{1}{120}, \ldots$

$a_n = \dfrac{1}{n!}$

55. $1, 3, 1, 3, 1, 3, \ldots$

$a_n = 2 + (-1)^n$

57. $a_1 = 28$ and $a_{k+1} = a_k - 4$

$a_1 = 28$

$a_2 = a_1 - 4 = 28 - 4 = 24$

$a_3 = a_2 - 4 = 24 - 4 = 20$

$a_4 = a_3 - 4 = 20 - 4 = 16$

$a_5 = a_4 - 4 = 16 - 4 = 12$

59. $a_1 = 3$ and $a_{k+1} = 2(a_k - 1)$

$a_1 = 3$

$a_2 = 2(a_1 - 1) = 2(3 - 1) = 4$

$a_3 = 2(a_2 - 1) = 2(4 - 1) = 6$

$a_4 = 2(a_3 - 1) = 2(6 - 1) = 10$

$a_5 = 2(a_4 - 1) = 2(10 - 1) = 18$

61. $a_1 = 6$ and $a_{k+1} = a_k + 2$

$a_1 = 6$

$a_2 = a_1 + 2 = 6 + 2 = 8$

$a_3 = a_2 + 2 = 8 + 2 = 10$

$a_4 = a_3 + 2 = 10 + 2 = 12$

$a_5 = a_4 + 2 = 12 + 2 = 14$

In general, $a_n = 2n + 4$.

63. $a_1 = 81$ and $a_{k+1} = \frac{1}{3}a_k$

$a_1 = 81$

$a_2 = \frac{1}{3}a_1 = \frac{1}{3}(81) = 27$

$a_3 = \frac{1}{3}a_2 = \frac{1}{3}(27) = 9$

$a_4 = \frac{1}{3}a_3 = \frac{1}{3}(9) = 3$

$a_5 = \frac{1}{3}a_4 = \frac{1}{3}(3) = 1$

In general, $a_n = 81\left(\frac{1}{3}\right)^{n-1} = 81(3)\left(\frac{1}{3}\right)^n = \frac{243}{3^n}$.

65. $a_n = \frac{1}{n!}$

$a_0 = \frac{1}{0!} = 1$

$a_1 = \frac{1}{1!} = 1$

$a_2 = \frac{1}{2}$

$a_3 = \frac{1}{3!} = \frac{1}{6}$

$a_4 = \frac{1}{4!} = \frac{1}{24}$

67. $a_n = \frac{n!}{2n + 1}$

$a_0 = \frac{0!}{1} = 1$

$a_1 = \frac{1!}{2 + 1} = \frac{1}{3}$

$a_2 = \frac{2!}{4 + 1} = \frac{2}{5}$

$a_3 = \frac{3!}{6 + 1} = \frac{6}{7}$

$a_4 = \frac{4!}{8 + 1} = \frac{24}{9} = \frac{8}{3}$

69. $a_n = \frac{(-1)^{2n}}{(2n)!}$

$a_0 = \frac{(-1)^0}{0!} = 1$

$a_1 = \frac{(-1)^2}{2!} = \frac{1}{2}$

$a_2 = \frac{(-1)^4}{4!} = \frac{1}{24}$

$a_3 = \frac{(-1)^6}{6!} = \frac{1}{720}$

$a_4 = \frac{(-1)^8}{8!} = \frac{1}{40,320}$

71. $\dfrac{2!}{4!} = \dfrac{2!}{4 \cdot 3 \cdot 2!} = \dfrac{1}{12}$

73. $\dfrac{12!}{4!8!} = \dfrac{12 \cdot 11 \cdot 10 \cdot 9 \cdot 8!}{4!8!}$

$= \dfrac{12 \cdot 11 \cdot 10 \cdot 9}{4 \cdot 3 \cdot 2} = 495$

75. $\dfrac{(n + 1)!}{n!} = \dfrac{(n + 1)n!}{n!} = n + 1$

77. $\dfrac{(2n - 1)!}{(2n + 1)!} = \dfrac{(2n - 1)!}{(2n + 1)(2n)(2n - 1)!}$

$= \dfrac{1}{2n(2n + 1)}$

79. $\displaystyle\sum_{i=1}^{5}(2i + 1) = (2 + 1) + (4 + 1) + (6 + 1) + (8 + 1) + (10 + 1) = 35$

81. $\displaystyle\sum_{k=1}^{4}10 = 10 + 10 + 10 + 10 = 40$

83. $\displaystyle\sum_{i=0}^{4}i^2 = 0^2 + 1^2 + 2^2 + 3^2 + 4^2 = 30$

85. $\displaystyle\sum_{k=0}^{3}\frac{1}{k^2 + 1} = \frac{1}{1} + \frac{1}{1 + 1} + \frac{1}{4 + 1} + \frac{1}{9 + 1} = \frac{9}{5}$

87. $\displaystyle\sum_{i=1}^{4}[(i-1)^2 + (i+1)^3] = [(0)^2 + (2)^3] + [(1)^2 + (3)^3] + [(2)^2 + (4)^3] + [(3)^2 + (5)^3] = 238$

89. $\displaystyle\sum_{i=1}^{4} 2^i = 2^1 + 2^2 + 2^3 + 2^4$
$= 30$

91. $\displaystyle\sum_{j=1}^{6}(24 - 3j) = 81$

93. $\displaystyle\sum_{k=0}^{4}\frac{(-1)^k}{k+1} = \frac{47}{60}$

95. $\displaystyle\frac{1}{3(1)} + \frac{1}{3(2)} + \frac{1}{3(3)} + \cdots + \frac{1}{3(9)} = \sum_{i=1}^{9}\frac{1}{3i} \approx 0.94299$

97. $\displaystyle\left[2\left(\frac{1}{8}\right) + 3\right] + \left[2\left(\frac{2}{8}\right) + 3\right] + \left[2\left(\frac{3}{8}\right) + 3\right] + \cdots + \left[2\left(\frac{8}{8}\right) + 3\right] = \sum_{i=1}^{8}\left[2\left(\frac{i}{8}\right) + 3\right] = 33$

99. $\displaystyle 3 - 9 + 27 - 81 + 243 - 729 = \sum_{i=1}^{6}(-1)^{i+1}3^i = -546$

101. $\displaystyle\frac{1}{1^2} - \frac{1}{2^2} + \frac{1}{3^2} - \frac{1}{4^2} + \cdots - \frac{1}{20^2} = \sum_{i=1}^{20}\frac{(-1)^{i+1}}{i^2} \approx 0.82128$

103. $\displaystyle\frac{1}{4} + \frac{3}{8} + \frac{7}{16} + \frac{15}{32} + \frac{31}{64} = \sum_{i=1}^{5}\frac{2^i - 1}{2^{i+1}} = \frac{129}{64} = 2.015625$

105. $\displaystyle\sum_{i=1}^{4} 5\left(\frac{1}{2}\right)^i = 4.6875 = \frac{75}{16}$

107. $\displaystyle\sum_{n=1}^{3} 4\left(-\frac{1}{2}\right)^n = -1.5 = -\frac{3}{2}$

109. (a) $\displaystyle\sum_{i=1}^{4} 6\left(\frac{1}{10}\right)^i = 6\left(\frac{1}{10}\right) + 6\left(\frac{1}{10}\right)^2 + 6\left(\frac{1}{10}\right)^3 + 6\left(\frac{1}{10}\right)^4$
$= 0.6666$
$= \dfrac{3333}{5000}$

(b) $\displaystyle\sum_{i=1}^{\infty} 6\left(\frac{1}{10}\right)^i = 6[0.1 + 0.01 + 0.001 + \ldots]$
$= 6[0.111\ldots]$
$= 0.666\ldots$
$= \dfrac{2}{3}$

111. (a) $\displaystyle\sum_{k=1}^{4}\left(\frac{1}{10}\right)^k = \frac{1}{10} + \frac{1}{100} + \frac{1}{1000} + \frac{1}{10,000}$
$= 0.1111$
$= \dfrac{1111}{10,000}$

(b) $\displaystyle\sum_{k=1}^{\infty}\left(\frac{1}{10}\right)^k = 0.1 + 0.01 + 0.001 + \ldots$
$= 0.111\ldots$
$= \dfrac{1}{9}$

113. $A_n = 5000\left(1 + \dfrac{0.03}{4}\right)^n$, $n = 1, 2, 3, \dots$

(a) $A_1 = 5000\left(1 + \dfrac{0.03}{4}\right)^1 = \$5037.50$

$A_2 \approx \$5075.28$ $A_3 \approx \$5113.35$

$A_4 \approx \$5151.70$ $A_5 \approx \$5190.33$

$A_6 \approx \$5229.26$ $A_7 \approx \$5268.48$

$A_8 \approx \$5307.99$

(b) $A_{40} \approx \$6741.74$

115. (a) $p_0 = 5500$ (year 2008)

$p_n = 0.75p_{n-1} + 500$

(b) $p_1 = 0.75p_0 + 500 = 4625$ (2009)

$p_2 = 0.75p_1 + 500 \approx 3969$ (2010)

$p_3 = 0.75p_2 + 500 \approx 3477$ (2011)

(Answers will vary slightly.)

(c) The population approaches 2000 trout because $0.75(2000) + 500 = 2000$.

117. (a) $a_0 = 50$ (end of January)

$a_n = \left(1 + \dfrac{0.06}{12}\right)a_{n-1} + 50 = 1.005a_{n-1} + 50$

(b) $a_1 = 1.005\,a_0 + 50 = 100.25$ (end of February)

$a_2 = 1.005\,a_1 + 50 = 150.75$

$a_3 = 201.51$ $a_4 = 252.51$

$a_5 = 303.78$ $a_6 = 355.29$

$a_7 = 407.07$ $a_8 = 459.11$

$a_9 = 511.40$ $a_{10} = 563.96$

$a_{11} = 616.78$ (end of December)

After one year, the IRA has \$616.78.

(c) After 50 deposits, $a_{49} \approx \$2832.26$.

119. (a)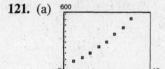

(b) For 2010, $n = 20$ and $r_{20} \approx \$12.25$.

For 2015, $n = 25$ and $r_{25} \approx \$12.64$.

(c) Answers will vary.

(d) $r_{18} = 11.97$ and $r_{19} = 12.12$. So, the average hourly wage reaches \$12 in 2008.

121. (a)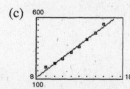

(b) Linear: $R_n = 54.58n - 336.3$

Quadratic: $R_n = 3.088n^2 - 22.62n + 130.0$

Coefficient of determination for linear model: 0.98656

Coefficient of determination for quadratic model: 0.99919.

(c)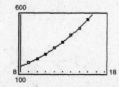

(d) The quadratic model is better.

The quadratic model is better because its coefficient of determination is closer to 1.

(e) For 2010, $n = 20$ and $R_{20} \approx 912.8$ million.

For 2015, $n = 25$ and $R_{25} \approx 1494.5$ million.

(f) $R_n = 1000$ when $n \approx 20.8$, or in 2010.

123. True

125. $a_0 = 1$, $a_1 = 1$, $a_{k+2} = a_{k+1} + a_k$

$a_0 = 1$	$b_0 = \frac{1}{1} = 1$	$a_6 = 8 + 5 = 13$	$b_6 = \frac{21}{13}$
$a_1 = 1$	$b_1 = \frac{2}{1} = 2$	$a_7 = 13 + 8 = 21$	$b_7 = \frac{34}{21}$
$a_2 = 1 + 1 = 2$	$b_2 = \frac{3}{2}$	$a_8 = 21 + 13 = 34$	$b_8 = \frac{55}{34}$
$a_3 = 2 + 1 = 3$	$b_3 = \frac{5}{3}$	$a_9 = 34 + 21 = 55$	$b_9 = \frac{89}{55}$
$a_4 = 3 + 2 = 5$	$b_4 = \frac{8}{5}$	$a_{10} = 55 + 34 = 89$	
$a_5 = 5 + 3 = 8$	$b_5 = \frac{13}{8}$	$a_{11} = 89 + 55 = 144$	

127. $a_n = \dfrac{\left(1 + \sqrt{5}\right)^n - \left(1 - \sqrt{5}\right)^n}{2^n \sqrt{5}}$

$a_1 = \dfrac{\left(1 + \sqrt{5}\right)^1 - \left(1 - \sqrt{5}\right)^1}{2^1 \sqrt{5}} = 1$

$a_2 = 1, \quad a_3 = 2$

$a_4 = 3, \quad a_5 = 5$

129. $a_{n+1} = \dfrac{\left(1 + \sqrt{5}\right)^{n+1} - \left(1 - \sqrt{5}\right)^{n+1}}{2^{n+1} \sqrt{5}}$

$a_{n+2} = \dfrac{\left(1 + \sqrt{5}\right)^{n+2} - \left(1 - \sqrt{5}\right)^{n+2}}{2^{n+2} \sqrt{5}}$

131. $a_n = \dfrac{x^n}{n!}$

$a_1 = \dfrac{x}{1} = x$

$a_2 = \dfrac{x^2}{2!} = \dfrac{x^2}{2}$

$a_3 = \dfrac{x^3}{3!} = \dfrac{x^3}{6}$

$a_4 = \dfrac{x^4}{4!} = \dfrac{x^4}{24}$

$a_5 = \dfrac{x^5}{5!} = \dfrac{x^5}{120}$

133. $a_n = \dfrac{(-1)^n x^{2n+1}}{2n + 1}$

$a_1 = \dfrac{-x^3}{3}$

$a_2 = \dfrac{x^5}{5}$

$a_3 = -\dfrac{x^7}{7}$

$a_4 = \dfrac{x^9}{9}$

$a_5 = \dfrac{-x^{11}}{11}$

135. $a_n = \dfrac{(-1)^n x^{2n}}{(2n)!}$

$a_1 = \dfrac{-x^2}{2}$

$a_2 = \dfrac{x^4}{4!} = \dfrac{x^4}{24}$

$a_3 = \dfrac{-x^6}{6!} = \dfrac{-x^6}{720}$

$a_4 = \dfrac{x^8}{8!} = \dfrac{x^8}{40,320}$

$a_5 = \dfrac{-x^{10}}{10!} = \dfrac{-x^{10}}{3,628,800}$

137. $a_n = \dfrac{(-1)^n x^n}{n!}$

$a_1 = -x$

$a_2 = \dfrac{x^2}{2}$

$a_3 = \dfrac{-x^3}{3!} = \dfrac{-x^3}{6}$

$a_4 = \dfrac{x^4}{4!} = \dfrac{x^4}{24}$

$a_5 = \dfrac{-x^5}{5!} = -\dfrac{x^5}{120}$

139. $a_n = \dfrac{(-1)^{n+1}(x + 1)^n}{n!}$

$a_1 = x + 1$

$a_2 = \dfrac{-(x + 1)^2}{2}$

$a_3 = \dfrac{(x + 1)^3}{6}$

$a_4 = -\dfrac{(x + 1)^4}{24}$

$a_5 = \dfrac{(x + 1)^5}{120}$

141. $a_n = \dfrac{1}{2n} - \dfrac{1}{2n + 2}$

$a_1 = \dfrac{1}{2} - \dfrac{1}{4} = \dfrac{1}{4}$

$a_2 = \dfrac{1}{4} - \dfrac{1}{6} = \dfrac{1}{12}$

$a_3 = \dfrac{1}{6} - \dfrac{1}{8} = \dfrac{1}{24}$

$a_4 = \dfrac{1}{8} - \dfrac{1}{10} = \dfrac{1}{40}$

$a_5 = \dfrac{1}{10} - \dfrac{1}{12} = \dfrac{1}{60}$

nth partial sum $= \left(\dfrac{1}{2} - \dfrac{1}{4}\right) + \left(\dfrac{1}{4} - \dfrac{1}{6}\right) + \cdots + \left(\dfrac{1}{2n} - \dfrac{1}{2n + 2}\right) = \dfrac{1}{2} - \dfrac{1}{2n + 2}$

143. $a_n = \dfrac{1}{n + 1} - \dfrac{1}{n + 2}$

$a_1 = \dfrac{1}{2} - \dfrac{1}{3} = \dfrac{1}{6}$

$a_2 = \dfrac{1}{3} - \dfrac{1}{4} = \dfrac{1}{12}$

$a_3 = \dfrac{1}{4} - \dfrac{1}{5} = \dfrac{1}{20}$

$a_4 = \dfrac{1}{5} - \dfrac{1}{6} = \dfrac{1}{30}$

$a_5 = \dfrac{1}{6} - \dfrac{1}{7} = \dfrac{1}{42}$

nth partial sum $= \left(\dfrac{1}{2} - \dfrac{1}{3}\right) + \left(\dfrac{1}{3} - \dfrac{1}{4}\right) + \cdots + \left(\dfrac{1}{n + 1} - \dfrac{1}{n + 2}\right) = \dfrac{1}{2} - \dfrac{1}{n + 2}$

145. $a_n = \ln n$

$a_1 = \ln 1 = 0$

$a_2 = \ln 2$

$a_3 = \ln 3$

$a_4 = \ln 4$

$a_5 = \ln 5$

nth partial sum $= \ln 2 + \ln 3 + \cdots + \ln n$

$\qquad\qquad\qquad = \ln(2 \cdot 3 \cdots n)$

$\qquad\qquad\qquad = \ln(n!)$

147. (a) $A - B = \begin{bmatrix} 8 & 1 \\ -3 & 7 \end{bmatrix}$

(b) $2B - 3A = \begin{bmatrix} -22 & -7 \\ 3 & -18 \end{bmatrix}$

(c) $AB = \begin{bmatrix} 18 & 9 \\ 18 & 0 \end{bmatrix}$

(d) $BA = \begin{bmatrix} 0 & 6 \\ 27 & 18 \end{bmatrix}$

149. (a) $A - B = \begin{bmatrix} -3 & -7 & 4 \\ 4 & 4 & 1 \\ 1 & 4 & 3 \end{bmatrix}$

(b) $2B - 3A = \begin{bmatrix} 8 & 17 & -14 \\ -12 & -13 & -9 \\ -3 & -15 & -10 \end{bmatrix}$

(c) $AB = \begin{bmatrix} -2 & 7 & -16 \\ 4 & 42 & 45 \\ 1 & 23 & 48 \end{bmatrix}$

(d) $BA = \begin{bmatrix} 16 & 31 & 42 \\ 10 & 47 & 31 \\ 13 & 22 & 25 \end{bmatrix}$

Section 8.2   Arithmetic Sequences and Partial Sums

■ You should be able to recognize an arithmetic sequence, find its common difference, and find its nth term.

■ You should be able to find the nth partial sum of an arithmetic sequence with common difference d using the formula

$$S_n = \frac{n}{2}(a_1 + a_n).$$

Vocabulary Check

1. arithmetic, common

2. $a_n = dn + c$

3. nth partial sum

1. 10, 8, 6, 4, 2, . . .

Arithmetic sequence, $d = -2$

3. $3, \frac{5}{2}, 2, \frac{3}{2}, 1, \ldots$

Arithmetic sequence, $d = -\frac{1}{2}$

5. $-24, -16, -8, 0, 8$

Arithmetic sequence, $d = 8$

7. 3.7, 4.3, 4.9, 5.5, 6.1, . . .

Arithmetic sequence, $d = 0.6$

9. $a_n = 8 + 13n$

21, 34, 47, 60, 73

Arithmetic sequence, $d = 13$

11. $a_n = \dfrac{1}{n + 1}$

$\dfrac{1}{2}, \dfrac{1}{3}, \dfrac{1}{4}, \dfrac{1}{5}, \dfrac{1}{6}$

Not an arithmetic sequence

13. $a_n = 150 - 7n$

143, 136, 129, 122, 115

Arithmetic sequence, $d = -7$

15. $a_n = 3 + 2(-1)^n$

1, 5, 1, 5, 1

Not an arithmetic sequence

17. $a_1 = 1$, $d = 3$

$a_n = a_1 + (n - 1)d = 1 + (n - 1)(3) = 3n - 2$

19. $a_1 = 100$, $d = -8$

$a_n = a_1 + (n - 1)d$

$\qquad = 100 + (n - 1)(-8) = 108 - 8n$

21. $4, \frac{3}{2}, -1, -\frac{7}{2}, \ldots, d = -\frac{5}{2}$

$a_n = a_1 + (n - 1)d = 4 + (n - 1)\left(-\frac{5}{2}\right) = \frac{13}{2} - \frac{5}{2}n$

23. $a_1 = 5$, $a_4 = 15$

$a_4 = a_1 + 3d \Rightarrow 15 = 5 + 3d \Rightarrow d = \frac{10}{3}$

$a_n = a_1 + (n - 1)d = 5 + (n - 1)\left(\frac{10}{3}\right) = \frac{10}{3}n + \frac{5}{3}$

25. $a_3 = 94, \ a_6 = 85$

$a_6 = a_3 + 3d \implies 85 = 94 + 3d \implies d = -3$

$a_1 = a_3 - 2d \implies a_1 = 94 - 2(-3) = 100$

$a_n = a_1 + (n-1)d$

$\quad = 100 + (n-1)(-3) = 103 - 3n$

27. $a_1 = 5, \ d = 6$

$a_1 = 5$

$a_2 = 5 + 6 = 11$

$a_3 = 11 + 6 = 17$

$a_4 = 17 + 6 = 23$

$a_5 = 23 + 6 = 29$

29. $a_1 = -10, \ d = -12$

$a_1 = -10$

$a_2 = -10 - 12 = -22$

$a_3 = -22 - 12 = -34$

$a_4 = -34 - 12 = -46$

$a_5 = -46 - 12 = -58$

31. $a_8 = 26, \ a_{12} = 42$

$26 = a_8 = a_1 + (n-1)d = a_1 + 7d$

$42 = a_{12} = a_1 + (n-1)d = a_1 + 11d$

Answer: $d = 4, \ a_1 = -2$

$a_1 = -2$

$a_2 = -2 + 4 = 2$

$a_3 = 2 + 4 = 6$

$a_4 = 6 + 4 = 10$

$a_5 = 10 + 4 = 14$

33. $a_3 = 19, a_{15} = -1.7$

$a_{15} = a_3 + 12d$

$-1.7 = 19 + 12d \implies d = -1.725$

$a_3 = a_1 + 2d \implies 19 = a_1 + 2(-1.725)$

$\qquad \implies a_1 = 22.45$

$a_2 = a_1 - 1.725 = 20.725$

$a_3 = 19$

$a_4 = 19 - 1.725 = 17.275$

$a_5 = 17.275 - 1.725 = 15.55$

35. $a_1 = 15, \ a_{k+1} = a_k + 4$

$a_2 = a_1 + 4 = 15 + 4 = 19$

$a_3 = 19 + 4 = 23$

$a_4 = 23 + 4 = 27$

$a_5 = 27 + 4 = 31$

$d = 4, \ a_n = 11 + 4n$

37. $a_1 = \frac{3}{5}, a_{k+1} = -\frac{1}{10} + a_k$

$a_2 = -\frac{1}{10} + \frac{3}{5} = \frac{5}{10} = \frac{1}{2}$

$a_3 = -\frac{1}{10} + \frac{1}{2} = \frac{4}{10} = \frac{2}{5}$

$a_4 = -\frac{1}{10} + \frac{2}{5} = \frac{3}{10}$

$a_5 = -\frac{1}{10} + \frac{3}{10} = \frac{1}{5}$

$d = -\frac{1}{10}$

$a_n = \frac{7}{10} - \frac{1}{10}n$

39. $a_1 = 5, a_2 = 11 \implies d = 6$

$a_{10} = a_1 + 9d = 5 + 9(6) = 59$

41. $a_1 = 4.2, a_2 = 6.6 \implies d = 2.4$

$a_7 = a_1 + 6d = 4.2 + 6(2.4) = 18.6$

43. $a_n = 15 - \frac{3}{2}n$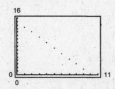

45. $a_n = 0.5n + 4$

47. $a_n = 4n - 5$

n	1	2	3	4	5	6	7	8	9	10
a_n	-1	3	7	11	15	19	23	27	31	35

49. $a_n = 20 - \frac{3}{4}n$

n	1	2	3	4	5	6	7	8	9	10
a_n	19.25	18.5	17.75	17	16.25	15.5	14.75	14	13.25	12.5

51. $a_n = 1.5 + 0.05n$

n	1	2	3	4	5	6	7	8	9	10
a_n	1.55	1.6	1.65	1.7	1.75	1.8	1.85	1.9	1.95	2.0

53. $S_{10} = \frac{10}{2}(2 + 20) = 110$

55. $S_5 = \frac{5}{2}(-1 + (-9)) = -25$

57. $S_{50} = \frac{50}{2}(2 + 100) = 2550$

59. $S_{131} = \frac{131}{2}(-100 + 30) = -4585$

61. $8, 20, 32, 44, \ldots n = 10$

$a_1 = 8, a_2 = 20 \implies d = 12$

$a_{10} = a_1 + 9d = 8 + 9(12) = 116$

$S_{10} = \frac{n}{2}[a_1 + a_{10}] = \frac{10}{2}[8 + 116] = 620$

63. $a_1 = 0.5, a_2 = 1.3 \implies d = 0.8$

$a_{10} = a_1 + 9d = 0.5 + 9(0.8) = 7.7$

$S_{10} = \frac{10}{2}(a_1 + a_{10}) = 5(0.5 + 7.7) = 41$

65. $a_1 = 100, a_{25} = 220$

$S_{25} = \frac{25}{2}(a_1 + a_{25}) = 12.5(100 + 220) = 4000$

67. $a_1 = 1, a_{50} = 50, n = 50$

$\sum_{n=1}^{50} n = \frac{50}{2}(1 + 50) = 1275$

69. $a_1 = 5, a_{100} = 500, n = 100$

$\sum_{n=1}^{100} 5n = \frac{100}{2}(5 + 500) = 25{,}250$

71. $\sum_{n=11}^{30} n - \sum_{n=1}^{10} n = \frac{20}{2}(11 + 30) - \frac{10}{2}(1 + 10)$

$= 410 - 55 = 355$

73. $\sum_{n=1}^{500} (n + 8) = \frac{500}{2}[9 + 508] = 129{,}250$

75. $\sum_{n=1}^{20} (2n + 1) = 440$

77. $\sum_{n=1}^{100} \frac{n + 1}{2} = 2575$

79. $\sum_{i=1}^{60} \left(250 - \frac{2}{5}i\right) = 14{,}268$

81. $a_1 = 14, a_{18} = 31$

$S_{18} = \frac{18}{2}(14 + 31) = 405$ bricks

83. $a_1 = 20,000$

$a_2 = 20,000 + 5000 = 25,000$

$d = 5000$

$a_5 = 20,000 + 4(5000) = 40,000$

$S_5 = \frac{5}{2}(20,000 + 40,000) = 150,000$

85. (a) $S_n = 0.91n + 5.7$

(b)

Year	1997	1998	1999	2000	2001	2002	2003	2004
Sales (Billions of $)	12.1	13.0	13.9	14.8	15.7	16.6	17.5	18.4

The model is a good fit.

(c) Total $= \frac{8}{2}(12.1 + 18.4) = \$122$ billion

(d) For 2005, $n = 15$ and $S_{15} = 19.35$.

For 2012, $n = 22$ and $S_{22} = 25.72$.

Total $= \frac{8}{2}(19.35 + 25.72) = \$180.3$ billion

Answers will vary.

87. True. Given a_1 and a_2, you know $d = a_2 - a_1$. Thus, $a_n = a_1 + (n - 1)d$.

89. $a_1 = x$ $a_6 = 11x$

$a_2 = x + 2x = 3x$ $a_7 = 13x$

$a_3 = 3x + 2x = 5x$ $a_8 = 15x$

$a_4 = 7x$ $a_9 = 17x$

$a_5 = 9x$ $a_{10} = 19x$

91. $a_{20} = a_1 + 19(3) = a_1 + 57$

$$S = \frac{n}{2}(a_1 + a_{20})$$

$$= \frac{20}{2}(a_1 + (a_1 + 57)) = 650$$

$$10(2a_1 + 57) = 650$$

$$20a_1 = 80$$

$$a_1 = 4$$

93. (a) $-7, -4, -1, 2, 5, 8, 11$

$a_{n+1} = a_n + 3, \; a_1 = -7$

(b) $17, 23, 29, 35, 41, 47, 53, 57$

$a_{n+1} = a_n + 6, \; a_1 = 17$

(c) Not arithmetic

(d) $4, 7.5, 11, 14.5, 18, 21.5, 25, 28.5$

$a_{n+1} = a_n + 3.5, \; a_1 = 4$

(e) Not arithmetic

95. $S = \dfrac{n(n + 1)}{2} = \dfrac{200(201)}{2} = 20,100$

97. $S = 1 + 3 + 5 + \cdots + 101$

$= (1 + 2 + 3 + \cdots + 101) - (2 + 4 + \cdots + 100)$

$= \dfrac{101(102)}{2} - 2\left(\dfrac{50(51)}{2}\right)$

$= 5151 - 2550 = 2601$

99. $\begin{bmatrix} 2 & -1 & 7 & \vdots & -10 \\ 3 & 2 & -4 & \vdots & 17 \\ 6 & -5 & 1 & \vdots & -20 \end{bmatrix}$ row reduces to $\begin{bmatrix} 1 & 0 & 0 & \vdots & 1 \\ 0 & 1 & 0 & \vdots & 5 \\ 0 & 0 & 1 & \vdots & -1 \end{bmatrix}$.

Answer: $(1, 5, -1)$

101. $\begin{vmatrix} 0 & 0 & 1 \\ 4 & -3 & 1 \\ 2 & 6 & 1 \end{vmatrix} = 30$

103. Answers will vary.

Area $= \frac{1}{2}(30) = 15$ square units

Section 8.3   Geometric Sequences and Series

- You should be able to identify a geometric sequence, find its common ratio, and find the nth term.
- You should be able to find the nth partial sum of a geometric sequence with common ratio r using the formula.

  $$S_n = a_1\left(\frac{1 - r^n}{1 - r}\right)$$

- You should know that if $|r| < 1$, then

  $$\sum_{n=1}^{\infty} a_1 r^{n-1} = \frac{a_1}{1 - r}.$$

Vocabulary Check

1. geometric, common

2. $a_n = a_1 r^{n-1}$

3. $S_n = \sum_{i=1}^{n} a_1 r^{i-1} = a_1\left(\frac{1 - r^n}{1 - r}\right)$

4. geometric series

5. $S = \sum_{i=0}^{\infty} a_1 r^i = \frac{a_1}{1 - r}$

---

1. $5, 15, 45, 135, \ldots$

Geometric sequence

$r = 3$

3. $6, 18, 30, 42, \ldots$

Not a geometric sequence

(**Note:** It is an arithmetic sequence with $d = 12$.)

5. $1, -\frac{1}{2}, \frac{1}{4}, -\frac{1}{8}, \ldots$

Geometric sequence

$r = -\frac{1}{2}$

7. $\frac{1}{8}, \frac{1}{4}, \frac{1}{2}, 1, \ldots$

Geometric sequence

$r = 2$

9. $1, \frac{1}{2}, \frac{1}{3}, \frac{1}{4}, \ldots$

Not a geometric sequence

11. $a_1 = 6, r = 3$

$a_2 = 6(3) = 18$

$a_3 = 18(3) = 54$

$a_4 = 54(3) = 162$

$a_5 = 162(3) = 486$

13. $a_1 = 1$, $r = \frac{1}{2}$

$a_1 = 1$

$a_2 = 1\left(\frac{1}{2}\right) = \frac{1}{2}$

$a_3 = \frac{1}{2}\left(\frac{1}{2}\right) = \frac{1}{4}$

$a_4 = \frac{1}{4}\left(\frac{1}{2}\right) = \frac{1}{8}$

$a_5 = \frac{1}{8}\left(\frac{1}{2}\right) = \frac{1}{16}$

15. $a_1 = 5$, $r = -\frac{1}{10}$

$a_1 = 5$

$a_2 = 5\left(-\frac{1}{10}\right) = -\frac{1}{2}$

$a_3 = \left(-\frac{1}{2}\right)\left(-\frac{1}{10}\right) = \frac{1}{20}$

$a_4 = \frac{1}{20}\left(-\frac{1}{10}\right) = -\frac{1}{200}$

$a_5 = \left(-\frac{1}{200}\right)\left(-\frac{1}{10}\right) = \frac{1}{2000}$

17. $a_1 = 1$, $r = e$

$a_1 = 1$

$a_2 = 1(e) = e$

$a_3 = (e)(e) = e^2$

$a_4 = (e^2)(e) = e^3$

$a_5 = (e^3)(e) = e^4$

19. $a_1 = 64$, $a_{k+1} = \frac{1}{2}a_k$

$a_1 = 64$

$a_2 = \frac{1}{2}(64) = 32$

$a_3 = \frac{1}{2}(32) = 16$

$a_4 = \frac{1}{2}(16) = 8$

$a_5 = \frac{1}{2}(8) = 4$

$r = \frac{1}{2}$, $a_n = 64\left(\frac{1}{2}\right)^{n-1} = 128\left(\frac{1}{2}\right)^n$

21. $a_1 = 9$, $a_{k+1} = 2a_k$

$a_2 = 2(9) = 18$

$a_3 = 2(18) = 36$

$a_4 = 2(36) = 72$

$a_5 = 2(72) = 144$

$r = 2$

$a_n = \left(\frac{9}{2}\right)2^n = 9(2^{n-1})$

23. $a_1 = 6$, $a_{k+1} = -\frac{3}{2}a_k$

$a_1 = 6$

$a_2 = -\frac{3}{2}(6) = -9$

$a_3 = -\frac{3}{2}(-9) = \frac{27}{2}$

$a_4 = -\frac{3}{2}\left(\frac{27}{2}\right) = -\frac{81}{4}$

$a_5 = -\frac{3}{2}\left(-\frac{81}{4}\right) = \frac{243}{8}$

$r = -\frac{3}{2}$, $a_n = 6\left(-\frac{3}{2}\right)^{n-1} = -4\left(-\frac{3}{2}\right)^n$

25. $a_1 = 4$, $a_4 = \frac{1}{2}$, $n = 10$

$a_1 r^3 = a_4$

$4r^3 = \frac{1}{2}$

$r^3 = \frac{1}{8}$

$r = \frac{1}{2}$

$a_{10} = a_4 r^6 = \frac{1}{2}\left(\frac{1}{2}\right)^6 = \frac{1}{2^7} = \frac{1}{128}$

27. $a_1 = 6$, $r = -\frac{1}{3}$, $n = 12$

$a_n = a_1 r^{n-1}$

$a_{12} = 6\left(-\frac{1}{3}\right)^{11} = \frac{-2}{3^{10}}$

29. $a_1 = 500$, $r = 1.02$, $n = 14$

$a_n = a_1 r^{n-1}$

$a_{14} = 500(1.02)^{13} \approx 646.8$

31. $a_2 = a_1 r = -18 \implies a_1 = \frac{-18}{r}$

$a_5 = a_1 r^4 = (a_1 r)r^3 = -18r^3 = \frac{2}{3} \implies r = -\frac{1}{3}$

$a_1 = \frac{-18}{r} = \frac{-18}{-1/3} = 54$

$a_6 = a_1 r^5 = 54\left(\frac{-1}{3}\right)^5 = \frac{-54}{243} = -\frac{2}{9}$

33. $7, 21, 63$

$r = 3$

$a_n = 7(3)^{n-1}$

$a_9 = 7(3)^{9-1} = 45{,}927$

35. 5, 30, 180

$r = \frac{30}{5} = 6$

$a_n = 5(6)^{n-1}$

$a_{10} = 5(6)^{10-1} = 50,388,480$

37. $a_n = 12(-0.75)^{n-1}$

39. $a_n = 2(1.3)^{n-1}$

41. 8, -4, 2, -1, $\frac{1}{2}$

$S_1 = 8$

$S_2 = 8 + (-4) = 4$

$S_3 = 8 + (-4) + 2 = 6$

$S_4 = 8 + (-4) + 2 + (-1) = 5$

43. $\displaystyle\sum_{n=1}^{\infty} 16\left(-\frac{1}{2}\right)^{n-1}$

n	1	2	3	4	5	6	7	8	9	10
S_n	16	24	28	30	31	31.5	31.75	31.875	31.9375	31.96875

45. $\displaystyle\sum_{n=1}^{9} 2^{n-1} \implies a_1 = 1, \ r = 2$

$S_9 = \dfrac{1(1 - 2^9)}{1 - 2} = 511$

47. $\displaystyle\sum_{i=1}^{7} 64\left(-\frac{1}{2}\right)^{i-1} \implies a_1 = 64, \ r = -\frac{1}{2}$

$S_7 = 64\left[\dfrac{1 - (-1/2)^7}{1 - (-1/2)}\right] = \dfrac{128}{3}\left[1 - \left(-\frac{1}{2}\right)^7\right] = 43$

49. $\displaystyle\sum_{n=0}^{20} 3\left(\frac{3}{2}\right)^n = \sum_{n=1}^{21} 3\left(\frac{3}{2}\right)^{n-1} \implies a_1 = 3, \ r = \frac{3}{2}$

$S_{21} = 3\left[\dfrac{1 - (3/2)^{21}}{1 - (3/2)}\right]$

$= -6\left[1 - \left(\frac{3}{2}\right)^{21}\right] \approx 29,921.31$

51. $\displaystyle\sum_{i=1}^{10} 8\left(-\frac{1}{4}\right)^{i-1} \implies a_1 = 8, \ r = -\frac{1}{4}$

$S_{10} = 8\left[\dfrac{1 - (-1/4)^{10}}{1 - (-1/4)}\right] = \dfrac{32}{5}\left[1 - \left(-\frac{1}{4}\right)^{10}\right] \approx 6.4$

53. $\displaystyle\sum_{n=0}^{5} 300(1.06)^n = \sum_{n=1}^{6} 300(1.06)^{n-1} \implies a_1 = 300, \ r = 1.06$

$S_6 = 300\left[\dfrac{1 - (1.06)^6}{1 - 1.06}\right] \approx 2092.60$

55. $5 + 15 + 45 + \cdots + 3645$

$r = 3$ and $3645 = 5(3)^{n-1} \implies n = 7$

Thus, the sum can be written as $\displaystyle\sum_{n=1}^{7} 5(3)^{n-1}$.

57. $2 - \frac{1}{2} + \frac{1}{8} - \cdots + \frac{1}{2048}$

$r = -\frac{1}{4}$ and $\frac{1}{2048} = 2\left(-\frac{1}{4}\right)^{n-1} \implies n = 7$

$\displaystyle\sum_{n=1}^{7} 2\left(-\frac{1}{4}\right)^{n-1}$

59. $a_1 = 10, \ r = \frac{4}{5}$

$\displaystyle\sum_{n=0}^{\infty} 10\left(\frac{4}{5}\right)^n = \frac{a_1}{1 - r} = \frac{10}{1 - \frac{4}{5}} = 50$

61. $a_1 = 5, \ r = -\frac{1}{2}$

$\displaystyle\sum_{n=0}^{\infty} 5\left(-\frac{1}{2}\right)^n = \frac{a_1}{1 - r} = \frac{5}{1 - \left(-\frac{1}{2}\right)} = \frac{5}{\left(\frac{3}{2}\right)} = \frac{10}{3}$

63. $\sum_{n=1}^{\infty} 2\left(\frac{7}{3}\right)^{n-1}$ does not have a finite sum $\left(\frac{7}{3} > 1\right)$.

65. $a_1 = 10, r = 0.11$

$$\sum_{n=0}^{\infty} 10(0.11)^n = \frac{a_1}{1-r} = \frac{10}{1-0.11} = \frac{10}{0.89}$$

$$= \frac{1000}{89} \approx 11.236$$

67. $a_1 = -3, r = -0.9$

$$\sum_{n=0}^{\infty} -3(-0.9)^n = \frac{a_1}{1-r} = \frac{-3}{1-(-0.9)}$$

$$= \frac{-3}{1.9} = \frac{-30}{19} \approx -1.579$$

69. $8 + 6 + \frac{9}{2} + \frac{27}{8} + \cdots = \sum_{n=0}^{\infty} 8\left(\frac{3}{4}\right)^n$

$$= \frac{8}{1-3/4} = 32$$

71. $3 - 1 + \frac{1}{3} - \frac{1}{9} + \cdots = \sum_{n=0}^{\infty} 3\left(-\frac{1}{3}\right)^n = \frac{a_1}{1-r} = \frac{3}{1-(-1/3)} = 3\left(\frac{3}{4}\right) = \frac{9}{4}$

73. $0.\overline{36} = \sum_{n=0}^{\infty} 0.36(0.01)^n$

$$= \frac{0.36}{1-0.01} = \frac{0.36}{0.99} = \frac{36}{99} = \frac{4}{11}$$

75. $1.2\overline{5} = 1.2 + \sum_{n=0}^{\infty} 0.05(0.1)^n$

$$= \frac{6}{5} + \frac{0.05}{1-0.1}$$

$$= \frac{6}{5} + \frac{0.05}{0.9}$$

$$= \frac{6}{5} + \frac{5}{90} = \frac{113}{90}$$

77. $A = P\left(1 + \frac{r}{n}\right)^{nt} = 1000\left(1 + \frac{0.03}{n}\right)^{n(10)}$

   (a) $n = 1$: $A = 1000(1 + 0.03)^{10} \approx 1343.92$

   (b) $n = 2$: $A = 1000\left(1 + \frac{0.03}{2}\right)^{2(10)} \approx 1346.86$

   (c) $n = 4$: $A = 1000\left(1 + \frac{0.03}{4}\right)^{4(10)} \approx 1348.35$

   (d) $n = 12$: $A = 1000\left(1 + \frac{0.03}{12}\right)^{12(10)} \approx 1349.35$

   (e) $n = 365$: $A = 1000\left(1 + \frac{0.03}{365}\right)^{365(10)} \approx 1349.84$

79. $A = \sum_{n=1}^{60} 100\left(1 + \frac{0.03}{12}\right)^n$

$$= 100\left(1 + \frac{0.03}{12}\right) \cdot \frac{[1 - (1 + 0.03/12)^{60}]}{[1 - (1 + 0.03/12)]}$$

$$= 100(1.0025) \cdot \left[\frac{1 - 1.0025^{60}}{1 - 1.0025}\right]$$

$$\approx \$6480.83$$

81. Let $N = 12t$ be the total number of deposits.

$$A = P\left(1 + \frac{r}{12}\right) + P\left(1 + \frac{r}{12}\right)^2 + \cdots + P\left(1 + \frac{r}{12}\right)^N$$

$$= \left(1 + \frac{r}{12}\right)\left[P + P\left(1 + \frac{r}{12}\right) + \cdots + P\left(1 + \frac{r}{12}\right)^{N-1}\right]$$

$$= P\left(1 + \frac{r}{12}\right)\sum_{n=1}^{N}\left(1 + \frac{r}{12}\right)^{n-1}$$

$$= P\left(1 + \frac{r}{12}\right)\frac{1 - \left(1 + \frac{r}{12}\right)^N}{1 - \left(1 + \frac{r}{12}\right)}$$

$$= P\left(1 + \frac{r}{12}\right)\left(-\frac{12}{r}\right)\left[1 - \left(1 + \frac{r}{12}\right)^N\right]$$

$$= P\left(\frac{12}{r} + 1\right)\left[-1 + \left(1 + \frac{r}{12}\right)^N\right]$$

$$= P\left[\left(1 + \frac{r}{12}\right)^N - 1\right]\left(1 + \frac{12}{r}\right)$$

$$= P\left[\left(1 + \frac{r}{12}\right)^{12t} - 1\right]\left(1 + \frac{12}{r}\right)$$

83. $P = \$50$, $r = 7\%$, $t = 20$ years

(a) Compounded monthly: $A = 50\left[\left(1 + \frac{0.07}{12}\right)^{12(20)} - 1\right]\left(1 + \frac{12}{0.07}\right) \approx \$26{,}198.27$

(b) Compounded continuously: $A = \dfrac{50e^{0.07/12}\left(e^{0.07(20)} - 1\right)}{e^{0.07/12} - 1} \approx \$26{,}263.88$

85. $P = 100$, $r = 5\% = 0.05$, $t = 40$

(a) Compounded monthly: $A = 100\left[\left(1 + \frac{0.05}{12}\right)^{12(40)} - 1\right]\left(1 + \frac{12}{0.05}\right) \approx \$153{,}237.86$

(b) Compounded continuously: $A = \dfrac{100e^{0.05/12}\left(e^{0.05(40)} - 1\right)}{e^{0.05/12} - 1} \approx \$153{,}657.02$

87. First shaded area: $\dfrac{16^2}{4}$

Second shaded area: $\dfrac{16^2}{4} + \dfrac{1}{2}\cdot\dfrac{16^2}{4}$

Third shaded area: $\dfrac{16^2}{4} + \dfrac{1}{2}\dfrac{16^2}{4} + \dfrac{1}{4}\dfrac{16^2}{4}$, etc

Total area of shaded region: $\dfrac{16^2}{4}\sum_{n=0}^{5}\left(\dfrac{1}{2}\right)^n = 64\left[\dfrac{1 - (1/2)^6}{1 - 1/2}\right] = 128\left(1 - \left(\dfrac{1}{2}\right)^6\right) = 126$ square units

89. (a) $a_0 = 70$ degrees

$a_1 = 0.8(70) = 56$ degrees

$\vdots$

$a_n = (0.8)^n(70)$

(b) $a_6 = (0.8)^6(70) \approx 18.35$ degrees

$a_{12} = (0.8)^{12}(70) \approx 4.81$ degrees

(c)

$a_3 \approx 35.8$

$a_4 \approx 28.7$

Thus, the water freezes between 3 and 4 hours, about 3.5 hours.

91. $400 + 0.75(400) + (0.75)^2(400) + \cdots = \displaystyle\sum_{n=0}^{\infty} 400(0.75)^n$

$$= \frac{400}{1 - 0.75} = \$1600$$

93. $250 + 0.80(250) + (0.80)^2(250) + \cdots = \displaystyle\sum_{n=0}^{\infty} 250(0.80)^2$

$$= \frac{250}{1 - 0.80} = \$1250$$

95. $600 + 0.725(600) + (0.725)^2(600) + \cdots = \displaystyle\sum_{n=0}^{\infty} 600(0.725)^n$

$$= \frac{600}{1 - 0.725} \approx \$2181.82$$

97. (a) Option 1: $30{,}000 + 1.025(30{,}000) + \cdots + (1.025)^4(30{,}000) = \displaystyle\sum_{n=0}^{4} 30{,}000(1.025)^n$

$$\approx \$157{,}689.86$$

Option 2: $32{,}500 + 1.02(32{,}500) + \cdots + (1.02)^4(32{,}500) = \displaystyle\sum_{n=0}^{4} 32{,}500(1.02)^n$

$$\approx \$169{,}131.31$$

Option 2 has the larger cumulative amount.

(b) Option 1: $(1.025)^4(30{,}000) \approx \$33{,}114.39$

Option 2: $(1.02)^4(32{,}500) \approx \$35{,}179.05$

Option 2 has the larger amount.

99. (a) Downward: $850 + 0.75(850) + (0.75)^2(850) + \cdots + (0.75)^9(850) = \displaystyle\sum_{n=0}^{9} 850(0.75)^n$

$$\approx 3208.53 \text{ feet}$$

Upward: $0.75(850) + (0.75)^2(850) + \cdots + (0.75)^{10}(850) = \displaystyle\sum_{n=0}^{9} (0.75)(850)(0.75)^n$

$$= \displaystyle\sum_{n=0}^{9} 637.5(0.75)^n \approx 2406.4 \text{ feet}$$

Total distance: $3208.53 + 2406.4 = 5614.93$ feet

(b) $\displaystyle\sum_{n=0}^{\infty} 850(0.75)^n + \sum_{n=0}^{\infty} 637.5(0.75)^n = \frac{850}{1 - 0.75} + \frac{637.5}{1 - 0.75} = 5950 \text{ feet}$

101. False. See definition page 535.

103. $a_1 = 3$, $r = \dfrac{x}{2}$

$$a_2 = 3\left(\frac{x}{2}\right) = \frac{3x}{2}$$

$$a_3 = \frac{3x}{2}\left(\frac{x}{2}\right) = \frac{3x^2}{4}$$

$$a_4 = \frac{3x^2}{4}\left(\frac{x}{2}\right) = \frac{3x^3}{8}$$

$$a_5 = \frac{3x^3}{8}\left(\frac{x}{2}\right) = \frac{3x^4}{16}$$

105. $a_1 = 100$, $r = e^x$, $n = 9$

$$a_n = a_1 r^{n-1}$$

$$a_9 = 100(e^x)^8 = 100e^{8x}$$

107. (a) $f(x) = 6\left[\dfrac{1 - 0.5^x}{1 - 0.5}\right]$

$$\sum_{n=0}^{\infty} 6\left(\frac{1}{2}\right)^n = \frac{6}{1 - 1/2} = 12$$

The horizontal asymptote of $f(x)$ is $y = 12$.
This corresponds to the sum of the series.

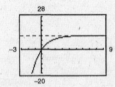

(b) $f(x) = 2\left[\dfrac{1 - 0.8^x}{1 - 0.8}\right]$

$$\sum_{n=0}^{\infty} 2\left(\frac{4}{5}\right)^n = \frac{2}{1 - 4/5} = 10$$

The horizontal asymptote of $f(x)$ is $y = 10$.
This corresponds to the sum of the series.

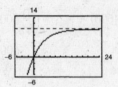

109. To use the first two terms of a geometric series to find the nth term, first divide the second term by the first term to obtain the constant ratio. The nth term is the first term multiplied by the common ratio raised to the $(n - 1)$ power.

$$r = \frac{a_2}{a_1}, \, a_n = a_1 r^{n-1}$$

111. $\text{Time} = \dfrac{\text{Distance}}{\text{Speed}} = \dfrac{200}{50} + \dfrac{200}{42} = 200\left[\dfrac{92}{2100}\right]$ hours

$\text{Speed} = \dfrac{\text{Distance}}{\text{Time}} = \dfrac{400}{200[92/2100]} = \dfrac{2(2100)}{92} \approx 45.65$ mph

113. $\det \begin{bmatrix} -1 & 3 & 4 \\ -2 & 8 & 0 \\ 2 & 5 & -1 \end{bmatrix} = 4(-10 - 16) - 1(-8 + 6)$

$$= -104 + 2 = -102$$

115. Answers will vary.

Section 8.4 Mathematical Induction

- You should be sure that you understand the principle of mathematical induction. If P_n is a statement involving the positive integer n, where P_1 is true and the truth of P_k implies the truth of P_{k+1}, then P_n is true for all positive integers n.
- You should be able to verify (by induction) the formulas for the sums of powers of integers and be able to use these formulas.
- You should be able to work with finite differences.

Vocabulary Check

1. mathematical induction

2. first

3. arithmetic

4. second

1. $P_k = \dfrac{5}{k(k+1)}$

$$P_{k+1} = \frac{5}{(k+1)[(k+1)+1]} = \frac{5}{(k+1)(k+2)}$$

3. $P_k = \dfrac{2^k}{(k+1)!}$

$$P_{k+1} = \frac{2^{k+1}}{((k+1)+1)!} = \frac{2^{k+1}}{(k+2)!}$$

5. $P_k = 1 + 6 + 11 + \cdots + [5(k-1) - 4] + [5k - 4]$

$P_{k+1} = 1 + 6 + 11 + \cdots + [5k - 4] + [5(k+1) - 4]$

$\quad\; = 1 + 6 + 11 + \cdots + [5k - 4] + [5k + 1]$

7. 1. When $n = 1$, $S_1 = 2 = 1(1 + 1)$.

   2. Assume that

$$S_k = 2 + 4 + 6 + 8 + \cdots + 2k = k(k+1).$$

Then,

$$S_{k+1} = 2 + 4 + 6 + 8 + \cdots + 2k + 2(k+1)$$
$$= S_k + 2(k+1) = k(k+1) + 2(k+1) = (k+1)(k+2).$$

Therefore, by mathematical induction, the formula is valid for all positive integer values of n.

9. 1. When $n = 1$, $S_1 = 3 = \dfrac{1}{2}(5(1) + 1)$

   2. Assume that $S_k = 3 + 8 + 13 + \cdots + (5k - 2) = \dfrac{k}{2}(5k + 1)$.

Then, $S_{k+1} = 3 + 8 + 13 + \cdots + (5k - 2) + [5(k+1) - 2]$

$$= S_k + [5k + 3] = \frac{k}{2}(5k + 1) + 5k + 3$$

$$= \frac{1}{2}[5k^2 + 11k + 6] = \frac{1}{2}(k + 1)(5k + 6)$$

$$= \frac{1}{2}(k + 1)(5(k + 1) + 1).$$

Therefore, by mathematical induction, the formula is valid for all positive integer values of n.

11. 1. When $n = 1$, $S_1 = 1 = 2^1 - 1$.

2. Assume that

$$S_k = 1 + 2 + 2^2 + 2^3 + \cdots + 2^{k-1} = 2^k - 1.$$

Then,

$$S_{k+1} = 1 + 2 + 2^2 + 2^3 + \cdots + 2^{k-1} + 2^k$$

$$= S_k + 2^k = 2^k - 1 + 2^k = 2(2^k) - 1 = 2^{k+1} - 1.$$

Therefore, by mathematical induction, the formula is valid for all positive integer values of n.

13. 1. When $n = 1$, $S_1 = 1 = \dfrac{1(1 + 1)}{2}$.

2. Assume that

$$S_k = 1 + 2 + 3 + 4 + \cdots + k = \frac{k(k + 1)}{2}.$$

Then,

$$S_{k+1} = 1 + 2 + 3 + 4 + \cdots + k + (k + 1)$$

$$= S_k + (k + 1) = \frac{k(k + 1)}{2} + \frac{2(k + 1)}{2} = \frac{(k + 1)(k + 2)}{2}.$$

Therefore, the formula is valid for all positive integer values of n.

15. 1. When $n = 1$,

$$S_1 = 1^4 = \frac{1(1 + 1)(2 \cdot 1 + 1)(3 \cdot 1^2 + 3 \cdot 1 - 1)}{30}.$$

2. Assume that $S_k = \displaystyle\sum_{i=1}^{k} i^4 = \frac{k(k + 1)(2k + 1)(3k^2 + 3k - 1)}{30}.$

Then, $S_{k+1} = S_k + (k + 1)^4$

$$= \frac{k(k + 1)(2k + 1)(3k^2 + 3k - 1)}{30} + (k + 1)^4 = \frac{k(k + 1)(2k + 1)(3k^2 + 3k - 1) + 30(k + 1)^4}{30}$$

$$= \frac{(k + 1)[k(2k + 1)(3k^2 + 3k - 1) + 30(k + 1)^3]}{30} = \frac{(k + 1)(6k^4 + 39k^3 + 91k^2 + 89k + 30)}{30}$$

$$= \frac{(k + 1)(k + 2)(2k + 3)(3k^2 + 9k + 5)}{30} = \frac{(k + 1)(k + 2)(2(k + 1) + 1)(3(k + 1)^2 + 3(k + 1) - 1)}{30}.$$

Therefore, the formula is valid for all positive integer values of n.

17. 1. When $n = 1$, $S_1 = 2 = \dfrac{1(2)(3)}{3}$.

2. Assume that $S_k = 1(2) + 2(3) + 3(4) + \cdots + k(k + 1) = \dfrac{k(k + 1)(k + 2)}{3}$.

Then,

$$S_{k+1} = 1(2) + 2(3) + 3(4) + \cdots + k(k + 1) + (k + 1)(k + 2)$$

$$= S_k + (k + 1)(k + 2)$$

$$= \frac{k(k + 1)(k + 2)}{3} + \frac{3(k + 1)(k + 2)}{3}$$

$$= \frac{(k + 1)(k + 2)(k + 3)}{3}.$$

Therefore, the formula is valid for all positive integer values of n.

19. 1. When $n = 1$, $S_1 = \dfrac{1}{1(1 + 1)} = \dfrac{1}{2}$

2. Assume $S_k = \displaystyle\sum_{i=1}^{k} \frac{1}{i(i + 1)} = \frac{k}{k + 1}$.

Thus,

$$S_{k+1} = \sum_{i=1}^{k+1} \frac{1}{i(i + 1)}$$

$$= \frac{1}{1(2)} + \frac{1}{2(3)} + \cdots + \frac{1}{k(k + 1)} + \frac{1}{(k + 1)(k + 2)}$$

$$= \frac{k}{k + 1} + \frac{1}{(k + 1)(k + 2)}$$

$$= \frac{k(k + 2) + 1}{(k + 2)(k + 2)}$$

$$= \frac{(k + 1)^2}{(k + 1)(k + 2)}$$

$$= \frac{k + 1}{k + 2}$$

$$= \frac{k + 1}{(k + 1) + 1}.$$

Therefore, the formula is valid for all positive integer values of n.

21. $\displaystyle\sum_{n=1}^{50} n^3 = \frac{50^2(50 + 1)^2}{4} = 1{,}625{,}625$

23. $\displaystyle\sum_{n=1}^{12} (n^2 - n) = \sum_{n=1}^{12} n^2 - \sum_{n=1}^{12} n$

$$= \frac{12(12 + 1)(2 \cdot 12 + 1)}{6} - \frac{12(12 + 1)}{2}$$

$$= 650 - 78 = 572$$

25. 1. When $n = 4$, $4! = 24$ and $2^4 = 16$, thus $4! > 2^4$.

    2. Assume $k! > 2^k$, $k > 4$. Then, $(k + 1)! = k!(k + 1) > 2^k(2)$ since $k + 1 > 2$. Thus, $(k + 1)! > 2^{k+1}$.

Therefore, by mathematical induction, the formula is valid for all integers n such that $n \geq 4$.

27. 1. When $n = 2$, $\dfrac{1}{\sqrt{1}} + \dfrac{1}{\sqrt{2}} \approx 1.707$ and $\sqrt{2} \approx 1.414$, thus $\dfrac{1}{\sqrt{1}} + \dfrac{1}{\sqrt{2}} > \sqrt{2}$.

    2. Assume $\dfrac{1}{\sqrt{1}} + \dfrac{1}{\sqrt{2}} + \dfrac{1}{\sqrt{3}} + \cdots + \dfrac{1}{\sqrt{k}} > \sqrt{k}$, $k > 2$.

Then, $\dfrac{1}{\sqrt{1}} + \dfrac{1}{\sqrt{2}} + \dfrac{1}{\sqrt{3}} + \cdots + \dfrac{1}{\sqrt{k}} + \dfrac{1}{\sqrt{k + 1}} > \sqrt{k} + \dfrac{1}{\sqrt{k + 1}}$.

Now we need to show that $\sqrt{k} + \dfrac{1}{\sqrt{k + 1}} > \sqrt{k + 1}$, $k > 2$.

This is true because $\sqrt{k(k + 1)} > k$

$$\sqrt{k(k + 1)} + 1 > k + 1$$

$$\frac{\sqrt{k(k + 1)} + 1}{\sqrt{k + 1}} > \frac{k + 1}{\sqrt{k + 1}}$$

$$\sqrt{k} + \frac{1}{\sqrt{k + 1}} > \sqrt{k + 1}.$$

Therefore, $\dfrac{1}{\sqrt{1}} + \dfrac{1}{\sqrt{2}} + \dfrac{1}{\sqrt{3}} + \cdots + \dfrac{1}{\sqrt{k}} + \dfrac{1}{\sqrt{k + 1}} > \sqrt{k + 1}$.

Therefore, by mathematical induction, the formula is valid for all integers n such that $n \geq 2$.

29. 1. When $n = 1$, $1 + a \geq a$ since $1 > 0$.

    2. Assume $(1 + a)^k \geq ka$.

Then, $(1 + a)^{k+1} = (1 + a)^k(1 + a) \geq ka(1 + a)$

$$= ka + ka^2 \geq ka + a \quad \text{(because } a > 1\text{)}$$

$$= (k + 1)a.$$

Therefore, by mathematical induction, the inequality is valid for all integers $n \geq 1$.

31. 1. When $n = 1$, $(ab)^1 = a^1 b^1 = ab$.

    2. Assume that $(ab)^k = a^k b^k$.

Then, $(ab)^{k+1} = (ab)^k(ab)$

$$= a^k b^k ab$$

$$= a^{k+1} b^{k+1}.$$

Thus, $(ab)^n = a^n b^n$.

33. 1. When $n = 1, (x_1)^{-1} = x_1^{-1}$.

2. Assume that

$$(x_1 x_2 x_3 \cdots x_k)^{-1} = x_1^{-1} x_2^{-1} x_3^{-1} \cdots x_k^{-1}.$$

Then,

$$\begin{aligned}
(x_1 x_2 x_3 \cdots x_k x_{k+1})^{-1} &= [(x_1 x_2 x_3 \cdots x_k) x_{k+1}]^{-1} \\
&= (x_1 x_2 x_3 \cdots x_k)^{-1} x_{k+1}^{-1} \\
&= x_1^{-1} x_2^{-1} x_3^{-1} \cdots x_k^{-1} x_{k+1}^{-1}.
\end{aligned}$$

Thus, the formula is valid.

35. 1. When $n = 1, x(y_1) = xy_1$.

2. Assume that $x(y_1 + y_2 + \cdots + y_k) = xy_1 + xy_2 + \cdots + xy_k$.

Then,

$$\begin{aligned}
xy_1 + xy_2 + \cdots + xy_k + xy_{k+1} &= x(y_1 + y_2 + \cdots + y_k) + xy_{k+1} \\
&= x[(y_1 + y_2 + \cdots + y_k) + y_{k+1}] \\
&= x(y_1 + y_2 + \cdots + y_k + y_{k+1}).
\end{aligned}$$

Hence, the formula holds.

37. 1. When $n = 1, [1^3 + 3(1)^2 + 2(1)] = 6$ and 3 is a factor.

2. Assume that 3 is a factor of $(k^3 + 3k^2 + 2k)$.

Then,

$$\begin{aligned}
[(k + 1)^3 + 3(k + 1)^2 + 2(k + 1)] &= k^3 + 3k^2 + 3k + 1 + 3k^2 + 6k + 3 + 2k + 2 \\
&= (k^3 + 3k^2 + 2k) + (3k^2 + 9k + 6) \\
&= (k^3 + 3k^2 + 2k) + 3(k^2 + 3k + 2).
\end{aligned}$$

Since 3 is a factor of $(k^3 + 3k^2 + 2k)$ by our assumption, and 3 is a factor of $3(k^2 + 3k + 2)$ then 3 is a factor of the whole sum.

Thus, 3 is a factor of $(n^3 + 3n^2 + 2n)$ for every positive integer n.

39. 1. When $n = 1, [1^3 - 1 + 3] = 3$, and 3 is a factor.

2. Assume that 3 is a factor of $k^3 - k + 3$. Then,

$$\begin{aligned}
[(k + 1)^3 - (k + 1) + 3] &= k^3 + 3k^2 + 3k + 1 - k - 1 + 3 \\
&= k^3 + 3k^2 + 2k + 3 \\
&= (k^3 - k + 3) + 3k^2 + 3k \\
&= (k^3 - k + 3) + 3(k^2 + k).
\end{aligned}$$

Since 3 is a factor of $k^3 - k + 3$ by our assumption, and 3 is a factor of $3(k^2 + k)$, then 3 is a factor of the whole sum.

Thus, 3 is a factor of $n^3 - n + 3$ for every positive integer n.

41. 1. When $n = 1$, $2^{2+1} + 1 = 9$, and 3 is a factor.

2. Assume that 3 is a factor of $2^{2k+1} + 1$.

Then,

$2^{2(k+1)+1} + 1 = 2^{2k+3} + 1$

$\qquad = 4 \cdot 2^{2k+1} + 1$

$\qquad = (3 + 1)2^{2k+1} + 1$

$\qquad = (2^{2k+1} + 1) + 3 \cdot 2^{2k+1}$.

Since 3 is a factor of $2^{2k+1} + 1$ by our assumption, and 3 is a factor of $3 \cdot 2^{2k+1}$, then 3 is a factor of the whole sum.

Thus, 3 is a factor of $2^{2n+1} + 1$ for every positive integer n.

45. $a_1 = 3, a_n = a_{n-1} - n$

$a_1 = 3$

$a_2 = a_1 - 2 = 3 - 2 = 1$

$a_3 = a_2 - 3 = 1 - 3 = -2$

$a_4 = a_3 - 4 = -2 - 4 = -6$

$a_5 = a_4 - 5 = -6 - 5 = -11$

First differences:

Second differences:

Since the second differences are all the same, the sequence has a quadratic model.

49. $a_1 = 2, a_n = a_{n-1} + 2$

$a_1 = 2$

$a_2 = a_1 + 2 = 2 + 2 = 4$

$a_3 = a_2 + 2 = 4 + 2 = 6$

$a_4 = a_3 + 2 = 6 + 2 = 8$

$a_5 = a_4 + 2 = 8 + 2 = 10$

First differences:

Second differences:

Since the first differences are equal, the sequence has a linear model.

43. $a_1 = 0, a_n = a_{n-1} + 3$

$a_1 = 0$

$a_2 = a_1 + 3 = 0 + 3 = 3$

$a_3 = a_2 + 3 = 3 + 3 = 6$

$a_4 = a_3 + 3 = 6 + 3 = 9$

$a_5 = a_4 + 3 = 9 + 3 = 12$

First differences:

Second differences:

Since the first differences are equal, the sequence has a linear model.

47. $a_0 = 0, a_n = a_{n-1} + n$

$a_0 = 0$

$a_1 = a_0 + 1 = 0 + 1 = 1$

$a_2 = a_1 + 2 = 1 + 2 = 3$

$a_3 = a_2 + 3 = 3 + 3 = 6$

$a_4 = a_3 + 4 = 6 + 4 = 10$

First differences:

Second differences:

Since the second differences are equal, the sequence has a quadratic model.

51. $a_1 = 3, a_2 = 3, a_3 = 5$

Let $a_n = an^2 + bn + c$.

$a_1 = a(1)^2 + b(1) + c = 3 \implies a + b + c = 3$

$a_2 = a(2)^2 + b(2) + c = 3 \implies 4a + 2b + c = 3$

$a_3 = a(3)^2 + b(3) + c = 5 \implies 9a + 3b + c = 5$

Solving the system, $a = 1, b = -3, c = 5$,

$a_n = n^2 - 3n + 5, n \geq 1$.

53. $a_0 = -3$, $a_2 = 1$, $a_4 = 9$

Let $a_n = an^2 + bn + c$. Then:

$a_0 = a(0)^2 + b(0) + c = -3 \implies c = -3$

$a_2 = a(2)^2 + b(2) + c = 1 \implies 4a + 2b + c = 1$

$$4a + 2b = 4$$
$$2a + b = 2$$

$a_4 = a(4)^2 + b(4) + c = 9 \implies 16a + 4b + c = 9$

$$16a + 4b = 12$$
$$4a + b = 3$$

By elimination: $-2a - b = -2$

$$\frac{4a + b = 3}{2a = 1}$$

$$a = \tfrac{1}{2} \implies b = 1$$

Thus, $a_n = \tfrac{1}{2}n^2 + n - 3$.

55. (a) $n = 1$: 3 sides

$n = 2$: $3 \cdot 4 = 12$ sides

$n = 3$: $3 \cdot 4^2 = 48$ sides

nth Koch snowflake: $3(4)^{n-1}$ sides

To prove this, use mathematical induction.

1. For $n - 1$, the number of sides is $3 \cdot 4^{1-1} = 3$.

2. Assume that the number of sides of the kth Koch snowflake is $3 \cdot 4^{k-1}$. When the $(k + 1)^{st}$ Koch snowflake is created, each side is replaced with 4 sides. That is, the number of sides is increased by a factor of 4:

   Number sides $= 4(3 \cdot 4^{k-1}) = 3 \cdot 4^k$.

Hence, the formula is valid for all positive integers n.

(b) $n = 1$: $A_1 = \dfrac{\sqrt{3}}{4}(1)^2 = \dfrac{\sqrt{3}}{4}$

$n = 2$: $A_2 = \dfrac{\sqrt{3}}{4}\left[1 + \dfrac{1}{3}\right]$

$n = 3$: $A_3 = \dfrac{\sqrt{3}}{4}\left[1 + \dfrac{1}{3} + \dfrac{1}{3}\left(\dfrac{4}{9}\right)\right]$

$n = 4$: $A_4 = \dfrac{\sqrt{3}}{4}\left[1 + \dfrac{1}{3} + \dfrac{1}{3}\left(\dfrac{4}{9}\right) + \dfrac{1}{3}\left(\dfrac{4}{9}\right)^2\right]$

$A_n = \dfrac{\sqrt{3}}{4}\left[1 + \displaystyle\sum_{k=2}^{n} \dfrac{1}{3}\left(\dfrac{4}{9}\right)^{k-2}\right]$, $n > 1$

(c) For the nth Koch snowflake, the length of a single side is $(1/3)^{n-1}$, and the number of sides is $3 \cdot 4^{n-1}$. Hence, the perimeter is

$$\left(\dfrac{1}{3}\right)^{n-1} 3 \cdot 4^{n-1} = 3\left(\dfrac{4}{3}\right)^{n-1}.$$

57. False. P_1 might not even be defined.

59. False. It has $n - 2$ second differences.

61. $(2x^2 - 1)^2 = 4x^4 - 4x^2 + 1$

63. $(5 - 4x)^3 = -64x^3 + 240x^2 - 300x + 125$

65. $3\sqrt{-27} - \sqrt{-12} = 3\sqrt{3 \cdot 3 \cdot (-3)} - \sqrt{2 \cdot 2(-3)} = 9\sqrt{3}i - 2\sqrt{3}i = 7\sqrt{3}i$

67. $10\left(\sqrt[3]{64} - 2\sqrt[3]{-16}\right) = 10\left(4 - 2^2\sqrt[3]{-2}\right)$

$$= 40 - 40\sqrt[3]{-2}$$
$$= 40\left(1 + \sqrt[3]{2}\right)$$

Section 8.5   The Binomial Theorem

- ■ You should be able to use the Binomial Theorem

$$(x + y)^n = x^n + nx^{n-1}y + \frac{n(n-1)}{2!}x^{n-2}y^2 + \cdots + {}_nC_r x^{n-r}y^r + \cdots + y^n$$

where ${}_nC_r = \dfrac{n!}{(n-r)!r!}$, to expand $(x + y)^n$.

- ■ You should be able to use Pascal's Triangle.

Vocabulary Check

1. binomial coefficients

2. Binomial Theorem, Pascal's Triangle

3. ${}_nC_r$ or $\left(\dfrac{n}{r}\right)$

4. expanding, binomial

---

1. ${}_7C_5 = \dfrac{7!}{2!5!} = \dfrac{7 \cdot 6 \cdot 5!}{2 \cdot 5!} = \dfrac{42}{2} = 21$

3. $\begin{pmatrix} 12 \\ 0 \end{pmatrix} = {}_{12}C_0 = \dfrac{12!}{0!12!} = 1$

5. ${}_{20}C_{15} = \dfrac{20!}{15!5!} = \dfrac{20 \cdot 19 \cdot 18 \cdot 17 \cdot 16}{5 \cdot 4 \cdot 3 \cdot 2 \cdot 1} = 15{,}504$

7. ${}_{14}C_1 = \dfrac{14!}{13!1!} = \dfrac{14 \cdot 13!}{13!} = 14$

9. $\begin{pmatrix} 100 \\ 98 \end{pmatrix} = {}_{100}C_{98} = \dfrac{100!}{98!2!} = \dfrac{100 \cdot 99}{2 \cdot 1} = 4950$

11. ${}_{41}C_{36} = 749{,}398$

13. ${}_{100}C_{98} = 4950$

15. ${}_{250}C_2 = 31{,}125$

17. $(x + 2)^4 = {}_4C_0 x^4 + {}_4C_1 x^3(2) + {}_4C_2 x^2(2)^2 + {}_4C_3 x(2)^3 + {}_4C_4(2)^4$

$\qquad = x^4 + 8x^3 + 24x^2 + 32x + 16$

19. $(a + 3)^3 = {}_3C_0 a^3 + {}_3C_1 a^2(3) + {}_3C_2 a(3)^2 + {}_3C_3(3)^3$

$\qquad = a^3 + 3a^2(3) + 3a(3)^2 + (3)^3 = a^3 + 9a^2 + 27a + 27$

21. $(y - 2)^4 = {}_4C_0 y^4 - {}_4C_1 y^3(2) + {}_4C_2 y^2(2)^2 - {}_4C_3 y(2)^3 + {}_4C_4(2)^4$

$\qquad = y^4 - 4y^3(2) + 6y^2(4) - 4y(8) + 16$

$\qquad = y^4 - 8y^3 + 24y^2 - 32y + 16$

23. $(x + y)^5 = {}_5C_0 x^5 + {}_5C_1 x^4 y + {}_5C_2 x^3 y^2 + {}_5C_3 x^2 y^3 + {}_5C_4 xy^4 + {}_5C_5 y^5$

$\qquad = x^5 + 5x^4 y + 10x^3 y^2 + 10x^2 y^3 + 5xy^4 + y^5$

25. $(3r + 2s)^6 = {}_6C_0(3r)^6 + {}_6C_1(3r)^5(2s) + {}_6C_2(3r)^4(2s)^2 + {}_6C_3(3r)^3(2s)^3 + {}_6C_4(3r)^2(2s)^4 + {}_6C_5(3r)(2s)^5 + {}_6C_6(2s)^6$

$\qquad = 729r^6 + 2916r^5 s + 4860r^4 s^2 + 4320r^3 s^3 + 2160r^2 s^4 + 576rs^5 + 64s^6$

27. $(x - y)^5 = {}_5C_0x^5 - {}_5C_1x^4y + {}_5C_2x^3y^2 - {}_5C_3x^2y^3 + {}_5C_4xy^4 - {}_5C_5y^5$

$\qquad = x^5 - 5x^4y + 10x^3y^2 - 10x^2y^3 + 5xy^4 - y^5$

29. $(1 - 4x)^3 = {}_3C_01^3 - {}_3C_11^2(4x) + {}_3C_21(4x)^2 - {}_3C_3(4x)^3$

$\qquad = 1 - 3(4x) + 3(4x)^2 - (4x)^3$

$\qquad = 1 - 12x + 48x^2 - 64x^3$

31. $(x^2 + 2)^4 = {}_4C_0(x^2)^4 + {}_4C_1(x^2)^3(2) + {}_4C_2(x^2)^22^2 + {}_4C_3(x^2)2^3 + {}_4C_4(2)^4$

$\qquad = x^8 + 8x^6 + 24x^4 + 32x^2 + 16$

33. $(x^2 - 5)^5 = {}_5C_0(x^2)^5 - {}_5C_1(x^2)^4(5) + {}_5C_2(x^2)^3(5^2) - {}_5C_3(x^2)^2(5^3) + {}_5C_4(x^2)(5^4) - {}_5C_5(5^5)$

$\qquad = x^{10} - 25x^8 + 250x^6 - 1250x^4 + 3125x^2 - 3125$

35. $(x^2 + y^2)^4 = {}_4C_0(x^2)^4 + {}_4C_1(x^2)^3(y^2) + {}_4C_2(x^2)^2(y^2)^2 + {}_4C_3(x^2)(y^2)^3 + {}_4C_4(y^2)^4$

$\qquad = x^8 + 4x^6y^2 + 6x^4y^4 + 4x^2y^6 + y^8$

37. $(x^3 - y)^6 = {}_6C_0(x^3)^6 - {}_6C_1(x^3)^5y + {}_6C_2(x^3)^4y^2 - {}_6C_3(x^3)^3y^3 + {}_6C_4(x^3)^2y^4 - {}_6C_5(x^3)y^5 + {}_6C_6y^6$

$\qquad = x^{18} - 6x^{15}y + 15x^{12}y^2 - 20x^9y^3 + 15x^6y^4 - 6x^3y^5 + y^6$

39. $\left(\dfrac{1}{x} + y\right)^5 = {}_5C_0\left(\dfrac{1}{x}\right)^5 + {}_5C_1\left(\dfrac{1}{x}\right)^4y + {}_5C_2\left(\dfrac{1}{x}\right)^3y^2 + {}_5C_3\left(\dfrac{1}{x}\right)^2y^3 + {}_5C_4\left(\dfrac{1}{x}\right)y^4 + {}_5C_5y^5$

$\qquad = \dfrac{1}{x^5} + \dfrac{5y}{x^4} + \dfrac{10y^2}{x^3} + \dfrac{10y^3}{x^2} + \dfrac{5y^4}{x} + y^5$

41. $\left(\dfrac{2}{x} - y\right)^4 = {}_4C_0\left(\dfrac{2}{x}\right)^4 - {}_4C_1\left(\dfrac{2}{x}\right)^3y + {}_4C_2\left(\dfrac{2}{x}\right)^2y^2 - {}_4C_3\left(\dfrac{2}{x}\right)y^3 + {}_4C_4y^4$

$\qquad = \dfrac{16}{x^4} - \dfrac{32}{x^3}y + \dfrac{24}{x^2}y^2 - \dfrac{8}{x}y^3 + y^4$

43. $(4x - 1)^3 - 2(4x - 1)^4 = (64x^3 - 48x^2 + 12x - 1) - 2(256x^4 - 256x^3 + 96x^2 - 16x + 1)$

$\qquad = -512x^4 + 576x^3 - 240x^2 + 44x - 3$

45. $2(x - 3)^4 + 5(x - 3)^2 = 2[x^4 - 4(x^3)(3) + 6(x^2)(3^2) - 4(x)(3^3) + 3^4] + 5[x^2 - 2(x)(3) + 3^2]$

$\qquad = 2(x^4 - 12x^3 + 54x^2 - 108x + 81) + 5(x^2 - 6x + 9)$

$\qquad = 2x^4 - 24x^3 + 113x^2 - 246x + 207$

47. $-3(x - 2)^3 - 4(x + 1)^6 = [-3x^3 + 18x^2 - 36x + 24] - [4x^6 + 24x^5 + 60x^4 + 80x^3 + 60x^2 + 24x + 4]$

$\qquad = -4x^6 - 24x^5 - 60x^4 - 83x^3 - 42x^2 - 60x + 20$

49. $(x + 8)^{10}, n = 4$

$\qquad {}_{10}C_3x^{10-3}(8)^3 = 120x^7(512) = 61{,}440x^7$

51. $(x - 6y)^5, n = 3$

$\qquad {}_5C_2x^{5-2}(-6y)^2 = 10x^3(36)y^2 = 360x^3y^2$

53. $(4x + 3y)^9, n = 8$

$$_9C_7(4x)^{9-7}(3y)^7 = 36(16)x^2(3^7)y^7$$
$$= 1,259,712x^2y^7$$

55. $(10x - 3y)^{12}, n = 9$

$$_{12}C_8(10x)^{12-8}(-3y)^8 = 495(10^4)(3^8)x^4y^8$$
$$= 32,476,950,000x^4y^8$$

57. The term involving x^4 in the expansion of $(x + 3)^{12}$ is $_{12}C_8x^4(3)^8 = 495x^4(3)^8 = 3,247,695x^4$. The coefficient is 3,247,695.

59. The term involving x^8y^2 in the expansion of $(x - 2y)^{10}$ is

$$_{10}C_2x^8(-2y)^2 = \frac{10!}{2!8!} \cdot 4x^8y^2 = 180x^8y^2.$$

The coefficient is 180.

61. The term involving x^6y^3 in $(3x - 2y)^9$ is

$$_9C_3(3x)^6(-2y)^3 = 84(3)^6(-2)^3x^6y^3$$
$$= -489,888x^6y^3.$$

The coefficient is $-489,888$.

63. The coefficient of $x^8y^6 = (x^2)^4y^6$ in the expansion of $(x^2 + y)^{10}$ is $_{10}C_6 = 210$.

65. 5th entry of 7th row: $_7C_5 = 21$

67. 5th entry of 6th row: $_6C_5 = 6$

69. 4th row of Pascal's Triangle: 1   4   6   4   1

$$(3t - 2v)^4 = 1(3t)^4 - 4(3t)^3(2v) + 6(3t)^2(2v)^2 - 4(3t)(2v)^3 + 1(2v)^4$$
$$= 81t^4 - 216t^3v + 216t^2v^2 - 96tv^3 + 16v^4$$

71. 5th row of Pascal's Triangle: 1   5   10   10   5   1

$$(2x - 3y)^5 = 1(2x)^5 - 5(2x)^4(3y) + 10(2x)^3(3y)^2 - 10(2x)^2(3y)^3 + 5(2x)(3y)^4 - (3y)^5$$
$$= 32x^5 - 240x^4y + 720x^3y^2 - 1080x^2y^3 + 810xy^4 - 243y^5$$

73. $\left(\sqrt{x} + 5\right)^4 = \left(\sqrt{x}\right)^4 + 4\left(\sqrt{x}\right)^3(5) + 6\left(\sqrt{x}\right)^2(5)^2 + 4\left(\sqrt{x}\right)(5^3) + 5^4$

$$= x^2 + 20x\sqrt{x} + 150x + 500\sqrt{x} + 625$$
$$= x^2 + 20x^{3/2} + 150x + 500x^{1/2} + 625$$

75. $(x^{2/3} - y^{1/3})^3 = (x^{2/3})^3 - 3(x^{2/3})^2(y^{1/3}) + 3(x^{2/3})(y^{1/3})^2 - (y^{1/3})^3$

$$= x^2 - 3x^{4/3}y^{1/3} + 3x^{2/3}y^{2/3} - y$$

77. $\dfrac{f(x + h) - f(x)}{h} = \dfrac{(x + h)^3 - x^3}{h}$

$$= \frac{x^3 + 3x^2h + 3xh^2 + h^3 - x^3}{h}$$
$$= \frac{h(3x^2 + 3xh + h^2)}{h}$$
$$= 3x^2 + 3xh + h^2, \ h \neq 0$$

79. $\dfrac{f(x + h) - f(x)}{h} = \dfrac{(x + h)^6 - x^6}{h}$

$$= \frac{(x^6 + 6x^5h + 15x^4h^2 + 20x^3h^3 + 15x^2h^4 + 6xh^5 + h^6) - x^6}{h}$$

$$= \frac{h(6x^5 + 15x^4h + 20x^3h^2 + 15x^2h^3 + 6xh^4 + h^5)}{h}$$

$$= 6x^5 + 15x^4h + 20x^3h^2 + 15x^2h^3 + 6xh^4 + h^5, h \neq 0$$

81. $\dfrac{f(x + h) - f(x)}{h} = \dfrac{\sqrt{x + h} - \sqrt{x}}{h}$

$$= \frac{\sqrt{x + h} - \sqrt{x}}{h} \cdot \frac{\sqrt{x + h} + \sqrt{x}}{\sqrt{x + h} + \sqrt{x}}$$

$$= \frac{(x + h) - x}{h\left[\sqrt{x + h} + \sqrt{x}\right]}$$

$$= \frac{1}{\sqrt{x + h} + \sqrt{x}}, h \neq 0$$

83. $(1 + i)^4 = {}_4C_0 1^4 + {}_4C_1(1)^3 i + {}_4C_2(1)^2 i^2 + {}_4C_3 1 \cdot i^3 + {}_4C_4 i^4$

$$= 1 + 4i - 6 - 4i + 1$$

$$= -4$$

85. $(4 + i)^4 = {}_4C_0(4)^4 + {}_4C_1(4^3)i + {}_4C_2(4^2)(i^2) + {}_4C_3(4)(i^3) + {}_4C_4 i^4$

$$= 256 + 256i - 96 - 16i + 1$$

$$= 161 + 240i$$

87. $(2 - 3i)^6 = {}_6C_0 2^6 - {}_6C_1 2^5(3i) + {}_6C_2 2^4(3i)^2 - {}_6C_3 2^3(3i)^3 + {}_6C_4 2^2(3i)^4 - {}_6C_5 2(3i)^5 + {}_6C_6(3i)^6$

$$= 64 - 576i - 2160 + 4320i + 4860 - 2916i - 729$$

$$= 2035 + 828i$$

89. $\left(5 + \sqrt{-16}\right)^3 = (5 + 4i)^3$

$$= 5^3 + 3(5^2)(4i) + 3(5)(4i)^2 + (4i)^3$$

$$= 125 + 300i - 240 - 64i$$

$$= -115 + 236i$$

91. $\left(4 + \sqrt{3}i\right)^4 = 4^4 + 4(4^3)\left(\sqrt{3}i\right) + 6(4^2)\left(\sqrt{3}i\right)^2 + 4(4)\left(\sqrt{3}i\right)^3 + \left(\sqrt{3}\right)^4$

$$= 256 + 256\sqrt{3}i - 288 - 48\sqrt{3}i + 9$$

$$= -23 + 208\sqrt{3}i$$

93. $\left(-\dfrac{1}{2} + \dfrac{\sqrt{3}}{2}i\right)^3 = \dfrac{1}{8}\left(-1 + \sqrt{3}i\right)^3$

$$= \dfrac{1}{8}\left[(-1)^3 + 3(-1)^2\left(\sqrt{3}i\right) + 3(-1)\left(\sqrt{3}i\right)^2 + \left(\sqrt{3}i\right)^3\right]$$

$$= \dfrac{1}{8}\left[-1 + 3\sqrt{3}i + 9 - 3\sqrt{3}i\right]$$

$$= 1$$

95. $\left(\dfrac{1}{4} - \dfrac{\sqrt{3}}{4}i\right)^3 = \left(\dfrac{1}{4}\right)^3 - 3\left(\dfrac{1}{4}\right)^2\left(\dfrac{\sqrt{3}}{4}i\right) + 3\left(\dfrac{1}{4}\right)\left(\dfrac{-\sqrt{3}}{4}i\right)^2 - \left(\dfrac{\sqrt{3}}{4}i\right)^3$

$$= \left[\dfrac{1}{64} - \dfrac{3}{4}\left(\dfrac{3}{16}\right)\right] + \left[\dfrac{-3}{16}\dfrac{\sqrt{3}}{4} + \dfrac{3\sqrt{3}}{64}\right]i$$

$$= -\dfrac{1}{8}$$

97. $(1.02)^8 = (1 + 0.02)^8 = 1 + 8(0.02) + 28(0.02)^2 + 56(0.02)^3 + 70(0.02)^4 + 56(0.02)^5$

$$+ 28(0.02)^6 + 8(0.02)^7 + (0.02)^8$$

$$= 1 + 0.16 + 0.0112 + 0.000448 + \cdots \approx 1.172$$

99. $(2.99)^{12} = (3 - 0.01)^{12}$

$$= 3^{12} - 12(3)^{11}(0.01) + 66(3)^{10}(0.01)^2 - 220(3)^9(0.01)^3 + 495(3)^8(0.01)^4$$

$$- 792(3)^7(0.01)^5 + 924(3)^6(0.01)^6 - 792(3)^5(0.01)^7 + 495(3)^4(0.01)^8$$

$$- 220(3)^3(0.01)^9 + 66(3)^2(0.01)^{10} - 12(3)(0.01)^{11} + (0.01)^{12}$$

$$\approx 510{,}568.785$$

101. $f(x) = x^3 - 4x$

$g(x) = f(x + 3)$

$$= (x + 3)^3 - 4(x + 3)$$

$$= x^3 + 9x^2 + 27x + 27 - 4x - 12$$

$$= x^3 + 9x^2 + 23x + 15$$

g is shifted three units to the left.

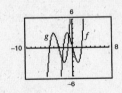

103. $f(x) = (1 - x)^3$

$g(x) = 1 - 3x$

$h(x) = 1 - 3x + 3x^2$

$p(x) = 1 - 3x + 3x^2 - x^3$

Since $p(x)$ is the expansion of $f(x)$, they have the same graph.

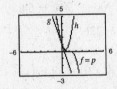

105. $_7C_4\left(\dfrac{1}{2}\right)^4\left(\dfrac{1}{2}\right)^3 = 35\left(\dfrac{1}{16}\right)\left(\dfrac{1}{8}\right) \approx 0.273$

107. $_8C_4\left(\dfrac{1}{3}\right)^4\left(\dfrac{2}{3}\right)^4 = 70\left(\dfrac{1}{81}\right)\left(\dfrac{16}{81}\right) \approx 0.171$

109. $f(t) = 0.064t^2 - 9.30t + 416.5, 5 \leq t \leq 23$

  (a) $g(t) = f(t + 20)$

$$= 0.064(t + 20)^2 - 9.30(t + 20) + 416.5$$

$$= 0.064t^2 - 6.74t + 256.1, -15 \leq t \leq 3$$

  (b)

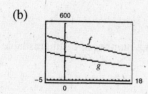

111. False. The x^4y^8 term is

$$_{12}C_4 x^4(-2y)^8 = 495x^4(-2)^8y^8 = 126{,}720x^4y^8.$$

[**Note:** 7920 is the coefficient of x^8y^4.]

113. Answers will vary. See page 557.

115. The expansions of $(x + y)^n$ and $(x - y)^n$ are almost the same except that the signs of the terms in the expansion of $(x - y)^n$ alternate from positive to negative.

117. $_nC_{n-r} = \dfrac{n!}{[n - (n - r)]!(n - r)!}$

$$= \dfrac{n!}{r!(n - r)!} = \dfrac{n!}{(n - r)!r!} = {_nC_r}$$

119. $_nC_r + {_nC_{r-1}} = \dfrac{n!}{(n - r)!r!} + \dfrac{n!}{(n - r + 1)!(r - 1)!}$

$$= \dfrac{n!(n - r + 1)}{(n - r)!r!(n - r + 1)} + \dfrac{n!}{(n - r + 1)!(r - 1)!} \cdot \dfrac{r}{r}$$

$$= \dfrac{n!(n - r + 1)}{(n - r + 1)!r!} + \dfrac{n!r}{(n - r + 1)!r!}$$

$$= \dfrac{n!(n - r + 1 + r)}{(n - r + 1)!r!}$$

$$= \dfrac{n!(n + 1)}{(n - r + 1)!r!}$$

$$= \dfrac{(n + 1)!}{(n + 1 - r)!r!} = {_{n+1}C_r}$$

121. $g(x) = f(x) + 8$

$g(x)$ is shifted eight units up from $f(x)$.

123. $g(x) = f(-x)$

$g(x)$ is the reflection of $f(x)$ in the y-axis.

125. $\begin{bmatrix} -6 & 5 \\ -5 & 4 \end{bmatrix}^{-1} = \dfrac{1}{-24 + 25}\begin{bmatrix} 4 & -5 \\ 5 & -6 \end{bmatrix} = \begin{bmatrix} 4 & -5 \\ 5 & -6 \end{bmatrix}$

Section 8.6   Counting Principles

- ■ You should know The Fundamental Counting Principle.

- ■ $_nP_r = \dfrac{n!}{(n-r)!}$ is the number of permutations of n elements taken r at a time.

- ■ Given a set of n objects that has n_1 of one kind, n_2 of a second kind, and so on, the number of distinguishable permutations is

$$\frac{n!}{n_1! n_2! \cdots n_k!}.$$

- ■ $_nC_r = \dfrac{n!}{(n-r)! r!}$ is the number of combinations of n elements taken r at a time.

Vocabulary Check

1. Fundamental Counting Principle

2. permutation

3. $_nP_r = \dfrac{n!}{(n-r)!}$

4. distinguishable permutations

5. combinations

---

1. Odd integers: 1, 3, 5, 7, 9, 11

6 ways

3. Prime integers: 2, 3, 5, 7, 11

5 ways

5. Divisible by 4: 4, 8, 12

3 ways

7. Sum is 8:

$1 + 7, 2 + 6, 3 + 5, 4 + 4, 5 + 3, 6 + 2, 7 + 1$

7 ways

9. Amplifiers: 4 choices

Compact disc players: 6 choices

Speakers: 5 choices

Total: $4 \cdot 6 \cdot 5 = 120$ ways

11. $2^{10} = 1024$ ways

13. (a) $9 \cdot 10 \cdot 10 = 900$

(b) $9 \cdot 9 \cdot 8 = 648$

15. $2(8 \cdot 10 \cdot 10)(10 \cdot 10 \cdot 10 \cdot 10) = 16{,}000{,}000$ numbers

17. (a) $26^3 + 26^3 = 35{,}152$

(b) There are $2 \cdot 25^3$ possibilities that don't have Q. Hence, $2 \cdot 26^3 - 2 \cdot 25^3 = 3902$ have at least one Q.

19. (a) $10^5 = 100{,}000$ zip codes

(b) $2 \cdot 10^4 = 20{,}000$ zip codes beginning with a one or a two

21. (a) $6 \cdot 5 \cdot 4 \cdot 3 \cdot 2 \cdot 1 = 720$

(b) $6 \cdot 1 \cdot 4 \cdot 1 \cdot 2 \cdot 1 = 48$

23. $_nP_r = \dfrac{n!}{(n-r)!}$

So, $_4P_4 = \dfrac{4!}{0!} = 4! = 24.$

25. $_8P_3 = \dfrac{8!}{5!} = 8 \cdot 7 \cdot 6 = 336$

27. $_5P_4 = \dfrac{5!}{1!} = 120$

29. $_{20}P_6 = 27,907,200$

31. $_{120}P_4 = 197,149,680$

33. $5! = 120$ ways

35. $9! = 362,880$ ways

37. $_{12}P_4 = \dfrac{12!}{8!}$

$= 12 \cdot 11 \cdot 10 \cdot 9$

$= 11,880$ ways

39. $37 \cdot 37 \cdot 37 = 50,653$

41.

ABCD	BACD	CABD	DABC
ABDC	BADC	CADB	DACB
ACBD	BCAD	CBAD	DBAC
ACDB	BCDA	CBDA	DBCA
ADBC	BDAC	CDAB	DCAB
ADCB	BDCA	CDBA	DCBA

43. $\dfrac{7!}{2!\,1!\,3!\,1!} = \dfrac{7!}{2!\,3!} = 420$

45. $\dfrac{7!}{2!\,1!\,1!\,1!\,1!\,1!\,1!} = \dfrac{7!}{2!}$

$= 7 \cdot 6 \cdot 5 \cdot 4 \cdot 3$

$= 2520$

47. $_5C_2 = \dfrac{5!}{2!\,3!} = \dfrac{5 \cdot 4}{2} = 10$

49. $_4C_1 = \dfrac{4!}{1!\,3!} = 4$

51. $_{25}C_0 = \dfrac{25!}{0!\,25!} = 1$

53. $_{20}C_4 = 4845$

55. $_{42}C_5 = 850,668$

57. AB, AC, AD, AE, AF,
BC, BD, BE, BF, CD
CE, CF, DE, DF, EF

$_6C_2 = 15$ ways

59. $_{100}C_{14} = \dfrac{100!}{14!\,86!} \approx 4.42 \times 10^{16}$ ways

61. $_{49}C_6 = 13,983,816$ ways

63. $_9C_2 = 36$ lines

65. Select type of card for three of a kind: $_{13}C_1$

Select three of four cards for three of a kind: $_4C_3$

Select type of card for pair: $_{12}C_1$

Select two of four cards for pair: $_4C_2$

$_{13}C_1 \cdot {}_4C_3 \cdot {}_{12}C_1 \cdot {}_4C_2 = 13 \cdot 4 \cdot 12 \cdot 6 = 3744$ ways to get a full house

67. (a) $_{12}C_4 = 495$ ways

(b) $(_5C_2)(_7C_2) = (10)(21) = 210$ ways

69. $(_7C_1)(_{12}C_3)(_{20}C_2) = 7 \cdot 220 \cdot 190$

$= 292,600$ ways

71. $_5C_2 - 5 = 10 - 5 = 5$ diagonals

73. $_8C_2 - 8 = 28 - 8 = 20$ diagonals

75. $$14 \cdot {}_nP_3 = {}_{n+2}P_4$$

Note: $n \geq 3$ for this to be defined.

$$14\left[\frac{n!}{(n-3)!}\right] = \frac{(n+2)!}{(n-2)!}$$

$$14n(n-1)(n-2) = (n+2)(n+1)n(n-1) \quad \text{(We can divide here by } n(n-1) \text{ since } n \neq 0, n \neq 1.)$$

$$14n - 28 = n^2 + 3n + 2$$

$$0 = n^2 - 11n + 30$$

$$0 = (n-5)(n-6)$$

$$n = 5 \quad \text{or} \quad n = 6$$

77. $${}_nP_4 = 10 \cdot {}_{n-1}P_3$$

$$\frac{n!}{(n-4)!} = 10\frac{(n-1)!}{(n-4)!}$$

$$n! = 10(n-1)!$$

$$n = 10$$

79. $${}_{n+1}P_3 = 4 \cdot {}_nP_2$$

$$\frac{(n+1)!}{(n-2)!} = 4\frac{n!}{(n-2)!}$$

$$(n+1)! = 4n!$$

$$n = 3$$

81. $$4 \cdot {}_{n+1}P_2 = {}_{n+2}P_3$$

$$4\frac{(n+1)!}{(n-1)!} = \frac{(n+2)!}{(n-1)!}$$

$$4(n+1)! = (n+2)!$$

$$n = 2$$

83. False

85. $${}_{100}P_{80} \approx 3.836 \times 10^{139}.$$

This number is too large for some calculators to evaluate.

87. $${}_nC_r = {}_nC_{n-r} = \frac{n!}{r!(n-r)!}$$

89. $${}_nP_{n-1} = \frac{n!}{(n-(n-1))!} = \frac{n!}{1!} = \frac{n!}{0!} = {}_nP_n$$

91. $${}_nC_{n-1} = \frac{n!}{[n-(n-1)]!(n-1)!}$$

$$= \frac{n!}{(1)!(n-1)!}$$

$$= \frac{n!}{(n-1)!1!} = {}_nC_1$$

93. From the graph of $y = \sqrt{x-3} - x + 6$, you see that there is one zero, $x \approx 8.303$. Analytically,

$$\sqrt{x-3} = x - 6$$

$$x - 3 = x^2 - 12x + 36$$

$$0 = x^2 - 13x + 39.$$

By the Quadratic Formula, $x = \dfrac{13 \pm \sqrt{(-13)^2 - 4(39)}}{2} = \dfrac{13 \pm \sqrt{13}}{2}.$

Selecting the larger solution, $x = \dfrac{13 + \sqrt{13}}{2} \approx 8.303.$ (The other solution is extraneous.)

95. $$\log_2(x-3) = 5$$

$$2^5 = x - 3$$

$$2^5 + 3 = x$$

$$x = 35$$

97. $x = \dfrac{\begin{vmatrix} -14 & 3 \\ 2 & -2 \end{vmatrix}}{\begin{vmatrix} -5 & 3 \\ 7 & -2 \end{vmatrix}} = \dfrac{22}{-11} = -2$

99. $x = \dfrac{\begin{vmatrix} -1 & -4 \\ -4 & 5 \end{vmatrix}}{\begin{vmatrix} -3 & -4 \\ 9 & 5 \end{vmatrix}} = \dfrac{-21}{21} = -1$

$y = \dfrac{\begin{vmatrix} -5 & -14 \\ 7 & 2 \end{vmatrix}}{\begin{vmatrix} -5 & 3 \\ 7 & -2 \end{vmatrix}} = \dfrac{88}{-11} = -8$

$y = \dfrac{\begin{vmatrix} -3 & -1 \\ 9 & -4 \end{vmatrix}}{\begin{vmatrix} -3 & -4 \\ 9 & 5 \end{vmatrix}} = \dfrac{21}{21} = 1$

Answer: $(-2, -8)$

Answer: $(-1, 1)$

Section 8.7   Probability

You should know the following basic principles of probability.

- If an event E has $n(E)$ equally likely outcomes and its sample space has $n(S)$ equally likely outcomes, then the probability of event E is

$$P(E) = \frac{n(E)}{n(S)}, \text{ where } 0 \leq P(E) \leq 1.$$

- If A and B are mutually exclusive events, then $P(A \cup B) = P(A) + P(B)$.

  If A and B are not mutually exclusive events, then $P(A \cup B) = P(A) + P(B) - P(A \cap B)$.

- If A and B are independent events, then the probability that both A and B will occur is $P(A)P(B)$.

- The probability of the complement of an event A is $P(A') = 1 - P(A)$.

Vocabulary Check

1. experiment, outcomes

2. sample space

3. probability

4. impossible, certain

5. mutually exclusive

6. independent

7. complement

8. (a) iii   (b) i   (c) iv   (d) ii

---

1. $\{(H, 1), (H, 2), (H, 3), (H, 4), (H, 5), (H, 6),$
$(T, 1), (T, 2), (T, 3), (T, 4), (T, 5), (T, 6)\}$

3. $\{ABC, ACB, BAC, BCA, CAB, CBA\}$

5. $\{(A, B), (A, C), (A, D), (A, E), (B, C),$
$(B, D), (B, E), (C, D), (C, E), (D, E)\}$

7. $E = \{HTT, THT, TTH\}$

$P(E) = \dfrac{n(E)}{n(S)} = \dfrac{3}{8}$

9. $E = \{HHH, HHT, HTH, HTT, THH, THT, TTH\}$

$P(E) = \dfrac{n(E)}{n(S)} = \dfrac{7}{8}$

11. $E = \{K, K, K, K, Q, Q, Q, Q, J, J, J, J\}$

$P(E) = \dfrac{n(E)}{n(S)} = \dfrac{12}{52} = \dfrac{3}{13}$

13. $E = \{A, A, A, A, K, K, K, K, Q, Q, Q, Q, J, J, J, J\}$

$P(E) = \dfrac{n(E)}{n(S)} = \dfrac{16}{52} = \dfrac{4}{13}$

15. $E = \{(1, 5), (2, 4), (3, 3), (4, 2), (5, 1)\}$

$P(E) = \dfrac{n(E)}{n(S)} = \dfrac{5}{36}$

17. not $E = \{(5, 6), (6, 5), (6, 6)\}$

$n(E) = n(S) - n(\text{not } E) = 36 - 3 = 33$

$P(E) = \dfrac{n(E)}{n(S)} = \dfrac{33}{36} = \dfrac{11}{12}$

19. $P(E) = \dfrac{{}_3C_2}{{}_6C_2} = \dfrac{3}{15} = \dfrac{1}{5}$

21. $P(E) = \dfrac{{}_4C_2}{{}_6C_2} = \dfrac{6}{15} = \dfrac{2}{5}$

23. $P(E') = 1 - P(E) = 1 - 0.75 = 0.25$

25. $P(E') = 1 - P(E) = 1 - \frac{2}{3} = \frac{1}{3}$

27. $P(E) = 1 - P(E') = 1 - p = 1 - 0.12 = 0.88$

29. $P(E) = 1 - P(E') = 1 - \frac{13}{20} = \frac{7}{20}$

31. (a) $0.15(8.15) \approx 1.22$ million

(b) $\dfrac{0.41}{1.0} = 0.41$

(c) $\dfrac{0.24}{1.0} = 0.24$

(d) $\dfrac{0.24 + 0.02}{1.0} = 0.26$

33. (a) $(0.128)(293.66) \approx 37.6$ million

(b) $\dfrac{0.01}{1.0} = 0.01$

(c) $\dfrac{0.01 + 0.002}{1.0} = 0.012$

35. (a) $\dfrac{34}{100} = 0.34$

(b) $\dfrac{45}{100} = 0.45$

(c) $\dfrac{23}{100} = 0.23$

37. (a) $\dfrac{672}{1254}$

(b) $\dfrac{582}{1254}$

(c) $\dfrac{672 - 124}{1254} = \dfrac{548}{1254}$

39. $p + p + 2p = 1$

$p = 0.25$

Taylor: $0.50 = \dfrac{1}{2}$

Perez: $0.25 = \dfrac{1}{4}$

Moore: $0.25 = \dfrac{1}{4}$

41. (a) $\dfrac{{}_{15}C_{10}}{{}_{20}C_{10}} = \dfrac{3003}{184{,}756} = \dfrac{21}{1292} \approx 0.016$

(b) $\dfrac{{}_{15}C_8 \cdot {}_5C_2}{{}_{20}C_{10}} = \dfrac{64{,}350}{184{,}756} = \dfrac{225}{646} \approx 0.348$

(c) $\dfrac{{}_{15}C_9 \cdot {}_5C_1}{{}_{20}C_{10}} + \dfrac{{}_{15}C_{10}}{{}_{20}C_{10}} = \dfrac{25{,}025 + 3003}{184{,}756}$

$= \dfrac{28{,}028}{184{,}756} = \dfrac{49}{323} \approx 0.152$

43. (a) $\dfrac{1}{{}_5P_5} = \dfrac{1}{120}$

(b) $\dfrac{1}{{}_4P_4} = \dfrac{1}{24}$

45. (a) There are three letters to be selected, and two must be Q and Y.

QY__, YQ__, Q__Y, Y__Q, __YQ, __QY

Thus, the probability is

$\dfrac{6(26)}{26^3} = \dfrac{6}{26^2} \approx 0.008876.$

(b) The three letters must be Q, Y, and X.

QYX, QXY, YQX, YXQ, XQY, XYQ

Thus, the probability is $\dfrac{6}{26^3} = \dfrac{3}{8778}.$

47. (a) $\dfrac{100}{(_{55}C_5)(_{42}C_1)} = \dfrac{100}{(3{,}478{,}761)(42)}$

(b) $\dfrac{1000}{(_{55}C_5)(_{42}C_1)} = \dfrac{1000}{(3{,}478{,}761)(42)}$

49. (a) $\dfrac{20}{52} = \dfrac{5}{13}$

(b) $\dfrac{13 + 13}{52} = \dfrac{1}{2}$

(c) $\dfrac{4 + 12}{52} = \dfrac{4}{13}$

51. (a) $\dfrac{_9C_4}{_{12}C_4} = \dfrac{126}{495} = \dfrac{14}{55}$ (4 good units)

(b) $\dfrac{(_9C_2)(_3C_2)}{_{12}C_4} = \dfrac{108}{495} = \dfrac{12}{55}$ (2 good units)

(c) $\dfrac{(_9C_3)(_3C_1)}{_{12}C_4} = \dfrac{252}{495} = \dfrac{28}{55}$ (3 good units)

At least 2 good units: $\dfrac{12}{55} + \dfrac{28}{55} + \dfrac{14}{55} = \dfrac{54}{55}$

53. $(0.32)^2 = 0.1024$

55. (a) $P(SS) = (0.985)^2 \approx 0.9702$

(b) $P(S) = 1 - P(FF) = 1 - (0.015)^2 \approx 0.9998$

(c) $P(FF) = (0.015)^2 \approx 0.0002$

57. (a) $\left(\dfrac{1}{5}\right)^6 = \dfrac{1}{15{,}625}$

(b) $\left(\dfrac{4}{5}\right)^6 = \dfrac{4096}{15{,}625} = 0.262144$

(c) $1 - 0.262144 = 0.737856 = \dfrac{11{,}529}{15{,}625}$

59. (a) If the *center* of the coin falls within the circle of radius $d/2$ around a vertex, the coin will cover the vertex.

$$P(\text{coin covers a vertex}) = \dfrac{\substack{\text{Area in which coin may fall} \\ \text{so that it covers a vertex}}}{\text{Total area}} = \dfrac{n\left[\pi\left(\dfrac{d}{2}\right)^2\right]}{nd^2} = \dfrac{1}{4}\pi$$

(b) Experimental results will vary.

61. True. $P(E) + P(E') = 1$

63. (a) As you consider successive people with distinct birthdays, the probabilities must decrease to take into account the birth dates already used. Since the birth dates of people are independent events, multiply the respective probabilities of distinct birthdays.

(b) $\dfrac{365}{365} \cdot \dfrac{364}{365} \cdot \dfrac{363}{365} \cdot \dfrac{362}{365}$

(c) $P_1 = \dfrac{365}{365} = 1$

$P_2 = \dfrac{365}{365} \cdot \dfrac{364}{365} = \dfrac{364}{365}P_1 = \dfrac{365 - (2 - 1)}{365}P_1$

$P_3 = \dfrac{365}{365} \cdot \dfrac{364}{365} \cdot \dfrac{363}{365} = \dfrac{363}{365}P_2 = \dfrac{365 - (3 - 1)}{365}P_2$

$P_n = \dfrac{365}{365} \cdot \dfrac{364}{365} \cdot \dfrac{363}{365} \cdot \ldots \cdot \dfrac{365 - (n - 1)}{365} = \dfrac{365 - (n - 1)}{365}P_{n-1}$

—CONTINUED—

63. **—CONTINUED—**

(d) Q_n is the probability that the birthdays are *not* distinct which is equivalent to at least

2 people having the same birthday.

(e)

n	10	15	20	23	30	40	50
P_n	0.88	0.75	0.59	0.49	0.29	0.11	0.03
Q_n	0.12	0.25	0.41	(0.51)	0.71	0.89	0.97

(f) 23, See the chart above.

65. $\dfrac{2}{x - 5} = 4$

$2 = 4(x - 5) = 4x - 20$

$4x = 22$

$x = \dfrac{11}{2}$

67. $\dfrac{3}{x - 2} + \dfrac{x}{x + 2} = 1$

$3(x + 2) + x(x - 2) = (x - 2)(x + 2)$

$3x + 6 + x^2 - 2x = x^2 - 4$

$x = -10$

69. $e^x + 7 = 35$

$e^x = 28$

$x = \ln(28) \approx 3.332$

71. $4 \ln 6x = 16$

$\ln 6x = 4$

$e^4 = 6x$

$x = \frac{1}{6}e^4 \approx 9.10$

73. $_5P_3 = \dfrac{5!}{(5 - 3)!} = \dfrac{120}{2} = 60$

75. $_{11}P_8 = \dfrac{11!}{(11 - 8)!} = \dfrac{11!}{3!} = 6,652,800$

77. $_6C_2 = \dfrac{6!}{4!2!} = \dfrac{6 \cdot 5 \cdot 4!}{4!2} = 15$

79. $_{11}C_8 = \dfrac{11!}{8!3!} = \dfrac{11 \cdot 10 \cdot 9 \cdot 8!}{8!6} = 165$

Review Exercises for Chapter 8

1. $a_n = \dfrac{2^n}{2^n + 1}$

$a_1 = \dfrac{2^1}{2^1 + 1} = \dfrac{2}{3}$

$a_2 = \dfrac{2^2}{2^2 + 1} = \dfrac{4}{5}$

$a_3 = \dfrac{2^3}{2^3 + 1} = \dfrac{8}{9}$

$a_4 = \dfrac{2^4}{2^4 + 1} = \dfrac{16}{17}$

$a_5 = \dfrac{2^5}{2^5 + 1} = \dfrac{32}{33}$

3. $a_n = \dfrac{(-1)^n}{n!}$

$a_1 = \dfrac{(-1)^1}{1!} = -1$

$a_2 = \dfrac{(-1)^2}{2!} = \dfrac{1}{2}$

$a_3 = \dfrac{(-1)^3}{3!} = -\dfrac{1}{6}$

$a_4 = \dfrac{(-1)^4}{4!} = \dfrac{1}{24}$

$a_5 = \dfrac{(-1)^5}{5!} = -\dfrac{1}{120}$

5. Common difference is 5.

$a_n = 5n, n = 1, 2, \ldots$

7. Denominators are successive odd numbers.

$$a_n = \frac{2}{2n-1}, n = 1, 2, 3, \ldots$$

9. $a_1 = 9, a_{k+1} = a_k - 4$

$a_2 = a_1 - 4 = 9 - 4 = 5$

$a_3 = 5 - 4 = 1$

$a_4 = 1 - 4 = -3$

$a_5 = -3 - 4 = -7$

11. $\dfrac{18!}{20!} = \dfrac{18!}{20 \cdot 19 \cdot 18!}$

$= \dfrac{1}{20 \cdot 19} = \dfrac{1}{380}$

13. $\dfrac{(n+1)!}{(n-1)!} = \dfrac{(n+1)n(n-1)!}{(n-1)!} = n(n+1)$

15. $\displaystyle\sum_{i=1}^{6} 5 = 6(5) = 30$

17. $\displaystyle\sum_{j=1}^{4} \frac{6}{j^2} = \frac{6}{1^2} + \frac{6}{2^2} + \frac{6}{3^2} + \frac{6}{4^2}$

$= 6 + \dfrac{3}{2} + \dfrac{2}{3} + \dfrac{3}{8} = \dfrac{205}{24}$

19. $\displaystyle\sum_{k=1}^{100} 2k^3 = 2 \cdot \frac{100^2(101)^2}{4} = 51,005,000$

21. $\displaystyle\sum_{n=0}^{50} (n^2 + 3) = \frac{50(51)(101)}{6} + 3(51) = 43,078$

23. $\dfrac{1}{2(1)} + \dfrac{1}{2(2)} + \dfrac{1}{2(3)} + \cdots + \dfrac{1}{2(20)} = \displaystyle\sum_{k=1}^{20} \frac{1}{2k}$

≈ 1.799

25. $\dfrac{1}{2} + \dfrac{2}{3} + \dfrac{3}{4} + \cdots + \dfrac{9}{10} = \displaystyle\sum_{k=1}^{9} \frac{k}{k+1} \approx 7.071$

27. (a) $\displaystyle\sum_{k=1}^{4} \frac{5}{10^k} = \frac{5}{10} + \frac{5}{100} + \frac{5}{1000} + \frac{5}{10,000} = 0.5 + 0.05 + 0.005 + 0.0005 = 0.5555 = \frac{1111}{2000}$

(b) $\displaystyle\sum_{k=1}^{\infty} \frac{5}{10^k} = \frac{5}{10} \sum_{k=0}^{\infty} \frac{1}{10^k} = \frac{5}{10} \cdot \frac{1}{1 - 1/10} = \frac{5}{10} \cdot \frac{10}{9} = \frac{5}{9}$

29. $\displaystyle\sum_{k=1}^{\infty} 2(0.5)^k$

(a) $\displaystyle\sum_{k=1}^{4} 2(0.5)^k = 2(0.5) + 2(0.5)^2 + 2(0.5)^3 + 2(0.5)^4$

$= 1.875 = \dfrac{15}{8}$

(b) $\displaystyle\sum_{k=1}^{\infty} 2(0.5)^k = 2(0.5)\frac{1}{1 - 0.5} = 2$

31. $a_n = 2500\left(1 + \dfrac{0.02}{4}\right)^n, n = 1, 2, 3$

(a) $a_1 = 2500\left(1 + \dfrac{0.02}{4}\right)^1 = 2512.5$

$a_2 = 2525.06 \qquad a_3 = 2537.69$

$a_4 = 2550.38 \qquad a_5 = 2563.13$

$a_6 = 2575.94 \qquad a_7 = 2588.82$

$a_8 = 2601.77$

(b) $a_{40} = 2500\left(1 + \dfrac{0.02}{4}\right)^{40} = \$3051.99$

33. Yes

$d = 3 - 5 = -2$

35. Yes

$d = 1 - \tfrac{1}{2} = \tfrac{1}{2}$

37. $a_1 = 3, d = 4$

$a_1 = 3$

$a_2 = 3 + 4 = 7$

$a_3 = 7 + 4 = 11$

$a_4 = 11 + 4 = 15$

$a_5 = 15 + 4 = 19$

39. $a_4 = 10, \ a_{10} = 28$

$a_{10} = a_4 + 6d$

$28 = 10 + 6d$

$18 = 6d$

$3 = d$

$a_1 = a_4 - 3d$

$a_1 = 10 - 3(3)$

$a_1 = 1$

$a_2 = 1 + 3 = 4$

$a_3 = 4 + 3 = 7$

$a_4 = 7 + 3 = 10$

$a_5 = 10 + 3 = 13$

41. $a_1 = 35, \ a_{k+1} = a_k - 3$

$a_1 = 35$

$a_2 = a_1 - 3 = 35 - 3 = 32$

$a_3 = a_2 - 3 = 32 - 3 = 29$

$a_4 = a_3 - 3 = 29 - 3 = 26$

$a_5 = a_4 - 3 = 26 - 3 = 23$

$a_n = 35 + (n-1)(-3) = 38 - 3n, d = -3$

43. $a_1 = 9, \ a_{k+1} = a_k + 7$

$a_1 = 9$

$a_2 = a_1 + 7 = 9 + 7 = 16$

$a_3 = a_2 + 7 = 16 + 7 = 23$

$a_4 = a_3 + 7 = 23 + 7 = 30$

$a_5 = a_4 + 7 = 30 + 7 = 37$

$a_n = 9 + (n-1)(7) = 2 + 7n, d = 7$

45. $a_n = 100 + (n-1)(-3) = 103 - 3n$

$$\sum_{n=1}^{20} (103 - 3n) = \sum_{n=1}^{20} 103 - 3\sum_{n=1}^{20} n = 20(103) - 3\left[\frac{(20)(21)}{2}\right] = 1430$$

47. $\displaystyle\sum_{j=1}^{10} (2j - 3) = 2\sum_{j=1}^{10} j - \sum_{j=1}^{10} 3$

$$= 2\left[\frac{10(11)}{2}\right] - 10(3) = 80$$

49. $\displaystyle\sum_{k=1}^{11} \left(\frac{2}{3}k + 4\right) = \frac{2}{3}\sum_{k=1}^{11} k + \sum_{k=1}^{11} 4$

$$= \frac{2}{3} \cdot \frac{(11)(12)}{2} + 11(4) = 88$$

51. $\displaystyle\sum_{k=1}^{100} 5k = 5\left[\frac{(100)(101)}{2}\right] = 25{,}250$

53. (a) $34{,}000 + 4(2250) = \$43{,}000$

(b) $\displaystyle\sum_{k=1}^{5} [34{,}000 + (k-1)(2250)]$

$$= \sum_{k=1}^{5} (31{,}750 + 2250k)$$

$$= \$192{,}500$$

55. $5, 10, 20, 40$

Geometric: $r = 2$

57. Geometric:

$r = -\dfrac{1}{3}$

59. $a_1 = 4$, $r = -\frac{1}{4}$

$a_1 = 4$

$a_2 = 4\left(-\frac{1}{4}\right) = -1$

$a_3 = -1\left(-\frac{1}{4}\right) = \frac{1}{4}$

$a_4 = \frac{1}{4}\left(-\frac{1}{4}\right) = -\frac{1}{16}$

$a_5 = -\frac{1}{16}\left(-\frac{1}{4}\right) = \frac{1}{64}$

61. $a_1 = 9$, $a_3 = 4$

$a_3 = a_1 r^2$

$4 = 9r^2$

$\frac{4}{9} = r^2 \implies r = \pm\frac{2}{3}$

$a_1 = 9$

$a_2 = 9\left(\frac{2}{3}\right) = 6$

$a_3 = 6\left(\frac{2}{3}\right) = 4$ or

$a_4 = 4\left(\frac{2}{3}\right) = \frac{8}{3}$

$a_5 = \frac{8}{3}\left(\frac{2}{3}\right) = \frac{16}{9}$

$a_1 = 9$

$a_2 = 9\left(-\frac{2}{3}\right) = -6$

$a_3 = -6\left(-\frac{2}{3}\right) = 4$

$a_4 = 4\left(-\frac{2}{3}\right) = -\frac{8}{3}$

$a_5 = -\frac{8}{3}\left(-\frac{2}{3}\right) = \frac{16}{9}$

63. $a_1 = 120$, $a_{k+1} = \frac{1}{3}a_k$

$a_1 = 120$

$a_2 = \frac{1}{3}(120) = 40$

$a_3 = \frac{1}{3}(40) = \frac{40}{3}$

$a_4 = \frac{1}{3}\left(\frac{40}{3}\right) = \frac{40}{9}$

$a_5 = \frac{1}{3}\left(\frac{40}{9}\right) = \frac{40}{27}$

$a_n = 120\left(\frac{1}{3}\right)^{n-1}$, $r = \frac{1}{3}$

65. $a_1 = 25$, $a_{k+1} = -\frac{3}{5}a_k$

$a_1 = 25$

$a_2 = -\frac{3}{5}(25) = -15$

$a_3 = -\frac{3}{5}(-15) = 9$

$a_4 = -\frac{3}{5}(9) = -\frac{27}{3}$

$a_5 = -\frac{3}{5}\left(-\frac{27}{5}\right) = \frac{81}{25}$

$a_n = 25\left(-\frac{3}{5}\right)^{n-1}$, $r = -\frac{3}{5}$

67. $a_2 = a_1 r$

$-8 = 16r$

$-\dfrac{1}{2} = r$

$a_n = 16\left(-\dfrac{1}{2}\right)^{n-1}$

$\displaystyle\sum_{n=1}^{20} 16\left(-\frac{1}{2}\right)^{n-1} = 16\left[\frac{1 - (-1/2)^{20}}{1 - (-1/2)}\right] \approx 10.67$

69. $a_1 = 100$, $r = 1.05$

$a_n = 100(1.05)^{n-1}$

$\displaystyle\sum_{n=1}^{20} 100(1.05)^{n-1} = 100\left[\frac{1 - (1.05)^{20}}{1 - 1.05}\right] \approx 3306.60$

71. $\displaystyle\sum_{i=1}^{7} 2^{i-1} = \frac{1 - 2^7}{1 - 2} = 127$

73. $\displaystyle\sum_{n=1}^{7} (-4)^{n-1} = \frac{1 - (-4)^7}{1 - (-4)} = 3277$

75. $\displaystyle\sum_{n=0}^{4} 250(1.02)^n = 250\left(\frac{1 - 1.02^5}{1 - 1.02}\right) = 1301.01004$

77. $\displaystyle\sum_{i=1}^{10} 10\left(\frac{3}{5}\right)^{i-1} \approx 24.849$

79. $\displaystyle\sum_{i=1}^{\infty} 4\left(\frac{7}{8}\right)^{i-1} = \sum_{i=0}^{\infty} 4\left(\frac{7}{8}\right)^{i} = \frac{4}{1 - 7/8} = 32$

81. $\displaystyle\sum_{k=1}^{\infty} 4\left(\frac{2}{3}\right)^{k-1} = \frac{4}{1 - 2/3} = 12$

83. (a) $a_t = 120,000(0.7)^t$

(b) $a_5 = 120,000(0.7)^5 = \$20,168.40$

85. 1. When $n = 1, 2 = \frac{1}{2}(5(1) - 1)$.

2. Assume that $S_k = 2 + 7 + \cdots + (5k - 3) = \frac{k}{2}(5k - 1)$. Then,

$$S_{k+1} = 2 + 7 + \cdots + (5k - 3) + [5(k + 1) - 3]$$

$$= S_k + 5k + 2$$

$$= \frac{k}{2}(5k - 1) + 5k + 2$$

$$= \frac{1}{2}[5k^2 + 9k + 4]$$

$$= \frac{1}{2}[(5k + 4)(k + 1)]$$

$$= \frac{k + 1}{2}(5(k + 1) - 1).$$

Therefore, by mathematical induction, the formula is true for all positive integers n.

87. 1. When $n = 1, a = a\left(\dfrac{1 - r}{1 - r}\right)$.

2. Assume that

$$S_k = \sum_{i=0}^{k-1} ar^i = \frac{a(1 - r^k)}{1 - r}.$$

Then,

$$S_{k+1} = \sum_{i=0}^{k} ar^i = \sum_{i=0}^{k-1} ar^i + ar^k = \frac{a(1 - r^k)}{1 - r} + ar^k$$

$$= \frac{a(1 - r^k + r^k - r^{k+1})}{1 - r} = \frac{a(1 - r^{k+1})}{1 - r}.$$

Therefore, by mathematical induction, the formula is valid for all positive integer values of n.

89. $\displaystyle\sum_{n=1}^{30} n = \frac{30(31)}{2} = 465$

91. $\displaystyle\sum_{n=1}^{7} (n^4 - n) = \sum_{n=1}^{7} n^4 - \sum_{n=1}^{7} n$

$$= \frac{7(8)(15)[3(7)^2 + 3(7) - 1]}{30} - \frac{7(8)}{2}$$

$$= \frac{840(167)}{30} - 28 = 4676 - 28 = 4648$$

93. $a_1 = f(1) = 5$

$a_2 = a_1 + 5 = 5 + 5 = 10$

$a_3 = a_2 + 5 = 15$

$a_4 = a_3 + 5 = 20$

$a_5 = a_4 + 5 = 25$

n:	1	2	3	4	5
a_n:	5	10	15	20	25

First differences: 5 5 5 5

Second difference: 0 0 0

Linear model: $a_n = 5n$

95. $a_1 = f(1) = 16$

$a_2 = a_1 - 1 = 16 - 1 = 15$

$a_3 = a_2 - 1 = 15 - 1 = 14$

$a_4 = 14 - 1 = 13$

$a_5 = 13 - 1 = 12$

n:	1	2	3	4	5
a_n:	16	15	14	13	12

First differences: -1 -1 -1 -1

Second difference: 0 0 0

Linear model: $a_n = 17 - n$

97. $_{10}C_8 = 45$

99. $\binom{9}{4} = {}_9C_4 = 126$

101. 4th number in 6th row is $_6C_3 = 20$.

103. 5th number in 8th row is $\binom{8}{4} = {}_8C_4 = 70$.

105. $(x + 5)^4 = x^4 + 4x^3(5) + 6x^2(5^2) + 4x(5^3) + 5^4$

$= x^4 + 20x^3 + 150x^2 + 500x + 625$

107. $(a - 4b)^5 = a^5 - 5a^4(4b) + 10a^3(4b)^2 - 10a^2(4b)^3 + 5a(4b)^4 - (4b)^5$

$= a^5 - 20a^4b + 160a^3b^2 - 640a^2b^3 + 1280ab^4 - 1024b^5$

109. $(7 + 2i)^4 = 7^4 + 4(7)^3(2i) + 6(7)^2(2i)^2 + 4(7)(2i)^3 + (2i)^4$

$= 2401 + 2744i - 1176 - 224i + 16$

$= 1241 + 2520i$

111. $E = \{(1, 11), (2, 10), (3, 9), (4, 8), (5, 7), (7, 5), (8, 4), (9, 3), (10, 2), (11, 1)\}$

$n(E) = 10$

113. (a) $(4)(3)(6)(3) = 216$ schedules

(b) $(2)(3)(6)(3) = 108$ schedules

(c) $(2)(3)(2)(3) = 36$ schedules

115. $_{10}C_8 = \dfrac{10!}{2!8!} = \dfrac{10 \cdot 9}{2} = 45$

117. $_{12}P_{10} = \dfrac{12!}{2!} = 239{,}500{,}800$

119. $_{100}C_{98} = \dfrac{100!}{2!98!} = \dfrac{100 \cdot 99}{2} = 4950$

121. $_{1000}P_2 = \dfrac{1000!}{998!} = 1000(999) = 999{,}000$

123. $\dfrac{8!}{2!2!2!1!1!} = \dfrac{8!}{8} = 7! = 5040$ permutations

125. $10! = 3{,}628{,}800$ ways

127. $_{20}C_{15} = 15{,}504$ ways

129. $_{n+1}P_2 = 4 \cdot {}_nP_1$

$\dfrac{(n + 1)!}{(n - 1)!} = 4 \cdot \dfrac{n!}{(n - 1)!}$

$(n + 1)! = 4 \cdot n!$

$n = 3$

131. $\frac{10}{10} \cdot \frac{1}{9} = \frac{1}{9}$

133. (a) $\frac{208}{500} = 0.416$

(b) $\frac{400}{500} = 0.8$

(c) $\frac{37}{500} = 0.074$

135. $P(2 \text{ pairs}) = \dfrac{(_{13}C_2)(_4C_2)(_4C_2)(_{44}C_1)}{(_{52}C_5)} = 0.0475$

137. True

$$\frac{(n + 2)!}{n!} = \frac{(n + 2)(n + 1)n!}{n!} = (n + 2)(n + 1)$$

139. Answers will vary. See pages 526 and 535.

141. (a) Arithmetic-linear model

(b) Geometric model

143. Answers will vary. See page 528. To define a sequence recursively, you need to be given one or more of the first few terms. All other terms are defined using previous terms.

145. If n is even, the expansion are the same. If n is odd, the expansion of $(-x + y)^n$ is the negative of that of $(x - y)^n$.

Chapter 8   Practice Test

1. Write out the first five terms of the sequence $a_n = \dfrac{2n}{(n + 2)!}$.

2. Write an expression for the nth term of the sequence $\left\{ \frac{4}{3}, \frac{5}{9}, \frac{6}{27}, \frac{7}{81}, \frac{8}{243}, \ldots \right\}$.

3. Find the sum $\displaystyle\sum_{i=1}^{6} (2i - 1)$.

4. Write out the first five terms of the arithmetic sequence where $a_1 = 23$ and $d = -2$.

5. Find a_{50} for the arithmetic sequence with $a_1 = 12$, $d = 3$, and $n = 50$.

6. Find the sum of the first 200 positive integers.

7. Write out the first five terms of the geometric sequence with $a_1 = 7$ and $r = 2$.

8. Evaluate $\displaystyle\sum_{n=0}^{9} 6\left(\frac{2}{3}\right)^n$.

9. Evaluate $\displaystyle\sum_{n=0}^{\infty} (0.03)^n$.

10. Use mathematical induction to prove that $1 + 2 + 3 + 4 + \cdots + n = \dfrac{n(n + 1)}{2}$.

11. Use mathematical induction to prove that $n! > 2^n$, $n \geq 4$.

12. Evaluate $_{13}C_4$. Verify with a graphing utility.

13. Expand $(x + 3)^5$.

14. Find the term involving x^7 in $(x - 2)^{12}$.

15. Evaluate $_{30}P_4$.

16. How many ways can six people sit at a table with six chairs?

17. Twelve cars run in a race. How many different ways can they come in first, second, and third place? (Assume that there are no ties.)

18. Two six-sided dice are tossed. Find the probability that the total of the two dice is less than 5.

19. Two cards are selected at random from a deck of 52 playing cards without replacement. Find the probability that the first card is a King and the second card is a black ten.

20. A manufacturer has determined that for every 1000 units it produces, 3 will be faulty. What is the probability that an order of 50 units will have one or more faulty units?

C H A P T E R   9
Topics in Analytic Geometry

Section 9.1   Circles and Parabolas **421**

Section 9.2   Ellipses **428**

Section 9.3   Hyperbolas **434**

Section 9.4   Rotation and Systems of Quadratic Equations **441**

Section 9.5   Parametric Equations **450**

Section 9.6   Polar Coordinates **455**

Section 9.7   Graphs of Polar Equations **461**

Section 9.8   Polar Equations of Conics **466**

Review Exercises **471**

Practice Test **484**

CHAPTER 9
Topics in Analytic Geometry

Section 9.1   Circles and Parabolas

- A **parabola** is the set of all points (x, y) that are equidistant from a fixed line (**directrix**) and a fixed point (**focus**) not on the line.
- The standard equation of a parabola with vertex (h, k) and
    - (a) Vertical axis $x = h$ and directrix $y = k - p$ is
      $(x - h)^2 = 4p(y - k)$, $p \neq 0$.
    - (b) Horizontal axis $y = k$ and directrix $x = h - p$ is
      $(y - k)^2 = 4p(x - h)$, $p \neq 0$.
- The tangent line to a parabola at a point P makes **equal angles** with
    - (a) the line through P and the focus.
    - (b) the axis of the parabola.

Vocabulary Check

1. conic section
2. locus
3. circle, center
4. parabola, directrix, focus
5. vertex
6. axis
7. tangent

---

1. $x^2 + y^2 = \left(\sqrt{18}\right)^2$

$x^2 + y^2 = 18$

3. Radius $= \sqrt{(3 - 1)^2 + (7 - 0)^2}$

$\quad\quad\quad = \sqrt{4 + 49} = \sqrt{53}$

$(x - h)^2 + (y - k)^2 = r^2$

$(x - 3)^2 + (y - 7)^2 = 53$

5. Diameter $= 2\sqrt{7} \implies$ radius $= \sqrt{7}$

$(x - h)^2 + (y - k)^2 = r^2$

$(x + 3)^2 + (y + 1)^2 = 7$

7. $x^2 + y^2 = 49$

Center: $(0, 0)$

Radius: 7

9. $(x + 2)^2 + (y - 7)^2 = 16$

Center: $(-2, 7)$

Radius: 4

11. $(x - 1)^2 + y^2 = 15$

Center: $(1, 0)$

Radius: $\sqrt{15}$

13. $\dfrac{1}{4}x^2 + \dfrac{1}{4}y^2 = 1$

$x^2 + y^2 = 4$

Center: $(0, 0)$

Radius: 2

15. $\dfrac{4}{3}x^2 + \dfrac{4}{3}y^2 = 1$

$x^2 + y^2 = \dfrac{3}{4}$

Center: $(0, 0)$

Radius: $\dfrac{\sqrt{3}}{2}$

17. $(x^2 - 2x + 1) + (y^2 + 6y + 9) = -9 + 1 + 9$

$(x - 1)^2 + (y + 3)^2 = 1$

Center: $(1, -3)$

Radius: 1

19. $4\left(x^2 + 3x + \dfrac{9}{4}\right) + 4(y^2 - 6y + 9) = -41 + 9 + 36$

$4\left(x + \dfrac{3}{2}\right)^2 + 4(y - 3)^2 = 4$

$\left(x + \dfrac{3}{2}\right)^2 + (y - 3)^2 = 1$

Center: $\left(-\dfrac{3}{2}, 3\right)$

Radius: 1

21. $x^2 = 16 - y^2$

$x^2 + y^2 = 16$

Center: $(0, 0)$

Radius: 4

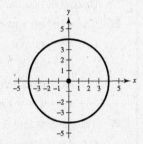

23. $x^2 + 4x + y^2 + 4y - 1 = 0$

$(x^2 + 4x + 4) + (y^2 + 4y + 4) = 1 + 4 + 4$

$(x + 2)^2 + (y + 2)^2 = 9$

Center: $(-2, -2)$

Radius: 3

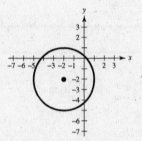

25. $x^2 - 14x + y^2 + 8y + 40 = 0$

$(x^2 - 14x + 4) + (y^2 + 8y + 16) = -40 + 49 + 16$

$(x - 7)^2 + (y + 4)^2 = 25$

Center: $(7, -4)$

Radius: 5

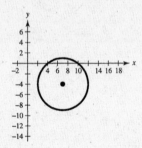

27. $x^2 + 2x + y^2 - 35 = 0$

$(x^2 + 2x + 1) + y^2 = 35 + 1$

$(x + 1)^2 + y^2 = 36$

Center: $(-1, 0)$

Radius: 6

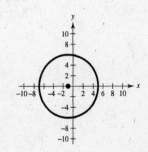

29. y-intercepts: $(0 - 2)^2 + (y + 3)^2 = 9$

$4 + (y + 3)^2 = 9$

$(y + 3)^2 = 5$

$y = -3 \pm \sqrt{5}$

$\left(0, -3 \pm \sqrt{5}\right)$

x-intercepts: $(x - 2)^2 + (0 + 3)^2 = 9$

$(x - 2)^2 = 0$

$x = 2$

$(2, 0)$

31. *y*-intercepts: Let $x = 0$.

$$y^2 - 6y - 27 = 0$$
$$y^2 - 6y + 9 = 27 + 9$$
$$(y - 3)^2 = 36$$
$$y - 3 = \pm 6$$
$$y = 9, -3$$
$$(0, 9), (0, -3)$$

x-intercepts: Let $y = 0$.

$$x^2 - 2x - 27 = 0$$
$$x^2 - 2x + 1 = 27 + 1$$
$$(x - 1)^2 = 28$$
$$x - 1 = \pm\sqrt{28}$$
$$x = 1 \pm 2\sqrt{7}$$
$$\left(1 \pm 2\sqrt{7}, 0\right)$$

33. *y*-intercepts: $(0 - 6)^2 + (y + 3)^2 = 16$

$$(y + 3)^2 = 16 - 36$$
$$= -20$$

No solution

No *y*-intercepts

x-intercepts: $(x - 6)^2 + (0 + 3)^2 = 16$

$$(x - 6)^2 = 7$$
$$x - 6 = \pm\sqrt{7}$$
$$x = 6 \pm \sqrt{7}$$
$$\left(6 \pm \sqrt{7}, 0\right)$$

35. (a) Radius: 81; Center: $(0, 0)$

$$x^2 + y^2 = 81^2 = 6561$$

(b) The distance from $(60, 45)$ to $(0, 0)$ is

$$\sqrt{60^2 + 45^2} = \sqrt{5625} = 75 \text{ miles.}$$

Yes, you would feel the earthquake.

(c)

You were $81 - 75 = 6$ miles from the outer boundary.

37. $y^2 = -4x$

Vertex: $(0, 0)$

Opens to the left since *p* is negative.

Matches graph (e).

39. $x^2 = -8y$

Vertex: $(0, 0)$

Opens downward since *p* is negative.

Matches graph (d).

41. $(y - 1)^2 = 4(x - 3)$

Vertex: $(3, 1)$

Opens to the right since *p* is positive.

Matches graph (a).

43. Vertex: $(0, 0) \implies h = 0, k = 0$

Graph opens upward.

$$x^2 = 4py$$

Point on graph: $(3, 6)$

$$3^2 = 4p(6)$$
$$9 = 24p$$
$$\tfrac{3}{8} = p$$

Thus, $x^2 = 4\left(\tfrac{3}{8}\right)y \implies y = \tfrac{2}{3}x^2$

$$\implies x^2 = \tfrac{3}{2}y.$$

45. Vertex: $(0, 0) \implies h = 0, k = 0$

Focus: $\left(0, -\tfrac{3}{2}\right) \implies p = -\tfrac{3}{2}$

$$(x - h)^2 = 4p(y - k)$$
$$x^2 = 4\left(-\tfrac{3}{2}\right)y$$
$$x^2 = -6y$$

47. Vertex:

$(0, 0) \implies h = 0, k = 0$

Focus: $(-2, 0) \implies p = -2$

$$(y - k)^2 = 4p(x - h)$$
$$y^2 = 4(-2)x$$
$$y^2 = -8x$$

49. Vertex: $(0, 0) \Rightarrow h = 0, k = 0$

Directrix: $y = -1 \Rightarrow p = 1$

$(x - h)^2 = 4p(y - k)$

$(x - 0)^2 = 4(1)(y - 0)$

$\quad x^2 = 4y$ or $y = \frac{1}{4}x^2$

51. Vertex: $(0, 0) \Rightarrow h = 0, k = 0$

Directrix: $x = 2 \Rightarrow p = -2$

$y^2 = 4px$

$y^2 = -8x$

53. Vertex: $(0, 0) \Rightarrow h = 0, k = 0$

Horizontal axis and passes through the point $(4, 6)$

$(y - k)^2 = 4p(x - h)$

$(y - 0)^2 = 4p(x - 0)$

$\quad y^2 = 4px$

$\quad 6^2 = 4p(4)$

$\quad 36 = 16p \Rightarrow p = \frac{9}{4}$

$\quad y^2 = 4\left(\frac{9}{4}\right)x$

$\quad y^2 = 9x$

55. $y = \frac{1}{2}x^2$

$x^2 = 2y = 4\left(\frac{1}{2}\right)y; \ p = \frac{1}{2}$

Vertex: $(0, 0)$

Focus: $\left(0, \frac{1}{2}\right)$

Directrix: $y = -\frac{1}{2}$

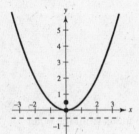

57. $y^2 = -6x$

$y^2 = 4\left(-\frac{3}{2}\right)x; \ p = -\frac{3}{2}$

Vertex: $(0, 0)$

Focus: $\left(-\frac{3}{2}, 0\right)$

Directrix: $x = \frac{3}{2}$

59. $x^2 + 8y = 0$

$\quad x^2 = 4(-2)y; \ p = -2$

Vertex: $(0, 0)$

Focus: $(0, -2)$

Directrix: $y = 2$

61. $(x + 1)^2 + 8(y + 3) = 0$

$\quad (x + 1)^2 = 4(-2)(y + 3)$

$h = -1, k = -3, p = -2$

Vertex: $(-1, -3)$

Focus: $(-1, -5)$

Directrix: $y = -1$

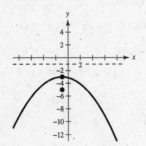

63. $y^2 + 6y + 8x + 25 = 0$

$\quad (y + 3)^2 = 4(-2)(x + 2); \ p = -2$

Vertex: $(-2, -3)$

Focus: $(-4, -3)$

Directrix: $x = 0$

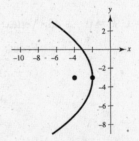

65. $\left(x + \frac{3}{2}\right)^2 = 4(y - 2) \implies h = -\frac{3}{2}, k = 2, p = 1$

Vertex: $\left(-\frac{3}{2}, 2\right)$

Focus: $\left(-\frac{3}{2}, 2 + 1\right) = \left(-\frac{3}{2}, 3\right)$

Directrix: $y = 1$

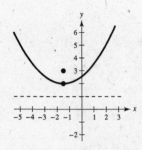

67. $\qquad y = \frac{1}{4}(x^2 - 2x + 5)$

$4y - 4 = (x - 1)^2$

$(x - 1)^2 = 4(1)(y - 1)$

$h = 1, k = 1, p = 1$

Vertex: $(1, 1)$

Focus: $(1, 2)$

Directrix: $y = 0$

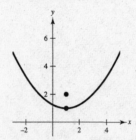

69. $x^2 + 4x + 6y - 2 = 0$

$\qquad x^2 + 4x + 4 = -6y + 2 + 4 = -6y + 6$

$\qquad (x + 2)^2 = -6(y - 1)$

$\qquad (x + 2)^2 = 4\left(-\frac{3}{2}\right)(y - 1)$

Vertex: $(-2, 1)$

Focus: $\left(-2, 1 - \frac{3}{2}\right) = \left(-2, -\frac{1}{2}\right)$

Directrix: $y = \frac{5}{2}$

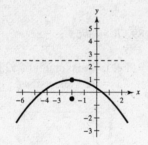

71. $y^2 + x + y = 0$

$y^2 + y + \frac{1}{4} = -x + \frac{1}{4}$

$\left(y + \frac{1}{2}\right)^2 = 4\left(-\frac{1}{4}\right)\left(x - \frac{1}{4}\right)$

$h = \frac{1}{4}, k = -\frac{1}{2}, p = -\frac{1}{4}$

Vertex: $\left(\frac{1}{4}, -\frac{1}{2}\right)$

Focus: $\left(0, -\frac{1}{2}\right)$

Directrix: $x = \frac{1}{2}$

To use a graphing
calculator, enter:

$y_1 = -\frac{1}{2} + \sqrt{\frac{1}{4} - x}$

$y_2 = -\frac{1}{2} - \sqrt{\frac{1}{4} - x}$

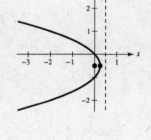

73. Vertex: $(3, 1)$, opens downward

Passes through: $(2, 0), (4, 0)$

$\qquad y = -(x - 2)(x - 4)$

$\qquad = -x^2 + 6x - 8$

$\qquad = -(x - 3)^2 + 1$

$(x - 3)^2 = -(y - 1)$

75. Vertex: $(-2, 0)$, opens to the right

Focus: $\left(-\frac{3}{2}, 0\right)$

$\frac{1}{2} = p$

$y^2 = 4\left(\frac{1}{2}\right)(x + 2)$

$y^2 = 2(x + 2)$

77. Vertex: $(5, 2)$

Focus: $(3, 2)$

Horizontal axis:

$p = 3 - 5 = -2$

$(y - 2)^2 = 4(-2)(x - 5)$

$(y - 2)^2 = -8(x - 5)$

79. Vertex: $(0, 4)$

Directrix: $y = 2$

Vertical axis

$p = 4 - 2 = 2$

$(x - 0)^2 = 4(2)(y - 4)$

$x^2 = 8(y - 4)$

81. Focus: $(2, 2)$

Directrix: $x = -2$

Horizontal axis

Vertex: $(0, 2)$

$p = 2 - 0 = 2$

$(y - 2)^2 = 4(2)(x - 0)$

$(y - 2)^2 = 8x$

83. $y^2 - 8x = 0$ and $x - y + 2 = 0$

$y^2 = 8x$ $\qquad y_3 = x + 2$

$y_1 = \sqrt{8x}$

$y_2 = -\sqrt{8x}$

The point of tangency is $(2, 4)$.

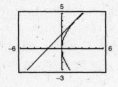

85. $x^2 = 2y$, $(4, 8)$, $p = \frac{1}{2}$, focus: $\left(0, \frac{1}{2}\right)$

Following Example 4, we find the y-intercept $(0, b)$.

$d_1 = \frac{1}{2} - b$

$d_2 = \sqrt{(4 - 0)^2 + \left(8 - \frac{1}{2}\right)^2} = \frac{17}{2}$

$d_1 = d_2 \implies \frac{1}{2} - b = \frac{17}{2} \implies b = -8$

$m = \frac{8 - (-8)}{4 - 0} = 4$

$y = 4x - 8$, Tangent line

Let $y = 0 \implies x = 2 \implies$ x-intercept $(2, 0)$.

87. $y = -2x^2 \implies x^2 = -\frac{1}{2}y = 4\left(-\frac{1}{8}\right)y$

$\implies p = -\frac{1}{8}$

Focus: $\left(0, -\frac{1}{8}\right)$

Following Example 4, we find the y-intercept $(0, b)$.

$d_1 = \frac{1}{8} + b$

$d_2 = \sqrt{(-1 - 0)^2 + \left(-2 + \frac{1}{8}\right)^2} = \frac{17}{8}$

$d_1 = d_2 \implies \frac{1}{8} + b = \frac{17}{8} \implies b = 2$

$m = \frac{-2 - 2}{-1 - 0} = 4$

$y = 4x + 2$

Let $y = 0 \implies x = -\frac{1}{2} \implies$ x-intercept $\left(-\frac{1}{2}, 0\right)$.

89. $R = 375x - \frac{3}{2}x^2$

R is a maximum of \$23,437.50 when $x = 125$ televisions.

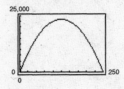

91. (a) $x^2 = 4py,\ p = \dfrac{3}{2}$

$x^2 = 4\left(\dfrac{3}{2}\right)y = 6y$

(or $y^2 = 6x$)

(b) When $x = 4$,

$6y = 16$

$y = \dfrac{16}{6} = \dfrac{8}{3}.$

Depth: $\dfrac{8}{3}$ inches

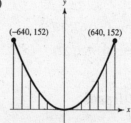

93. (a)

(b) $x^2 = 4py$

$640^2 = 4p(152)$

$p = \dfrac{12{,}800}{19}$

$y = \dfrac{19}{51{,}200}x^2$

(c)

x	0	200	400	500	600
y	0	14.84	59.38	92.77	133.59

95. Vertex: $(0, 0]$

$y^2 = 4px$

Point: $(1000, 800)$

$800^2 = 4p(1000] \implies p = 160$

$y^2 = 4(160)x$

$y^2 = 640x$

97. $-12.5(y - 7.125) = (x - 6.25)^2$

$-12.5y + 89.0625 = x^2 - 12.5x + 39.0625$

$y = -0.08x^2 + x + 4$

(a)

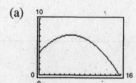

(b) The highest point is at $(6.25, 7.125)$. The distance is the x-intercept of ≈ 15.69 feet.

99. The slope of the line joining $(3, -4)$ and the center is $-\dfrac{4}{3}$. The slope of the tangent line at $(3, -4)$ is $\dfrac{3}{4}$. Thus,

$y + 4 = \dfrac{3}{4}(x - 3)$

$4y + 16 = 3x - 9$

$3x - 4y = 25,$ tangent line.

101. The slope of the line joining $\left(2, -2\sqrt{2}\right)$ and the center is $\left(-2\sqrt{2}\right)/2 = -\sqrt{2}$. The slope of the tangent line is $1/\sqrt{2} = \sqrt{2}/2$. Thus,

$y + 2\sqrt{2} = \dfrac{\sqrt{2}}{2}(x - 2)$

$2y + 4\sqrt{2} = \sqrt{2}x - 2\sqrt{2}$

$\sqrt{2}x - 2y = 6\sqrt{2},$ tangent line.

103. False. The center is $(0, -5)$.

105. False. A circle is a conic section.

107. True

109. Answers will vary. See the reflective property of parabolas, page 599.

111. $(y - 3)^2 = 6(x + 1)$

For the upper half of the parabola,

$y - 3 = \sqrt{6(x + 1)}$

$y = \sqrt{6(x + 1)} + 3.$

113. $f(x) = 3x^3 - 4x + 2$

Relative maximum: $(-0.67, 3.78)$

Relative minimum: $(0.67, 0.22)$

115. $f(x) = x^4 + 2x + 2$

Relative minimum: $(-0.79, 0.81)$

Section 9.2   Ellipses

- An **ellipse** is the set of all points (x, y) the sum of whose distances from two distinct fixed points (**foci**) is constant.
- The standard equation of an ellipse with center (h, k) and major and minor axes of lengths $2a$ and $2b$ is

  (a) $\dfrac{(x - h)^2}{a^2} + \dfrac{(y - k)^2}{b^2} = 1$ if the major axis is horizontal.

  (b) $\dfrac{(x - h)^2}{b^2} + \dfrac{(y - k)^2}{a^2} = 1$ if the major axis is vertical.

- $c^2 = a^2 - b^2$ where c is the distance from the center to a focus.
- The eccentricity of an ellipse is $e = \dfrac{c}{a}$.

Vocabulary Check

1. ellipse

2. major axis, center

3. minor axis

4. eccentricity

1. $\dfrac{x^2}{4} + \dfrac{y^2}{9} = 1$

Center: $(0, 0)$

$a = 3, b = 2$

Vertical major axis

Matches graph (b).

3. $\dfrac{x^2}{4} + \dfrac{y^2}{25} = 1$

Center: $(0, 0)$

$a = 5, b = 2$

Vertical major axis

Matches graph (d).

5. $\dfrac{(x - 2)^2}{16} + (y + 1)^2 = 1$

Center: $(2, -1)$

$a = 4, b = 1$

Horizontal major axis

Matches graph (a).

7. $\dfrac{x^2}{64} + \dfrac{y^2}{9} = 1$

Center: $(0, 0)$

$a = 8, b = 3,$

$c = \sqrt{64 - 9} = \sqrt{55}$

Vertices: $(\pm 8, 0)$

Foci: $\left(\pm\sqrt{55}, 0\right)$

$e = \dfrac{c}{a} = \dfrac{\sqrt{55}}{8}$

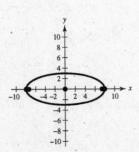

9. $\dfrac{(x - 4)^2}{16} + \dfrac{(y + 1)^2}{25} = 1$

Center: $(4, -1)$

$a = 5, b = 4, c = 3$

Vertices: $(4, -1 \pm 5)$; $(4, -6), (4, 4)$

Foci: $(4, -1 \pm 3)$; $(4, -4), (4, 2)$

$e = \dfrac{c}{a} = \dfrac{3}{5}$

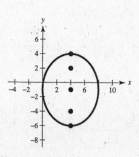

11. $\dfrac{(x + 5)^2}{9/4} + (y - 1)^2 = 1$

Center: $(-5, 1)$

$a = \dfrac{3}{2}, b = 1, c = \sqrt{\dfrac{9}{4} - 1} = \dfrac{\sqrt{5}}{2}$

Foci: $\left(-5 + \dfrac{\sqrt{5}}{2}, 1\right), \left(-5 - \dfrac{\sqrt{5}}{2}, 1\right)$

Vertices: $\left(-5 + \dfrac{3}{2}, 1\right) = \left(-\dfrac{7}{2}, 1\right), \left(-5 - \dfrac{3}{2}, 1\right) = \left(-\dfrac{13}{2}, 1\right)$

$e = \dfrac{\sqrt{5}/2}{3/2} = \dfrac{\sqrt{5}}{3}$

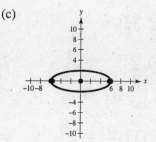

13. (a) $x^2 + 9y^2 = 36$

$\dfrac{x^2}{36} + \dfrac{y^2}{4} = 1$

(b) $a = 6, b = 2, c = \sqrt{36 - 4} = \sqrt{32} = 4\sqrt{2}$

Center: $(0, 0)$

Vertices: $(\pm 6, 0)$

Foci: $\left(\pm 4\sqrt{2}, 0\right)$

$e = \dfrac{c}{a} = \dfrac{4\sqrt{2}}{6} = \dfrac{2\sqrt{2}}{3}$

(c)

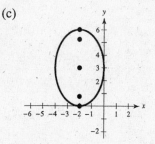

15. (a) $9x^2 + 4y^2 + 36x - 24y + 36 = 0$

$9(x^2 + 4x + 4) + 4(y^2 - 6y + 9) = -36 + 36 + 36$

$\dfrac{(x + 2)^2}{4} + \dfrac{(y - 3)^2}{9} = 1$

(b) $a = 3, b = 2, c = \sqrt{5}$

Center: $(-2, 3)$

Foci: $\left(-2, 3 \pm \sqrt{5}\right)$

Vertices: $(-2, 6), (-2, 0)$

$e = \dfrac{\sqrt{5}}{3}$

(c)

17. (a)
$$6x^2 + 2y^2 + 18x - 10y + 2 = 0$$

$$6\left(x^2 + 3x + \frac{9}{4}\right) + 2\left(y^2 - 5y + \frac{25}{4}\right) = -2 + \frac{27}{2} + \frac{25}{2}$$

$$6\left(x + \frac{3}{2}\right)^2 + 2\left(y - \frac{5}{2}\right)^2 = 24$$

$$\frac{\left(x + \frac{3}{2}\right)^2}{4} + \frac{\left(y - \frac{5}{2}\right)^2}{12} = 1$$

(b) $a = 2\sqrt{3}, b = 2, c = 2\sqrt{2}$

Center: $\left(-\frac{3}{2}, \frac{5}{2}\right)$

Foci: $\left(-\frac{3}{2}, \frac{5}{2} \pm 2\sqrt{2}\right)$

Vertices: $\left(-\frac{3}{2}, \frac{5}{2} \pm 2\sqrt{3}\right)$

$e = \dfrac{\sqrt{2}}{\sqrt{3}} = \dfrac{\sqrt{6}}{3}$

(c)

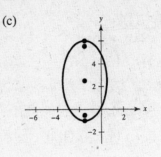

19. (a)
$$16x^2 + 25y^2 - 32x + 50y + 16 = 0$$

$$16(x^2 - 2x + 1) + 25(y^2 + 2y + 1) = -16 + 16 + 25$$

$$\frac{(x - 1)^2}{25/16} + (y + 1)^2 = 1$$

(b) $a = \dfrac{5}{4}, b = 1, c = \dfrac{3}{4}$

Center: $(1, -1)$

Foci: $\left(\frac{7}{4}, -1\right), \left(\frac{1}{4}, -1\right)$

Vertices: $\left(\frac{9}{4}, -1\right), \left(-\frac{1}{4}, -1\right)$

$e = \dfrac{3}{5}$

(c)

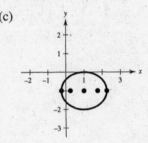

21. (a) $12x^2 + 20y^2 - 12x + 40y - 37 = 0$

$$12\left(x^2 - 1 + \frac{1}{4}\right) + 20(y^2 + 2y + 1) = 37 + 3 + 20$$

$$12\left(x - \frac{1}{2}\right)^2 + 20(y + 1)^2 = 60$$

$$\frac{\left(x - \frac{1}{2}\right)^2}{5} + \frac{(y + 1)^2}{3} = 1$$

(b) $a = \sqrt{5}, b = \sqrt{3}, c = \sqrt{5 - 3} = \sqrt{2}$

Center: $\left(\frac{1}{2}, -1\right)$

Vertices: $\left(\frac{1}{2} \pm \sqrt{5}, -1\right)$

Foci: $\left(\frac{1}{2} \pm \sqrt{2}, -1\right)$

Eccentricity: $\dfrac{c}{a} = \dfrac{\sqrt{2}}{\sqrt{5}} = \dfrac{\sqrt{10}}{5}$

(c)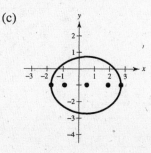

23. Center: $(0, 0)$

$a = 4, b = 2$

Vertical major axis

$$\frac{x^2}{4} + \frac{y^2}{16} = 1$$

25. Center: $(0, 0)$

$a = 3,$

$c = 2 \implies b = \sqrt{9 - 4} = \sqrt{5}$

Horizontal major axis

$$\frac{x^2}{9} + \frac{y^2}{5} = 1$$

27. Center: $(0, 0)$

$c = 3$

$a = 4 \implies b = \sqrt{16 - 9} = \sqrt{7}$

Horizontal major axis

$$\frac{x^2}{16} + \frac{y^2}{7} = 1$$

29. Vertices: $(0, \pm5) \implies a = 5$

Center: $(0, 0)$

Vertical major axis

$$\frac{(x - h)^2}{b^2} + \frac{(y - k)^2}{a^2} = 1$$

$$\frac{x^2}{b^2} + \frac{y^2}{25} = 1$$

Point: $(4, 2)$

$$\frac{4^2}{b^2} + \frac{2^2}{25} = 1$$

$$\frac{16}{b^2} = 1 - \frac{4}{25} = \frac{21}{25}$$

$$400 = 21b^2$$

$$\frac{400}{21} = b^2$$

$$\frac{x^2}{400/21} + \frac{y^2}{25} = 1$$

$$\frac{21x^2}{400} + \frac{y^2}{25} = 1$$

31. Center: $(2, 3)$

$a = 3, b = 1$

Vertical major axis

$\dfrac{(x - h)^2}{b^2} + \dfrac{(y - k)^2}{a^2} = 1$

$\dfrac{(x - 2)^2}{1} + \dfrac{(y - 3)^2}{9} = 1$

33. Center: $(4, 2)$

$a = 4, b = 1 \implies c = \sqrt{16 - 1} = \sqrt{15}$

Horizontal major axis

$\dfrac{(x - 4)^2}{16} + \dfrac{(y - 2)^2}{1} = 1$

35. Center: $(0, 4)$

$c = 4, a = 18 \implies b^2 = a^2 - c^2 = 324 - 16 = 308$

Vertical major axis

$\dfrac{x^2}{308} + \dfrac{(y - 4)^2}{324} = 1$

37. Vertices: $(3, 1), (3, 9) \implies a = 4$

Center: $(3, 5)$

Minor axis of length $6 \implies b = 3$

Vertical major axis

$\dfrac{(x - h)^2}{b^2} + \dfrac{(y - k)^2}{a^2} = 1$

$\dfrac{(x - 3)^2}{9} + \dfrac{(y - 5)^2}{16} = 1$

39. Center: $(0, 4)$

Vertices: $(-4, 4), (4, 4) \implies a = 4$

$a = 2c \implies 4 = 2c \implies c = 2$

$2^2 = 4^2 - b^2 \implies b^2 = 12$

Horizontal major axis

$\dfrac{(x - h)^2}{a^2} + \dfrac{(y - k)^2}{b^2} = 1$

$\dfrac{x^2}{16} + \dfrac{(y - 4)^2}{12} = 1$

41. $\dfrac{x^2}{4} + \dfrac{y^2}{9} = 1$

$a = 3, b = 2,$

$c = \sqrt{9 - 4} = \sqrt{5}$

$e = \dfrac{c}{a} = \dfrac{\sqrt{5}}{3}$

43. $\qquad x^2 + 9y^2 - 10x + 36y + 52 = 0$

$(x^2 - 10x + 25) + 9(y^2 + 4y + 4) = -52 + 25 + 36$

$\qquad (x - 5)^2 + 9(y + 2)^2 = 9$

$\qquad \dfrac{(x - 5)^2}{9} + \dfrac{(y + 2)^2}{1} = 1$

$a = 3, b = 1, c = \sqrt{9 - 1} = 2\sqrt{2}$

$e = \dfrac{c}{a} = \dfrac{2\sqrt{2}}{3}$

45. Vertices: $(\pm 5, 0) \implies a = 5$

Eccentricity: $\dfrac{4}{5} = \dfrac{c}{a} \implies c = \dfrac{4}{5}a = 4$

$b^2 = a^2 - c^2 = 25 - 16 = 9$

Center: $(0, 0)$

Horizontal major axis

$\dfrac{x^2}{25} + \dfrac{y^2}{9} = 1$

47. (a)

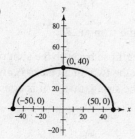

(b) Vertices: $(\pm 50, 0) \implies a = 50$

Height at center:

$40 \implies b = 40$

Horizontal major axis

$\dfrac{x^2}{a^2} + \dfrac{y^2}{b^2} = 1$

$\dfrac{x^2}{2500} + \dfrac{y^2}{1600} = 1, \; y \geq 0$

(c) For $x = 45$, $\dfrac{45^2}{2500} + \dfrac{y^2}{1600} = 1$.

$y^2 = 1600\left(1 - \dfrac{45^2}{2500}\right)$

$y^2 = 304$

$y \approx 17.44$

The height five feet from the edge of the tunnel is approximately 17.44 feet.

49. Let $\dfrac{x^2}{a^2} + \dfrac{y^2}{b^2} = 1$ be the equation of the ellipse. Then $b = 2$ and $a = 3 \implies c^2 = a^2 - b^2 = 9 - 4 = 5$.

Thus, the tacks are placed at $\left(\pm\sqrt{5}, 0\right)$. The string has a length of $2a = 6$ feet.

51. Area of ellipse $= 2$(area of circle)

$\pi ab = 2\pi r^2$

$\pi a(10) = 2\pi(10)^2$

$\pi a(10) = 200$

$a = 20$

Length of major axis: $2a = 2(20) = 40$ units

53. $a + c = 4.08$

$a - c = 0.34$

$2a = 4.42 \implies a = 2.21 \implies c = 1.87$

$b^2 = a^2 - c^2 \implies b^2 = 1.3872$

$\dfrac{x^2}{4.8841} + \dfrac{y^2}{1.3872} = 1$

55. For $\dfrac{x^2}{a^2} + \dfrac{y^2}{b^2} = 1$, we have $c^2 = a^2 - b^2$.

When $x = c$,

$\dfrac{c^2}{a^2} + \dfrac{y^2}{b^2} = 1 \implies y^2 = b^2\left(1 - \dfrac{a^2 - b^2}{a^2}\right)$

$\implies y^2 = \dfrac{b^4}{a^2}$

$\implies 2y = \dfrac{2b^2}{a}.$

57. $\dfrac{x^2}{9} + \dfrac{y^2}{16} = 1$

$a = 4, b = 3, c = \sqrt{7}$

Points on the ellipse:

$(\pm 3, 0), (0, \pm 4)$

Length of latus recta:

$\dfrac{2b^2}{a} = \dfrac{2(3)^2}{4} = \dfrac{9}{2}$

Additional points: $\left(\pm\dfrac{9}{4}, -\sqrt{7}\right), \left(\pm\dfrac{9}{4}, \sqrt{7}\right)$

59. $5x^2 + 3y^2 = 15$

$$\frac{x^2}{3} + \frac{y^2}{5} = 1$$

$a = \sqrt{5}, b = \sqrt{3}, c = \sqrt{2}$

Points on the ellipse: $\left(\pm\sqrt{3}, 0\right), \left(0, \pm\sqrt{5}\right)$

Length of latus recta: $\dfrac{2b^2}{a} = \dfrac{2 \cdot 3}{\sqrt{5}} = \dfrac{6\sqrt{5}}{5}$

Additional points: $\left(\pm\dfrac{3\sqrt{5}}{5}, -\sqrt{2}\right), \left(\pm\dfrac{3\sqrt{5}}{5}, \sqrt{2}\right)$

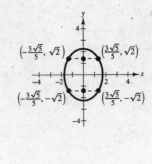

61. True. If $e \approx 1$ then the ellipse is elongated, not circular.

63. (a) The length of the string is $2a$.

   (b) The path is an ellipse because the sum of the distances from the two thumbtacks is always the length of the string, that is, it is constant.

65. Center: $(6, 2)$

Foci: $(2, 2), (10, 2) \implies c = 4$

$(a + c) + (a - c) = 2a = 36 \implies a = 18$

$b^2 = a^2 - b^2 \implies b = \sqrt{18^2 - 16} = \sqrt{308}$

Horizontal major axis

$$\frac{(x - 6)^2}{324} + \frac{(y - 2)^2}{308} = 1$$

67. Arithmetic: $d = -11$ **69.** Geometric: $r = 2$ **71.** $\displaystyle\sum_{n=0}^{6} 3^n = 1093$ **73.** $\displaystyle\sum_{n=1}^{10} 4\left(\tfrac{3}{4}\right)^{n-1} \approx 15.099$

Section 9.3   Hyperbolas

- A **hyperbola** is the set of all points (x, y) the difference of whose distances from two distinct fixed points **(foci)** is constant.

- The standard equation of a hyperbola with center (h, k) and transverse and conjugate axes of lengths $2a$ and $2b$ is:

  (a) $\dfrac{(x - h)^2}{a^2} - \dfrac{(y - k)^2}{b^2} = 1$ if the transverse axis is horizontal.

  (b) $\dfrac{(y - k)^2}{a^2} - \dfrac{(x - h)^2}{b^2} = 1$ if the transverse axis is vertical.

- $c^2 = a^2 + b^2$ where c is the distance from the center to a focus.

- The asymptotes of a hyperbola are:

  (a) $y = k \pm \dfrac{b}{a}(x - h)$ if the transverse axis is horizontal.

  (b) $y = k \pm \dfrac{a}{b}(x - h)$ the transverse axis is vertical.

—CONTINUED—

Section 9.3   —CONTINUED—

■ The eccentricity of a hyperbola is $e = \dfrac{c}{a}$.

■ To classify a nondegenerate conic from its general equation $Ax^2 + Cy^2 + Dx + Ey + F = 0$:
  (a) If $A = C \ (A \neq 0, C \neq 0)$, then it is a circle.
  (b) If $AC = 0 \ (A = 0 \text{ or } C = 0, \text{ but not both})$, then it is a parabola.
  (c) If $AC > 0$, then it is an ellipse.
  (d) If $AC < 0$, then it is a hyperbola.

Vocabulary Check

1. hyperbola

2. branches

3. transverse axis, center

4. asymptotes

5. $Ax^2 + Cy^2 + Dx + Ey + F = 0$

---

1. Center: $(0, 0)$

$a = 3, b = 5, c = \sqrt{34}$

Vertical transverse axis

Matches graph (b).

3. Center: $(1, 0)$

$a = 4, b = 2$

Horizontal transverse axis

Matches graph (a).

5. $x^2 - y^2 = 1$

$a = 1, b = 1, c = \sqrt{2}$

Center: $(0, 0)$

Vertices: $(\pm 1, 0)$

Foci: $\left(\pm\sqrt{2}, 0\right)$

Asymptotes: $y = \pm x$

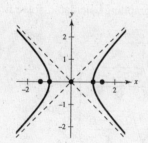

7. $\dfrac{y^2}{1} - \dfrac{x^2}{4} = 1$

$a = 1, b = 2, c = \sqrt{5}$

Center: $(0, 0)$

Vertices: $(0, \pm 1)$

Foci: $\left(0, \pm\sqrt{5}\right)$

Asymptotes: $y = \pm\dfrac{1}{2}x$

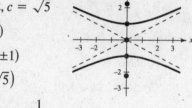

9. $\dfrac{y^2}{25} - \dfrac{x^2}{81} = 1$

$a = 5, b = 9, c = \sqrt{a^2 + b^2} = \sqrt{106}$

Center: $(0, 0)$

Vertices: $(0, \pm 5)$

Foci: $\left(0, \pm\sqrt{106}\right)$

Asymptotes:

$y = \pm\dfrac{a}{b}x = \pm\dfrac{5}{9}x$

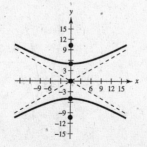

11. $\dfrac{(x - 1)^2}{4} - \dfrac{(y + 2)^2}{1} = 1$

$a = 2, b = 1, c = \sqrt{5}$

Center: $(1, -2)$

Vertices:

$(-1, -2), (3, -2)$

Foci: $\left(1 \pm \sqrt{5}, -2\right)$

Asymptotes: $y = -2 \pm \dfrac{1}{2}(x - 1)$

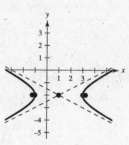

13. $\dfrac{(y + 5)^2}{1/9} - \dfrac{(x - 1)^2}{1/4} = 1$

$a = \dfrac{1}{3},\, b = \dfrac{1}{2},\, c = \sqrt{\dfrac{1}{9} + \dfrac{1}{4}} = \dfrac{\sqrt{13}}{6}$

Center: $(1, -5)$

Vertices: $\left(1, -5 \pm \dfrac{1}{3}\right)$: $\left(1, -\dfrac{16}{3}\right), \left(1, -\dfrac{14}{3}\right)$

Foci: $\left(1, -5 \pm \dfrac{\sqrt{13}}{6}\right)$

Asymptotes: $y = k \pm \dfrac{a}{b}(x - h)$

$$y = -5 \pm \dfrac{2}{3}(x - 1)$$

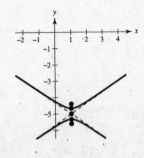

15. (a) $4x^2 - 9y^2 = 36$

$$\dfrac{x^2}{9} - \dfrac{y^2}{4} = 1$$

(b) Center: $(0, 0)$

$a = 3,\, b = 2,\, c = \sqrt{9 + 4} = \sqrt{13}$

Vertices: $(\pm 3, 0)$

Foci: $\left(\pm\sqrt{13}, 0\right)$

Asymptotes: $y = \pm\dfrac{b}{a}x = \pm\dfrac{2}{3}x$

(c)

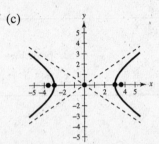

17. (a) $2x^2 - 3y^2 = 6$

$$\dfrac{x^2}{3} - \dfrac{y^2}{2} = 1$$

(b) $a = \sqrt{3},\, b = \sqrt{2},\, c = \sqrt{5}$

Center: $(0, 0)$

Vertices: $\left(\pm\sqrt{3}, 0\right)$

Foci: $\left(\pm\sqrt{5}, 0\right)$

Asymptotes: $y = \pm\sqrt{\dfrac{2}{3}}\,x$

$$= \pm\dfrac{\sqrt{6}}{3}x$$

(c) To use a graphing calculator, solve first for y.

$$y^2 = \dfrac{2x^2 - 6}{3}$$

$\left. \begin{array}{l} y_1 = \sqrt{\dfrac{2x^2 - 6}{3}} \\[3mm] y_2 = -\sqrt{\dfrac{2x^2 - 6}{3}} \end{array} \right\}$ Hyperbola

$\left. \begin{array}{l} y_3 = \sqrt{\dfrac{2}{3}}\,x \\[3mm] y_4 = -\sqrt{\dfrac{2}{3}}\,x \end{array} \right\}$ Asymptotes

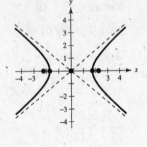

19. (a) $9x^2 - y^2 - 36x - 6y + 18 = 0$

$9(x^2 - 4x + 4) - (y^2 + 6y + 9) = -18 + 36 - 9$

$$\dfrac{(x - 2)^2}{1} - \dfrac{(y + 3)^2}{9} = 1$$

(b) $a = 1,\, b = 3,\, c = \sqrt{10}$

Center: $(2, -3)$

Vertices: $(1, -3), (3, -3)$

Foci: $\left(2 \pm \sqrt{10}, -3\right)$

Asymptotes: $y = -3 \pm 3(x - 2)$

(c)

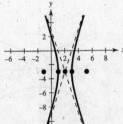

21. (a) $x^2 - 9y^2 + 2x - 54y - 80 = 0$

$(x^2 + 2x + 1) - 9(y^2 + 6y + 9) = 80 + 1 - 81$

$(x + 1)^2 - 9(y + 3)^2 = 0$

$y + 3 = \pm\frac{1}{3}(x + 1)$

(b) Degenerate hyperbola is two lines intersecting at $(-1, -3)$.

(c)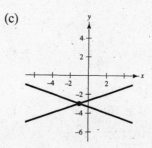

23. (a) $9y^2 - x^2 + 2x + 54y + 62 = 0$

$9(y^2 + 6y + 9) - (x^2 - 2x + 1) = -62 - 1 + 81$

$$\frac{(y + 3)^2}{2} - \frac{(x - 1)^2}{18} = 1$$

(b) $a = \sqrt{2}, b = 3\sqrt{2}, c = 2\sqrt{5}$

Center: $(1, -3)$

Vertices: $\left(1, -3 \pm \sqrt{2}\right)$

Foci: $\left(1, -3 \pm 2\sqrt{5}\right)$

Asymptotes: $y = -3 \pm \frac{1}{3}(x - 1)$

(c) To use a graphing calculator, solve for y first.

$9(y + 3)^2 = 18 + (x - 1)^2$

$$y = -3 \pm \sqrt{\frac{18 + (x - 1)^2}{9}}$$

$\left.\begin{array}{l} y_1 = -3 + \frac{1}{3}\sqrt{18 + (x - 1)^2} \\[2mm] y_2 = -3 - \frac{1}{3}\sqrt{18 + (x - 1)^2} \end{array}\right\}$ Hyperbola

$\left.\begin{array}{l} y_3 = -3 + \frac{1}{3}(x - 1) \\[2mm] y_4 = -3 - \frac{1}{3}(x - 1) \end{array}\right\}$ Asymptotes

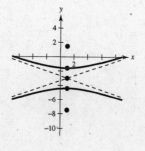

25. Vertices: $(0, \pm 2) \implies a = 2$

Foci: $(0, \pm 4) \implies c = 4$

$b^2 = c^2 - a^2 = 16 - 4 = 12$

Center: $(0, 0) = (h, k)$

$$\frac{(y - k)^2}{a^2} - \frac{(x - h)^2}{b^2} = 1$$

$$\frac{y^2}{4} - \frac{x^2}{12} = 1$$

27. Vertices: $(\pm 1, 0) \implies a = 1$

Asymptotes:

$y = \pm 5x \implies \frac{b}{a} = 5$

$\implies b = 5$

Center: $(0, 0)$

$$\frac{x^2}{1} - \frac{y^2}{25} = 1$$

29. Foci: $(0, \pm 8) \implies c = 8$

Asymptotes: $y = \pm 4x \implies \dfrac{a}{b} = 4 \implies a = 4b$

Center: $(0, 0) = (h, k)$

$c^2 = a^2 + b^2 \implies 64 = 16b^2 + b^2$

$\dfrac{64}{17} = b^2 \implies a^2 = \dfrac{1024}{17}$

$\dfrac{(y - k)^2}{a^2} - \dfrac{(x - h)^2}{b^2} = 1$

$\dfrac{y^2}{1024/17} - \dfrac{x^2}{64/17} = 1$

$\dfrac{17y^2}{1024} - \dfrac{17x^2}{64} = 1$

31. Vertices: $(2, 0), (6, 0) \implies a = 2$

Foci: $(0, 0), (8, 0) \implies c = 4$

$b^2 = c^2 - a^2 = 16 - 4 = 12$

Center: $(4, 0) = (h, k)$

$\dfrac{(x - h)^2}{a^2} - \dfrac{(y - k)^2}{b^2} = 1$

$\dfrac{(x - 4)^2}{4} - \dfrac{y^2}{12} = 1$

33. Vertices: $(4, 1), (4, 9) \implies a = 4$

Foci: $(4, 0), (4, 10) \implies c = 5$

$b^2 = c^2 - a^2 = 25 - 16 = 9$

Center: $(4, 5) = (h, k)$

$\dfrac{(y - k)^2}{a^2} - \dfrac{(x - h)^2}{b^2} = 1$

$\dfrac{(y - 5)^2}{16} - \dfrac{(x - 4)^2}{9} = 1$

35. Vertices: $(2, 3), (2, -3) \implies a = 3$

Solution point: $(0, 5)$

Center: $(2, 0) = (h, k)$

$\dfrac{(y - k)^2}{a^2} - \dfrac{(x - h)^2}{b^2} = 1$

$\dfrac{y^2}{9} - \dfrac{(x - 2)^2}{b^2} = 1 \implies$

$b^2 = \dfrac{9(x - 2)^2}{y^2 - 9}$

$= \dfrac{9(-2)^2}{25 - 9} = \dfrac{36}{16} = \dfrac{9}{4}$

$\dfrac{y^2}{9} - \dfrac{(x - 2)^2}{9/4} = 1$

37. Vertices: $(0, 4), (0, 0)$

Center: $(0, 2), a = 2$

$\dfrac{(y - 2)^2}{4} - \dfrac{x^2}{b^2} = 1$

Passes through $\left(\sqrt{5}, -1\right)$

$\dfrac{(-1 - 2)^2}{4} - \dfrac{5}{b^2} = 1$

$\dfrac{9}{4} - 1 = \dfrac{5}{b^2}$

$b^2 = 4 \implies b = 2$

$\dfrac{(y - 2)^2}{4} - \dfrac{x^2}{4} = 1$

39. Vertices: $(1, 2), (3, 2) \implies a = 1$

Center: $(2, 2)$

Asymptotes: $y = x, y = 4 - x$

$\dfrac{b}{a} = 1 \implies b = 1$

$\dfrac{(x - 2)^2}{1} - \dfrac{(y - 2)^2}{1} = 1$

41. Vertices: $(0, 2), (6, 2) \implies a = 3$

Asymptotes: $y = \frac{2}{3}x, \; y = 4 - \frac{2}{3}x$

$\frac{b}{a} = \frac{2}{3} \implies b = 2$

Center: $(3, 2) = (h, k)$

$\frac{(x - h)^2}{a^2} - \frac{(y - k)^2}{b^2} = 1$

$\frac{(x - 3)^2}{9} - \frac{(y - 2)^2}{4} = 1$

43. F_1: Friend's location $(-10,560, 0)$

F_2: Your location $(10,560, 0)$

$P(x, y)$: Location of lightning strike

$(1100)(18) = 19,800$

$\frac{x^2}{a^2} - \frac{y^2}{b^2} = 1$

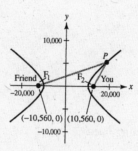

$c = 10,560, \; a = \frac{19,800}{2} = 9900 \implies a^2 = 98,010,000$

$b^2 = c^2 - a^2 = 13,503,600$

$\frac{x^2}{98,010,000} - \frac{y^2}{13,503,600}$

45. (a) $\frac{x^2}{a^2} - \frac{y^2}{b^2} = 1$

$a = 1; \; (2, 9)$ is on the curve, so

$\frac{4}{1} - \frac{81}{b^2} = 1 \implies \frac{81}{b^2} = 3$

$\implies b^2 = \frac{81}{3} \implies b = 3\sqrt{3}.$

$\frac{x^2}{1} - \frac{y^2}{27} = 1, \quad -9 \leq y \leq 9$

(b) Because each unit is $\frac{1}{2}$ foot, 4 inches is $\frac{2}{3}$ of a unit. The base is 9 units from the origin, so

$y = 9 - \frac{2}{3} = 8\frac{1}{3}.$

When $y = \frac{25}{3}$,

$x^2 = 1 + \frac{(25/3)^2}{27} \implies x \approx 1.88998.$

So the width is $2x \approx 3.779956$ units, or 22.68 inches, or 1.88998 feet.

47. Center: $(0, 0)$

Focus: $(24, 0)$

$b^2 = c^2 - a^2 = 24^2 - a^2 = 576 - a^2$

$\frac{x^2}{a^2} - \frac{y^2}{576 - a^2} = 1$

$\frac{24^2}{a^2} - \frac{24^2}{576 - a^2} = 1$

$\frac{576}{a^2} - \frac{576}{576 - a^2} = 1$

$576(576 - a^2) - 576a^2 = a^2(576 - a^2)$

$a^4 - 1728a^2 + 331,776 = 0$

$a \approx \pm 38.83 \;\; \text{or} \;\; a \approx \pm 14.83$

Since $a < c$ and $c = 24$, we choose $a = 14.83$. The vertex is approximate at $(14.83, 0)$. [**Note:** By the Quadratic Formula, the exact value of a is $a = 12(\sqrt{5} - 1)$.]

49. $9x^2 + 4y^2 - 18x + 16y - 119 = 0$

$A = 9, C = 4$

$AC = 36 > 0$, Ellipse

51. $16x^2 - 9y^2 + 32x + 54y - 209 = 0$

$A = 16, C = -9$

$AC = 16(-9) < 0$, Hyperbola

53. $y^2 + 12x + 4y + 28 = 0$

$C = 1, A = 0$

$AC = 0$, Parabola

55. $x^2 + y^2 + 2x - 6y = 0$

$A = C = 1$, Circle

57. $x^2 - 6x - 2y + 7 = 0$

$A = 1, C = 0, D = -6, E = -2, F = 7$

$AC = 0 \implies$ Parabola

59. True. $e = \dfrac{c}{a} = \dfrac{\sqrt{a^2 + b^2}}{a}$

61. False. For example,

$x^2 - y^2 - 2x + 2y = 0$

$(x - 1)^2 - (y - 1)^2 = 0$

is the graph of two intersecting lines.

63. Let (x, y) be such that the difference of the distances from $(c, 0)$ and $(-c, 0)$ is $2a$ (again only deriving one of the forms).

$$2a = \left| \sqrt{(x + c)^2 + y^2} - \sqrt{(x - c) + y^2} \right|$$
$$2a + \sqrt{(x - c)^2 + y^2} = \sqrt{(x + c)^2 + y^2}$$
$$4a^2 + 4a\sqrt{(x - c)^2 + y^2} + (x - c)^2 + y^2 = (x + c)^2 + y^2$$
$$4a\sqrt{(x - c)^2 + y^2} = 4cx - 4a^2$$
$$a\sqrt{(x - c)^2 + y^2} = cx - a^2$$
$$a^2(x^2 - 2cx + c^2 + y^2) = c^2x^2 - 2a^2cx + a^4$$
$$a^2(c^2 - a^2) = (c^2 - a^2)x^2 - a^2y^2$$

Let $b^2 = c^2 - a^2$. Then $a^2b^2 = b^2x^2 - a^2y^2 \implies 1 = \dfrac{x^2}{a^2} - \dfrac{y^2}{b^2}$.

65. $|d_2 - d_1| =$ constant by definition of hyperbola

At the point $(a, 0)$,

$|d_2 - d_1| = |(a + c) - (c - a)| = 2a$.

67. At the point $(a, 0)$, the difference of the distances to the foci $(\pm c, 0)$ is $(c + a) - (c - a) = 2a$. Let (x, y) be a point on the hyperbola.

$$2a = \sqrt{(x + c)^2 + y^2} - \sqrt{(x - c)^2 + y^2}$$

$$2a + \sqrt{(x - c)^2 + y^2} = \sqrt{(x + c)^2 + y^2}$$

$$4a^2 + 4a\sqrt{(x - c)^2 + y^2} + (x - c)^2 + y^2 = (x + c)^2 + y^2$$

$$4a\sqrt{(x - c)^2 + y^2} = 4cx - 4a^2$$

$$a\sqrt{(x - c)^2 + y^2} = cx - a^2$$

$$a^2(x^2 - 2cx + c^2 + y^2) = c^2x^2 - 2a^2cx + a^4$$

$$a^2(c^2 - a^2) = (c^2 - a^2)x^2 - a^2y^2$$

$$1 = \frac{x^2}{a^2} - \frac{y^2}{c^2 - a^2}$$

Thus, $c^2 - a^2 = b^2$, as desired.

69. $(x^3 - 3x^2) - (6 - 2x - 4x^2) = x^3 + x^2 + 2x - 6$

71.

$$
\begin{array}{r|rrrr}
-2 & 1 & 0 & -3 & 4 \\
   &   & -2 & 4 & -2 \\
\hline
   & 1 & -2 & 1 & 2
\end{array}
$$

$$\frac{x^3 - 3x + 4}{x + 2} = x^2 - 2x + 1 + \frac{2}{x + 2}$$

73. $x^3 - 16x = x(x^2 - 16) = x(x - 4)(x + 4)$

75. $2x^3 - 24x^2 + 72x = 2x(x^2 - 12x + 36)$

$$= 2x(x - 6)^2$$

77. $16x^3 + 54 = 2(8x^3 + 27)$

$$= 2(2x + 3)(4x^2 - 6x + 9)$$

Section 9.4   Rotation and Systems of Quadratic Equations

- The general second-degree equation $Ax^2 + Bxy + Cy^2 + Dx + Ey + F = 0$ can be rewritten as $A'(x')^2 + C'(y')^2 + D'x' + E'y' + F' = 0$ by rotating the coordinate axes through the angle θ where $\cot 2\theta = (A - C)/B$.

- $x = x'\cos\theta - y'\sin\theta$
  $y = x'\sin\theta + y'\cos\theta$

- The graph of the nondegenerate equation $Ax^2 + Bxy + Cy^2 + Dx + Ey + F = 0$ is:

  (a) An ellipse or circle if $B^2 - 4AC < 0$.

  (b) A parabola if $B^2 - 4AC = 0$.

  (c) A hyperbola if $B^2 - 4AC > 0$.

Vocabulary Check

1. rotation, axes

2. invariant under rotation

3. discriminant

1. $\theta = 90°$; Point: $(0, 3)$

$x = x' \cos \theta - y' \sin \theta$

$y = x' \sin \theta + y' \cos \theta$

$0 = x' \cos 90° - y' \sin 90°$

$3 = x' \sin 90° + y' \cos 90°$

$0 = y'$

$3 = x'$

Thus, $(x', y') = (3, 0)$.

3. $xy + 1 = 0$

$A = 0, B = 1, C = 0$

$\cot 2\theta = \dfrac{A - C}{B} = 0 \implies 2\theta = \dfrac{\pi}{2} \implies \theta = \dfrac{\pi}{4}$

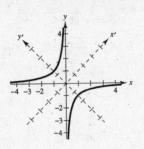

$x = x' \cos \dfrac{\pi}{4} - y' \sin \dfrac{\pi}{4} = x'\left(\dfrac{\sqrt{2}}{2}\right) - y'\left(\dfrac{\sqrt{2}}{2}\right) = \dfrac{x' - y'}{\sqrt{2}}$

$y = x' \sin \dfrac{\pi}{4} + y' \cos \dfrac{\pi}{4} = x'\left(\dfrac{\sqrt{2}}{2}\right) + y'\left(\dfrac{\sqrt{2}}{2}\right) = \dfrac{x' + y'}{\sqrt{2}}$

$xy + 1 = 0$

$\left(\dfrac{x' - y'}{\sqrt{2}}\right)\left(\dfrac{x' + y'}{\sqrt{2}}\right) + 1 = 0$

$\dfrac{(y')^2}{2} - \dfrac{(x')^2}{2} = 1$, Hyperbola

5. $x^2 - 4xy + y^2 + 1 = 0$

$A = 1, B = -4, C = 1$

$\cot 2\theta = \dfrac{A - C}{B} = 0 \implies 2\theta = \dfrac{\pi}{2} \implies \theta = \dfrac{\pi}{4}$

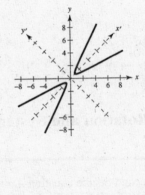

$x = x' \cos \dfrac{\pi}{4} - y' \sin \dfrac{\pi}{4}$

$y = x' \sin \dfrac{\pi}{4} + y' \cos \dfrac{\pi}{4}$

$= x'\left(\dfrac{\sqrt{2}}{2}\right) - y'\left(\dfrac{\sqrt{2}}{2}\right)$

$= x'\left(\dfrac{\sqrt{2}}{2}\right) + y'\left(\dfrac{\sqrt{2}}{2}\right)$

$= \dfrac{\sqrt{2}}{2}(x' - y')$

$= \dfrac{\sqrt{2}}{2}(x' + y')$

$x^2 - 4xy + y^2 + 1 = 0$

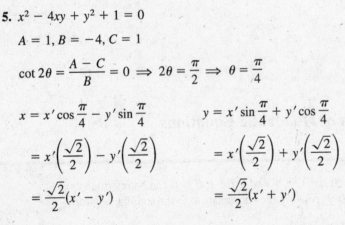

$\left[\dfrac{\sqrt{2}}{2}(x' - y')\right]^2 - 4\left[\dfrac{\sqrt{2}}{2}(x' - y')\dfrac{\sqrt{2}}{2}(x' + y')\right] + \left[\dfrac{\sqrt{2}}{2}(x' + y')\right]^2 + 1 = 0$

$\dfrac{1}{2}(x')^2 - x'y' + \dfrac{1}{2}(y')^2 - 2[(x')^2 - (y')^2] + \dfrac{1}{2}(x')^2 + x'y' + \dfrac{1}{2}(y')^2 + 1 = 0$

$-(x')^2 + 3(y')^2 = -1$

$(x')^2 - \dfrac{(y')^2}{1/3} = 1$, Hyperbola

7. $xy - 2y - 4x = 0$

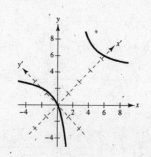

$A = 0, B = 1, C = 0$

$\cot 2\theta = \dfrac{A - C}{B} = 0 \Rightarrow 2\theta = \dfrac{\pi}{2} \Rightarrow \theta = \dfrac{\pi}{4}$

$x = x' \cos \dfrac{\pi}{4} - y' \sin \dfrac{\pi}{4}$ $\qquad\qquad$ $y = x' \sin \dfrac{\pi}{4} + y' \cos \dfrac{\pi}{4}$

$\qquad = x'\left(\dfrac{\sqrt{2}}{2}\right) - y'\left(\dfrac{\sqrt{2}}{2}\right)$ $\qquad\qquad = x'\left(\dfrac{\sqrt{2}}{2}\right) + y'\left(\dfrac{\sqrt{2}}{2}\right)$

$\qquad = \dfrac{x' - y'}{\sqrt{2}}$ $\qquad\qquad\qquad\qquad = \dfrac{x' + y'}{\sqrt{2}}$

$$xy - 2y - 4x = 0$$

$$\left(\dfrac{x' - y'}{\sqrt{2}}\right)\left(\dfrac{x' + y'}{\sqrt{2}}\right) - 2\left(\dfrac{x' + y'}{\sqrt{2}}\right) - 4\left(\dfrac{x' - y'}{\sqrt{2}}\right) = 0$$

$$\dfrac{(x')^2}{2} - \dfrac{(y')^2}{2} - \sqrt{2}x' - \sqrt{2}y' - 2\sqrt{2}x' + 2\sqrt{2}y' = 0$$

$$\left[(x')^2 - 6\sqrt{2}x' + \left(3\sqrt{2}\right)^2\right] - \left[(y')^2 - 2\sqrt{2}y' + \left(\sqrt{2}\right)^2\right] = 0 + \left(3\sqrt{2}\right)^2 - \left(\sqrt{2}\right)^2$$

$$\left(x' - 3\sqrt{2}\right)^2 - \left(y' - \sqrt{2}\right)^2 = 16$$

$$\dfrac{\left(x' - 3\sqrt{2}\right)^2}{16} - \dfrac{\left(y' - \sqrt{2}\right)^2}{16} = 1, \quad \text{Hyperbola}$$

9. $5x^2 - 6xy + 5y^2 - 12 = 0$

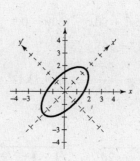

$A = 5, B = -6, C = 5$

$\cot 2\theta = \dfrac{A - C}{B} = 0 \Rightarrow 2\theta = \dfrac{\pi}{2} \Rightarrow \theta = \dfrac{\pi}{4}$

$x = x' \cos \dfrac{\pi}{4} - y' \sin \dfrac{\pi}{4} = \dfrac{\sqrt{2}}{2}(x' - y')$

$y = x' \sin \dfrac{\pi}{4} + y' \cos \dfrac{\pi}{4} = \dfrac{\sqrt{2}}{2}(x' + y')$

$$5x^2 - 6xy + 5y^2 - 12 = 0$$

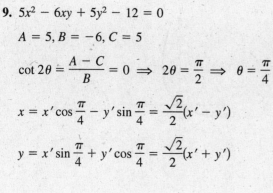

$$5\left[\dfrac{\sqrt{2}}{2}(x' - y')\right]^2 - 6\left[\dfrac{\sqrt{2}}{2}(x' - y')\dfrac{\sqrt{2}}{2}(x' + y')\right] + 5\left[\dfrac{\sqrt{2}}{2}(x' + y')\right]^2 = 12$$

$$\dfrac{5}{2}(x')^2 - 5x'y' + \dfrac{5}{2}(y')^2 - 3(x')^2 + 3(y')^2 + \dfrac{5}{2}(x')^2 + 5x'y' + \dfrac{5}{2}(y')^2 = 12$$

$$2(x')^2 + 8(y')^2 = 12$$

$$\dfrac{(x')^2}{6} + \dfrac{(y')^2}{3/2} = 1, \quad \text{Ellipse}$$

11. $3x^2 - 2\sqrt{3}xy + y^2 + 2x + 2\sqrt{3}y = 0$

$A = 3, B = -2\sqrt{3}, C = 1$

$\cot 2\theta = \dfrac{A - C}{B} = -\dfrac{1}{\sqrt{3}} \implies \theta = 60°$

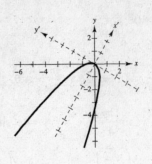

$x = x' \cos 60° - y' \sin 60°$

$\quad = x'\left(\dfrac{1}{2}\right) - y'\left(\dfrac{\sqrt{3}}{2}\right) = \dfrac{x' - \sqrt{3}y'}{2}$

$y = x' \sin \theta + y' \cos \theta = \dfrac{\sqrt{3}x' + y'}{2}$

$$3x^2 - 2\sqrt{3}xy + y^2 + 2x + 2\sqrt{3}y = 0$$

$$3\left(\dfrac{x' - \sqrt{3}y'}{2}\right)^2 - 2\sqrt{3}\left(\dfrac{x' - \sqrt{3}y'}{2}\right)\left(\dfrac{\sqrt{3}x' + y'}{2}\right) + \left(\dfrac{\sqrt{3}x' + y'}{2}\right)^2 + 2\left(\dfrac{x' - \sqrt{3}y'}{2}\right)$$

$$+ 2\sqrt{3}\left(\dfrac{\sqrt{3}x' + y'}{2}\right) = 0$$

$$\dfrac{3(x')^2}{4} - \dfrac{6\sqrt{3}x'y'}{4} + \dfrac{9(y')^2}{4} - \dfrac{6(x')^2}{4} + \dfrac{4\sqrt{3}x'y'}{4} + \dfrac{6(y')^2}{4} + \dfrac{3(x')^2}{4} + \dfrac{2\sqrt{3}x'y'}{4} + \dfrac{(y')^2}{4}$$

$$+ x' - \sqrt{3}y' + 3x' + \sqrt{3}y' = 0$$

$$4(y')^2 + 4x' = 0$$

$$x' = -(y')^2, \quad \text{Parabola}$$

13. $9x^2 + 24xy + 16y^2 + 90x - 130y = 0$

$A = 9, B = 24, C = 16$

$\cot 2\theta = \dfrac{A - C}{B} = -\dfrac{7}{24} \implies \theta \approx 53.13°$

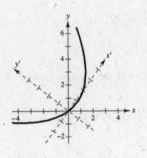

$\cos 2\theta = -\dfrac{7}{25}$

$\sin \theta = \sqrt{\dfrac{1 - \cos 2\theta}{2}} = \sqrt{\dfrac{1 - (-7/25)}{2}} = \dfrac{4}{5}$

$\cos \theta = \sqrt{\dfrac{1 + \cos 2\theta}{2}} = \sqrt{\dfrac{1 + (-7/25)}{2}} = \dfrac{3}{5}$

$x = x' \cos \theta - y' \sin \theta \qquad y = x' \sin \theta + y' \cos \theta$

$\quad = x'\left(\dfrac{3}{5}\right) - y'\left(\dfrac{4}{5}\right) \qquad\quad = x'\left(\dfrac{4}{5}\right) + y'\left(\dfrac{3}{5}\right)$

$\quad = \dfrac{3x' - 4y'}{5} \qquad\qquad\quad = \dfrac{4x' + 3y'}{5}$

—CONTINUED—

13. —CONTINUED—

$$9x^2 + 24xy + 16y^2 + 90x - 130y = 0$$

$$9\left(\frac{3x' - 4y'}{5}\right)^2 + 24\left(\frac{3x' - 4y'}{5}\right)\left(\frac{4x' + 3y'}{5}\right) + 16\left(\frac{4x' + 3y'}{5}\right)^2 + 90\left(\frac{3x' - 4y'}{5}\right)$$

$$- 130\left(\frac{4x' + 3y'}{5}\right) = 0$$

$$\frac{81(x')^2}{25} - \frac{216x'y'}{25} + \frac{144(y')^2}{25} + \frac{288(x')^2}{25} - \frac{168x'y'}{25} - \frac{288(y')^2}{25} + \frac{256(x')^2}{25} + \frac{384x'y'}{25}$$

$$+ \frac{144(y')^2}{25} + 54x' - 72y' - 104x' - 78y' = 0$$

$$25(x')^2 - 50x' - 150y' = 0$$

$$(x')^2 - 2x' + 1 = 6y' + 1$$

$$(x' - 1)^2 = 6\left(y' + \frac{1}{6}\right), \text{ Parabola}$$

15. $x^2 + 3xy + y^2 = 20$

$$\cot 2\theta = \frac{A - C}{B} = \frac{1 - 1}{3} = 0 \implies \theta = \frac{\pi}{4} = 45°$$

Solve for y in terms of x:

$$y^2 + 3xy = 20 - x^2$$

$$y^2 + 3xy + \frac{9x^2}{4} = 20 - x^2 + \frac{9x^2}{4}$$

$$\left(y + \frac{3}{2}x\right)^2 = 20 + \frac{5x^2}{4} = \frac{80 + 5x^2}{4}$$

$$y = -\frac{3}{2}x \pm \frac{\sqrt{80 + 5x^2}}{2}$$

Graph $y_1 = -\frac{3x}{2} + \frac{\sqrt{80 + 5x^2}}{2}$

and $y_2 = -\frac{3x}{2} - \frac{\sqrt{80 + 5x^2}}{2}$.

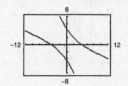

17. $17x^2 + 32xy - 7y^2 = 75$

$$\cot 2\theta = \frac{A - C}{B} = \frac{17 + 7}{32} = \frac{24}{32} = \frac{3}{4} \implies \theta \approx 26.57°$$

Solve for y in terms of x by completing the square.

$$-7y^2 + 32xy = -17x^2 + 75$$

$$y^2 - \frac{32}{7}xy = \frac{17}{7}x^2 - \frac{75}{7}$$

$$y^2 - \frac{32}{7}xy + \frac{256}{49}x^2 = \frac{119}{49}x^2 - \frac{525}{49} + \frac{256}{49}x^2$$

$$\left(y - \frac{16}{7}x\right)^2 = \frac{375x^2 - 525}{49}$$

$$y = \frac{16}{7}x \pm \sqrt{\frac{375x^2 - 525}{49}}$$

$$y = \frac{16x \pm 5\sqrt{15x^2 - 21}}{7}$$

Graph $y_1 = \frac{16x + 5\sqrt{15x^2 - 21}}{7}$ and

$$y_2 = \frac{16x - 5\sqrt{15x^2 - 21}}{7}.$$

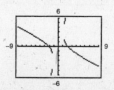

19. $32x^2 + 48xy + 8y^2 = 50$

$\cot 2\theta = \dfrac{A - C}{B} = \dfrac{32 - 8}{48} = \dfrac{1}{2} \implies \theta \approx 31.72°$

Solve for y in terms of x:

$$8y^2 + 48xy = -32x^2 + 50$$

$$y^2 + 6xy = -4x^2 + \dfrac{25}{4}$$

$$y^2 + 6xy + 9x^2 = -4x^2 + \dfrac{25}{4} + 9x^2$$

$$(y + 3x)^2 = 5x^2 + \dfrac{25}{4} = \dfrac{20x^2 + 25}{4}$$

$$y = -3x \pm \dfrac{\sqrt{20x^2 + 25}}{2}$$

Graph $y_1 = -3x + \dfrac{\sqrt{20x^2 + 25}}{2}$ and

$y_2 = -3x - \dfrac{\sqrt{20x^2 + 25}}{2}$.

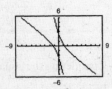

21. $xy + 4 = 0$

$B^2 - 4AC = 1 \implies$ The graph is a hyperbola.

$\cot 2\theta = \dfrac{A - C}{B} = 0 \implies \theta = 45°$

Matches graph (e).

23. $-2x^2 + 3xy + 2y^2 + 3 = 0$

$B^2 - 4AC = (3)^2 - 4(-2)(2)$

$\qquad = 25 \implies$ The graph is a hyperbola.

$\cot 2\theta = \dfrac{A - C}{B} = -\dfrac{4}{3} \implies \theta \approx -18.43°$

Matches graph (f).

25. $3x^2 + 2xy + y^2 - 10 = 0$

$B^2 - 4AC = (2)^2 - 4(3)(1)$

$\qquad = -8 \implies$ The graph is an ellipse or circle.

$\cot 2\theta = \dfrac{A - C}{B} = 1 \implies \theta = 22.5°$

Matches graph (d).

27. $16x^2 - 24xy + 9y^2 - 30x - 40y = 0$

(a) $B^2 - 4AC = (-24)^2 - 4(16)(9) = 0 \implies$ Parabola

(b) $9y^2 - (24x + 40)y + (16x^2 - 30x) = 0$

$$y = \dfrac{(24x + 40) \pm \sqrt{(24x + 40)^2 - 4(9)(16x^2 - 30x)}}{2(9)}$$

$$= \dfrac{24x + 40 \pm \sqrt{3000x + 1600}}{18}$$

(c)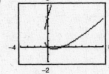

29. $15x^2 - 8xy + 7y^2 - 45 = 0$

   (a) $B^2 - 4AC = (-8)^2 - 4(15)(7)$

$$= -356 \implies \text{Ellipse or circle}$$

   (b) $7y^2 - 8xy + (15x^2 - 45) = 0$

$$y = \frac{8x \pm \sqrt{(-8x)^2 - 4(7)(15x^2 - 45)}}{14}$$

$$= \frac{8x \pm \sqrt{1260 - 356x^2}}{14}$$

   (c)

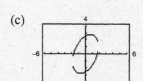

31. $x^2 - 6xy - 5y^2 + 4x - 22 = 0$

   (a) $B^2 - 4AC = (-6)^2 - 4(1)(-5)$

$$= 56 \implies \text{Hyperbola}$$

   (b) $-5y^2 - 6xy + (x^2 + 4x - 22) = 0$

$$y = \frac{6x \pm \sqrt{(-6x)^2 - 4(-5)(x^2 + 4x - 22)}}{-10}$$

$$= \frac{6x \pm \sqrt{56x^2 + 80x - 440}}{-10}$$

   (c)

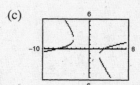

33. $x^2 + 4xy + 4y^2 - 5x - y - 3 = 0$

   (a) $B^2 - 4AC = 4^2 - 4(1)(4) = 0 \implies \text{Parabola}$

   (b) $4y^2 + (4x - 1)y + (x^2 - 5x - 3) = 0$

$$y = \frac{(1 - 4x) \pm \sqrt{(4x - 1)^2 - 4(4)(x^2 - 5x - 3)}}{8}$$

$$= \frac{1 - 4x \pm \sqrt{72x + 49}}{8}$$

   (c)

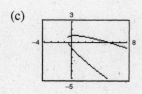

35. $y^2 - 16x^2 = 0$

$$y^2 = 16x^2$$

$$y = \pm 4x$$

Two intersecting lines

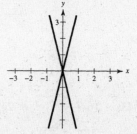

37. $x^2 + 2xy + y^2 - 1 = 0$

$$(x + y)^2 - 1 = 0$$

$$(x + y)^2 = 1$$

$$x + y = \pm 1$$

$$y = -x \pm 1$$

Two parallel lines

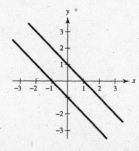

39. $x^2 + y^2 = 4$

$3x - y^2 = 0$

Adding:

$$x^2 + 3x - 4 = 0$$

$$(x + 4)(x - 1) = 0 \implies x = 1, -4$$

For $x = 1, y = \pm\sqrt{3}$.

$x = -4$ is impossible.

Solutions: $\left(1, \sqrt{3}\right), \left(1, -\sqrt{3}\right)$

41. $-4x^2 - y^2 - 16x + 24y - 16 = 0$

$\underline{4x^2 + y^2 + 40x - 24y + 208 = 0}$

$24x \qquad + 192 = 0$

$24x = -192$

$x = -8$

For $x = -8$:

$-4(64) - y^2 - 16(-8) + 24y - 16 = 0$

$-y^2 + 24y - 144 = 0$

$y^2 - 24y + 144 = 0$

$(y - 12)^2 = 0$

$\implies y = 12$

Solution: $(-8, 12)$

43. $x^2 - y^2 - 12x + 16y - 64 = 0$

$\underline{x^2 + y^2 - 12x - 16y + 64 = 0}$

$2x^2 \qquad - 24x \qquad = 0$

$x^2 - 12x = 0$

$x(x - 12) = 0 \implies x = 0, 12$

For $x = 0$:

$-y^2 + 16y - 64 = 0$

$y^2 - 16y + 64 = 0$

$(y - 8)^2 = 0 \implies y = 8$

For $x = 12$:

$144 - y^2 - 12(12) + 16y - 64 = 0$

$-y^2 + 16y - 64 = 0 \implies y = 8$

Solutions: $(0, 8), (12, 8)$

45. $-16x^2 - y^2 + 24y - 80 = 0$

$\underline{16x^2 + 25y^2 \qquad - 400 = 0}$

$24y^2 + 24y - 480 = 0$

$24(y + 5)(y - 4) = 0$

$y = -5 \text{ or } y = 4$

When $y = -5$:

$16x^2 + 25(-5)^2 - 400 = 0$

$16x^2 = -225$

No real solution

When $y = 4$:

$16x^2 + 25(4)^2 - 400 = 0$

$16x^2 = 0$

$x = 0$

The point of intersection is $(0, 4)$. In standard form the equations are:

$$\frac{x^2}{4} + \frac{(y - 12)^2}{64} = 1$$

$$\frac{x^2}{25} + \frac{y^2}{16} = 1$$

47. $2x^2 - y^2 + 6 = 0$

$2x + y = 0 \implies y = -2x$

$2x^2 - (-2x)^2 + 6 = 0$

$-2x^2 + 6 = 0$

$x^2 = 3 \implies x = \pm\sqrt{3}$

Two solutions: $\left(\sqrt{3}, -2\sqrt{3}\right), \left(-\sqrt{3}, 2\sqrt{3}\right)$

49. $10x^2 - 25y^2 - 100x + 160 = 0$

$$y^2 - 2x + 16 = 0 \Rightarrow y^2 = 2x - 16$$

$$10x^2 - 25(2x - 16) - 100x + 160 = 0$$

$$10x^2 - 150x + 560 = 0$$

$$x^2 - 15x + 56 = 0$$

$$(x - 8)(x - 7) = 0$$

$x = 8 \Rightarrow y^2 = 0 \Rightarrow (8, 0)$

$x = 7 \Rightarrow y^2 = -2$ impossible

One solution: $(8, 0)$

51.
$$xy + x - 2y + 3 = 0 \Rightarrow y = \frac{-x - 3}{x - 2}$$

$$x^2 + 4y^2 - 9 = 0$$

$$x^2 + 4\left(\frac{-x - 3}{x - 2}\right)^2 = 9$$

$$x^2(x - 2)^2 + 4(-x - 3)^2 = 9(x - 2)^2$$

$$x^2(x^2 - 4x + 4) + 4(x^2 + 6x + 9) = 9(x^2 - 4x + 4)$$

$$x^4 - 4x^3 + 4x^2 + 4x^2 + 24x + 36 = 9x^2 - 36x + 36$$

$$x^4 - 4x^3 - x^2 + 60x = 0$$

$$x(x + 3)(x^2 - 7x + 20) = 0$$

$$x = 0 \text{ or } x = -3$$

Note: $x^2 - 7x + 20 = 0$ has no real solution.

When $x = 0$: $y = \dfrac{-0 - 3}{0 - 2} = \dfrac{3}{2}$

When $x = -3$: $y = \dfrac{-(-3) - 3}{-3 - 2} = 0$

The points of intersection are $\left(0, \frac{3}{2}\right), (-3, 0)$.

53. True. $B^2 - 4AC = 1 - 4k$

If $k < \frac{1}{4}$, then $B^2 - 4AC > 0$.

55. $g(x) = \dfrac{2}{2 - x}$

Asymptotes:
$x = 2, y = 0$

Intercepts: $(0, 1)$

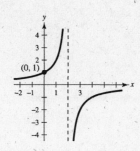

57. $h(t) = \dfrac{t^2}{2 - t} = -t - 2 + \dfrac{4}{2 - t}$

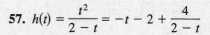

Slant asymptote: $y = -t - 2$

Vertical asymptote: $t = 2$

Intercept: $(0, 0)$

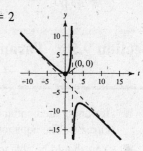

59. (a) $AB = \begin{bmatrix} 1 & -3 \\ 2 & 5 \end{bmatrix}\begin{bmatrix} 0 & 6 \\ 5 & -1 \end{bmatrix} = \begin{bmatrix} -15 & 9 \\ 25 & 7 \end{bmatrix}$

   (b) $BA = \begin{bmatrix} 0 & 6 \\ 5 & -1 \end{bmatrix}\begin{bmatrix} 1 & -3 \\ 2 & 5 \end{bmatrix} = \begin{bmatrix} 12 & 30 \\ 3 & -20 \end{bmatrix}$

   (c) $A^2 = \begin{bmatrix} 1 & -3 \\ 2 & 5 \end{bmatrix}\begin{bmatrix} 1 & -3 \\ 2 & 5 \end{bmatrix} = \begin{bmatrix} -5 & -18 \\ 12 & 19 \end{bmatrix}$

61. (a) $AB = \begin{bmatrix} 4 & -2 & 5 \end{bmatrix} \begin{bmatrix} 3 \\ -4 \\ 5 \end{bmatrix} = [12 + 8 + 25] = [45]$

(b) $BA = \begin{bmatrix} 3 \\ -4 \\ 5 \end{bmatrix} \begin{bmatrix} 4 & -2 & 5 \end{bmatrix} = \begin{bmatrix} 12 & -6 & 15 \\ -16 & 8 & -20 \\ 20 & -10 & 25 \end{bmatrix}$

(c) A^2 does not exist.

63. $f(x) = |x + 3|$

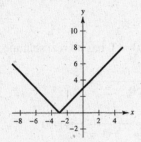

65. $g(x) = \sqrt{4 - x^2}$

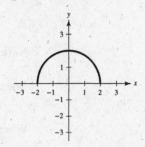

67. $h(t) = -(t - 2)^3 + 3$

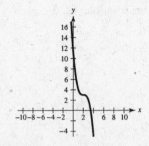

69. $f(t) = [\![t - 5]\!] + 1$

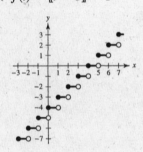

71. Area $= \frac{1}{2}ab \sin C$

$= \frac{1}{2}(8)(12) \sin 110°$

≈ 45.11

73. $s = \frac{1}{2}(11 + 18 + 10) = \frac{39}{2}$

Area $= \sqrt{s(s - a)(s - b)(s - c)}$

$= \sqrt{\frac{39}{2}\left(\frac{17}{2}\right)\left(\frac{3}{2}\right)\left(\frac{19}{2}\right)}$

≈ 48.60

Section 9.5 Parametric Equations

■ If f and g are continuous functions of t on an interval I, then the set of ordered pairs $(f(t), g(t))$ is a *plane curve C*. The equations $x = f(t)$ and $y = g(t)$ are *parametric equations* for C and t is the *parameter.*

■ You should be able to graph plane curves with your graphing utility.

■ To eliminate the parameter:

Solve for t in one equation and substitute into the second equation.

■ You should be able to find the parametric equations for a graph.

Vocabulary Check

1. plane curve, parametric equations, parameter

2. orientation

3. eliminating, parameter

---

1. $x = t$

$y = t + 2$

$y = x + 2$, line

Matches (c).

3. $x = \sqrt{t}$

$y = t$

$y = x^2$, parabola, $x \geq 0$

Matches (b).

5. $x = \ln t \iff t = e^x$

$y = \dfrac{1}{2}t - 2$

$y = \dfrac{1}{2}e^x - 2$

Matches (f).

7. $x = \sqrt{t},\ y = 2 - t$

(a)

t	0	1	2	3	4
x	0	1	$\sqrt{2}$	$\sqrt{3}$	2
y	2	1	0	-1	-2

(c)

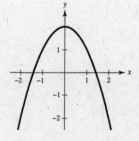

(b) Graph by hand.

   Note: $x \geq 0$

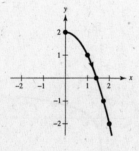

(d) $y = 2 - t = 2 - x^2$,

   Parabola

   In part (c), $x \geq 0$.

9. The graph opens upward, contains $(1, 0)$, and is oriented left to right.
   Matches (b).

11. $x = t,\ y = -4t$

$y = -4x$

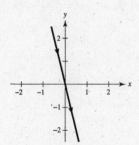

13. $x = 3t - 3,\ y = 2t + 1$

$t = \dfrac{x + 3}{3}$

$y = 2\left(\dfrac{x + 3}{3}\right) + 1$

$y = \dfrac{2}{3}x + 3$

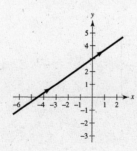

15. $x = \dfrac{1}{4}t,\ y = t^2$

$y = (4x)^2$

$y = 16x^2$

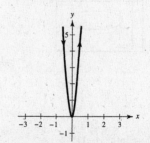

17. $x = t + 2, y = t^2$

$t = x - 2$

$y = (x - 2)^2$

19. $x = 2t, y = |t - 2|$

$t = \dfrac{x}{2} \implies y = |t - 2|$

$= \left|\dfrac{x}{2} - 2\right|$

$= \dfrac{1}{2}|x - 4|$

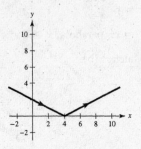

21. $x = 2\cos\theta, y = 3\sin\theta$

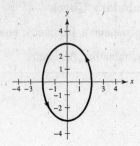

$\left(\dfrac{x}{2}\right)^2 = \cos^2\theta, \left(\dfrac{y}{3}\right)^2 = \sin^2\theta$

$\dfrac{x^2}{4} + \dfrac{y^2}{9} = \cos^2\theta + \sin^2\theta = 1$

$\dfrac{x^2}{4} + \dfrac{y^2}{9} = 1, \quad \text{ellipse}$

23. $x = e^{-t} \implies \dfrac{1}{x} = e^t$

$y = e^{3t} \implies y = (e^t)^3$

$y = \left(\dfrac{1}{x}\right)^3$

$y = \dfrac{1}{x^3}, \; x > 0, \; y > 0$

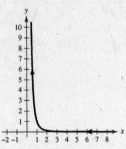

25. $x = t^3 \implies x^{1/3} = t$

$y = 3\ln t \implies y = \ln t^3$

$y = \ln(x^{1/3})^3$

$y = \ln x$

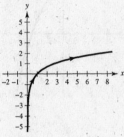

27. $x = 4 + 3\cos\theta, y = -2 + \sin\theta$

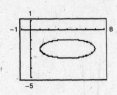

29. $x = 4\sec\theta, y = 2\tan\theta$

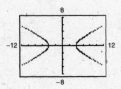

31. $x = \dfrac{t}{2}$

$y = \ln(t^2 + 1)$

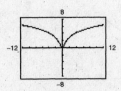

33. By eliminating the parameters in (a)–(d), we get $y = 2x + 1$. They differ from each other in restricted domain and in orientation.

(a) Domain: $-\infty < x < \infty$

Orientation: Left to right

(b) Domain: $-1 \le x \le 1$

Orientation: Depends on θ

(c) Domain: $0 < x < \infty$

Orientation: Right to left

(d) Domain: $0 < x < \infty$

Orientation: Left to right

35. $t = \dfrac{(x - x_1)}{(x_2 - x_1)}$

$y = y_1 + \left(\dfrac{x - x_1}{x_2 - x_1}\right)(y_2 - y_1)$

$\Rightarrow y - y_1 = \left(\dfrac{y_2 - y_1}{x_2 - x_1}\right)(x - x_1)$

37. $x = h + a\cos\theta$

$y = k + b\sin\theta$

$\dfrac{x - h}{a} = \cos\theta, \ \dfrac{y - k}{b} = \sin\theta$

$\dfrac{(x - h)^2}{a^2} + \dfrac{(y - k)^2}{b^2} = 1$

39. $x = x_1 + t(x_2 - x_1) = 1 + t(6 - 1) = 1 + 5t$

$y = y_1 + t(y_2 - y_1) = 4 + t(-3 - 4) = 4 - 7t$

41. $a = 5$, $c = 4$, and $b = \sqrt{a^2 - c^2} = 3$.

The center is $(0, 0)$, so $h = 0$ and $k = 0$.

$\cos^2\theta + \sin^2\theta = 1 = \dfrac{x^2}{5^2} + \dfrac{y^2}{3^2}$, so $x = 5\cos\theta$ and $y = 3\sin\theta$. This solution is not unique.

43. $y = 5x - 3$

Answers will vary.

$x = t, y = 5t - 3$

$x = \dfrac{1}{5}t, y = t - 3$

45. $y = \dfrac{1}{x}$

Sample answers:

$x = t, y = \dfrac{1}{t}$

$x = t^3, y = \dfrac{1}{t^3}$

47. $y = 6x^2 - 5$

Sample answers:

$x = t, y = 6t^2 - 5$

$x = 2t, y = 24t^2 - 5$

49. $x = 2\cot\theta, \ y = 2\sin^2\theta$

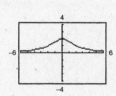

51. Matches (b).

53. Matches (d).

55. $x = (v_0 \cos \theta)t$, $y = h + (v_0 \sin \theta)t - 16t^2$

(a) $100 \text{ miles/hour} = \dfrac{100 \text{ mi/hr} \cdot 5280 \text{ ft/mi}}{3600 \text{ sec/hr}}$

$$= 146.67 \text{ ft/sec}$$

$x = (146.67 \cos \theta)t$

$y = 3 + (146.67 \sin \theta)t - 16t^2$

(b) $\theta = 15°$

$x = (146.67 \cos 15°)t = 141.7t$

$y = 3 + (146.67 \sin 15°)t - 16t^2$

$$= 3 + 38.0t - 16t^2$$

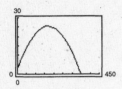

It is not a home run because $y < 10$ when $x = 400$.

(c) $\theta = 23°$

$x = (146.67 \cos 23°)t = 135.0t$

$y = 3 + (146.67 \sin 23°)t - 16t^2$

$$= 3 + 57.3t - 16t^2$$

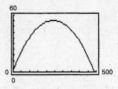

Yes, it is a home run because $y > 10$ when $x = 400$.

(d) $\theta \approx 19.4°$ is the minimum angle.

57. (a) $x = (\cos 35°)v_0 t$

$y = 7 + (\sin 35°)v_0 t - 16t^2$

(c)

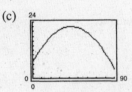

Maximum height ≈ 22 feet

(d) From part (b), $t_1 \approx 2.03$ seconds.

(b) If the ball is caught at time t_1, then:

$90 = (\cos 35°)v_0 t_1$

$4 = 7 + (\sin 35°)v_0 t_1 - 16t_1^2$.

$v_0 t_1 = \dfrac{90}{\cos 35°} \Rightarrow -3 = (\sin 35°)\dfrac{90}{\cos 35°} - 16t_1^2$

$\Rightarrow 16t_1^2 = 90 \tan 35° + 3$

$\Rightarrow t_1 \approx 2.03$ seconds

$\Rightarrow v_0 = \dfrac{90}{t_1 \cos 35°} \approx 54.09$ ft/sec

59. True

$x = t$ first set

$y = t^2 + 1 = x^2 + 1$

$x = 3t$ second set

$y = 9t^2 + 1 = (3t)^2 + 1 = x^2 + 1$

61. False. For example, $x = t^2$ and $y = t$ does not represent y as a function of x.

63. Sample answer: $x = \cos \theta$

$$y = -2 \sin \theta$$

65. $f(-x) = \dfrac{4(-x)^2}{(-x)^2 + 1} = \dfrac{4x^2}{x^2 + 1} = f(x)$

Symmetric about the y-axis

Even function

67. $y = e^x \neq e^{-x}; e^{-x} \neq -e^x$

No symmetry

Neither even nor odd

Section 9.6   Polar Coordinates

- In polar coordinates you do not have unique representation of points. The point (r, θ) can be represented by $(r, \theta \pm 2n\pi)$ or by $(-r, \theta \pm (2n + 1)\pi)$ where n is any integer. The pole is represented by $(0, \theta)$ where θ is any angle.

- To convert from polar coordinates to rectangular coordinates, use the following relationships.
  $$x = r \cos \theta$$
  $$y = r \sin \theta$$

- To convert from rectangular coordinates to polar coordinates, use the following relationships.
  $$r = \pm\sqrt{x^2 + y^2}$$
  $$\tan \theta = y/x$$

  If θ is in the same quadrant as the point (x, y), then r is positive. If θ is in the opposite quadrant as the point (x, y), then r is negative.

- You should be able to convert rectangular equations to polar form and vice versa.

Vocabulary Check

1. pole   **2.** directed distance, directed angle   **3.** polar

---

1. Polar coordinates: $\left(4, \dfrac{\pi}{2}\right)$

$x = 4 \cos\left(\dfrac{\pi}{2}\right) = 0$

$y = 4 \sin\left(\dfrac{\pi}{2}\right) = 4$

Rectangular coordinates: $(0, 4)$

3. Polar coordinates: $\left(-1, \dfrac{5\pi}{4}\right)$

$x = -1 \cos\left(\dfrac{5\pi}{4}\right) = \dfrac{\sqrt{2}}{2}$

$y = -1 \sin\left(\dfrac{5\pi}{4}\right) = \dfrac{\sqrt{2}}{2}$

Rectangular coordinates: $\left(\dfrac{\sqrt{2}}{2}, \dfrac{\sqrt{2}}{2}\right)$

5.

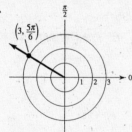

Three additional representations:

$\left(3, \dfrac{5\pi}{6} - 2\pi\right) = \left(3, -\dfrac{7\pi}{6}\right)$

$\left(-3, \dfrac{5\pi}{6} + \pi\right) = \left(-3, \dfrac{11\pi}{6}\right)$

$\left(-3, \dfrac{5\pi}{6} - \pi\right) = \left(-3, -\dfrac{\pi}{6}\right)$

7.

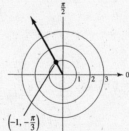

Three additional representations:

$\left(-1, -\dfrac{\pi}{3} + 2\pi\right) = \left(-1, \dfrac{5\pi}{3}\right)$

$\left(1, -\dfrac{\pi}{3} + \pi\right) = \left(1, \dfrac{2\pi}{3}\right)$

$\left(1, -\dfrac{\pi}{3} - \pi\right) = \left(1, -\dfrac{4\pi}{3}\right)$

9.

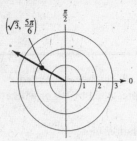

Three additional representations:

$$\left(\sqrt{3}, -\frac{7\pi}{6}\right), \left(-\sqrt{3}, -\frac{\pi}{6}\right), \left(-\sqrt{3}, \frac{11\pi}{6}\right)$$

13. Polar coordinates: $\left(4, -\frac{\pi}{3}\right)$

$$x = 4\cos\left(-\frac{\pi}{3}\right) = 2$$

$$y = 4\sin\left(-\frac{\pi}{3}\right) = -2\sqrt{3}$$

Rectangular coordinates: $\left(2, -2\sqrt{3}\right)$

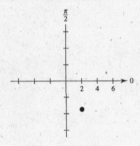

17. Polar coordinates: $\left(0, -\frac{7\pi}{6}\right)$ (origin!)

$$x = 0\cos\left(-\frac{7\pi}{6}\right) = 0$$

$$y = 0\sin\left(-\frac{7\pi}{6}\right) = 0$$

Rectangular coordinates: $(0, 0)$

11.

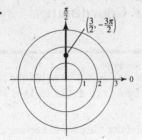

Three additional representations:

$$\left(\frac{3}{2}, \frac{\pi}{2}\right), \left(-\frac{3}{2}, \frac{3\pi}{2}\right), \left(-\frac{3}{2}, -\frac{\pi}{2}\right)$$

15. Polar coordinates: $\left(-1, \frac{-3\pi}{4}\right)$

$$x = -1\cos\left(\frac{-3\pi}{4}\right) = \frac{\sqrt{2}}{2}$$

$$y = -1\sin\left(\frac{-3\pi}{4}\right) = \frac{\sqrt{2}}{2}$$

Rectangular coordinates: $\left(\frac{\sqrt{2}}{2}, \frac{\sqrt{2}}{2}\right)$

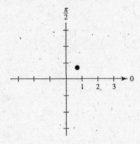

19. Polar coordinates: $\left(\sqrt{2}, 2.36\right)$

$$x = \sqrt{2}\cos(2.36) \approx -1.004$$

$$y = \sqrt{2}\sin(2.36) \approx 0.996$$

Rectangular coordinates: $(-1.004, 0.996)$

21. $(r, \theta) = \left(2, \dfrac{2\pi}{9}\right) \Rightarrow (x, y) = (1.53, 1.29)$

23. $(r, \theta) = (-4.5, 1.3) \Rightarrow (x, y) = (-1.204, -4.336)$

25. $(r, \theta) = (2.5, 1.58) \Rightarrow (x, y) = (-0.02, 2.50)$

27. $(r, \theta) = (-4.1, -0.5) \Rightarrow (x, y) = (-3.60, 1.97)$

29. Rectangular coordinates: $(-7, 0)$

$r = 7$, $\tan \theta = 0$, $\theta = 0$

Polar coordinates: $(7, \pi)$, $(-7, 0)$

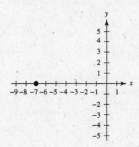

31. Rectangular coordinates: $(1, 1)$

$r = \sqrt{2}$, $\tan \theta = 1$, $\theta = \dfrac{\pi}{4}$

Polar coordinates: $\left(\sqrt{2}, \dfrac{\pi}{4}\right)$, $\left(-\sqrt{2}, \dfrac{5\pi}{4}\right)$

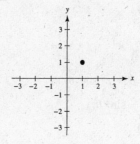

33. Rectangular coordinates: $\left(-\sqrt{3}, -\sqrt{3}\right)$

$r = \sqrt{3+3} = \sqrt{6}$, $\tan \theta = 1$, $\theta = \dfrac{\pi}{4}$

Polar coordinates: $\left(\sqrt{6}, \dfrac{5\pi}{4}\right)$, $\left(-\sqrt{6}, \dfrac{\pi}{4}\right)$

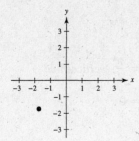

35. $(x, y) = (6, 9)$

$r = \sqrt{6^2 + 9^2} = \sqrt{117} \approx 10.8$

$\tan \theta = \dfrac{9}{6} = \dfrac{3}{2} \Rightarrow \theta \approx 0.983$

Polar coordinates: $(10.8, 0.983)$, $(-10.8, 4.124)$

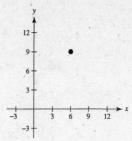

37. $(x, y) = (3, -2) \Rightarrow r = \sqrt{3^2 + (-2)^2} = \sqrt{13}$

$\theta = \arctan\left(-\dfrac{2}{3}\right) \approx -0.588$

$(r, \theta) \approx \left(\sqrt{13}, -0.588\right)$

39. $(x, y) = \left(\sqrt{3}, 2\right) \Rightarrow r = \sqrt{3 + 2^2} = \sqrt{7}$

$\theta = \arctan\left(\dfrac{2}{\sqrt{3}}\right) \approx 0.857$

$(r, \theta) \approx \left(\sqrt{7}, 0.857\right)$

41. $(x, y) = \left(\dfrac{5}{2}, \dfrac{4}{3}\right) \Rightarrow r = \sqrt{\left(\dfrac{5}{2}\right)^2 + \left(\dfrac{4}{3}\right)^2} = \dfrac{17}{6}$

$\theta = \arctan\left(\dfrac{4/3}{5/2}\right) \approx 0.490$

$(r, \theta) \approx \left(\dfrac{17}{6}, 0.490\right)$

43. $x^2 + y^2 = 9$

$r^2 = 9$

$r = 3$

45. $y = 4$

$r \sin \theta = 4$

$r = 4 \csc \theta$

47. $x = 8$

$r \cos \theta = 8$

$r = 8 \sec \theta$

49. $3x - 6y + 2 = 0$

$3r \cos \theta - 6r \sin \theta = -2$

$r(3 \cos \theta - 6 \sin \theta) = -2$

$$r = \frac{2}{6 \sin \theta - 3 \cos \theta}$$

51. $xy = 4$

$(r \cos \theta)(r \sin \theta) = 4$

$r^2 \cos \theta \sin \theta = 4$

$r^2(2 \cos \theta \sin \theta) = 8$

$r^2 \sin 2\theta = 8$

$r^2 = 8 \csc 2\theta$

53. $(x^2 + y^2)^2 = 9(x^2 - y^2)$

$(r^2)^2 = 9(r^2 \cos^2 \theta - r^2 \sin^2 \theta)$

$r^2 = 9(\cos^2 \theta - \sin^2 \theta)$

$r^2 = 9 \cos(2\theta)$

55. $x^2 + y^2 - 6x = 0$

$r^2 - 6r \cos \theta = 0$

$r^2 = 6r \cos \theta$

$r = 6 \cos \theta$

57. $x^2 + y^2 - 2ax = 0$

$r^2 - 2ar \cos \theta = 0$

$r(r - 2a \cos \theta) = 0$

$r = 2a \cos \theta$

59. $y^2 = x^3$

$(r \sin \theta)^2 = (r \cos \theta)^3$

$\sin^2 \theta = r \cos^3 \theta$

$$r = \frac{\sin^2 \theta}{\cos^3 \theta}$$

$= \tan^2 \theta \sec \theta$

61. $r = 6 \sin \theta$

$r^2 = 6r \sin \theta$

$x^2 + y^2 = 6y$

$x^2 + y^2 - 6y = 0$

63. $\theta = \dfrac{4\pi}{3}$

$\tan \theta = \tan \dfrac{4\pi}{3} = \dfrac{y}{x}$

$\sqrt{3} = \dfrac{y}{x}$

$y = \sqrt{3}x$

65. $\theta = \dfrac{5\pi}{6}$

$\tan \theta = \tan \dfrac{5\pi}{6} = \dfrac{y}{x}$

$\dfrac{-\sqrt{3}}{3} = \dfrac{y}{x}$

$y = \dfrac{-\sqrt{3}}{3}x$

67. $\theta = \dfrac{\pi}{2}$, vertical line

$x = 0$

69. $r = 4$

$r^2 = 16$

$x^2 + y^2 = 16$

71. $r = -3 \csc \theta$

$r \sin \theta = -3$

$y = -3$

73. $r^2 = \cos \theta$

$r^3 = r \cos \theta$

$(x^2 + y^2)^{3/2} = x$

$x^2 + y^2 = x^{2/3}$

$(x^2 + y^2)^3 = x^2$

75.

$$r = 2 \sin 3\theta$$
$$r = 2(3 \sin \theta - 4 \sin^3 \theta)$$
$$r^4 = 6r^3 \sin \theta - 8r^3 \sin^3 \theta$$
$$(x^2 + y^2)^2 = 6(x^2 + y^2)y - 8y^3$$
$$(x^2 + y^2)^2 = 6x^2y - 2y^3$$

77.

$$r = \frac{1}{1 - \cos \theta}$$
$$r - r \cos \theta = 1$$
$$\sqrt{x^2 + y^2} - x = 1$$
$$x^2 + y^2 = 1 + 2x + x^2$$
$$y^2 = 2x + 1$$

79.

$$r = \frac{6}{2 - 3 \sin \theta}$$
$$r(2 - 3 \sin \theta) = 6$$
$$2r = 6 + 3r \sin \theta$$
$$2\left(\pm\sqrt{x^2 + y^2}\right) = 6 + 3y$$
$$4(x^2 + y^2) = (6 + 3y)^2$$
$$4x^2 + 4y^2 = 36 + 36y + 9y^2$$
$$4x^2 - 5y^2 - 36y - 36 = 0$$

81.

$$r = 7$$
$$r^2 = 49$$
$$x^2 + y^2 = 49$$

The graph is a circle centered at the origin with radius 7.

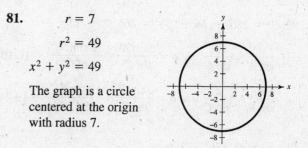

83.

$$\theta = \frac{\pi}{4}$$
$$\tan \theta = \tan \frac{\pi}{4} = 1 = \frac{y}{x}$$
$$y = x$$

The graph is the line $y = x$, which makes an angle of $\theta = \pi/4$ with the positive x-axis.

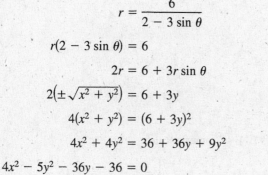

85.

$$r = 3 \sec \theta$$
$$r \cos \theta = 3$$
$$x = 3$$
$$x - 3 = 0$$

Vertical line

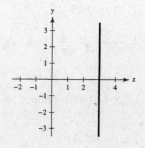

87. True, the distances from the origin are the same.

89. (a) $(r_1, \theta_1) = (x_1, y_1)$ where $x_1 = r_1 \cos \theta_1$ and $y_1 = r_1 \sin \theta_1$.

$(r_2, \theta_2) = (x_2, y_2)$ where $x_2 = r_2 \cos \theta_2$ and $y_2 = r_2 \sin \theta_2$.

Then $x_1^2 + y_1^2 = r_1^2 \cos^2 \theta_1 + r_1^2 \sin^2 \theta_1 = r_1^2$ and $x_2^2 + y_2^2 = r_2^2$. Thus,

$$d = \sqrt{(x_1 - x_2)^2 + (y_1 - y_2)^2}$$
$$= \sqrt{x_1^2 - 2x_1x_2 + x_2^2 + y_1^2 - 2y_1y_2 + y_2^2}$$
$$= \sqrt{(x_1^2 + y_1^2) + (x_2^2 + y_2^2) - 2(x_1x_2 + y_1y_2)}$$
$$= \sqrt{r_1^2 + r_2^2 - 2(r_1r_2 \cos \theta_1 \cos \theta_2 + r_1r_2 \sin \theta_1 \sin \theta_2)}$$
$$= \sqrt{r_1^2 + r_2^2 - 2r_1r_2 \cos(\theta_1 - \theta_2)}.$$

(b) If $\theta_1 = \theta_2$, the points are on the same line through the origin. In this case,

$$d = \sqrt{r_1^2 + r_2^2 - 2r_1r_2 \cos(0)} = \sqrt{(r_1 - r_2)^2} = |r_1 - r_2|.$$

(c) If $\theta_1 - \theta_2 = 90°$, $d = \sqrt{r_1^2 + r_2^2}$, the Pythagorean Theorem.

(d) For instance, $\left(3, \frac{\pi}{6}\right)$, $\left(4, \frac{\pi}{3}\right)$ gives $d \approx 2.053$ and $\left(-3, \frac{7\pi}{6}\right)$, $\left(-4, \frac{4\pi}{3}\right)$ gives $d \approx 2.053$. (Same!)

91. $\cos A = \dfrac{b^2 + c^2 - a^2}{2bc} = \dfrac{19^2 + 25^2 - 13^2}{2(19)(25)} = 0.86$

$A \approx 30.7°$

$\cos B = \dfrac{a^2 + c^2 - b^2}{2ac} = \dfrac{13^2 + 25^2 - 19^2}{2(13)(25)} = 0.66615$

$B \approx 48.2°$

$C \approx 180° - 30.7° - 48.2° \approx 101.1°$

93. $B = 180° - 56° - 38° = 86°$

$\dfrac{a}{\sin A} = \dfrac{c}{\sin C} \Rightarrow a = \dfrac{c \sin A}{\sin C} = \dfrac{12 \sin(56°)}{\sin(38°)} \approx 16.16$

$\dfrac{b}{\sin B} = \dfrac{c}{\sin C} \Rightarrow b = \dfrac{c \sin B}{\sin C} = \dfrac{12 \sin(86°)}{\sin(38°)} \approx 19.44$

95. $c^2 = a^2 + b^2 - 2ab \cos C$

$= 8^2 + 4^2 - 2(8)(4) \cos(35°)$

≈ 27.57

$c \approx 5.25$

$\dfrac{b}{\sin B} = \dfrac{c}{\sin C} \Rightarrow \sin B = \dfrac{b \sin C}{c} = \dfrac{4 \sin(35°)}{5.25}$

$\Rightarrow B \approx 25.9°$

$A = 180° - B - C = 119.1°$

97. By Cramer's Rule,

$D = \begin{vmatrix} 5 & -7 \\ -3 & 1 \end{vmatrix} = 5 - 21 = -16$

$D_x = \begin{vmatrix} -11 & -7 \\ -3 & 1 \end{vmatrix} = -11 - 21 = -32$

$D_y = \begin{vmatrix} 5 & -11 \\ -3 & -3 \end{vmatrix} = -15 - 33 = -48$

$x = \dfrac{D_x}{D} = \dfrac{-32}{-16} = 2, \ y = \dfrac{D_y}{D} = \dfrac{-48}{-16} = 3$

Solution: $(2, 3)$

99. By Cramer's Rule,

$D = \begin{vmatrix} 3 & -2 & 1 \\ 2 & 1 & -3 \\ 1 & -3 & 9 \end{vmatrix} = 35$

$D_a = \begin{vmatrix} 0 & -2 & 1 \\ 0 & 1 & -3 \\ 0 & -3 & 9 \end{vmatrix} = 0$

$D_b = \begin{vmatrix} 3 & 0 & 1 \\ 2 & 0 & -3 \\ 1 & 0 & 9 \end{vmatrix} = 0$

$D_c = \begin{vmatrix} 3 & -2 & 0 \\ 2 & 1 & 0 \\ 1 & -3 & 0 \end{vmatrix} = 0$

$a = \dfrac{D_a}{D} = 0, \ b = \dfrac{D_b}{D} = 0, \ c = \dfrac{D_c}{D} = 0$

Solution: $(0, 0, 0)$

101. By Cramer's Rule,

$D = \begin{vmatrix} -1 & 1 & 2 \\ 2 & 3 & 1 \\ 5 & 4 & 2 \end{vmatrix} = -15$

$D_x = \begin{vmatrix} 1 & 1 & 2 \\ -2 & 3 & 1 \\ 4 & 4 & 2 \end{vmatrix} = -30$

$D_y = \begin{vmatrix} -1 & 1 & 2 \\ 2 & -2 & 1 \\ 5 & 4 & 2 \end{vmatrix} = 45$

$D_z = \begin{vmatrix} -1 & 1 & 1 \\ 2 & 3 & -2 \\ 5 & 4 & 4 \end{vmatrix} = -45$

$x = \dfrac{D_x}{D} = \dfrac{-30}{-15} = 2, \ y = \dfrac{D_y}{D} = \dfrac{45}{-15} = -3,$

$z = \dfrac{D_z}{D} = \dfrac{-45}{-15} = 3$

Solution: $(2, -3, 3)$

103. Points: $(4, -3), (6, -7), (-2, -1)$

$$\begin{vmatrix} 4 & -3 & 1 \\ 6 & -7 & 1 \\ -2 & -1 & 1 \end{vmatrix} = -20 \neq 0$$

The points are not collinear.

105. Points: $(-6, -4), (-1, -3), (1.5, -2.5)$

$$\begin{vmatrix} -6 & -4 & 1 \\ -1 & -3 & 1 \\ 1.5 & -2.5 & 1 \end{vmatrix} = 0$$

The points are collinear.

Section 9.7   Graphs of Polar Equations

- When graphing polar equations:
  1. Test for symmetry
     (a) $\theta = \pi/2$: Replace (r, θ) by $(r, \pi - \theta)$ or $(-r, -\theta)$.
     (b) Polar axis: Replace (r, θ) by $(r, -\theta)$ or $(-r, \pi - \theta)$.
     (c) Pole: Replace (r, θ) by $(r, \pi + \theta)$ or $(-r, \theta)$.
     (d) $r = f(\sin \theta)$ is symmetric with respect to the line $\theta = \pi/2$.
     (e) $r = f(\cos \theta)$ is symmetric with respect to the polar axis.
  2. Find the θ values for which $|r|$ is maximum.
  3. Find the θ values for which $r = 0$.
  4. Know the different types of polar graphs.
     (a) Limaçons   (b) Rose curves, $n \geq 2$ (c) Circles   (d) Lemniscates
     $r = a \pm b \cos \theta$ $r = a \cos n\theta$ $r = a \cos \theta$ $r^2 = a^2 \cos 2\theta$
     $r = a \pm b \sin \theta$ $r = a \sin n\theta$ $r = a \sin \theta$ $r^2 = a^2 \sin 2\theta$

                                                              $r = a$

- You should be able to graph polar equations of the form $r = f(\theta)$ with your graphing utility. If your utility does not have a polar mode, use
  $$x = f(t) \cos t$$
  $$y = f(t) \sin t$$
  in parametric mode.

Vocabulary Check

1. $\theta = \dfrac{\pi}{2}$ **2.** polar axis   **3.** convex limaçon

4. circle   **5.** lemniscate   **6.** cardioid

1. $r = 3 \cos 2\theta$ is a rose curve.   **3.** Lemniscate   **5.** $r = 6 \sin 2\theta$ is a rose curve.

7. The graph is symmetric about the line $\theta = \pi/2$, and passes through $(r, \theta) = (3, 3\pi/2)$. Matches (a).

9. The graph has four leaves. Matches (c).

11. $r = 14 + 4 \cos \theta$

$\theta = \dfrac{\pi}{2}$: $\quad -r = 14 + 4 \cos(-\theta)$

$\qquad\qquad -r = 14 + 4 \cos \theta$

Not an equivalent equation

$\qquad\qquad r = 14 + 4 \cos(\pi - \theta)$

$\qquad\qquad r = 14 + 4(\cos \pi \cos \theta + \sin \pi \sin \theta)$

$\qquad\qquad r = 14 - 4 \cos \theta$

Not an equivalent equation

Polar axis: $r = 14 + 4 \cos(-\theta)$

$\qquad\qquad r = 14 + 4 \cos \theta$

Equivalent equation

Pole: $\quad -r = 14 + 4 \cos \theta$

Not an equivalent equation

$\qquad\qquad r = 14 + 4 \cos(\pi + \theta)$

$\qquad\qquad r = 14 - 4 \cos \theta$

Not an equivalent equation

Answer: Symmetric with respect to polar axis

13. $r = \dfrac{4}{1 + \sin \theta}$

$\theta = \dfrac{\pi}{2}$: $\quad r = \dfrac{4}{1 + \sin(\pi - \theta)}$

$\qquad\qquad r = \dfrac{4}{1 + \sin \pi \cos \theta - \cos \pi \sin \theta}$

$\qquad\qquad r = \dfrac{4}{1 + \sin \theta}$

Equivalent equation

Polar axis: $r = \dfrac{4}{1 + \sin(-\theta)}$

$\qquad\qquad r = \dfrac{4}{1 - \sin \theta}$

Not an equivalent equation

$\qquad\qquad -r = \dfrac{4}{1 + \sin(\pi - \theta)}$

$\qquad\qquad -r = \dfrac{4}{1 + \sin \theta}$

Not an equivalent equation

Pole: $-r = \dfrac{4}{1 + \sin \theta}$

Not an equivalent equation

$\qquad\qquad r = \dfrac{4}{1 + \sin(\pi + \theta)}$

$\qquad\qquad r = \dfrac{4}{1 - \sin \theta}$

Not an equivalent equation

Answer: Symmetric with respect to $\theta = \dfrac{\pi}{2}$

15. $r = 6 \sin \theta$

$\theta = \dfrac{\pi}{2}$: $\quad -r = 6 \sin(-\theta)$

$\qquad\qquad r = 6 \sin \theta$

Equivalent equation

Polar axis: $r = 6 \sin(-\theta)$

$\qquad\qquad r = -6 \sin \theta$

Not an equivalent equation

$\qquad\qquad -r = 6 \sin(\pi - \theta)$

$\qquad\qquad -r = 6(\sin \pi \cos \theta - \cos \pi \sin \theta)$

$\qquad\qquad -r = 6 \sin \theta$

Not an equivalent equation

Pole: $\quad -r = 6 \sin \theta$

Not an equivalent equation

$\qquad\qquad r = 6 \sin(\pi + \theta)$

$\qquad\qquad r = -6 \sin \theta$

Not an equivalent equation

Answer: Symmetric with respect to $\theta = \dfrac{\pi}{2}$

17. $r^2 = 16 \sin 2\theta$

$\theta = \dfrac{\pi}{2}: \quad (-r)^2 = 16 \sin(2(-\theta))$

$\qquad\qquad r^2 = -16 \sin 2\theta$

Not an equivalent equation

$\qquad r^2 = 16 \sin(2(\pi - \theta))$

$\qquad r^2 = 16 \sin(2\pi - 2\theta)$

$\qquad r^2 = -16 \sin 2\theta$

Not an equivalent equation

Polar axis: $r^2 = 16 \sin(2(-\theta))$

$\qquad\qquad r^2 = -16 \sin 2\theta$

Not an equivalent equation

$\qquad (-r)^2 = 16 \sin(2(\pi - \theta))$

$\qquad\quad r^2 = -16 \sin 2\theta$

Not an equivalent equation

Pole: $\qquad (-r)^2 = 16 \sin(2\theta)$

$\qquad\qquad r^2 = 16 \sin 2\theta$

Equivalent equation

Answer: Symmetric with respect to pole

19. $|r| = |10(1 - \sin \theta)|$

$\qquad\quad = 10|1 - \sin \theta| \le 10(2) = 20$

$|1 - \sin \theta| = 2$

$1 - \sin \theta = 2 \quad$ or $\quad 1 - \sin \theta = -2$

$\quad \sin \theta = -1 \qquad\qquad \sin \theta = 3$

$\quad \theta = \dfrac{3\pi}{2} \qquad\qquad$ Not possible

Maximum: $|r| = 20$ when $\theta = \dfrac{3\pi}{2}$

$r = 0$ when $1 - \sin \theta = 0$

$\qquad\qquad\qquad \sin \theta = 1$

$\qquad\qquad\qquad \theta = \dfrac{\pi}{2}$

21. $\quad |r| = |4 \cos 3\theta| = 4 |\cos 3\theta| \le 4$

$|\cos 3\theta| = 1$

$\cos 3\theta = \pm 1$

$\theta = 0, \dfrac{\pi}{3}, \dfrac{2\pi}{3}, \pi$

Maximum: $|r| = 4$ when $\theta = 0, \dfrac{\pi}{3}, \dfrac{2\pi}{3}, \pi$

$r = 0$ when $\cos 3\theta = 0$

$\theta = \dfrac{\pi}{6}, \dfrac{\pi}{2}, \dfrac{5\pi}{6}$

23. $r = 5$

Circle

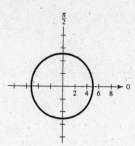

25. $r = 3 \sin \theta$

Symmetric with respect to $\theta = \pi/2$

Circle with radius of 3/2

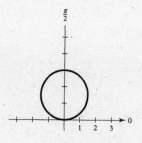

27. $r = 3(1 - \cos \theta)$

Cardioid

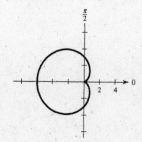

29. $r = 3 - 4 \cos \theta$

Limaçon

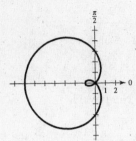

31. $r = 4 + 5 \sin \theta$

Limaçon

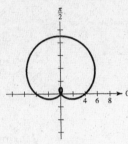

33. $r = 5 \cos 3\theta$

Rose curve

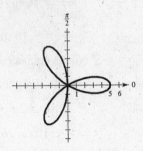

35. $r = 7 \sin 2\theta$

Rose curve, four petals

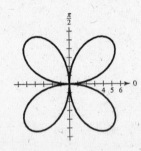

37. $r = 8 \cos 2\theta$

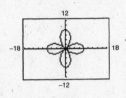

$0 \le \theta < 2\pi$

39. $r = 2(5 - \sin \theta)$

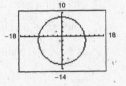

$0 \le \theta < 2\pi$

41. $r = 3 - 6 \cos \theta, \ 0 \le \theta \le 2\pi$

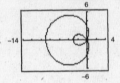

43. $r = \dfrac{3}{\sin \theta - 2 \cos \theta}, \ 0 \le \theta \le \dfrac{\pi}{2}$

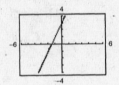

45. $r^2 = 4 \cos 2\theta, \ -2\pi \le \theta \le 2\pi$

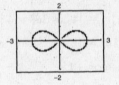

47. $r = 4 \sin \theta \cos^2 \theta, \ 0 \le \theta \le \pi$

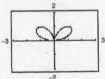

49. $r = 2 \csc \theta + 6$

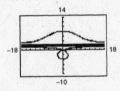

$0 \le \theta < 2\pi$

51. $r = e^{2\theta}$

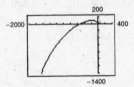

Answers will vary.

53. $r = 3 - 2 \cos \theta, \ 0 \le \theta < 2\pi$

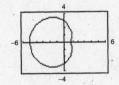

55. $r = 2 \cos\left(\dfrac{3\theta}{2}\right), \ 0 \le \theta < 4\pi$

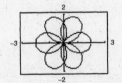

57. $r^2 = \sin 2\theta$, $0 \le \theta < \dfrac{\pi}{2}$

$\left(\text{Use } r_1 = \sqrt{\sin 2\theta} \text{ and}\right.$

$\left. r_2 = -\sqrt{\sin 2\theta}.\right)$

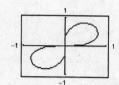

59. $r = 2 - \sec \theta$

$x = -1$ is an asymptote.

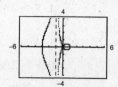

61. $r = \dfrac{2}{\theta}$

$y = 2$ is an asymptote.

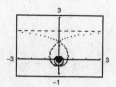

63. True. It has five petals.

65. $r = \cos(5\theta) + n \cos \theta$, $0 \le \theta < \pi$; Answers will vary.

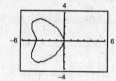

$n = -5$

$n = -4$

$n = -3$

$n = -2$

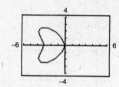

$n = -1$

$n = 0$

$n = 1$

$n = 2$

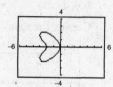

$n = 3$

$n = 4$

$n = 5$

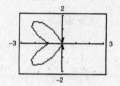
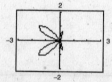
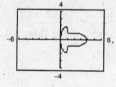

67. Use the result of Exercise 66.

(a) Rotation: $\phi = \dfrac{\pi}{2}$

Original graph: $r = f(\sin \theta)$

Rotated graph: $r = f\left(\sin\left(\theta - \dfrac{\pi}{2}\right)\right) = f(-\cos \theta)$

(b) Rotation: $\phi = \pi$

Original graph: $r = f(\sin \theta)$

Rotated graph: $r = f(\sin(\theta - \pi)) = f(-\sin \theta)$

(c) Rotation: $\phi = \dfrac{3\pi}{2}$

Original graph: $r = f(\sin \theta)$

Rotated graph: $r = f\left(\sin\left(\theta - \dfrac{3\pi}{2}\right)\right) = f(\cos \theta)$

69. (a) $r = 2 \sin\left[2\left(\theta - \dfrac{\pi}{6}\right)\right]$

$= 2 \sin\left(2\theta - \dfrac{\pi}{3}\right)$

$= \sin 2\theta - \sqrt{3} \cos 2\theta$

(b) $r = 2 \sin\left[2\left(\theta - \dfrac{\pi}{2}\right)\right]$

$= 2 \sin(2\theta - \pi)$

$= -2 \sin 2\theta$

$= -4 \sin \theta \cos \theta$

(ç) $r = 2 \sin\left[2\left(\theta - \dfrac{2\pi}{3}\right)\right]$

$= 2 \sin\left(2\theta - \dfrac{4\pi}{3}\right)$

$= \sqrt{3} \cos 2\theta - \sin 2\theta$

(d) $r = 2 \sin[2(\theta - \pi)]$

$= 2 \sin(2\theta - 2\pi)$

$= 2 \sin 2\theta$

$= 4 \sin \theta \cos \theta$

71. $r = 2 + k \cos \theta$

$k = 0$	$k = 1$	$k = 2$	$k = 3$
Circle	Convex limaçon	Cardioid	Limaçon with inner loop

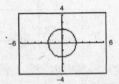 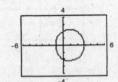 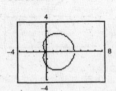 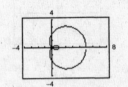

Section 9.8 Polar Equations of Conics

■ The graph of a polar equation of the form

$$r = \frac{ep}{1 \pm e \cos \theta} \quad \text{or} \quad r = \frac{ep}{1 \pm e \sin \theta}$$

is a conic, where $e > 0$ is the eccentricity and $|p|$ is the distance between the focus (pole) and the directrix.

(a) If $e < 1$, the graph is an ellipse.

(b) If $e = 1$, the graph is a parabola.

(c) If $e > 1$, the graph is a hyperbola.

■ Guidelines for finding polar equations of conics:

(a) Horizontal directrix above the pole: $r = \dfrac{ep}{1 + e \sin \theta}$

(b) Horizontal directrix below the pole: $r = \dfrac{ep}{1 - e \sin \theta}$

(c) Vertical directrix to the right of the pole: $r = \dfrac{ep}{1 + e \cos \theta}$

(d) Vertical directrix to the left of the pole: $r = \dfrac{ep}{1 - e \cos \theta}$

Vocabulary Check

1. conic   **2.** eccentricity, e **3.** (a) i (b) iii (c) ii

1. $r = \dfrac{2e}{1 + e \cos \theta}$

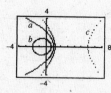

(a) Parabola

(b) Ellipse

(c) Hyperbola

3. $r = \dfrac{2e}{1 - e \sin \theta}$

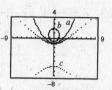

(a) Parabola

(b) Ellipse

(c) Hyperbola

5. $r = \dfrac{4}{1 - \cos \theta}$

$e = 1 \implies$ parabola

Vertical directrix to left of pole

Matches (b).

7. $r = \dfrac{3}{2 + \cos \theta} = \dfrac{3/2}{1 + (1/2) \cos \theta}$

$e = \dfrac{1}{2} \implies$ ellipse

Vertical directrix to right of pole

Matches (f).

9. $r = \dfrac{3}{1 + 2 \sin \theta}$

$e = 2 \implies$ hyperbola

Horizontal directrix above the pole.

Matches (d).

11. $r = \dfrac{2}{1 - \cos \theta}$

$e = 1 \implies$ parabola

Vertex: $(r, \theta) = (1, \pi)$

13. $r = \dfrac{4}{4 - \cos \theta} = \dfrac{1}{1 - (1/4) \cos \theta}$

$e = \dfrac{1}{4}, p = 4,$ ellipse

Vertices:

$(r, \theta) = \left(\dfrac{4}{3}, 0\right), \left(\dfrac{4}{5}, \pi\right)$

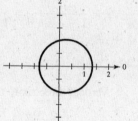

15. $r = \dfrac{8}{4 + 3 \sin \theta} = \dfrac{2}{1 + (3/4) \sin \theta}$

$e = \dfrac{3}{4} \implies$ ellipse

Vertices:

$(r, \theta) = \left(\dfrac{8}{7}, \dfrac{\pi}{2}\right), \left(8, \dfrac{3\pi}{2}\right)$

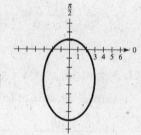

17. $r = \dfrac{6}{2 + \sin \theta} = \dfrac{(1/2)(6)}{1 + (1/2) \sin \theta}$

$e = \dfrac{1}{2} \implies$ ellipse

Vertices:

$\left(2, \dfrac{\pi}{2}\right), \left(6, \dfrac{3\pi}{2}\right)$

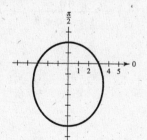

19. $r = \dfrac{3}{4 - 8 \cos \theta} = \dfrac{3/4}{1 - 2 \cos \theta}$

$e = 2 \implies$ hyperbola

Hyperbola

Vertices:

$(r, \theta) = \left(-\dfrac{3}{4}, 0\right), \left(\dfrac{1}{4}, \pi\right)$

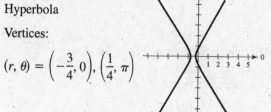

21. $r = \dfrac{-5}{1 - \sin \theta}$

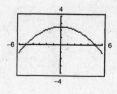

Parabola

23. $r = \dfrac{14}{14 + 17 \sin \theta}$

$\quad = \dfrac{1}{1 + (17/14) \sin \theta}$

Hyperbola

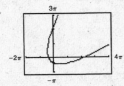

25.

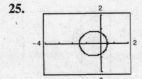

Ellipse

27.

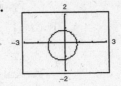

29.

31.

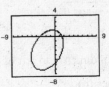

33. $e = 1, x = -1, p = 1$

Vertical directrix to the left of the pole

$r = \dfrac{1(1)}{1 - 1\cos \theta} = \dfrac{1}{1 - \cos \theta}$

35. $e = \dfrac{1}{2}, y = 1, p = 1$

Horizontal directrix above the pole

$r = \dfrac{(1/2)(1)}{1 + (1/2)\sin \theta} = \dfrac{1}{2 + \sin \theta}$

37. $e = 2, x = 1, p = 1$

Vertical directrix to the right of the pole

$r = \dfrac{2(1)}{1 + 2\cos \theta} = \dfrac{2}{1 + 2\cos \theta}$

39. Vertex: $\left(1, -\dfrac{\pi}{2}\right) \Rightarrow e = 1, p = 2$

Horizontal directrix below the pole

$r = \dfrac{1(2)}{1 - 1\sin \theta} = \dfrac{2}{1 - \sin \theta}$

41. Vertex: $(5, \pi) \Rightarrow e = 1, p = 10$

Vertical directrix to left of pole

$r = \dfrac{1(10)}{1 - 1\cos \theta} = \dfrac{10}{1 - \cos \theta}$

43. Center: $(4, \pi), c = 4, a = 6, e = \dfrac{2}{3}$

Vertical directrix to the right of the pole

$r = \dfrac{(2/3)p}{1 + (2/3)\cos \theta} = \dfrac{2p}{3 + 2\cos \theta}$

$2 = \dfrac{2p}{3 + 2\cos 0} = \dfrac{2p}{5} \Rightarrow p = 5$

$r = \dfrac{10}{3 + 2\cos \theta}$

45. Center: $(8, 0), c = 8, a = 12, e = \dfrac{c}{a} = \dfrac{2}{3}$

Vertical directrix to left of pole

$r = \dfrac{(2/3)p}{1 - (2/3)\cos \theta} = \dfrac{2p}{3 - 2\cos \theta}$

$20 = \dfrac{2p}{3 - 2} = 2p \Rightarrow p = 10$

$r = \dfrac{20}{3 - 2\cos \theta}$

47. Center: $\left(\dfrac{5}{2}, \dfrac{\pi}{2}\right)$, $c = \dfrac{5}{2}$, $a = \dfrac{3}{2}$, $e = \dfrac{5}{3}$

Horizontal directrix above the pole

$$r = \frac{(5/3)p}{1 + (5/3)\sin\theta} = \frac{5p}{3 + 5\sin\theta}$$

Substitute the point $\left(1, \dfrac{-3\pi}{2}\right)$ rather than $\left(-1, \dfrac{3\pi}{2}\right)$

in order to get a directrix between the vertices.

$$1 = \frac{5p}{3 + 5\sin(-3\pi/2)}$$

$$p = \frac{8}{5}$$

$$r = \frac{5(8/5)}{3 + 5\sin\theta} = \frac{8}{3 + 5\sin\theta}$$

49. When $\theta = 0$, $r = c + a = ea + a = a(1 + e)$.

Therefore,

$$a(1 + e) = \frac{ep}{1 - e\cos 0}$$

$$a(1 + e)(1 - e) = ep$$

$$a(1 - e^2) = ep.$$

Thus, $r = \dfrac{ep}{1 - e\cos\theta} = \dfrac{(1 - e^2)a}{1 - e\cos\theta}.$

51. $r = \dfrac{[1 - (0.0167)^2](92.956 \times 10^6)}{1 - 0.0167\cos\theta}$

$\approx \dfrac{9.2930 \times 10^7}{1 - 0.0167\cos\theta}$

Perihelion distance:

$r = 92.956 \times 10^6(1 - 0.0167) \approx 9.1404 \times 10^7$

Aphelion distance:

$r = 92.956 \times 10^6(1 + 0.0167) \approx 9.4508 \times 10^7$

53. $r = \dfrac{(1 - 0.0484^2)77.841 \times 10^7}{1 - 0.0484\cos\theta}$

$= \dfrac{7.7659 \times 10^8}{1 - 0.0484\cos\theta}$

Perihelion:

$r = 77.841 \times 10^7(1 - 0.0484) \approx 7.4073 \times 10^8$ km

Aphelion:

$r = 77.841 \times 10^7(1 + 0.0484) \approx 8.1609 \times 10^8$ km

55. $a = 4.498 \times 10^9$, $e = 0.0086$, Neptune

$a = 5.906 \times 10^9$, $e = 0.2488$, Pluto

(a) Neptune: $r = \dfrac{(1 - 0.0086^2)4.498 \times 10^9}{1 - 0.0086\cos\theta} = \dfrac{4.4977 \times 10^9}{1 - 0.0086\cos\theta}$

Pluto: $r = \dfrac{(1 - 0.2488^2)5.906 \times 10^9}{1 - 0.2488\cos\theta} = \dfrac{5.5404 \times 10^9}{1 - 0.2488\cos\theta}$

(b) Neptune: Perihelion: $4.498 \times 10^9(1 - 0.0086) \approx 4.4593 \times 10^9$ km

Aphelion: $4.498 \times 10^9(1 + 0.0086) \approx 4.5367 \times 10^9$ km

Pluto: Perihelion: $5.906 \times 10^9(1 - 0.2488) \approx 4.4366 \times 10^9$ km

Aphelion: $5.906 \times 10^9(1 + 0.2488) \approx 7.3754 \times 10^9$ km

(c)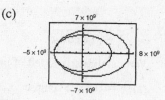

(d) Yes. Pluto is closer to the sun for just a very short time. Pluto was considered the ninth planet because its mean distance from the sun is larger than that of Neptune.

(e) Although the graphs intersect, the orbits do not, and the planets won't collide.

57. $r = \dfrac{4}{-3 - 3\sin\theta} = \dfrac{-4/3}{1 + \sin\theta}$

False. The directrix is below the pole.

59.
$$\frac{x^2}{a^2} + \frac{y^2}{b^2} = 1$$

$$\frac{r^2 \cos^2 \theta}{a^2} + \frac{r^2 \sin^2 \theta}{b^2} = 1$$

$$\frac{r^2 \cos^2 \theta}{a^2} + \frac{r^2(1 - \cos^2 \theta)}{b^2} = 1$$

$$r^2 b^2 \cos^2 \theta + r^2 a^2 - r^2 a^2 \cos^2 \theta = a^2 b^2$$

$$r^2(b^2 - a^2) \cos^2 \theta + r^2 a^2 = a^2 b^2$$

For an ellipse, $b^2 - a^2 = -c^2$. Hence,

$$-r^2 c^2 \cos^2 \theta + r^2 a^2 = a^2 b^2$$

$$-r^2 \left(\frac{c}{a}\right)^2 \cos^2 \theta + r^2 = b^2, \ e = \frac{c}{a}$$

$$-r^2 e^2 \cos^2 \theta + r^2 = b^2$$

$$r^2(1 - e^2 \cos^2 \theta) = b^2$$

$$r^2 = \frac{b^2}{1 - e^2 \cos^2 \theta}.$$

61. $\frac{x^2}{169} + \frac{y^2}{144} = 1$

$$a = 13, b = 12, c = 5, e = \frac{5}{13}$$

$$r^2 = \frac{144}{1 - (25/169) \cos^2 \theta} = \frac{24{,}336}{169 - 25 \cos^2 \theta}$$

63. $\frac{x^2}{25} + \frac{y^2}{16} = 1$

$$a = 5, b = 4, c = 3, e = \frac{3}{5}$$

$$r^2 = \frac{b^2}{1 - e^2 \cos^2 \theta} = \frac{16}{1 - (9/25) \cos^2 \theta} = \frac{400}{25 - 9 \cos^2 \theta}$$

65. Center: $(x, y) = (0, 0), c = 5, a = 4, e = \frac{5}{4}$

$$b^2 = c^2 - a^2 = 25 - 16 = 9 \implies b = 3$$

$$r^2 = \frac{-b^2}{1 - e^2 \cos^2 \theta} = \frac{-9}{1 - (25/16) \cos^2 \theta} = \frac{144}{25 \cos^2 \theta - 16}$$

67. $r = \dfrac{4}{1 - 0.4 \cos \theta}$

Vertical directrix to left of pole

(a) $e = 0.4 \implies$ ellipse

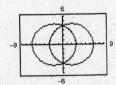 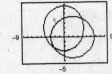

(b) $r = \dfrac{4}{1 + 0.4 \cos \theta}$

Vertical directrix to right of pole

Graph is reflected in line $\theta = \pi/2$.

$$r = \frac{4}{1 - 0.4 \sin \theta}$$

Horizontal directrix below pole

$90°$ rotation counterclockwise

69. Answers will vary.

71. $4\sqrt{3}\tan\theta - 3 = 1$

$$\tan\theta = \frac{1}{\sqrt{3}} = \frac{\sqrt{3}}{3}$$

$$\theta = \frac{\pi}{6} + n\pi$$

73. $12\sin^2\theta = 9$

$$\sin^2\theta = \frac{3}{4}$$

$$\sin\theta = \pm\frac{\sqrt{3}}{2}$$

$$\theta = \frac{\pi}{3} + n\pi, \frac{2\pi}{3} + n\pi$$

75. $2\cot x = 5\cos\dfrac{\pi}{2}$

$$2\cot x = 0$$

$$\cot x = 0$$

$$x = \frac{\pi}{2} + n\pi$$

For Exercises 77 and 79: $\sin u = -\dfrac{3}{5}$, $\cos u = \dfrac{4}{5}$, $\cos v = \dfrac{1}{\sqrt{2}}$, $\sin v = -\dfrac{1}{\sqrt{2}}$

77. $\cos(u + v) = \cos u\cos v - \sin u\sin v$

$$= \frac{4}{5}\left(\frac{1}{\sqrt{2}}\right) - \left(-\frac{3}{5}\right)\left(-\frac{1}{\sqrt{2}}\right)$$

$$= \frac{1}{5\sqrt{2}} = \frac{\sqrt{2}}{10}$$

79. $\sin(u - v) = \sin u\cos v - \sin v\cos u$

$$= \left(-\frac{3}{5}\right)\left(\frac{1}{\sqrt{2}}\right) - \left(-\frac{1}{\sqrt{2}}\right)\left(\frac{4}{5}\right)$$

$$= \frac{1}{5\sqrt{2}} = \frac{\sqrt{2}}{10}$$

81. $_{12}C_9 = 220$

83. $_{10}P_3 = 720$

Review Exercises for Chapter 9

1. Radius $= \sqrt{(-3 - 0)^2 + (-4 - 0)^2}$

$$= \sqrt{9 + 16} = \sqrt{25} = 5$$

$x^2 + y^2 = 25$

3. Radius $= \dfrac{1}{2}\sqrt{(5 - (-1))^2 + (6 - 2)^2}$

$$= \frac{1}{2}\sqrt{36 + 16} = \frac{1}{2}\sqrt{52} = \sqrt{13}$$

Center $= \left(\dfrac{5 + (-1)}{2}, \dfrac{6 + 2}{2}\right) = (2, 4)$

$(x - 2)^2 + (y - 4)^2 = 13$

5. $\dfrac{1}{2}x^2 + \dfrac{1}{2}y^2 = 18$

$$x^2 + y^2 = 36$$

Center: $(0, 0)$

Radius: 6

7. $16x^2 + 16y^2 - 16x + 24y - 3 = 0$

$$16\left(x^2 - x + \frac{1}{4}\right) + 16\left(y^2 + \frac{3}{2}y + \frac{9}{16}\right) = 3 + 4 + 9$$

$$16\left(x - \frac{1}{2}\right)^2 + 16\left(y + \frac{3}{4}\right)^2 = 16$$

$$\left(x - \frac{1}{2}\right)^2 + \left(y + \frac{3}{4}\right)^2 = 1$$

Center: $\left(\dfrac{1}{2}, -\dfrac{3}{4}\right)$

Radius: 1

9. $(x^2 + 4x + 4) + (y^2 + 6y + 9) = 3 + 4 + 9$

$$(x + 2)^2 + (y + 3)^2 = 16$$

Center: $(-2, -3)$

Radius: 4

11. x-intercepts: $(x - 3)^2 + (0 + 1)^2 = 7$

$$(x - 3)^2 = 6$$

$$x - 3 = \pm\sqrt{6}$$

$$x = 3 \pm \sqrt{6}$$

$$\left(3 \pm \sqrt{6}, 0\right)$$

y-intercepts: $(0 - 3)^2 + (y + 1)^2 = 7$

$$(y + 1)^2 = -2, \text{ impossible}$$

No y-intercepts

13. $4x - y^2 = 0$

$$y^2 = 4(1)x, \ p = 1$$

Vertex: $(0, 0)$

Focus: $(1, 0)$

Directrix: $x = -1$

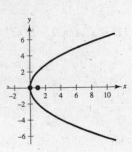

15. $\frac{1}{2}y^2 + 18x = 0$

$$\frac{1}{2}y^2 = -18x$$

$$y^2 = -36x = 4(-9)x, \ p = -9$$

Vertex: $(0, 0)$

Focus: $(-9, 0)$

Directrix: $x = 9$

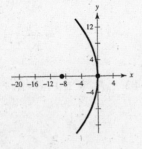

17. Vertex: $(0, 0)$

Focus: $(-6, 0)$

Parabola opens to left.

$$y^2 = 4px$$

$$y^2 = 4(-6)x$$

$$y^2 = -24x$$

19. Vertex: $(-6, 4)$

Passes through $(0, 0)$

Vertical axis

$$(x + 6)^2 = 4p(y - 4)$$

$$(0 + 6)^2 = 4p(0 - 4)$$

$$36 = -16p$$

$$-\frac{9}{4} = p$$

$$(x + 6)^2 = 4\left(-\frac{9}{4}\right)(y - 4)$$

$$(x + 6)^2 = -9(y - 4)$$

21. $x^2 = -2y = 4\left(-\frac{1}{2}\right)y, p = -\frac{1}{2}$

Focus: $\left(0, -\frac{1}{2}\right)$

$$d_1 = \frac{1}{2} + b$$

$$d_2 = \sqrt{(2 - 0)^2 + \left(-2 + \frac{1}{2}\right)^2} = \frac{5}{2}$$

$$d_1 = d_2 \implies \frac{1}{2} + b = \frac{5}{2} \implies b = 2$$

Slope of tangent line: $\dfrac{b + 2}{0 - 2} = \dfrac{4}{-2} = -2$

Equation: $y + 2 = -2(x - 2)$

$$y = -2x + 2$$

x-intercept: $(1, 0)$

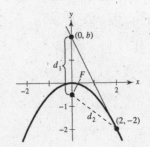

23. $x^2 = 4p(y - 12)$

$(4, 10)$ on curve:

$16 = 4p(10 - 12) = -8p \implies p = -2$

$x^2 = 4(-2)(y - 12) = -8y + 96$

$y = \dfrac{-x^2 + 96}{8}$

$y = 0$ if $x^2 = 96 \implies x = 4\sqrt{6} \implies$ width is $8\sqrt{6}$ meters.

25. $\dfrac{x^2}{4} + \dfrac{y^2}{16} = 1$

$a = 4, b = 2, c = \sqrt{16 - 4} = \sqrt{12} = 2\sqrt{3}$

Center: $(0, 0)$

Vertices: $(0, \pm4)$

Foci: $\left(0, \pm2\sqrt{3}\right)$

Eccentricity $= \dfrac{c}{a}$

$\qquad = \dfrac{2\sqrt{3}}{4}$

$\qquad = \dfrac{\sqrt{3}}{2}$

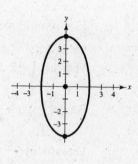

27. $\dfrac{(x - 4)^2}{6} + \dfrac{(y + 4)^2}{9} = 1$

$a = 3, b = \sqrt{6}, c = \sqrt{9 - 6} = \sqrt{3}$

Center: $(4, -4)$

Vertices: $(4, -1), (4, -7)$

Foci: $\left(4, -4 \pm \sqrt{3}\right)$

Eccentricity $= \dfrac{c}{a} = \dfrac{\sqrt{3}}{3}$

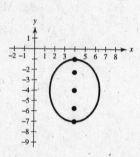

29. (a) $16(x^2 - 2x + 1) + 9(y^2 + 8y + 16) = -16 + 16 + 144$

$\qquad 16(x - 1)^2 + 9(y + 4)^2 = 144$

$\qquad \dfrac{(x - 1)^2}{9} + \dfrac{(y + 4)^2}{16} = 1$

(b) Center: $(1, -4)$

$a = 4, b = 3, c = \sqrt{16 - 9} = \sqrt{7}$

Vertices: $(1, 0), (1, -8)$

Foci: $\left(1, -4 \pm \sqrt{7}\right)$

$e = \dfrac{c}{a} = \dfrac{\sqrt{7}}{4}$

(c)

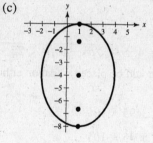

31. (a) $3(x^2 + 4x + 4) + 8(y^2 - 14y + 49) = -403 + 12 + 392$

$\qquad 3(x + 2)^2 + 8(y - 7)^2 = 1$

$\qquad \dfrac{(x + 2)^2}{1/3} + \dfrac{(y - 7)^2}{1/8} = 1$

—**CONTINUED**—

31. —CONTINUED—

(b) Center: $(-2, 7)$

$$a = \frac{\sqrt{3}}{3}, b = \frac{\sqrt{2}}{4}$$

$$c^2 = a^2 - b^2 = \frac{1}{3} - \frac{1}{8} = \frac{5}{24} \Rightarrow c = \frac{\sqrt{30}}{12}$$

Vertices: $\left(-2 \pm \frac{\sqrt{3}}{3}, 7\right)$

Foci: $\left(-2 \pm \frac{\sqrt{30}}{12}, 7\right)$

Eccentricity: $\dfrac{c}{a} = \dfrac{\sqrt{30}/12}{\sqrt{3}/3} = \dfrac{\sqrt{10}}{4}$

(c)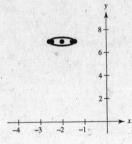

33. Vertices: $(\pm 5, 0)$

Foci: $(\pm 4, 0)$

$a = 5, c = 4 \Rightarrow b = 3$

$$\frac{x^2}{25} + \frac{y^2}{9} = 1$$

35. Vertices: $(-3, 0), (7, 0)$

Foci: $(0, 0), (4, 0)$

Horizontal major axis

Center: $(2, 0)$

$a = 5, c = 2,$

$b = \sqrt{25 - 4} = \sqrt{21}$

$$\frac{(x - h)^2}{a^2} + \frac{(y - k)^2}{b^2} = 1$$

$$\frac{(x - 2)^2}{25} + \frac{y^2}{21} = 1$$

37. $a = 5, b = 4, c = \sqrt{a^2 - b^2} = \sqrt{25 - 16} = 3$

The foci should be placed 3 feet on either side of the center and have the same height as the pillars.

39. $a - c = 1.3495 \times 10^9$

$a + c = 1.5045 \times 10^9$

Adding, $2a = 2.854 \times 10^9 \Rightarrow a = 1.427 \times 10^9$.
Then

$c = 1.5045 \times 10^9 - 1.427 \times 10^9 = 0.0775 \times 10^9$

$e = \dfrac{c}{a} \approx 0.0543.$

41. (a) $5y^2 - 4x^2 = 20$

$$\frac{y^2}{4} - \frac{x^2}{5} = 1$$

(b) $a = 2, b = \sqrt{5},$

$c = \sqrt{4 + 5} = 3$

Center: $(0, 0)$

Vertices: $(0, \pm 2)$

Foci: $(0, \pm 3)$

Eccentricity $= \dfrac{c}{a} = \dfrac{3}{2}$

(c)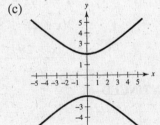

43. (a) $9(x^2 - 2x + 1) - 16(y^2 + 2y + 1) = 151 + 9 - 16$

$$9(x - 1)^2 - 16(y + 1)^2 = 144$$

$$\frac{(x - 1)^2}{16} - \frac{(y + 1)^2}{9} = 1$$

(b) Center: $(1, -1)$, $a = 4$, $b = 3$, $c = 5$

Vertices: $(5, -1)$, $(-3, -1)$

Foci: $(6, -1)$, $(-4, -1)$

Eccentricity: $\dfrac{5}{4}$

(c)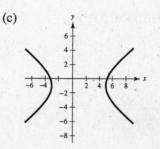

45. (a) $(y^2 - 2y + 1) - 4(x^2 + 12x + 36) = -59 + 1 - 144$

$$(y - 1)^2 - 4(x + 6)^2 = -202$$

$$\frac{(x + 6)^2}{(101/2)} - \frac{(y - 1)^2}{202} = 1$$

(b) Center: $(-6, 1)$

$$a^2 = \frac{101}{2},\ b^2 = 202,\ c^2 = \frac{101}{2} + 202 = \frac{505}{2}$$

Vertices: $\left(-6 \pm \sqrt{\dfrac{101}{2}}, 1\right)$

Foci: $\left(-6 \pm \sqrt{\dfrac{505}{2}}, 1\right)$

Eccentricity: $e = \dfrac{c}{a} = \dfrac{\sqrt{505}}{\sqrt{101}} = \sqrt{5}$

(c)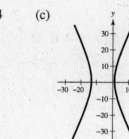

47. $\dfrac{x^2}{a^2} - \dfrac{y^2}{b^2} = 1$

$a = 4$

$c^2 = a^2 + b^2 \Rightarrow 36 = 16 + b^2$

$\Rightarrow b = \sqrt{20} = 2\sqrt{5}$

$\dfrac{x^2}{16} - \dfrac{y^2}{20} = 1$

49. Foci: $(0, 0)$, $(8, 0) \Rightarrow c = 4$

Center: $(4, 0)$

Asymptotes:

$y = \pm 2(x - 4) \Rightarrow \dfrac{b}{a} = 2 \Rightarrow b = 2a$

$c^2 = a^2 + b^2$

$16 = a^2 + (2a)^2 = 5a^2 \Rightarrow a = \dfrac{4}{\sqrt{5}}, b = \dfrac{8}{\sqrt{5}}$

$\dfrac{(x - 4)^2}{16/5} - \dfrac{y^2}{64/5} = 1$

51. $d_2 - d_1 = 186{,}000(0.0005)$

$2a = 93$

$a = 46.5$

$c = 100$

$b = \sqrt{c^2 - a^2}$

$\dfrac{x^2}{a^2} - \dfrac{y^2}{b^2} = 1$

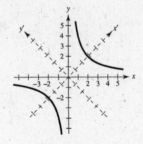

$x = 60 \Rightarrow y^2 = b^2\left(\dfrac{x^2}{a^2} - 1\right) = (100^2 - 46.5^2)\left(\dfrac{60^2}{46.5^2} - 1\right) \approx 5211.57 \Rightarrow y \approx 72.2$

72.2 miles north

53. $3x^2 + 2y^2 - 12x + 12y + 29 = 0$

$3(x^2 - 4x + 4) + 2(y^2 + 6y + 9) = -29 + 12 + 18$

$3(x - 2)^2 + 2(y + 3)^2 = 1$

Ellipse

55. $5x^2 - 2y^2 + 10x - 4y + 17 = 0$

$5(x^2 + 2x + 1) - 2(y^2 + 2y + 1) = -17 + 5 - 2$

$5(x + 1)^2 - 2(y + 1)^2 = -14$

$\dfrac{(y + 1)^2}{7} - \dfrac{(x + 1)^2}{(14/5)} = 1$

Hyperbola

57. $xy - 4 = 0$

$A = 0, B = 1, C = 0$

$\cot 2\theta = \dfrac{A - C}{B} = 0 \Rightarrow \theta = \dfrac{\pi}{4}$

$x = \dfrac{\sqrt{2}}{2}(x' - y'),\, y = \dfrac{\sqrt{2}}{2}(x' + y')$

$xy = 4$

$\dfrac{\sqrt{2}}{2}(x' - y')\,\dfrac{\sqrt{2}}{2}(x' + y') = 4$

$\dfrac{1}{2}(x')^2 - \dfrac{1}{2}(y')^2 = 4$

$\dfrac{(x')^2}{8} - \dfrac{(y')^2}{8} = 1$

Hyperbola

59. $5x^2 - 2xy + 5y^2 - 12 = 0$

$A = 5, B = -2, C = 5$

$\cot 2\theta = 0 \implies \theta = \dfrac{\pi}{4}$

$x = \dfrac{\sqrt{2}}{2}(x' - y'),\ y = \dfrac{\sqrt{2}}{2}(x' + y')$

$$5x^2 - 2xy + 5y^2 = 12$$

$$5\left[\frac{\sqrt{2}}{2}(x' - y')\right]^2 - 2\left[\frac{\sqrt{2}}{2}(x' - y')\right]\left[\frac{\sqrt{2}}{2}(x' + y')\right] + 5\left[\frac{\sqrt{2}}{2}(x' + y')\right]^2 = 12$$

$$5\left[\frac{1}{2}(x')^2 - x'y' + \frac{1}{2}(y')^2\right] - (x')^2 + (y')^2 + 5\left[\frac{1}{2}(x')^2 + x'y' + \frac{1}{2}(y')^2\right] = 12$$

$$4(x')^2 + 6(y')^2 = 12$$

$$\frac{(x')^2}{3} + \frac{(y')^2}{2} = 1$$

Ellipse

61. (a) $B^2 - 4AC = (-8)^2 - 4(16)(1) = 0$

Parabola

(b) $y^2 + (5 - 8x)y + (16x^2 - 10x) = 0$

$y = \dfrac{(8x - 5) \pm \sqrt{(5 - 8x)^2 - 4(16x^2 - 10x)}}{2}$

(c)

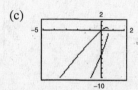

63. (a) $B^2 - 4AC = (2)^2 - 4(1)(1) = 0$

Parabola

(b) $y^2 + \left(2x - 2\sqrt{2}\right)y + \left(x^2 + 2\sqrt{2}x + 2\right) = 0$

$y = \dfrac{\left(2\sqrt{2} - 2x\right) \pm \sqrt{\left(2x - 2\sqrt{2}\right)^2 - 4\left(x^2 + 2\sqrt{2}x + 2\right)}}{2}$

(c)

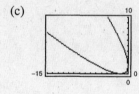

65. Adding the equations,

$24x + 240 = 0 \implies x = -10$. Then:

$4(100) + y^2 - 560 - 24y + 304 = 0$

$\qquad\qquad y^2 - 24y + 144 = 0$

$\qquad\qquad\quad (y - 12)^2 = 0 \implies y = 12$

Solution: $(-10, 12)$

67.

t	-2	-1	0	1	2	3
x	-8	-5	-2	1	4	7
y	15	11	7	3	-1	-5

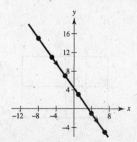

69. $x = 5t - 1, y = 2t + 5$

$t = \frac{1}{5}(x + 1) \implies$

$y = \frac{2}{5}(x + 1) + 5 = \frac{2}{5}x + \frac{27}{5}$, line

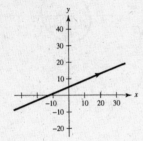

71. $x = t^2 + 2, y = 4t^2 - 3$

$t^2 = x - 2 \implies$

$y = 4(x - 2) - 3 = 4x - 11, \quad x \geq 2$

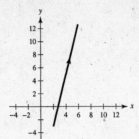

73. $x = t^3, y = \frac{1}{2}t^2$

$t = x^{1/3} \implies y = \frac{1}{2}x^{2/3}$

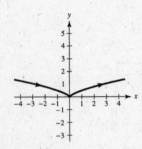

75. $x = \sqrt[3]{t}$

$y = t$

$t = x^3 \implies y = t = x^3$

$y = x^3$

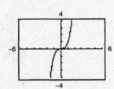

77. $x = \frac{1}{t}$

$y = t$

$y = t = \frac{1}{x}$

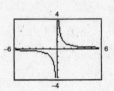

79. $x = 2t$

$y = 4t$

$y = 2(2t) = 2x$

Line

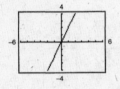

81. $x = 1 + 4t$

$y = 2 - 3t$

$t = \frac{x - 1}{4} \implies y = 2 - 3\left(\frac{x - 1}{4}\right) = 2 - \frac{3}{4}x + \frac{3}{4}$

$y = \frac{11}{4} - \frac{3}{4}x$

$3x + 4y - 11 = 0$

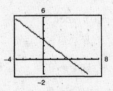

83. $x = 3$

$y = t$

Vertical line: $x = 3$

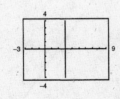

85. $x = 6 \cos \theta, y = 6 \sin \theta$

$\cos \theta = \frac{x}{6}, \sin \theta = \frac{y}{6}$

$\frac{x^2}{36} + \frac{y^2}{36} = 1$

$x^2 + y^2 = 36$

87. $y = 6x + 2$

$x = t, y = 6t + 2$

$x = -t, y = -6t + 2$

Other answers possible

89. $y = x^2 + 2$

$x = t, y = t^2 + 2$

$x = t + 1, y = (t + 1)^2 + 2 = t^2 + 2t + 3$

Other answers possible

91. $x = x_1 + t(x_2 - x_1) = 3 + t(8 - 3) = 5t + 3$

$y = y_1 + t(y_2 - y_1) = 5 + t(0 - 0) = 5$

or $x = t, \; y = 5$

93. $x = x_1 + t(x_2 - x_1)$

$\quad = -1 + t[10 - (-1)] = 11t - 1$

$y = y_1 + t(y_2 - y_1) = 6 + t(0 - 6) = -6t + 6$

95. $(90, 4)$ is on the curve:

$$90 = 0.82v_0 t \Rightarrow v_0 = \frac{90}{0.82t}$$

$$4 = 7 + 0.57\left[\frac{90}{0.82t}\right]t - 16t^2 \Rightarrow$$

$$16t^2 = 3 + \frac{0.57(90)}{0.82} \Rightarrow t \approx 2.024$$

Hence, $v_0 \approx \dfrac{90}{0.82(2.024)} \approx 54.23$ ft/sec.

97. From Exercise 96:

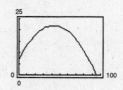

The maximum height is approximately 21.9 feet for $t \approx 0.97$.

99. $\left(1, \dfrac{\pi}{4}\right)$

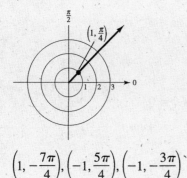

$\left(1, -\dfrac{7\pi}{4}\right), \left(-1, \dfrac{5\pi}{4}\right), \left(-1, -\dfrac{3\pi}{4}\right)$

101. $(r, \theta) = \left(-2, -\dfrac{11\pi}{6}\right)$

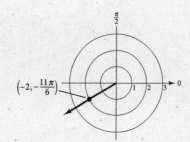

$\left(-2, \dfrac{\pi}{6}\right), \left(2, \dfrac{7\pi}{6}\right), \left(2, -\dfrac{5\pi}{6}\right)$

103. $\left(\sqrt{5}, -\dfrac{4\pi}{3}\right)$

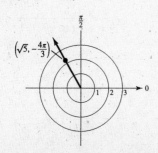

$\left(\sqrt{5}, \dfrac{2\pi}{3}\right), \left(-\sqrt{5}, -\dfrac{\pi}{3}\right), \left(-\sqrt{5}, \dfrac{5\pi}{3}\right)$

105. $(r, \theta) = \left(5, -\dfrac{7\pi}{6}\right)$

$(x, y) = \left(-\dfrac{5\sqrt{3}}{2}, \dfrac{5}{2}\right)$

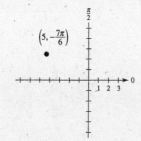

107. $\left(2, -\dfrac{5\pi}{3}\right)$

$x = r \cos \theta = 2\left(\dfrac{1}{2}\right) = 1$

$y = r \sin \theta = 2\left(\dfrac{\sqrt{3}}{2}\right) = \sqrt{3}$

$(x, y) = \left(1, \sqrt{3}\right)$

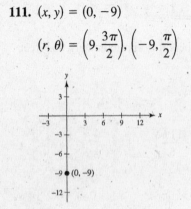

109. $(r, \theta) = \left(3, \dfrac{3\pi}{4}\right)$

$(x, y) = \left(3 \cos \dfrac{3\pi}{4}, 3 \sin \dfrac{3\pi}{4}\right) = \left(\dfrac{-3\sqrt{2}}{2}, \dfrac{3\sqrt{2}}{2}\right)$

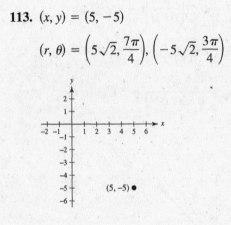

111. $(x, y) = (0, -9)$

$(r, \theta) = \left(9, \dfrac{3\pi}{2}\right), \left(-9, \dfrac{\pi}{2}\right)$

113. $(x, y) = (5, -5)$

$(r, \theta) = \left(5\sqrt{2}, \dfrac{7\pi}{4}\right), \left(-5\sqrt{2}, \dfrac{3\pi}{4}\right)$

115. $x^2 + y^2 = 9$

$r^2 = 9$

$r = 3$

117. $x^2 + y^2 - 4x = 0$

$r^2 - 4r \cos \theta = 0$

$r = 4 \cos \theta$

119. $xy = 5$

$(r \cos \theta)(r \sin \theta) = 5$

$r^2 = 5 \csc \theta \cdot \sec \theta$

121. $4x^2 + y = 1$

$4(r \cos \theta)^2 + (r \sin \theta)^2 = 1$

$4r^2 \cos^2 \theta + r^2(1 - \cos^2 \theta) = 1$

$r^2[3 \cos^2 \theta + 1] = 1$

$r^2 = \dfrac{1}{3 \cos^2 \theta + 1}$

123. $r = 5$

$x^2 + y^2 = 5^2 = 25$

Circle

125. $r = 3 \cos \theta$

$r^2 = 3r \cos \theta$

$x^2 + y^2 = 3x$

127.
$$r^2 = \cos 2\theta$$
$$r^2 = 1 - 2\sin^2 \theta$$
$$r^4 = r^2 - 2r^2 \sin^2 \theta$$
$$(x^2 + y^2)^2 = x^2 + y^2 - 2y^2$$
$$(x^2 + y^2)^2 - x^2 + y^2 = 0$$

129. $\theta = \dfrac{5\pi}{6}$

$\tan \theta = \dfrac{y}{x} = -\dfrac{1}{\sqrt{3}}$

$y = -\dfrac{\sqrt{3}}{3}x,$ line

131. $r = 5$, circle

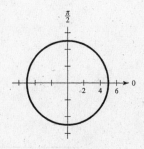

133. $\theta = \dfrac{\pi}{2}$, y-axis

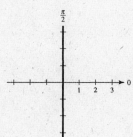

135. $r = 5\cos \theta$, circle

137. $r = 5 + 4\cos \theta$

Dimpled limaçon

Symmetric with respect to polar axis

r is maximum at $\theta = 0$: $(r, \theta) = (9, 0)$

$r \neq 0$ (No zeros)

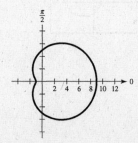

139. $r = 3 - 5\sin \theta$

Limaçon with loop

Symmetry: line $\theta = \dfrac{\pi}{2}$

Maximum $|r|$-value: $|r| = 8$ when $\theta = \dfrac{3\pi}{2}$

Zeros: $r = 0$ when $\theta \approx 0.6435, 2.4981$ $\left(\sin \theta = \dfrac{3}{5}\right)$

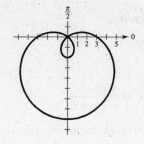

141. $r = -3\cos 2\theta$, $0 \le \theta \le 2\pi$

Four-leaved rose

Symmetric with respect to $\theta = \dfrac{\pi}{2}$, polar axis, and pole

The value of $|r|$ is a maximum (3) at $\theta = 0, \dfrac{\pi}{2}, \pi, \dfrac{3\pi}{2}$.

$r = 0$ for $\theta = \dfrac{\pi}{4}, \dfrac{3\pi}{4}, \dfrac{5\pi}{4}, \dfrac{7\pi}{4}$

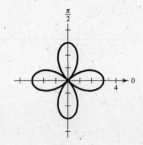

143. $r^2 = 5 \sin 2\theta$

Lemniscate

Symmetry with respect to pole

Maximum $|r|$-value: $\sqrt{5}$ when $\theta = \dfrac{\pi}{4}, \dfrac{5\pi}{4}$

Zeros: $r = 0$ when $\theta = 0, \dfrac{\pi}{2}, \pi, \dfrac{3\pi}{2}$

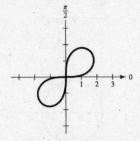

145. $r = \dfrac{2}{1 - \sin \theta}$

$e = 1$

Parabola

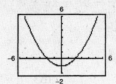

147. $r = \dfrac{4}{5 - 3 \cos \theta}$

$\quad = \dfrac{4/5}{1 - (3/5) \cos \theta}$

$e = \dfrac{3}{5}$

Ellipse

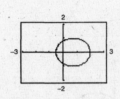

149. $r = \dfrac{5}{6 + 2 \sin \theta}$

$\quad = \dfrac{5/6}{1 + (1/3) \sin \theta}$

$e = \dfrac{1}{3}$

Ellipse

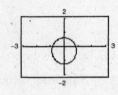

151. $e = 1$

$\quad r = \dfrac{4}{1 - \cos \theta}$

Vertical directrix: $x = -4$

153. Ellipse: $r = \dfrac{ep}{1 - e \cos \theta}$

Vertices: $(5, 0), (1, \pi) \implies a = 3$

One focus: $(0, 0) \implies c = 2$

$e = \dfrac{c}{a} = \dfrac{2}{3}, \quad 5 = \dfrac{2/3\, p}{1 - (2/3) \cos 0} \implies p = \dfrac{5}{2}$

$r = \dfrac{(2/3)(5/2)}{1 - (2/3) \cos \theta} = \dfrac{5/3}{1 - (2/3) \cos \theta} = \dfrac{5}{3 - 2 \cos \theta}$

155. $e = 0.093$

Use $r = \dfrac{ep}{1 - e \cos \theta}$.

$2a = \dfrac{0.093p}{1 - 0.093 \cos 0} + \dfrac{0.093p}{1 - 0.093 \cos \pi} = 0.1876p = 3.05 \Rightarrow p \approx 16.258, \; ep \approx 1.512$

$r = \dfrac{1.512}{1 - 0.093 \cos \theta}$

Perihelion: $\dfrac{1.512}{1 + 0.093} \approx 1.383$ astronomical units

Aphelion: $\dfrac{1.512}{1 - 0.093} \approx 1.667$ astronomical units

157. False. The y^4-term is not second degree.

159. (a) Vertical translation

 (b) Horizontal translation

 (c) Reflection in the y-axis

 (d) Parabola opens more slowly.

161. The number b must be less than 5. The ellipse becomes more circular and approaches a circle of radius 5.

163. (a) The speed would double.

 (b) The elliptical orbit would be flatter. The length of the major axis is greater.

Chapter 9   Practice Test

1. Find the vertex, focus and directrix of the parabola $x^2 - 6x - 4y + 1 = 0$.

2. Find an equation of the parabola with its vertex at $(2, -5)$ and focus at $(2, -6)$.

3. Find the center, foci, vertices, and eccentricity of the ellipse $x^2 + 4y^2 - 2x + 32y + 61 = 0$.

4. Find an equation of the ellipse with vertices $(0, \pm 6)$ and eccentricity $e = \frac{1}{2}$.

5. Find the center, vertices, foci, and asymptotes of the hyperbola $16y^2 - x^2 - 6x - 128y + 231 = 0$.

6. Find an equation of the hyperbola with vertices at $(\pm 3, 2)$ and foci at $(\pm 5, 2)$.

7. Rotate the axes to eliminate the xy-term. Sketch the graph of the resulting equation, showing both sets of axes.

   $5x^2 + 2xy + 5y^2 - 10 = 0$

8. Use the discriminant to determine whether the graph of the equation is a parabola, ellipse, or hyperbola.

   (a) $6x^2 - 2xy + y^2 = 0$ (b) $x^2 + 4xy + 4y^2 - x - y + 17 = 0$

For Exercises 9 and 10, eliminate the parameter and write the corresponding rectangular equation.

9. $x = 3 - 2 \sin \theta, y = 1 + 5 \cos \theta$ 10. $x = e^{2t}, y = e^{4t}$

11. Convert the polar point $\left(\sqrt{2}, (3\pi)/4 \right)$ to rectangular coordinates.

12. Convert the rectangular point $\left(\sqrt{3}, -1 \right)$ to polar coordinates.

13. Convert the rectangular equation $4x - 3y = 12$ to polar form.

14. Convert the polar equation $r = 5 \cos \theta$ to rectangular form.

15. Sketch the graph of $r = 1 - \cos \theta$.

16. Sketch the graph of $r = 5 \sin 2\theta$.

17. Sketch the graph of $r = \dfrac{3}{6 - \cos \theta}$.

18. Find a polar equation of the parabola with its vertex at $(6, \pi/2)$ and focus at $(0, 0)$.

C H A P T E R   1 0
Analytic Geometry in Three Dimensions

Section 10.1   The Three-Dimensional Coordinate System **486**

Section 10.2   Vectors in Space **491**

Section 10.3   The Cross Product of Two Vectors **496**

Section 10.4   Lines and Planes in Space **500**

Review Exercises **504**

Practice Test **508**

CHAPTER 10
Analytic Geometry in Three Dimensions

Section 10.1   The Three-Dimensional Coordinate System

■ You should be able to plot points in the three-dimensional coordinate system.

■ The distance between the points (x_1, y_1, z_1) and (x_2, y_2, z_2) is
$$d = \sqrt{(x_2 - x_1)^2 + (y_2 - y_1)^2 + (z_2 - z_1)^2}.$$

■ The midpoint of the line segment joining the points (x_1, y_1, z_1) and (x_2, y_2, z_2) is
$$\left(\frac{x_1 + x_2}{2}, \frac{y_1 + y_2}{2}, \frac{z_1 + z_2}{2}\right).$$

■ The equation of the sphere with center (h, k, j) and radius r is
$$(x - h)^2 + (y - k)^2 + (z - j)^2 = r^2.$$

■ You should be able to find the trace of a surface in space.

Vocabulary Check

1. three-dimensional

2. xy-plane, xz-plane, yz-plane

3. octants

4. Distance Formula

5. $\left(\dfrac{x_1 + x_2}{2}, \dfrac{y_1 + y_2}{2}, \dfrac{z_1 + z_2}{2}\right)$

6. sphere

7. surface, space

8. trace

---

1. $A(-1, 4, 3), B(1, 3, -2), C(-3, 0, -2)$

3. $A(-2, -1, 4), B(3, -2, 0), C(-2, 2, -3)$

5.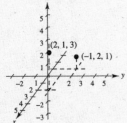

7.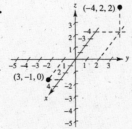

9.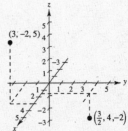

11. $x = -3, y = 3, z = 4$: $(-3, 3, 4)$

13. $y = z = 0, x = 10$: $(10, 0, 0)$

15. Octant IV

17. Octants I, II, III, IV
(above the xy-plane)

19. Octants II, IV, VI, VIII

21. $d = \sqrt{(7 - 3)^2 + (4 - 2)^2 + (8 - (-5))^2}$

$= \sqrt{4^2 + 2^2 + 13^2}$

$= \sqrt{16 + 4 + 169}$

$= \sqrt{189}$

$= 3\sqrt{21} \approx 13.748$

23. $d = \sqrt{[6 - (-1)]^2 + [0 - 4]^2 + [-9 - (-2)]^2}$

$= \sqrt{7^2 + 4^2 + 7^2}$

$= \sqrt{49 + 16 + 49}$

$= \sqrt{114}$

≈ 10.677

25. $d = \sqrt{(1 - 0)^2 + [0 - (-3)]^2 + (-10 - 0)^2}$

$= \sqrt{1 + 9 + 100}$

$= \sqrt{110} \approx 10.488$

27. $d_1 = \sqrt{(0 - 0)^2 + (0 - 4)^2 + (2 - 0)^2} = \sqrt{20} = 2\sqrt{5}$

$d_2 = \sqrt{(0 - (-2))^2 + (0 - 5)^2 + (2 - 2)^2} = \sqrt{29}$

$d_3 = \sqrt{(-2 - 0)^2 + (5 - 4)^2 + (2 - 0)^2} = 3$

$d_1{}^2 + d_3{}^2 = 20 + 9 = 29 = d_2{}^2$

29. $d_1 = \sqrt{(2 - 0)^2 + (2 - 0)^2 + (1 - 0)^2} = \sqrt{9} = 3$

$d_2 = \sqrt{(2 - 2)^2 + (-4 - 2)^2 + (4 - 1)^2} = \sqrt{45} = 3\sqrt{5}$

$d_3 = \sqrt{(2 - 0)^2 + (-4 - 0)^2 + (4 - 0)^2} = \sqrt{36} = 6$

$d_1{}^2 + d_3{}^2 = 9 + 36 = 45 = d_2{}^2$

31. $d_1 = \sqrt{(5 - 1)^2 + (-1 + 3)^2 + (2 + 2)^2} = \sqrt{16 + 4 + 16} = \sqrt{36} = 6$

$d_2 = \sqrt{(5 + 1)^2 + (-1 - 1)^2 + (2 - 2)^2} = \sqrt{36 + 4} = \sqrt{40} = 2\sqrt{10}$

$d_3 = \sqrt{(-1 - 1)^2 + (1 + 3)^2 + (2 + 2)^2} = \sqrt{4 + 16 + 16} = \sqrt{36} = 6$

$d_1 = d_3$, Isosceles triangle

33. $d_1 = \sqrt{(8 - 4)^2 + (1 + 1)^2 + (2 + 2)^2} = \sqrt{36} = 6$

$d_2 = \sqrt{(8 - 2)^2 + (1 - 3)^2 + (2 - 2)^2} = \sqrt{40} = 2\sqrt{10}$

$d_3 = \sqrt{(4 - 2)^2 + (-1 - 3)^2 + (-2 - 2)^2} = \sqrt{36} = 6$

Since $d_1 = d_3$, the triangle is isosceles.

35. Midpoint: $\left(\dfrac{3 - 3}{2}, \dfrac{-6 + 4}{2}, \dfrac{10 + 4}{2}\right) = (0, -1, 7)$

37. Midpoint: $\left(\dfrac{6 - 4}{2}, \dfrac{-2 + 2}{2}, \dfrac{5 + 6}{2}\right) = \left(1, 0, \dfrac{11}{2}\right)$

39. Midpoint: $\left(\dfrac{-2 + 7}{2}, \dfrac{8 - 4}{2}, \dfrac{10 + 2}{2}\right) = \left(\dfrac{5}{2}, 2, 6\right)$

41. $(x - 3)^2 + (y - 2)^2 + (z - 4)^2 = 16$

43. $(x + 1)^2 + (y - 2)^2 + z^2 = 3$

45. $(x - 0)^2 + (y - 4)^2 + (z - 3)^2 = 3^2$

$x^2 + (y - 4)^2 + (z - 3)^2 = 9$

47. Radius $= \dfrac{\text{Diameter}}{2} = 5$

$(x + 3)^2 + (y - 7)^2 + (z - 5)^2 = 5^2 = 25$

49. Center: $\left(\dfrac{3+0}{2}, \dfrac{0+0}{2}, \dfrac{0+6}{2} \right) = \left(\dfrac{3}{2}, 0, 3 \right)$

Radius: $\sqrt{\left(3 - \dfrac{3}{2} \right)^2 + (0-0)^2 + (0-3)^2} = \sqrt{\dfrac{9}{4} + 9} = \sqrt{\dfrac{45}{4}}$

Sphere: $\left(x - \dfrac{3}{2} \right)^2 + (y-0)^2 + (z-3)^2 = \dfrac{45}{4}$

51. $\left(x^2 - 5x + \dfrac{25}{4} \right) + y^2 + z^2 = \dfrac{25}{4}$

$\left(x - \dfrac{5}{2} \right)^2 + y^2 + z^2 = \dfrac{25}{4}$

Center: $\left(\dfrac{5}{2}, 0, 0 \right)$

Radius: $\dfrac{5}{2}$

53. $(x^2 - 4x + 4) + (y^2 + 2y + 1) + z^2 = 4 + 1$

$(x-2)^2 + (y+1)^2 + z^2 = 5$

Center: $(2, -1, 0)$

Radius: $\sqrt{5}$

55. $(x^2 - 4x + 4) + (y^2 + 2y + 1) + (z^2 - 6z + 9) = -10 + 4 + 1 + 9$

$(x-2)^2 + (y+1)^2 + (z-3)^2 = 4$

Center: $(2, -1, 3)$

Radius: 2

57. $(x^2 + 4x + 4) + y^2 + (z^2 - 8z + 16) = -19 + 4 + 16$

$(x+2)^2 + y^2 + (z-4)^2 = 1$

Center: $(-2, 0, 4)$

Radius: 1

59. $$x^2 + y^2 + z^2 - 2x - \dfrac{2}{3}y - 8z = -\dfrac{73}{9}$$

$(x^2 - 2x + 1) + \left(y^2 - \dfrac{2}{3}y + \dfrac{1}{9} \right) + (z^2 - 8z + 16) = -\dfrac{73}{9} + 1 + \dfrac{1}{9} + 16$

$(x-1)^2 + \left(y - \dfrac{1}{3} \right)^2 + (z-4)^2 = 9$

Center: $\left(1, \dfrac{1}{3}, 4 \right)$

Radius: 3

61. $4(x^2 - 2x + 1) + 4(y^2 + 4y + 4) + 4z^2 = 4 + 16 + 1$

$(x-1)^2 + (y+2)^2 + z^2 = \dfrac{21}{4}$

Center: $(1, -2, 0)$

Radius: $\dfrac{\sqrt{21}}{2}$

63. $9x^2 - 6x + 9y^2 + 18y + 9z^2 = -1$

$x^2 - \dfrac{2}{3}x + \dfrac{1}{9} + y^2 + 2y + 1 + z^2 = -\dfrac{1}{9} + \dfrac{1}{9} + 1$

$\left(x - \dfrac{1}{3} \right)^2 + (y+1)^2 + z^2 = 1$

Center: $\left(\dfrac{1}{3}, -1, 0 \right)$

Radius: 1

65. xz-trace $(y = 0)$: $(x - 1)^2 + z^2 = 36$, Circle

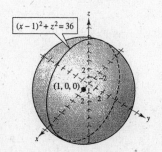

67. yz-trace $(x = 0)$: $(y - 3)^2 + z^2 = 9 - 4 = 5$,

Circle

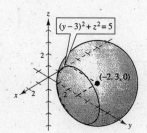

69. $(x^2 - 2x + 1) + y^2 + (z^2 - 4z + 4) = -1 + 1 + 4$

$(x - 1)^2 + y^2 + (z - 2)^2 = 4$

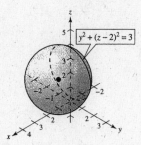

$yz -$ trace: $x = 0$: $y^2 + (z - 2)^2 = 3$

71.

73.

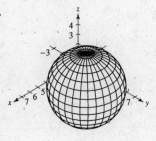

75. The length of each side is 3.
Thus, $(x, y, z) = (3, 3, 3)$.

77. $d = 165 \implies r = \frac{165}{2} = 82.5$

$x^2 + y^2 + z^2 = \left(\frac{165}{2}\right)^2$

79. False. x is the directed distance from the yz-plane to P.

81. In the xy-plane, the z-coordinate is 0.
In the xz-plane, the y-coordinate is 0.
In the yz-plane, the x-coordinate is 0.

83. The trace is a circle, or a single point.

85. $x_m = \frac{x_2 + x_1}{2} \implies x_2 = 2x_m - x_1$

Similarly for y_2 and z_2,

$(x_2, y_2, z_2) = (2x_m - x_1, 2y_m - y_1, 2z_m - z_1)$.

87. $v^2 + 3v + \dfrac{9}{4} = 2 + \dfrac{9}{4}$

$\left(v + \dfrac{3}{2}\right)^2 = \dfrac{17}{4}$

$v + \dfrac{3}{2} = \pm\dfrac{\sqrt{17}}{2}$

$v = -\dfrac{3}{2} \pm \dfrac{\sqrt{17}}{2}$

89. $x^2 - 5x + \dfrac{25}{4} = -5 + \dfrac{25}{4}$

$\left(x - \dfrac{5}{2}\right)^2 = \dfrac{5}{4}$

$x - \dfrac{5}{2} = \pm\dfrac{\sqrt{5}}{2}$

$x = \dfrac{5}{2} \pm \dfrac{\sqrt{5}}{2}$

91. $4y^2 + 4y = 9$

$y^2 + y + \dfrac{1}{4} = \dfrac{9}{4} + \dfrac{1}{4}$

$\left(y + \dfrac{1}{2}\right)^2 = \dfrac{10}{4}$

$y + \dfrac{1}{2} = \pm\dfrac{\sqrt{10}}{2}$

$y = -\dfrac{1}{2} \pm \dfrac{\sqrt{10}}{2}$

93. $\mathbf{v} = 3\mathbf{i} - 3\mathbf{j}$, Quadrant IV

$\|\mathbf{v}\| = \sqrt{3^2 + (-3)^2}$

$= \sqrt{18}$

$= 3\sqrt{2}$

$\tan\theta = -\dfrac{3}{3} = -1 \Longrightarrow$

$\theta = -45° \text{ or } 315°$

95. $\mathbf{v} = 4\mathbf{i} + 5\mathbf{j}$, Quadrant I

$\|\mathbf{v}\| = \sqrt{16 + 25} = \sqrt{41}$

$\tan\theta = \dfrac{5}{4} \Longrightarrow \theta \approx 51.34°$

97. $\mathbf{u} \cdot \mathbf{v} = \langle -4, 1 \rangle \cdot \langle 3, 5 \rangle$

$= -4(3) + 1(5)$

$= -7$

99. $a_0 = 1, a_n = a_{n-1} + n^2$

$a_1 = 1 + 1^2 = 2$

$a_2 = 2 + 2^2 = 6$

$a_3 = 6 + 3^2 = 15$

$a_4 = 15 + 4^2 = 31$

	1		2		6		15		31
First differences:		1		4		9		16		
Second differences:			3		5		7			

Neither model

101. $a_1 = -1, a_n = a_{n-1} + 3$

$a_2 = -1 + 3 = 2$

$a_3 = 2 + 3 = 5$

$a_4 = 5 + 3 = 8$

$a_5 = 8 + 3 = 11$

	−1		2		5		8		11
First differences:		3		3		3		3		
Second differences:			0		0		0			

Linear model

103. $(x + 5)^2 + (y - 1)^2 = 49$

105. $(y - 1)^2 = 4p(x - 4), \ p = -3$

$(y - 1)^2 = 4(-3)(x - 4)$

$(y - 1)^2 = -12(x - 4)$

107. $a = 3, b = 2,$ center: $(3, 3),$ horizontal major axis

$$\frac{(x - 3)^2}{9} + \frac{(y - 3)^2}{4} = 1$$

109. Center: $(6, 0),$ horizontal transverse axis

$a = 2, c = 6, b^2 = c^2 - a^2 = 36 - 4 = 32$

$$\frac{(x - 6)^2}{4} - \frac{y^2}{32} = 1$$

Section 10.2   Vectors in Space

- Vectors in space $\mathbf{v} = \langle v_1, v_2, v_3 \rangle$ have many of the same properties as vectors in the plane.
- The dot product of two vectors $\mathbf{u} = \langle u_1, u_2, u_3 \rangle$ and $\mathbf{v} = \langle v_1, v_2, v_3 \rangle$ in space is $\mathbf{u} \cdot \mathbf{v} = u_1 v_1 + u_2 v_2 + u_3 v_3$.
- Two nonzero vectors $\mathbf{u}$ and $\mathbf{v}$ are said to be parallel if there is some scalar c such that $\mathbf{u} = c\mathbf{v}$.
- You should be able to use vectors to solve real life problems.

Vocabulary Check

1. zero

2. $\mathbf{v} = v_1 \mathbf{i} + v_2 \mathbf{j} + v_3 \mathbf{k}$

3. component form

4. orthogonal

5. parallel

1. $\mathbf{v} = \langle 0 - 2, 3 - 0, 2 - 1 \rangle = \langle -2, 3, 1 \rangle$

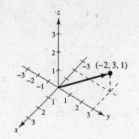

3. (a) $\mathbf{v} = \langle 1 - 1, 4 - 4, 0 - 4 \rangle = \langle 0, 0, -4 \rangle$

(b)

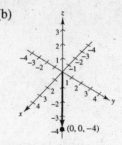

5. (a) $\mathbf{v} = \langle 1 - (-6), -1 - 4, 3 - (-2) \rangle$

$= \langle 7, -5, 5 \rangle$

(b) $\|\mathbf{v}\| = \sqrt{7^2 + (-5)^2 + 5^2}$

$= \sqrt{49 + 25 + 25}$

$= \sqrt{99}$

$= 3\sqrt{11}$

(c) $\dfrac{\mathbf{v}}{\|\mathbf{v}\|} = \dfrac{1}{3\sqrt{11}} \langle 7, -5, 5 \rangle = \dfrac{\sqrt{11}}{33} \langle 7, -5, 5 \rangle$

7. (a) $\mathbf{v} = \langle 1 - (-1), 4 - 2, -4 - (-4) \rangle = \langle 2, 2, 0 \rangle$

(b) $\|\mathbf{v}\| = \sqrt{2^2 + 2^2 + 0^2} = \sqrt{8} = 2\sqrt{2}$

(c) Unit vector: $\dfrac{1}{2\sqrt{2}} \langle 2, 2, 0 \rangle = \left\langle \dfrac{\sqrt{2}}{2}, \dfrac{\sqrt{2}}{2}, 0 \right\rangle$

9. (a)

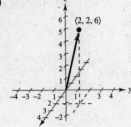

(b)

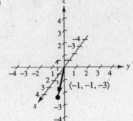

(c)

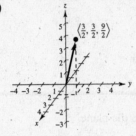

(d)

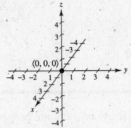

11. $\mathbf{v} = 2\mathbf{i} + 2\mathbf{j} - \mathbf{k}$

(a) $2\mathbf{v} = 4\mathbf{i} + 4\mathbf{j} - 2\mathbf{k}$

(b) $-\mathbf{v} = -2\mathbf{i} - 2\mathbf{j} + \mathbf{k}$

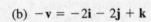

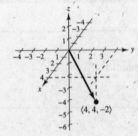

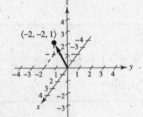

(c) $\dfrac{5}{2}\mathbf{v} = 5\mathbf{i} + 5\mathbf{j} - \dfrac{5}{2}\mathbf{k}$

(d) $0\mathbf{v} = \mathbf{0}$

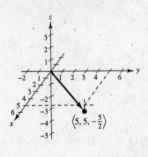

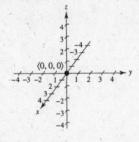

13. $\mathbf{z} = \mathbf{u} - 2\mathbf{v} = \langle -1, 3, 2 \rangle - 2\langle 1, -2, -2 \rangle = \langle -3, 7, 6 \rangle$

15. $2\mathbf{z} - 4\mathbf{u} = \mathbf{w} \implies \mathbf{z} = \frac{1}{2}(4\mathbf{u} + \mathbf{w}) = \frac{1}{2}(4\langle -1, 3, 2 \rangle + \langle 5, 0, -5 \rangle) = \left\langle \frac{1}{2}, 6, \frac{3}{2} \right\rangle$

17. $\mathbf{z} = 2\langle -1, 3, 2 \rangle - 3\langle 1, -2, -2 \rangle + \frac{1}{2}\langle 5, 0, -5 \rangle = \left\langle -\frac{5}{2}, 12, \frac{15}{2} \right\rangle$

19. $4\mathbf{z} = 4\langle 5, 0, -5 \rangle - \langle -1, 3, 2 \rangle + \langle 1, -2, -2 \rangle = \langle 22, -5, -24 \rangle$

$\mathbf{z} = \left\langle \frac{11}{2}, -\frac{5}{4}, -6 \right\rangle$

21. $\|\mathbf{v}\| = \|\langle 7, 8, 7 \rangle\|$
$= \sqrt{49 + 64 + 49} = \sqrt{162} = 9\sqrt{2}$

23. $\|\mathbf{v}\| = \sqrt{1^2 + (-2)^2 + 4^2} = \sqrt{21}$

25. $\|\mathbf{v}\| = \sqrt{2^2 + (-4)^2 + 1^2} = \sqrt{21}$

27. $\|\mathbf{v}\| = \sqrt{4^2 + (-3)^2 + (-7)^2}$
$= \sqrt{16 + 9 + 49} = \sqrt{74}$

29. $\mathbf{v} = \langle 1 - 1, 0 - (-3), -1 - 4 \rangle = \langle 0, 3, -5 \rangle$
$\|\mathbf{v}\| = \sqrt{0 + 3^2 + (-5)^2} = \sqrt{34}$

31. $\|\mathbf{u}\| = \sqrt{5^2 + (-12)^2} = \sqrt{169} = 13$
(a) $\frac{1}{13}(5\mathbf{i} - 12\mathbf{k})$
(b) $-\frac{1}{13}(5\mathbf{i} - 12\mathbf{k})$

33. (a) $\dfrac{\mathbf{u}}{\|\mathbf{u}\|} = \dfrac{\langle 8, 3, -1 \rangle}{\sqrt{74}}$
$= \dfrac{1}{\sqrt{74}}(8\mathbf{i} + 3\mathbf{j} - \mathbf{k}) = \dfrac{\sqrt{74}}{74}\langle 8, 3, -1 \rangle$

(b) $-\dfrac{1}{\sqrt{74}}(8\mathbf{i} + 3\mathbf{j} - \mathbf{k}) = -\dfrac{\sqrt{74}}{74}\langle 8, 3, -1 \rangle$

35. $6\mathbf{u} - 4\mathbf{v} = 6\langle -1, 3, 4 \rangle - 4\langle 5, 4.5, -6 \rangle = \langle -6, 18, 24 \rangle + \langle -20, -18, 24 \rangle = \langle -26, 0, 48 \rangle$

37. $\mathbf{u} + \mathbf{v} = \langle -1, 3, 4 \rangle + \langle 5, 4.5, -6 \rangle = \langle 4, 7.5, -2 \rangle$
$\|\mathbf{u} + \mathbf{v}\| = \sqrt{4^2 + 7.5^2 + (-2)^2} = \frac{1}{2}\sqrt{305} \approx 8.73$

39. $\mathbf{u} \cdot \mathbf{v} = \langle 4, 4, -1 \rangle \cdot \langle 2, -5, -8 \rangle$
$= 8 - 20 + 8 = -4$

41. $\mathbf{u} \cdot \mathbf{v} = \langle 2, -5, 3 \rangle \cdot \langle 9, 3, -1 \rangle$
$= 18 - 15 - 3 = 0$

43. $\cos \theta = \dfrac{\mathbf{u} \cdot \mathbf{v}}{\|\mathbf{u}\| \, \|\mathbf{v}\|} = \dfrac{-8}{\sqrt{8}\sqrt{25}} \implies \theta \approx 124.45°$

45. $\cos \theta = \dfrac{\mathbf{u} \cdot \mathbf{v}}{\|\mathbf{u}\| \, \|\mathbf{v}\|} = \dfrac{-120}{\sqrt{1700}\sqrt{73}} \implies \theta \approx 109.92°$

47. $-\frac{3}{2}\langle 8, -4, -10 \rangle = \langle -12, 6, 15 \rangle \implies$ parallel

49. $\mathbf{u} \cdot \mathbf{v} = 3 - 5 + 2 = 0 \implies$ orthogonal

51. $\mathbf{u} \neq c\mathbf{v}$
$\mathbf{u} \cdot \mathbf{v} = -2 - 6 \neq 0$
Neither parallel nor orthogonal

53. $\mathbf{u} \cdot \mathbf{v} = -4 + 3 + 1 = 0$
Orthogonal

55. $\mathbf{v} = \langle 7 - 5, 3 - 4, -1 - 1 \rangle = \langle 2, -1, -2 \rangle$
$\mathbf{u} = \langle 4 - 7, 5 - 3, 3 - (-1) \rangle = \langle -3, 2, 4 \rangle$
Since $\mathbf{u}$ and $\mathbf{v}$ are not parallel, the points are not collinear.

57. $\mathbf{v} = \langle -1 - 1, 2 - 3, 5 - 2 \rangle = \langle -2, -1, 3 \rangle$
$\mathbf{u} = \langle 3 - (-1), 4 - 2, -1 - 5 \rangle = \langle 4, 2, -6 \rangle$
Since $\mathbf{u} = -2\mathbf{v}$, the points are collinear.

59. The vector $\langle 1, 2, 0 \rangle$ joining $(1, 2, 0)$ and $(0, 0, 0)$ is perpendicular to the vector $\langle -2, 1, 0 \rangle$ joining $(-2, 1, 0)$ and $(0, 0, 0)$:
$\langle 1, 2, 0 \rangle \cdot \langle -2, 1, 0 \rangle = -2 + 2 = 0$
The triangle is a right triangle.

61. The three sides of the triangle are given by the vectors:

$\mathbf{u} = \langle -2, 4, -2 \rangle$

$\mathbf{v} = \langle -3, 5, -4 \rangle$

$\mathbf{w} = \langle -1, 1, -2 \rangle$

$\mathbf{u} \cdot \mathbf{v} = 34 > 0$

$\mathbf{u} \cdot \mathbf{w} = 10 > 0$

$\mathbf{v} \cdot \mathbf{w} = 16 > 0$

The triangle has three acute angles.

Acute triangle

63. $\mathbf{v} = \langle 2, -4, 7 \rangle = \langle q_1 - 1, q_2 - 5, q_3 - 0 \rangle \implies$

$\left. \begin{array}{l} 2 = q_1 - 1 \\ -4 = q_2 - 5 \\ 7 = q_3 \end{array} \right\} \implies \left. \begin{array}{l} q_1 = 3 \\ q_2 = 1 \\ q_3 = 7 \end{array} \right\} \implies$ Terminal point is $(3, 1, 7)$.

65. $\mathbf{v} = \langle 4, \frac{3}{2}, -\frac{1}{4} \rangle = \langle q_1 - 2, q_2 - 1, q_3 + \frac{3}{2} \rangle$

$4 = q_1 - 2 \implies q_1 = 6$

$\frac{3}{2} = q_2 - 1 \implies q_2 = \frac{5}{2}$

$-\frac{1}{4} = q_3 + \frac{3}{2} \implies q_3 = -\frac{7}{4}$

Terminal point: $\left(6, \frac{5}{2}, -\frac{7}{4} \right)$

67. $c\mathbf{u} = c\mathbf{i} + 2c\mathbf{j} + 3c\mathbf{k}$

$\|c\mathbf{u}\| = \sqrt{c^2 + 4c^2 + 9c^2} = |c|\sqrt{14} = 3 \implies$

$c = \pm \dfrac{3}{\sqrt{14}} = \pm \dfrac{3\sqrt{14}}{14}$

69. $\mathbf{v} = \langle q_1, q_2, q_3 \rangle$

Since $\mathbf{v}$ lies in the yz-plane, $q_1 = 0$. Since $\mathbf{v}$ makes an angle of $45°$, $|q_2| = |q_3|$.
Finally, $\|\mathbf{v}\| = 4$ implies that $q_2^2 + q_3^2 = 16$. Thus, $q_2 = q_3 = 2\sqrt{2}$ and
$\mathbf{v} = \langle 0, 2\sqrt{2}, 2\sqrt{2} \rangle$, or $q_2 = 2\sqrt{2}$ and $q_3 = -2\sqrt{2}$ and $\mathbf{v} = \langle 0, 2\sqrt{2}, -2\sqrt{2} \rangle$.

71. $\overrightarrow{PQ_1} = \langle 0, -24, -12\sqrt{21} \rangle$

$\overrightarrow{PQ_2} = \langle 12\sqrt{3}, 12, -12\sqrt{21} \rangle$

$\overrightarrow{PQ_3} = \langle -12\sqrt{3}, 12, -12\sqrt{21} \rangle$

Let $\mathbf{F}_1$, $\mathbf{F}_2$, and $\mathbf{F}_3$ be the tension on each wire. Since $\|\mathbf{F}_1\| = \|\mathbf{F}_2\| = \|\mathbf{F}_3\|$,
there exists a constant c such that

$\mathbf{F}_1 = c\langle 0, -24, -12\sqrt{21} \rangle$

$\mathbf{F}_2 = c\langle 12\sqrt{3}, 12, -12\sqrt{21} \rangle$

$\mathbf{F}_3 = c\langle -12\sqrt{3}, 12, -12\sqrt{21} \rangle$

The total force is $-30\mathbf{k} = \mathbf{F}_1 + \mathbf{F}_2 + \mathbf{F}_3 \implies$ the vertical $(\mathbf{k})$ component satisfies

$-10 = -12\sqrt{21}\,c \implies c = \dfrac{5}{6\sqrt{21}}$.

—CONTINUED—

71. —CONTINUED—

Hence,

$$\mathbf{F}_1 = \left\langle 0, \frac{-20}{\sqrt{21}}, -10 \right\rangle$$

$$\mathbf{F}_2 = \left\langle \frac{10}{\sqrt{7}}, \frac{10}{\sqrt{21}}, -10 \right\rangle$$

$$\mathbf{F} = \left\langle \frac{-10}{\sqrt{7}}, \frac{10}{\sqrt{21}}, -10 \right\rangle$$

$$\|\mathbf{F}_1\| = \|\mathbf{F}_2\| = \|\mathbf{F}_3\| \approx 10.91 \text{ pounds.}$$

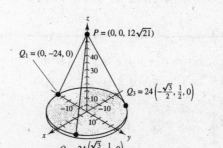

$$Q_1 = (0, -24, 0)$$

$$Q_2 = (20.8, 12, 0)$$

$$Q_3 = (-20.8, 12, 0)$$

$$P = (0, 0, 55)$$

73. True. $\cos \theta = 0 \implies \theta = 90°$

75. (a)

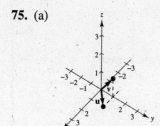

(c) $\mathbf{w} = \langle 1, 2, 1 \rangle = a\langle 1, 1, 0 \rangle + b\langle 0, 1, 1 \rangle$

$$1 = a$$

$$2 = a + b$$

$$1 = b$$

Hence, $a = b = 1$.

(b) $\mathbf{w} = a\mathbf{u} + b\mathbf{v} = a\langle 1, 1, 0 \rangle + b\langle 0, 1, 1 \rangle$

$$\mathbf{0} = \langle a, a + b, b \rangle \implies a = b = 0$$

(d) $\mathbf{w} = \langle 1, 2, 3 \rangle = a\langle 1, 1, 0 \rangle + b\langle 0, 1, 1 \rangle$

$$1 = a$$

$$2 = a + b$$

$$3 = b$$

Impossible

77. If $\mathbf{u} \cdot \mathbf{v} < 0$, then $\cos \theta < 0$ and the angle between $\mathbf{u}$ and $\mathbf{v}$ is obtuse, $180° > \theta > 90°$.

79. (a) $x = t$

$y = 3t + 2$

(b) $x = t - 1$

$y = 3(t - 1) + 2 = 3t - 1$

81. (a) $x = t$

$y = t^2 - 8$

(b) $x = t - 1$

$y = (t - 1)^2 - 8 = t^2 - 2t - 7$

Section 10.3   The Cross Product of Two Vectors

■ The cross product of two vectors $\mathbf{u} = u_1\mathbf{i} + u_2\mathbf{j} + u_3\mathbf{k}$ and $\mathbf{v} = v_1\mathbf{i} + v_2\mathbf{j} + v_3\mathbf{k}$ is given by

$$\mathbf{u} \times \mathbf{v} = (u_2v_3 - u_3v_2)\mathbf{i} - (u_1v_3 - u_3v_1)\mathbf{j} + (u_1v_2 - u_2v_1)\mathbf{k}$$

$$= \begin{vmatrix} \mathbf{i} & \mathbf{j} & \mathbf{k} \\ u_1 & u_2 & u_3 \\ v_1 & v_2 & v_3 \end{vmatrix}.$$

■ The cross product satisfies the following algebraic properties.

(a) $\mathbf{u} \times \mathbf{v} = -(\mathbf{v} \times \mathbf{u})$

(b) $\mathbf{u} \times (\mathbf{v} + \mathbf{w}) = (\mathbf{u} \times \mathbf{v}) + (\mathbf{u} \times \mathbf{w})$

(c) $c(\mathbf{u} \times \mathbf{v}) = (c\mathbf{u}) \times \mathbf{v} = \mathbf{u} \times (c\mathbf{v})$

(d) $\mathbf{u} \times \mathbf{0} = \mathbf{0} \times \mathbf{u} = \mathbf{0}$

(e) $\mathbf{u} \times \mathbf{u} = \mathbf{0}$

(f) $\mathbf{u} \cdot (\mathbf{v} \times \mathbf{w}) = (\mathbf{u} \times \mathbf{v}) \cdot \mathbf{w}$

■ The following geometric properties of the cross product are valid, where θ is the angle between the vectors $\mathbf{u}$ and $\mathbf{v}$:

(a) $\mathbf{u} \times \mathbf{v}$ is orthogonal to both $\mathbf{u}$ and $\mathbf{v}$.

(b) $\|\mathbf{u} \times \mathbf{v}\| = \|\mathbf{u}\| \|\mathbf{v}\| \sin \theta$

(c) $\mathbf{u} \times \mathbf{v} = \mathbf{0}$ if and only if $\mathbf{u}$ and $\mathbf{v}$ are scalar multiples.

(d) $\|\mathbf{u} \times \mathbf{v}\|$ is the area of the parallelogram having $\mathbf{u}$ and $\mathbf{v}$ as sides.

■ The absolute value of the triple scalar product is the volume of the parallelepiped having $\mathbf{u}$, $\mathbf{v}$, and $\mathbf{w}$ as sides.

$$\mathbf{u} \cdot (\mathbf{v} \times \mathbf{w}) = \begin{vmatrix} u_1 & u_2 & u_3 \\ v_1 & v_2 & v_3 \\ w_1 & w_2 & w_3 \end{vmatrix}$$

Vocabulary Check

1. cross product

2. $\mathbf{0}$

3. $\|\mathbf{u}\| \|\mathbf{v}\| \sin \theta$

4. triple scalar product

---

1. $\mathbf{j} \times \mathbf{i} = \begin{vmatrix} \mathbf{i} & \mathbf{j} & \mathbf{k} \\ 0 & 1 & 0 \\ 1 & 0 & 0 \end{vmatrix} = -\mathbf{k}$

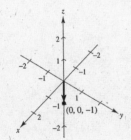

3. $\mathbf{i} \times \mathbf{k} = \begin{vmatrix} \mathbf{i} & \mathbf{j} & \mathbf{k} \\ 1 & 0 & 0 \\ 0 & 0 & 1 \end{vmatrix} = -\mathbf{j}$

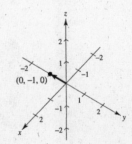

5. $\mathbf{u} \times \mathbf{v} = \begin{vmatrix} \mathbf{i} & \mathbf{j} & \mathbf{k} \\ 1 & -1 & 0 \\ 0 & 1 & -1 \end{vmatrix} = \mathbf{i} + \mathbf{j} + \mathbf{k} = \langle 1, 1, 1 \rangle$

7. $\mathbf{u} \times \mathbf{v} = \begin{vmatrix} \mathbf{i} & \mathbf{j} & \mathbf{k} \\ 3 & -2 & 5 \\ 0 & -1 & 1 \end{vmatrix} = \langle 3, -3, -3 \rangle$

$(\mathbf{u} \times \mathbf{v}) \cdot \mathbf{u} = \langle 3, -3, -3 \rangle \cdot \langle 3, -2, 5 \rangle = 0$

$(\mathbf{u} \times \mathbf{v}) \cdot \mathbf{v} = \langle 3, -3, -3 \rangle \cdot \langle 0, -1, 1 \rangle = 0$

9. $\mathbf{u} \times \mathbf{v} = \begin{vmatrix} \mathbf{i} & \mathbf{j} & \mathbf{k} \\ -10 & 0 & 6 \\ 7 & 0 & 0 \end{vmatrix} = \langle 0, 42, 0 \rangle$

$(\mathbf{u} \times \mathbf{v}) \cdot \mathbf{u} = \langle 0, 42, 0 \rangle \cdot \langle -10, 0, 6 \rangle = 0$

$(\mathbf{u} \times \mathbf{v}) \cdot \mathbf{v} = \langle 0, 42, 0 \rangle \cdot \langle 7, 0, 0 \rangle = 0$

11. $\mathbf{u} \times \mathbf{v} = \begin{vmatrix} \mathbf{i} & \mathbf{j} & \mathbf{k} \\ 6 & 2 & 1 \\ 1 & 3 & -2 \end{vmatrix} = \langle -7, 13, 16 \rangle$

$= -7\mathbf{i} + 13\mathbf{j} + 16\mathbf{k}$

13. $\mathbf{u} \times \mathbf{v} = \begin{vmatrix} \mathbf{i} & \mathbf{j} & \mathbf{k} \\ 2 & 4 & 3 \\ -1 & 3 & -2 \end{vmatrix} = -17\mathbf{i} + \mathbf{j} + 10\mathbf{k}$

15. $\mathbf{u} \times \mathbf{v} = \begin{vmatrix} \mathbf{i} & \mathbf{j} & \mathbf{k} \\ \frac{1}{2} & -\frac{2}{3} & 1 \\ -\frac{3}{4} & 1 & \frac{1}{4} \end{vmatrix} = -\frac{7}{6}\mathbf{i} - \frac{7}{8}\mathbf{j}$

17. $\mathbf{u} \times \mathbf{v} = \begin{vmatrix} \mathbf{i} & \mathbf{j} & \mathbf{k} \\ 0 & 0 & 6 \\ -1 & 3 & 1 \end{vmatrix} = \langle -18, -6, 0 \rangle$

$= -18\mathbf{i} - 6\mathbf{j}$

19. $\mathbf{u} \times \mathbf{v} = \begin{vmatrix} \mathbf{i} & \mathbf{j} & \mathbf{k} \\ -1 & 0 & 1 \\ 0 & 1 & -2 \end{vmatrix} = \langle -1, -2, -1 \rangle$

$= -\mathbf{i} - 2\mathbf{j} - \mathbf{k}$

21. $\mathbf{u} \times \mathbf{v} = \begin{vmatrix} \mathbf{i} & \mathbf{j} & \mathbf{k} \\ 2 & 4 & 3 \\ 0 & -2 & 1 \end{vmatrix} = \langle 10, -2, -4 \rangle$

23. $\mathbf{u} \times \mathbf{v} = \begin{vmatrix} \mathbf{i} & \mathbf{j} & \mathbf{k} \\ 1 & -2 & 4 \\ -4 & 2 & -1 \end{vmatrix} = -6\mathbf{i} - 15\mathbf{j} - 6\mathbf{k}$

25. $\mathbf{u} \times \mathbf{v} = \begin{vmatrix} \mathbf{i} & \mathbf{j} & \mathbf{k} \\ 6 & -5 & 1 \\ \frac{1}{2} & -\frac{3}{4} & \frac{2}{10} \end{vmatrix} = \langle -0.25, -0.7, -2 \rangle$

27. $\mathbf{u} \times \mathbf{v} = \begin{vmatrix} \mathbf{i} & \mathbf{j} & \mathbf{k} \\ 1 & 2 & 3 \\ 2 & -3 & 0 \end{vmatrix} = \langle 9, 6, -7 \rangle$

$\|\mathbf{u} \times \mathbf{v}\| = \sqrt{166}$

Unit vector: $\dfrac{\sqrt{166}}{166} \langle 9, 6, -7 \rangle$

29. $\mathbf{u} \times \mathbf{v} = \begin{vmatrix} \mathbf{i} & \mathbf{j} & \mathbf{k} \\ 3 & 1 & 0 \\ 0 & 1 & 1 \end{vmatrix} = \mathbf{i} - 3\mathbf{j} + 3\mathbf{k}$

$\|\mathbf{u} \times \mathbf{v}\| = \sqrt{19}$

Unit vector $= \dfrac{\mathbf{u} \times \mathbf{v}}{\|\mathbf{u} \times \mathbf{v}\|} = \dfrac{1}{\sqrt{19}}(\mathbf{i} - 3\mathbf{j} + 3\mathbf{k})$

$= \dfrac{\sqrt{19}}{19} \langle 1, -3, 3 \rangle$

31. $\mathbf{u} \times \mathbf{v} = \begin{vmatrix} \mathbf{i} & \mathbf{j} & \mathbf{k} \\ -3 & 2 & -5 \\ \frac{1}{2} & -\frac{3}{4} & \frac{1}{10} \end{vmatrix} = \left\langle -\dfrac{71}{20}, -\dfrac{11}{5}, \dfrac{5}{4} \right\rangle$

Consider the parallel vector $\langle -71, -44, 25 \rangle = \mathbf{w}$.

$\|\mathbf{w}\| = \sqrt{71^2 + 44^2 + 25^2} = \sqrt{7602}$

Unit vector $= \dfrac{1}{\sqrt{7602}} \langle -71, -44, 25 \rangle$

$= \dfrac{\sqrt{7602}}{7602} \langle -71, -44, 25 \rangle$

33. $\mathbf{u} \times \mathbf{v} = \begin{vmatrix} \mathbf{i} & \mathbf{j} & \mathbf{k} \\ 1 & 1 & -1 \\ 1 & 1 & 1 \end{vmatrix} = 2\mathbf{i} - 2\mathbf{j}$

$\|\mathbf{u} \times \mathbf{v}\| = 2\sqrt{2}$

Unit vector $= \dfrac{\mathbf{u} \times \mathbf{v}}{\|\mathbf{u} \times \mathbf{v}\|} = \dfrac{1}{2\sqrt{2}}(2\mathbf{i} - 2\mathbf{j})$

$= \dfrac{1}{\sqrt{2}}\mathbf{i} - \dfrac{1}{\sqrt{2}}\mathbf{j}$

$= \dfrac{\sqrt{2}}{2}\mathbf{i} - \dfrac{\sqrt{2}}{2}\mathbf{j}$

35. $\mathbf{u} \times \mathbf{v} = \begin{vmatrix} \mathbf{i} & \mathbf{j} & \mathbf{k} \\ 0 & 0 & 1 \\ 1 & 0 & 1 \end{vmatrix} = \mathbf{j}$

Area $= \|\mathbf{u} \times \mathbf{v}\| = \|\mathbf{j}\| = 1$ square unit

37. $\mathbf{u} \times \mathbf{v} = \begin{vmatrix} \mathbf{i} & \mathbf{j} & \mathbf{k} \\ 3 & 4 & 6 \\ 2 & -1 & 5 \end{vmatrix} = 26\mathbf{i} - 3\mathbf{j} - 11\mathbf{k}$

Area $= \|\mathbf{u} \times \mathbf{v}\| = \sqrt{26^2 + (-3)^2 + (-11)^2}$

$= \sqrt{806}$ square units

39. $\mathbf{u} \times \mathbf{v} = \begin{vmatrix} \mathbf{i} & \mathbf{j} & \mathbf{k} \\ 2 & 2 & -3 \\ 0 & 2 & 3 \end{vmatrix} = \langle 12, -6, 4 \rangle$

Area $= \|\mathbf{u} \times \mathbf{v}\| = \sqrt{12^2 + (-6)^2 + 4^2}$

$= 14$ square units

41. (a) $\overrightarrow{AB} = \langle 3 - 2, 1 - (-1), 2 - 4 \rangle = \langle 1, 2, -2 \rangle$
is parallel to

$\overrightarrow{DC} = \langle 0 - (-1), 5 - 3, 6 - 8 \rangle = \langle 1, 2, -2 \rangle$.

$\overrightarrow{AD} = \langle -3, 4, 4 \rangle$ is parallel to $\overrightarrow{BC} = \langle -3, 4, 4 \rangle$.

(c) $\overrightarrow{AB} \cdot \overrightarrow{AD} = \langle 1, 2, -2 \rangle \cdot \langle -3, 4, 4 \rangle$

$\neq 0 \implies$ not a rectangle

(b) $\overrightarrow{AB} \times \overrightarrow{AD} = \begin{vmatrix} \mathbf{i} & \mathbf{j} & \mathbf{k} \\ 1 & 2 & -2 \\ -3 & 4 & 4 \end{vmatrix} = \langle 16, 2, 10 \rangle$

Area $= \|\overrightarrow{AB} \times \overrightarrow{AD}\|$

$= \sqrt{16^2 + 2^2 + 10^2}$

$= \sqrt{360} = 6\sqrt{10}$ square units

43. $\mathbf{u} = \langle 1, 2, 3 \rangle, \mathbf{v} = \langle -3, 0, 0 \rangle$

$\mathbf{u} \times \mathbf{v} = \begin{vmatrix} \mathbf{i} & \mathbf{j} & \mathbf{k} \\ 1 & 2 & 3 \\ -3 & 0 & 0 \end{vmatrix} = \langle 0, -9, 6 \rangle$

Area $= \frac{1}{2}\|\mathbf{u} \times \mathbf{v}\| = \frac{1}{2}\sqrt{81 + 36} = \frac{3}{2}\sqrt{13}$

45. $\mathbf{u} = \langle -2 - 2, -2 - 3, 0 - (-5) \rangle = \langle -4, -5, 5 \rangle$

$\mathbf{v} = \langle 3 - 2, 0 - 3, 6 - (-5) \rangle = \langle 1, -3, 11 \rangle$

$\mathbf{u} \times \mathbf{v} = \begin{vmatrix} \mathbf{i} & \mathbf{j} & \mathbf{k} \\ -4 & -5 & 5 \\ 1 & -3 & 11 \end{vmatrix} = \langle -40, 49, 17 \rangle$

Area $= \frac{1}{2}\|\mathbf{u} \times \mathbf{v}\| = \frac{1}{2}\sqrt{(-40)^2 + 49^2 + 17^2}$

$= \frac{1}{2}\sqrt{4290}$ square units

47. $\mathbf{u} \cdot (\mathbf{v} \times \mathbf{w}) = \begin{vmatrix} 2 & 3 & 3 \\ 4 & 4 & 0 \\ 0 & 0 & 4 \end{vmatrix} = 2(16) - 3(16) + 3(0) = -16$

49. $\mathbf{u} \cdot (\mathbf{v} \times \mathbf{w}) = \begin{vmatrix} 2 & 3 & 1 \\ 1 & -1 & 0 \\ 4 & 3 & 1 \end{vmatrix} = 2(-1) - 3(1) + 1(7) = 2$

51. $\mathbf{u} \cdot (\mathbf{v} \times \mathbf{w}) = \begin{vmatrix} 1 & 1 & 0 \\ 0 & 1 & 1 \\ 1 & 0 & 1 \end{vmatrix} = 1 + 1 = 2$

Volume $= \left| \mathbf{u} \cdot (\mathbf{v} \times \mathbf{w}) \right| = 2$ cubic units

53. $\mathbf{u} \cdot (\mathbf{v} \times \mathbf{w}) = \begin{vmatrix} 0 & 2 & 2 \\ 0 & 0 & -2 \\ 3 & 0 & 2 \end{vmatrix} = 0 - 2(6) + 2(0) = -12$

Volume $= \left| \mathbf{u} \cdot (\mathbf{v} \times \mathbf{w}) \right| = 12$ cubic units

55. $\mathbf{u} = \langle 4, 0, 0 \rangle, \quad \mathbf{v} = \langle 0, -2, 3 \rangle, \quad \mathbf{w} = \langle 0, 5, 3 \rangle$

$\mathbf{u} \cdot (\mathbf{v} \times \mathbf{w}) = \begin{vmatrix} 4 & 0 & 0 \\ 0 & -2 & 3 \\ 0 & 5 & 3 \end{vmatrix} = 4(-21) = -84$

Volume $= |-84| = 84$ cubic units

57. $\mathbf{V} \times \mathbf{F} = \begin{vmatrix} \mathbf{i} & \mathbf{j} & \mathbf{k} \\ 0 & -\frac{1}{2}\cos 40^\circ & -\frac{1}{2}\sin 40^\circ \\ 0 & 0 & -p \end{vmatrix} = \left(\frac{p}{2}\cos 40^\circ \right)\mathbf{i}$

(a) $T(p) = \|\mathbf{V} \times \mathbf{F}\| = \dfrac{p}{2}\cos 40^\circ$

(b)

p	15	20	25	30	35	40	45
T	5.75	7.66	9.58	11.49	13.41	15.32	17.24

59. True. The cross product is defined for vectors in three-dimensional space.

61. If $\mathbf{u}$ and $\mathbf{v}$ are orthogonal, then $\sin \theta = 1$ and hence, $\|\mathbf{u} \times \mathbf{v}\| = \|\mathbf{u}\| \, \|\mathbf{v}\| \sin \theta = \|\mathbf{u}\| \, \|\mathbf{v}\|$.

63. $\mathbf{v} \times \mathbf{w} = \begin{vmatrix} \mathbf{i} & \mathbf{j} & \mathbf{k} \\ v_1 & v_2 & v_3 \\ w_1 & w_2 & w_3 \end{vmatrix} = (v_2 w_3 - w_2 v_3)\mathbf{i} - (v_1 w_3 - w_1 v_3)\mathbf{j} + (v_1 w_2 - v_2 w_1)\mathbf{k}$

Hence,

$\mathbf{u} \cdot (\mathbf{v} \times \mathbf{w}) = u_1(v_2 w_3 - w_2 v_3) - u_2(v_1 w_3 - w_1 v_3) + u_3(v_1 w_2 - v_2 w_1)$

$= \begin{vmatrix} u_1 & u_2 & u_3 \\ v_1 & v_2 & v_3 \\ w_1 & w_2 & w_3 \end{vmatrix}.$

65. $\cos 480^\circ = \cos 120^\circ = -\frac{1}{2}$

67. $\sin 690^\circ = \sin 330^\circ = -\frac{1}{2}$

69. $\sin \dfrac{19\pi}{6} = \sin\left(\dfrac{7\pi}{6}\right) = -\dfrac{1}{2}$

71. $\tan \dfrac{15\pi}{4} = \tan \dfrac{7\pi}{4} = -1$

Section 10.4   Lines and Planes in Space

■ The parametric equations of the line in space parallel to the vector $\langle a, b, c \rangle$ and passing through the point (x_1, y_2, z_3) are

$$x = x_1 + at, \quad y = y_1 + bt, \quad z = z_1 + ct.$$

■ The standard equation of the plane in space containing the point (x_1, y_1, z_1) and having normal vector (a, b, c) is

$$a(x - x_1) + b(y - y_1) + c(z - z_1) = 0.$$

■ You should be able to find the angle between two planes by calculating the angle between their normal vectors.

■ You should be able to sketch a plane in space.

■ The distance between a point Q and a plane having normal $\mathbf{n}$ is

$$D = \|\text{proj}_{\mathbf{n}}\, \overrightarrow{PQ}\| = \frac{|\overrightarrow{PQ} \cdot \mathbf{n}|}{\|\mathbf{n}\|}$$

where P is a point in the plane.

Vocabulary Check

1. direction, $\dfrac{\overrightarrow{PQ}}{t}$

2. parametric equations

3. symmetric equations

4. normal

5. $a(x - x_1) + b(y - y_1) + c(z - z_1) = 0$

---

1. $x = x_1 + at = 0 + t$

   $y = y_1 + bt = 0 + 2t$

   $z = z_1 + ct = 0 + 3t$

   (a) Parametric equations: $x = t, y = 2t, z = 3t$ (b) Symmetric equations: $\dfrac{x}{1} = \dfrac{y}{2} = \dfrac{z}{3}$

3. $x = x_1 + at = -4 + \dfrac{1}{2}t, \quad y = y_1 + bt = 1 + \dfrac{4}{3}t, \quad z = z_1 + ct = 0 - t$

   (a) Parametric equations: $x = -4 + \dfrac{1}{2}t, y = 1 + \dfrac{4}{3}t, z = -t$

      Equivalently: $x = -4 + 3t, y = 1 + 8t, z = -6t$

   (b) Symmetric equations: $\dfrac{x + 4}{3} = \dfrac{y - 1}{8} = \dfrac{z}{-6}$

5. $x = x_1 + at = 2 + 2t,$

   $y = y_1 + bt = -3 - 3t,$

   $z = z_1 + ct = 5 + t$

   (a) Parametric equations:
      $x = 2 + 2t, y = -3 - 3t, z = 5 + t$

   (b) Symmetric equations: $\dfrac{x - 2}{2} = \dfrac{y + 3}{-3} = z - 5$

7. $\mathbf{v} = \langle 1 - 2, 4 - 0, -3 - 2 \rangle = \langle -1, 4, -5 \rangle$

   Point: $(2, 0, 2)$

   (a) $x = 2 - t, y = 4t, z = 2 - 5t$

   (b) $\dfrac{x - 2}{-1} = \dfrac{y}{4} = \dfrac{z - 2}{-5}$

9. $\mathbf{v} = \langle 1 - (-3), -2 - 8, 16 - 15 \rangle = \langle 4, -10, 1 \rangle$

Point: $(-3, 8, 15)$

(a) $x = -3 + 4t, y = 8 - 10t, z = 15 + t$

(b) $\dfrac{x + 3}{4} = \dfrac{y - 8}{-10} = \dfrac{z - 15}{1}$

11. $\mathbf{v} = \langle -1 - 3, 1 - 1, 5 - 2 \rangle = \langle -4, 0, 3 \rangle$

Point: $(3, 1, 2)$

(a) $x = 3 - 4t, y = 1, z = 2 + 3t$

(b) $\dfrac{x - 3}{-4} = \dfrac{z - 2}{3}, y = 1$

Not possible

13. $\mathbf{v} = \left\langle 1 + \dfrac{1}{2}, -\dfrac{1}{2} - 2, 0 - \dfrac{1}{2} \right\rangle = \left\langle \dfrac{3}{2}, -\dfrac{5}{2}, -\dfrac{1}{2} \right\rangle$ or $\langle 3, -5, -1 \rangle$

Point: $\left(-\dfrac{1}{2}, 2, \dfrac{1}{2} \right)$

(a) $x = -\dfrac{1}{2} + 3t, y = 2 - 5t, z = \dfrac{1}{2} - t$

(b) $\dfrac{x + \frac{1}{2}}{3} = \dfrac{y - 2}{-5} = \dfrac{z - \frac{1}{2}}{-1}$

15.

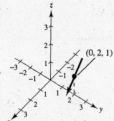

17. $a(x - x_1) + b(y - y_1) + c(z - z_1) = 0$

$1(x - 2) + 0(y - 1) + 0(z - 2) = 0$

$x - 2 = 0$

19. $-2(x - 5) + 1(y - 6) - 2(z - 3) = 0$

$-2x + y - 2z + 10 = 0$

21. $\mathbf{n} = \langle -1, -2, 1 \rangle \implies$

$-1(x - 2) - 2(y - 0) + 1(z - 0) = 0$

$-x - 2y + z + 2 = 0$

23. $\mathbf{u} = \langle 1 - 0, 2 - 0, 3 - 0 \rangle = \langle 1, 2, 3 \rangle$

$\mathbf{v} = \langle -2 - 0, 3 - 0, 3 - 0 \rangle = \langle -2, 3, 3 \rangle$

$\mathbf{n} = \mathbf{u} \times \mathbf{v} = \begin{vmatrix} \mathbf{i} & \mathbf{j} & \mathbf{k} \\ 1 & 2 & 3 \\ -2 & 3 & 3 \end{vmatrix} = \langle -3, -9, 7 \rangle$

$-3(x - 0) - 9(y - 0) + 7(z - 0) = 0$

$-3x - 9y + 7z = 0$

$3x + 9y - 7z = 0$

25. $\mathbf{u} = \langle 3 - 2, 4 - 3, 2 + 2 \rangle = \langle 1, 1, 4 \rangle$

$\mathbf{v} = \langle 1 - 2, -1 - 3, 0 + 2 \rangle = \langle -1, -4, 2 \rangle$

$\mathbf{n} = \mathbf{u} \times \mathbf{v} = \begin{vmatrix} \mathbf{i} & \mathbf{j} & \mathbf{k} \\ 1 & 1 & 4 \\ -1 & -4 & 2 \end{vmatrix} = \langle 18, -6, -3 \rangle$

$18(x - 2) - 6(y - 3) - 3(z + 2) = 0$

$18x - 6y - 3z - 24 = 0$

$6x - 2y - z - 8 = 0$

27. $\mathbf{n} = \mathbf{j}: 0(x - 2) + 1(y - 5) + 0(z - 3) = 0$

$y - 5 = 0$

29. $\langle 0 - (-1), 2 - (-2), 4 - 0 \rangle = \langle 1, 4, 4 \rangle$ and $\langle 1, 0, 0 \rangle$ are parallel to the plane.

$\mathbf{n} = \begin{vmatrix} \mathbf{i} & \mathbf{j} & \mathbf{k} \\ 1 & 4 & 4 \\ 1 & 0 & 0 \end{vmatrix} = \langle 0, 4, -4 \rangle$

$0(x - 0) + 4(y - 2) - 4(z - 4) = 0$

$4y - 4z + 8 = 0$

$y - z + 2 = 0$

31. $\langle -1 - 2, 1 - 2, -1 - 1 \rangle = \langle -3, -1, -2 \rangle$ and
$\langle 2, -3, 1 \rangle$ are parallel to plane.

$$\mathbf{n} = \begin{vmatrix} \mathbf{i} & \mathbf{j} & \mathbf{k} \\ -3 & -1 & -2 \\ 2 & -3 & 1 \end{vmatrix} = \langle -7, -1, 11 \rangle$$

$-7(x - 2) - 1(y - 2) + 11(z - 1) = 0$

$\qquad -7x - y + 11z + 5 = 0$

33. $\mathbf{v} = \langle 0, 0, 1 \rangle$ and $P = (2, 3, 4)$

$x = 2$

$y = 3$

$z = 4 + t$

35. $\mathbf{v} = \langle 3, 2, -1 \rangle$
and $P = (2, 3, 4)$

$x = 2 + 3t$

$y = 3 + 2t$

$z = 4 - t$

37. $\mathbf{v} = \langle 2, -1, 3 \rangle$ and
$P = (5, -3, -4)$

$x = 5 + 2t$

$y = -3 - t$

$z = -4 + 3t$

39. $\mathbf{v} = \langle -1, 1, 1 \rangle$
and $P = (2, 1, 2)$

$x = 2 - t$

$y = 1 + t$

$z = 2 + t$

41. $\mathbf{n}_1 = \langle 5, -3, 1 \rangle, \mathbf{n}_2 = \langle 1, 4, 7 \rangle$

$\mathbf{n}_1 \cdot \mathbf{n}_2 = 5 - 12 + 7 = 0$; orthogonal

43. $\mathbf{n}_1 = \langle 2, 0, -1 \rangle, \mathbf{n}_2 = \langle 4, 1, 8 \rangle$

$\mathbf{n}_1 \cdot \mathbf{n}_2 = 8 - 8 = 0$; orthogonal

45. (a) $\mathbf{n}_1 = \langle 3, -4, 5 \rangle, \mathbf{n}_2 = \langle 1, 1, -1 \rangle$; normal vectors to planes

$$\cos \theta = \frac{|\mathbf{n}_1 \cdot \mathbf{n}_2|}{\|\mathbf{n}_1\| \|\mathbf{n}_2\|} = \frac{|-6|}{\sqrt{50}\sqrt{3}} = \frac{6}{\sqrt{150}} \implies \theta \approx 60.67°$$

(b) $3x - 4y + 5z = 6 \qquad$ Equation 1

$\quad x + y - z = 2 \qquad$ Equation 2

(-3) times Equation 2 added to Equation 1 gives

$-7y + 8z = 0$

$\qquad y = \frac{8}{7}z.$

Substituting back into Equation 2, $x = 2 - y + z = 2 - \frac{8}{7}z + z = 2 - \frac{1}{7}z.$

Letting $t = z/7$, we obtain $x = 2 - t, y = 8t, z = 7t.$

47. (a) $\mathbf{n}_1 = \langle 1, 1, -1 \rangle, \mathbf{n}_2 = \langle 2, -5, -1 \rangle$; normal vectors to planes

$$\cos \theta = \frac{|\mathbf{n}_1 \cdot \mathbf{n}_2|}{\|\mathbf{n}_1\| \|\mathbf{n}_2\|} = \frac{|-2|}{\sqrt{3}\sqrt{30}} = \frac{2}{\sqrt{90}} \implies \theta \approx 77.83°$$

(b) $\quad x + y - z = 0 \qquad$ Equation 1

$\quad 2x - 5y - z = 1 \qquad$ Equation 2

(-2) times Equation 1 added to Equation 2 gives

$-7y + z = 1$

$\qquad y = \frac{z - 1}{7}.$

Substituting back into Equation 1, $x = z - y = z - \frac{z - 1}{7} = \frac{6z}{7} + \frac{1}{7} = \frac{1}{7}(6z + 1).$

Letting $z = t, x = \frac{6t + 1}{7}, y = \frac{t - 1}{7}$. Equivalently, let $y = t, z = 7t + 1$ and $x = 6t + 1.$

49. $x + 2y + 3z = 6$

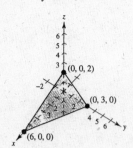

51. $x + 2y = 4$

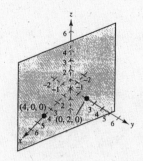

53. $3x + 2y - z = 6$

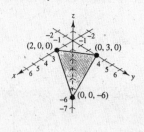

55. $D = \dfrac{|\overrightarrow{PQ} \cdot \mathbf{n}|}{\|\mathbf{n}\|}$

$P = (1, 0, 0)$ on plane, $Q = (0, 0, 0)$,

$\mathbf{n} = \langle 8, -4, 1 \rangle$, $\overrightarrow{PQ} = \langle -1, 0, 0 \rangle$

$D = \dfrac{|\langle -1, 0, 0 \rangle \cdot \langle 8, -4, 1 \rangle|}{\sqrt{64 + 16 + 1}} = \dfrac{|-8|}{\sqrt{81}} = \dfrac{8}{9}$

57. $D = \dfrac{|\overrightarrow{PQ} \cdot \mathbf{n}|}{\|\mathbf{n}\|}$

$P = (2, 0, 0)$ on plane, $Q = (4, -2, -2)$,

$\mathbf{n} = \langle 2, -1, 1 \rangle$, $\overrightarrow{PQ} = \langle 2, -2, -2 \rangle$

$D = \dfrac{|\langle 2, -2, -2 \rangle \cdot \langle 2, -1, 1 \rangle|}{\sqrt{6}} = \dfrac{4}{\sqrt{6}} = \dfrac{2\sqrt{6}}{3}$

59. The normal vector to plane containing $(0, 0, 0)$, $(2, 2, 12)$ and $(10, 0, 0)$ is obtained as follows.

$\mathbf{v}_1 = \langle 2, 2, 12 \rangle$, $\mathbf{v}_2 = \langle 10, 0, 0 \rangle$

$\mathbf{v}_1 \times \mathbf{v}_2 = \begin{vmatrix} \mathbf{i} & \mathbf{j} & \mathbf{k} \\ 2 & 2 & 12 \\ 10 & 0 & 0 \end{vmatrix} = \langle 0, 120, -20 \rangle$

$\mathbf{n}_1 = \langle 0, 6, -1 \rangle$

The normal vector to the plane containing $(0, 0, 0)$, $(2, 2, 12)$ and $(0, 10, 0)$ is obtained as follows.

$\mathbf{u}_1 = \langle 2, 2, 12 \rangle$, $\mathbf{u}_2 = \langle 0, 10, 0 \rangle$

$\mathbf{u}_1 \times \mathbf{u}_2 = \begin{vmatrix} \mathbf{i} & \mathbf{j} & \mathbf{k} \\ 2 & 2 & 12 \\ 0 & 10 & 0 \end{vmatrix} = \langle -120, 0, 20 \rangle$

$\mathbf{n}_2 = \langle -6, 0, 1 \rangle$

The angle θ between two adjacent sides is given by

$\cos \theta = \dfrac{|\mathbf{n}_1 \cdot \mathbf{n}_2|}{\|\mathbf{n}_1\| \|\mathbf{n}_2\|} = \dfrac{|-1|}{\sqrt{37}\sqrt{37}} = \dfrac{1}{37} \implies \theta \approx 88.45°.$

61. False. They might be skew lines, such as:

$L_1: x = t, y = 0, z = 0$ (x-axis)

and $L_2: x = 0, y = t, z = 1$

63. The lines are parallel:

$-\frac{3}{2}\langle 10, -18, 20 \rangle = \langle -15, 27, -30 \rangle$

65. $x^2 + y^2 = 10^2 = 100$

67. $r = 3 \cos \theta$, $r^2 = 3r \cos \theta$, $x^2 + y^2 = 3x$

69. $r^2 = 49$

$r = 7$

71. $y = 5$

$r \sin \theta = 5$

$r = 5 \csc \theta$

Review Exercises for Chapter 10

1. (a) and (b)

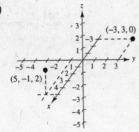

3. $(-5, 4, 0)$

5. $d = \sqrt{(5-4)^2 + (2-0)^2 + (1-7)^2}$

$ = \sqrt{1 + 4 + 36}$

$ = \sqrt{41}$

7. $d_1 = \sqrt{(3-0)^2 + (-2-3)^2 + (0-2)^2} = \sqrt{9 + 25 + 4} = \sqrt{38}$

$d_2 = \sqrt{(0-0)^2 + (5-3)^2 + (-3-2)^2} = \sqrt{4 + 25} = \sqrt{29}$

$d_3 = \sqrt{(0-3)^2 + (5-(-2))^2 + (-3-0)^2} = \sqrt{9 + 49 + 9} = \sqrt{67}$

$d_1^2 + d_2^2 = 38 + 29 = 67 = d_3^2$

9. Midpoint: $\left(\dfrac{-2+2}{2}, \dfrac{3-5}{2}, \dfrac{2+(-2)}{2} \right) = (0, -1, 0)$

11. Midpoint: $\left(\dfrac{10-8}{2}, \dfrac{6-2}{2}, \dfrac{-12-6}{2} \right) = (1, 2, -9)$

13. $(x-2)^2 + (y-3)^2 + (z-5)^2 = 1$

15. Radius: 6

$(x-1)^2 + (y-5)^2 + (z-2)^2 = 36$

17. $(x^2 - 4x + 4) + (y^2 - 6y + 9) + z^2 = -4 + 4 + 9$

$(x-2)^2 + (y-3)^2 + z^2 = 9$

Center: $(2, 3, 0)$

Radius: 3

19. (a) xz-trace $(y = 0)$: $x^2 + z^2 = 7$, circle

(b) yz-trace $(x = 0)$: $(y-3)^2 + z^2 = 16$, circle

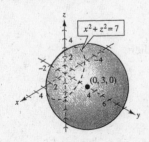

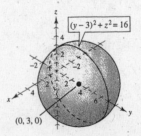

21. (a) $\mathbf{v} = \langle 3 - 2, 3 - (-1), 0 - 4 \rangle = \langle 1, 4, -4 \rangle$

(b) $\|\mathbf{v}\| = \sqrt{1^2 + 4^2 + (-4)^2} = \sqrt{33}$

(c) Unit vector: $\dfrac{\sqrt{33}}{33} \langle 1, 4, -4 \rangle$

23. (a) $\mathbf{v} = \langle -3 - 7, 2 - (-4), 10 - 3 \rangle = \langle -10, 6, 7 \rangle$

(b) $\|\mathbf{v}\| = \sqrt{(-10)^2 + 6^2 + 7^2} = \sqrt{185}$

(c) Unit vector: $\dfrac{\sqrt{185}}{185} \langle -10, 6, 7 \rangle$

25. $\mathbf{u} \cdot \mathbf{v} = -1(0) + 4(-6) + 3(5) = -9$

27. $\mathbf{u} \cdot \mathbf{v} = 2(1) - 1(0) + 1(-1) = 1$

29. $\cos \theta = \dfrac{\mathbf{u} \cdot \mathbf{v}}{\|\mathbf{u}\| \, \|\mathbf{v}\|} = \dfrac{12 - 2 - 10}{\sqrt{42}\sqrt{17}} = 0$

$\theta \approx 90°$ The vectors are orthogonal.

31. Since $\mathbf{u} \cdot \mathbf{v} = 0$, the angle is $90°$.

33. $\mathbf{u} \cdot \mathbf{v} = 7(-1) + (-2)(4) + 3(5) = 0$

Orthogonal

35. Since $-\frac{2}{3}\langle 39, -12, 21 \rangle = \langle -26, 8, -14 \rangle$, the vectors are parallel.

37. First two points: $\mathbf{u} = \langle -3, 4, 1 \rangle$

Last two points: $\mathbf{v} = \langle 0, -2, 6 \rangle$

Since $\mathbf{u} \neq c\mathbf{v}$, the points are not collinear.

39. First two points: $\langle 4, -2, -10 \rangle$

First and third points: $\langle 2, -1, -5 \rangle$

Since $\langle 4, -2, -10 \rangle = 2\langle 2, -1, -5 \rangle$, the three points are collinear.

41. Let $\mathbf{a}$, $\mathbf{b}$, and $\mathbf{c}$ be the three force vectors determined by $A(0, 10, 10)$, $B(-4, -6, 10)$, and $C(4, -6, 10)$.

$$\mathbf{a} = \|\mathbf{a}\|\frac{\langle 0, 10, 10 \rangle}{10\sqrt{2}} = \|\mathbf{a}\|\left\langle 0, \frac{1}{\sqrt{2}}, \frac{1}{\sqrt{2}} \right\rangle$$

$$\mathbf{b} = \|\mathbf{b}\|\frac{\langle -4, -6, 10 \rangle}{\sqrt{152}} = \|\mathbf{b}\|\left\langle \frac{-2}{\sqrt{38}}, \frac{-3}{\sqrt{38}}, \frac{5}{\sqrt{38}} \right\rangle$$

$$\mathbf{c} = \|\mathbf{c}\|\frac{\langle 4, -6, 10 \rangle}{\sqrt{152}} = \|\mathbf{c}\|\left\langle \frac{2}{\sqrt{38}}, \frac{-3}{\sqrt{38}}, \frac{5}{\sqrt{38}} \right\rangle$$

Must have $\mathbf{a} + \mathbf{b} + \mathbf{c} = 300\mathbf{k}$. Thus,

$$\frac{-2}{\sqrt{38}}\|\mathbf{b}\| + \frac{2}{\sqrt{38}}\|\mathbf{c}\| = 0$$

$$\frac{1}{\sqrt{2}}\|\mathbf{a}\| - \frac{3}{\sqrt{38}}\|\mathbf{b}\| - \frac{3}{\sqrt{38}}\|\mathbf{c}\| = 0$$

$$\frac{1}{\sqrt{2}}\|\mathbf{a}\| + \frac{5}{\sqrt{38}}\|\mathbf{b}\| + \frac{5}{\sqrt{38}}\|\mathbf{c}\| = 300.$$

From the first equation, $\|\mathbf{b}\| = \|\mathbf{c}\|$. From the second equation, $\dfrac{1}{\sqrt{2}}\|\mathbf{a}\| = \dfrac{6}{\sqrt{38}}\|\mathbf{b}\|$.

From the third equation, $\dfrac{1}{\sqrt{2}}\|\mathbf{a}\| = 300 - \dfrac{10}{\sqrt{38}}\|\mathbf{b}\|$. Thus,

$$\frac{6}{\sqrt{38}}\|\mathbf{b}\| = 300 - \frac{10}{\sqrt{38}}\|\mathbf{b}\| \implies \frac{16}{\sqrt{38}}\|\mathbf{b}\| = 300 \text{ and } \|\mathbf{b}\| = \|\mathbf{c}\| = \frac{75\sqrt{38}}{4} \approx 115.58.$$

Finally, $\|\mathbf{a}\| = \sqrt{2}\left(\dfrac{6}{\sqrt{38}}\right)\left(\dfrac{75\sqrt{38}}{4}\right) = \dfrac{225\sqrt{2}}{2} \approx 159.10.$

43. $\mathbf{u} \times \mathbf{v} = \begin{vmatrix} \mathbf{i} & \mathbf{j} & \mathbf{k} \\ -2 & 8 & 2 \\ 1 & 1 & -1 \end{vmatrix} = \langle -10, 0, -10 \rangle$

45. $\mathbf{u} \times \mathbf{v} = \begin{vmatrix} \mathbf{i} & \mathbf{j} & \mathbf{k} \\ -3 & 2 & -5 \\ 10 & -15 & 2 \end{vmatrix} = \langle -71, -44, 25 \rangle$

$\|\mathbf{u} \times \mathbf{v}\| = \sqrt{7602}$

Unit vector: $\dfrac{1}{\sqrt{7602}}\langle -71, -44, 25 \rangle$

47. First two points: $\langle 3, 2, 3 \rangle$

Last two points: $\langle 3, 2, 3 \rangle$

First and third points: $\langle -2, 2, 0 \rangle$

$$\begin{vmatrix} \mathbf{i} & \mathbf{j} & \mathbf{k} \\ 3 & 2 & 3 \\ -2 & 2 & 0 \end{vmatrix} = \langle -6, -6, 10 \rangle$$

Area $= \|\langle -6, -6, 10 \rangle\|$

$= \sqrt{36 + 36 + 100}$

$= \sqrt{172}$

$= 2\sqrt{43}$ square units

49. The parallelogram is determined by the three vectors with initial point $(0, 0, 0)$.

$\mathbf{u} = \langle 3, 0, 0 \rangle$, $\mathbf{v} = \langle 2, 0, 5 \rangle$, $\mathbf{w} = \langle 0, 5, 1 \rangle$

$$\mathbf{u} \cdot (\mathbf{v} \times \mathbf{w}) = \begin{vmatrix} 3 & 0 & 0 \\ 2 & 0 & 5 \\ 0 & 5 & 1 \end{vmatrix} = -75$$

Volume $= |-75| = 75$ cubic units

51. $\mathbf{v} = \langle 9 - 3, 11 - 0, 6 - 2 \rangle = \langle 6, 11, 4 \rangle$

Point: $(3, 0, 2)$

(a) $x = 3 + 6t$, $y = 11t$, $z = 2 + 4t$

(b) $\dfrac{x - 3}{6} = \dfrac{y}{11} = \dfrac{z - 2}{4}$

53. $\mathbf{v} = \langle 3 + 1, 6 - 3, -1 - 5 \rangle = \langle 4, 3, -6 \rangle$, point: $(-1, 3, 5)$

(a) Parametric equations: $x = -1 + 4t$, $y = 3 + 3t$, $z = 5 - 6t$

(b) Symmetric equations: $\dfrac{x + 1}{4} = \dfrac{y - 3}{3} = \dfrac{z - 5}{-6}$

55. Use $2\mathbf{v} = \langle -4, 5, 2 \rangle$, point: $(0, 0, 0)$.

(a) Parametric equations: $x = -4t$, $y = 5t$, $z = 2t$

(b) Symmetric equations: $\dfrac{x}{-4} = \dfrac{y}{5} = \dfrac{z}{2}$

57. $\mathbf{u} = \langle 5, 0, 2 \rangle$, $\mathbf{v} = \langle 2, 3, 8 \rangle$

$$\mathbf{u} \times \mathbf{v} = \begin{vmatrix} \mathbf{i} & \mathbf{j} & \mathbf{k} \\ 5 & 0 & 2 \\ 2 & 3 & 8 \end{vmatrix} = \langle -6, -36, 15 \rangle$$

$\mathbf{n} = \langle 2, 12, -5 \rangle$

$a(x - x_0) + b(y - y_0) + c(z - z_0) = 0$

$2(x - 0) + 12(y - 0) - 5(z - 0) = 0$

$2x + 12y - 5z = 0$

59. $\mathbf{n} = \mathbf{k}$, normal vector

Plane: $0(x - 5) + 0(y - 3) + 1(z - 2) = 0$

$z - 2 = 0$

61. $3x - 2y + 3z = 6$

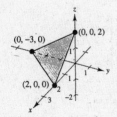

63. $2x - 3z = 6$

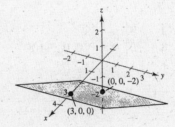

65. $\mathbf{n} = \langle 2, -20, 6 \rangle$, $P = (0, 0, 1)$ in plane, $Q = (2, 3, 10)$, $\overrightarrow{PQ} = \langle 2, 3, 9 \rangle$

$$D = \frac{|\overrightarrow{PQ} \cdot \mathbf{n}|}{\|\mathbf{n}\|} = \frac{|-2|}{\sqrt{440}} = \frac{1}{\sqrt{110}} = \frac{\sqrt{110}}{110} \approx 0.0953$$

67. $\mathbf{n} = \langle 1, -10, 3 \rangle$, $P = (2, 0, 0)$ in plane, $Q = (0, 0, 0)$, $\overrightarrow{PQ} = \langle -2, 0, 0 \rangle$

$$D = \frac{|\overrightarrow{PQ} \cdot \mathbf{n}|}{\|\mathbf{n}\|} = \frac{|-2|}{\sqrt{1 + 100 + 9}} = \frac{2}{\sqrt{110}} = \frac{2\sqrt{110}}{110} = \frac{\sqrt{110}}{55} \approx 0.191$$

69. False. $\mathbf{a} \times \mathbf{b} = -(\mathbf{b} \times \mathbf{a})$

71. $\mathbf{u} \cdot \mathbf{u} = \langle 3, -2, 1 \rangle \cdot \langle 3, -2, 1 \rangle$

$$= 9 + 4 + 1$$
$$= 14$$
$$= \|\mathbf{u}\|^2$$

73. $\mathbf{u} \cdot (\mathbf{v} + \mathbf{w}) = \langle 3, -2, 1 \rangle \cdot \langle 1, -2, -1 \rangle = 6$

$\mathbf{u} \cdot \mathbf{v} + \mathbf{u} \cdot \mathbf{w} = 11 + (-5) = 6$

75. $\mathbf{u} \times \mathbf{v} = \begin{vmatrix} \mathbf{i} & \mathbf{j} & \mathbf{k} \\ u_1 & u_2 & u_3 \\ v_1 & v_2 & v_3 \end{vmatrix} = (u_2 v_3 - u_3 v_2)\mathbf{i} - (u_1 v_3 - u_3 v_1)\mathbf{j} + (u_1 v_2 - u_2 v_1)\mathbf{k}$

77. The magnitude will increase by a factor of 4.

Chapter 10   Practice Test

1. Find the lengths of the sides of the triangle with vertices $(0, 0, 0)$, $(1, 2, -4)$, and $(0, -2, -1)$. Show that the triangle is a right triangle.

2. Find the standard form of the equation of a sphere having center $(0, 4, 1)$ and radius 5.

3. Find the center and radius of the sphere $x^2 + y^2 + z^2 + 2x - 4z - 11 = 0$.

4. Find the vector $\mathbf{u} - 3\mathbf{v}$ given $\mathbf{u} = \langle 1, 0, -1 \rangle$ and $\mathbf{v} = \langle 4, 3, -6 \rangle$.

5. Find the length of $\frac{1}{2}\mathbf{v}$ if $\mathbf{v} = \langle 2, 4, -6 \rangle$.

6. Find the dot product of $\mathbf{u} = \langle 2, 1, -3 \rangle$ and $\mathbf{v} = \langle 1, 1, -2 \rangle$.

7. Determine whether $\mathbf{u} = \langle 1, 1, -1 \rangle$ and $\mathbf{v} = \langle -3, -3, 3 \rangle$ are orthogonal, parallel, or neither.

8. Find the cross product of $\mathbf{u} = \langle -1, 0, 2 \rangle$ and $\mathbf{v} = \langle 1, -1, 3 \rangle$. What is $\mathbf{v} \times \mathbf{u}$?

9. Use the triple scalar product to find the volume of the parallelepiped having adjacent edges $\mathbf{u} = \langle 1, 1, 1 \rangle$, $\mathbf{v} = \langle 0, -1, 1 \rangle$, and $\mathbf{w} = \langle 1, 0, 4 \rangle$.

10. Find a set of parametric equations for the line through the points $(0, -3, 3)$ and $(2, -3, 4)$.

11. Find an equation of the plane passing through $(1, 2, 3)$ and perpendicular to the vector $\mathbf{n} = \langle 1, -1, 0 \rangle$.

12. Find an equation of the plane passing through the three points $A = (0, 0, 0)$, $B = (1, 1, 1)$, and $C = (1, 2, 3)$.

13. Determine whether the planes $x + y - z = 12$ and $3x - 4y - z = 9$ are parallel, orthogonal, or neither.

14. Find the distance between the point $(1, 1, 1)$ and the plane $x + 2y + z = 6$.

C H A P T E R 1 1
Limits and an Introduction to Calculus

Section 11.1   Introduction to Limits 510

Section 11.2   Techniques for Evaluating Limits 514

Section 11.3   The Tangent Line Problem 520

Section 11.4   Limits at Infinity and Limits of Sequences 526

Section 11.5   The Area Problem 530

Review Exercises 535

Practice Test 542

CHAPTER 11
Limits and an Introduction to Calculus

Section 11.1   Introduction to Limits

- If $f(x)$ becomes arbitrarily close to a unique number L as x approaches c from either side, then the limit of $f(x)$ as x approaches c is L:

  $$\lim_{x \to c} f(x) = L.$$

- You should be able to use a calculator to find a limit.
- You should be able to use a graph to find a limit.
- You should understand how limits can fail to exist:
  - (a) $f(x)$ approaches a different number from the right of c than it approaches from the left of c.
  - (b) $f(x)$ increases or decreases without bound as x approaches c.
  - (c) $f(x)$ oscillates between two fixed values as x approaches c.
- You should know and be able to use the elementary properties of limits.

Vocabulary Check

1. limit

2. oscillates

3. direct substitution

---

1. (a)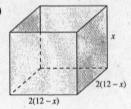

(b) $V = (\text{base})\text{height} = (24 - 2x)^2 x = 4x(12 - x)^2$

(d)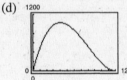

Maximum at $x = 4$

(c) $\displaystyle\lim_{x \to 4} V = 1024$

x	3	3.5	3.9	4	4.1	4.5	5
V	972.0	1011.5	1023.5	1024.0	1023.5	1012.5	980.0

3. $\displaystyle\lim_{x \to 2} (5x + 4) = 14$

x	1.9	1.99	1.999	2	2.001	2.01	2.1
$f(x)$	13.5	13.95	13.995	14	14.005	14.05	14.5

The limit is reached.

5. $\lim\limits_{x \to 3} \dfrac{x - 3}{x^2 - 9} = \dfrac{1}{6}$

x	2.9	2.99	2.999	3	3.001	3.01	3.1
$f(x)$	0.1695	0.1669	0.16669	Error	0.16664	0.1664	0.1639

The limit is not reached.

7. $\lim\limits_{x \to 0} \dfrac{\sin 2x}{x} = 2$

x	-0.1	-0.01	-0.001	0	0.001	0.01	0.1
$f(x)$	1.9867	1.99987	1.9999987	Error	1.9999987	1.99987	1.987

The limit is not reached.

9. $\lim\limits_{x \to 0} \dfrac{e^{2x} - 1}{x} = 2$

x	-0.1	-0.01	-0.001	0	0.001	0.01	0.1
$f(x)$	1.8127	1.9801	1.9980	Error	2.0020	2.0201	2.2140

The limit is not reached.

11. $\lim\limits_{x \to 1} \dfrac{x - 1}{x^2 + 2x - 3} = \dfrac{1}{4}$

x	0.9	0.99	0.999	1.0	1.001	1.01	1.1
$f(x)$	0.2564	0.2506	0.2501	Error	0.2499	0.2494	0.2439

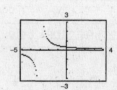

13. $\lim\limits_{x \to 0} \dfrac{\sqrt{x + 5} - \sqrt{5}}{x} \approx 0.2236$ $\left(\text{Actual limit is } \dfrac{1}{2\sqrt{5}}. \right)$

x	-0.1	-0.01	-0.001	0	0.001	0.01	0.1
$f(x)$	0.2247	0.2237	0.2236	Error	0.2236	0.2235	0.2225

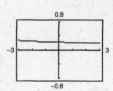

15. $\lim\limits_{x \to -4} \dfrac{[x/(x + 2)] - 2}{x + 4} = \dfrac{1}{2}$

x	-4.1	-4.01	-4.001	-4.0	-3.999	-3.99	-3.9
$f(x)$	0.4762	0.4975	0.4998	Error	0.5003	0.5025	0.5263

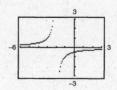

17. Make sure your calculator is set in radian mode.

$$\lim_{x \to 0} \frac{\sin x}{x} = 1$$

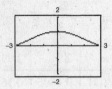

x	−0.1	−0.01	−0.001	0	0.001	0.01	0.1
f(x)	0.9983	0.99998	0.9999998	Error	0.9999998	0.99998	0.9983

19. $\displaystyle\lim_{x \to 0} \frac{\sin^2 x}{x} = 0$

x	−0.1	−0.01	−0.001	0	0.001	0.01	0.1
f(x)	−0.0997	−0.0100	−0.0010	Error	0.0010	0.0100	0.0997

21. $\displaystyle\lim_{x \to 0} \frac{e^{2x} - 1}{2x} = 1.0$

x	−0.1	−0.01	−0.001	0	0.001	0.01	0.1
f(x)	0.9063	0.9901	0.9990	Error	1.0010	1.0101	1.1070

23. $\displaystyle\lim_{x \to 1} \frac{\ln(2x - 1)}{x - 1} = 2$

x	0.9	0.99	0.999	1	1.001	1.01	1.1
f(x)	2.2314	2.0203	2.0020	Error	1.9980	1.9803	1.8232

25. $f(x) = \begin{cases} 2x + 1, & x < 2 \\ x + 3, & x \geq 2 \end{cases}$

The limit exists as x approaches 2:

$$\lim_{x \to 2} f(x) = 5$$

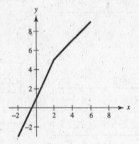

27. $\displaystyle\lim_{x \to 2} f(x)$ does not exist.

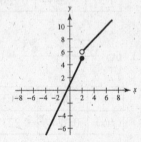

29. $\displaystyle\lim_{x \to -4} (x^2 - 3) = 13$

31. $\displaystyle\lim_{x \to -2} \frac{|x + 2|}{x + 2}$ does not exist. $f(x) = \dfrac{|x + 2|}{x + 2}$ equals -1 to the left of -2, and equals 1 to the right of -2.

33. The limit does not exist because $f(x)$ oscillates between 2 and -2.

35. $\displaystyle\lim_{x \to \pi/2} \tan x$ does not exist.

37. $\lim\limits_{x \to 0} \dfrac{5}{2 + e^{1/x}}$ does not exist.

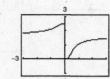

39. $\lim\limits_{x \to 0} \cos \dfrac{1}{x}$ does not exist.

The graph oscillates between -1 and 1.

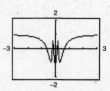

41. $\lim\limits_{x \to 4} \dfrac{\sqrt{x + 3} - 1}{x - 4}$ does not exist.

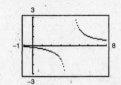

43. $\lim\limits_{x \to 1} \dfrac{x - 1}{x^2 - 4x + 3} = -\dfrac{1}{2}$

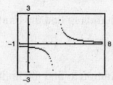

45. $\lim\limits_{x \to 4} \ln(x + 3) \approx 1.946$ (Exact limit is $\ln 7$.)

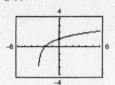

47. (a) $\lim\limits_{x \to c} [-2g(x)] = -2(6) = -12$

(b) $\lim\limits_{x \to c} [f(x) + g(x)] = 3 + 6 = 9$

(c) $\lim\limits_{x \to c} \dfrac{f(x)}{g(x)} = \dfrac{3}{6} = \dfrac{1}{2}$

(d) $\lim\limits_{x \to c} \sqrt{f(x)} = \sqrt{3}$

49. (a) $\lim\limits_{x \to 2} f(x) = 2^3 = 8$

(b) $\lim\limits_{x \to 2} g(x) = \dfrac{\sqrt{2^2 + 5}}{2(2^2)} = \dfrac{3}{8}$

(c) $\lim\limits_{x \to 2} [f(x)g(x)] = 8\left(\dfrac{3}{8}\right) = 3$

(d) $\lim\limits_{x \to 2} [g(x) - f(x)] = \dfrac{3}{8} - 8 = -\dfrac{61}{8}$

51. $\lim\limits_{x \to 5} (10 - x^2) = 10 - 5^2 = -15$

53. $\lim\limits_{x \to -3} (2x^2 + 4x + 1) = 2(-3)^2 + 4(-3) + 1 = 7$

55. $\lim\limits_{x \to 3} \left(-\dfrac{9}{x}\right) = -\dfrac{9}{3} = -3$

57. $\lim\limits_{x \to -3} \dfrac{3x}{x^2 + 1} = -\dfrac{9}{10}$

59. $\lim\limits_{x \to -2} \dfrac{5x + 3}{2x - 9} = \dfrac{5(-2) + 3}{2(-2) - 9} = \dfrac{-7}{-13} = \dfrac{7}{13}$

61. $\lim\limits_{x \to -1} \sqrt{x + 2} = \sqrt{-1 + 2} = 1$

63. $\lim\limits_{x \to 7} \dfrac{5x}{\sqrt{x + 2}} = \dfrac{5(7)}{\sqrt{7 + 2}} = \dfrac{35}{3}$

65. $\lim\limits_{x \to 3} e^x = e^3 \approx 20.0855$

67. $\lim\limits_{x \to \pi} \sin 2x = \sin 2\pi = 0$

69. $\lim\limits_{x \to 1/2} \arcsin x = \arcsin \dfrac{1}{2} = \dfrac{\pi}{6} \approx 0.5236$

71. True

73. Answers will vary.

75. (a) No. The limit may or may not exist, and if it does exist, it may not equal 4.

(b) No. $f(2)$ may or may not exist, and if $f(2)$ exists, it may not equal 4.

77. $\dfrac{5 - x}{3x - 15} = \dfrac{5 - x}{-3(5 - x)} = -\dfrac{1}{3}, \; x \neq 5$

79. $\dfrac{15x^2 + 7x - 4}{15x^2 + x - 2} = \dfrac{(3x - 1)(5x + 4)}{(3x - 1)(5x + 2)}$

$\qquad = \dfrac{5x + 4}{5x + 2}, \; x \neq \dfrac{1}{3}$

81. $\dfrac{x^2 + 27}{x^2 + x - 6} = \dfrac{(x + 3)(x^2 - 3x + 9)}{(x + 3)(x - 2)}$

$\qquad = \dfrac{x^2 - 3x + 9}{x - 2}, \; x \neq -3$

Section 11.2   Techniques for Evaluating Limits

- You can use direct substitution to find the limit of a polynomial function $p(x)$:

  $$\lim_{x \to c} p(x) = p(c).$$

- You can use direct substitution to find the limit of a rational function $r(x) = \dfrac{p(x)}{q(x)}$, as long as $q(c) \neq 0$:

  $$\lim_{x \to c} r(x) = r(c) = \dfrac{p(c)}{q(c)}, q(c) \neq 0.$$

- You should be able to use cancellation techniques to find a limit.

- You should know how to use rationalization techniques to find a limit.

- You should know how to use technology to find a limit.

- You should be able to calculate one-sided limits.

Vocabulary Check

1. dividing out technique   **2.** indeterminate form   **3.** one-sided limit   **4.** difference quotient

1. $g(x) = \dfrac{-2x^2 + x}{x}, \; g_2(x) = -2x + 1$

   (a) $\displaystyle\lim_{x \to 0} g(x) = 1$

   (b) $\displaystyle\lim_{x \to -1} g(x) = 3$

   (c) $\displaystyle\lim_{x \to -2} g(x) = 5$

3. $g(x) = \dfrac{x^3 - x}{x - 1}, \; g_2(x) = x^2 + x = x(x + 1)$

   (a) $\displaystyle\lim_{x \to 1} g(x) = 2$

   (b) $\displaystyle\lim_{x \to -1} g(x) = 0$

   (c) $\displaystyle\lim_{x \to 0} g(x) = 0$

5. $\displaystyle\lim_{x \to 6} \dfrac{x - 6}{x^2 - 36} = \lim_{x \to 6} \dfrac{x - 6}{(x - 6)(x + 6)}$

$\qquad = \displaystyle\lim_{x \to 6} \dfrac{1}{x + 6} = \dfrac{1}{12}$

7. $\displaystyle\lim_{x \to -1} \dfrac{1 - 2x - 3x^2}{1 + x} = \lim_{x \to -1} \dfrac{(1 + x)(1 - 3x)}{1 + x}$

$\qquad = \displaystyle\lim_{x \to -1} (1 - 3x) = 4$

9. $\displaystyle\lim_{t \to 2} \dfrac{t^3 - 8}{t - 2} = \lim_{t \to 2} \dfrac{(t - 2)(t^2 + 2t + 4)}{t - 2}$

$\qquad = \displaystyle\lim_{t \to 2} (t^2 + 2t + 4)$

$\qquad = 4 + 4 + 4 = 12$

11. $\displaystyle\lim_{x \to 1} \dfrac{x^4 - 1}{x^4 - 3x^2 - 4} = \dfrac{0}{-6} = 0$

13. $\lim\limits_{x \to -1} \dfrac{x^3 + 2x^2 - x - 2}{x^3 + 4x^2 - x - 4} = \lim\limits_{x \to -1} \dfrac{(x-1)(x+1)(x+2)}{(x-1)(x+1)(x+4)}$

$= \lim\limits_{x \to -1} \dfrac{(x-1)(x+2)}{(x-1)(x+4)}$

$= \dfrac{(-2)(1)}{(-2)(3)} = \dfrac{1}{3}$

15. $\lim\limits_{x \to 2} \dfrac{x^3 + 2x^2 - 5x - 6}{x^3 - 7x + 6} = \lim\limits_{x \to 2} \dfrac{(x-2)(x+1)(x+3)}{(x-2)(x-1)(x+3)}$

$= \lim\limits_{x \to 2} \dfrac{x+1}{x-1}$

$= \dfrac{3}{1} = 3$

17. $\lim\limits_{y \to 0} \dfrac{\sqrt{5+y} - \sqrt{5}}{y} = \lim\limits_{y \to 0} \dfrac{\sqrt{5+y} - \sqrt{5}}{y} \cdot \dfrac{\sqrt{5+y} + \sqrt{5}}{\sqrt{5+y} + \sqrt{5}}$

$= \lim\limits_{y \to 0} \dfrac{(5+y) - 5}{y\left(\sqrt{5+y} + \sqrt{5}\right)}$

$= \lim\limits_{y \to 0} \dfrac{1}{\sqrt{5+y} + \sqrt{5}}$

$= \dfrac{1}{2\sqrt{5}} = \dfrac{\sqrt{5}}{10}$

19. $\lim\limits_{x \to -3} \dfrac{\sqrt{x+7} - 2}{x+3} = \lim\limits_{x \to -3} \dfrac{\sqrt{x+7} - 2}{x+3} \cdot \dfrac{\sqrt{x+7} + 2}{\sqrt{x+7} + 2}$

$= \lim\limits_{x \to -3} \dfrac{(x+7) - 4}{(x+3)\left(\sqrt{x+7} + 2\right)}$

$= \lim\limits_{x \to -3} \dfrac{1}{\sqrt{x+7} + 2} = \dfrac{1}{4}$

21. $\lim\limits_{x \to 0} \dfrac{1/(1+x) - 1}{x} = \lim\limits_{x \to 0} \dfrac{1 - (1+x)}{(1+x)x}$

$= \lim\limits_{x \to 0} \dfrac{-1}{1+x} = -1$

23. $\lim\limits_{x \to 0} \dfrac{\sec x}{\tan x} = \lim\limits_{x \to 0} \dfrac{1}{\cos x} \cdot \dfrac{\cos x}{\sin x}$

$= \lim\limits_{x \to 0} \dfrac{1}{\sin x}$, does not exist

25. $\lim\limits_{x \to 0} \dfrac{\cos 2x}{\cot 2x} = \lim\limits_{x \to 0} \dfrac{\cos 2x}{(\cos 2x)/\sin(2x)}$

$= \lim\limits_{x \to 0} \sin 2x = 0$

27. $\lim\limits_{x \to \pi/2} \dfrac{\sin x - 1}{x} = \dfrac{1 - 1}{\pi/2} = 0$

29. $f(x) = \dfrac{\sqrt{x+3} - \sqrt{3}}{x}$

$\lim\limits_{x \to 0} f(x) \approx 0.2887$

$\left(\text{Exact limit: } \dfrac{1}{2\sqrt{3}}\right)$

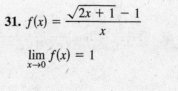

31. $f(x) = \dfrac{\sqrt{2x+1} - 1}{x}$

$\lim\limits_{x \to 0} f(x) = 1$

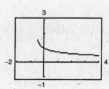

33. $\lim\limits_{x \to 2} \dfrac{x^5 - 32}{x - 2} = 80$

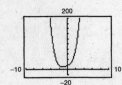

35. $f(x) = \dfrac{1/(x + 4) - (1/4)}{x}$

$$\lim\limits_{x \to 0} f(x) = -\dfrac{1}{16}, \ (-0.0625)$$

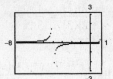

37. $f(x) = \dfrac{e^{2x} - 1}{x}$

$$\lim\limits_{x \to 0} f(x) = 2$$

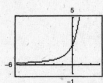

39. $\lim\limits_{x \to 0^+} x \ln x = 0$

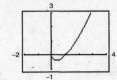

41. $\lim\limits_{x \to 0} \dfrac{\sin 2x}{x} = 2$

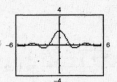

43. $\lim\limits_{x \to 0} \dfrac{\tan x}{x} = 1$

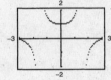

45. $\lim\limits_{x \to 1} \dfrac{1 - \sqrt[3]{x}}{1 - x} = \dfrac{1}{3} \approx 0.333$

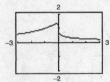

47. $f(x) = (1 - x)^{2/x}$

$$\lim\limits_{x \to 0} f(x) \approx 0.135$$

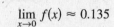

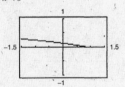

49. $f(x) = \dfrac{x - 1}{x^2 - 1}$

(a) Graphically, $\lim\limits_{x \to 1^-} \dfrac{x - 1}{x^2 - 1} = \dfrac{1}{2}$.

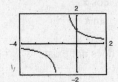

(b)

x	0.5	0.9	0.99	0.999	1
$f(x)$	0.6667	0.5263	0.5025	0.5003	Error

Numerically, $\lim\limits_{x \to 1^-} \dfrac{x - 1}{x^2 - 1} = \dfrac{1}{2}$.

(c) Algebraically, $\lim\limits_{x \to 1^-} \dfrac{x - 1}{x^2 - 1} = \lim\limits_{x \to 1^-} \dfrac{x - 1}{(x - 1)(x + 1)} = \lim\limits_{x \to 1^-} \dfrac{1}{x + 1} = \dfrac{1}{2}$.

51. $f(x) = \dfrac{4 - \sqrt{x}}{x - 16}$

(a) Graphically, $\displaystyle\lim_{x \to 16^+} \dfrac{4 - \sqrt{x}}{x - 16} = -\dfrac{1}{8}$.

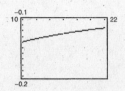

(b)

x	16	16.001	16.01	16.1	16.5
$f(x)$	Error	-0.1250	-0.1250	-0.1248	-0.1240

Numerically, $\displaystyle\lim_{x \to 16^+} \dfrac{4 - \sqrt{x}}{x - 16} = -0.125$.

(c) Algebraically,

$$\lim_{x \to 16^+} \dfrac{4 - \sqrt{x}}{x - 16} = \lim_{x \to 16^+} \dfrac{4 - \sqrt{x}}{(\sqrt{x} - 4)(\sqrt{x} + 4)}$$

$$= \lim_{x \to 16^+} \dfrac{-1}{\sqrt{x} + 4} = \dfrac{-1}{4 + 4} = -\dfrac{1}{8}.$$

53. $f(x) = \dfrac{|x - 6|}{x - 6}$

$\displaystyle\lim_{x \to 6^+} f(x) = 1$

$\displaystyle\lim_{x \to 6^-} f(x) = -1$

Limit does not exist.

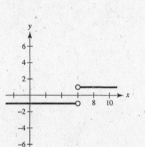

55. $f(x) = \dfrac{1}{x^2 + 1}$

$$\lim_{x \to 1^-} \dfrac{1}{x^2 + 1} = \lim_{x \to 1^+} \dfrac{1}{x^2 + 1}$$

$$= \lim_{x \to 1} \dfrac{1}{x^2 + 1}$$

$$= \dfrac{1}{2}$$

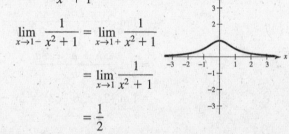

57. $\displaystyle\lim_{x \to 2^-} f(x) = 2 - 1 = 1$

$\displaystyle\lim_{x \to 2^+} f(x) = 2(2) - 3 = 1$

$\displaystyle\lim_{x \to 2} f(x) = 1$

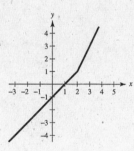

59. $f(x) = \begin{cases} 4 - x^2, & x \le 1 \\ 3 - x, & x > 1 \end{cases}$

$\displaystyle\lim_{x \to 1^-} f(x) = 4 - 1 = 3$

$\displaystyle\lim_{x \to 1^+} f(x) = 3 - 1 = 2$

$\displaystyle\lim_{x \to 1} f(x)$ does not exist.

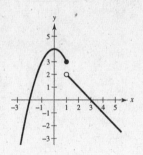

61.

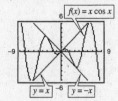

$\displaystyle\lim_{x \to 0} f(x) = 0$

63.

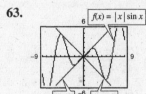

$\displaystyle\lim_{x \to 0} f(x) = 0$

65.

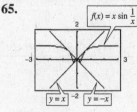

$\displaystyle\lim_{x \to 0} f(x) = 0$

67. (a) Can be evaluated by direct substitution: $\displaystyle\lim_{x \to 0} x^2 \sin x^2 = 0^2 \sin 0^2 = 0$

(b) Cannot be evaluated by direct substitution:

$$\lim_{x \to 0} \dfrac{\sin x^2}{x^2} = 1$$

69. $\lim\limits_{h\to 0} \dfrac{f(x+h)-f(x)}{h} = \lim\limits_{h\to 0} \dfrac{3(x+h)-1-(3x-1)}{h}$

$$= \lim\limits_{h\to 0} \dfrac{3x+3h-1-3x+1}{h}$$

$$= \lim\limits_{h\to 0} \dfrac{3h}{h} = 3$$

71. $\lim\limits_{h\to 0} \dfrac{f(x+h)-f(x)}{h} = \lim\limits_{h\to 0} \dfrac{\sqrt{x+h}-\sqrt{x}}{h} \cdot \left(\dfrac{\sqrt{x+h}+\sqrt{x}}{\sqrt{x+h}+\sqrt{x}}\right)$

$$= \lim\limits_{h\to 0} \dfrac{(x+h)-x}{h\left(\sqrt{x+h}+\sqrt{x}\right)}$$

$$= \lim\limits_{h\to 0} \dfrac{1}{\sqrt{x+h}+\sqrt{x}} = \dfrac{1}{2\sqrt{x}}$$

73. $\lim\limits_{h\to 0} \dfrac{f(x+h)-f(x)}{h} = \lim\limits_{h\to 0} \dfrac{((x+h)^2-3(x+h))-(x^2-3x)}{h}$

$$= \lim\limits_{h\to 0} \dfrac{x^2+2xh+h^2-3x-3h-x^2+3x}{h}$$

$$= \lim\limits_{h\to 0} \dfrac{2xh+h^2-3h}{h}$$

$$= \lim\limits_{h\to 0} (2x+h-3) = 2x-3$$

75. $\lim\limits_{h\to 0} \dfrac{f(x+h)-f(x)}{h} = \lim\limits_{h\to 0} \dfrac{1/(x+h+2)-1/(x+2)}{h}$

$$= \lim\limits_{h\to 0} \dfrac{(x+2)-(x+h+2)}{h(x+h+2)(x+2)}$$

$$= \lim\limits_{h\to 0} \dfrac{-h}{h(x+h+2)(x+2)}$$

$$= \lim\limits_{h\to 0} \dfrac{-1}{(x+h+2)(x+2)}$$

$$= \dfrac{-1}{(x+2)^2}$$

77. $\lim\limits_{t\to 1} \dfrac{(-16(1)+128)-(-16t^2+128)}{1-t} = \lim\limits_{t\to 1} \dfrac{16t^2-16}{1-t}$

$$= \lim\limits_{t\to 1} \dfrac{16(t-1)(t+1)}{1-t}$$

$$= \lim\limits_{t\to 1} -16(t+1)$$

$$= -32\, \dfrac{\text{ft}}{\text{sec}}$$

79. $C(t) = 1.00 - 0.25[\![-(t-1)]\!]$

(a)

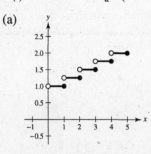

(b)

t	3	3.3	3.4	3.5	3.6	3.7	4
C	1.50	1.75	1.75	1.75	1.75	1.75	1.75

$\displaystyle\lim_{t\to3.5} C(t) = 1.75$

(c)

t	2	2.5	2.9	3	3.1	3.5	4
C	1.25	1.50	1.50	1.50	1.75	1.75	1.75

$\displaystyle\lim_{t\to3} C(t)$ does not exist. The one-sided limits do not agree.

81. Answers will vary. As $t \to 2$ from the left, $f(t) \to 39.00$. As $t \to 2$ from the right, $f(t) \to 46.80$.

83. True

85. Many answers possible

(a)

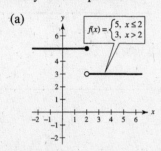

$f(x) = \begin{cases} 5, & x \le 2 \\ 3, & x > 2 \end{cases}$

(b)

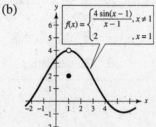

$f(x) = \begin{cases} \dfrac{4\sin(x-1)}{x-1}, & x \ne 1 \\ 2, & x = 1 \end{cases}$

87. Slope of line through $(4, -6)$ and $(3, -4)$:

$$\frac{-6+4}{4-3} = -2$$

Slope of perpendicular line: $\dfrac{1}{2}$

Equation: $y + 10 = \dfrac{1}{2}(x - 6)$

$$2y - x + 26 = 0$$

89. $r = \dfrac{3}{1 + \cos\theta}$, $e = 1$, Parabola

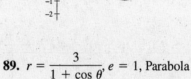

91. $r = \dfrac{9}{2 + 3\cos\theta} = \dfrac{9/2}{1 + (3/2)\cos\theta}$, $e = \dfrac{3}{2}$,

Hyperbola

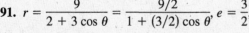

93. $r = \dfrac{5}{1 - \sin\theta}$, $e = 1$, Parabola

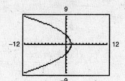

95. $\langle 7, -2, 3\rangle \cdot \langle -1, 4, 5\rangle = -7 - 8 + 15$

$$= 0 \Rightarrow \text{orthogonal}$$

97. $-3\langle -4, 3, -6\rangle = \langle 12, -9, 18\rangle \Rightarrow$ parallel

Section 11.3   The Tangent Line Problem

■ You should be able to visually approximate the slope of a graph.

■ The slope m of the graph of f at the point $(x, f(x))$ is given by

$$m = \lim_{h \to 0} \frac{f(x + h) - f(x)}{h}$$

provided this limit exists.

■ You should be able to use the limit definition to find the slope of a graph.

■ The derivative of f at x is given by

$$f'(x) = \lim_{h \to 0} \frac{f(x + h) - f(x)}{h}$$

provided this limit exists. Notice that this is the same limit as that for the tangent line slope.

■ You should be able to use the limit definition to find the derivative of a function.

Vocabulary Check

1. Calculus   **2.** tangent line   **3.** secant line

4. difference quotient   **5.** derivative

---

1. Slope is 0 at (x, y).   **3.** Slope is $\frac{1}{2}$ at (x, y).

5. $m_{\text{sec}} = \dfrac{g(3 + h) - g(3)}{h} = \dfrac{(3 + h)^2 - 4(3 + h) - (-3)}{h} = \dfrac{h^2 + 2h}{h}$

$\qquad m = \lim\limits_{h \to 0} \dfrac{h^2 + 2h}{h} = \lim\limits_{h \to 0} \dfrac{h(h + 2)}{h} = \lim\limits_{h \to 0} (h + 2) = 2$

7. $m_{\text{sec}} = \dfrac{g(1 + h) - g(1)}{h} = \dfrac{5 - 2(1 + h) - 3}{h} = \dfrac{-2h}{h}$

$\qquad m = \lim\limits_{h \to 0} \dfrac{-2h}{h} = -2$

9. $m_{\text{sec}} = \dfrac{g(2 + h) - g(2)}{h} = \dfrac{[4/(2 + h)] - 2}{h} = \dfrac{4 - 2(2 + h)}{(2 + h)h} = \dfrac{-2}{2 + h},\ h \neq 0$

$\qquad m = \lim\limits_{h \to 0} \left(\dfrac{-2}{2 + h} \right) = -1$

11. $m_{\text{sec}} = \dfrac{h(9 + k) - h(9)}{k} = \dfrac{\sqrt{9 + k} - 3}{k} \cdot \dfrac{\sqrt{9 + k} + 3}{\sqrt{9 + k} + 3} = \dfrac{(9 + k) - 9}{k[\sqrt{9 + k} + 3]} = \dfrac{1}{\sqrt{9 + k} + 3},\ k \neq 0$

$\qquad m = \lim\limits_{k \to 0} \dfrac{1}{\sqrt{9 + k} + 3} = \dfrac{1}{6}$

13. $m_{\text{sec}} = \dfrac{g(x+h) - g(x)}{h} = \dfrac{4 - (x+h)^2 - (4 - x^2)}{h} = \dfrac{-2xh - h^2}{h} = -2x - h, \ h \neq 0$

$m = \lim\limits_{h \to 0}(-2x - h) = -2x$

(a) At $(0, 4)$, $m = -2(0) = 0$. (b) At $(-1, 3)$, $m = -2(-1) = 2$.

15. $m_{\text{sec}} = \dfrac{g(x+h) - g(x)}{h} = \dfrac{\dfrac{1}{x+h+4} - \dfrac{1}{x+4}}{h} = \dfrac{(x+4) - (x+4+h)}{(x+h+4)(x+4)(h)}$

$\quad = \dfrac{-h}{(x+h+4)(x+4)h} = \dfrac{-1}{(x+h+4)(x+4)}, \ h \neq 0$

$m = \lim\limits_{h \to 0} \dfrac{-1}{(x+h+4)(x+4)} = \dfrac{-1}{(x+4)^2}$

(a) At $\left(0, \dfrac{1}{4}\right)$, $m = \dfrac{-1}{(0+4)^2} = \dfrac{-1}{16}$. (b) At $\left(-2, \dfrac{1}{2}\right)$, $m = \dfrac{-1}{(-2+4)^2} = \dfrac{-1}{4}$.

17. $m_{\text{sec}} = \dfrac{g(x+h) - g(x)}{h} = \dfrac{\sqrt{x+h-1} - \sqrt{x-1}}{h} \cdot \dfrac{\sqrt{x+h+1} + \sqrt{x-1}}{\sqrt{x+h-1} + \sqrt{x-1}}$

$\quad = \dfrac{(x+h-1) - (x-1)}{h\left(\sqrt{x+h-1} + \sqrt{x-1}\right)} = \dfrac{1}{\sqrt{x+h-1} + \sqrt{x-1}}, \ h \neq 0$

$m = \lim\limits_{h \to 0}\left(\dfrac{1}{\sqrt{x+h-1} + \sqrt{x-1}}\right) = \dfrac{1}{2\sqrt{x-1}}$

(a) At $(5, 2)$, $m = \dfrac{1}{2\sqrt{5-1}} = \dfrac{1}{4}$. (b) At $(10, 3)$, $m = \dfrac{1}{2\sqrt{10-1}} = \dfrac{1}{6}$.

19.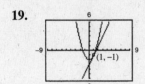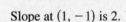

Slope at $(1, -1)$ is 2.

21.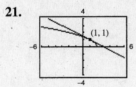

Slope at $(1, 1)$ is $-\dfrac{1}{2}$.

23.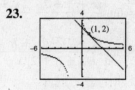

Slope at $(1, 2)$ is -1.

25. $f'(x) = \lim\limits_{h \to 0} \dfrac{f(x+h) - f(x)}{h} = \lim\limits_{h \to 0} \dfrac{5 - 5}{h} = 0$

27. $g'(x) = \lim\limits_{h \to 0} \dfrac{g(x+h) - g(x)}{h} = \lim\limits_{h \to 0} \dfrac{\left[9 - \frac{1}{3}(x+h)\right] - \left[9 - \frac{1}{3}x\right]}{h} = \lim\limits_{h \to 0} \dfrac{-\frac{1}{3}h}{h} = -\dfrac{1}{3}$

29. $f'(x) = \lim\limits_{h \to 0} \dfrac{f(x+h) - f(x)}{h} = \lim\limits_{h \to 0} \dfrac{\left[4 - 3(x+h)^2\right] - (4 - 3x^2)}{h}$

$\quad = \lim\limits_{h \to 0} \dfrac{-3(x^2 + 2xh + h^2) + 3x^2}{h} = \lim\limits_{h \to 0} \dfrac{-6xh - 3h^2}{h} = \lim\limits_{h \to 0}(-6x - 3h) = -6x$

31. $f'(x) = \lim\limits_{h \to 0} \dfrac{f(x + h) - f(x)}{h}$

$= \lim\limits_{h \to 0} \dfrac{\dfrac{1}{(x + h)^2} - \dfrac{1}{x^2}}{h}$

$= \lim\limits_{h \to 0} \dfrac{x^2 - (x^2 + 2xh + h^2)}{(x + h)^2 x^2 h}$

$= \lim\limits_{h \to 0} \dfrac{-2x - h}{(x + h)^2 x^2} = -\dfrac{2x}{x^4} = -\dfrac{2}{x^3}$

33. $f'(x) = \lim\limits_{h \to 0} \dfrac{f(x + h) - f(x)}{h}$

$= \lim\limits_{h \to 0} \dfrac{\sqrt{x + h - 4} - \sqrt{x - 4}}{h} \cdot \dfrac{\sqrt{x + h - 4} + \sqrt{x - 4}}{\sqrt{x + h - 4} + \sqrt{x - 4}}$

$= \lim\limits_{h \to 0} \dfrac{(x + h - 4) - (x - 4)}{h\left[\sqrt{x + h - 4} + \sqrt{x - 4}\right]}$

$= \lim\limits_{h \to 0} \dfrac{1}{\sqrt{x + h - 4} + \sqrt{x - 4}}$

$= \dfrac{1}{2\sqrt{x - 4}}$

35. $f'(x) = \lim\limits_{h \to 0} \dfrac{f(x + h) - f(x)}{h}$

$= \lim\limits_{h \to 0} \dfrac{\dfrac{1}{(x + h + 2)} - \dfrac{1}{x + 2}}{h}$

$= \lim\limits_{h \to 0} \dfrac{(x + 2) - (x + h + 2)}{h(x + h + 2)(x + 2)}$

$= \lim\limits_{h \to 0} \dfrac{-1}{(x + h + 2)(x + 2)}$

$= \dfrac{-1}{(x + 2)^2}$

37. $f'(x) = \lim\limits_{h \to 0} \dfrac{f(x + h) - f(x)}{h} = \lim\limits_{h \to 0} \dfrac{\dfrac{1}{\sqrt{x + h - 9}} - \dfrac{1}{\sqrt{x - 9}}}{h} \cdot \dfrac{\dfrac{1}{\sqrt{x + h - 9}} + \dfrac{1}{\sqrt{x - 9}}}{\dfrac{1}{\sqrt{x + h - 9}} + \dfrac{1}{\sqrt{x - 9}}}$

$= \lim\limits_{h \to 0} \dfrac{\dfrac{1}{(x + h - 9)} - \dfrac{1}{(x - 9)}}{h\left[\dfrac{1}{\sqrt{x + h - 9}} + \dfrac{1}{\sqrt{x - 9}}\right]} = \lim\limits_{h \to 0} \dfrac{(x - 9) - (x + h - 9)}{h(x + h - 9)(x - 9)\left[\dfrac{1}{\sqrt{x + h - 9}} + \dfrac{1}{\sqrt{x - 9}}\right]}$

$= \lim\limits_{h \to 0} \dfrac{-1}{(x + h - 9)(x - 9)\left[\dfrac{1}{\sqrt{x + h - 9}} + \dfrac{1}{\sqrt{x - 9}}\right]} = \dfrac{-1}{(x - 9)^2\left[\dfrac{2}{\sqrt{x - 9}}\right]} = \dfrac{-1}{2(x - 9)^{3/2}}$

39. (a) $m_{sec} = \dfrac{f(2 + h) - f(2)}{h}$

$= \dfrac{(2 + h)^2 - 1 - 3}{h}$

$= \dfrac{4 + 4h + h^2 - 4}{h}$

$= 4 + h, \ h \neq 0$

$m = \lim\limits_{h \to 0} (4 + h) = 4$

(b) $y - 3 = 4(x - 2)$

$y = 4x - 5$

(c)

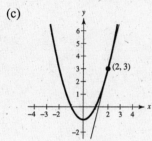

41. (a) $m_{sec} = \dfrac{f(1 + h) - f(1)}{h}$

$= \dfrac{(1 + h)^3 - 2(1 + h) - (-1)}{h}$

$= \dfrac{1 + 3h + 3h^2 + h^3 - 2 - 2h + 1}{h}$

$= \dfrac{h^3 + 3h^2 + h}{h} = h^2 + 3h + 1, \ h \neq 0$

$m = \lim\limits_{h \to 0} (h^2 + 3h + 1) = 1$

(b) $y - (-1) = 1(x - 1)$

$y = x - 2$

(c)

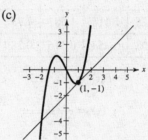

43. (a) $m_{sec} = \dfrac{f(3 + h) - f(3)}{h} = \dfrac{\sqrt{3 + h + 1} - 2}{h}$

$= \dfrac{\sqrt{4 + h} - 2}{h} \cdot \dfrac{\sqrt{4 + h} + 2}{\sqrt{4 + h} + 2}$

$= \dfrac{4 + h - 4}{h\left[\sqrt{4 + h} + 2\right]}$

$= \dfrac{1}{\sqrt{4 + h} + 2}$

$m = \lim\limits_{h \to 0} \dfrac{1}{\sqrt{4 + h} + 2} = \dfrac{1}{4}$

(b) $y - 2 = \dfrac{1}{4}(x - 3)$

$y = \dfrac{1}{4}x + \dfrac{5}{4}$

(c)

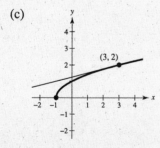

45. (a) $m_{sec} = \dfrac{f(-4 + h) - f(-4)}{h}$

$= \dfrac{\dfrac{1}{-4 + h + 5} - 1}{h}$

$= \dfrac{1 + 4 - h - 5}{h(-4 + h + 5)} = \dfrac{-1}{h + 1}, \ h \neq 0$

$m = \lim\limits_{h \to 0} \dfrac{-1}{h + 1} = -1$

(b) $y - 1 = -1(x + 4)$

$y = -x - 3$

(c)

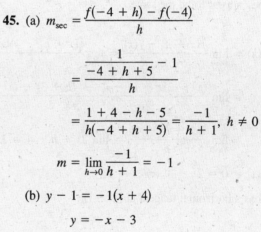

47.

x	−2	−1.5	−1	−0.5	0	0.5	1	1.5	2
f(x)	2	1.125	0.5	0.125	0	0.125	0.5	1.125	2
f'(x)	−2	−1.5	−1	−0.5	0	0.5	1	1.5	2

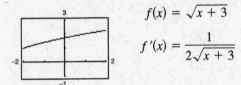

$$f(x) = \tfrac{1}{2}x^2$$
$$f'(x) = x$$

They appear to be the same.

49.

x	−2	−1.5	−1	−0.5	0	0.5	1	1.5	2
f(x)	1	1.225	1.414	1.581	1.732	1.871	2	2.121	2.236
f'(x)	0.5	0.408	0.354	0.316	0.289	0.267	0.25	0.236	0.224

$$f(x) = \sqrt{x + 3}$$
$$f'(x) = \frac{1}{2\sqrt{x + 3}}$$

They appear to be the same.

51. $f'(x) = \lim\limits_{h \to 0} \dfrac{f(x + h) - f(x)}{h} = \lim\limits_{h \to 0} \dfrac{[(x + h)^2 - 4(x + h) + 3] - [x^2 - 4x + 3]}{h}$

$= \lim\limits_{h \to 0} \dfrac{(x^2 + 2xh + h^2 - 4x - 4h + 3) - (x^2 - 4x + 3)}{h}$

$= \lim\limits_{h \to 0} \dfrac{2xh + h^2 - 4h}{h} = \lim\limits_{h \to 0} 2x + h - 4 = 2x - 4$

$f'(x) = 0 = 2x - 4 \implies x = 2$

f has a horizontal tangent at $(2, -1)$.

53. $f'(x) = \lim\limits_{h \to 0} \dfrac{f(x + h) - f(x)}{h} = \lim\limits_{h \to 0} \dfrac{[3(x + h)^3 - 9(x + h)] - [3x^3 - 9x]}{h}$

$= \lim\limits_{h \to 0} \dfrac{3(x^3 + 3x^2h + 3xh^2 + h^3) - 9x - 9h - 3x^3 + 9x}{h}$

$= \lim\limits_{h \to 0} \dfrac{9x^2h + 9xh^2 + 3h^3 - 9h}{h}$

$= \lim\limits_{h \to 0} (9x^2 + 9xh + 3h - 9) = 9x^2 - 9$

$f'(x) = 0 = 9x^2 - 9 = 9(x + 1)(x - 1) \implies x = \pm1$

Horizontal tangents at $(1, -6)$ and $(-1, 6)$.

55.
$$f'(x) = 4x^3 - 4x = 0$$
$$4x(x - 1)(x + 1) = 0$$
$$x = 0, 1, -1$$
$$(0, 0), (1, -1), (-1, -1)$$

57. $f'(x) = -2 \sin x + 1 = 0$
$$\sin x = \frac{1}{2}$$
$$x = \frac{\pi}{6}, \frac{5\pi}{6}$$
$$\left(\frac{\pi}{6}, \sqrt{3} + \frac{\pi}{6} \right), \left(\frac{5\pi}{6}, \frac{5\pi}{6} - \sqrt{3} \right)$$

59. $f'(x) = x^2 e^x + 2xe^x = 0$
$$xe^x(x + 2) = 0$$
$$x = 0, -2$$
$$(0, 0), (-2, 4e^{-2})$$

61. $f'(x) = \ln x + 1 = 0$
$$\ln x = -1$$
$$x = e^{-1}$$
$$(e^{-1}, -e^{-1})$$

63. (a) $P(t) = -0.63t^2 + 63.3t + 8448$

   (c) Using the limit definition,
$$P'(t) = -1.26t + 63.3$$
$$P'(20) = -1.26(20) + 63.3 = 38.1.$$

   (d) Answers will vary.

(b)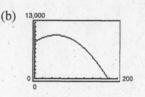

$P'(20) \approx 38$

At time 2020, the population is increasing at approximately 38,000 per year.

65. (a) $V = \frac{4}{3}\pi r^3$

$$V'(r) = \lim_{h \to 0} \frac{V(r + h) - V(r)}{h}$$
$$= \lim_{h \to 0} \frac{(4/3)\pi(r + h)^3 - (4/3)\pi r^3}{h}$$
$$= \lim_{h \to 0} \left(\frac{4}{3}\pi \right) \frac{r^3 + 3r^2h + 3rh^2 + h^3 - r^3}{h}$$
$$= \lim_{h \to 0} \frac{4}{3}\pi(3r^2 + 3rh + h^2) = 4\pi r^2$$

   (b) $V'(4) = 4\pi(4)^2 \approx 201.06$

   (c) Cubic inches per inch; Answers will vary.

67. $s(t) = -16t^2 + 64t + 80$

   (a) Using the limit definition, $s'(t) = -32t + 64.$

   (b) $s(0) = 80, s(3) = 128$

$$\text{Average rate of change} = \frac{128 - 80}{3}$$
$$= \frac{48}{3} = 16 \text{ ft/sec}$$

   (c) $s'(t) = -32t + 64 = 0 \implies t = 2$ seconds

     Answers will vary.

   (d) $s(t) = -16t^2 + 64t + 80$
$$= 0 \implies t = 5 \text{ seconds}$$
$$s'(5) = -32(5) + 64 = -96 \text{ ft/sec}$$

   (e)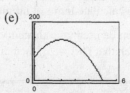

69. True. The slope is $2x$, which is different for all x.

71. Matches (b).

   (Derivative is always positive, but decreasing.)

73. Matches (d).

(Derivative is -1 for $x < 0$, 1 for $x > 0$.)

75. Answers will vary.

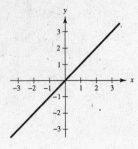

77. $f(x) = \dfrac{1}{x^2 - x - 2} = \dfrac{1}{(x + 1)(x - 2)}$

Intercept: $\left(0, -\dfrac{1}{2}\right)$

Vertical asymptotes: $x = -1, x = 2$

Horizontal asymptote: $y = 0$

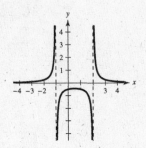

79. $f(x) = \dfrac{x^2 - x - 2}{x - 2} = \dfrac{(x - 2)(x + 1)}{x - 2} = x + 1, \ x \neq 2$

Line with hole at $(2, 3)$

Intercepts: $(0, 1), (-1, 0)$

Slant asymptote: $y = x + 1$

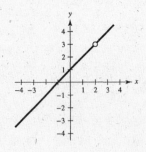

81. $\langle 1, 1, 1 \rangle \times \langle 2, 1, -1 \rangle = \begin{vmatrix} \mathbf{i} & \mathbf{j} & \mathbf{k} \\ 1 & 1 & 1 \\ 2 & 1 & -1 \end{vmatrix}$

$\qquad\qquad = \langle -2, 3, -1 \rangle$

83. $\langle -4, 10, 0 \rangle \times \langle 4, -1, 0 \rangle = \begin{vmatrix} \mathbf{i} & \mathbf{j} & \mathbf{k} \\ -4 & 10 & 0 \\ 4 & -1 & 0 \end{vmatrix}$

$\qquad\qquad = \langle 0, 0, -36 \rangle$

Section 11.4 Limits at Infinity and Limits of Sequences

- The limit at infinity $\lim\limits_{x \to \infty} f(x) = L$ means that $f(x)$ get arbitrarily close to L as x increases without bound.

- Similarly, the limit at infinity $\lim\limits_{x \to -\infty} f(x) = L$ means that $f(x)$ get arbitrarily close to L as x decreases without bound.

- You should be able to calculate limits at infinity, especially those arising from rational functions.

- Limits of functions can be used to evaluate limits of sequences. If f is a function such that $\lim\limits_{x \to \infty} f(x) = L$ and if a_n is a sequence such that $f(n) = a_n$, then $\lim\limits_{n \to \infty} a_n = L$.

Vocabulary Check

1. limit, infinity

2. converge

3. diverge

1. Intercept: $(0, 0)$

Horizontal asymptote: $y = 4$

Matches (c).

3. Horizontal asymptote: $y = 4$

Vertical asymptote: $x = 0$

Matches (d).

5. Vertical asymptotes: $x = \pm 1$

Horizontal asymptote: $y = 1$

Matches (f).

7. Vertical asymptote: $x = 2$

Horizontal asymptote: $y = -2$

Matches (h).

9. $\lim\limits_{x \to \infty} \dfrac{3}{x^2} = 0$

11. $\lim\limits_{x \to \infty} \dfrac{3 + x}{3 - x} = -1$

13. $\lim\limits_{x \to -\infty} \dfrac{5x - 2}{6x + 1} = \dfrac{5}{6}$

15. $\lim\limits_{x \to -\infty} \dfrac{4x^2 - 3}{2 - x^2} = \dfrac{4}{-1} = -4$

17. $\lim\limits_{t \to \infty} \dfrac{t^2}{t + 3}$ does not exist.

19. $\lim\limits_{t \to \infty} \dfrac{4t^2 + 3t - 1}{3t^2 + 2t - 5} = \dfrac{4}{3}$

21. $\lim\limits_{y \to -\infty} \dfrac{3 + 8y - 4y^2}{3 - y - 2y^2} = \dfrac{-4}{-2} = 2$

23. $\lim\limits_{x \to -\infty} \dfrac{-(x^2 + 3)}{(2 - x)^2} = \lim\limits_{x \to -\infty} \dfrac{-x^2 - 3}{x^2 - 4x + 4} = -1$

25. $\lim\limits_{x \to -\infty} \left[\dfrac{x}{(x + 1)^2} - 4 \right] = 0 - 4 = -4$

27. $\lim\limits_{t \to \infty} \left(\dfrac{1}{3t^2} - \dfrac{5t}{t + 2} \right) = 0 - 5 = -5$

29. $y = \dfrac{3x}{1 - x}$

Horizontal asymptote: $y = -3$

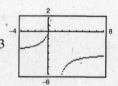

31. $y = \dfrac{2x}{1 - x^2}$

Horizontal asymptote: $y = 0$

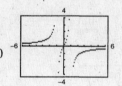

33. $y = 1 - \dfrac{3}{x^2}$

Horizontal asymptote: $y = 1$

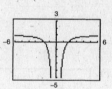

35. $f(x) = x - \sqrt{x^2 + 2}$

(a)

x	10^0	10^1	10^2	10^3	10^4	10^5	10^6
$f(x)$	-0.7321	-0.0995	-0.0100	-0.0010	-1.0×10^{-4}	-1.0×10^{-5}	-1.0×10^{-6}

$\lim\limits_{x \to \infty} f(x) = 0$

(b)

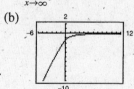

$\lim\limits_{x \to \infty} f(x) = 0$

37. $f(x) = 3\left(2x - \sqrt{4x^2 + x}\right)$

(a)

x	10^0	10^1	10^2	10^3	10^4	10^5	10^6
$f(x)$	-0.7082	-0.7454	-0.7495	-0.74995	-0.749995	-0.7499995	-0.75

$$\lim_{x \to \infty} f(x) = -0.75$$

(b)

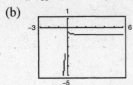

$$\lim_{x \to \infty} f(x) = -0.75$$

39. $a_n = \dfrac{n + 1}{n^2 + 1}$

$1, \dfrac{3}{5}, \dfrac{2}{5}, \dfrac{5}{17}, \dfrac{3}{13}$

$$\lim_{n \to \infty} a_n = 0$$

41. $a_n = \dfrac{n}{2n + 1}$

$\dfrac{1}{3}, \dfrac{2}{5}, \dfrac{3}{7}, \dfrac{4}{9}, \dfrac{5}{11}$

$$\lim_{n \to \infty} a_n = \dfrac{1}{2}$$

43. $a_n = \dfrac{n^2}{3n + 2}$

$\dfrac{1}{5}, \dfrac{1}{2}, \dfrac{9}{11}, \dfrac{8}{7}, \dfrac{25}{17}$

$$\lim_{n \to \infty} \dfrac{n^2}{3n + 2} \text{ does not exist.}$$

45. $a_n = \dfrac{(n + 1)!}{n!}$

$2, 3, 4, 5, 6$

$$\lim_{n \to \infty} \dfrac{(n + 1)!}{n!} = \lim_{n \to \infty} (n + 1) \text{ does not exist.}$$

47. $a_n = \dfrac{(-1)^n}{n}$

$-1, \dfrac{1}{2}, -\dfrac{1}{3}, \dfrac{1}{4}, -\dfrac{1}{5}$

$$\lim_{n \to \infty} \dfrac{(-1)^n}{n} = 0$$

49.

n	10^0	10^1	10^2	10^3	10^4	10^5	10^6
a_n	2	1.55	1.505	1.5005	1.5001	1.500	1.500

$$\lim_{n \to \infty} a_n = 1.5$$

$$a_n = \frac{1}{n}\left(n + \frac{1}{n}\left[\frac{n(n + 1)}{2}\right]\right) = 1 + \frac{1}{n^2}\left[\frac{n^2 + n}{2}\right] = 1 + \frac{n^2 + n}{2n^2}$$

$$\lim_{n \to \infty} a_n = 1 + \frac{1}{2} = \frac{3}{2}$$

51.

n	10^0	10^1	10^2	10^3	10^4	10^5	10^6
a_n	16	6.16	5.4136	5.3413	5.3341	5.3334	5.3333

$$\lim_{n \to \infty} a_n = 5.33$$

$$a_n = \frac{16}{n^3}\left[\frac{n(n + 1)(2n + 1)}{6}\right] = \left(\frac{8}{3}\right)\frac{n(n + 1)(2n + 1)}{n^3}$$

$$\lim_{n \to \infty} a_n = \frac{16}{3}$$

53. (a) Average cost $= \overline{C} = \dfrac{C}{x} = 13.50 + \dfrac{45{,}750}{x}$

(b) $\overline{C}(100) = \$471$

$\overline{C}(1000) = \$59.25$

(c) $\lim\limits_{x \to \infty} C(x) = 13.50$

As more units are produced, the fixed costs (45,750) become less dominant.

55. (a)

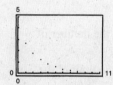

(b) For 2004, $t = 14$ and $E(14) \approx 72.0$ million.

For 2008, $t = 18$ and $E(18) \approx 73.8$ million.

(c) $\lim\limits_{t \to \infty} E(t) = \dfrac{0.702}{0.009} = 78$

The enrollment approaches 78 million.

(d) Answers will vary.

57. False. $f(x) = \dfrac{x^2 + 1}{1}$ does not have a horizontal asymptote.

59. True

61. For example, let $f(x) = \dfrac{1}{x^2}$ and $g(x) = \dfrac{1}{x^2}$.

Then, $\lim\limits_{x \to 0} \dfrac{1}{x^2}$ increases without bound, but $\lim\limits_{x \to 0} \left[f(x) - g(x) \right] = 0$.

63. Converges to 0

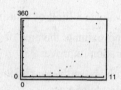

65. Diverges

67. $y = x^4$

(a) $f(x) = (x + 3)^4$

(b) $f(x) = x^4 - 1$

(c) $f(x) = -2 + x^4$

(d) $f(x) = \frac{1}{2}(x - 4)^4$

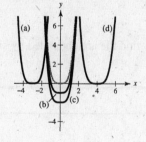

69.

$$
\begin{array}{r}
x^2 + 2x + 1 \\
x^2 - 4 \overline{)\, x^4 + 2x^3 - 3x^2 - 8x - 4} \\
\underline{x^4 \qquad\;\; - 4x^2} \\
2x^3 + \;\; x^2 \\
\underline{2x^3 \qquad\;\; - 8x} \\
x^2 \qquad\;\; - 4
\end{array}
$$

$x^4 + 2x^3 - 3x^2 - 8x - 4 = (x^2 - 4)(x^2 + 2x + 1)$

71.

$$
\begin{array}{r}
x^3 + 5x^2 \qquad\;\; - 3 \\
3x + 2 \overline{)\, 3x^4 + 17x^3 + 10x^2 - 9x - 8} \\
\underline{3x^4 + \;\; 2x^3} \\
15x^3 + 10x^2 \\
\underline{15x^3 + 10x^2} \\
-9x - 8 \\
\underline{-9x - 6} \\
-2
\end{array}
$$

$\dfrac{3x^4 + 17x^3 + 10x^2 - 9x - 8}{3x + 2} = x^3 + 5x^2 - 3 + \dfrac{-2}{3x + 2}$

73. $f(x) = x^4 - x^3 - 20x^2$

$\quad = x^2(x^2 - x - 20)$

$\quad = x^2(x - 5)(x + 4)$

Real zeros: $0, 0, 5, -4$

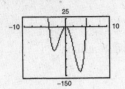

75. $f(x) = x^3 - 3x^2 + 2x - 6$

$\quad = x^2(x - 3) + 2(x - 3)$

$\quad = (x - 3)(x^2 + 2)$

Real zero: 3

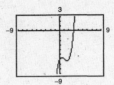

77. $\displaystyle\sum_{i=1}^{6} (2i + 3) = 5 + 7 + 9 + 11 + 13 + 15 = 60$

79. $\displaystyle\sum_{k=1}^{10} 15 = 10(15) = 150$

Section 11.5   The Area Problem

■ You should know the following summation formulas and properties.

(a) $\displaystyle\sum_{i=1}^{n} c = cn$

(b) $\displaystyle\sum_{i=1}^{n} i = \frac{n(n + 1)}{2}$

(c) $\displaystyle\sum_{i=1}^{n} i^2 = \frac{n(n + 1)(2n + 1)}{6}$

(d) $\displaystyle\sum_{i=1}^{n} i^3 = \frac{n^2(n + 1)^2}{4}$

(e) $\displaystyle\sum_{i=1}^{n} (a_i \pm b_i) = \sum_{i=1}^{n} a_i \pm \sum_{i=1}^{n} b_i$

(f) $\displaystyle\sum_{i=1}^{n} ka_i = k\sum_{i=1}^{n} a_i$

■ You should be able to evaluate a limit of a summation, $\displaystyle\lim_{n\to\infty} S(n)$.

■ You should be able to approximate the area of a region using rectangles. By increasing the number of rectangles, the approximation improves.

■ The area of a plane region above the x-axis bounded by f between $x = a$ and $x = b$ is the limit of the sum of the approximating rectangles:

$$A = \lim_{n\to\infty} \sum_{i=1}^{n} f\left(a + \frac{(b - a)i}{n}\right)\left(\frac{b - a}{n}\right)$$

■ You should be able to use the limit definition of area to find the area bounded by simple functions in the plane.

Vocabulary Check

1. $\dfrac{n(n + 1)}{2}$

2. $\dfrac{n^2(n + 1)^2}{4}$

3. area

1. $\displaystyle\sum_{i=1}^{60} 7 = 7(60) = 420$

3. $\displaystyle\sum_{i=1}^{20} i^3 = \frac{20^2(21)^2}{4} = 44{,}100$

5. $\displaystyle\sum_{k=1}^{20} (k^3 + 2) = \frac{20^2(21)^2}{4} + 2(20)$

$\quad = 44{,}100 + 40$

$\quad = 44{,}140$

7. $\displaystyle\sum_{j=1}^{25} (j^2 + j) = \frac{25(26)(51)}{6} + \frac{25(26)}{2} = 5850$

9. (a) $S(n) = \sum_{i=1}^{n} \dfrac{i^3}{n^4} = \dfrac{1}{n^4}\left[\dfrac{n^2(n+1)^2}{4}\right] = \dfrac{n^2 + 2n + 1}{4n^2}$

(b)

n	10^0	10^1	10^2	10^3	10^4
$S(n)$	1	0.3025	0.255025	0.25050025	0.25005

(c) $\lim\limits_{n\to\infty} S(n) = \dfrac{1}{4}$

11. (a) $S(n) = \sum_{i=1}^{n} \dfrac{3}{n^3}(1 + i^2) = \dfrac{3}{n^3}\left[n + \dfrac{n(n+1)(2n+1)}{6}\right] = \dfrac{3}{n^2} + \dfrac{6n^2 + 9n + 3}{6n^2} = \dfrac{2n^2 + 3n + 7}{2n^2}$

(b)

n	10^0	10^1	10^2	10^3	10^4
$S(n)$	6	1.185	1.0154	1.0015	1.00015

(c) $\lim\limits_{n\to\infty} S(n) = 1$

13. (a) $S(n) = \sum_{i=1}^{n} \left(\dfrac{i^2}{n^3} + \dfrac{2}{n}\right)\left(\dfrac{1}{n}\right) = \dfrac{1}{n}\left[\dfrac{n(n+1)(2n+1)}{6n^3} + \dfrac{2n}{n}\right] = \dfrac{1}{6n^3}(2n^2 + 3n + 1) + \dfrac{2}{n} = \dfrac{14n^2 + 3n + 1}{6n^3}$

(b)

n	10^0	10^1	10^2	10^3	10^4
$S(n)$	3	0.2385	0.02338	0.00233	0.0002333

(c) $\lim\limits_{n\to\infty} S(n) = 0$

15. (a) $S(n) = \sum_{i=1}^{n} \left[1 - \left(\dfrac{i}{n}\right)^2\right]\left(\dfrac{1}{n}\right) = \dfrac{1}{n}\left[n - \dfrac{1}{n^2}\left(\dfrac{n(n+1)(2n+1)}{6}\right)\right] = 1 - \dfrac{2n^2 + 3n + 1}{6n^2} = \dfrac{4n^2 - 3n - 1}{6n^2}$

(b)

n	10^0	10^1	10^2	10^3	10^4
$S(n)$	0	0.615	0.66165	0.66617	0.666617

(c) $\lim\limits_{n\to\infty} S(n) = \dfrac{2}{3}$

17. $f(x) = x + 4, [-1, 2], n = 6,$ width $= \frac{1}{2}$

Area $\approx \frac{1}{2}[3.5 + 4 + 4.5 + 5 + 5.5 + 6] = 14.25$ square units

19. The width of each rectangle is $\frac{1}{4}$. The height is obtained by evaluating f at the right-hand endpoint of each interval.

$A \approx \sum_{i=1}^{8} f\left(\dfrac{i}{4}\right)\left(\dfrac{1}{4}\right) = \sum_{i=1}^{8} \dfrac{1}{4}\left(\dfrac{i}{4}\right)^3\left(\dfrac{1}{4}\right) = 1.265625$ square units

21. Width of each rectangle is $12/n$. The height is

$f\left(\dfrac{12}{n}i\right) = -\dfrac{1}{3}\left(\dfrac{12}{n}i\right) + 4.$

$A = \sum_{i=1}^{n} \left[-\dfrac{1}{3}\left(\dfrac{12i}{n}\right) + 4\right]\left(\dfrac{12}{n}\right)$

n	4	8	20	50
Approximate area	18	21	22.8	23.52

Note: Exact area is 24.

23. The width of each rectangle is $3/n$. The height is

$\dfrac{1}{9}\left(\dfrac{3i}{n}\right)^3.$

$A \approx \sum_{i=1}^{n} \dfrac{1}{9}\left(\dfrac{3i}{n}\right)^3\left(\dfrac{3}{n}\right)$

n	4	8	20	50
Approximate area	3.52	2.85	2.48	2.34

25. $f(x) = 2x + 5$, $[0, 4]$

The width of each rectangle is $4/n$. The height is

$$f\left(\frac{4i}{n}\right) = 2\left(\frac{4i}{n}\right) + 5 = \frac{8i}{n} + 5.$$

$$A \approx \sum_{i=1}^{n}\left(\frac{8i}{n} + 5\right)\left(\frac{4}{n}\right)$$

n	4	8	20	50	100	∞
Area	40	38	36.8	36.32	36.16	36

$$A \approx \sum_{i=1}^{n}\left(\frac{8i}{n} + 5\right)\left(\frac{4}{n}\right) = \sum_{i=1}^{n}\left(\frac{20}{n} + \frac{32}{n^2}i\right) \approx \frac{20}{n}(n) + \frac{32}{n^2}\left(\frac{n(n + 1)}{2}\right) = 20 + 16\left(\frac{n^2 + n}{n^2}\right)$$

$$A = \lim_{n\to\infty}\left[20 + 16\left(\frac{n^2 + n}{n^2}\right)\right] = 20 + 16 = 36$$

27. $f(x) = 16 - 2x$, $[1, 5]$

The width of each rectangle is $4/n$. The height is

$$f\left(1 + \frac{4i}{n}\right) = 16 - 2\left(1 + \frac{4i}{n}\right) = 14 - \frac{8i}{n}.$$

$$A \approx \sum_{i=1}^{n}\left(14 - \frac{8i}{n}\right)\frac{4}{n}$$

n	4	8	20	50	100	∞
Area	36	38	39.2	39.68	39.84	40

$$A \approx \sum_{i=1}^{n}\left(\frac{56}{n} - \frac{32i}{n^2}\right) = \frac{56}{n}(n) - \frac{32}{n^2}\left(\frac{n(n + 1)}{2}\right) = 56 - 16\left(\frac{n(n + 1)}{n^2}\right)$$

$$A = \lim_{n\to\infty}\left[56 - 16\left(\frac{n(n + 1)}{n^2}\right)\right] = 56 - 16 = 40$$

29. $f(x) = 9 - x^2$, $[0, 2]$

The width of each rectangle is $2/n$. The height is

$$f\left(\frac{2i}{n}\right) = 9 - \left(\frac{2i}{n}\right)^2 = 9 - \frac{4i^2}{n^2}.$$

$$A \approx \sum_{i=1}^{n}\left(9 - \frac{4i^2}{n^2}\right)\left(\frac{2}{n}\right)$$

n	4	8	20	50	100	∞
Area	14.25	14.8125	15.13	15.2528	15.2932	$\frac{46}{3}$

$$A \approx \sum_{i=1}^{n}\left(\frac{18}{n} - \frac{8i^2}{n^3}\right) = \left(\frac{18}{n}\right)n - \frac{8}{n^3}\left[\frac{n(n + 1)(2n + 1)}{6}\right] = 18 - \frac{4}{3}\left[\frac{n(n + 1)(2n + 1)}{n^3}\right]$$

$$A = \lim_{n\to\infty}\left[18 - \frac{4}{3}\left(\frac{n(n + 1)(2n + 1)}{n^3}\right)\right] = 18 - \frac{8}{3} = \frac{46}{3}$$

31. $f(x) = \dfrac{1}{2}x + 4, \ [-1, 3]$

The width of each rectangle is $4/n$. The height is

$$f\left(-1 + \frac{4i}{n}\right) = \frac{1}{2}\left(-1 + \frac{4i}{n}\right) + 4 = \frac{7}{2} + \frac{2i}{n}.$$

$$A \approx \sum_{i=1}^{n}\left(\frac{7}{2} + \frac{2i}{n}\right)\left(\frac{4}{n}\right)$$

$$A \approx \sum_{i=1}^{n}\left(\frac{14}{n} + \frac{8i}{n^2}\right) = \left(\frac{14}{n}\right)n + \frac{8}{n^2}\left(\frac{n(n+1)}{2}\right)$$

$$A = \lim_{n\to\infty}\left[14 + \frac{4}{n^2}\left(\frac{n(n+1)}{1}\right)\right] = 14 + 4 = 18$$

n	4	8	20	50	100	∞
Area	19	18.5	18.2	18.08	18.04	18

33. $A \approx \displaystyle\sum_{i=1}^{n} f\left(\frac{i}{n}\right)\left(\frac{1}{n}\right)$

$$= \sum_{i=1}^{n}\left[4\left(\frac{i}{n}\right) + 1\right]\left(\frac{1}{n}\right)$$

$$= \frac{1}{n}\sum_{i=1}^{n}\left[\frac{4}{n}i + 1\right]$$

$$= \frac{1}{n}\left[\frac{4}{n}\frac{n(n+1)}{2} + n\right]$$

$$= \frac{1}{n}[2(n+1) + n]$$

$$= \frac{3n+2}{n}$$

$$A = \lim_{n\to\infty}\frac{3n+2}{n} = 3 \text{ square units}$$

35. $A \approx \displaystyle\sum_{i=1}^{n} f\left(\frac{i}{n}\right)\left(\frac{1}{n}\right)$

$$= \sum_{i=1}^{n}\left[-2\left(\frac{i}{n}\right) + 3\right]\left(\frac{1}{n}\right)$$

$$= \frac{1}{n}\sum_{i=1}^{n}\left[-\frac{2i}{n} + 3\right]$$

$$= \frac{1}{n}\left[-\frac{2}{n}\frac{n(n+1)}{2} + 3n\right]$$

$$= \frac{1}{n}[2n - 1]$$

$$A = \lim_{n\to\infty}\frac{2n-1}{n} = 2 \text{ square units}$$

37. $A \approx \displaystyle\sum_{i=1}^{n} f\left(-1 + \frac{2i}{n}\right)\left(\frac{2}{n}\right)$

$$= \sum_{i=1}^{n}\left[2 - \left(-1 + \frac{2i}{n}\right)^2\right]\frac{2}{n}$$

$$= \sum_{i=1}^{n}\left[2 - 1 + \frac{4i}{n} - \frac{4i^2}{n^2}\right]\left(\frac{2}{n}\right)$$

$$= \frac{2}{n}\sum_{i=1}^{n}1 + \frac{8}{n^2}\sum_{i=1}^{n}i - \frac{8}{n^3}\sum_{i=1}^{n}i^2$$

$$= \frac{2}{n}(n) + \frac{8}{n^2}\frac{n(n+1)}{2} - \frac{8}{n^3}\frac{n(n+1)(2n+1)}{6}$$

$$A = \lim_{n\to\infty}\left[2 + 4\frac{n(n+1)}{n^2} - \frac{4}{3}\frac{n(n+1)(2n+1)}{n^3}\right]$$

$$= 2 + 4 - \frac{8}{3} = \frac{10}{3} \text{ square units}$$

39. $A \approx \sum_{i=1}^{n} g\left(1 + \frac{i}{n}\right)\left(\frac{1}{n}\right)$

$= \sum_{i=1}^{n} \left[8 - \left(1 + \frac{i}{n}\right)^3\right]\frac{1}{n}$

$= \sum_{i=1}^{n} \left[7 - \frac{3i}{n} - \frac{3i^2}{n^2} - \frac{i^3}{n^3}\right]\frac{1}{n}$

$= \frac{7}{n}\sum_{i=1}^{n} 1 - \frac{3}{n^2}\sum_{i=1}^{n} i - \frac{3}{n^3}\sum_{i=1}^{n} i^2 - \frac{1}{n^4}\sum_{i=1}^{n} i^3$

$= \frac{7}{n}(n) - \frac{3}{n^2}\frac{n(n+1)}{2} - \frac{3}{n^3}\frac{n(n+1)(2n+1)}{6} - \frac{1}{n^4}\frac{n^2(n+1)^2}{4}$

$A = \lim_{n\to\infty}\left[7 - \frac{3}{2}\frac{n(n+1)}{n^2} - \frac{1}{2n^3}n(n+1)(2n+1) - \frac{1}{n^4}\frac{n^2(n+1)^2}{4}\right] = 7 - \frac{3}{2} - 1 - \frac{1}{4} = \frac{17}{4}$ square units

41. $A \approx \sum_{i=1}^{n} g\left(\frac{i}{n}\right)\left(\frac{1}{n}\right)$

$= \sum_{i=1}^{n} \left[2\left(\frac{i}{n}\right) - \left(\frac{i}{n}\right)^3\right]\left(\frac{1}{n}\right)$

$= \frac{1}{n}\sum_{i=1}^{n} \left[\frac{2}{n}i - \frac{1}{n^3}i^3\right]$

$= \frac{1}{n}\left[\frac{2}{n}\frac{n(n+1)}{2} - \frac{1}{n^3}\frac{n^2(n+1)^2}{4}\right]$

$= \frac{n+1}{n} - \frac{(n+1)^2}{4n^2}$

$A = \lim_{n\to\infty}\left[\frac{n+1}{n} - \frac{(n+1)^2}{4n^2}\right]$

$= 1 - \frac{1}{4} = \frac{3}{4}$ square units

43. $A \approx \sum_{i=1}^{n} f\left(1 + \frac{3i}{n}\right)\left(\frac{3}{n}\right)$

$= \sum_{i=1}^{n} \left[\frac{1}{4}\left(1 + \frac{3i}{n}\right)^2 + \left(1 + \frac{3i}{n}\right)\right]\left(\frac{3}{n}\right)$

$= \sum_{i=1}^{n} \left(\frac{1}{4} + \frac{3}{2}\frac{i}{n} + \frac{9}{4}\frac{i^2}{n^2} + 1 + \frac{3i}{n}\right)\left(\frac{3}{n}\right)$

$= \frac{15}{4n}\sum_{i=1}^{n} 1 + \frac{27}{2n^2}\sum_{i=1}^{n} i + \frac{27}{4n^3}\sum_{i=1}^{n} i^2$

$= \frac{15}{4n}(n) + \frac{27}{2n^2}\left(\frac{n(n+1)}{2}\right) + \frac{27}{4n^3}\frac{n(n+1)(2n+1)}{6}$

$A = \lim_{n\to\infty}\left[\frac{15}{4} + \frac{27}{4}\frac{n(n+1)}{n^2} + \frac{9}{8n^3}n(n+1)(2n+1)\right]$

$= \frac{15}{4} + \frac{27}{4} + \frac{9}{4} = \frac{51}{4}$ square units

45. $y = (-3.0 \cdot 10^{-6})x^3 + 0.002x^2 - 1.05x + 400$

Note that $y = 0$ when $x = 500$.

Area $\approx 105{,}208.33$ square feet ≈ 2.4153 acres

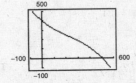

47. True. See Formula 2, page 820.

49. Answers will vary.

51. $2\tan x = \tan 2x = \frac{2\tan x}{1 - \tan^2 x}$

$\tan x = 0 \implies x = n\pi$

53. $(\mathbf{u} \cdot \mathbf{v})\mathbf{u} = (\langle 4, -5\rangle \cdot \langle -1, -2\rangle)\langle 4, -5\rangle$

$= 6\langle 4, -5\rangle$

$= \langle 24, -30\rangle$

55. $\|\mathbf{v}\| - 2 = \sqrt{5} - 2$

Review Exercises for Chapter 11

1. $\lim\limits_{x \to 3} (6x - 1)$

The limit (17) can be reached.

x	2.9	2.99	2.999	3	3.001	3.01	3.1
$f(x)$	16.4	16.94	16.994	17	17.006	17.06	17.6

3. $f(x) = \dfrac{1 - e^{-x}}{x}$

$\lim\limits_{x \to 0} \dfrac{1 - e^{-x}}{x} = 1$

The limit cannot be reached.

x	-0.1	-0.01	-0.001	0	0.001	0.01	0.1
$f(x)$	1.0517	1.0050	1.0005	Error	0.9995	0.9950	0.9516

5. $\lim\limits_{x \to 1} (3 - x) = 2$

7. $\lim\limits_{x \to 1} \dfrac{x^2 - 1}{x - 1} = 2$

9. (a) $\lim\limits_{x \to c} [f(x)]^3 = 4^3 = 64$

(b) $\lim\limits_{x \to c} [3f(x) - g(x)] = 3(4) - 5 = 7$

(c) $\lim\limits_{x \to c} [f(x)g(x)] = (4)(5) = 20$

(d) $\lim\limits_{x \to c} \dfrac{f(x)}{g(x)} = \dfrac{4}{5}$

11. $\lim\limits_{x \to 4} \left(\tfrac{1}{2}x + 3\right) = \tfrac{1}{2}(4) + 3 = 5$

13. $\lim\limits_{x \to 2} (5x - 3)(3x + 5) = (5(2) - 3)(3(2) + 5)$
$$= (7)(11) = 77$$

15. $\lim\limits_{t \to 3} \dfrac{t^2 + 1}{t} = \dfrac{9 + 1}{3} = \dfrac{10}{3}$

17. $\lim\limits_{x \to -2} \sqrt[3]{4x} = (-8)^{1/3} = -2$

19. $\lim\limits_{x \to \pi} \sin 3x = \sin 3\pi = 0$

21. $\lim\limits_{x \to -1} 2e^x = 2e^{-1} = \dfrac{2}{e}$

23. $\lim\limits_{x \to -1/2} \arcsin x = \arcsin\left(-\dfrac{1}{2}\right) = -\dfrac{\pi}{6}$

25. $\lim\limits_{t \to -2} \dfrac{t + 2}{t^2 - 4} = \lim\limits_{t \to -2} \dfrac{t + 2}{(t + 2)(t - 2)}$
$$= \lim\limits_{t \to -2} \dfrac{1}{t - 2} = -\dfrac{1}{4}$$

27. $\lim\limits_{x \to 5} \dfrac{x - 5}{x^2 + 5x - 50} = \lim\limits_{x \to 5} \dfrac{x - 5}{(x - 5)(x + 10)}$
$$= \lim\limits_{x \to 5} \dfrac{1}{x + 10} = \dfrac{1}{15}$$

29. $\lim\limits_{x \to -2} \dfrac{x^2 - 4}{x^3 + 8} = \lim\limits_{x \to -2} \dfrac{(x + 2)(x - 2)}{(x + 2)(x^2 - 2x + 4)}$
$$= \lim\limits_{x \to -2} \dfrac{x - 2}{x^2 - 2x + 4}$$
$$= \dfrac{-4}{12} = \dfrac{-1}{3}$$

31. $\lim\limits_{x \to -1} \dfrac{1/(x + 2) - 1}{x + 1} = \lim\limits_{x \to -1} \dfrac{1 - (x + 2)}{(x + 2)(x + 1)}$
$$= \lim\limits_{x \to -1} \dfrac{-(x + 1)}{(x + 2)(x + 1)}$$
$$= \lim\limits_{x \to -1} \dfrac{-1}{(x + 2)} = -1$$

33. $\lim\limits_{u \to 0} \dfrac{\sqrt{4+u}-2}{u} = \lim\limits_{u \to 0} \dfrac{\sqrt{4+u}-2}{u} \cdot \dfrac{\sqrt{4+u}+2}{\sqrt{4+u}+2}$

$\qquad = \lim\limits_{u \to 0} \dfrac{(4+u)-4}{u\left(\sqrt{4+u}+2\right)}$

$\qquad = \lim\limits_{u \to 0} \dfrac{1}{\sqrt{4+u}+2} = \dfrac{1}{4}$

35. $\lim\limits_{x \to 5} \dfrac{\sqrt{x-1}-2}{x-5} = \lim\limits_{x \to 5} \dfrac{\sqrt{x-1}-2}{x-5} \cdot \dfrac{\sqrt{x-1}+2}{\sqrt{x-1}+2}$

$\qquad = \lim\limits_{x \to 5} \dfrac{(x-1)-4}{(x-5)\left(\sqrt{x-1}+2\right)}$

$\qquad = \lim\limits_{x \to 5} \dfrac{1}{\sqrt{x-1}+2} = \dfrac{1}{2+2} = \dfrac{1}{4}$

37. (a)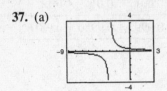

(b)

x	2.9	2.99	3	3.01	3.1
$f(x)$	0.1695	0.1669	Error	0.1664	0.1639

$\lim\limits_{x \to 3} \dfrac{x-3}{x^2-9} = \dfrac{1}{6}$

39. (a)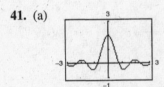

$\lim\limits_{x \to 0} e^{-2/x}$ does not exist.

(b) Answers will vary.

x	-0.1	-0.01	-0.001	0	0.001	0.01	0.1
y_1	4.85 E 8	7.2 E 86	Error	Error	0	1 E -87	2.1 E -9

41. (a)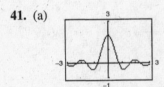

$\lim\limits_{x \to 0} \dfrac{\sin 4x}{2x} = 2$

(b)

x	-0.1	-0.01	-0.001	0	0.001	0.01	0.1
y_1	1.9471	1.9995	1.999995	error	1.999995	1.995	1.9471

43. (a)

$\lim\limits_{x \to 1^+} \dfrac{\sqrt{2x+1}-\sqrt{3}}{x-1} \approx 0.577$

$\left(\text{Exact value: } \dfrac{\sqrt{3}}{3}\right)$

(b)

x	1.1	1.01	1.001	1.0001
$f(x)$	0.5680	0.5764	0.5773	0.5773

45. $f(x) = \dfrac{|x-3|}{x-3}$

Limit does not exist because

$\lim\limits_{x \to 3^+} f(x) = 1$ and

$\lim\limits_{x \to 3^-} f(x) = -1$.

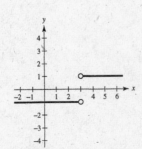

47. $f(x) = \dfrac{2}{x^2-4}$

Limit does not exist.

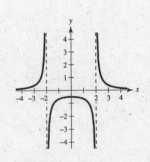

49. $\displaystyle\lim_{x\to5}\frac{|x-5|}{x-5}$ does not exist.

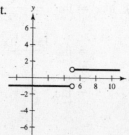

51. $\displaystyle\lim_{x\to2} f(x)$ does not exist.

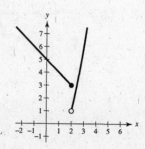

53. $\displaystyle\lim_{h\to0}\frac{f(x+h)-f(x)}{h}=\lim_{h\to0}\frac{3(x+h)-(x+h)^2-(3x-x^2)}{h}$

$$=\lim_{h\to0}\frac{3x+3h-x^2-2xh-h^2-3x+x^2}{h}=\lim_{h\to0}\frac{3h-2xh-h^2}{h}$$

$$=\lim_{h\to0}(3-2x-h)=3-2x$$

55. Slope ≈ 2 Answers will vary.

57.

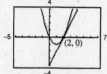

Slope at $(2, f(2))$ is approximately 2.

59.

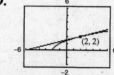

Slope is $\frac{1}{4}$ at $(2, 2)$.

61.

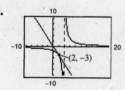

At $(2, f(2)) = (2, -3)$, the slope is approximately -1.5.

63. $m = \displaystyle\lim_{h\to0}\frac{f(x+h)-f(x)}{h}$

$$= \lim_{h\to0}\frac{(x+h)^2-4(x+h)-(x^2-4x)}{h}$$

$$= \lim_{h\to0}\frac{x^2+2xh+h^2-4x-4h-x^2-4x}{h}$$

$$= \lim_{h\to0}\frac{2xh+h^2-4h}{h}$$

$$= \lim_{h\to0}(2x+h-4) = 2x-4$$

(a) At $(0, 0)$, $m = 2(0) - 4 = -4$.

(b) At $(5, 5)$, $m = 2(5) - 4 = 6$.

65. $m = \displaystyle\lim_{h\to0}\frac{f(x+h)-f(x)}{h}=\lim_{h\to0}\frac{\dfrac{4}{x+h-6}-\dfrac{4}{x-6}}{h}$

$$= \lim_{h\to0}\frac{4(x-6)-4(x+h-6)}{(x+h-6)(x-6)h}$$

$$= \lim_{h\to0}\frac{-4h}{(x+h-6)(x-6)h}$$

$$= \lim_{h\to0}\frac{-4}{(x+h-6)(x-6)}=\frac{-4}{(x-6)^2}$$

(a) At $(7, 4)$, $m = \dfrac{-4}{(7-6)^2} = -4$.

(b) At $(8, 2)$, $m = \dfrac{-4}{(8-6)^2} = -1$.

67. $f'(x) = \lim_{h \to 0} \dfrac{f(x + h) - f(x)}{h} = \lim_{h \to 0} \dfrac{5 - 5}{h} = 0$

69. $h'(x) = \lim_{k \to 0} \dfrac{h(x + k) - h(x)}{k}$

$= \lim_{k \to 0} \dfrac{\left[5 - \frac{1}{2}(x + k)\right] - \left[5 - \frac{1}{2}x\right]}{k}$

$= \lim_{k \to 0} \dfrac{-\frac{1}{2}k}{k} = -\dfrac{1}{2}$

71. $g'(x) = \lim_{h \to 0} \dfrac{g(x + h) - g(x)}{h}$

$= \lim_{h \to 0} \dfrac{2(x + h)^2 - 1 - (2x^2 - 1)}{h}$

$= \lim_{h \to 0} \dfrac{2x^2 + 4xh + 2h^2 - 2x^2}{h}$

$= \lim_{h \to 0} (4x + 2h)$

$= 4x$

73. $f'(t) = \lim_{h \to 0} \dfrac{f(t + h) - f(t)}{h}$

$= \lim_{h \to 0} \dfrac{\sqrt{t + h + 5} - \sqrt{t + 5}}{h} \cdot \dfrac{\sqrt{t + h + 5} + \sqrt{t + 5}}{\sqrt{t + h + 5} + \sqrt{t + 5}}$

$= \lim_{h \to 0} \dfrac{(t + h + 5) - (t + 5)}{h\left(\sqrt{t + h + 5} + \sqrt{t + 5}\right)}$

$= \lim_{h \to 0} \dfrac{1}{\sqrt{t + h + 5} + \sqrt{t + 5}}$

$= \dfrac{1}{2\sqrt{t + 5}}$

75. $g'(s) = \dfrac{g(s + h) - g(s)}{h}$

$= \lim_{h \to 0} \dfrac{\dfrac{4}{s + h + 5} - \dfrac{4}{s + 5}}{h}$

$= \lim_{h \to 0} \dfrac{4s + 20 - 4s - 4h - 20}{(s + h + 5)(s + 5)h}$

$= \lim_{h \to 0} \dfrac{-4h}{(s + h + 5)(s + 5)h}$

$= \lim_{h \to 0} \dfrac{-4}{(s + h + 5)(s + 5)}$

$= \dfrac{-4}{(s + 5)^2}$

77. $g'(x) = \lim_{h \to 0} \dfrac{g(x+h) - g(x)}{h}$

$= \lim_{h \to 0} \dfrac{\dfrac{1}{\sqrt{x+h+4}} - \dfrac{1}{\sqrt{x+4}}}{h}$

$= \lim_{h \to 0} \dfrac{\sqrt{x+4} - \sqrt{x+h+4}}{h\sqrt{x+h+4}\sqrt{x+4}} \cdot \dfrac{\sqrt{x+4} + \sqrt{x+h+4}}{\sqrt{x+4} + \sqrt{x+h+4}}$

$= \lim_{h \to 0} \dfrac{(x+4) - (x+h+4)}{h\sqrt{x+h+4}\sqrt{x+4}\left[\sqrt{x+4} + \sqrt{x+h+4}\right]}$

$= \lim_{h \to 0} \dfrac{-1}{\sqrt{x+h+4}\sqrt{x+4}\left[\sqrt{x+4} + \sqrt{x+h+4}\right]}$

$= \dfrac{-1}{(x+4)2\sqrt{x+4}}$

$= \dfrac{-1}{2(x+4)^{3/2}}$

79. $\lim\limits_{x \to \infty} \dfrac{4x}{2x-3} = \dfrac{4}{2} = 2$

81. $\lim\limits_{x \to -\infty} \dfrac{2x}{x^2 - 25} = 0$

83. $\lim\limits_{x \to \infty} \dfrac{x^2}{2x+3}$ does not exist.

85. $\lim\limits_{x \to \infty}\left[\dfrac{x}{(x-2)^2} + 3\right] = 0 + 3 = 3$

87. $a_n = \dfrac{2n-3}{5n+4}$

$a_1 = -\dfrac{1}{9} \qquad a_4 = \dfrac{5}{24}$

$a_2 = \dfrac{1}{14} \qquad a_5 = \dfrac{7}{29}$

$a_3 = \dfrac{3}{19}$

$\lim\limits_{n \to \infty} a_n = \dfrac{2}{5}$

89. $a_n = \dfrac{(-1)^n}{n^3}$

$a_1 = -1 \qquad a_4 = \dfrac{1}{64}$

$a_2 = \dfrac{1}{8} \qquad a_5 = -\dfrac{1}{125}$

$a_3 = -\dfrac{1}{27}$

$\lim\limits_{n \to \infty} a_n = 0$

91. $a_n = \dfrac{1}{2n^2}[3 - 2n(n+1)] = \dfrac{3}{2n^2} - \dfrac{n+1}{n}$

$-0.5, -1.125, -1.16\overline{6}, -1.15625, -1.14$

$\lim\limits_{n \to \infty} a_n = 0 - 1 = -1$

93. (a) $\sum_{i=1}^{n}\left(\dfrac{4i^2}{n^2}-\dfrac{i}{n}\right)\dfrac{1}{n}=\dfrac{4}{n^3}\sum_{i=1}^{n}i^2-\dfrac{1}{n^2}\sum_{i=1}^{n}i$

(b)

n	10^0	10^1	10^2	10^3	10^4
$S(n)$	3	0.99	0.8484	0.8348	0.8335

$$=\dfrac{4}{n^3}\dfrac{n(n+1)(2n+1)}{6}-\dfrac{1}{n^2}\dfrac{n(n+1)}{2}$$

$$=\dfrac{4n(n+1)(2n+1)-3n^2(n+1)}{6n^3}$$

$$=\dfrac{n(n+1)(8n+4-3n)}{6n^3}$$

$$=\dfrac{(n+1)(5n+4)}{6n^2}$$

(c) $\lim\limits_{n\to\infty}S(n)=\dfrac{5}{6}$

95. Area $\approx\dfrac{1}{2}\left(\dfrac{7}{2}+3+\dfrac{5}{2}+2+\dfrac{3}{2}+1\right)$

$\qquad=\dfrac{1}{2}\left(\dfrac{27}{2}\right)=\dfrac{27}{4}=6.75$

97. $f(x)=\dfrac{1}{4}x^2,\ b-a=4-0=4$

$A\approx\sum_{i=1}^{n}f\left(\dfrac{4i}{n}\right)\left(\dfrac{4}{n}\right)$

n	4	8	20	50
Approximate area	7.5	6.375	5.74	5.4944

$=\sum_{i=1}^{n}\dfrac{1}{4}\left(\dfrac{4i}{n}\right)^2\left(\dfrac{4}{n}\right)$

$\left(\text{Exact area is }\dfrac{16}{3}\approx 5.33.\right)$

$=\dfrac{1}{n}\sum_{i=1}^{n}\dfrac{16}{n^2}i^2$

$=\dfrac{16}{n^3}\dfrac{n(n+1)(2n+1)}{6}$

$=\dfrac{8(n+1)(2n+1)}{3n^2}$

99. $A=\lim\limits_{n\to\infty}\sum_{i=1}^{n}\left(10-\dfrac{10i}{n}\right)\left(\dfrac{10}{n}\right)$

$=\lim\limits_{n\to\infty}\left[\dfrac{100}{n}\sum_{i=1}^{n}1-\dfrac{100}{n^2}\sum_{i=1}^{n}i\right]$

$=\lim\limits_{n\to\infty}\left[\dfrac{100}{n}(n)-\dfrac{100}{n^2}\left(\dfrac{n(n+1)}{2}\right)\right]$

$=\lim\limits_{n\to\infty}\left[100-50\dfrac{n(n+1)}{n^2}\right]$

$=100-50=50,\ \text{exact area}$

101. $A=\lim\limits_{n\to\infty}\sum_{i=1}^{n}\left[\left(-1+\dfrac{3i}{n}\right)^2+4\right]\left(\dfrac{3}{n}\right)$

$=\lim\limits_{n\to\infty}\sum_{i=1}^{n}\left[5-\dfrac{6i}{n}+\dfrac{9i^2}{n^2}\right]\dfrac{3}{n}$

$=\lim\limits_{n\to\infty}\left[\dfrac{15}{n}\sum_{i=1}^{n}1-\dfrac{18}{n^2}\sum_{i=1}^{n}i+\dfrac{27}{n^3}\sum_{i=1}^{n}i^2\right]$

$=\lim\limits_{n\to\infty}\left[\dfrac{15}{n}(n)-\dfrac{18}{n^2}\dfrac{n(n+1)}{2}+\dfrac{27}{n^3}\dfrac{n(n+1)(2n+1)}{6}\right]$

$=15-9+9=15,\ \text{exact area}$

103. $f(x) = x^3 + 1$, $[0, 2]$

The width of each rectangle is $2/n$. The height is

$$f\left(\frac{2i}{n}\right) = \left(\frac{2i}{n}\right)^3 + 1.$$

$$A \approx \sum_{i=1}^{n}\left[\left(\frac{2i}{n}\right)^3 + 1\right]\left(\frac{2}{n}\right)$$

$$= \sum_{i=1}^{n}\left[\frac{16i^3}{n^4} + \frac{2}{n}\right]$$

$$= \left(\frac{16}{n^4}\right)\frac{n^2(n+1)^2}{4} + \left(\frac{2}{n}\right)n$$

$$= \frac{4}{n^2}(n+1)^2 + 2$$

$$A = \lim_{n\to\infty}\left[\frac{4}{n^2}(n+1)^2 + 2\right] = 4 + 2 = 6$$

105. $f(x) = 3(x^3 - x^2)$, $[1, 3]$

The width of each rectangle is $2/n$. The height is

$$f\left(1 + \frac{2i}{n}\right) = 3\left[\left(1 + \frac{2i}{n}\right)^3 - \left(1 + \frac{2i}{n}\right)^2\right]$$

$$= \frac{6i}{n} + \frac{24i^2}{n^2} + \frac{24i^3}{n^3}.$$

$$A \approx \sum_{i=1}^{n}\left[\frac{6i}{n} + \frac{24i^2}{n^2} + \frac{24i^3}{n^3}\right]\left(\frac{2}{n}\right)$$

$$= \frac{12}{n^2}\frac{n(n+1)}{2} + \frac{48}{n^3}\frac{n(n+1)(2n+1)}{6} + \frac{48}{n^4}\frac{n^2(n+1)^2}{4}$$

$$A = \lim_{n\to\infty}\left[6 + 8\frac{n(n+1)(2n+1)}{n^3} + 12\frac{n^2(n+1)^2}{n^4}\right]$$

$$= 6 + 16 + 12 = 34$$

107. (a) $y = -3.376 \times 10^{-7}x^3 + 3.753 \times 10^{-4}x^2 - 0.168x + 132.168$

(c) Area $\approx 88{,}868$ square feet; answers will vary.

(b)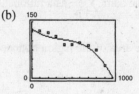

109. False. The limit does not exist.

Chapter 11   Practice Test

1. Use a graphing utility to complete the table and use the result to estimate the limit

$$\lim_{x \to 3} \frac{x - 3}{x^2 - 9}.$$

x	2.9	2.99	3	3.01	3.1
$f(x)$			?		

2. Graph the function

$$f(x) = \frac{\sqrt{x + 4} - 2}{x}$$

and estimate the limit

$$\lim_{x \to 0} \frac{\sqrt{x + 4} - 2}{x}.$$

3. Find the limit $\lim_{x \to 2} e^{x-2}$ by direct substitution.

4. Find the limit $\lim_{x \to 1} \frac{x^3 - 1}{x - 1}$ analytically.

5. Use a graphing utility to estimate the limit

$$\lim_{x \to 0} \frac{\sin 5x}{2x}.$$

6. Find the limit

$$\lim_{x \to -2} \frac{|x + 2|}{x + 2}.$$

7. Use the limit process to find the slope of the graph of $f(x) = \sqrt{x}$ at the point $(4, 2)$.

8. Find the derivative of the function $f(x) = 3x - 1$.

9. Find the limits.

   (a) $\lim_{x \to \infty} \frac{3}{x^4}$

   (b) $\lim_{x \to -\infty} \frac{x^2}{x^2 + 3}$

   (c) $\lim_{x \to \infty} \frac{|x|}{1 - x}$

10. Write the first four terms of the sequence $a_n = \dfrac{1 - n^2}{2n^2 + 1}$ and find the limit of the sequence.

11. Find the sum $\displaystyle\sum_{i=1}^{25} (i^2 + i) \cdot \frac{1}{2}$

12. Write the sum $\displaystyle\sum_{i=1}^{n} \frac{i^2}{n^3}$ as a rational function $S(n)$, and find $\lim_{n \to \infty} S(n)$.

13. Find the area of the region bounded by $f(x) = 1 - x^2$ over the interval $0 \le x \le 1$.

APPENDICES

Appendix B Prerequisites **544**

    B.1 The Cartesian Plane **544**

    B.2 Graphs of Equations **549**

    B.3 Solving Equations Algebraically and Graphically **554**

    B.4 Solving Inequalities Algebraically and Graphically **568**

    B.5 Representing Data Graphically **574**

Appendix C Concepts in Statistics **576**

    C.1 Measures of Central Tendency and Dispersion **576**

    C.2 Least Squares Regression **578**

Appendix D Variation **578**

Appendix E Solving Linear Equations and Inequalities **581**

Appendix F Systems of Inequalities **582**

    F.1 Solving Systems of Inequalities **582**

    F.2 Linear Programming **588**

APPENDIX B
Prerequisites

Appendix B.1   The Cartesian Plane

- ■ You should be able to plot points.
- ■ You should know that the distance between (x_1, y_1) and (x_2, y_2) in the plane is
  $$d = \sqrt{(x_2 - x_1)^2 + (y_2 - y_1)^2}.$$
- ■ You should know that the midpoint of the line segment joining (x_1, y_1) and (x_2, y_2) is
  $$\left(\frac{x_1 + x_2}{2}, \frac{y_1 + y_2}{2}\right).$$
- ■ You should know the equation of a circle: $(x - h)^2 + (y - k)^2 = r^2$.
- ■ You should be able to translate points in the plane.

Vocabulary Check

1. (a) iii   (b) vi   (c) i   (d) iv   (e) v   (f) ii   **2.** Cartesian

3. Distance Formula   **4.** Midpoint Formula

5. $(x - h)^2 + (y - k)^2 = r^2$, center, radius

---

1. A: $(2, 6)$, B: $(-6, -2)$, C: $(4, -4)$, D: $(-3, 2)$

3.

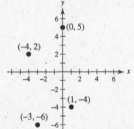

5.

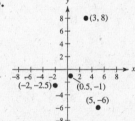

7. $(-5, 4)$

9. $(-6, -6)$

11. $x > 0 \implies$ The point lies in Quadrant I or in Quadrant IV.

$y < 0 \implies$ The point lies in Quadrant III or in Quadrant IV.

$x > 0$ and $y < 0 \implies (x, y)$ lies in Quadrant IV.

13. $x = -4 \implies$ x is negative $\implies$ The point lies in Quadrant II or in Quadrant III.

$y > 0 \implies$ The point lies in Quadrant I or Quadrant II.

$x = -4$ and $y > 0 \implies (x, y)$ lies in Quadrant II.

15. $y < -5 \Rightarrow y$ is negative $\Rightarrow$ The point lies in either Quadrant III or Quadrant IV.

17. If $-y > 0$, then $y < 0$.

$x < 0 \Rightarrow$ The point lies in Quadrant II or in Quadrant III.

$y < 0 \Rightarrow$ The point lies in Quadrant III or in Quadrant IV.

$x < 0$ and $y < 0 \Rightarrow (x, y)$ lies in Quadrant III.

19. If $xy > 0$, then either x and y are both positive, or both negative. Hence, (x, y) lies in either Quadrant I or Quadrant III.

21.

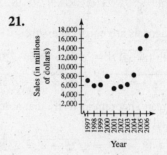

23. $(6, -3), (6, 5)$

$d = \sqrt{(6 - 6)^2 + (5 - (-3))^2} = \sqrt{64} = 8$

25. $(-3, -1), (2, -1)$

$d = \sqrt{(2 - (-3))^2 + (-1 - (-1))^2} = \sqrt{25} = 5$

27. $d = \sqrt{(3 - (-2))^2 + (-6 - 6)^2} = \sqrt{5^2 + (-12)^2}$

$= \sqrt{25 + 144} = \sqrt{169} = 13$

29. $\left(\frac{1}{2}, \frac{4}{3}\right), (2, -1)$

$d = \sqrt{\left(\frac{1}{2} - 2\right)^2 + \left(\frac{4}{3} + 1\right)^2}$

$= \sqrt{\frac{9}{4} + \frac{49}{9}}$

$= \sqrt{\frac{277}{36}} = \frac{\sqrt{277}}{6} \approx 2.77$

31. $(-4.2, 3.1), (-12.5, 4.8)$

$d = \sqrt{(-4.2 + 12.5)^2 + (3.1 - 4.8)^2}$

$= \sqrt{68.89 + 2.89}$

$= \sqrt{71.78} \approx 8.47$

33. (a) The distance between $(0, 2)$ and $(4, 2)$ is 4.

The distance between $(4, 2)$ and $(4, 5)$ is 3.

The distance between $(0, 2)$ and $(4, 5)$ is

$\sqrt{(4 - 0)^2 + (5 - 2)^2} = \sqrt{16 + 9}$

$= \sqrt{25} = 5.$

(b) $4^2 + 3^2 = 16 + 9 = 25 = 5^2$

35. (a) The distance between $(-1, 1)$ and $(9, 1)$ is 10.

The distance between $(9, 1)$ and $(9, 4)$ is 3.

The distance between $(-1, 1)$ and $(9, 4)$ is

$\sqrt{(9 - (-1))^2 + (4 - 1)^2} = \sqrt{100 + 9}$

$= \sqrt{109}.$

(b) $10^2 + 3^2 = 109 = \left(\sqrt{109}\right)^2$

37. Find distances between pairs of points.

$d_1 = \sqrt{(4 - 2)^2 + (0 - 1)^2} = \sqrt{5}$

$d_2 = \sqrt{(4 + 1)^2 + (0 + 5)^2} = \sqrt{50}$

$d_3 = \sqrt{(2 + 1)^2 + (1 + 5)^2} = \sqrt{45}$

$\left(\sqrt{5}\right)^2 + \left(\sqrt{45}\right)^2 = \left(\sqrt{50}\right)^2$

Because $d_1{}^2 + d_3{}^2 = d_2{}^2$, the triangle is a right triangle.

39. $d_1 = \sqrt{(1 - 3)^2 + (-3 - 2)^2} = \sqrt{4 + 25} = \sqrt{29}$

$d_2 = \sqrt{(3 + 2)^2 + (2 - 4)^2} = \sqrt{25 + 4} = \sqrt{29}$

$d_3 = \sqrt{(1 + 2)^2 + (-3 - 4)^2} = \sqrt{9 + 49} = \sqrt{58}$

$d_1 = d_2$. Triangle is isosceles.

41. Find distances between pairs of points.

$$d_1 = \sqrt{(0-2)^2 + (9-5)^2} = \sqrt{4+16} = \sqrt{20} = 2\sqrt{5}$$

$$d_2 = \sqrt{(-2-0)^2 + (0-9)^2} = \sqrt{4+81} = \sqrt{85}$$

$$d_3 = \sqrt{(0-(-2))^2 + (-4-0)^2} = \sqrt{4+16} = \sqrt{20} = 2\sqrt{5}$$

$$d_4 = \sqrt{(0-2)^2 + (-4-5)^2} = \sqrt{4+81} = \sqrt{85}$$

Opposite sides have equal lengths of $2\sqrt{5}$ and $\sqrt{85}$, so the figure is a parallelogram.

43. First show that the diagonals are equal in length.

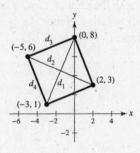

$$d_1 = \sqrt{(0-(-3))^2 + (8-1)^2} = \sqrt{9+49} = \sqrt{58}$$

$$d_2 = \sqrt{(2-(-5))^2 + (3-6)^2} = \sqrt{49+9} = \sqrt{58}$$

Now use the Pythagorean Theorem to verify that at least one angle is 90°
(and, hence, they are all right angles).

$$d_3 = \sqrt{(0-(-5))^2 + (8-6)^2} = \sqrt{25+4} = \sqrt{29}$$

$$d_4 = \sqrt{(-3-(-5))^2 + (1-6)^2} = \sqrt{4+25} = \sqrt{29}$$

Thus, $d_3{}^2 + d_4{}^2 = d_1{}^2$.

45. (a)

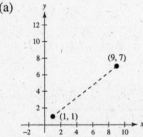

(b) $d = \sqrt{(9-1)^2 + (7-1)^2}$

$$= \sqrt{64+36} = 10$$

(c) $\left(\dfrac{9+1}{2}, \dfrac{7+1}{2}\right) = (5, 4)$

47. (a)

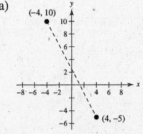

(b) $d = \sqrt{(4+4)^2 + (-5-10)^2}$

$$= \sqrt{64+225} = 17$$

(c) $\left(\dfrac{4-4}{2}, \dfrac{-5+10}{2}\right) = \left(0, \dfrac{5}{2}\right)$

49. (a)

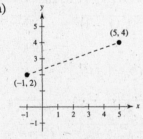

(b) $d = \sqrt{(5+1)^2 + (4-2)^2}$

$$= \sqrt{36+4} = \sqrt{40} = 2\sqrt{10}$$

(c) $\left(\dfrac{-1+5}{2}, \dfrac{2+4}{2}\right) = (2, 3)$

51. (a)

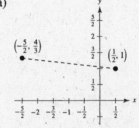

(b) $d = \sqrt{\left(\dfrac{1}{2} + \dfrac{5}{2}\right)^2 + \left(1 - \dfrac{4}{3}\right)^2}$

$$d = \sqrt{9 + \dfrac{1}{9}} = \dfrac{\sqrt{82}}{3}$$

(c) $\left(\dfrac{-\frac{5}{2} + \frac{1}{2}}{2}, \dfrac{\frac{4}{3} + 1}{2}\right) = \left(-1, \dfrac{7}{6}\right)$

53. (a)

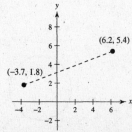

(b) $d = \sqrt{(6.2 + 3.7)^2 + (5.4 - 1.8)^2}$

$= \sqrt{98.01 + 12.96} = \sqrt{110.97}$

(c) $\left(\dfrac{6.2 - 3.7}{2}, \dfrac{5.4 + 1.8}{2}\right) = (1.25, 3.6)$

55. Calculate the midpoint:

$$\left(\dfrac{2000 + 2006}{2}, \dfrac{2237 + 3950}{2}\right) = (2003, 3093.5)$$

The sales in 2003 are \$3093.5 million.

57. Since $x_m = \dfrac{x_1 + x_2}{2}$ and $y_m = \dfrac{y_1 + y_2}{2}$ we have:

$2x_m = x_1 + x_2 \qquad 2y_m = y_1 + y_2$

$2x_m - x_1 = x_2 \qquad 2y_m - y_1 = y_2$

So, $(x_2, y_2) = (2x_m - x_1, 2y_m - y_1)$.

(a) $(x_2, y_2) = (2x_m - x_1, 2y_m - y_1) = (2(4) - 1, 2(-1) - (-2)) = (7, 0)$

(b) $(x_2, y_2) = (2x_m - x_1, 2y_m - y_1) = (2(2) - (-5), 2(4) - 11) = (9, -3)$

59. $(x - 0)^2 + (y - 0)^2 = 3^2$

$x^2 + y^2 = 9$

61. $(x - 2)^2 + (y + 1)^2 = 4^2$

$(x - 2)^2 + (y + 1)^2 = 16$

63. $(x + 1)^2 + (y - 2)^2 = r^2$

$(0 + 1)^2 + (0 - 2)^2 = r^2 \implies r^2 = 5$

$(x + 1)^2 + (y - 2)^2 = 5$

65. $r = \dfrac{1}{2}\sqrt{(6 - 0)^2 + (8 - 0)^2} = \dfrac{1}{2}\sqrt{100} = 5$

Center: $\left(\dfrac{0 + 6}{2}, \dfrac{0 + 8}{2}\right) = (3, 4)$

$(x - 3)^2 + (y - 4)^2 = 25$

67. Because the circle is tangent to the x-axis, the radius is 1.

$(x + 2)^2 + (y - 1)^2 = 1$

69. The center is the midpoint of one of the diagonals of the square.

Center: $\left(\dfrac{7 + (-1)}{2}, \dfrac{-2 + (-10)}{2}\right) = (3, -6)$

The radius is one half the length of a side of the square.

Radius: $\dfrac{1}{2}(7 - (-1)) = 4$

Circle: $(x - 3)^2 + (y + 6)^2 = 16$

71. $(x - 2)^2 + (y + 1)^2 = 16$

73. $x^2 + y^2 = 25$

Center: $(0, 0)$

Radius: 5

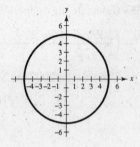

75. Center: $(1, -3)$

Radius: 2

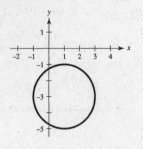

77. Center: $\left(\frac{1}{2}, \frac{1}{2}\right)$

Radius: $\frac{3}{2}$

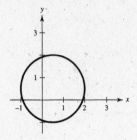

79. The x-coordinates are increased by 2, and the y-coordinates are increased by 5.

Old vertex	Shifted vertex
$(-1, -1)$	$(1, 4)$
$(-2, -4)$	$(0, 1)$
$(2, -3)$	$(4, 2)$

81.

Old vertex	Shifted vertex
$(0, 2)$	$(-1, 5)$
$(3, 5)$	$(2, 8)$
$(5, 2)$	$(4, 5)$
$(2, -1)$	$(1, 2)$

83. The point $(65, 83)$ represents an entrance exam score of 65.

85. (a) Sample answer: The number of artists inducted each year seems to be nearly steady except for the first few years. Estimate: Between 5 and 7 new members

(b) Sample answer: The Rock and Roll Hall of Fame was opened in 1986.

87. $d = \sqrt{(45 - 10)^2 + (40 - 15)^2} = \sqrt{35^2 + 25^2} = \sqrt{1850} = 5\sqrt{74} \approx 43$ yards

89. (a)

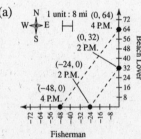

(b) Distance at 2 P.M.:

$$\sqrt{(-24 - 0)^2 + (0 - 32)^2} = \sqrt{1600} = 40 \text{ miles}$$

Distance at 4 P.M.:

$$\sqrt{(-48 - 0)^2 + (0 - 64)^2} = \sqrt{6400} = 80 \text{ miles}$$

Yes, the yachts are twice as far from each other.

91. Find the distances between pairs of points.

$d_1 = \sqrt{(2 + 2\sqrt{3} - 2)^2 + (0 - 6)^2} = \sqrt{12 + 36} = \sqrt{48} = 4\sqrt{3}$

$d_2 = \sqrt{((2 + 2\sqrt{3}) - (2 - 2\sqrt{3}))^2 + (0 - 0)^2} = 4\sqrt{3}$

$d_3 = \sqrt{(2 - 2\sqrt{3} - 2)^2 + (0 - 6)^2} = \sqrt{12 + 36} = 4\sqrt{3}$

Because $d_1 = d_2 = d_3$, the triangle is equilateral.

93. False. It would be sufficient to use the midpoint formula 15 times.

95. False. The polygon could be a rhombus. For example, consider the points $(4, 0)$, $(0, 6)$, $(-4, 0)$ and $(0, -6)$.

97. No, the scales can be different. The scales depend on the magnitude of the coordinates. See Figure P.13.

Appendix B.2   Graphs of Equations

> ■ You should be able to use the point-plotting method of graphing.
>
> ■ You should be able to find x- and y-intercepts.
>
>   (a) To find the x-intercepts, let $y = 0$ and solve for x.
>
>   (b) To find the y-intercepts, let $x = 0$ and solve for y.
>
> ■ You should know how to graph an equation with a graphing utility. You should be able to determine an appropriate viewing rectangle.
>
> ■ You should be able to use the zoom and trace features of a graphing utility.

Vocabulary Check

1. solution point   **2.** graph   **3.** intercepts

---

1. $y = \sqrt{x + 4}$

  (a) $(0, 2)$: $2 \overset{?}{=} \sqrt{0 + 4}$

  $2 = 2$ ✓

  Yes, the point *is* on the graph.

  (b) $(5, 3)$: $3 \overset{?}{=} \sqrt{5 + 4}$

  $3 = \sqrt{9}$ ✓

  Yes, the point *is* on the graph.

3. $y = 4 - |x - 2|$

  (a) $(1, 5)$: $5 \overset{?}{=} 4 - |1 - 2|$

  $5 \neq 4 - 1$

  No, the point *is not* on the graph.

  (b) $(1.2, 3.2)$: $3.2 \overset{?}{=} 4 - |1.2 - 2|$

  $3.2 \overset{?}{=} 4 - |-0.8|$

  $3.2 \overset{?}{=} 4 - 0.8$

  $3.2 \overset{?}{=} 3.2$ ✓

  Yes, the point *is* on the graph.

5. $x^2 + y^2 = 20$

  (a) $(3, -2)$: $3^2 + (-2)^2 \overset{?}{=} 20$

  $9 + 4 \overset{?}{=} 20$

  $13 \neq 20$

  No, the point *is not* on the graph.

  (b) $(-4, 2)$: $(-4)^2 + 2^2 \overset{?}{=} 20$

  $16 + 4 \overset{?}{=} 20$

  $20 = 20$

  Yes, the point *is* on the graph.

7. $y = \frac{3}{2}x - 1$

x	-2	0	$\frac{2}{3}$	1	2
y	-4	-1	0	$\frac{1}{2}$	2
Solution point	$(-2, -4)$	$(0, -1)$	$\left(\frac{3}{2}, 0\right)$	$\left(1, \frac{1}{2}\right)$	$(2, 2)$

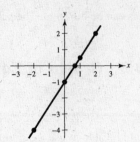

9. (a) $y = \frac{1}{4}x - 3$

x	-2	-1	0	1	2
y	$-\frac{7}{2}$	$-\frac{13}{4}$	-3	$-\frac{11}{4}$	$-\frac{5}{2}$

(b)

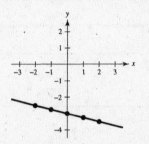

(c) $y = -\frac{1}{4}x - 3$

x	-2	-1	0	1	2
y	$-\frac{5}{2}$	$-\frac{11}{4}$	-3	$-\frac{13}{4}$	$-\frac{7}{2}$

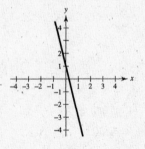

Both graphs are lines. The first graph rises to the right, whereas the second falls. Both pass through $(0, -3)$.

11. $y = 2x + 3$ has intercepts $(0, 3)$ and $\left(-\frac{3}{2}, 0\right)$.

Matches graph (e).

13. $y = x^2 - 2x$ has intercepts $(0, 0)$ and $(2, 0)$.

Matches graph (b).

15. $y = 2\sqrt{x}$ has one intercept $(0, 0)$.

Matches graph (c).

17. $y = -4x + 1$

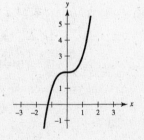

19. $y = 2 - x^2$

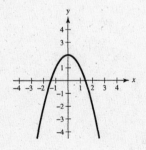

21. $y = x^2 - 3x$

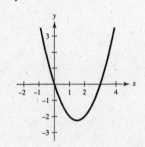

23. $y = x^3 + 2$

25. $y = \sqrt{x - 3}$

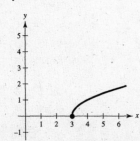

27. $y = |x - 2|$

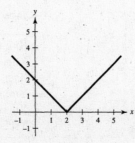

29. $x = y^2 - 1$

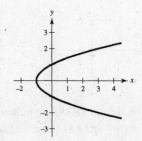

31. $y = x - 7$

Intercepts: $(0, -7), (7, 0)$

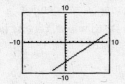

33. $y = 3 - \dfrac{1}{2}x$

Intercepts: $(6, 0), (0, 3)$

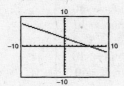

35. $y = \dfrac{2x}{x - 1}$

Intercepts: $(0, 0)$

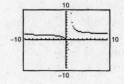

37. $y = x\sqrt{x + 3}$

Intercepts: $(0, 0), (-3, 0)$

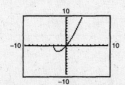

39. $y = \sqrt[3]{x - 8}$

Intercepts: $(8, 0), (0, -2)$

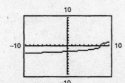

41. $y = x^2 - 4x + 3$

Intercepts: $(3, 0), (1, 0), (0, 3)$

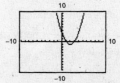

43. $y = x^2(x - 4) + 4x$

$= x^3 - 4x^2 + 4x$

Intercepts: $(0, 0), (2, 0)$

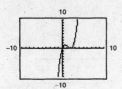

45. $y = \frac{5}{2}x + 5$

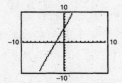

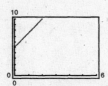

The first setting shows the line and its intercepts.
The first setting is better.

The second setting does not show the x-intercept
$(-2, 0)$.

47. $y = -x^2 + 10x - 5$

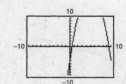 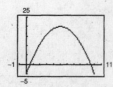

The second viewing window is better because it shows more of the essential features of the function.

49. $y = -10x + 50$

Range/Window

Xmin = -10
Xmax = 10
Xscl = 2
Ymin = -50
Ymax = 100
Yscl = 25

51. $y = \sqrt{x + 2} - 1$

Range/Window

Xmin = -5
Xmax = 1
Xscl = 1
Ymin = -3
Ymax = 1
Yscl = 1

53. $y = |x| + |x - 10|$

Range/Window

Xmin = -30
Xmax = 30
Xscl = 5
Ymin = -10
Ymax = 50
Yscl = 5

55. $y_1 = \frac{1}{4}(x^2 - 8)$

$y_2 = \frac{1}{4}x^2 - 2$

The graphs are identical.
The Distributive Property is illustrated.

57. $y_1 = \frac{1}{5}[10(x^2 - 1)]$

$y_2 = 2(x^2 - 1)$

The graphs are identical.

The Associative Property of Multiplication is illustrated.

59. $y = \sqrt{5 - x}$

(a) $(2, y) \approx (2, 1.73)$

(b) $(x, 3) = (-4, 3)$

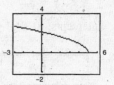

61. $y = x^5 - 5x$

(a) $(-0.5, y) \approx (-0.5, 2.47)$

(b) $(x, -4) = (1, -4)$ or $(x, -4) \approx (-1.65, -4)$

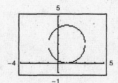

63. $x^2 + y^2 = 16$

$$y^2 = 16 - x^2$$
$$y = \pm\sqrt{16 - x^2}$$

Use $y_1 = \sqrt{16 - x^2}$

$$y_2 = -\sqrt{16 - x^2}.$$

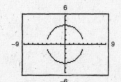

65. $(x - 1)^2 + (y - 2)^2 = 4$

$$(y - 2)^2 = 4 - (x - 1)^2$$
$$y - 2 = \pm\sqrt{4 - (x - 1)^2}$$
$$y = 2 \pm \sqrt{4 - (x - 1)^2}$$

Use $y_1 = 2 + \sqrt{4 - (x - 1)^2}$

$$y_2 = 2 - \sqrt{4 - (x - 1)^2}.$$

67. The center is in the first quadrant and the circle is tangent to the *x*-axis. Matches (a).

69. $(x - 1)^2 + (y - 2)^2 = 25$

    (a) $(1 - 1)^2 + (2 - 2)^2 = 0 \neq 25$ No

    (b) $(-2 - 1)^2 + (6 - 2)^2 = 9 + 16 = 25$ Yes

    (c) $(5 - 1)^2 + (-1 - 2)^2 = 16 + 9 = 25$ Yes

    (d) $(0 - 1)^2 + \left(2 + 2\sqrt{6} - 2\right)^2 = 1 + 24 = 25$ Yes

71. (a) $y = 225{,}000 - 20{,}000t$, $0 \le t \le 8$

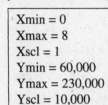
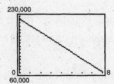

    Xmin = 0
    Xmax = 8
    Xscl = 1
    Ymin = 60,000
    Ymax = 230,000
    Yscl = 10,000

    (b) When $t = 5.8$, $y = 109{,}000$. Algebraically, $225{,}000 - 20{,}000(5.8) = \$109{,}000$.

    (c) When $t = 2.35$, $y = 178{,}000$. Algebraically, $225{,}000 - 20{,}000(2.35) = \$178{,}000$.

73. (a) Model: $y = -0.0049t^3 + 0.443t^2 - 0.75t + 116.7$, $5 \le t \le 14$

t	5	6	7	8	9	10	11	12	13	14
Model	123.4	127.1	131.5	136.5	142.3	148.6	155.5	163.0	171.1	179.6

    The model is a good fit.

    (b)

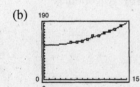

    The model is a good fit.

    (c) For 2008, $t = 18$ and $y \approx 218.2$ thousand dollars. For 2010, $t = 20$ and $y \approx 239.7$ thousand dollars. These values seem reasonable.

    (d) For $y = 150$, $t \approx 10.2$, or 2000.

75. (a)

    (b) Perimeter: $12 = 2x + 2w$

                     $12 = 2(x + w)$

                       $6 = x + w$

        Thus, $w = 6 - x$.

        Area: $xw = x(6 - x) \implies A = x(6 - x)$

    (c)

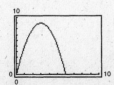

    (d) When $w = 4.9$, $x = 1.1$ and Area $= 5.39$ square meters.

        Algebraically, Area $= xw = (1.1)(4.9) = 5.39$ square meters.

    (e) The maximum area corresponds to the highest point on the graph, which appears to be $(3, 9)$. Thus, $x = 3$ and $w = 3$, and the rectangle is a square.

77. False. $y = 1 - x^2$ has two x-intercepts, $(1, 0)$ and $(-1, 0)$. Also, $y = x^2 + 1$ has no x-intercepts.

79. Answers will vary.

81. Answers will vary. Sample answer: $y = 250x + 1000$ could represent the amount of money in someone's checking account after x months if they deposited an initial $1000 and added $250 per month.

Appendix B.3   Solving Equations Algebraically and Graphically

■ You should know how to solve linear equations: $ax + b = 0$.

■ An identity is an equation whose solution consists of every real number in its domain.

■ To solve an equation you can:
   (a) Add or subtract the same quantity from both sides.
   (b) Multiply or divide both sides by the same nonzero quantity.

■ To solve an equation that can be simplified to a linear equation:
   (a) Remove all symbols of grouping and all fractions.   (b) Combine like terms.
   (c) Solve by algebra.   (d) Check the answer.

■ A "solution" that does not satisfy the original equation is called an extraneous solution.

■ You should be able to set up mathematical models to solve problems.

■ You should be able to translate key words and phrases.
   (a) Equality:
       Equals, equal to, is, are, was, will be, represents
   (b) Addition:
       Sum, plus, greater, increased by, more than, exceeds, total of
   (c) Subtraction:
       Difference, minus, less than, decreased by, subtracted from, reduced by, the remainder
   (d) Multiplication:
       Product, multiplied by, twice, times, percent of
   (e) Division:
       Quotient, divided by, ratio, per
   (f) Consecutive:
       Next, subsequent

■ You should know the following formulas:
   (a) Perimeter:
       1. Square: $P = 4s$
       2. Rectangle: $P = 2L + 2W$
       3. Circle: $C = 2\pi r$
   (b) Area:
       1. Square: $A = s^2$
       2. Rectangle: $A = LW$
       3. Circle: $A = \pi r^2$
       4. Triangle: $A = \left(\dfrac{1}{2}\right)bh$
   (c) Volume
       1. Cube: $V = s^3$
       2. Rectangular solid: $V = LWH$
       3. Cylinder: $V = \pi r^2 h$
       4. Sphere: $V = \left(\dfrac{4}{3}\right)\pi r^3$
   (d) Simple Interest: $I = Prt$
   (e) Compound Interest: $A = P\left(1 + \dfrac{r}{n}\right)^{nt}$
   (f) Distance: $D = r \cdot t$
   (g) Temperature: $F = \dfrac{9}{5}C + 32$

■ You should be able to solve word problems. Study the examples in the text carefully.

1. $\dfrac{5}{2x} - \dfrac{4}{x} = 3$

(a) $\dfrac{5}{2(-1/2)} - \dfrac{4}{(-1/2)} \stackrel{?}{=} 3$

$3 = 3$

$x = -\dfrac{1}{2}$ *is* a solution.

(c) $\dfrac{5}{2(0)} - \dfrac{4}{0}$ is undefined.

$x = 0$ *is not* a solution.

(b) $\dfrac{5}{2(4)} - \dfrac{4}{4} \stackrel{?}{=} 3$

$-\dfrac{3}{8} \neq 3$

$x = 4$ *is not* a solution.

(d) $\dfrac{5}{2(1/4)} - \dfrac{4}{1/4} \stackrel{?}{=} 3$

$-6 \neq 3$

$x = \dfrac{1}{4}$ *is not* a solution.

3. $3 + \dfrac{1}{x+2} = 4$

(a) $3 + \dfrac{1}{(-1)+2} \stackrel{?}{=} 4$

$4 = 4$

$x = -1$ *is* a solution.

(c) $3 + \dfrac{1}{0+2} \stackrel{?}{=} 4$

$\dfrac{7}{2} \neq 4$

$x = 0$ *is not* a solution.

(b) $3 + \dfrac{1}{(-2)+2} = 3 + \dfrac{1}{0}$ is undefined.

$x = -2$ *is not* a solution.

(d) $3 + \dfrac{1}{5+2} \stackrel{?}{=} 4$

$\dfrac{22}{7} = 4$

$x = 5$ *is not* a solution.

5. $\dfrac{\sqrt{x+4}}{6} + 3 = 4$

(a) $\dfrac{\sqrt{-3+4}}{6} + 3 \stackrel{?}{=} 4$

$\dfrac{19}{6} \neq 4$

$x = -3$ *is not* a solution.

(c) $\dfrac{\sqrt{21+4}}{6} + 3 \stackrel{?}{=} 4$

$\dfrac{23}{6} \neq 4$

$x = 21$ *is not* a solution.

(b) $\dfrac{\sqrt{0+4}}{6} + 3 \stackrel{?}{=} 4$

$\dfrac{10}{3} \neq 4$

$x = 0$ *is not* a solution.

(d) $\dfrac{\sqrt{32+4}}{6} + 3 \stackrel{?}{=} 4$

$4 = 4$

$x = 32$ *is* a solution.

7. $2(x - 1) = 2x - 2$ is an *identity* by the Distributive Property. It is true for all real values of x.

9. $x^2 - 8x + 5 = (x - 4)^2 - 11$ is an *identity* since

$$(x - 4)^2 - 11 = x^2 - 8x + 16 - 11$$
$$= x^2 - 8x + 5.$$

11. $3 + \dfrac{1}{x + 1} = \dfrac{4x}{x + 1}$ is *conditional*. There are real values of x for which the equation is not true.

13. *Method 1:*
$$\frac{3x}{8} - \frac{4x}{3} = 4$$
$$\frac{9x - 32x}{24} = 4$$
$$-23x = 96$$
$$x = -\frac{96}{23}$$

Method 2: Graph $y_1 = \dfrac{3x}{8} - \dfrac{4x}{3}$ and $y_2 = 4$ in the same viewing window. These lines intersect at

$$x \approx -4.1739 \approx -\frac{96}{23}.$$

15. *Method 1:*
$$\frac{2x}{5} + 5x = \frac{4}{3}$$
$$\frac{2x + 25x}{5} = \frac{4}{3}$$
$$27x = \frac{20}{3}$$
$$x = \frac{20}{3(27)} = \frac{20}{81}$$

Method 2: Graph $y_1 = \dfrac{2x}{5} + 5x$ and $y_2 = \dfrac{4}{3}$ in the same viewing window. These lines intersect at $x \approx 0.2469 \approx \dfrac{20}{81}$.

17. $3x - 5 = 2x + 7$
$3x - 2x = 7 + 5$
$x = 12$

19. $4y + 2 - 5y = 7 - 6y$
$-y + 2 = 7 - 6y$
$6y - y = 7 - 2$
$5y = 5$
$y = 1$

21. $3(y - 5) = 3 + 5y$
$3y - 15 = 3 + 5y$
$-18 = 2y$
$y = -9$

23. $\dfrac{x}{5} - \dfrac{x}{2} = 3$
$\dfrac{2x - 5x}{10} = 3$
$-3x = 30$
$x = -10$

25. $\dfrac{3}{2}(z + 5) - \dfrac{1}{4}(z + 24) = 0$
$4\left(\dfrac{3}{2}\right)(z + 5) - 4\left(\dfrac{1}{4}\right)(z + 24) = 4(0)$
$6(z + 5) - (z + 24) = 0$
$6z + 30 - z - 24 = 0$
$5z = -6$
$z = -\dfrac{6}{5}$

27. $\dfrac{2(z-4)}{5} + 5 = 10z$

$\dfrac{(2z-8)+25}{5} = 10z$

$2z + 17 = 50z$

$17 = 48z$

$z = \dfrac{17}{48}$

29. $\dfrac{100 - 4u}{3} = \dfrac{5u+6}{4} + 6$

$12\left(\dfrac{100-4u}{3}\right) = 12\left(\dfrac{5u+6}{4}\right) + 12(6)$

$4(100 - 4u) = 3(5u + 6) + 72$

$400 - 16u = 15u + 18 + 72$

$-31u = -310$

$u = 10$

31. $\dfrac{5x-4}{5x+4} = \dfrac{2}{3}$

$3(5x - 4) = 2(5x + 4)$

$15x - 12 = 10x + 8$

$5x = 20$

$x = 4$

33. $\dfrac{1}{x-3} + \dfrac{1}{x+3} = \dfrac{10}{x^2-9}$

$\dfrac{(x+3)+(x-3)}{x^2-9} = \dfrac{10}{x^2-9}$

$2x = 10$

$x = 5$

35. $\dfrac{7}{2x+1} - \dfrac{8x}{2x-1} = -4$

$7(2x - 1) - 8x(2x + 1) = -4(2x + 1)(2x - 1)$

$14x - 7 - 16x^2 - 8x = -16x^2 + 4$

$6x = 11$

$x = \dfrac{11}{6}$

37. $\dfrac{1}{x} + \dfrac{2}{x-5} = 0$

$1(x - 5) + 2x = 0$

$3x - 5 = 0$

$3x = 5$

$x = \dfrac{5}{3}$

39. $\dfrac{3}{x(x-3)} + \dfrac{4}{x} = \dfrac{1}{x-3}$

$3 + 4(x - 3) = x$

$3 + 4x - 12 = x$

$3x = 9$

$x = 3$

A check reveals that $x = 3$ is an extraneous solution, so there is no solution.

41. $y = x - 5$

Let $y = 0$: $0 = x - 5 \implies x = 5 \implies (5, 0)$ x-intercept

Let $x = 0$: $y = 0 - 5 \implies y = -5 \implies (0, -5)$ y-intercept

43. $y = x^2 + x - 2$

Let $y = 0$: $(x^2 + x - 2) = (x + 2)(x - 1) = 0 \implies x = -2, 1 \implies (-2, 0), (1, 0)$ x-intercepts

Let $x = 0$: $y = 0^2 + 0 - 2 = -2 \implies (0, -2)$ y-intercept

45. $y = x\sqrt{x + 2}$

Let $y = 0$: $0 = x\sqrt{x + 2} \implies x = 0, -2 \implies (0, 0), (-2, 0)$ x-intercepts

Let $x = 0$: $y = 0\sqrt{0 + 2} = 0 \implies (0, 0)$ y-intercept

47. $xy = 4$

If $x = 0$, then $0y = 0 = 4$, which is impossible. Similarly, $y = 0$ is impossible.
Hence there are no intercepts.

49. $y = |x - 2| - 4$

Let $y = 0$: $|x - 2| - 4 = 0 \implies |x - 2| = 4 \implies x = -2, 6 \implies (-2, 0), (6, 0)$ x-intercepts

Let $x = 0$: $|0 - 2| - 4 = |-2| - 4 = 2 - 4 = -2 = y \implies (0, -2)$ y-intercept

51. $xy - 2y - x + 1 = 0$

Let $y = 0$:

$-x + 1 = 0 \implies x = 1 \implies (1, 0)$ x-intercept

Let $x = 0$:

$-2y + 1 = 0 \implies y = \frac{1}{2} \implies \left(0, \frac{1}{2}\right)$ y-intercept

53.

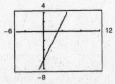

$y = 0 = 2(x - 1) - 4$

$\qquad = 2x - 2 - 4$

$\qquad = 2x - 6 \implies 2x = 6 \implies x = 3$

$(3, 0)$

55.

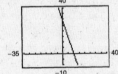

$y = 0 = 20 - (3x - 10)$

$\qquad = 20 - 3x + 10$

$\qquad = 30 - 3x \implies 3x = 30 \implies x = 10$

$(10, 0)$

57. $f(x) = 5(4 - x)$

$\qquad 5(4 - x) = 0$

$\qquad 4 - x = 0$

$\qquad x = 4$

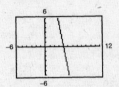

59. $f(x) = x^3 - 6x^2 + 5x$

$x^3 - 6x^2 + 5x = 0$

$x(x^2 - 6x + 5) = 0$

$x(x - 5)(x - 1) = 0$

$x = 0, 5, 1$

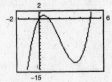

61.

$$f(x) = \frac{x+2}{3} - \frac{x-1}{5} - 1$$

$$\frac{x+2}{3} - \frac{x-1}{5} - 1 = 0$$

$$5(x+2) - 3(x-1) - 15 = 0$$

$$2x = 2$$

$$x = 1$$

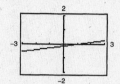

63. $2.7x - 0.4x = 1.2$

$$2.3x = 1.2$$

$$x = \frac{1.2}{2.3} \approx 0.522$$

$$f(x) = 2.7x - 0.4x - 1.2 = 0$$

$$x \approx 0.522$$

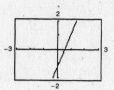

65. $25(x-3) = 12(x+2) - 10$

$$25x - 75 = 12x + 24 - 10$$

$$13x - 89 = 0$$

$$x = \frac{89}{13}$$

$$f(x) = 25(x-3) - 12(x+2) + 10 = 0$$

$$x = 6.846$$

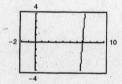

67. $\frac{3x}{2} + \frac{1}{4}(x-2) = 10$

$$\frac{6x}{4} + \frac{x}{4} = 10 + \frac{1}{2}$$

$$\frac{7x}{4} = \frac{21}{2}$$

$$x = 6$$

$$f(x) = \frac{3x}{2} + \frac{1}{4}(x-2) - 10 = 0$$

$$x = 6.0$$

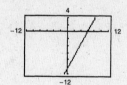

69. $0.60x + 0.40(100 - x) = 1.2$

$$0.60x + 40 - 0.40x = 1.2$$

$$0.20x = -38.8$$

$$x = -194$$

$$f(x) = 0.60x + 0.40(100 - x) - 1.2 = 0$$

$$x = -194$$

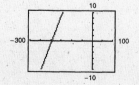

71.

$$\frac{2x}{3} = 10 - \frac{24}{x}$$

$$\frac{2x}{3}(3x) = 10(3x) - \frac{24}{x}(3x)$$

$$2x^2 = 30x - 72$$

$$2x^2 - 30x + 72 = 0$$

$$x^2 - 15x + 36 = 0$$

$$(x-3)(x-12) = 0$$

$$x = 3, 12$$

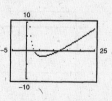

$$f(x) = \frac{2x}{3} - 10 + \frac{24}{x}$$

$$x = 3, 12$$

73.
$$\frac{3}{x + 2} - \frac{4}{x - 2} = 5$$

$$3(x - 2) - 4(x + 2) = 5(x + 2)(x - 2)$$

$$3x - 6 - 4x - 8 = 5(x^2 - 4)$$

$$0 = 5x^2 + x - 6$$

$$0 = (x - 1)(5x + 6)$$

$$x = 1, -\frac{6}{5}$$

$$f(x) = \frac{3}{x + 2} - \frac{4}{x - 2} - 5$$

$$= 0$$

$$x = 1.0, -1.2$$

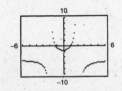

75.
$$(x + 2)^2 = x^2 - 6x + 1$$

$$x^2 + 4x + 4 = x^2 - 6x + 1$$

$$10x = -3$$

$$x = -\frac{3}{10}$$

$$f(x) = (x + 2)^2 - x^2 + 6x - 1$$

$$x = -\frac{3}{10}$$

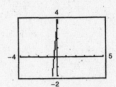

77. $2x^3 - x^2 - 18x + 9 = 0$

$$x = -3.0, 0.5, 3.0$$

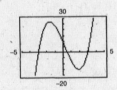

79.
$$x^4 = 2x^3 + 1$$

$$x^4 - 2x^3 - 1 = 0$$

$$x \approx -0.717, 2.107$$

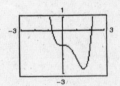

81.
$$\frac{2}{x + 2} = 3$$

$$\frac{2}{x + 2} - 3 = 0$$

$$x = -\frac{4}{3}$$

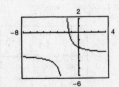

83.
$$|x - 3| = 4$$

$$|x - 3| - 4 = 0$$

$$x = -1, 7$$

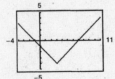

85.
$$\sqrt{x - 2} = 3$$

$$\sqrt{x - 2} - 3 = 0$$

$$x = 11$$

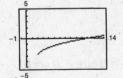

87.
$$y = 2 - x$$
$$y = 2x - 1$$
$$2 - x = 2x - 1$$
$$3 = 3x$$
$$x = 1, y = 2 - 1 = 1$$
$$(x, y) = (1, 1)$$

89.
$$x - y = -4 \implies y = x + 4$$
$$x^2 - y = -2 \implies y = x^2 + 2$$
$$x^2 + 2 = x + 4$$
$$x^2 - x - 2 = 0$$
$$(x - 2)(x + 1) = 0$$
$$x = 2, y = 6$$
$$x = -1, y = 3$$
$$(2, 6), (-1, 3)$$

91.
$$y = x^2 - x + 1$$
$$y = x^2 + 2x + 4$$
$$x^2 - x + 1 = x^2 + 2x + 4$$
$$-3 = 3x$$
$$x = -1$$
$$y = (-1)^2 - (-1) + 1 = 3$$
$$(x, y) = (-1, 3)$$

93. $y = 9 - 2x$
$$y = x - 3$$
$$(4, 1)$$

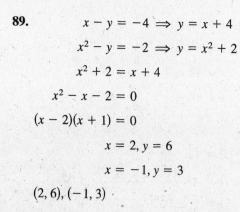

95. $y = 4 - x^2$
$$y = 2x - 1$$
$$(x, y) = (1.449, 1.898), (-3.449, -7.899)$$

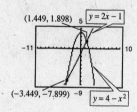

97. $y = 2x^2$
$$y = x^4 - 2x^2$$
$$(x, y) = (0, 0), (2, 8), (-2, 8)$$

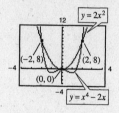

99.
$$6x^2 + 3x = 0$$
$$3x(2x + 1) = 0$$
$$3x = 0 \text{ or } 2x + 1 = 0$$
$$x = 0 \text{ or } \quad x = -\tfrac{1}{2}$$

101.
$$x^2 - 2x - 8 = 0$$
$$(x - 4)(x + 2) = 0$$
$$x - 4 = 0 \text{ or } x + 2 = 0$$
$$x = 4 \text{ or } \quad x = -2$$

103.
$$3 + 5x - 2x^2 = 0$$
$$(3 - x)(1 + 2x) = 0$$
$$3 - x = 0 \text{ or } 1 + 2x = 0$$
$$x = 3 \text{ or } \quad x = -\tfrac{1}{2}$$

105.
$$x^2 + 4x = 12$$
$$x^2 + 4x - 12 = 0$$
$$(x + 6)(x - 2) = 0$$
$$x + 6 = 0 \quad \text{or } x - 2 = 0$$
$$x = -6 \text{ or } \quad x = 2$$

107.
$$(x + a)^2 - b^2 = 0$$
$$[(x + a) + b][(x + a) - b] = 0$$
$$x + a + b = 0 \implies x = -a - b$$
$$x + a - b = 0 \implies x = -a + b$$

109. $x^2 = 49$
$$x = \pm\sqrt{49} = \pm 7$$

111. $(x - 12)^2 = 16$

$$x - 12 = \pm\sqrt{16} = \pm 4$$
$$x = 12 \pm 4$$
$$x = 16, 8$$

113. $(3x - 1)^2 + 6 = 0$

$$(3x - 1)^2 = -6$$
$$3x - 1 = \pm\sqrt{-6} = \pm\sqrt{6}i$$
$$x = \frac{1}{3} \pm \frac{\sqrt{6}}{3}i \approx 0.33 \pm 0.82i$$

115. $(2x - 1)^2 = 12$

$$2x - 1 = \pm\sqrt{12} = \pm 2\sqrt{3}$$
$$2x = 1 \pm 2\sqrt{3}$$
$$x = \frac{1}{2} \pm \sqrt{3}$$
$$x = 2.23, -1.23$$

117. $(x - 7)^2 = (x + 3)^2$

$$x - 7 = \pm(x + 3)$$
$$x - 7 = x + 3, \quad \text{impossible}$$
$$x - 7 = -(x + 3) \implies 2x = 4$$
$$\implies \quad x = 2$$

119. $\qquad x^2 + 4x = 32$

$$x^2 + 4x + 4 = 32 + 4$$
$$(x + 2)^2 = 36$$
$$x + 2 = \pm 6$$
$$x = -2 \pm 6$$
$$x = -8, 4$$

121. $\quad x^2 + 6x + 2 = 0$

$$x^2 + 6x = -2$$
$$x^2 + 6x + 3^2 = -2 + 3^2$$
$$(x + 3)^2 = 7$$
$$x + 3 = \pm\sqrt{7}$$
$$x = -3 \pm \sqrt{7}$$

123. $9x^2 - 18x + 3 = 0$

$$x^2 - 2x + \frac{1}{3} = 0$$
$$x^2 - 2x = -\frac{1}{3}$$
$$x^2 - 2x + 1^2 = -\frac{1}{3} + 1^2$$
$$(x - 1)^2 = \frac{2}{3}$$
$$x - 1 = \pm\sqrt{\frac{2}{3}}$$
$$x = 1 \pm \sqrt{\frac{2}{3}}$$
$$x = 1 \pm \frac{\sqrt{6}}{3}$$

125. $-6 + 2x - x^2 = 0$

$$(x^2 - 2x + 1) = -6 + 1$$
$$(x - 1)^2 = -5$$
$$x - 1 = \pm\sqrt{-5}$$
$$= \pm\sqrt{5}i$$
$$x = 1 \pm \sqrt{5}i$$

127. $2x^2 + 5x - 8 = 0$

$$x^2 + \frac{5}{2}x - 4 = 0$$
$$x^2 + \frac{5}{2}x + \frac{25}{16} = 4 + \frac{25}{16}$$
$$\left(x + \frac{5}{4}\right)^2 = \frac{89}{16}$$
$$x + \frac{5}{4} = \pm\frac{\sqrt{89}}{4}$$
$$x = \frac{-5}{4} \pm \frac{\sqrt{89}}{4}$$

129. $-x^2 + 2x + 2 = 0$

$$x = \frac{-b \pm \sqrt{b^2 - 4ac}}{2a}$$
$$= \frac{-2 \pm \sqrt{2^2 - 4(-1)(2)}}{2(-1)}$$
$$= \frac{-2 \pm 2\sqrt{3}}{-2} = 1 \pm \sqrt{3}$$

131. $x^2 + 8x - 4 = 0$

$$x = \frac{-b \pm \sqrt{b^2 - 4ac}}{2a}$$

$$= \frac{-8 \pm \sqrt{8^2 - 4(1)(-4)}}{2(1)}$$

$$= \frac{-8 \pm 4\sqrt{5}}{2} = -4 \pm 2\sqrt{5}$$

133. $x^2 + 3x + 8 = 0$

$$x = \frac{-3 \pm \sqrt{9 - 4(8)}}{2}$$

$$= \frac{-3 \pm \sqrt{-23}}{2}$$

$$= -\frac{3}{2} \pm \frac{\sqrt{23}i}{2}$$

135. $28x - 49x^2 = 4$

$-49x^2 + 28x - 4 = 0$

$$x = \frac{-b \pm \sqrt{b^2 - 4ac}}{2a}$$

$$= \frac{-28 \pm \sqrt{28^2 - 4(-49)(-4)}}{2(-49)}$$

$$= \frac{-28 \pm 0}{-98} = \frac{2}{7}$$

137. $4x^2 + 16x + 17 = 0$

$$x = \frac{-b \pm \sqrt{b^2 - 4ac}}{2a}$$

$$= \frac{-16 \pm \sqrt{16^2 - 4(4)(17)}}{2(4)}$$

$$= \frac{-16 \pm \sqrt{-16}}{8}$$

$$= \frac{-16 \pm 4i}{8}$$

$$= -2 \pm \frac{1}{2}i$$

139. $x^2 - 2x - 1 = 0$

$x^2 - 2x = 1$

$x^2 - 2x + 1^2 = 1 + 1^2$

$(x - 1)^2 = 2$

$x - 1 = \pm\sqrt{2}$

$x = 1 \pm \sqrt{2}$

141. $(x + 3)^2 = 81$

$x + 3 = \pm 9$

$x + 3 = 9$ or $x + 3 = -9$

$x = 6$ or $x = -12$

143. $x^2 - 14x + 49 = 0$

$(x - 7)^2 = 0$

$x = 7$

145. $x^2 - x - \frac{11}{4} = 0$

$$x^2 - x + \frac{1}{4} = \frac{11}{4} + \frac{1}{4}$$

$$\left(x - \frac{1}{2}\right)^2 = 3$$

$$x - \frac{1}{2} = \pm\sqrt{3}$$

$$x = \frac{1}{2} \pm \sqrt{3}$$

$$x = \frac{1}{2} \pm \sqrt{3}$$

147. $(x + 1)^2 = x^2$

$(x + 1)^2 - x^2 = 0$

$(x + 1 - x)(x + 1 + x) = 0$

$2x + 1 = 0$

$2x = -1$

$$x = -\frac{1}{2}$$

149. $4x^4 - 16x^2 = 0$

$4x^2(x^2 - 4) = 0$

$4x^2(x - 2)(x + 2) = 0$

$x = 0, \pm 2$

151. $5x^3 + 30x^2 + 45x = 0$

$5x(x^2 + 6x + 9) = 0$

$5x(x + 3)^2 = 0$

$5x = 0 \implies x = 0$

$x + 3 = 0 \implies x = -3$

153. $4x^4 - 18x^2 = 0$

$2x^2(2x^2 - 9) = 0$

$2x^2 = 0 \implies x = 0$

$2x^2 - 9 = 0 \implies x = \pm\dfrac{3\sqrt{2}}{2}$

155.
$$x^4 - 4x^2 + 3 = 0$$

$$(x^2 - 3)(x^2 - 1) = 0$$

$$\left(x + \sqrt{3}\right)\left(x - \sqrt{3}\right)(x + 1)(x - 1) = 0$$

$x + \sqrt{3} = 0 \implies x = -\sqrt{3}$

$x - \sqrt{3} = 0 \implies x = \sqrt{3}$

$x + 1 = 0 \implies x = -1$

$x - 1 = 0 \implies x = 1$

157. $x^3 - 3x^2 - x + 3 = 0$

$x^2(x - 3) - (x - 3) = 0$

$(x - 3)(x^2 - 1) = 0$

$(x - 3)(x + 1)(x - 1) = 0$

$x - 3 = 0 \implies x = 3$

$x + 1 = 0 \implies x = -1$

$x - 1 = 0 \implies x = 1$

159.
$$4x^4 - 65x^2 + 16 = 0$$

$$(4x^2 - 1)(x^2 - 16) = 0$$

$$(2x + 1)(2x - 1)(x + 4)(x - 4) = 0$$

$2x + 1 = 0 \implies x = -\frac{1}{2}$

$2x - 1 = 0 \implies x = \frac{1}{2}$

$x + 4 = 0 \implies x = -4$

$x - 4 = 0 \implies x = 4$

161. $\dfrac{1}{t^2} + \dfrac{8}{t} + 15 = 0$

$1 + 8t + 15t^2 = 0$

$(1 + 3t)(1 + 5t) = 0$

$1 + 3t = 0 \implies t = -\dfrac{1}{3}$

$1 + 5t = 0 \implies t = -\dfrac{1}{5}$

163. $6\left(\dfrac{s}{s + 1}\right)^2 + 5\left(\dfrac{s}{s + 1}\right) - 6 = 0$

Let $u = \dfrac{s}{(s + 1)}$.

$6u^2 + 5u - 6 = 0$

$(3u - 2)(2u + 3) = 0$

$3u - 2 = 0 \implies u = \dfrac{2}{3}$

$2u + 3 = 0 \implies u = -\dfrac{3}{2}$

$\dfrac{s}{s + 1} = \dfrac{2}{3} \implies s = 2$

$\dfrac{s}{s + 1} = -\dfrac{3}{2} \implies s = -\dfrac{3}{5}$

165.
$$2x + 9\sqrt{x} - 5 = 0$$
$$(2\sqrt{x} - 1)(\sqrt{x} + 5) = 0$$
$$\sqrt{x} = \tfrac{1}{2} \implies x = \tfrac{1}{4}$$
$(\sqrt{x} = -5$ is not possible.$)$

Note: You can see graphically that there is only one solution.

167.
$$\sqrt{x - 10} - 4 = 0$$
$$\sqrt{x - 10} = 4$$
$$x - 10 = 16$$
$$x = 26$$

169.
$$\sqrt{x + 1} - 3x = 1$$
$$\sqrt{x + 1} = 3x + 1$$
$$x + 1 = 9x^2 + 6x + 1$$
$$0 = 9x^2 + 5x$$
$$0 = x(9x + 5)$$
$$x = 0$$

$$9x + 5 = 0 \implies x = -\frac{5}{9}, \text{extraneous}$$

171.
$$\sqrt[3]{2x + 1} + 8 = 0$$
$$\sqrt[3]{2x + 1} = -8$$
$$2x + 1 = -512$$
$$2x = -513$$
$$x = -\frac{513}{2} = -256.5$$

173.
$$\sqrt{x} - \sqrt{x - 5} = 1$$
$$\sqrt{x} = 1 + \sqrt{x - 5}$$
$$(\sqrt{x})^2 = (1 + \sqrt{x - 5})^2$$
$$x = 1 + 2\sqrt{x - 5} + x - 5$$
$$4 = 2\sqrt{x - 5}$$
$$2 = \sqrt{x - 5}$$
$$4 = x - 5$$
$$9 = x$$

175.
$$(x - 5)^{2/3} = 16$$
$$x - 5 = \pm 16^{3/2}$$
$$x - 5 = \pm 64$$
$$x = 69, -59$$

177. $3x(x - 1)^{1/2} + 2(x - 1)^{3/2} = 0$
$$(x - 1)^{1/2}[3x + 2(x - 1)] = 0$$
$$(x - 1)^{1/2}(5x - 2) = 0$$
$$(x - 1)^{1/2} = 0 \implies x - 1 = 0 \implies x = 1$$
$$5x - 2 = 0 \implies x = \tfrac{2}{5} \text{ which is extraneous.}$$

179.
$$\frac{1}{x} - \frac{1}{x + 1} = 3$$
$$x(x + 1)\frac{1}{x} - x(x + 1)\frac{1}{x + 1} = x(x + 1)(3)$$
$$x + 1 - x = 3x(x + 1)$$
$$1 = 3x^2 + 3x$$
$$0 = 3x^2 + 3x - 1; \ a = 3, \ b = 3, \ c = -1$$
$$x = \frac{-3 \pm \sqrt{(3)^2 - 4(3)(-1)}}{2(3)} = \frac{-3 \pm \sqrt{21}}{6}$$

181.
$$x = \frac{3}{x} + \frac{1}{2}$$

$$(2x)(x) = (2x)\left(\frac{3}{x}\right) + (2x)\left(\frac{1}{2}\right)$$

$$2x^2 = 6 + x$$

$$2x^2 - x - 6 = 0$$

$$(2x + 3)(x - 2) = 0$$

$$2x + 3 = 0 \Rightarrow x = -\frac{3}{2}$$

$$x - 2 = 0 \Rightarrow x = 2$$

183. $|2x - 1| = 5$

$$2x - 1 = 5 \Rightarrow x = 3$$

$$-(2x - 1) = 5 \Rightarrow x = -2$$

185. $|x| = x^2 + x - 3$

$x = x^2 + x - 3$ OR $-x = x^2 + x - 3$

$x^2 - 3 = 0$ $x^2 + 2x - 3 = 0$

$x = \pm\sqrt{3}$ $(x - 1)(x + 3) = 0$

$$x - 1 = 0 \Rightarrow x = 1$$

$$x + 3 = 0 \Rightarrow x = -3$$

Only $x = \sqrt{3}$, and $x = -3$ are solutions to the original equation. $x = -\sqrt{3}$ and $x = 1$ are extraneous. Note that the graph of $y = x^2 + x - 3 - |x|$ has two x-intercepts.

187. $y = x^3 - 2x^2 - 3x$

(a)

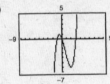

(b) x-intercepts: $(-1, 0), (0, 0), (3, 0)$

(c) $0 = x^3 - 2x^2 - 3x$

$0 = x(x + 1)(x - 3)$

$x = 0$

$x + 1 = 0 \Rightarrow x = -1$

$x - 3 = 0 \Rightarrow x = 3$

(d) The x-intercepts are the same as the solutions.

189. $y = \sqrt{11x - 30} - x$

(a)

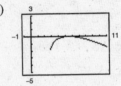

(b) x-intercepts: $(5, 0), (6, 0)$

(c)
$$0 = \sqrt{11x - 30} - x$$

$$x = \sqrt{11x - 30}$$

$$x^2 = 11x - 30$$

$$x^2 - 11x + 30 = 0$$

$$(x - 5)(x - 6) = 0$$

$$x - 5 = 0 \Rightarrow x = 5$$

$$x - 6 = 0 \Rightarrow x = 6$$

(d) The x-intercepts and the solutions are the same.

191. $y = \dfrac{1}{x} - \dfrac{4}{x-1} - 1$

(a)

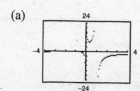

(b) x-intercept: $(-1, 0)$

(c) $\quad 0 = \dfrac{1}{x} - \dfrac{4}{x-1} - 1$

$\quad 0 = (x-1) - 4x - x(x-1)$

$\quad 0 = x - 1 - 4x - x^2 + x$

$\quad 0 = -x^2 - 2x - 1$

$\quad 0 = x^2 + 2x + 1$

$\quad x + 1 = 0 \implies x = -1$

(d) The x-intercepts and the solutions are the same.

195. (a)

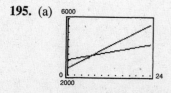

Intersection: $(6.7, 3388.7)$

(c) The slopes indicate the change in population per year. Arizona's population is growing faster.

197. $C = 0.45x^2 - 1.65x + 50.75, \ 10 \le x \le 25$

(a)

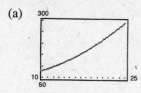

193. $y = |x + 1| - 2$

(a)

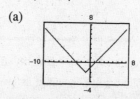

(b) x-intercept: $(1, 0), (-3, 0)$

(c) $0 = |x + 1| - 2$

$2 = |x + 1|$

$x + 1 = 2$ or $-(x + 1) = 2$

$x = 1$ or $\quad -x - 1 = 2$

$-x = 3$

$x = -3$

(d) The x-intercepts and the solutions are the same.

(b) $45.2t + 3087 = 128.2t + 2533$

$83t = 554$

$t \approx 6.7$

$A = S \approx 3388.7$

The point $(6.7, 3388.7)$ indicates the year, 1986, in which the two populations were the same, about 3388.7 thousand.

(d) For 2010, $t = 30$ and $S \approx 4443$ thousand and $A \approx 6379$ thousand. Answers will vary.

(b) If $C = 150$, then $x = 16.797$ degrees.

(c) If the temperature is increased $10°$ to $20°$, then C increases from 79.25 to 197.75, a factor of 2.5.

199. False. Two linear equations could have an infinite number of points of intersection. For example, $x + y = 1$ and $2x + 2y = 2$.

201. $2x - 5c = 10 + 3c - 3x, \quad x = 3$

$2(3) - 5c = 10 + 3c - 3(3)$

$6 - 5c = 1 + 3c$

$5 = 8c$

$c = \frac{5}{8}$

203. (a) $ax^2 + bx = 0$

$x(ax + b) = 0$

$x = 0$

$x = -b/a$

(b) $ax^2 - ax = 0$

$ax(x - 1) = 0$

$x = 0$

$x = 1$

Appendix B.4 Solving Inequalities Algebraically and Graphically

- ■ You should be able to solve an inequality algebraically using the Properties of Inequalities.
- ■ You should be able to solve inequalities involving absolute values.
- ■ You should be able to solve polynomial inequalities using critical numbers and test intervals.
- ■ You should be able to solve rational inequalities.
- ■ You should be able to solve inequalities using a graphing utility.

Vocabulary Check

1. negative

2. double

3. $-a \leq x \leq a$

4. $x \leq -a, x \geq a$

5. zeros, undefined values

---

1. $x < 3$

Matches (f).

3. $-3 < x \leq 4$

Matches (d).

5. $-1 \leq x \leq \frac{5}{2}$

Matches (e).

7. (a) $x = 3$

$$5(3) - 12 \overset{?}{>} 0$$

$$3 > 0$$

Yes, $x = 3$ is a solution.

(b) $x = -3$

$$5(-3) - 12 \overset{?}{>} 0$$

$$-27 \not> 0$$

No, $x = -3$ is not a solution.

(c) $x = \frac{5}{2}$

$$5\left(\tfrac{5}{2}\right) - 12 \overset{?}{>} 0$$

$$\tfrac{1}{2} > 0$$

Yes, $x = \frac{5}{2}$ is a solution.

(d) $x = \frac{3}{2}$

$$5\left(\tfrac{3}{2}\right) - 12 \overset{?}{>} 0$$

$$-\tfrac{9}{2} \not> 0$$

No, $x = \frac{3}{2}$ is not a solution.

9. $-1 < \dfrac{3 - x}{2} \leq 1$

(a) $x = 0$

$$-1 \overset{?}{<} \frac{3 - 0}{2} \overset{?}{\leq} 1$$

$$-1 \overset{?}{<} \frac{3}{2} \overset{?}{\leq} 1$$

No, $x = 0$ is not a solution.

(b) $x = \sqrt{5}$

$$-1 \overset{?}{<} \frac{3 - \sqrt{5}}{2} \overset{?}{\leq} 1$$

$$-1 \overset{?}{<} 0.382 \overset{?}{\leq} 1$$

Yes, $x = \sqrt{5}$ is a solution.

(c) $x = 1$

$$-1 \overset{?}{<} \frac{3 - 1}{2} \overset{?}{\leq} 1$$

$$-1 \overset{?}{<} 1 \overset{?}{\leq} 1$$

Yes, $x = 1$ is a solution.

(d) $x = 5$

$$-1 \overset{?}{<} \frac{3 - 5}{2} \leq 1$$

$$-1 \overset{?}{<} -1 \overset{?}{\leq} 1$$

No, $x = 5$ is not a solution.

11. $-10x < 40$

$-\frac{1}{10}(-10x) > -\frac{1}{10}(40)$

$x > -4$

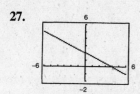

13. $4(x + 1) < 2x + 3$

$4x + 4 < 2x + 3$

$2x < -1$

$x < -\frac{1}{2}$

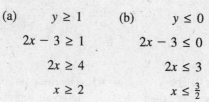

15. $\frac{3}{4}x - 6 \leq x - 7$

$1 \leq \frac{1}{4}x$

$4 \leq x$

$x \geq 4$

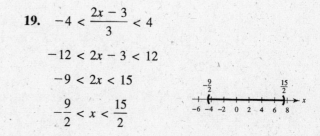

17. $-8 \leq 1 - 3(x - 2) < 13$

$-8 \leq 1 - 3x + 6 < 13$

$-8 \leq -3x + 7 < 13$

$-15 \leq -3x < 6$

$5 \geq x > -2 \implies -2 < x \leq 5$

19. $-4 < \dfrac{2x - 3}{3} < 4$

$-12 < 2x - 3 < 12$

$-9 < 2x < 15$

$-\dfrac{9}{2} < x < \dfrac{15}{2}$

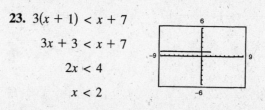

21. $5 - 2x \geq 1$

$-2x \geq -4$

$x \leq 2$

23. $3(x + 1) < x + 7$

$3x + 3 < x + 7$

$2x < 4$

$x < 2$

25.

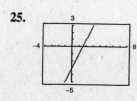

Using the graph, (a) $y \geq 1$ for $x \geq 2$ and (b) $y \leq 0$ for $x \leq \frac{3}{2}$. Algebraically:

(a)　　$y \geq 1$

$2x - 3 \geq 1$

$2x \geq 4$

$x \geq 2$

(b)　　$y \leq 0$

$2x - 3 \leq 0$

$2x \leq 3$

$x \leq \frac{3}{2}$

27.

Using the graph, (a) $0 \leq y \leq 3$ for $-2 \leq x \leq 4$ and (b) $y \geq 0$ for $x \leq 4$. Algebraically:

(a)　$0 \leq y \leq 3$

$0 \leq -\frac{1}{2}x + 2 \leq 3$

$-2 \leq -\frac{1}{2}x \leq 1$

$4 \geq x \geq -2$

(b)　　$y \geq 0$

$-\frac{1}{2}x + 2 \geq 0$

$2 \geq \frac{1}{2}x$

$4 \geq x$

29. $|5x| > 10$

$5x < -10$ or $5x > 10$

$x < -2$ or $x > 2$

31. $|x - 7| < 6$

$$-6 < x - 7 < 6$$

$$1 < x < 13$$

33. $|x + 14| + 3 > 17$

$$|x + 14| > 14$$

$$x + 14 < -14 \text{ or } x + 14 > 14$$

$$x < -28 \text{ or } \qquad x > 0$$

35. $10|1 - 2x| < 5$

$$|1 - 2x| < \tfrac{1}{2}$$

$$-\tfrac{1}{2} < 1 - 2x < \tfrac{1}{2}$$

$$-\tfrac{3}{2} < -2x < -\tfrac{1}{2}$$

$$\tfrac{3}{4} > x > \tfrac{1}{4}$$

$$\tfrac{1}{4} < x < \tfrac{3}{4}$$

37. $y = |x - 3|$

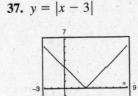

Graphically, (a) $y \leq 2$ for $1 \leq x \leq 5$ and (b) $y \geq 4$ for $x \leq -1$ or $x \geq 7$. Algebraically:

(a) $\qquad y \leq 2$

$$|x - 3| \leq 2$$

$$-2 \leq x - 3 \leq 2$$

$$1 \leq x \leq 5$$

(b) $\qquad y \geq 4$

$$|x - 3| \geq 4$$

$$x - 3 \leq -4 \text{ or } \quad x - 3 \geq 4$$

$$x \leq -1 \text{ or } \qquad x \geq 7$$

39. The midpoint of the interval $[-3, 3]$ is 0. The interval represents all real numbers x no more than three units from 0.

$$|x - 0| \leq 3$$

$$|x| \leq 3$$

41. The midpoint of the interval $[-3, 3]$ is 0. The two intervals represent all numbers x more than three units from 0.

$$|x - 0| > 3$$

$$|x| > 3$$

43. All real numbers within 10 units of 7

$$|x - 7| \leq 10$$

45. All real numbers at least five units from 3

$$|x - 3| \geq 5$$

47. $x^2 - 4x - 5 > 0$

$$(x - 5)(x + 1) > 0$$

Critical numbers: $-1, 5$

Testing the intervals $(-\infty, -1)$, $(-1, 5)$ and $(5, \infty)$, we have $x^2 - 4x - 5 > 0$ on $(-\infty, -1)$ and $(5, \infty)$. Similarly, $x^2 - 4x - 5 < 0$ on $(-1, 5)$.

49. $2x^2 - 4x - 3 = 0$

$$x = \frac{4 \pm \sqrt{16 + 24}}{4} = 1 \pm \frac{\sqrt{10}}{2}$$

Entirely negative:

$$\left(1 - \frac{\sqrt{10}}{2}, 1 + \frac{\sqrt{10}}{2}\right) \approx (-0.581, 2.581)$$

Entirely positive:

$$\left(-\infty, 1 - \frac{\sqrt{10}}{2}\right) \cup \left(1 + \frac{\sqrt{10}}{2}, \infty\right)$$

51. $x^2 - 4x + 5 > 0$ for all x. There are no critical numbers because $x^2 - 4x + 5 \neq 0$. The only test interval is $(-\infty, \infty)$.

53.

$$(x + 2)^2 < 25$$

$$x^2 + 4x + 4 < 25$$

$$x^2 + 4x - 21 < 0$$

$$(x + 7)(x - 3) < 0$$

Critical numbers: $x = -7, x = 3$

Test intervals: $(-\infty, -7), (-7, 3), (3, \infty)$

Test: Is $(x + 7)(x - 3) < 0$?

Solution set: $(-7, 3)$

55.

$$x^2 + 4x + 4 \geq 9$$

$$x^2 + 4x - 5 \geq 0$$

$$(x + 5)(x - 1) \geq 0$$

Critical numbers: $x = -5, x = 1$

Test intervals: $(-\infty, -5), (-5, 1), (1, \infty)$

Test: Is $(x + 5)(x - 1) \geq 0$?

Solution set: $(-\infty, -5] \cup [1, \infty)$

57.

$$x^3 - 4x \geq 0$$

$$x(x + 2)(x - 2) \geq 0$$

Critical number: $x = 0, x = \pm 2$

Test intervals: $(-\infty, -2), (-2, 0), (0, 2), (2, \infty)$

Test: Is $x(x + 2)(x - 2) \geq 0$?

Solution set: $[-2, 0] \cup [2, \infty)$

59. $3x^2 - 11x + 16 \leq 0$

Since $b^2 - 4ac = -71 < 0$, there are no real solutions to $3x^2 - 11x + 16 = 0$. In fact, $3x^2 - 11x + 16 > 0$ for all x.

No solution

61.

$$2x^3 + 5x^2 - 6x - 9 > 0$$

$$(x + 1)(x + 3)(2x - 3) > 0$$

Critical numbers: $-3, -1, \frac{3}{2}$

Testing the four intervals, we see that $2x^3 + 5x^2 - 6x - 9 > 0$ on $(-3, -1)$ and $\left(\frac{3}{2}, \infty\right)$.

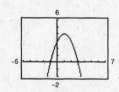

63. (a) $f(x) = g(x)$ when $x = 1$.

(b) $f(x) \geq g(x)$ when $x \geq 1$.

(c) $f(x) > g(x)$ when $x > 1$.

65. $y = -x^2 + 2x + 3$

(a) $y \leq 0$ when $x \leq -1$ or $x \geq 3$.

(b) $y \geq 3$ when $0 \leq x \leq 2$.

Algebraically,

$$-x^2 + 2x + 3 \leq 0$$

$$x^2 - 2x - 3 \geq 0$$

$$(x - 3)(x + 1) \geq 0$$

Critical numbers: $x = -1, x = 3$

Testing the intervals $(-\infty, -1)$, $(-1, 3)$, and $(3, \infty)$, you obtain $x \leq -1$ or $x \geq 3$.

$$-x^2 + 2x + 3 \geq 3$$

$$-x^2 + 2x \geq 0$$

$$x^2 - 2x \leq 0$$

$$x(x - 2) \leq 0$$

Critical numbers: $x = 0, x = 2$

Testing the intervals $(-\infty, 0)$, $(0, 2)$, and $(2, \infty)$, you obtain $0 \leq x \leq 2$.

67. $\dfrac{1}{x} - x > 0$

$\dfrac{1 - x^2}{x} > 0$

Critical numbers: $x = 0, x = \pm 1$

Test intervals: $(-\infty, -1), (-1, 0), (0, 1), (1, \infty)$

Test: Is $\dfrac{1 - x^2}{x} > 0$?

Solution set: $(-\infty, -1) \cup (0, 1)$

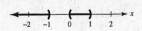

69. $\dfrac{x + 6}{x + 1} - 2 < 0$

$\dfrac{x + 6 - 2(x + 1)}{x + 1} < 0$

$\dfrac{4 - x}{x + 1} < 0$

Critical numbers: $x = -1, x = 4$

Test intervals: $(-\infty, -1), (-1, 4), (4, \infty)$

Test: Is $\dfrac{4 - x}{x + 1} < 0$?

Solution set: $(-\infty, -1) \cup (4, \infty)$

71. $y = \dfrac{3x}{x - 2}$

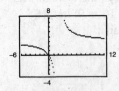

(a) $y \le 0$ when $0 \le x < 2$.

(b) $y \ge 6$ when $2 < x \le 4$.

73. $\sqrt{x - 5}$

Need: $x - 5 \ge 0$

$x \ge 5$

Domain: $[5, \infty)$

75. $\sqrt[3]{6 - x}$

Domain: all real x

77. $\sqrt{x^2 - 4}$

Need: $x^2 - 4 \ge 0$

$(x + 2)(x - 2) \ge 0$

$x \le -2$ or $x \ge 2$

Domain: $(-\infty, -2] \cup [2, \infty)$

79. (a) $P(t) = 1000$

This occurs at the point of intersection, $t \approx 4$, or 1994.

(b) Less than one million: $P(t) < 1000$
This occurs for $t < 4$, or before 1994.

Greater than one million: $P(t) > 1000$
This occurs for $t > 4$, or after 1994.

81. (a) $s = -16t^2 + v_0 t + s_0$

$s = -16t^2 + 160t$

$s = 16t(10 - t)$

$s = 0$ when $t = 10$ seconds.

(b) $s = -16t^2 + 160t > 384$

$16t^2 - 160t + 384 < 0$

$16(t - 6)(t - 4) < 0$

$s > 384$ when $4 < t < 6$.

83. (a)

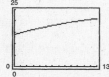

(b) $15 < D < 20$ for $1.28 < t < 10.09$, or between 1991 and 2000

(c) $15 < D < 20$

$$15 < -0.0165t^2 + 0.755t + 14.06 < 20$$

To solve these inequalities, find the critical numbers.

$$0.0165t^2 - 0.755t + 0.94 = 0$$

$$t = \frac{0.755 \pm \sqrt{(-0.755)^2 - 4(0.0165)(0.94)}}{2(0.0165)}$$

$$= \frac{0.755 \pm \sqrt{0.507985}}{0.033}$$

Because $0 < t < 13$, select the negative sign, $t \approx 1.28$. Hence, $15 < D$ for $1.28 < t$.
Similarly, $D < 20$ for $t < 10.09$.

(d) No. $D(t) < 30$ for all t.

85. $V(t) \geq 65$

$$3.37t + 57.9 \geq 65$$

$$3.37t \geq 7.1$$

$$t \geq 2.11$$

The number of hours playing video games exceeded 65 in 2002.

87. $V(t) = N(t)$

$$3.37t + 57.9 = -2.51t + 179.6$$

$$5.88t = 121.7$$

$$t \approx 20.7$$

According to these models, the number of hours reading daily newspapers and playing video games will be the same in 2020.

89. When $t = 2$, $v \approx 333$ vibrations per second.

91. When $200 \leq v \leq 400$, $1.2 < t < 2.4$.

93. (a) Option A: $A(t) = 0.15t + 12$

Option B: $B(t) = 0.20t$

(b)

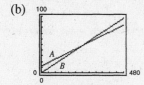

(c) $A(t) = B(t)$ when $t = 240$. $B(t)$ is the better choice if you use less than 240 minutes.
$A(t)$ is the better choice if you use more than 240 minutes.

(d) Answers will vary.

95. False. If $-10 \leq x \leq 8$, then $10 \geq -x$ and $-x \geq -8$.

97. The polynomial $f(x) = (x - a)(x - b)$ is zero at $x = a$ and $x = b$.

99. (iv) $a < b$

   (ii) $2a < 2b$

   (iii) $2a < a + b < 2b$

   (i) $a < \dfrac{a + b}{2} < b$

Appendix B.5   Representing Data Graphically

■ You should be able to construct line plots.

■ You should be able to construct histograms or frequency distributions.

■ You should be able to construct bar graphs.

■ You should be able to construct line graphs.

Vocabulary Check

1. Statistics   **2.** Line plots   **3.** histogram

4. frequency distribution   **5.** bar graph   **6.** Line graphs

1. (a) The price 2.569 occurred with the greatest frequency (6).

   (b) The prices range from 2.459 to 2.649. The range is $2.649 - 2.459 = 0.19$.

3.

```
        ×
        ×      × × ×
    × × ×      × × × ×
    × × ×      × × × ×
  × ×  × × ×   × × × × ×     ×
  +--+--+--+--+--+--+--+--+--+
  10  12  14  16  18  20  22  24
              Quiz Scores
```

The score of 15 occurred with the greatest frequency.

5. (Answers will depend on intervals selected.)

Interval	Tally
$[0, 25)$	ⅢⅢⅢⅢⅢⅢ |||
$[25, 50)$	ⅢⅢⅢⅢⅢⅢ	
$[50, 75)$	ⅢⅢ	
$[75, 100)$	ⅢⅢ |||
$[100, 125)$		
$[125, 150)$	
$[150, 175)$	
$[175, 200)$	
$[200, 225)$	
$[225, 250)$		

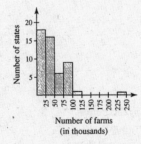

7.

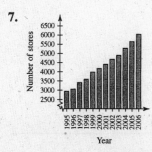

From 1995 to 2006, the number of Wal-Mart stores increases at a fairly constant rate.

9. 1999: 13,428 − 2430 = $10,998

2000: 14,081 − 2506 = $11,575

2001: 15,000 − 2562 = $12,438

2002: 15,742 − 2700 = $13,042

2003: 16,383 − 2903 = $13,480

2004: 17,442 − 3313 = $14,129

11.

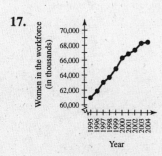

13. From 1996 to 2005:

$$\frac{2400 - 1100}{1100} \approx 1.18, \text{ or } 118\%$$

15. Highest price was $2.59 in January.

17.

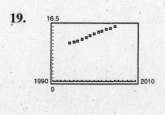

From 1995 to 2004, the total number of women in the work force increases at a fairly constant rate.

19.

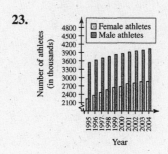

21.

Average monthly bill (in dollars) vs. Year

Answers will vary.

23.

Number of athletes (in thousands) — Female athletes, Male athletes — vs. Year

Answers will vary.

25. Answers will vary. Line plots are useful for ordering small sets of data. Histograms or bar graphs can be used to organize larger sets.

A P P E N D I X   C
Concepts in Statistics

Appendix C.1   Measures of Central Tendency and Dispersion

Vocabulary Check

1. measure, central tendency

2. modes, bimodal

3. variance, standard deviation

4. Quartiles

1. Mean $= \dfrac{5 + 12 + 7 + 14 + 8 + 9 + 7}{7}$

   $= \dfrac{62}{7} \approx 8.86$

   Median: 8

   Mode: 7

3. Mean $= \dfrac{5 + 12 + 7 + 24 + 8 + 9 + 7}{7}$

   $= \dfrac{72}{7} \approx 10.29$

   Median: 8

   Mode: 7

5. Mean $= \dfrac{5 + 12 + 7 + 14 + 9 + 7}{6} = \dfrac{54}{6} = 9$

   Median: $\dfrac{7 + 9}{2} = 8$

   Mode: 7

7. (a) The mean is sensitive to extreme values

   (b) Mean: 14.86

   Median: 14

   Mode: 13

   Each is increased by 6.

   (c) Each will increase by k.

9. Mean $= \dfrac{410 + 260 + 320 + 320 + 460 + 150}{6}$

   $= \dfrac{1920}{6} = 320$

   Median: 320

   Mode: 320

11. (a) Jay: $\dfrac{181 + 222 + 196}{3} = 199\frac{2}{3}$

    Hank: $\dfrac{199 + 195 + 205}{3} = 199\frac{2}{3}$

    Buck: $\dfrac{202 + 251 + 235}{3} = 229\frac{1}{3}$

    (b) Adding all nine numbers, you obtain

    $$\text{Mean} = \dfrac{1886}{9} = 209\frac{5}{9}.$$

    (c) Median = 202 (four scores below 202 and four scores above 202)

13. There are many possible answers. For example: $\{4, 4, 10\}$

15. The mean is 76.55 and the median is 82. The median is the best description.

17. (a) Mean = 12, $\sigma \approx 2.83$

    (c) Mean = 12, $\sigma \approx 1.41$

    (b) Mean = 20, $\sigma \approx 2.83$

    (d) Mean = 9, $\sigma \approx 1.41$

19. $\bar{x} = 6$

$v = 10$

$\sigma \approx 3.16$

21. $\bar{x} = 2$

$v = \frac{4}{3}$

$\sigma \approx 1.15$

23. $\bar{x} = 4$

$v = 4$

$\sigma \approx 2$

25. $\bar{x} = 47$

$v = 226$

$\sigma \approx 15.03$

27. $\bar{x} = 6$

$$\sigma = \sqrt{\frac{2^2 + 4^2 + 6^2 + 6^2 + 13^2 + 5^2}{6} - 6^2} = \sqrt{\frac{286}{6} - 36} = \sqrt{\frac{35}{3}} \approx 3.42$$

29. $\bar{x} = 5.8$

$$\sigma = \sqrt{\frac{8.1^2 + 6.9^2 + 3.7^2 + 4.2^2 + 6.1^2}{5} - 5.8^2} = \sqrt{2.712} \approx 1.65$$

31. $\bar{x} = 12$ and $|x_i - 12| = 8$ for all x_i. Hence, $\sigma = 8$.

33. The mean will increase by 5. The standard deviation will not change.

35. $\bar{x} = 235$

$\sigma = 28$

$n = 600$

$1 - \dfrac{1}{2^2} = \dfrac{3}{4}$ lies within two standard deviations:

    $[235 - 2(28), 235 + 2(28)] = [179, 291]$.

$1 - \dfrac{1}{3^2} = \dfrac{8}{9}$ lies within three standard deviations:

    $[235 - 3(28), 235 + 3(28)] = [151, 319]$.

If $\sigma = 16$, then

    $[235 - 2(16), 235 + 2(16)] = [203, 267]$

    $[235 - 3(16), 235 + 3(16)] = [187, 283]$.

37. (a) 12, 13, 13, 14, 14, 15, 20, 23, 23

Median: 14

Lower quartile is median of $\{12, 13, 13\} = 13$.

Upper quartile is median of $\{15, 20, 23, 23\} = 21.5$.

(b)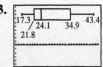

39. (a) 46, 47, 47, 48, 48, 49, 50, 51, 52, 53

Median: $\dfrac{48 + 49}{2} = 48.5$

Lower quartile is median of $\{46, 47, 47, 48, 48\} = 47$.

Upper quartile is median of $\{49, 50, 51, 52, 53\} = 51$.

(b)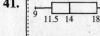

41.

43.

45.

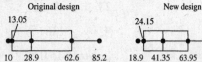

From the plots, you can see that the lifetimes of the units in the new design are greater than the original design. The median increased by over 12 months.

Appendix C.2   Least Squares Regression

1.

x	y	xy	x^2
-4	1	-4	16
-3	3	-9	9
-2	4	-8	4
-1	6	-6	1
Total	-10	14	-27	30

$n = 4$

$4b + (-10)a = 14$

$(-10)b + 30a = -27$

Solving this system, $a = 1.6$ and $b = 7.5$.

Answer: $y = 1.6x + 7.5$

3.

x	y	xy	x^2
-3	1	-3	9
-1	2	-2	1
1	2	2	1
4	3	12	16
Total	1	8	9	27

$n = 4$

$4b + a = 8$

$b + 27a = 9$

Solving this system, $a \approx 0.262$ and $b \approx 1.93$.

Answer: $y = 0.262x + 1.93$

A P P E N D I X   D
Variation

Vocabulary Check

1. directly proportional

2. constant, variation

3. directly proportional

4. inverse

5. combined

6. jointly proportional

1. $y = kx$

$12 = k(5)$

$\frac{12}{5} = k$

$y = \frac{12}{5}x$

3. $y = kx$

$2050 = k(10)$

$205 = k$

$y = 205x$

5. $y = kx$

$33 = k(13)$

$\frac{33}{13} = k$

$y = \frac{33}{13}x$

When $x = 10$ inches, $y \approx 25.4$ centimeters.

When $x = 20$ inches, $y \approx 50.8$ centimeters.

7. $y = kx$

$5520 = k(150,000)$

$0.0368 = k$

$y = 0.0368x$

$y = 0.0368(200,000)$

$= \$7360$

The property tax is $7360.

9. $d = kF$

$0.15 = k(265)$

$\frac{3}{5300} = k$

$d = \frac{3}{5300}F$

(a) $d = \frac{3}{5300}(90) \approx 0.05$ meter

(b) $0.1 = \frac{3}{5300}F$

$\frac{530}{3} = F$

$F = 176\frac{2}{3}$ newtons

11. $k = 1$

x	2	4	6	8	10
$y = kx^2$	4	16	36	64	100

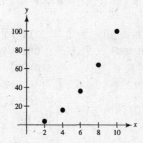

13. $k = \frac{1}{2}$

x	2	4	6	8	10
$y = kx^2$	2	8	18	32	50

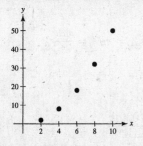

15. $d = kv^2$

$0.02 = k\left(\frac{1}{4}\right)^2$

$k = 0.32$

$d = 0.32v^2$

$0.12 = 0.32v^2$

$v^2 = \frac{0.12}{0.32} = \frac{3}{8}$

$v = \frac{\sqrt{3}}{2\sqrt{2}} = \frac{\sqrt{6}}{4} \approx 0.61$ mi/hr

17. $k = 2$

x	2	4	6	8	10
$y = \dfrac{k}{x^2}$	$\frac{1}{2}$	$\frac{1}{8}$	$\frac{1}{18}$	$\frac{1}{32}$	$\frac{1}{50}$

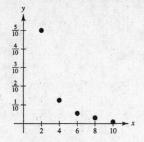

19. $k = 10$

x	2	4	6	8	10
$y = \dfrac{k}{x^2}$	$\frac{5}{2}$	$\frac{5}{8}$	$\frac{5}{18}$	$\frac{5}{32}$	$\frac{1}{10}$

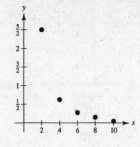

21. The table represents the equation $y = 5/x$.

23. $y = kx$

$-7 = k(10)$

$-\frac{7}{10} = k$

$y = -\frac{7}{10}x$

This equation checks with the other points given in the table.

25. $A = kr^2$

27. $y = \dfrac{k}{x^2}$

29. $F = \dfrac{kg}{r^2}$

31. $P = \dfrac{k}{V}$

33. $R = k(T - T_e)$

35. $A = \dfrac{1}{2}bh$

The area of a triangle is jointly proportional to its base and height.

37. $V = \dfrac{4}{3}\pi r^3$

The volume of a sphere varies directly as the cube of its radius.

39. $r = \dfrac{d}{t}$

Average speed is directly proportional to the distance and inversely proportional to the time.

41. $A = kr^2$

$9\pi = k(3)^2$

$\pi = k$

$A = \pi r^2$

43. $y = \dfrac{k}{x}$

$7 = \dfrac{k}{4}$

$28 = k$

$y = \dfrac{28}{x}$

45. $F = krs^3$

$4158 = k(11)(3)^3$

$k = 14$

$F = 14rs^3$

47. $z = \dfrac{kx^2}{y}$

$6 = \dfrac{k(6)^2}{4}$

$\dfrac{24}{36} = k$

$\dfrac{2}{3} = k$

$z = \dfrac{2/3\,x^2}{y} = \dfrac{2x^2}{3y}$

49. $r = \dfrac{kl}{A}, \quad A = \pi r^2 = \dfrac{\pi d^2}{4}$

$r = \dfrac{4kl}{\pi d^2}$

$66.17 = \dfrac{4(1000)k}{\pi\left(\frac{0.0126}{12}\right)^2}$

$k \approx 5.73 \times 10^{-8}$

$r = \dfrac{4(5.73 \times 10^{-8})l}{\pi\left(\frac{0.0126}{12}\right)^2}$

$33.5 = \dfrac{4(5.73 \times 10^{-8})l}{\pi\left(\frac{0.0126}{12}\right)^2}$

$\dfrac{33.5\,\pi\left(\frac{0.0126}{12}\right)^2}{4(5.73 \times 10^{-8})} = l$

$l \approx 506 \text{ feet}$

51. $W = kmh$

$2116.8 = k(120)(1.8)$

$k = \dfrac{2116.8}{(120)(1.8)} = 9.8$

$W = 9.8mh$

When $m = 100$ kilograms and $h = 1.5$ meters, we have $W = 9.8(100)(1.5) = 1470$ joules.

53. $v = \dfrac{k}{A}$

$v = \dfrac{k}{0.75A} = \dfrac{4}{3}\left(\dfrac{k}{A}\right)$

The velocity is increased by one-third.

55. (a)

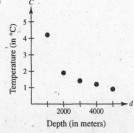

(b) Yes, the data appears to be modeled (approximately) by the inverse proportion model.

$$4.2 = \frac{k_1}{1000} \qquad 1.9 = \frac{k_2}{2000} \qquad 1.4 = \frac{k_3}{3000} \qquad 1.2 = \frac{k_4}{4000} \qquad 0.9 = \frac{k_5}{5000}$$

$$4200 = k_1 \qquad 3800 = k_2 \qquad 4200 = k_3 \qquad 4800 = k_4 \qquad 4500 = k_5$$

(c) Mean: $k = \dfrac{4200 + 3800 + 4200 + 4800 + 4500}{5} = 4300$, Model: $C = \dfrac{4300}{d}$

(d)

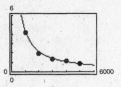

(e) $3 = \dfrac{4300}{d}$

$$d = \frac{4300}{3} = 1433\frac{1}{3} \text{ meters}$$

57. False. y will increase if k is positive and y will decrease if k is negative.

59. The graph appears to represent $y = 4/x$, so y varies inversely as x.

A P P E N D I X E
Solving Linear Equations and Inequalities

Vocabulary Check

1. linear

2. equivalent inequalities

---

1. $x + 11 = 15$
$x = 15 - 11$
$x = 4$

3. $x - 2 = 5$
$x = 5 + 2$
$x = 7$

5. $3x = 12$
$x = \dfrac{12}{3}$
$x = 4$

7. $\dfrac{x}{5} = 4$
$x = 4(5)$
$x = 20$

9. $8x + 7 = 39$
$8x = 32$
$x = 4$

11. $24 - 7x = 3$
$-7x = -21$
$x = 3$

13. $8x - 5 = 3x + 20$
$5x = 25$
$x = 5$

15. $-2(x + 5) = 10$

$-2x - 10 = 10$

$-2x = 20$

$x = -10$

17. $2x + 3 = 2x - 2$

$3 = -2$

No solution

19. $\frac{3}{2}(x + 5) - \frac{1}{4}(x + 24) = 0$

$\frac{3}{2}(x + 5) = \frac{1}{4}(x + 24)$

$12(x + 5) = 2(x + 24)$

$12x + 60 = 2x + 48$

$10x = -12$

$x = -\frac{12}{10}$

$x = -\frac{6}{5}$

21. $0.25x + 0.75(10 - x) = 3$

$25x + 75(10 - x) = 300$

$25x + 750 - 75x = 300$

$-50x = -450$

$x = 9$

23. $x + 6 < 8$

$x < 8 - 6$

$x < 2$

25. $-x - 8 > -17$

$17 - 8 > x$

$9 > x$

$x < 9$

27. $6 + x \le -8$

$x \le -8 - 6$

$x \le -14$

29. $\frac{4}{5}x > 8$

$x > \frac{5}{4}(8)$

$x > 10$

31. $-\frac{3}{4}x > -3$

$\frac{3}{4}x < 3$

$x < 4$

33. $4x < 12$

$x < 3$

35. $-11x \le -22$

$11x \ge 22$

$x \ge 2$

37. $x - 3(x + 1) \ge 7$

$x - 3x - 3 \ge 7$

$-2x \ge 10$

$x \le -5$

39. $7x - 12 < 4x + 6$

$3x < 18$

$x < 6$

41. $\frac{3}{4}x - 6 \le x - 7$

$1 \le \frac{1}{4}x$

$4 \le x$

$x \ge 4$

43. $3.6x + 11 \ge -3.4$

$3.6x \ge -14.4$

$x \ge \frac{-14.4}{3.6}$

$x \ge -4$

APPENDIX F
Systems of Inequalities

Appendix F.1   Solving Systems of Inequalities

Vocabulary Check

1. solution

2. graph

3. linear

4. point, equilibrium

---

1. $x < 2$

Vertical boundary

Matches graph (g).

3. $2x + 3y \ge 6$

$y \ge -\frac{2}{3}x + 2$

Line with negative slope

Matches (a).

5. $x^2 + y^2 < 9$

Circular boundary

Matches (e).

7. $xy > 1$ or $y > \dfrac{1}{x}$

Matches (f).

9. $y < 2 - x^2$

Graph the parabola $y = 2 - x^2$. The region lies below the parabola.

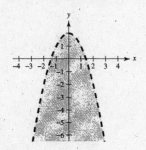

11. $y^2 + 1 \geq x$

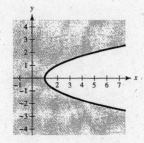

13. $x \geq 4$

Using a solid line, graph the vertical line $x = 4$ and shade to the right of this line.

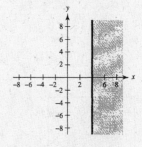

15. $y \geq -1$

Using a solid line, graph the horizontal line $y = -1$ and shade above this line.

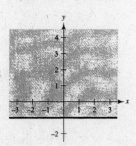

17. $2y - x \geq 4$

Using a solid line, graph $2y - x = 4$, and then shade above the line. (Use $(0, 0)$ as a test point.)

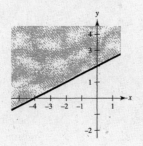

19. $2x + 3y < 6$

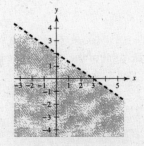

21. $4x - 3y \leq 24$

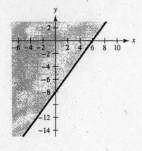

23. $y > 3x^2 + 1$

Sketch the parabola $y = 3x^2 + 1$. The region lies above the parabola.

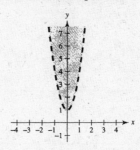

25. $2x - y^2 > 0$

$2x > y^2$

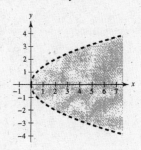

27. $(x + 1)^2 + y^2 < 9$

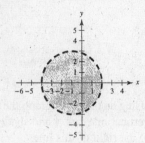

29. $y \geq \frac{2}{3}x - 1$

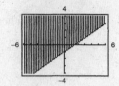

31. $y < -3.8x + 1.1$

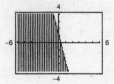

33. $x^2 + 5y - 10 \leq 0$

$$y \leq 2 - \frac{x^2}{5}$$

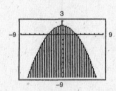

35. $y \leq \dfrac{1}{1 + x^2}$

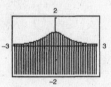

37. $y < \ln x$

Using a dashed line, graph $y = \ln x$, and shade to the right of the curve. (Use $(2, 0)$ as a test point.)

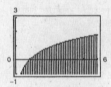

39. $y > 3^{-x-4}$

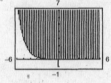

41. The line through $(0, 2)$ and $(3, 0)$ is $y = -\frac{2}{3}x + 2$. For the shaded region above the line, we have:

$$y > -\frac{2}{3}x + 2$$

$$3y > -2x + 6$$

$$2x + 3y > 6$$

$$\frac{x}{3} + \frac{y}{2} > 1$$

43. The circle shown is $x^2 + y^2 = 9$. For the shaded region inside the circle, we have $x^2 + y^2 \leq 9$.

45. (a) $(0, 2)$ is a solution: $-2(0) + 5(2) \geq 3$

$$2 < 4$$

$$-4(0) + 2(2) < 7$$

(b) $(-6, 4)$ is not a solution: $4 \not< 4$

(c) $(-8, -2)$ is not a solution: $-4(-8) + 2(-2) \not< 7$

(d) $(-3, 2)$ is not a solution: $-4(-3) + 2(2) \not< 7$

47. $\begin{cases} x + y \le 1 \\ -x + y \le 1 \\ y \ge 0 \end{cases}$

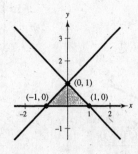

First, find the points of intersection of each pair of equations.

Vertex A	***Vertex B***	***Vertex C***
$\begin{cases} x + y = 1 \\ -x + y = 1 \end{cases}$	$\begin{cases} x + y = 1 \\ y = 0 \end{cases}$	$\begin{cases} -x + y = 1 \\ y = 0 \end{cases}$
$(0, 1)$	$(1, 0)$	$(-1, 0)$

49. $\begin{cases} -3x + 2y < 6 \\ x - 4y > -2 \\ 2x + y < 3 \end{cases}$

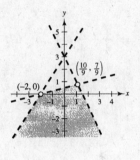

First, find the points of intersection of each pair of equations.

Vertex A	***Vertex B***	***Vertex C***
$\begin{cases} -3x + 2y = 6 \\ x - 4y = -2 \end{cases}$	$\begin{cases} -3x + 2y = 6 \\ 2x + y = 3 \end{cases}$	$\begin{cases} x - 4y = -2 \\ 2x + y = 3 \end{cases}$
$(-2, 0)$	$(0, 3)$	$\left(\frac{10}{9}, \frac{7}{9}\right)$

51. $3x + y \le y^2$

$x - y > 0$

The curves given by $3x + y = y^2$ and $x - y = 0$ intersect as follows:

$3x + x = x^2$

$4x = x^2$

$x = 0, 4$

Intersection points:
$(0, 0), (4, 4)$

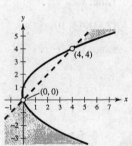

53. $2x + y < 2 \implies y < 2 - 2x$

$x + 3y > 2 \implies y > \frac{1}{3}(2 - x)$

$2 - 2x = \frac{1}{3}(2 - x)$

$6 - 6x = (2 - x)$

$4 = 5x$

$x = \frac{4}{5}$

Intersection: $\left(\frac{4}{5}, \frac{2}{5}\right)$

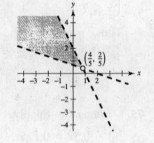

55. $\begin{cases} x < y^2 \\ x > y + 2 \end{cases}$

Points of intersection:

$y^2 = y + 2$

$y^2 - y - 2 = 0$

$(y + 1)(y - 2) = 0$

$y = -1, 2$

$(1, -1), (4, 2)$

57. $\begin{cases} x^2 + y^2 \le 9 \\ x^2 + y^2 \ge 1 \end{cases}$

There are no points of intersection. The region in common to both inequalities is the region between the circles.

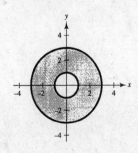

59. $\begin{cases} y \le \sqrt{3x} + 1 \\ y \ge x^2 + 1 \end{cases}$

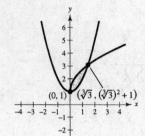

61. $\begin{cases} y < x^3 - 2x + 1 \\ y > -2x \\ x \le 1 \end{cases}$

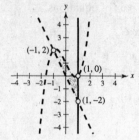

63. $\begin{cases} x^2 y \ge 1 \\ 0 < x \le 4 \\ y \le 4 \end{cases}$

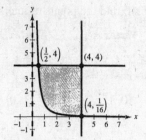

65. $\begin{cases} y < -x + 4 \implies \dfrac{x}{4} + \dfrac{y}{4} < 1 \\ x \ge 0 \qquad\qquad\quad x \ge 0 \\ y \ge 0 \qquad\qquad\quad y \ge 0 \end{cases}$

67. $(0, 4), (4, 0)$

Line: $y \le 4 - x$

$(0, 2); (8, 0)$

Line: $y \le -\dfrac{1}{4}x + 2$

$x \ge 0, \ y \ge 0$

69. Circle of radius 2 and center $(0, 2)$

$x^2 + (y - 2)^2 \le 4$

71. $\begin{cases} x \ge 2 \\ x \le 5 \\ y \ge 1 \\ y \le 7 \end{cases}$

Thus,
$2 \le x \le 5, 1 \le y \le 7.$

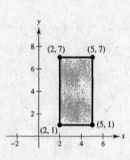

73. $(0, 0), (5, 0)$

Line: $y \ge 0$

$(0, 0), (2, 3)$

Line: $y \le \dfrac{3}{2}x$

$(2, 3), (5, 0)$

Line: $y \le -x + 5$

75. Demand = Supply

$50 - 0.5x = 0.125x$

$50 = 0.625x$

$x = 80$

$p = 10$

Point of equilibrium: $(80, 10)$

Consumer surplus $= \frac{1}{2}(40)(80) = 1600$

Producer surplus $= \frac{1}{2}(10)(80) = 400$

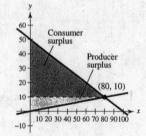

77.   Demand = Supply

$$300 - 0.0002x = 225 + 0.0005x$$

$$75 = 0.0007x$$

$$x = \frac{75}{0.0007} = \frac{750,000}{7}$$

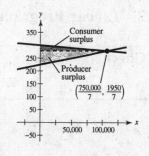

Equilibrium point: $\left(\dfrac{750,000}{7}, \dfrac{1950}{7}\right) \approx (107,142.86, 278.57)$

Consumer surplus: $\dfrac{(107,142.86)(300 - 278.57)}{2} \approx 1,148,036$

Producer surplus: $\dfrac{(107,142.86)(278.57 - 225)}{2} \approx 2,869,822$

79. $x + y \le 30,000$

   $x \ge 7500$

   $y \ge 7500$

   $x \ge 2y$

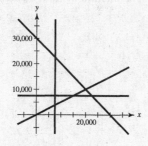

81. (a) Let x = number of ounces of food X.

   Let y = number of ounces of food Y.

   Calcium: $20x + 10y \ge 280$

   Iron: $15x + 10y \ge 160$

   Vitamin B: $10x + 20y \ge 180$

   $x \ge 0$

   $y \ge 0$

   (b)

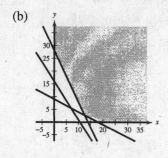

83. (a) $xy \ge 500$ Body-building space

   $2x + \pi y \ge 125$ Track (two semi-circles and two lengths)

   $x \ge 0$ Physical constraint

   $y \ge 0$ Physical constraint

   (b)

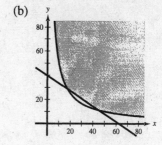

85. Area $= 9 \cdot 11 = 99$ square units

   True

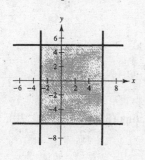

87. Test a point on either side of the boundary.

Appendix F.2   Linear Programming

Vocabulary Check

1. optimization   **2.** objective function   **3.** constraints, feasible solutions

1. $z = 3x + 5y$

At $(0, 6)$: $z = 3(0) + 5(6) = 30$

At $(0, 0)$: $z = 3(0) + 5(0) = 0$

At $(6, 0)$: $z = 3(6) + 5(0) = 18$

The minimum value is 0 at $(0, 0)$.

The maximum value is 30 at $(0, 6)$.

3. $z = 10x + 7y$

At $(0, 6)$: $z = 10(0) + 7(6) = 42$

At $(0, 0)$: $z = 10(0) + 7(0) = 0$

At $(6, 0)$: $z = 10(6) + 7(0) = 60$

The minimum value is 0 at $(0, 0)$.

The maximum value is 60 at $(6, 0)$.

5. $z = 3x + 2y$

$x + 3y = 15 \implies y = \frac{1}{3}(15 - x)$

$4x + y = 16 \implies y = (16 - 4x)$

$\frac{1}{3}(15 - x) = 16 - 4x$

$(15 - x) = 48 - 12x$

$11x = 33$

$x = 3$

$y = 4$

At $(0, 0)$: $z = 0$

At $(0, 5)$: $z = 10$

At $(4, 0)$: $z = 12$

At $(3, 4)$: $z = 17$

The minimum value is 0 at $(0, 0)$.

The maximum value is 17 at $(3, 4)$.

7. $z = 5x + 0.5y$

At $(0, 0)$: $z = 0$

At $(0, 5)$: $z = 2.5$

At $(4, 0)$: $z = 20$

At $(3, 4)$: $z = 17$

The minimum value is 0 at $(0, 0)$.

The maximum value is 20 at $(4, 0)$.

9. $z = 10x + 7y$

At $(0, 45)$: $z = 10(0) + 7(45) = 315$

At $(30, 45)$: $z = 10(30) + 7(45) = 615$

At $(60, 20)$: $z = 10(60) + 7(20) = 740$

At $(60, 0)$: $z = 10(60) + 7(0) = 600$

At $(0, 0)$: $z = 10(0) + 7(0) = 0$

The minimum value is 0 at $(0, 0)$.

The maximum value is 740 at $(60, 20)$.

11. $z = 25x + 30y$

At $(0, 45)$: $z = 25(0) + 30(45) = 1350$

At $(30, 45)$: $z = 25(30) + 30(45) = 2100$

At $(60, 20)$: $z = 25(60) + 30(20) = 2100$

At $(60, 0)$: $z = 25(60) + 30(0) = 1500$

At $(0, 0)$: $z = 25(0) + 30(0) = 0$

The minimum value is 0 at $(0, 0)$.

The maximum value is 2100 at any point along the line segment connecting $(30, 45)$ and $(60, 20)$.

13. $z = 6x + 10y$

At $(0, 2)$: $z = 6(0) + 10(2) = 20$

At $(5, 0)$: $z = 6(5) + 10(0) = 30$

At $(0, 0)$: $z = 6(0) + 10(0) = 0$

The minimum value is 0 at $(0, 0)$.

The maximum value is 30 at $(5, 0)$.

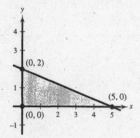

15. $z = 3x + 4y$

At $(0, 0)$: $z = 0$

At $(7, 0)$: $z = 21$

At $(0, 10)$: $z = 40$

At $(5, 8)$: $z = 47$

The minimum value is 0 at $(0, 0)$.

The maximum value is 47 at $(5, 8)$.

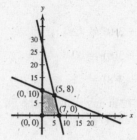

17. $z = x + 2y$

At $(0, 0)$: $z = 0 + 2(0) = 0$

At $(0, 10)$: $z = 0 + 2(10) = 20$

At $(5, 8)$: $z = 5 + 2(8) = 21$

At $(7, 0)$: $z = 7 + 2(0) = 7$

The minimum value is 0 at $(0, 0)$.

The maximum value is 21 at $(5, 8)$.

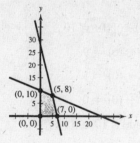

19. $z = 2x$

At $(0, 0)$: $z = 2(0) = 0$

At $(0, 10)$: $z = 2(0) = 0$

At $(5, 8)$: $z = 2(5) = 10$

At $(7, 0)$: $z = 2(7) = 14$

The maximum value is 14 at $(7, 0)$.

The minimum value is 0 along the line segment joining $(0, 0)$ and $(0, 10)$.

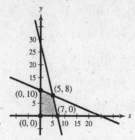

21. $z = 4x + y$

At $(36, 0)$: $z = 4(36) + 0 = 144$

At $(40, 0)$: $z = 4(40) + 0 = 160$

At $(24, 8)$: $z = 4(24) + 8 = 104$

The minimum value is 104 at $(24, 8)$.

The maximum value is 160 at $(40, 0)$.

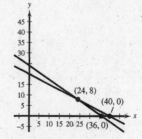

23. $z = x + 4y$

At $(36, 0)$: $z = 36 + 4(0) = 36$

At $(40, 0)$: $z = 40 + 4(0) = 40$

At $(24, 8)$: $z = 24 + 4(8) = 56$

The minimum value is 36 at $(36, 0)$.

The maximum value is 56 at $(24, 8)$.

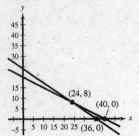

25. $z = 2x + 3y$

At $(36, 0)$: $z = 2(36) + 3(0) = 72$

At $(40, 0)$: $z = 2(40) + 3(0) = 80$

At $(24, 8)$: $z = 2(24) + 3(8) = 72$

The minimum value is 72 at any point on the line segment joining $(36, 0)$ and $(24, 8)$.

The maximum value is 80 at $(40, 0)$.

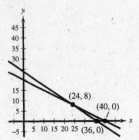

27. $z = 2x + y$

(a), (b)

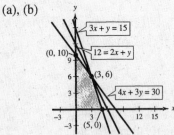

(c) At $(0, 10)$: $z = 2(0) + (10) = 10$

At $(3, 6)$: $z = 2(3) + (6) = 12$

At $(5, 0)$: $z = 2(5) + (0) = 10$

At $(0, 0)$: $z = 2(0) + (0) = 0$

The maximum value is 12 at $(3, 6)$.

29. $z = x + y$

(a), (b)

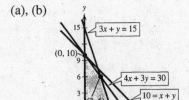

(c) At $(0, 10)$: $z = (0) + (10) = 10$

At $(3, 6)$: $z = (3) + (6) = 9$

At $(5, 0)$: $z = (5) + (0) = 5$

At $(0, 0)$: $z = (0) + (0) = 0$

The maximum value is 10 at $(0, 10)$.

31. $-x + y \leq 1 \implies y \leq x + 1$

$-x + 2y \leq 4 \implies y \leq \frac{1}{2}x + 2$

Intersection: $(2, 3)$

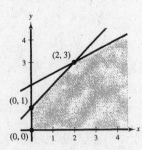

The constraints do not form a closed set of points. Therefore, $z = x + y$ is unbounded.

33. $-x + y \leq 0 \implies y \leq x$

$-3x + y \geq 3 \implies y \geq 3x + 3$

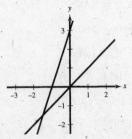

The feasible set is empty.

35. Let x = number of audits.

Let y = number of tax returns.

Constraints: $100x + 12.5y \leq 800$

$$8x + 2y \leq 96$$

$$x \geq 0$$

$$y \geq 0$$

Objective function: $R = 2000x + 300y$

Vertices of feasible region:
$(0, 0), (8, 0), (0, 48), (4, 32)$

At $(0, 0)$: $R = 0$

At $(8, 0)$: $R = 16{,}000$

At $(0, 48)$: $R = 14{,}400$

At $(4, 32)$: $R = 17{,}600$

4 audits, 32 tax returns yields maximum revenue of $17,600.

37. x = number of bags of Brand X
y = number of bags of Brand Y

Constraints: $2x + y \geq 12$

$$2x + 9y \geq 36$$

$$2x + 3y \geq 24$$

$$x \geq 0$$

$$y \geq 0$$

Objective function: $C = 25x + 20y$

Vertices: $(0, 12), (3, 6), (9, 2), (18, 0)$

At $(0, 12)$: $C = 25(0) + 20(12) = 240$

At $(3, 6)$: $C = 25(3) + 20(6) = 195$

At $(9, 2)$: $C = 25(9) + 20(2) = 265$

At $(18, 0)$: $C = 25(18) + 20(0) = 450$

To minimize cost, use three bags of Brand X and six bags of Brand Y for a total cost of $195.

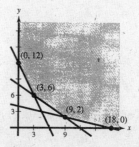

39. True, the maximum value is attained at all points in the segment joining these two vertices.

41. There are an infinite number of objective functions that would have a maximum at $(0, 4)$. One such objective function is $z = x + 5y$.

43. There are an infinite number of objective functions that would have a maximum at $(5, 0)$. One such objective function is $z = 4x + y$.

45. Constraints: $x \geq 0, y \geq 0, x + 3y \leq 15, 4x + y \leq 16$

Vertex	Value of $z = 3x + ty$
$(0, 0)$	$z = 0$
$(0, 5)$	$z = 5t$
$(3, 4)$	$z = 9 + 4t$
$(4, 0)$	$z = 12$

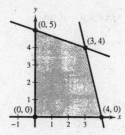

(a) For the maximum value to be at $(0, 5)$, $z = 5t$ must be greater than

$z = 9 + 4t$ and $z = 12$.

$5t > 9 + 4t$ and $5t > 12$

$\quad t > 9 \qquad\qquad t > \frac{12}{5}$

Thus, $t > 9$.

(b) For the maximum value to be at $(3, 4)$, $z = 9 + 4t$ must be greater than $z = 5t$ and $z = 12$.

$9 + 4t > 5t$ and $9 + 4t > 12$

$\quad 9 > t \qquad\qquad t > 3$

$\qquad\qquad\qquad\qquad t > \frac{3}{4}$

Thus, $\frac{3}{4} < t < 9$.

Chapter 1   Practice Test Solutions

1. $m = \dfrac{3 - 2}{1 - (-2)} = \dfrac{1}{3}$

2. Slope $= \dfrac{-2 - (-5)}{3 - 4} = \dfrac{3}{-1} = -3$

$y + 2 = -3(x - 3)$

$y + 2 = -3x + 9$

$y + 3x = 7 \ $ or $\ y = -3x + 7$

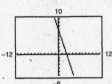

3. $y - 5 = -3(x + 1)$

$y - 5 = -3x - 3$

$y + 3x = 2 \ $ or $\ y = -3x + 2$

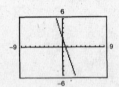

4. $3x + 5y = 7$

$5y = -3x + 7$

$y = -\tfrac{3}{5}x + \tfrac{7}{5}$

Slope of perpendicular line is $m = \tfrac{5}{3}$.

$y - 2 = \tfrac{5}{3}(x + 3)$

$y = \tfrac{5}{3}x + 7$

5. No, y is not a function of x. For example, $(0, 2)$ and $(0, -2)$ both satisfy the equation.

6. $f(0) = \dfrac{|0 - 2|}{(0 - 2)} = \dfrac{2}{-2} = -1$

$f(2)$ is not defined.

$f(4) = \dfrac{|4 - 2|}{(4 - 2)} = \dfrac{2}{2} = 1$

7. The domain of $f(x) = \dfrac{5}{x^2 - 16}$ is all $x \neq \pm 4$.

8. The domain of $g(t) = \sqrt{4 - t}$ consists of all t satisfying

$4 - t \geq 0$ or $t \leq 4$.

9.

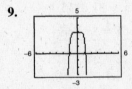

$f(x) = 3 - x^6$ is even.

10. $f(x) = 12x - x^3$

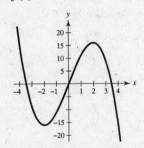

f is increasing on $(-2, 2)$.

11.

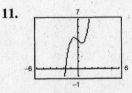

Relative minimum: $(0.577, 3.615)$

Relative maximum: $(-0.577, 4.385)$

12. $f(x) = x^3 - 3$ is a vertical shift of three units downward of $y = x^3$.

13. $f(x) = \sqrt{x - 6}$ is a horizontal shift six units to the right of $y = \sqrt{x}$.

14. $(g \circ f)(x) = g(f(x))$
$$= g(\sqrt{x}) = (\sqrt{x})^2 - 2 = x - 2$$

Domain: $x \geq 0$

15. $\left(\dfrac{f}{g}\right)(x) = \dfrac{f(x)}{g(x)} = \dfrac{3x^2}{16 - x^4}$

The domain is all $x \neq \pm 2$.

16. $(f \circ g)(x) = f\left(\dfrac{x - 1}{3}\right)$
$$= 3\left(\dfrac{x - 1}{3}\right) + 1 = (x - 1) + 1 = x$$

$(g \circ f)(x) = g(3x + 1) = \dfrac{(3x + 1) - 1}{3} = \dfrac{3x}{3} = x$

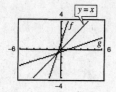

17. $y = \sqrt{9 - x^2}, \quad 0 \leq x \leq 3$
$$x = \sqrt{9 - y^2}$$
$$x^2 = 9 - y^2$$
$$y^2 = 9 - x^2$$
$$y = \sqrt{9 - x^2}$$

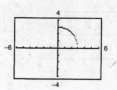

18. $y = 0.882 + 0.912x$

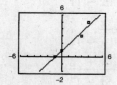

Chapter 2   Practice Test Solutions

1. x-intercepts: $(1, 0), (5, 0)$

y-intercept: $(0, 5)$

Vertex: $(3, -4)$

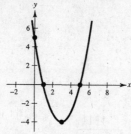

2. $a = 0.01, b = -90$
$$\dfrac{-b}{2a} = \dfrac{90}{2(.01)} = 4500 \text{ units}$$

3. Vertex: $(1, 7)$

Opening downward through $(2, 5)$

$y = a(x - 1)^2 + 7$ Standard form

$5 = a(2 - 1)^2 + 7$

$5 = a + 7$

$a = -2$

$y = -2(x - 1)^2 + 7 = -2(x^2 - 2x + 1) + 7 = -2x^2 + 4x + 5$

4. $y = \pm a(x - 2)(3x - 4)$ where a is any real number.

$$y = \pm(3x^2 - 10x + 8)$$

5. Leading coefficient: -3

Degree: 5

Moves down to the right and up to the left.

6. $0 = x^5 - 5x^3 + 4x$

$\quad = x(x^4 - 5x^2 + 4)$

$\quad = x(x^2 - 1)(x^2 - 4)$

$\quad = x(x + 1)(x - 1)(x + 2)(x - 2)$

$x = 0, x = \pm 1, x = \pm 2$

7. $f(x) = x(x - 3)(x + 2)$

$\quad = x(x^2 - x - 6)$

$\quad = x^3 - x^2 - 6x$

8. Intercepts: $(0, 0), \left(\pm 2\sqrt{3}, 0\right)$

Moves up to the right.

Moves down to the left.

x	-2	-1	0	1	2
y	16	11	0	-11	-16

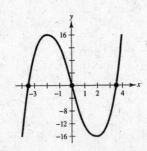

9.

$$
\begin{array}{r}
3x^3 + 9x^2 + 20x + 62 + \dfrac{176}{x - 3} \\
\end{array}
$$

$$
\begin{array}{l}
x - 3 \overline{)\, 3x^4 + 0x^3 - 7x^2 + 2x - 10} \\
\quad\ \underline{3x^4 - 9x^3} \\
\qquad\quad 9x^3 - 7x^2 \\
\qquad\quad \underline{9x^3 - 27x^2} \\
\qquad\qquad\quad 20x^2 + 2x \\
\qquad\qquad\quad \underline{20x^2 - 60x} \\
\qquad\qquad\qquad\quad 62x - 10 \\
\qquad\qquad\qquad\quad \underline{62x - 186} \\
\qquad\qquad\qquad\qquad\quad 176
\end{array}
$$

10.

$$
x - 2 + \dfrac{5x - 13}{x^2 + 2x - 1}
$$

$$
\begin{array}{l}
x^2 + 2x - 1 \overline{)\, x^3 + 0x^2 + 0x - 11} \\
\qquad\qquad\ \underline{x^3 + 2x^2 - x} \\
\qquad\qquad\quad -2x^2 + x - 11 \\
\qquad\qquad\quad \underline{-2x^2 - 4x + 2} \\
\qquad\qquad\qquad\qquad 5x - 13
\end{array}
$$

11.

$$
\begin{array}{r|rrrrrr}
-5 & 3 & 13 & 0 & 0 & 12 & -1 \\
   &   & -15 & 10 & -50 & 250 & -1310 \\
\hline
   & 3 & -2 & 10 & -50 & 262 & -1311
\end{array}
$$

$$\frac{3x^5 + 13x^4 + 12x - 1}{x + 5} = 3x^4 - 2x^3 + 10x^2 - 50x + 262 - \frac{1311}{x + 5}$$

12.

$$
\begin{array}{r|rrrr}
-6 & 7 & 40 & -12 & 15 \\
   &   & -42 & 12 & 0 \\
\hline
   & 7 & -2 & 0 & 15
\end{array}
$$

$f(-6) = 15$

13. $0 = x^3 - 19x - 30$

Possible rational roots:

$\pm 1, \pm 2, \pm 3, \pm 5, \pm 6, \pm 10, \pm 15, \pm 30$

$$
\begin{array}{r|rrrr}
-2 & 1 & 0 & -19 & -30 \\
   &   & -2 & 4 & 30 \\
\hline
   & 1 & -2 & -15 & 0
\end{array}
$$

-2 is a zero.

$0 = (x + 2)(x^2 - 2x - 15)$

$0 = (x + 2)(x + 3)(x - 5)$

Zeros: $x = -2, x = -3, x = 5$

14. $0 = x^4 + x^3 - 8x^2 - 9x - 9$

Possible rational roots: $\pm 1, \pm 3, \pm 9$

$$
\begin{array}{r|rrrrr}
3 & 1 & 1 & -8 & -9 & -9 \\
  &   & 3 & 12 & 12 & 9 \\
\hline
  & 1 & 4 & 4 & 3 & 0
\end{array}
$$

$x = 3$ is a zero.

$0 = (x - 3)(x^3 + 4x^2 + 4x + 3)$

Possible rational roots of $x^3 + 4x^2 + 4x + 3$: $\pm 1, \pm 3$

$$
\begin{array}{r|rrrr}
-3 & 1 & 4 & 4 & 3 \\
   &   & -3 & -3 & -3 \\
\hline
   & 1 & 1 & 1 & 0
\end{array}
$$

$x = -3$ is a zero.

$0 = (x - 3)(x + 3)(x^2 + x + 1)$

The zeros of $x^2 + x + 1$ are $x = \dfrac{-1 \pm \sqrt{3}i}{2}$.

Zeros: $x = 3, x = -3,$

$$x = -\frac{1}{2} + \frac{\sqrt{3}}{2}i, x = -\frac{1}{2} - \frac{\sqrt{3}}{2}i$$

15. $0 = 6x^3 - 5x^2 + 4x - 15$

Possible rational roots: $\pm 1, \pm 3, \pm 5, \pm 15, \pm\frac{1}{2}, \pm\frac{3}{2}, \pm\frac{5}{2}, \pm\frac{15}{2}, \pm\frac{1}{3}, \pm\frac{5}{3}, \pm\frac{1}{6}, \pm\frac{5}{6}$

16. $0 = x^3 - \frac{20}{3}x^2 + 9x - \frac{10}{3}$

$0 = 3x^3 - 20x^2 + 27x - 10$

Possible rational roots:

$\pm 1, \pm 2, \pm 5, \pm 10, \pm\frac{1}{3}, \pm\frac{2}{3}, \pm\frac{5}{3}, \pm\frac{10}{3}$

$$
\begin{array}{r|rrrr}
1 & 3 & -20 & 27 & -10 \\
  &   & 3 & -17 & 10 \\
\hline
  & 3 & -17 & 10 & 0
\end{array}
$$

$x = 1$ is a zero.

$0 = (x - 1)(3x^2 - 17x + 10)$

$0 = (x - 1)(3x - 2)(x - 5)$

Zeros: $x = 1, x = \frac{2}{3}, x = 5$

17. $f(x) = x^4 + x^3 + 3x^2 + 5x - 10$

Possible rational roots: $\pm 1, \pm 2, \pm 5, \pm 10$

$$
\begin{array}{r|rrrrr}
1 & 1 & 1 & 3 & 5 & -10 \\
  &   & 1 & 2 & 5 & 10 \\
\hline
  & 1 & 2 & 5 & 10 & 0
\end{array}
$$

$x = 1$ is a zero.

$$
\begin{array}{r|rrrr}
-2 & 1 & 2 & 5 & 10 \\
   &   & -2 & 0 & -10 \\
\hline
   & 1 & 0 & 5 & 0
\end{array}
$$

$x = -2$ is a zero.

$f(x) = (x - 1)(x + 2)(x^2 + 5)$

$= (x - 1)(x + 2)(x + \sqrt{5}i)(x - \sqrt{5}i)$

18. $\dfrac{2}{1 + i} = \dfrac{2}{1 + i} \cdot \dfrac{1 - i}{1 - i}$

$= \dfrac{2 - 2i}{1 + 1}$

$= 1 - i$

19. $\dfrac{3 + i}{2} - \dfrac{i + 1}{4} = \dfrac{6 + 2i - i - 1}{4} = \dfrac{5}{4} + \dfrac{1}{4}i$

20. $f(x) = (x - 2)[x - (3 + i)][x - (3 - i)][x - (3 - 2i)][x - (3 + 2i)]$

$\quad\quad = (x - 2)[(x - 3)^2 + 1][(x - 3)^2 + 4]$

$\quad\quad = (x - 2)(x^2 - 6x + 10)(x^2 - 6x + 13)$

$\quad\quad = x^5 - 14x^4 + 83x^3 - 256x^2 + 406x - 260$

21.

$$
\begin{array}{r|rrrr}
3i & 1 & 4 & 9 & 36 \\
   &   & 3i & 12i - 9 & -36 \\
\hline
   & 1 & 4 + 3i & 12i & 0
\end{array}
$$

22. $z = \dfrac{kx^2}{\sqrt{y}}$

23. $f(x) = \dfrac{x - 1}{2x}$

Vertical asymptote: $x = 0$

Horizontal asymptote: $y = \dfrac{1}{2}$

x-intercept: $(1, 0)$

24. $f(x) = \dfrac{3x^2 - 4}{x}$

Vertical asymptote: $x = 0$

Slant asymptote: $y = 3x$

x-intercepts: $\left(\pm\dfrac{2}{\sqrt{3}}, 0\right)$

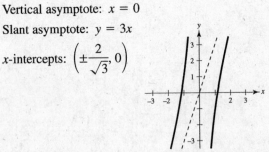

25. $y = 8$ is a horizontal asymptote since the degree of the numerator equals the degree of the denominator. There are no vertical asymptotes.

26. $x = 1$ is a vertical asymptote.

$\dfrac{4x^2 - 2x + 7}{x - 1} = 4x + 2 + \dfrac{9}{x - 1}$

so $y = 4x + 2$ is a slant asymptote.

27. $f(x) = \dfrac{x - 5}{(x - 5)^2} = \dfrac{1}{x - 5}$

Vertical asymptote: $x = 5$

Horizontal asymptote: $y = 0$

y-intercept: $\left(0, -\dfrac{1}{5}\right)$

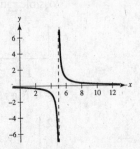

Chapter 3   Practice Test Solutions

1. $x^{3/5} = 8$

$\quad x = 8^{5/3}$

$\quad\quad = \left(\sqrt[3]{8}\right)^5 = 2^5 = 32$

2. $3^{x-1} = \dfrac{1}{81}$

$\quad 3^{x-1} = 3^{-4}$

$\quad x - 1 = -4$

$\quad\quad x = -3$

3. $f(x) = 2^{-x} = \left(\frac{1}{2}\right)^x$

x	-2	-1	0	1	2.
$f(x)$	4	2	1	$\frac{1}{2}$	$\frac{1}{4}$

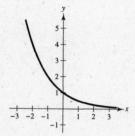

4. $g(x) = e^x + 1$

x	-2	-1	0	1	2
$g(x)$	1.14	1.37	2	3.72	8.39

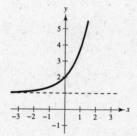

5. $A = P\left(1 + \dfrac{r}{n}\right)^{nt}$

   (a) $A = 5000\left(1 + \dfrac{0.09}{12}\right)^{12(3)} \approx \$6543.23$

   (b) $A = 5000\left(1 + \dfrac{0.09}{4}\right)^{4(3)} \approx \$6530.25$

   (c) $A = 5000e^{(0.09)(3)} \approx \$6549.82$

6. $\quad 7^{-2} = \dfrac{1}{49}$

$\log_7 \dfrac{1}{49} = -2$

7. $x - 4 = \log_2 \dfrac{1}{64}$

$\quad\quad 2^{x-4} = \dfrac{1}{64}$

$\quad\quad 2^{x-4} = 2^{-6}$

$\quad\quad x - 4 = -6$

$\quad\quad\quad\; x = -2$

8. $\log_b \sqrt[4]{\dfrac{8}{25}} = \dfrac{1}{4} \log_b \dfrac{8}{25}$

$\quad\quad = \dfrac{1}{4}[\log_b 8 - \log_b 25]$

$\quad\quad = \dfrac{1}{4}[\log_b 2^3 - \log_b 5^2]$

$\quad\quad = \dfrac{1}{4}[3 \log_b 2 - 2 \log_b 5]$

$\quad\quad = \dfrac{1}{4}[3(0.3562) - 2(0.8271)]$

$\quad\quad = -0.1464$

9. $5 \ln x - \dfrac{1}{2} \ln y + 6 \ln z = \ln x^5 - \ln \sqrt{y} + \ln z^6 = \ln\left(\dfrac{x^5 z^6}{\sqrt{y}}\right)$

10. $\log_9 28 = \dfrac{\log 28}{\log 9} \approx 1.5166$

11. $\log_{10} N = 0.6646$

$\quad\quad\quad N = 10^{0.6646} \approx 4.62$

12.

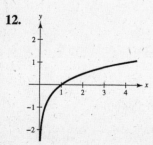

13. Domain:

$\quad\quad\quad x^2 - 9 > 0$

$\quad\quad (x + 3)(x - 3) > 0$

$\quad\quad x < -3 \text{ or } x > 3$

14.

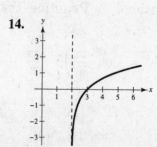

15. $\dfrac{\ln x}{\ln y} \neq \ln(x - y)$ since $\dfrac{\ln x}{\ln y} = \log_y x$.

16. $5^x = 41$

$$x = \log_5 41 = \frac{\ln 41}{\ln 5} \approx 2.3074$$

17. $x - x^2 = \log_5 \frac{1}{25}$

$5^{x - x^2} = \frac{1}{25}$

$5^{x - x^2} = 5^{-2}$

$x - x^2 = -2$

$0 = x^2 - x - 2$

$0 = (x + 1)(x - 2)$

$x = -1 \text{ or } x = 2$

18. $\log_2 x + \log_2(x - 3) = 2$

$\log_2[x(x - 3)] = 2$

$x(x - 3) = 2^2$

$x^2 - 3x = 4$

$x^2 - 3x - 4 = 0$

$(x + 1)(x - 4) = 0$

$x = 4$

$x = -1 \quad \text{(extraneous solution)}$

19. $\dfrac{e^x + e^{-x}}{3} = 4$

$e^x(e^x + e^{-x}) = 12e^x$

$e^{2x} + 1 = 12e^x$

$e^{2x} - 12e^x + 1 = 0$

$$e^x = \frac{12 \pm \sqrt{144 - 4}}{2}$$

$e^x \approx 11.9161 \qquad \text{or} \qquad e^x \approx 0.08392$

$x \approx \ln 11.9161 \qquad\qquad x \approx \ln 0.08392$

$x \approx 2.4779 \qquad\qquad\quad x \approx -2.4779$

20. $A = Pe^{rt}$

$12{,}000 = 6000e^{0.13t}$

$2 = e^{0.13t}$

$\ln 2 = 0.13t$

$\dfrac{\ln 2}{0.13} = t$

$t \approx 5.3319 \text{ yr or } 5 \text{ yr } 4 \text{ mo}$

21. There are two points of intersection:

$(0.0169, -2.983)$,

$(1.731, 1.647)$

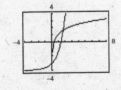

22. $y = 1.0597x^{1.9792}$

Chapter 4   Practice Test Solutions

1. $350° = 350°\left(\dfrac{\pi}{180°}\right) = \dfrac{35\pi}{18}$

2. $\dfrac{5\pi}{9} = \dfrac{5\pi}{9} \cdot \dfrac{180°}{\pi} = 100°$

3. $135° \, 14' \, 12'' = \left(135 + \frac{14}{60} + \frac{12}{3600}\right)°$

$\approx 135.2367°$

4. $-22.569° = -(22° + 0.569(60)')$

$= -22° \, 34.14'$

$= -(22° \, 34' + 0.14(60)'')$

$\approx -22° \, 34' \, 8''$

5. $\cos \theta = \dfrac{2}{3}$

$x = 2,\ r = 3,\ y = \pm\sqrt{9 - 4} = \pm\sqrt{5}$

$\tan \theta = \dfrac{y}{x} = \pm\dfrac{\sqrt{5}}{2}$

6. $\sin \theta = 0.9063$

$\theta = \arcsin 0.9063$

$\theta \approx 65°$ or $\dfrac{13\pi}{36}$

7. $\tan 20° = \dfrac{35}{x}$

$x = \dfrac{35}{\tan 20°} \approx 96.1617$

8. $\theta = \dfrac{6\pi}{5}$, θ is in Quadrant III.

Reference angle: $\dfrac{6\pi}{5} - \pi = \dfrac{\pi}{5}$ or $36°$

9. $\csc 3.92 = \dfrac{1}{\sin 3.92} \approx -1.4242$

10. $\tan \theta = 6 = \dfrac{6}{1}$, θ lies in Quadrant III.

$y = -6,\ x = -1,\ r = \sqrt{36 + 1} = \sqrt{37}$, so

$\sec \theta = \dfrac{\sqrt{37}}{-1} \approx -6.0828.$

11. Period: 4π

Amplitude: 3

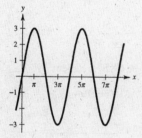

12. Period: 2π

Amplitude: 2

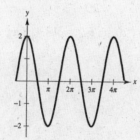

13. Period: $\dfrac{\pi}{2}$

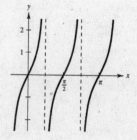

14. Period: 2π

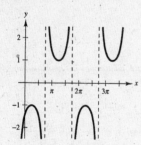

15.

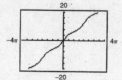

16.

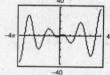

17. $\theta = \arcsin 1$

$\sin \theta = 1$

$\theta = \dfrac{\pi}{2}$

18. $\theta = \arctan(-3)$

$\tan \theta = -3$

$\theta \approx -1.249$ or $-71.565°$

19. $\sin\left(\arccos\dfrac{4}{\sqrt{35}}\right)$

$\sin\theta = \dfrac{\sqrt{19}}{\sqrt{35}} \approx 0.7368$

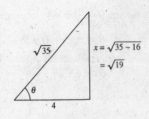

$x = \sqrt{35-16}$
$= \sqrt{19}$

20. $\cos\left(\arcsin\dfrac{x}{4}\right)$

$\cos\theta = \dfrac{\sqrt{16-x^2}}{4}$

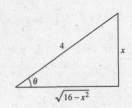

21. Given $A = 40°$, $c = 12$

$B = 90° - 40° = 50°$

$\sin 40° = \dfrac{a}{12}$

$a = 12\sin 40° \approx 7.713$

$\cos 40° = \dfrac{b}{12}$

$b = 12\cos 40° \approx 9.193$

22. Given $B = 6.84°$, $a = 21.3$

$A = 90° - 6.84° = 83.16°$

$\sin 83.16° = \dfrac{21.3}{c}$

$c = \dfrac{21.3}{\sin 83.16°} \approx 21.453$

$\tan 83.16° = \dfrac{21.3}{b}$

$b = \dfrac{21.3}{\tan 83.16°} \approx 2.555$

23. Given $a = 5$, $b = 9$

$c = \sqrt{25+81} = \sqrt{106}$

≈ 10.296

$\tan A = \dfrac{5}{9}$

$A = \arctan\dfrac{5}{9} \approx 29.055°$

$B = 90° - 29.055° = 60.945°$

24. $\sin 67° = \dfrac{x}{20}$

$x = 20\sin 67° \approx 18.41$ feet

25. $\tan 5° = \dfrac{250}{x}$

$x = \dfrac{250}{\tan 5°}$

≈ 2857.513 feet

≈ 0.541 mi

Chapter 5   Practice Test Solutions

1. $\tan x = \dfrac{4}{11}$, $\sec x < 0 \Rightarrow x$ is in Quadrant III.

$y = -4$, $\bar{x} = -11$, $r = \sqrt{16+121} = \sqrt{137}$

$\sin x = -\dfrac{4}{\sqrt{137}} = -\dfrac{4\sqrt{137}}{137}$ $\csc x = -\dfrac{\sqrt{137}}{4}$

$\cos x = -\dfrac{11}{\sqrt{137}} = -\dfrac{11\sqrt{137}}{137}$ $\sec x = -\dfrac{\sqrt{137}}{11}$

$\tan x = \dfrac{4}{11}$ $\cot x = \dfrac{11}{4}$

2. $\dfrac{\sec^2 x + \csc^2 x}{\csc^2 x(1+\tan^2 x)} = \dfrac{\sec^2 x + \csc^2 x}{\csc^2 x + (\csc^2 x)\tan^2 x}$

$= \dfrac{\sec^2 x + \csc^2 x}{\csc^2 x + \dfrac{1}{\sin^2 x}\cdot\dfrac{\sin^2 x}{\cos^2 x}}$

$= \dfrac{\sec^2 x + \csc^2 x}{\csc^2 x + \dfrac{1}{\cos^2 x}}$

$= \dfrac{\sec^2 x + \csc^2 x}{\csc^2 x + \sec^2 x} = 1$

3. $\ln|\tan \theta| - \ln|\cot \theta| = \ln\dfrac{|\tan \theta|}{|\cot \theta|}$

$\qquad = \ln\left|\dfrac{\sin \theta/\cos \theta}{\cos \theta/\sin \theta}\right| = \ln\left|\dfrac{\sin^2 \theta}{\cos^2 \theta}\right|$

$\qquad = \ln|\tan^2 \theta| = 2\ln|\tan \theta|$

4. $\cos\left(\dfrac{\pi}{2} - x\right) = \dfrac{1}{\csc x}$ is true since

$\qquad \cos\left(\dfrac{\pi}{2} - x\right) = \sin x = \dfrac{1}{\csc x}.$

5. $\sin^4 x + (\sin^2 x)\cos^2 x = \sin^2 x(\sin^2 x + \cos^2 x)$

$\qquad = \sin^2 x(1) = \sin^2 x$

6. $(\csc x + 1)(\csc x - 1) = \csc^2 x - 1 = \cot^2 x$

7. $\dfrac{\cos^2 x}{1 - \sin x} \cdot \dfrac{1 + \sin x}{1 + \sin x} = \dfrac{\cos^2 x(1 + \sin x)}{1 - \sin^2 x} = \dfrac{\cos^2 x(1 + \sin x)}{\cos^2 x} = 1 + \sin x$

8. $\dfrac{1 + \cos \theta}{\sin \theta} + \dfrac{\sin \theta}{1 + \cos \theta} = \dfrac{(1 + \cos \theta)^2 + \sin^2 \theta}{\sin \theta(1 + \cos \theta)}$

$\qquad = \dfrac{1 + 2\cos \theta + \cos^2 \theta + \sin^2 \theta}{\sin \theta(1 + \cos \theta)}$

$\qquad = \dfrac{2 + 2\cos \theta}{\sin \theta(1 + \cos \theta)} = \dfrac{2}{\sin \theta} = 2\csc \theta$

9. $\tan^4 x + 2\tan^2 x + 1 = (\tan^2 x + 1)^2 = (\sec^2 x)^2 = \sec^4 x$

10. (a) $\sin 105° = \sin(60° + 45°)$

$\qquad = \sin 60° \cos 45° + \cos 60° \sin 45°$

$\qquad = \dfrac{\sqrt{3}}{2} \cdot \dfrac{\sqrt{2}}{2} + \dfrac{1}{2} \cdot \dfrac{\sqrt{2}}{2}$

$\qquad = \dfrac{\sqrt{2}}{4}(\sqrt{3} + 1)$

(b) $\tan 15° = \tan(60° - 45°) = \dfrac{\tan 60° - \tan 45°}{1 + \tan 60° \tan 45°}$

$\qquad = \dfrac{\sqrt{3} - 1}{1 + \sqrt{3}} \cdot \dfrac{1 - \sqrt{3}}{1 - \sqrt{3}} = \dfrac{2\sqrt{3} - 1 - 3}{1 - 3}$

$\qquad = \dfrac{2\sqrt{3} - 4}{-2} = 2 - \sqrt{3}$

11. $(\sin 42°)\cos 38° - (\cos 42°)\sin 38° = \sin(42° - 38°) = \sin 4°$

12. $\tan\left(\theta + \dfrac{\pi}{4}\right) = \dfrac{\tan \theta + \tan(\pi/4)}{1 - (\tan \theta)\tan(\pi/4)} = \dfrac{\tan \theta + 1}{1 - \tan \theta(1)} = \dfrac{1 + \tan \theta}{1 - \tan \theta}$

13. $\sin(\arcsin x - \arccos x) = \sin(\arcsin x)\cos(\arccos x) - \cos(\arcsin x)\sin(\arccos x)$

$\qquad = (x)(x) - \left(\sqrt{1 - x^2}\right)\left(\sqrt{1 - x^2}\right) = x^2 - (1 - x^2) = 2x^2 - 1$

14. (a) $\cos(120°) = \cos[2(60°)] = 2\cos^2 60° - 1 = 2\left(\dfrac{1}{2}\right)^2 - 1 = -\dfrac{1}{2}$

(b) $\tan(300°) = \tan[2(150°)] = \dfrac{2\tan 150°}{1 - \tan^2 150°} = \dfrac{-2\sqrt{3}/3}{1 - (1/3)} = -\sqrt{3}$

15. (a) $\sin 22.5° = \sin \dfrac{45°}{2} = \sqrt{\dfrac{1 - \cos 45°}{2}} = \sqrt{\dfrac{1 - \sqrt{2}/2}{2}} = \dfrac{\sqrt{2 - \sqrt{2}}}{2}$

(b) $\tan \dfrac{\pi}{12} = \tan \dfrac{\pi/6}{2} = \dfrac{\sin(\pi/6)}{1 + \cos(\pi/6)} = \dfrac{1/2}{1 + \sqrt{3}/2} = \dfrac{1}{2 + \sqrt{3}} = 2 - \sqrt{3}$

16. $\sin \theta = \dfrac{4}{5}$, θ lies in Quadrant II $\Rightarrow \cos \theta = -\dfrac{3}{5}$.

$\cos \dfrac{\theta}{2} = \sqrt{\dfrac{1 + \cos \theta}{2}} = \sqrt{\dfrac{1 - (3/5)}{2}}$

$= \sqrt{\dfrac{2}{10}} = \dfrac{1}{\sqrt{5}} = \dfrac{\sqrt{5}}{5}$

17. $(\sin^2 x) \cos^2 x = \dfrac{1 - \cos 2x}{2} \cdot \dfrac{1 + \cos 2x}{2}$

$= \dfrac{1}{4}[1 - \cos^2 2x]$

$= \dfrac{1}{4}\left[1 - \dfrac{1 + \cos 4x}{2}\right]$

$= \dfrac{1}{8}[2 - (1 + \cos 4x)]$

$= \dfrac{1}{8}[1 - \cos 4x]$

18. $6(\sin 5\theta) \cos 2\theta = 6\left\{\dfrac{1}{2}[\sin(5\theta + 2\theta) + \sin(5\theta - 2\theta)]\right\} = 3[\sin 7\theta + \sin 3\theta]$

19. $\sin(x + \pi) + \sin(x - \pi) = 2\left(\sin \dfrac{[(x + \pi) + (x - \pi)]}{2}\right) \cos \dfrac{[(x + \pi) - (x - \pi)]}{2} = 2 \sin x \cos \pi = -2 \sin x$

20. $\dfrac{\sin 9x + \sin 5x}{\cos 9x - \cos 5x} = \dfrac{2 \sin 7x \cos 2x}{-2 \sin 7x \sin 2x} = -\dfrac{\cos 2x}{\sin 2x} = -\cot 2x$

21. $\dfrac{1}{2}[\sin(u + v) - \sin(u - v)] = \dfrac{1}{2}\{(\sin u) \cos v + (\cos u) \sin v - [(\sin u) \cos v - (\cos u) \sin v]\}$

$= \dfrac{1}{2}[2(\cos u) \sin v] = (\cos u) \sin v$

22. $4 \sin^2 x = 1$

$\sin^2 x = \dfrac{1}{4}$

$\sin x = \pm\dfrac{1}{2}$

$\sin x = \dfrac{1}{2}$ or $\sin x = -\dfrac{1}{2}$

$x = \dfrac{\pi}{6}$ or $\dfrac{5\pi}{6}$ $x = \dfrac{7\pi}{6}$ or $\dfrac{11\pi}{6}$

23. $\tan^2 \theta + \left(\sqrt{3} - 1\right) \tan \theta - \sqrt{3} = 0$

$(\tan \theta - 1)(\tan \theta + \sqrt{3}) = 0$

$\tan \theta = 1$ or $\tan \theta = -\sqrt{3}$

$\theta = \dfrac{\pi}{4}$ or $\dfrac{5\pi}{4}$ $\theta = \dfrac{2\pi}{3}$ or $\dfrac{5\pi}{3}$

24. $\sin 2x = \cos x$

$2(\sin x) \cos x - \cos x = 0$

$\cos x(2 \sin x - 1) = 0$

$\cos x = 0$ or $\sin x = \dfrac{1}{2}$

$x = \dfrac{\pi}{2}$ or $\dfrac{3\pi}{2}$ $x = \dfrac{\pi}{6}$ or $\dfrac{5\pi}{6}$

25. $\tan^2 x - 6 \tan x + 4 = 0$

$$\tan x = \frac{-(-6) \pm \sqrt{(-6)^2 - 4(1)(4)}}{2(1)}$$

$$\tan x = \frac{6 \pm \sqrt{20}}{2} = 3 \pm \sqrt{5}$$

$$\tan x = 3 + \sqrt{5} \qquad \text{or} \qquad \tan x = 3 - \sqrt{5}$$

$$x \approx 1.3821 \text{ or } 4.5237 \qquad\qquad x = 0.6524 \text{ or } 3.7940$$

Chapter 6   Practice Test Solutions

1. $C = 180° - (40° + 12°) = 128°$

$a = \sin 40°\left(\dfrac{100}{\sin 12°}\right) \approx 309.164$

$c = \sin 128°\left(\dfrac{100}{\sin 12°}\right) \approx 379.012$

2. $\sin A = 5\left(\dfrac{\sin 150°}{20}\right) = 0.125$

$A \approx 7.181°$

$B \approx 180° - (150° + 7.181°) = 22.819°$

$b = \sin 22.819°\left(\dfrac{20}{\sin 150°}\right) \approx 15.513$

3. Area $= \frac{1}{2}ab \sin C$

$= \frac{1}{2}(3)(6) \sin 130°$

≈ 6.894 square units

4. $h = b \sin A = 35 \sin 22.5° \approx 13.394$

$a = 10$

Since $a < h$ and A is acute, the triangle has no solution.

5. $\cos A = \dfrac{(53)^2 + (38)^2 - (49)^2}{2(53)(38)} \approx 0.4598$

$A \approx 62.627°$

$\cos B = \dfrac{(49)^2 + (38)^2 - (53)^2}{2(49)(38)} \approx 0.2782$

$B \approx 73.847°$

$C \approx 180° - (62.627° + 73.847°) = 43.526°$

6. $c^2 = (100)^2 + (300)^2 - 2(100)(300) \cos 29°$

$\approx 47{,}522.8176$

$c \approx 218$

$\cos A = \dfrac{(300)^2 + (218)^2 - (100)^2}{2(300)(218)} \approx 0.97495$

$A \approx 12.85°$

$B \approx 180° - (12.85° + 29°) = 138.15°$

7. $s = \dfrac{a + b + c}{2} = \dfrac{4.1 + 6.8 + 5.5}{2} = 8.2$

Area $= \sqrt{s(s - a)(s - b)(s - c)}$

$= \sqrt{8.2(8.2 - 4.1)(8.2 - 6.8)(8.2 - 5.5)}$

$= 11.273$ square units

8. $x^2 = (40)^2 + (70)^2 - 2(40)(70) \cos 168°$

$\approx 11{,}977.6266$

$x \approx 190.442$ miles

9. $\mathbf{w} = 4(3\mathbf{i} + \mathbf{j}) - 7(-\mathbf{i} + 2\mathbf{j}) = 19\mathbf{i} - 10\mathbf{j}$

10. $\dfrac{\mathbf{v}}{\|\mathbf{v}\|} = \dfrac{5\mathbf{i} - 3\mathbf{j}}{\sqrt{25 + 9}} = \dfrac{5}{\sqrt{34}}\mathbf{i} - \dfrac{3}{\sqrt{34}}\mathbf{j}$

$= \dfrac{5\sqrt{34}}{34}\mathbf{i} - \dfrac{3\sqrt{34}}{34}\mathbf{j}$

11. $\mathbf{u} = 6\mathbf{i} + 5\mathbf{j}, \ \mathbf{v} = 2\mathbf{i} - 3\mathbf{j}$

$\mathbf{u} \cdot \mathbf{v} = 6(2) + 5(-3) = -3$

$\|\mathbf{u}\| = \sqrt{61}, \ \|\mathbf{v}\| = \sqrt{13}$

$\cos \theta = \dfrac{-3}{\sqrt{61}\sqrt{13}}$

$\theta \approx 96.116°$

12. $4(\mathbf{i} \cos 30° + \mathbf{j} \sin 30°) = 4\left(\dfrac{\sqrt{3}}{2}\mathbf{i} + \dfrac{1}{2}\mathbf{j}\right) = \langle 4\sqrt{3}, 2 \rangle$

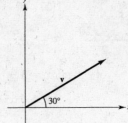

13. $\operatorname{proj}_{\mathbf{v}}\mathbf{u} = \left(\dfrac{\mathbf{u} \cdot \mathbf{v}}{\|\mathbf{v}\|^2}\right)\mathbf{v} = \dfrac{-10}{20}\langle -2, 4 \rangle = \langle 1, -2 \rangle$

14. $r = \sqrt{25 + 25} = \sqrt{50} = 5\sqrt{2}$

$\tan \theta = \dfrac{-5}{5} = -1$

Since z is in Quadrant IV,

$\theta = 315°$

$z = 5\sqrt{2}(\cos 315° + i \sin 315°).$

15. $\cos 225° = -\dfrac{\sqrt{2}}{2} \ \sin 225° = -\dfrac{\sqrt{2}}{2}$

$z = 6\left(-\dfrac{\sqrt{2}}{2} - i\dfrac{\sqrt{2}}{2}\right)$

$= -3\sqrt{2} - 3\sqrt{2}i$

16. $[7(\cos 23° + i \sin 23°)][4(\cos 7° + i \sin 7°)] = 7(4)[\cos(23° + 7°) + i \sin(23° + 7°)]$

$= 28(\cos 30° + i \sin 30°) = 14\sqrt{3} + 14i$

17. $\dfrac{9\left(\cos \dfrac{5\pi}{4} + i \sin \dfrac{5\pi}{4}\right)}{3(\cos \pi + i \sin \pi)} = \dfrac{9}{3}\left[\cos\left(\dfrac{5\pi}{4} - \pi\right) + i \sin\left(\dfrac{5\pi}{4} - \pi\right)\right] = 3\left(\cos \dfrac{\pi}{4} + i \sin \dfrac{\pi}{4}\right) = \dfrac{3\sqrt{2}}{2} + \dfrac{3\sqrt{2}}{2}i$

18. $(2 + 2i)^8 = [2\sqrt{2}(\cos 45° + i \sin 45°)]^8 = \left(2\sqrt{2}\right)^8[\cos(8)(45°) + i \sin (8)(45°)]$

$= 4096[\cos 360° + i \sin 360°] = 4096$

19. $z = 8\left(\cos \dfrac{\pi}{3} + i \sin \dfrac{\pi}{3}\right), \ n = 3$

The cube roots of z are:

For $k = 0, \ \sqrt[3]{8}\left[\cos \dfrac{\pi/3}{3} + i \sin \dfrac{\pi/3}{3}\right] = 2\left(\cos \dfrac{\pi}{9} + i \sin \dfrac{\pi}{9}\right).$

For $k = 1, \ \sqrt[3]{8}\left[\cos \dfrac{\pi/3 + 2\pi}{3} + i \sin \dfrac{\pi/3 + 2\pi}{3}\right] = 2\left(\cos \dfrac{7\pi}{9} + i \sin \dfrac{7\pi}{9}\right).$

For $k = 2, \ \sqrt[3]{8}\left[\cos \dfrac{\pi/3 + 4\pi}{3} + i \sin \dfrac{\pi/3 + 4\pi}{3}\right] = 2\left(\cos \dfrac{13\pi}{9} + i \sin \dfrac{13\pi}{9}\right).$

20. $x^4 = -i = 1\left(\cos\dfrac{3\pi}{2} + i\sin\dfrac{3\pi}{2}\right)$

For $k = 0$, $\cos\dfrac{3\pi/2}{4} + i\sin\dfrac{3\pi/2}{4} = \cos\dfrac{3\pi}{8} + i\sin\dfrac{3\pi}{8}$.

For $k = 1$, $\cos\dfrac{3\pi/2 + 2\pi}{4} + i\sin\dfrac{3\pi/2 + 2\pi}{4} = \cos\dfrac{7\pi}{8} + i\sin\dfrac{7\pi}{8}$.

For $k = 2$, $\cos\dfrac{3\pi/2 + 4\pi}{4} + i\sin\dfrac{3\pi/2 + 4\pi}{4} = \cos\dfrac{11\pi}{8} + i\sin\dfrac{11\pi}{8}$.

For $k = 3$, $\cos\dfrac{3\pi/2 + 6\pi}{4} + i\sin\dfrac{3\pi/2 + 6\pi}{4} = \cos\dfrac{15\pi}{8} + i\sin\dfrac{15\pi}{8}$.

Chapter 7   Practice Test Solutions

1. $\begin{cases} x + y = 1 \\ 3x - y = 15 \end{cases} \implies y = 3x - 15$

$x + (3x - 15) = 1$

$\qquad\qquad 4x = 16$

$\qquad\qquad\; x = 4$

$\qquad\qquad\; y = -3$

2. $\begin{cases} x - 3y = -3 \implies x = 3y - 3 \\ x^2 + 6y = 5 \end{cases}$

$\qquad\quad (3y - 3)^2 + 6y = 5$

$\qquad 9y^2 - 18y + 9 + 6y = 5$

$\qquad\qquad 9y^2 - 12y + 4 = 0$

$\qquad\qquad\quad (3y - 2)^2 = 0$

$\qquad\qquad\qquad\qquad y = \tfrac{2}{3}$

$\qquad\qquad\qquad\qquad x = -1$

3. $\begin{cases} x + y + z = 6 \implies z = 6 - x - y \\ 2x - y + 3z = 0 \qquad 2x - y + 3(6 - x - y) = 0 \implies -x - 4y = -18 \\ 5x + 2y - z = -3 \qquad 5x + 2y - (6 - x - y) = -3 \implies 6x + 3y = 3 \end{cases}$

$\qquad\qquad\qquad x = 18 - 4y$

$6(18 - 4y) + 3y = 3$

$\qquad\quad -21y = -105$

$\qquad\qquad\; y = 5$

$\qquad\qquad\; x = 18 - 4y = -2$

$\qquad\qquad\; z = 6 - x - y = 3$

4. $\begin{cases} x + y = 110 \implies y = 110 - x \\ xy = 2800 \end{cases}$

$x(110 - x) = 2800$

$\quad 0 = x^2 - 110x + 2800$

$\quad 0 = (x - 40)(x - 70)$

$\quad x = 40 \;$ or $\; x = 70$

$\quad y = 70 \qquad\quad y = 40$

5. $\begin{cases} 2x + 2y = 170 \implies y = \dfrac{170 - 2x}{2} = 85 - x \\ xy = 1500 \end{cases}$

$x(85 - x) = 1500$

$\quad 0 = x^2 - 85x + 1500$

$\quad 0 = (x - 25)(x - 60)$

$\quad x = 25 \;$ or $\; x = 60$

$\quad y = 60 \qquad\quad y = 25$

Dimensions: $60' \times 25'$

6. $\begin{cases} 2x + 15y = 4 \\ x - 3y = 23 \end{cases} \Rightarrow \begin{array}{r} 2x + 15y = 4 \\ 5x - 15y = 115 \\ \hline 7x = 119 \end{array}$

$$x = 17$$

$$y = \frac{x - 23}{3} = -2$$

7. $\begin{cases} x + y = 2 \\ 38x - 19y = 7 \end{cases} \Rightarrow \begin{array}{r} 19x + 19y = 38 \\ 38x - 19y = 7 \\ \hline 57x = 45 \end{array}$

$$x = \frac{45}{57} = \frac{15}{19}$$

$$y = 2 - x$$

$$= \frac{38}{19} - \frac{15}{19}$$

$$= \frac{23}{19}$$

8. $y_1 = 2(0.112 - 0.4x)$

$y_2 = \dfrac{(0.131 + 0.3x)}{0.7}$

$\begin{cases} 0.4x + 0.5y = 0.112 \\ 0.3x - 0.7y = -0.131 \end{cases} \Rightarrow \begin{array}{r} 0.28x + 0.35y = 0.0784 \\ 0.15x - 0.35y = -0.0655 \\ \hline 0.43x = 0.0129 \end{array}$

$$x = \frac{0.0129}{0.43} = 0.03$$

$$y = (2)(0.112 - 0.4x) = 0.20$$

9. Let x = amount in 11% fund and y = amount in 13% fund.

$$\begin{cases} x + y = 17{,}000 \Rightarrow y = 17{,}000 - x \\ 0.11x + 0.13y = 2080 \end{cases}$$

$$0.11x + 0.13(17{,}000 - x) = 2080$$

$$-0.02x = -130$$

$$x = \$6500$$

$$y = \$10{,}500$$

10. Using a graphing utility, you obtain $y = 0.7857x - 0.1429$. Analytically, $(4, 3)$, $(1, 1)$, $(-1, -2)$, $(-2, -1)$.

$$n = 4, \quad \sum_{i=1}^{4} x_i = 2, \quad \sum_{i=1}^{4} y_i = 1, \quad \sum_{i=1}^{4} x_i^2 = 22, \quad \sum_{i=1}^{4} x_i y_i = 17$$

$\begin{array}{r} 4b + 2a = 1 \Rightarrow 4b + 2a = 1 \\ 2b + 22a = 17 \Rightarrow -4b - 44a = -34 \\ \hline -42a = -33 \end{array}$

$$a = \frac{33}{42} = \frac{11}{14}$$

$$b = \frac{1}{4}\left(1 - 2\left(\frac{33}{42}\right)\right) = -\frac{1}{7}$$

$$y = ax + b = \frac{11}{14}x - \frac{1}{7}$$

11. $\begin{array}{ll} x + y = -2 & \text{Equation 1} \\ 2x - y + z = 11 & \text{Equation 2} \\ 4y - 3z = -20 & \text{Equation 3} \end{array}$

$\begin{cases} x + y = -2 \\ -3y + z = 15 \quad -2\text{Eq.1} + \text{Eq.2} \\ 4y - 3z = -20 \end{cases}$

$\begin{cases} x + y = -2 \\ -3y + z = 15 \\ -5y = 25 \quad 3\text{Eq.2} + \text{Eq.3} \end{cases}$

Answer: $y = -5$

$ x = 3$

$ z = 0$

12. $4x - y + 5z = 4$ Equation 1
$2x + y - z = 0$ Equation 2
$2x + 4y + 8z = 0$ Equation 3

$$\begin{cases} 4x - y + 5z = 4 \\ \quad -3y + 7z = 4 \quad \text{Eq.1} - 2\text{Eq.2} \\ \quad 3y + 9z = 0 \quad -\text{Eq.2} + \text{Eq.3} \end{cases}$$

$$\begin{cases} 4x - y + 5z = 4 \\ \quad -3y + 7z = 4 \\ \quad 16z = 4 \quad \text{Eq.2} + \text{Eq.3} \end{cases}$$

Answer: $z = \frac{1}{4}$
$y = -\frac{3}{4}$
$x = \frac{1}{2}$

13. $\begin{cases} 3x + 2y - z = 5 \implies \quad 6x + 4y - 2z = 10 \\ 6x - y + 5z = 2 \implies \underline{-6x + y - 5z = -2} \\ \qquad\qquad\qquad\qquad\qquad 5y - 7z = 8 \end{cases}$

$$y = \frac{8 + 7z}{5}$$

$3x + 2y - z = 5$
$\underline{12x - 2y + 10z = 4}$
$15x \qquad + 9z = 9$

$$x = \frac{9 - 9z}{15} = \frac{3 - 3z}{5}$$

Let $z = a$, then $x = \dfrac{3 - 3a}{5}$ and $y = \dfrac{8 + 7a}{5}$.

14. $y = ax^2 + bx + c$ passes through $(0, -1)$, $(1, 4)$, and $(2, 13)$.

At $(0, -1)$: $-1 = a(0)^2 + b(0) + c \implies c = -1$

At $(1, 4)$: $\quad 4 = a(1)^2 + b(1) - 1 \implies 5 = a + b \implies 5 = a + b$

At $(2, 13)$: $\quad 13 = a(2)^2 + b(2) - 1 \implies 14 = 4a + 2b \implies \underline{-7 = -2a - b}$

$\qquad\qquad\qquad\qquad\qquad\qquad\qquad\qquad\qquad\qquad -2 = -a$

$\qquad\qquad\qquad\qquad\qquad\qquad\qquad\qquad\qquad\qquad\quad a = 2$

$\qquad\qquad\qquad\qquad\qquad\qquad\qquad\qquad\qquad\qquad\quad b = 3$

Thus, $y = 2x^2 + 3x - 1$.

15. $s = \frac{1}{2}at^2 + v_0 t + s_0$ passes through $(1, 12)$, $(2, 5)$, and $(3, 4)$.

At $(1, 12)$: $12 = \frac{1}{2}a + v_0 + s_0 \implies$
At $(2, 5)$: $\quad 5 = 2a + 2v_0 + s_0 \implies$
At $(3, 4)$: $\quad 4 = \frac{9}{2}a + 3v_0 + s_0 \implies$

$\begin{cases} \frac{1}{2}a + v_0 + s_0 = 12 \\ -a \qquad + s_0 = 19 \quad 2\text{Eq.1} - \text{Eq.2} \\ -3a \qquad + s_0 = 7 \quad 3\text{Eq.2} - 2\text{Eq.3} \end{cases}$

$a = 6$
$s_0 = 25$
$v_0 = -16$

$\begin{cases} \frac{1}{2}a + v_0 + s_0 = 12 \\ -a \qquad + s_0 = 19 \\ -2a \qquad\qquad = -12 \quad -\text{Eq.2} + \text{Eq.3} \end{cases}$

Thus,
$s = \frac{1}{2}(6)t^2 - 16t + 25 = 3t^2 - 16t + 25$.

16. $\begin{bmatrix} 1 & -2 & 4 \\ 3 & -5 & 9 \end{bmatrix}$

$-3R_1 + R_2 \rightarrow \begin{bmatrix} 1 & -2 & 4 \\ 0 & 1 & -3 \end{bmatrix}$

$2R_2 + R_1 \rightarrow \begin{bmatrix} 1 & 0 & -2 \\ 0 & 1 & -3 \end{bmatrix}$

17. $3x + 5y = 3$

$2x - y = -11$

$$\begin{bmatrix} 3 & 5 & \vdots & 3 \\ 2 & -1 & \vdots & -11 \end{bmatrix}$$

$-R_2 + R_1 \rightarrow \begin{bmatrix} 1 & 6 & \vdots & 14 \\ 2 & -1 & \vdots & -11 \end{bmatrix}$

$-2R_1 + R_2 \rightarrow \begin{bmatrix} 1 & 6 & \vdots & 14 \\ 0 & -13 & \vdots & -39 \end{bmatrix}$

$-\frac{1}{13}R_2 \rightarrow \begin{bmatrix} 1 & 6 & \vdots & 14 \\ 0 & 1 & \vdots & 3 \end{bmatrix}$

$-6R_2 + R_1 \rightarrow \begin{bmatrix} 1 & 0 & \vdots & -4 \\ 0 & 1 & \vdots & 3 \end{bmatrix}$

Answer: $x = -4, y = 3$

18. $\begin{cases} 2x + 3y = -3 \\ 3x + 2y = 8 \\ x + y = 1 \end{cases}$

$$\begin{bmatrix} 2 & 3 & \vdots & -3 \\ 3 & 2 & \vdots & 8 \\ 1 & 1 & \vdots & 1 \end{bmatrix}$$

$\begin{matrix} R_3 \\ \\ R_1 \end{matrix} \begin{bmatrix} 1 & 1 & \vdots & 1 \\ 3 & 2 & \vdots & 8 \\ 2 & 3 & \vdots & -3 \end{bmatrix}$

$\begin{matrix} -3R_1 + R_2 \rightarrow \\ -2R_1 + R_3 \rightarrow \end{matrix} \begin{bmatrix} 1 & 1 & \vdots & 1 \\ 0 & -1 & \vdots & 5 \\ 0 & 1 & \vdots & -5 \end{bmatrix}$

$\begin{matrix} R_2 + R_1 \rightarrow \\ -R_2 \rightarrow \\ -R_2 + R_3 \rightarrow \end{matrix} \begin{bmatrix} 1 & 0 & \vdots & 6 \\ 0 & 1 & \vdots & -5 \\ 0 & 0 & \vdots & 0 \end{bmatrix}$

Answer: $x = 6, y = -5$

19. $\begin{cases} x + 3z = -5 \\ 2x + y = 0 \\ 3x + y - z = 3 \end{cases}$

$$\begin{bmatrix} 1 & 0 & 3 & \vdots & -5 \\ 2 & 1 & 0 & \vdots & 0 \\ 3 & 1 & -1 & \vdots & 3 \end{bmatrix}$$

$\begin{matrix} -2R_1 + R_2 \rightarrow \\ -3R_1 + R_3 \rightarrow \end{matrix} \begin{bmatrix} 1 & 0 & 3 & \vdots & -5 \\ 0 & 1 & -6 & \vdots & 10 \\ 0 & 1 & -10 & \vdots & 18 \end{bmatrix}$

$-R_2 + R_3 \rightarrow \begin{bmatrix} 1 & 0 & 3 & \vdots & -5 \\ 0 & 1 & -6 & \vdots & 10 \\ 0 & 0 & -4 & \vdots & 8 \end{bmatrix}$

$\begin{matrix} -3R_3 + R_1 \rightarrow \\ 6R_3 + R_2 \rightarrow \\ -\frac{1}{4}R_3 \rightarrow \end{matrix} \begin{bmatrix} 1 & 0 & 0 & \vdots & 1 \\ 0 & 1 & 0 & \vdots & -2 \\ 0 & 0 & 1 & \vdots & -2 \end{bmatrix}$

Answer: $x = 1, y = -2, z = -2$

20. $\begin{bmatrix} 1 & 4 & 5 \\ 2 & 0 & -3 \end{bmatrix} \begin{bmatrix} 1 & 6 \\ 0 & -7 \\ -1 & 2 \end{bmatrix} = \begin{bmatrix} -4 & -12 \\ 5 & 6 \end{bmatrix}$

21. $3A - 5B = 3\begin{bmatrix} 9 & 1 \\ -4 & 8 \end{bmatrix} - 5\begin{bmatrix} 6 & -2 \\ 3 & 5 \end{bmatrix}$

$= \begin{bmatrix} -3 & 13 \\ -27 & -1 \end{bmatrix}$

22. $f(A) = \begin{bmatrix} 3 & 0 \\ 7 & 1 \end{bmatrix}^2 - 7\begin{bmatrix} 3 & 0 \\ 7 & 1 \end{bmatrix} + 8\begin{bmatrix} 1 & 0 \\ 0 & 1 \end{bmatrix}$

$= \begin{bmatrix} 3 & 0 \\ 7 & 1 \end{bmatrix}\begin{bmatrix} 3 & 0 \\ 7 & 1 \end{bmatrix} - \begin{bmatrix} 21 & 0 \\ 49 & 7 \end{bmatrix} + \begin{bmatrix} 8 & 0 \\ 0 & 8 \end{bmatrix}$

$= \begin{bmatrix} 9 & 0 \\ 28 & 1 \end{bmatrix} - \begin{bmatrix} 21 & 0 \\ 49 & 7 \end{bmatrix} + \begin{bmatrix} 8 & 0 \\ 0 & 8 \end{bmatrix}$

$= \begin{bmatrix} -4 & 0 \\ -21 & 2 \end{bmatrix}$

23. False

$$(A + B)(A + 3B) = A(A + 3B) + B(A + 3B)$$
$$= A^2 + 3AB + BA + 3B^2$$

24.

$$\begin{bmatrix} 1 & 2 & \vdots & 1 & 0 \\ 3 & 5 & \vdots & 0 & 1 \end{bmatrix}$$

$$-3R_1 + R_2 \rightarrow \begin{bmatrix} 1 & 2 & \vdots & 1 & 0 \\ 0 & -1 & \vdots & -3 & 1 \end{bmatrix}$$

$$\begin{array}{c} 2R_2 + R_1 \rightarrow \\ -R_2 \rightarrow \end{array} \begin{bmatrix} 1 & 0 & \vdots & -5 & 2 \\ 0 & 1 & \vdots & 3 & -1 \end{bmatrix}$$

$$A^{-1} = \begin{bmatrix} -5 & 2 \\ 3 & -1 \end{bmatrix}$$

25. $\begin{bmatrix} 1 & 1 & 1 & \vdots & 1 & 0 & 0 \\ 3 & 6 & 5 & \vdots & 0 & 1 & 0 \\ 6 & 10 & 8 & \vdots & 0 & 0 & 1 \end{bmatrix}$

$$\begin{array}{c} -3R_1 + R_2 \rightarrow \\ -6R_1 + R_3 \rightarrow \end{array} \begin{bmatrix} 1 & 1 & 1 & \vdots & 1 & 0 & 0 \\ 0 & 3 & 2 & \vdots & -3 & 1 & 0 \\ 0 & 4 & 2 & \vdots & -6 & 0 & 1 \end{bmatrix}$$

$$\begin{array}{c} -\frac{1}{3}R_2 + R_1 \rightarrow \\ \frac{1}{3}R_2 \rightarrow \\ -4R_2 + R_3 \rightarrow \end{array} \begin{bmatrix} 1 & 0 & \frac{1}{3} & \vdots & 2 & -\frac{1}{3} & 0 \\ 0 & 1 & \frac{2}{3} & \vdots & -1 & \frac{1}{3} & 0 \\ 0 & 0 & -\frac{2}{3} & \vdots & -2 & -\frac{4}{3} & 1 \end{bmatrix}$$

$$\begin{array}{c} \frac{1}{2}R_3 + R_1 \rightarrow \\ R_3 + R_2 \rightarrow \\ -\frac{3}{2}R_3 \rightarrow \end{array} \begin{bmatrix} 1 & 0 & 0 & \vdots & 1 & -1 & \frac{1}{2} \\ 0 & 1 & 0 & \vdots & -3 & -1 & 1 \\ 0 & 0 & 1 & \vdots & 3 & 2 & -\frac{3}{2} \end{bmatrix}$$

$$A^{-1} = \begin{bmatrix} 1 & -1 & \frac{1}{2} \\ -3 & -1 & 1 \\ 3 & 2 & -\frac{3}{2} \end{bmatrix}$$

26. (a) $x + 2y = 4$

$3x + 5y = 1$

$$\begin{bmatrix} 1 & 2 & \vdots & 1 & 0 \\ 3 & 5 & \vdots & 0 & 1 \end{bmatrix}$$

$$-3R_1 + R_2 \rightarrow \begin{bmatrix} 1 & 2 & \vdots & 1 & 0 \\ 0 & -1 & \vdots & -3 & 1 \end{bmatrix}$$

$$\begin{array}{c} -2R_2 + R_1 \rightarrow \\ -R_2 \rightarrow \end{array} \begin{bmatrix} 1 & 0 & \vdots & -5 & 2 \\ 0 & 1 & \vdots & 3 & -1 \end{bmatrix}$$

$$X = A^{-1}B = \begin{bmatrix} -5 & 2 \\ 3 & -1 \end{bmatrix}\begin{bmatrix} 4 \\ 1 \end{bmatrix} = \begin{bmatrix} -18 \\ 11 \end{bmatrix}$$

$x = -18, y = 11$

(b) $x + 2y = 3$

$3x + 5y = -2$

$$X = A^{-1}B = \begin{bmatrix} -5 & 2 \\ 3 & -1 \end{bmatrix}\begin{bmatrix} 3 \\ -2 \end{bmatrix} = \begin{bmatrix} -19 \\ 11 \end{bmatrix}$$

$x = -19, y = 11$

27. $\begin{vmatrix} 6 & -1 \\ 3 & 4 \end{vmatrix} = 24 - (-3) = 27$

28. $\begin{vmatrix} 1 & 3 & -1 \\ 5 & 9 & 0 \\ 6 & 2 & -5 \end{vmatrix} = 1(-45) + (-3)(-25) + (-1)(-44)$

$$= 74$$

29. $\begin{vmatrix} 1 & 4 & 2 & 3 \\ 0 & 1 & -2 & 0 \\ 3 & 5 & -1 & 1 \\ 2 & 0 & 6 & 1 \end{vmatrix} = -7$

30. $\begin{vmatrix} 6 & 4 & 3 & 0 & 6 \\ 0 & 5 & 1 & 4 & 8 \\ 0 & 0 & 2 & 7 & 3 \\ 0 & 0 & 0 & 9 & 2 \\ 0 & 0 & 0 & 0 & 1 \end{vmatrix} = 6(5)(2)(9)(1) = 540$

31. Area $= \frac{1}{2}\begin{vmatrix} 0 & 7 & 1 \\ 5 & 0 & 1 \\ 3 & 9 & 1 \end{vmatrix} = \frac{1}{2}(31)$

$= 15.5$ square units

32. $\begin{vmatrix} x & y & 1 \\ 2 & 7 & 1 \\ -1 & 4 & 1 \end{vmatrix} = 3x - 3y + 15 = 0$

or $x - y + 5 = 0$

33. $x = \dfrac{\begin{vmatrix} 4 & -7 \\ 11 & 5 \end{vmatrix}}{\begin{vmatrix} 6 & -7 \\ 2 & 5 \end{vmatrix}} = \dfrac{97}{44}$

34. $z = \dfrac{\begin{vmatrix} 3 & 0 & 1 \\ 0 & 1 & 3 \\ 1 & -1 & 2 \end{vmatrix}}{\begin{vmatrix} 3 & 0 & 1 \\ 0 & 1 & 4 \\ 1 & -1 & 0 \end{vmatrix}} = \dfrac{14}{11}$

35. $y = \dfrac{\begin{vmatrix} 721.4 & 33.77 \\ 45.9 & 19.85 \end{vmatrix}}{\begin{vmatrix} 721.4 & -29.1 \\ 45.9 & 105.6 \end{vmatrix}}$

$= \dfrac{12{,}769.747}{77{,}515.530} \approx 0.1647$

Chapter 8   Practice Test Solutions

1. $a_n = \dfrac{2n}{(n+2)!}$

$a_1 = \dfrac{2(1)}{3!} = \dfrac{2}{6} = \dfrac{1}{3}$

$a_2 = \dfrac{2(2)}{4!} = \dfrac{4}{24} = \dfrac{1}{6}$

$a_3 = \dfrac{2(3)}{5!} = \dfrac{6}{120} = \dfrac{1}{20}$

$a_4 = \dfrac{2(4)}{6!} = \dfrac{8}{720} = \dfrac{1}{90}$

$a_5 = \dfrac{2(5)}{7!} = \dfrac{10}{5040} = \dfrac{1}{504}$

Terms: $\dfrac{1}{3}, \dfrac{1}{6}, \dfrac{1}{20}, \dfrac{1}{90}, \dfrac{1}{504}$

2. $a_n = \dfrac{n+3}{3^n}$

3. $\displaystyle\sum_{i=1}^{6}(2i - 1) = 1 + 3 + 5 + 7 + 9 + 11 = 36$

4. $a_1 = 23,\ d = -2$

$a_2 = a_1 + d = 21$

$a_3 = a_2 + d = 19$

$a_4 = a_3 + d = 17$

$a_5 = a_4 + d = 15$

Terms: 23, 21, 19, 17, 15

5. $a_1 = 12,\ d = 3,\ n = 50$

$a_n = a_1 + (n-1)d$

$a_{50} = 12 + (50 - 1)3 = 159$

6. $a_1 = 1$

$a_{200} = 200$

$S_n = \dfrac{n}{2}(a_1 + a_n)$

$S_{200} = \dfrac{200}{2}(1 + 200) = 20{,}100$

7. $a_1 = 7,\ r = 2$

$a_2 = a_1 r = 14$

$a_3 = a_1 r^2 = 28$

$a_4 = a_1 r^3 = 56$

$a_5 = a_1 r^4 = 112$

Terms: 7, 14, 28, 56, 112

8. $\sum_{n=0}^{9} 6\left(\frac{2}{3}\right)^n$, $a_1 = 6$, $r = \frac{2}{3}$, $n = 10$

$S_n = \frac{a_1(1 - r^n)}{1 - r} = \frac{6\left(1 - (2/3)^{10}\right)}{1 - (2/3)} \approx 17.6879$

9. $\sum_{n=0}^{\infty} (0.03)^n$, $a_1 = 1$, $r = 0.03$

$S = \frac{a_1}{1 - r} = \frac{1}{1 - 0.03} = \frac{1}{0.97} = \frac{100}{97} \approx 1.0309$

10. For $n = 1$, $1 = \frac{1(1 + 1)}{2}$. Assume that $1 + 2 + 3 = 4 + \cdots + k = \frac{k(k + 1)}{2}$. Now for $n = k + 1$,

$1 + 2 + 3 + 4 + \cdots + k + (k + 1) = \frac{k(k + 1)}{2} + k + 1$

$= \frac{k(k + 1)}{2} + \frac{2(k + 1)}{2}$

$= \frac{(k + 1)(k + 2)}{2}$.

Thus, $1 + 2 + 3 + 4 + \cdots + n = \frac{n(n + 1)}{2}$ for all integers $n \geq 1$.

11. For $n = 4$, $4! > 2^4$. Assume that $k! > 2^k$. Then
$(k + 1)! = (k + 1)(k!) > (k + 1)2^k > 2 \cdot 2^k$
$= 2^{k+1}$.

Thus, $n! > 2^n$ for all integers $n \geq 4$.

12. $_{13}C_4 = \frac{13!}{(13 - 4)!4!} = 715$

13. $(x + 3)^5 = x^5 + 5x^4(3) + 10x^3(3)^2 + 10x^2(3)^3 + 5x(3)^4 + (3)^5$
$= x^5 + 15x^4 + 90x^3 + 270x^2 + 405x + 243$

14. $_{12}C_5 x^7(-2)^5 = -25{,}344x^7$

15. $_{30}P_4 = \frac{30!}{(30 - 4)!} = 657{,}720$

16. $6! = 720$ ways

17. $_{12}P_3 = 1320$

18. $P(2) + P(3) + P(4) = \frac{1}{36} + \frac{2}{36} + \frac{3}{36}$
$= \frac{6}{36} = \frac{1}{6}$

19. $P(K, B10) = \frac{4}{52} \cdot \frac{2}{51} = \frac{2}{663}$

20. Let A = probability of no faulty units.
$P(A) = \left(\frac{997}{1000}\right)^{50} \approx 0.8605$
$P(A') = 1 - P(A) \approx 0.1395$

Chapter 9 Practice Test Solutions

1. $x^2 - 6x - 4y + 1 = 0$

$$x^2 - 6x + 9 = 4y - 1 + 9$$

$$(x - 3)^2 = 4y + 8$$

$$(x - 3)^2 = 4(1)(y + 2) \implies p = 1$$

Vertex: $(3, -2)$

Focus: $(3, -1)$

Directrix: $y = -3$

2. Vertex: $(2, -5)$

Focus: $(2, -6)$

Vertical axis; opens downward with $p = -1$

$$(x - h)^2 = 4p(y - k)$$

$$(x - 2)^2 = 4(-1)(y + 5)$$

$$x^2 - 4x + 4 = -4y - 20$$

$$x^2 - 4x + 4y + 24 = 0$$

3. $x^2 + 4y^2 - 2x + 32y + 61 = 0$

$$(x^2 - 2x + 1) + 4(y^2 + 8y + 16) = -61 + 1 + 64$$

$$(x - 1)^2 + 4(y + 4)^2 = 4$$

$$\frac{(x - 1)^2}{4} + \frac{(y + 4)^2}{1} = 1$$

$a = 2, b = 1, c = \sqrt{3}$

Horizontal major axis

Center: $(1, -4)$

Foci: $\left(1 \pm \sqrt{3}, -4\right)$

Vertices: $(3, -4), (-1, -4)$

Eccentricity: $e = \dfrac{\sqrt{3}}{2}$

4. Vertices: $(0, \pm 6)$

Eccentricity: $e = \dfrac{1}{2}$

Center: $(0, 0)$

Vertical major axis

$a = 6, e = \dfrac{c}{a} = \dfrac{c}{6} = \dfrac{1}{2} \implies c = 3$

$b^2 = (6)^2 - (3)^2 = 27$

$$\frac{x^2}{27} + \frac{y^2}{36} = 1$$

5. $16y^2 - x^2 - 6x - 128y + 231 = 0$

$$16(y^2 - 8y + 16) - (x^2 + 6x + 9) = -231 + 256 - 9$$

$$16(y - 4)^2 - (x + 3)^2 = 16$$

$$\frac{(y - 4)^2}{1} - \frac{(x + 3)^2}{16} = 1$$

$a = 1, b = 4, c = \sqrt{17}$

Center: $(-3, 4)$

Vertical transverse axis

Vertices: $(-3, 5), (-3, 3)$

Foci: $\left(-3, 4 \pm \sqrt{17}\right)$

Asymptotes: $y = 4 \pm \dfrac{1}{4}(x + 3)$

6. Vertices: $(\pm 3, 2)$

Foci: $(\pm 5, 2)$

Center: $(0, 2)$

Horizontal transverse axis

$a = 3, c = 5, b = 4$

$$\frac{(x - 0)^2}{9} - \frac{(y - 2)^2}{16} = 1$$

$$\frac{x^2}{9} - \frac{(y - 2)^2}{16} = 1$$

7. $5x^2 + 2xy + 5y^2 - 10 = 0$

$A = 5, B = 2, C = 5$

$\cot 2\theta = \dfrac{5-5}{2} = 0$

$2\theta = \dfrac{\pi}{2} \Rightarrow \theta = \dfrac{\pi}{4}$

$x = x' \cos \dfrac{\pi}{4} - y' \sin \dfrac{\pi}{4} = \dfrac{x' - y'}{\sqrt{2}}$

$y = x' \sin \dfrac{\pi}{4} + y' \cos \dfrac{\pi}{4} = \dfrac{x' + y'}{\sqrt{2}}$

$5\left(\dfrac{x'-y'}{\sqrt{2}}\right)^2 + 2\left(\dfrac{x'-y'}{\sqrt{2}}\right)\left(\dfrac{x'+y'}{\sqrt{2}}\right) + 5\left(\dfrac{x'+y'}{\sqrt{2}}\right)^2 - 10 = 0$

$\dfrac{5(x')^2}{2} - \dfrac{10x'y'}{2} + \dfrac{5(y')^2}{2} + (x')^2 - (y')^2 + \dfrac{5(x')^2}{2} + \dfrac{10x'y'}{2} + \dfrac{5(y')^2}{2} - 10 = 0$

$6(x')^2 + 4(y')^2 - 10 = 0$

$\dfrac{3(x')^2}{5} + \dfrac{2(y')^2}{5} = 1$

$\dfrac{(x')^2}{5/3} + \dfrac{(y')^2}{5/2} = 1$

Ellipse centered at the origin

8. (a) $6x^2 - 2xy + y^2 = 0$

$A = 6, B = -2, C = 1$

$B^2 - 4AC = (-2)^2 - 4(6)(1) = -20 < 0$

Ellipse

(b) $x^2 + 4xy + 4y^2 - x - y + 17 = 0$

$A = 1, B = 4, C = 4$

$B^2 - 4AC = (4)^2 - 4(1)(4) = 0$

Parabola

9. $x = 3 - 2 \sin \theta, y = 1 + 5 \cos \theta$

$\dfrac{x-3}{-2} = \sin \theta, \dfrac{y-1}{5} = \cos \theta$

$\left(\dfrac{x-3}{-2}\right)^2 + \left(\dfrac{y-1}{5}\right)^2 = 1$

$\dfrac{(x-3)^2}{4} + \dfrac{(y-1)^2}{25} = 1$

10. $x = e^{2t}, y = e^{4t}$

$x > 0, y > 0$

$y = (e^{2t})^2 = (x)^2 = x^2, x > 0, y > 0$

11. Polar: $\left(\sqrt{2}, \dfrac{3\pi}{4}\right)$

$x = \sqrt{2} \cos \dfrac{3\pi}{4} = \sqrt{2}\left(-\dfrac{1}{\sqrt{2}}\right) = -1$

$y = \sqrt{2} \sin \dfrac{3\pi}{4} = \sqrt{2}\left(\dfrac{1}{\sqrt{2}}\right) = 1$

Rectangular: $(-1, 1)$

12. Rectangular: $\left(\sqrt{3}, -1\right)$

$r = \pm\sqrt{(\sqrt{3})^2 + (-1)^2} = \pm 2$

$\tan \theta = \dfrac{-1}{\sqrt{3}} = -\dfrac{\sqrt{3}}{3}$

$\theta = \dfrac{5\pi}{6}$ or $\theta = \dfrac{11\pi}{6}$

Polar: $\left(-2, \dfrac{5\pi}{6}\right)$ or $\left(2, \dfrac{11\pi}{6}\right)$

13. Rectangular: $4x - 3y = 12$

Polar: $4r \cos \theta - 3r \sin \theta = 12$

$r(4 \cos \theta - 3 \sin \theta) = 12$

$$r = \frac{12}{4 \cos \theta - 3 \sin \theta}$$

14. Polar: $r = 5 \cos \theta$

$r^2 = 5r \cos \theta$

Rectangular: $x^2 + y^2 = 5x$

$x^2 + y^2 - 5x = 0$

15. $r = 1 - \cos \theta$

Cardioid

Symmetry: Polar axis

Maximum value of $|r|$: $r = 2$ when $\theta = \pi$.

Zero of r: $r = 0$ when $\theta = 0$.

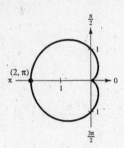

θ	0	$\frac{\pi}{2}$	π	$\frac{3\pi}{2}$
r	0	1	2	1

16. $r = 5 \sin 2\theta$

Rose curve with four petals

Symmetry: Polar axis, $\theta = \dfrac{\pi}{2}$, and pole

Maximum value of $|r|$: $|r| = 5$ when $\theta = \dfrac{\pi}{4}, \dfrac{3\pi}{4}, \dfrac{5\pi}{4}, \dfrac{7\pi}{4}$.

Zeros of r: $r = 0$ when $\theta = 0, \dfrac{\pi}{2}, \pi, \dfrac{3\pi}{2}$.

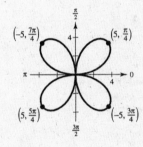

17. $r = \dfrac{3}{6 - \cos \theta}$

$r = \dfrac{1/2}{1 - (1/6) \cos \theta}$

$e = \dfrac{1}{6} < 1$, so the graph is an ellipse.

θ	0	$\frac{\pi}{2}$	π	$\frac{3\pi}{2}$
r	$\frac{3}{5}$	$\frac{1}{2}$	$\frac{3}{7}$	$\frac{1}{2}$

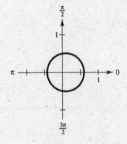

18. Parabola

Vertex: $\left(6, \dfrac{\pi}{2} \right)$

Focus: $(0, 0)$

$e = 1$

$r = \dfrac{ep}{1 + e \sin \theta}$

$r = \dfrac{p}{1 + \sin \theta}$

$6 = \dfrac{p}{1 + \sin(\pi/2)}$

$6 = \dfrac{p}{2}$

$12 = p$

$r = \dfrac{12}{1 + \sin \theta}$

Chapter 10   Practice Test Solutions

1. Let $A = (0, 0, 0)$, $B = (1, 2, -4)$, $C = (0, -2, -1)$.

   Side AB: $\sqrt{1^2 + 2^2 + 4^2} = \sqrt{21}$

   Side AC: $\sqrt{0^2 + 2^2 + 1^2} = \sqrt{5}$

   Side BC: $\sqrt{(-1)^2 + (-2-2)^2 + (-1+4)^2} = \sqrt{1 + 16 + 9} = \sqrt{26}$

   $BC^2 = AB^2 + AC^2$

   $26 = 21 + 5$

2. $(x - 0)^2 + (y - 4)^2 + (z - 1)^2 = 5^2$

   $x^2 + (y - 4)^2 + (z - 1)^2 = 25$

3. $(x^2 + 2x + 1) + y^2 + (z^2 - 4z + 4) = 1 + 4 + 11$

   $(x + 1)^2 + y^2 + (z - 2)^2 = 16$

   Center: $(-1, 0, 2)$

   Radius: 4

4. $\mathbf{u} - 3\mathbf{v} = \langle 1, 0, -1 \rangle - 3\langle 4, 3, -6 \rangle$

   $= \langle 1, 0, -1 \rangle - \langle 12, 9, -18 \rangle$

   $= \langle -11, -9, 17 \rangle$

5. $\frac{1}{2}\mathbf{v} = \frac{1}{2}\langle 2, 4, -6 \rangle = \langle 1, 2, -3 \rangle$

   $\left\| \frac{1}{2}\mathbf{v} \right\| = \sqrt{1^2 + 2^2 + (-3)^2} = \sqrt{14}$

6. $\mathbf{u} \cdot \mathbf{v} = \langle 2, 1, -3 \rangle \cdot \langle 1, 1, -2 \rangle$

   $= 2 + 1 + 6 = 9$

7. Because $\mathbf{v} = \langle -3, -3, 3 \rangle = -3\langle 1, 1, -1 \rangle = -3\mathbf{u}$, $\mathbf{u}$ and $\mathbf{v}$ are parallel.

8. $\mathbf{u} \times \mathbf{v} = \begin{vmatrix} \mathbf{i} & \mathbf{j} & \mathbf{k} \\ -1 & 0 & 2 \\ 1 & -1 & 3 \end{vmatrix} = \langle 2, 5, 1 \rangle$

   $\mathbf{v} \times \mathbf{u} = -(\mathbf{u} \times \mathbf{v}) = \langle -2, -5, -1 \rangle$

9. $\mathbf{u} \cdot (\mathbf{v} \times \mathbf{w}) = \begin{vmatrix} 1 & 1 & 1 \\ 0 & -1 & 1 \\ 1 & 0 & 4 \end{vmatrix}$

   $= 1(-4) - 1(-1) + 1(1)$

   $= -4 + 1 + 1 = -2$

   Volume $= |\mathbf{u} \cdot (\mathbf{v} \times \mathbf{w})| = |-2| = 2$

10. $\mathbf{v} = \langle (2 - 0), -3 - (-3), 4 - 3 \rangle = \langle 2, 0, 1 \rangle$

    $x = 2 + 2t, y = -3, z = 4 + t$

11. $1(x - 1) - 1(y - 2) + 0(z - 3) = 0$

    $x - 1 - y + 2 = 0$

    $x - y + 1 = 0$

12. $\overrightarrow{AB} = \langle 1, 1, 1 \rangle, \overrightarrow{AC} = \langle 1, 2, 3 \rangle$

    $\mathbf{n} = \overrightarrow{AB} \times \overrightarrow{AC} = \begin{vmatrix} \mathbf{i} & \mathbf{j} & \mathbf{k} \\ 1 & 1 & 1 \\ 1 & 2 & 3 \end{vmatrix} = \langle 1, -2, 1 \rangle$

    Plane: $1(x - 0) - 2(y - 0) + (z - 0) = 0$

    $x - 2y + z = 0$

13. $\mathbf{n}_1 = \langle 1, 1, -1 \rangle, \mathbf{n}_2 = \langle 3, -4, -1 \rangle$

    $\mathbf{n}_1 \cdot \mathbf{n}_2 = 3 - 4 + 1 = 0 \implies$ Orthogonal planes

14. $\mathbf{n} = \langle 1, 2, 1 \rangle, Q = (1, 1, 1), P = (0, 0, 6)$ on plane, $\overrightarrow{PQ} = \langle 1, 1, -5 \rangle$

    $D = \frac{|\overrightarrow{PQ} \cdot \mathbf{n}|}{\|\mathbf{n}\|} = \frac{|1 + 2 - 5|}{\sqrt{1 + 4 + 1}} = \frac{2}{\sqrt{6}} = \frac{\sqrt{6}}{3}$

Chapter 11   Practice Test Solutions

1.

x	2.9	2.99	3	3.01	3.1
$f(x)$	0.1695	0.1669	?	0.1664	0.1639

$$\lim_{x \to 3} \frac{x - 3}{x^2 - 9} \approx 0.1667$$

2. $\lim_{x \to 0} \dfrac{\sqrt{x + 4} - 2}{x} \approx \dfrac{1}{4}$

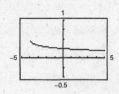

3. $\lim_{x \to 2} e^{x-2} = e^{2-2} = e^0 = 1$

4. $\lim_{x \to 1} \dfrac{x^3 - 1}{x - 1} = \lim_{x \to 1} \dfrac{(x - 1)(x^2 + x + 1)}{x - 1}$

$$= \lim_{x \to 1} (x^2 + x + 1) = 3$$

5. $\lim_{x \to 0} \dfrac{\sin 5x}{2x} \approx 2.5$

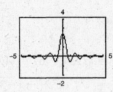

6. The limit does not exist. If

$$f(x) = \frac{|x + 2|}{x + 2},$$

then $f(x) = 1$ for $x > -2$, and $f(x) = -1$ for $x < -2$.

7. $m_{\text{sec}} = \dfrac{f(4 + h) - f(4)}{h}$

$$= \frac{\sqrt{4 + h} - 2}{h}$$

$$= \frac{\sqrt{4 + h} - 2}{h} \cdot \frac{\sqrt{4 + h} + 2}{\sqrt{4 + h} + 2}$$

$$= \frac{(4 + h) - 4}{h\left[\sqrt{4 + h} + 2\right]}$$

$$= \frac{h}{h\left[\sqrt{4 + h} + 2\right]}$$

$$= \frac{1}{\sqrt{4 + h} + 2}, \ h \neq 0$$

$$m = \lim_{h \to 0} \frac{1}{\sqrt{4 + h} + 2} = \frac{1}{\sqrt{4} + 2} = \frac{1}{4}$$

8. $f'(x) = \lim_{h \to 0} \dfrac{f(x + h) - f(x)}{h}$

$$= \lim_{h \to 0} \frac{[3(x + h) - 1] - [3x - 1]}{h}$$

$$= \lim_{h \to 0} \frac{3x + 3h - 1 - 3x + 1}{h}$$

$$= \lim_{h \to 0} \frac{3h}{h} = \lim_{h \to 0} 3 = 3$$

9. (a) $\lim_{x \to \infty} \dfrac{3}{x^4} = 0$

(b) $\lim_{x \to -\infty} \dfrac{x^2}{x^2 + 3} = 1$

(c) $\lim_{x \to \infty} \dfrac{|x|}{1 - x} = -1$

10. $a_1 = 0, a_2 = \dfrac{1 - 4}{8 + 1} = -\dfrac{1}{3}, a_3 = \dfrac{1 - 9}{18 + 1} = -\dfrac{8}{19},$

$$a_4 = \frac{1 - 16}{33} = -\frac{15}{33}$$

$$\lim_{n \to \infty} a_n = \lim_{n \to \infty} \frac{1 - n^2}{2n^2 + 1} = -\frac{1}{2}$$

11. $\displaystyle\sum_{i=1}^{25} i^2 + \sum_{i=1}^{25} i = \dfrac{25(26)(51)}{6} + \dfrac{25(26)}{2} = \dfrac{25(26)}{6}[51 + 3] = \dfrac{25(26)(54)}{6} = 5850$

12. $\displaystyle\sum_{i=1}^{n} \frac{i^2}{n^3} = \frac{1}{n^3} \sum_{i=1}^{n} i^2 = \frac{1}{n^3}\left[\frac{n(n+1)(2n+1)}{6}\right] = \frac{2n^2 + 3n + 1}{6n^2} = S(n)$

$\displaystyle\lim_{n\to\infty} S(n) = \frac{1}{3}$

13. Width of rectangles: $\dfrac{b-a}{n} = \dfrac{1}{n}$

Height: $f\left(a + \dfrac{(b-a)i}{n}\right) = f\left(\dfrac{i}{n}\right) = 1 - \left(\dfrac{i}{n}\right)^2$

$\displaystyle A_n \approx \sum_{i=1}^{n}\left[1 - \frac{i^2}{n^2}\right]\frac{1}{n} = \sum_{i=1}^{n}\frac{1}{n} - \sum_{i=1}^{n}\frac{i^2}{n^3} = 1 - \frac{1}{n^2}\frac{n(n+1)(2n+1)}{6}$

$\displaystyle A = \lim_{n\to\infty} A_n = 1 - \frac{1}{3} = \frac{2}{3}$

PART II

Chapter 1 Chapter Test Solutions

1. $5x + 2y = 3$

$2y = -5x + 3$

$y = -\frac{5}{2}x + \frac{3}{2}$

Slope $= -\frac{5}{2}$

(a) Parallel line

$y - 4 = -\frac{5}{2}(x - 0)$

$y = -\frac{5}{2}x + 4$

$5x + 2y - 8 = 0$

(b) Perpendicular line

$y - 4 = \frac{2}{5}(x - 0)$

$y = \frac{2}{5}x + 4$

$2x - 5y + 20 = 0$

2. Slope $= \dfrac{4 - (-1)}{-3 - 2} = \dfrac{5}{-5} = -1$

$y + 1 = -1(x - 2)$

$y = -x + 1$

3. No, for some x there corresponds more than one value of y. For instance, if $x = 1$, $y = \pm 1/\sqrt{3}$.

4. $f(x) = |x + 2| - 15$

(a) $f(-8) = |-8 + 2| - 15 = 6 - 15 = -9$

(b) $f(14) = |14 + 2| - 15 = 16 - 15 = 1$

(c) $f(t - 6) = |t - 6 + 2| - 15 = |t - 4| - 15$

5. $3 - x \geq 0 \implies$ domain is all $x \leq 3$.

6. $C = 5.60x + 24{,}000$

$P = R - C$

$= 99.50x - (5.60x + 24{,}000)$

$= 93.9x - 24{,}000$

7. $f(-x) = 2(-x)^3 - 3(-x)$

$= -2x^3 + 3x = -f(x)$

Odd

8. $f(-x) = 3(-x)^4 + 5(-x)^2$

$= 3x^4 + 5x^2 = f(x)$

Even

9. $h(x) = \frac{1}{4}x^4 - 2x^2 = \frac{1}{4}x^2(x^2 - 8)$

By graphing h, you see that the graph is increasing on $(-2, 0)$ and $(2, \infty)$ and decreasing on $(-\infty, -2)$ and $(0, 2)$.

10. $g(t) = |t + 2| - |t - 2|$

By graphing g, you see that the graph is increasing on $(-2, 2)$, and constant on $(-\infty, -2)$ and $(2, \infty)$.

11. Relative minimum: $(-3.33, -6.52)$

Relative maximum: $(0, 12)$

12. Relative minimum: $(0.77, 1.81)$

Relative maximum: $(-0.77, 2.19)$

13. (a) Parent function $f(x) = x^3$

(b) g is obtained from f by a horizontal shift five units to the right, a vertical stretch of 2, a reflection in the x-axis, and a vertical shift three units upward.

(c)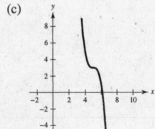

14. (a) Parent function $f(x) = \sqrt{x}$

(b) g is obtained from f by a reflection in the y-axis, and a horizontal shift seven units to the left.

(c)

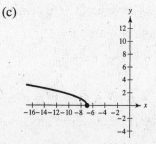

15. (a) Parent function $f(x) = |x|$

(b) $g(x) = 4|-x| - 7 = 4|x| - 7$ is obtained from f by a vertical stretch of 4 followed by a vertical shift seven units downward.

(c)

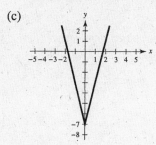

16. (a) $(f - g)(x) = x^2 - \sqrt{2 - x}$

Domain: $x \le 2$

(b) $\left(\dfrac{f}{g}\right)(x) = \dfrac{x^2}{\sqrt{2 - x}}$

Domain: $x < 2$

(c) $(f \circ g)(x) = f\left(\sqrt{2 - x}\right) = 2 - x$

Domain: $x \le 2$

(d) $(g \circ f)x = g(x^2) = \sqrt{2 - x^2}$

Domain: $-\sqrt{2} \le x \le \sqrt{2}$

17. $f(x) = x^3 + 8$

Yes, f is one-to-one and has an inverse function.

$$y = x^3 + 8$$
$$x = y^3 + 8$$
$$x - 8 = y^3$$
$$\sqrt[3]{x - 8} = y$$
$$f^{-1}(x) = \sqrt[3]{x - 8}$$

18. $f(x) = x^2 + 6$

No, f is not one-to-one, and does not have an inverse function.

19. $f(x) = \dfrac{3x\sqrt{x}}{8}$

Yes, f is one-to-one and has an inverse function.

$$y = \tfrac{3}{8}x^{3/2}, \ x \ge 0, y \ge 0$$
$$x = \tfrac{3}{8}y^{3/2}, \ y \ge 0, \ x \ge 0$$
$$\tfrac{8}{3}x = y^{3/2}$$
$$\left(\tfrac{8}{3}x\right)^{2/3} = y$$
$$f^{-1}(x) = \left(\tfrac{8}{3}x\right)^{2/3}, \ \ x \ge 0$$

20. $y = 18.30t - 76.2$, $r \approx 0.99622$

$y = 200$ for $t \approx 15$, or 2005

Chapter 2   Chapter Test Solutions

1. $y = x^2 + 4x + 3 = x^2 + 4x + 4 - 1 = (x + 2)^2 - 1$

Vertex: $(-2, -1)$

$x = 0 \implies y = 3$

$y = 0 \implies x^2 + 4x + 3 = 0 \implies (x + 3)(x + 1) = 0 \implies x = -1, -3$

Intercepts: $(0, 3), (-1, 0), (-3, 0)$

2. Let $y = a(x - h)^2 + k$. The vertex $(3, -6)$ implies that $y = a(x - 3)^2 - 6$. For $(0, 3)$ you obtain

$3 = a(0 - 3)^2 - 6 = 9a - 6 \Rightarrow a = 1$.

Thus, $y = (x - 3)^2 - 6 = x^2 - 6x + 3$.

3. $f(x) = 4x^3 + 4x^2 + x = x(4x^2 + 4x + 1) = x(2x + 1)^2$

Zeros: 0   (multiplicity 1)

$-\frac{1}{2}$ (multiplicity 2)

4. $f(x) = -x^3 + 7x + 6$

5.

$$
\begin{array}{r}
3x \\
x^2 + 1 \overline{)\ 3x^3 + 0x^2 + 4x - 1} \\
\underline{3x^3 \qquad\quad + 3x} \\
x - 1
\end{array}
$$

$$3x + \frac{x - 1}{x^2 + 1}$$

6.

$$
\begin{array}{r|rrrrr}
2 & 2 & 0 & -5 & 0 & -3 \\
  &   & 4 & 8 & 6 & 12 \\
\hline
  & 2 & 4 & 3 & 6 & 9
\end{array}
$$

$$2x^3 + 4x^2 + 3x + 6 + \frac{9}{x - 2}$$

7.

$$
\begin{array}{r|rrrrr}
-2 & 3 & 0 & -6 & 5 & -1 \\
   &   & -6 & 12 & -12 & 14 \\
\hline
   & 3 & -6 & 6 & -7 & 13
\end{array}
$$

$f(-2) = 13$

8. Possible rational zeros:

$\pm 24, \pm 12, \pm 8, \pm 6, \pm 4, \pm 3, \pm 2, \pm 1, \pm\frac{3}{2}, \pm\frac{1}{2}$

Rational zeros: $-2, \frac{3}{2}$

9. Possible rational zeros: $\pm 2, \pm 1, \pm\frac{2}{3}, \pm\frac{1}{3}$

Rational zeros: $\pm 1, -\frac{2}{3}$

10. $f(x) = x^3 - 7x^2 + 11x + 19$

$= (x + 1)(x^2 - 8x + 19)$

For the quadratic,

$$x = \frac{8 \pm \sqrt{64 - 4(19)}}{2} = 4 \pm \sqrt{3}\,i.$$

Zeros: $-1, 4 \pm \sqrt{3}\,i$

$f(x) = (x + 1)(x - 4 + \sqrt{3}\,i)(x - 4 - \sqrt{3}\,i)$

11. $(-8 - 3i) + (-1 - 15i) = -9 - 18i$

12. $\left(10 + \sqrt{-20}\right) - \left(4 - \sqrt{-14}\right) = 6 + 2\sqrt{5}i + \sqrt{14}i = 6 + \left(2\sqrt{5} + \sqrt{14}\right)i$

13. $(2 + i)(6 - i) = 12 + 6i - 2i + 1 = 13 + 4i$

14. $(4 + 3i)^2 - (5 + i)^2 = (16 + 24i - 9) - (25 + 10i - 1) = -17 + 14i$

15. $\dfrac{8 + 5i}{6 - i} \cdot \dfrac{6 + i}{6 + i} = \dfrac{48 + 30i + 8i - 5}{36 + 1} = \dfrac{43}{37} + \dfrac{38}{37}i$

16. $\dfrac{5i}{2 + i} \cdot \dfrac{2 - i}{2 - i} = \dfrac{10i + 5}{4 + 1} = 1 + 2i$

17. $\dfrac{(2i - 1)}{(3i + 2)} \cdot \dfrac{2 - 3i}{2 - 3i} = \dfrac{6 - 2 + 4i + 3i}{4 + 9}$

$\qquad = \dfrac{4}{13} + \dfrac{7}{13}i$

18.

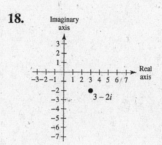

19.

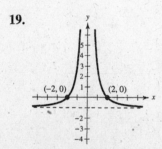

Vertical asymptote: $x = 0$

Intercepts: $(2, 0), (-2, 0)$

Symmetry: y-axis

Horizontal asymptote: $y = -1$

20. $g(x) = \dfrac{x^2 + 2}{x - 1} = x + 1 + \dfrac{3}{x - 1}$

Vertical asymptote: $x = 1$

Intercept: $(0, -2)$

Slant asymptote: $y = x + 1$

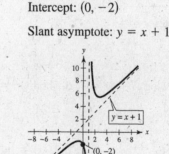

21. $f(x) = \dfrac{2x^2 + 9}{5x^2 + 2}$

Horizontal asymptote: $y = \dfrac{2}{5}$

y-axis symmetry

Intercept: $\left(0, \dfrac{9}{2}\right)$

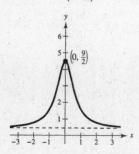

22. (a)

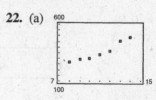

(b) $y = 5.582t^2 - 85.53t + 602.0$

(c)

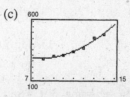

Yes, the model is a good fit.

(d) For 2005, $t = 15$ and $y \approx \$575$ billion.

For 2010, $t = 20$ and $y \approx \$1124$ billion.

(e) Answers will vary.

Chapter 3   Chapter Test Solutions

1.

x	-2	-1	0	1	2
$f(x)$	100	10	1	0.1	0.01

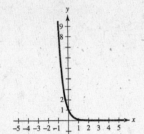

$f(x) = 10^{-x}$

Horizontal asymptote: $y = 0$

Intercept: $(0, 1)$

2.

x	0	2	3	4
$f(x)$	-0.03	-1	-6	-36

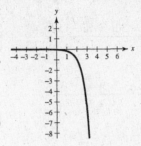

$f(x) = -6^{x-2}$

Horizontal asymptote: $y = 0$

Intercept: $\left(0, -\frac{1}{36}\right)$

3.

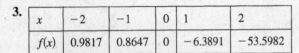

x	-2	-1	0	1	2
$f(x)$	0.9817	0.8647	0	-6.3891	-53.5982

$f(x) = 1 - e^{2x}$

Horizontal asymptote: $y = 1$

Intercept: $(0, 0)$

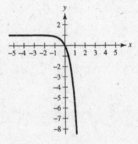

4. $\log_7 7^{-0.89} = -0.89 \log_7 7$
$$= -0.89$$

5. $4.6 \ln e^2 = 4.6(2) \ln e = 9.2$

6. $2 - \log_{10} 100 = 2 - 2 = 0$

7. $f(x) = -\log_{10} x - 6$

Domain: $x > 0$

Vertical asymptote: $x = 0$

x-intercept: $(10^{-6}, 0) \approx (0, 0)$

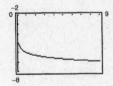

8. $f(x) = \ln(x - 4)$

Domain: $x > 4$

Vertical asymptote: $x = 4$

x-intercept: $(5, 0)$

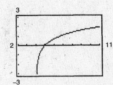

9. $f(x) = 1 + \ln(x + 6)$

Domain: $x > -6$

Vertical asymptote: $x = -6$

x-intercept: $(-5.632, 0)$

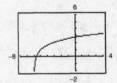

10. $\log_7 44 = \dfrac{\ln 44}{\ln 7} \approx 1.945$

11. $\log_{2/5}(0.9) = \dfrac{\ln(0.9)}{\ln(2/5)} \approx 0.115$

12. $\log_{24} 68 = \dfrac{\ln 68}{\ln 24} \approx 1.328$

13. $\log_2 3a^4 = \log_2 3 + \log_2 a^4 = \log_2 3 + 4 \log_2 a$

14. $\ln \dfrac{5\sqrt{x}}{6} = \ln 5 + \ln\sqrt{x} - \ln 6$

$\qquad = \ln 5 + \dfrac{1}{2}\ln x - \ln 6$

15. $\ln \dfrac{x\sqrt{x+1}}{2e^4} = \ln x + \ln\sqrt{x+1} - \ln(2) - \ln e^4$

$\qquad = \ln x + \dfrac{1}{2}\ln(x+1) - \ln 2 - 4$

16. $\log_3 13 + \log_3 y = \log_3(13y)$

17. $4\ln x - 4\ln y = \ln x^4 - \ln y^4 = \ln\left(\dfrac{x^4}{y^4}\right) = \ln\left(\dfrac{x}{y}\right)^4$

18. $\ln x - \ln(x+2) + \ln(2x-3) = \ln\left[\dfrac{x(2x-3)}{x+2}\right]$

19. $3^x = 81 = 3^4$

$\qquad x = 4$

20. $\qquad 5^{2x} = 2500$

$\qquad 2x\ln 5 = \ln 2500$

$\qquad x = \dfrac{1}{2}\dfrac{\ln 2500}{\ln 5} \approx 2.431$

21. $\log_7 x = 3$

$\qquad 7^3 = x$

$\qquad x = 343$

22. $\log_{10}(x-4) = 5$

$\qquad 10^5 = x - 4$

$\qquad x = 10^5 + 4$

$\qquad = 100{,}004$

23. $\dfrac{1025}{8 + e^{4x}} = 5$

$\qquad 1025 = 40 + 5e^{4x}$

$\qquad 985 = 5e^{4x}$

$\qquad e^{4x} = 197$

$\qquad 4x = \ln(197)$

$\qquad x = \dfrac{1}{4}\ln(197) \approx 1.321$

24. $-xe^{-x} + e^{-x} = 0$

$\qquad e^{-x}(1 - x) = 0$

$\qquad x = 1$

25. $\log_{10} x - \log_{10}(8 - 5x) = 2$

$\qquad \log_{10}\left(\dfrac{x}{8 - 5x}\right) = 2$

$\qquad 10^2 = \dfrac{x}{8 - 5x}$

$\qquad 800 - 500x = x$

$\qquad 800 = 501x$

$\qquad x = \dfrac{800}{501} \approx 1.597$

26. $2x\ln x - x = 0$

$\qquad 2\ln x = 1, \quad (x \neq 0)$

$\qquad \ln x = \dfrac{1}{2}$

$\qquad x = e^{1/2} \approx 1.649$

27. $\qquad \dfrac{1}{2} = 1e^{k(22)}$ (half-life is 22 years)

$\ln\dfrac{1}{2} = 22k$

$k = \dfrac{1}{22}\ln\dfrac{1}{2} = -\dfrac{1}{22}\ln 2 \approx -0.03151$

$A = e^{-0.03151(19)} \approx 0.54953$ or   55% remains

28. (a) Quadratic model: $R = -0.092t^2 + 3.29t + 38.1$

Exponential model: $R = 46.99(1.026)^t$

Power model: $R = 36.00t^{0.233}$

(b)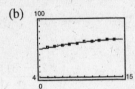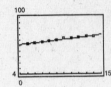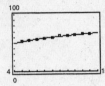

(c) The quadratic model is best for 2010, $t = 20$, and $R \approx 67.1$ billion dollars.

Chapters 1–3   Cumulative Test Solutions

1. (a)   Slope $= \dfrac{8-4}{-5-(-1)} = \dfrac{4}{-4} = -1$

$y - 8 = -1(x - (-5)) = -x - 5$

$x + y - 3 = 0$

(b) Sample answers: $(0, 3), (1, 2), (2, 1)$

2. (a) $y - 1 = -2\left(x + \dfrac{1}{2}\right)$

$y - 1 = -2x - 1$

$y = -2x$

$y + 2x = 0$

(b) Three additional points: $(0, 0), (1, -2), (2, -4)$

3. (a) Vertical line: $x = -\dfrac{3}{7}$ or $x + \dfrac{3}{7} = 0$

(b) Three additional points:

$\left(-\dfrac{3}{7}, 0\right), \left(-\dfrac{3}{7}, 1\right), \left(-\dfrac{3}{7}, 2\right)$

4. $f(x) = \dfrac{x}{x - 2}$

(a) $f(5) = \dfrac{5}{5 - 2} = \dfrac{5}{3}$

(b) $f(2)$ is undefined (division by 0).

(c) $f(5 + 4s) = \dfrac{5 + 4s}{(5 + 4s) - 2} = \dfrac{5 + 4s}{3 + 4s}$

5. $f(x) = \begin{cases} 3x - 8, & x < 0 \\ x^2 + 4, & x \geq 0 \end{cases}$

(a) $f(-8) = 3(-8) - 8 = -32$

(b) $f(0) = 0^2 + 4 = 4$

(c) $f(4) = 4^2 + 4 = 20$

6. No, for some x there corresponds two values of y.

7.
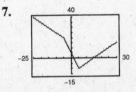

Decreasing on $(-\infty, 5)$,
Increasing on $(5, \infty)$

8. (a) $r(x) = \frac{1}{2}\sqrt[3]{x}$ is a vertical shrink of $y = \sqrt[3]{x}$.

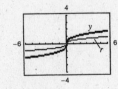

(b) $h(x) = \sqrt[3]{x} + 2$ is a vertical shift two units upward.

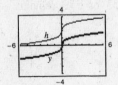

—CONTINUED—

8. —CONTINUED—

(c) $f(x) = \sqrt[3]{x}$

$g(x) = -\sqrt[3]{x + 2}$

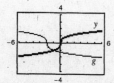

g is a horizontal shift two units to the left followed by a reflection in the x-axis.

9. $(f + g)(-4) = f(-4) + g(-4) = [-(-4)^2 + 3(-4) - 10] + [4(-4) + 1]$

$$= -38 - 15 = -53$$

10. $(g - f)\left(\tfrac{3}{4}\right) = \left[4\left(\tfrac{3}{4}\right) + 1\right] - \left[-\left(\tfrac{3}{4}\right)^2 + 3\left(\tfrac{3}{4}\right) - 10\right]$

$$= 4 - (-8.3125) = 12.3125 = \tfrac{197}{16}$$

11. $(g \circ f)(-2) = g(f(-2)) = g(-20) = 4(-20) + 1 = -79$

12. $(fg)(-1) = f(-1)g(-1) = (-14)(-3) = 42$

13. Yes, $h(x) = 5x - 2$ has an inverse function.

$y = 5x - 2$

$x = 5y - 2$

$x + 2 = 5y$

$\dfrac{x + 2}{5} = y$

$h^{-1}(x) = \dfrac{x + 2}{5}$

14. $f(x) = -\dfrac{1}{2}(x^2 + 4x)$

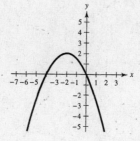

15. $f(x) = \tfrac{1}{4}x(x - 2)^2$

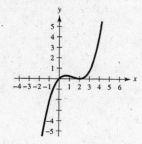

16.

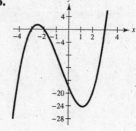

17. $x^3 + 2x^2 + 4x + 8 = (x + 2)(x^2 + 4)$

Zeros: $-2, \pm 2i$

18. Using a graphing utility, $x \approx 1.424$.

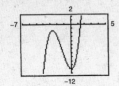

19.

$$\begin{array}{r} 4x + 2 \\ x + 3 \overline{) 4x^2 + 14x - 9} \\ \underline{4x^2 + 12x} \\ 2x - 9 \\ \underline{2x + 6} \\ -15 \end{array}$$

$$\frac{4x^2 + 14x - 9}{x + 3} = 4x + 2 - \frac{15}{x + 3}$$

20.

$$\begin{array}{r} 6 \ \big| \ \begin{array}{rrrr} 2 & -5 & 6 & -20 \\ & 12 & 42 & 288 \end{array} \\ \hline \begin{array}{rrrr} 2 & 7 & 48 & 268 \end{array} \end{array}$$

$$\frac{2x^3 - 5x^2 + 6x - 20}{x - 6} = 2x^2 + 7x + 48 + \frac{268}{x - 6}$$

21.

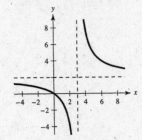

22. $f(x) = (x - 0)(x + 3)\left[x - \left(1 + \sqrt{5}i\right)\right]\left[x - \left(1 - \sqrt{5}i\right)\right]$

$= x(x + 3)[(x - 1)^2 + 5]$

$= (x^2 + 3x)(x^2 - 2x + 6)$

$= x^4 + x^3 + 18x$

23. $f(x) = \dfrac{2x}{x - 3}$

Vertical asymptote: $x = 3$

Horizontal asymptote: $y = 2$

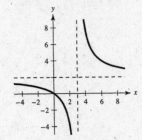

24. $f(x) = \dfrac{5x}{x^2 + x - 6} = \dfrac{5x}{(x + 3)(x - 2)}$

Vertical asymptotes: $x = -3, 2$

Horizontal asymptote: $y = 0$

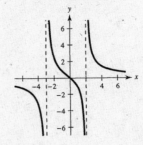

25. $f(x) = \dfrac{x^2 - 3x + 8}{x - 2} = x - 1 + \dfrac{6}{x - 2}$

Vertical asymptote: $x = 2$

Slant asymptote: $y = x - 1$

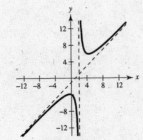

26. $(1.85)^{3.1} \approx 6.733$

27. $58^{\sqrt{5}} \approx 8772.934$

28. $e^{-20/11} \approx 0.162$

29. $4e^{2.56} \approx 51.743$

30. $f(x) = -3^{x+4} - 5$

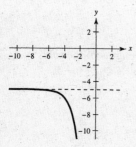

31. $f(x) = -\left(\frac{1}{2}\right)^{-x} - 3$

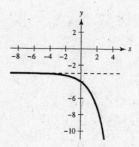

32. $f(x) = 4 + \log_{10}(x - 3)$

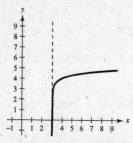

33. $f(x) = \ln(4 - x)$

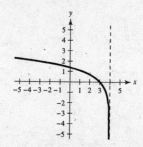

34. $\log_5 21 = \dfrac{\ln 21}{\ln 5} \approx 1.892$

35. $\log_9 6.8 = \dfrac{\ln 6.8}{\ln 9} \approx 0.872$

36. $\log_2\left(\dfrac{3}{2}\right) = \dfrac{\ln\left(\frac{3}{2}\right)}{\ln 2} \approx 0.585$

37. $\ln\left(\dfrac{x^2 - 4}{x^2 + 1}\right) = \ln[(x - 2)(x + 2)] - \ln(x^2 + 1)$

$$= \ln(x - 2) + \ln(x + 2) - \ln(x^2 + 1)$$

38. $2 \ln x - \ln(x - 1) + \ln(x + 1) = \ln\left[x^2\left(\dfrac{x + 1}{x - 1}\right)\right]$

39. $6e^{2x} = 72$

$e^{2x} = 12$

$2x = \ln 12$

$x = \dfrac{1}{2} \ln 12 \approx 1.242$

40. $4^{x-5} + 21 = 30$

$4^{x-5} = 9$

$(x - 5) \ln 4 = \ln 9$

$x - 5 = \dfrac{\ln 9}{\ln 4}$

$x = 5 + \dfrac{\ln 9}{\ln 4} \approx 6.585$

41. $\log_2 x + \log_2 5 = 6$

$\log_2 5x = 6$

$5x = 2^6 = 64$

$x = \dfrac{64}{5} = 12.8$

42. $250e^{0.05x} = 500,000$

$\qquad e^{0.05x} = 2000$

$\qquad 0.05x = \ln 2000$

$\qquad x = 20 \ln 2000$

$\qquad \approx 152.018$

43. $2x^2e^{2x} - 2xe^{2x} = 0$

$\qquad (2x^2 - 2x)e^{2x} = 0$

$\qquad 2x^2 - 2x = 0$

$\qquad 2x(x - 1) = 0$

$\qquad x = 0, 1$

44. $\ln(2x - 5) - \ln x = 1$

$$\ln \frac{2x - 5}{x} = 1$$

$$e = \frac{2x - 5}{x}$$

$$ex = 2x - 5$$

$$x(e - 2) = -5$$

$$x = \frac{-5}{e - 2} < 0$$

No solution because $\ln\left(\dfrac{-5}{e - 2}\right)$ does not exist.

45. (a) Let x and y be the lengths of the sides
$2x + 2y = 546 \implies y = 273 - x$.

$A = xy = x(273 - x)$

(b)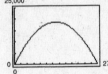

Domain: $0 < x < 273$

(c) If $A = 15,000$, then $x = 76.23$ or 196.77.

Dimensions in feet:

76.23×196.77 or 196.77×76.23

46. (a) Quadratic model: $y = 0.0707x^2 - 0.183x + 1.45$, Coefficient of determination: 0.99871

Exponential model: $y = 0.8915(1.2106)^x$, Coefficient of determination: 0.99862

Power model: $y = 0.7865x^{0.6762}$, Coefficient of determination: 0.93130

(b) Quadratic model: Exponential model: Power model:

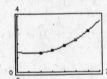   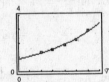   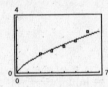

(c) The quadratic model is best because its coefficient of determination is closest to 1.

(d) For 2008, $x = 8$ and $y \approx \$4.51$. For 2010, $x = 10$ and $y \approx \$6.69$. Answers will vary.

Chapter 4   Chapter Test Solutions

1. (a)

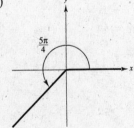

(b) $\dfrac{5\pi}{4} + 2\pi = \dfrac{13\pi}{4}$; $\dfrac{5\pi}{4} - 2\pi = -\dfrac{3\pi}{4}$

(c) $\dfrac{5\pi}{4} \cdot \dfrac{180}{\pi} = 225°$

2. $90 \text{ km/hr} = \dfrac{90 \text{ km/hr}}{60 \text{ min/h}} \cdot 1000 \text{ m/km}$

$\quad\quad\quad\quad = 1500 \text{ m/min}$

$1500 = \dfrac{s}{t} = \dfrac{r\theta}{t} \implies$ angular speed $= \dfrac{\theta}{t} = \dfrac{1500}{r} = \dfrac{1500}{1.25/2} = 2400 \text{ rad/min}$

3. $\sin\theta = \dfrac{4}{\sqrt{17}} = \dfrac{4\sqrt{17}}{17}$ $\csc\theta = \dfrac{\sqrt{17}}{4}$

$\cos\theta = -\dfrac{1}{\sqrt{17}} = -\dfrac{\sqrt{17}}{17}$ $\sec\theta = -\sqrt{17}$

$\tan\theta = -4$ $\cot\theta = -\dfrac{1}{4}$

4. $\tan\theta = \dfrac{7}{2}$

$\tan^2\theta + 1 = \sec^2\theta \implies \sec\theta = \sqrt{\dfrac{49}{4}+1} = \dfrac{\sqrt{53}}{2}$

$\cos\theta = \dfrac{2}{\sqrt{53}} = \dfrac{2\sqrt{53}}{53}$

$\sin\theta = \tan\theta\cos\theta = \dfrac{7\sqrt{53}}{53}$

$\csc\theta = \dfrac{53}{7\sqrt{53}} = \dfrac{\sqrt{53}}{7}$

$\cot\theta = \dfrac{2}{7}$

$\sec\theta = \dfrac{\sqrt{53}}{2}$

5. $\theta = 255°$

$\theta' = 255° - 180° = 75°$

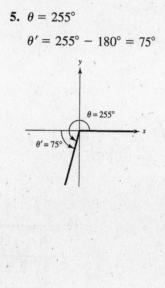

6. $\sec\theta = \dfrac{1}{\cos\theta} < 0 \implies$ Quadrants II or III

$\tan\theta > 0 \implies$ Quadrants I or III

Hence, Quadrant III.

7. $\cos\theta = -\dfrac{\sqrt{2}}{2}$

$\theta = 135°, 225°$

8. $\csc\theta = \dfrac{1}{\sin\theta} = 1.030 \implies \sin\theta = \dfrac{1}{1.030}$ and θ in Quadrant I or II.

Using a calculator, $\theta = 1.33, 1.81$ radians.

9. $\cos\theta = -\dfrac{3}{5}, \sin\theta > 0$, Quadrant II

$\sin\theta = \dfrac{4}{5}$ $\tan\theta = -\dfrac{4}{3}$

$\sec\theta = -\dfrac{5}{3}$ $\csc\theta = \dfrac{5}{4}$

$\cot\theta = -\dfrac{3}{4}$

10. $g(x) = -2\sin\left(x - \dfrac{\pi}{4}\right)$

Amplitude: 2, shifted $\pi/4$ to the right

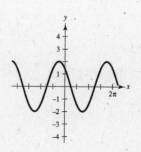

11. $f(x) = \frac{1}{2} \tan 4x$

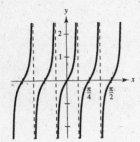

12. $f(x) = \frac{1}{2} \sec(x - \pi)$ is $y = \frac{1}{2} \sec x$ shifted π to the right.

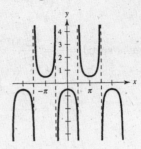

13. $f(x) = 2\cos(\pi - 2x) + 3 = 2\cos(2x - \pi) + 3$

Amplitude: 2

Shifted $\dfrac{\pi}{2}$ to the right, period π

Shifted vertically upward 3

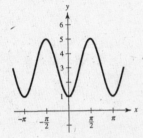

14. $f(x) = 2\csc\left(x + \dfrac{\pi}{2}\right)$

Shifted $\dfrac{\pi}{2}$ to the left

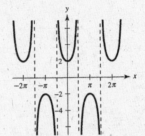

15. $f(x) = 2\cot\left(x - \dfrac{\pi}{2}\right)$

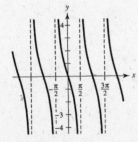

16.

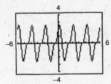

Period is 2.

17.

Not periodic

18. Amplitude: 2

Reflected in x-axis $\implies a = -2$

Period 4π and shifted to the right:

$$y = -2\sin\left(\dfrac{x}{2} - \dfrac{\pi}{4}\right)$$

19. Let $u = \arccos \dfrac{2}{3} \implies \cos u = \dfrac{2}{3}$.

Then $\tan\left(\arccos \dfrac{2}{3}\right) = \tan u = \dfrac{\sqrt{5}}{2}$.

20. $f(x) = 2 \arcsin\left(\frac{1}{2}x\right)$

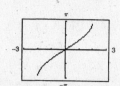

21. $f(x) = 2 \arccos x$

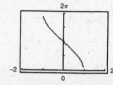

22. $f(x) = \arctan\left(\frac{x}{2}\right)$

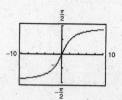

23. $\tan \theta = \frac{110}{160} \Rightarrow \theta \approx 34.5°$

Bearing: 214.5°

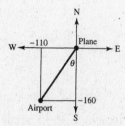

Chapter 5   Chapter Test Solutions

1. $\tan \theta = \frac{3}{2}$, $\cos \theta < 0 \Rightarrow \theta$ in Quadrant III

$\sec^2 \theta = \tan^2 \theta + 1 = \frac{9}{4} + 1 = \frac{13}{4} \Rightarrow \sec \theta = -\frac{\sqrt{13}}{2}$

$\cos \theta = -\frac{2}{\sqrt{13}} = -\frac{2\sqrt{13}}{13}$

$\sin \theta = \tan \theta \cos \theta = \frac{3}{2}\left(-\frac{2\sqrt{13}}{13}\right) = -\frac{3\sqrt{13}}{13}$

$\csc \theta = -\frac{13}{3\sqrt{13}} = -\frac{\sqrt{13}}{3}$

$\cot \theta = \frac{2}{3}$

2. $\csc^2 \beta(1 - \cos^2 \beta) = \frac{1}{\sin^2 \beta} \cdot \sin^2 \beta = 1$

3. $\frac{\sec^4 x - \tan^4 x}{\sec^2 x + \tan^2 x} = \frac{[(\sec^2 x) + (\tan^2 x)][\sec^2 x - \tan^2 x]}{\sec^2 x + \tan^2 x} = \sec^2 x - \tan^2 x = 1$

4. $\frac{\cos \theta}{\sin \theta} + \frac{\sin \theta}{\cos \theta} = \frac{\cos^2 \theta + \sin^2 \theta}{\sin \theta \cos \theta} = \frac{1}{\sin \theta \cos \theta} = \csc \theta \sec \theta$

5. Since $\tan^2 \theta = \sec^2 \theta - 1$ for all θ, then $\tan \theta = -\sqrt{\sec^2 \theta - 1}$ in Quadrants II and IV. Thus, $\frac{\pi}{2} < \theta \leq \pi$ and $\frac{3\pi}{2} < \theta < 2\pi$.

6. Conjecture: $y_1 = y_2$

Algebraically,

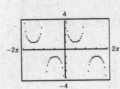

$$y_1 = \sin x + \cos x \cot x = \sin x + \frac{\cos^2 x}{\sin x}$$

$$= \frac{\sin^2 x + \cos^2 x}{\sin x}$$

$$= \frac{1}{\sin x} = \csc x.$$

7. $\sin \theta \cdot \sec \theta = \sin \theta \, \dfrac{1}{\cos \theta} = \tan \theta$

8. $\sec^2 x \tan^2 x + \sec^2 x = \sec^2 x(\tan^2 x + 1) = \sec^4 x$

9. $\dfrac{\csc \alpha + \sec \alpha}{\sin \alpha + \cos \alpha} = \dfrac{\dfrac{1}{\sin \alpha} + \dfrac{1}{\cos \alpha}}{\sin \alpha + \cos \alpha} = \dfrac{\dfrac{\cos \alpha + \sin \alpha}{\sin \alpha \cdot \cos \alpha}}{(\sin \alpha + \cos \alpha)}$

$$= \dfrac{1}{\sin \alpha \cos \alpha} = \dfrac{\cos^2 \alpha + \sin^2 \alpha}{\sin \alpha \cos \alpha}$$

$$= \dfrac{\cos \alpha}{\sin \alpha} + \dfrac{\sin \alpha}{\cos \alpha} = \cot \alpha + \tan \alpha$$

10. $\cos\!\left(x + \dfrac{\pi}{2}\right) = \cos x \cos \dfrac{\pi}{2} - \sin x \sin \dfrac{\pi}{2}$

$$= 0 - \sin x = -\sin x$$

11. $\sin(n\pi + \theta) = \sin n\pi \cos \theta + \cos n\pi \sin \theta$

$$= 0 + (-1)^n \sin \theta$$

$$= (-1)^n \sin \theta$$

12. $(\sin x + \cos x)^2 = \sin^2 x + \cos^2 x + 2 \sin x \cos x$

$$= 1 + \sin 2x$$

13. $\tan 105° = \tan(45° + 60°)$

$$= \dfrac{\tan 45° + \tan 60°}{1 - \tan 45° \tan 60°}$$

$$= \dfrac{1 + \sqrt{3}}{1 - \sqrt{3}} = -2 - \sqrt{3}$$

14. $\sin^4 x \tan^2 x = \dfrac{\sin^6 x}{\cos^2 x}$

$$= \dfrac{1}{32} \cdot \dfrac{(10 - 15 \cos 2x + 6 \cos 4x - \cos 6x)}{(1 + \cos 2x)/2}$$

$$= \dfrac{1}{16}\!\left[\dfrac{10 - 15 \cos 2x + 6 \cos 4x - \cos 6x}{1 + \cos 2x}\right]$$

15. $\dfrac{\sin 4\theta}{1 + \cos 4\theta} = \tan \dfrac{4\theta}{2} = \tan 2\theta$

16. $4 \cos 2\theta \sin 4\theta = 4\!\left(\dfrac{1}{2}\right)\![\sin(2\theta + 4\theta) - \sin(2\theta - 4\theta)]$

$$= 2[\sin 6\theta - \sin(-2\theta)]$$

$$= 2[\sin 6\theta + \sin 2\theta]$$

17. $\sin 3\theta - \sin 4\theta = 2\cos\left(\dfrac{3\theta + 4\theta}{2}\right)\sin\left(\dfrac{3\theta - 4\theta}{2}\right)$

$$= 2\cos\left(\dfrac{7\theta}{2}\right)\sin\left(\dfrac{-\theta}{2}\right)$$

$$= -2\cos\dfrac{7\theta}{2}\sin\dfrac{\theta}{2}$$

18. $\tan^2 x + \tan x = 0$

$\tan x(\tan x + 1) = 0$

$\tan x = 0 \Rightarrow x = 0, \pi$

$\tan x + 1 = 0 \Rightarrow \tan x = -1 \Rightarrow x = \dfrac{3\pi}{4}, \dfrac{7\pi}{4}$

19. $\sin 2\alpha - \cos\alpha = 0$

$2\sin\alpha\cos\alpha - \cos\alpha = 0$

$\cos\alpha(2\sin\alpha - 1) = 0$

$\cos\alpha = 0 \Rightarrow a = \dfrac{\pi}{2}, \dfrac{3\pi}{2}$

$2\sin\alpha - 1 = 0 \Rightarrow \sin a = \dfrac{1}{2} \Rightarrow \alpha = \dfrac{\pi}{6}, \dfrac{5\pi}{6}$

20. $4\cos^2 x - 3 = 0$

$$\cos^2 x = \dfrac{3}{4}$$

$$\cos x = \pm\dfrac{\sqrt{3}}{2}$$

$$x = \dfrac{\pi}{6}, \dfrac{5\pi}{6}, \dfrac{7\pi}{6}, \dfrac{11\pi}{6}$$

21. $\csc^2 x - \csc x - 2 = 0$

$(\csc x - 2)(\csc x + 1) = 0$

$\csc x - 2 = 0 \Rightarrow \csc x = 2 \Rightarrow \sin x = \dfrac{1}{2} \Rightarrow x = \dfrac{\pi}{6}, \dfrac{5\pi}{6}$

$\csc x + 1 = 0 \Rightarrow \csc x = -1 \Rightarrow \sin x = -1 \Rightarrow x = \dfrac{3\pi}{2}$

22. $3\cos x - x = 0$

$x \approx 1.170, -2.663, -2.938$

23. $\sin 2u = 2\sin u\cos u = 2\dfrac{2}{\sqrt{5}} \cdot \dfrac{1}{\sqrt{5}} = \dfrac{4}{5}$

$$\tan 2u = \dfrac{2\tan u}{1 - \tan^2 u} = \dfrac{2(2)}{1 - 2^2} = -\dfrac{4}{3}$$

$$\cos 2u = 1 - 2\sin^2 u = 1 - 2\left(\dfrac{2}{\sqrt{5}}\right)^2 = -\dfrac{3}{5}$$

24. $n = \dfrac{\sin[(\theta/2) + (\alpha/2)]}{\sin(\theta/2)}$

$\dfrac{3}{2} = \dfrac{\sin[(\theta/2) + 30°]}{\sin(\theta/2)}$

$3\sin\dfrac{\theta}{2} = 2\left[\sin\dfrac{\theta}{2}\cos 30° + \cos\dfrac{\theta}{2}\sin 30°\right]$

$3\sin\dfrac{\theta}{2} = \sqrt{3}\sin\dfrac{\theta}{2} + \cos\dfrac{\theta}{2}$

$\left(3 - \sqrt{3}\right)\sin\dfrac{\theta}{2} = \cos\dfrac{\theta}{2}$

$\tan\dfrac{\theta}{2} = \dfrac{1}{3 - \sqrt{3}}$

$\dfrac{\theta}{2} = \arctan\left(\dfrac{1}{3 - \sqrt{3}}\right) \approx 38.26°$

$\theta \approx 76.52°$

Chapter 6   Chapter Test Solutions

1. $A = 36°, B = 98°, c = 16$

$C = 180° - 36° - 98° = 46°$

$a = \dfrac{c}{\sin C} \sin A \approx 13.07$

$b = \dfrac{a}{\sin A} \sin B \approx 22.03$

2. $a = 4, b = 8, c = 10$

$\cos C = \dfrac{a^2 + b^2 - c^2}{2ab} = \dfrac{-20}{64} \Rightarrow C \approx 108.21°$

$\cos B = \dfrac{a^2 + c^2 - b^2}{2ac} = \dfrac{52}{80} \Rightarrow B \approx 49.46°$

$A = 180° - C - B \approx 22.33°$

3. $A = 35°, b = 8, c = 12$

$a^2 = b^2 + c^2 - 2bc \cdot \cos A$

$\quad = 64 + 144 - 2(8)(12) \cos 35° \approx 50.7228$

$a \approx 7.12$

$\sin B = \dfrac{\sin A}{a} b \approx 0.6443 \Rightarrow B \approx 40.11°$

$C = 180° - A - B \approx 104.89°$

4. $A = 25°, b = 28, a = 18$

$\sin B = \dfrac{\sin A}{a} b = \dfrac{(\sin 25°)28}{18} \approx 0.6574$

$B_1 \approx 41.10°$ or $B_2 \approx 180° - 41.1° \approx 138.90°$

Case 1: $B_1 = 41.10°$

$\qquad C = 180° - A - B_1 \approx 113.90°$

$\qquad c = \dfrac{a}{\sin A} \sin C \approx 38.94$

Case 2: $B_2 = 138.90°$

$\qquad C = 180° - A - B_2 \approx 16.10°$

$\qquad c = \dfrac{a}{\sin A} \sin C \approx 11.81$

5. No triangle possible $(5.2 \le 10.1)$

6. $\sin B = \dfrac{\sin A}{a} b = \dfrac{\sin 150°}{9.4} 4.8$

$\qquad \approx 0.2553 \Rightarrow B \approx 14.8°$

$C = 180° - A - B = 15.2°$

$c = \dfrac{a}{\sin A} \sin C \approx 4.9$

7. Law of Cosines:

$a^2 = b^2 + c^2 - 2bc \cos \theta$

$\quad = 480^2 + 565^2 - 2(480)(565) \cos 80°$

$\quad = 455,438.2 \Rightarrow a \approx 674.9 \text{ ft}$

8. $s = \dfrac{a + b + c}{2} = \dfrac{55 + 85 + 100}{2} = 120$

$A = \sqrt{s(s - a)(s - b)(s - c)}$

$\quad = \sqrt{120(65)(35)(20)}$

$\quad \approx 2336.7 \text{ square meters}$

9. $\mathbf{w} = \langle 4 - (-8), 1 - (-12) \rangle = \langle 12, 13 \rangle$

$\|\mathbf{w}\| = \sqrt{12^2 + 13^2} = \sqrt{313} \approx 17.69$

10. (a) $2\mathbf{v} + \mathbf{u} = 2\langle -2, -8 \rangle + \langle 0, -4 \rangle = \langle -4, -20 \rangle$

(b) $\mathbf{u} - 3\mathbf{v} = \langle 0, -4 \rangle - 3\langle -2, -8 \rangle = \langle 6, 20 \rangle$

(c) $5\mathbf{u} - \mathbf{v} = 5\langle 0, -4 \rangle - \langle -2, -8 \rangle = \langle 2, -12 \rangle$

11. (a) $2\mathbf{v} + \mathbf{u} = 2\langle -1, -10 \rangle + \langle -2, -3 \rangle$

$\qquad = \langle -4, -23 \rangle$

(b) $\mathbf{u} - 3\mathbf{v} = \langle -2, -3 \rangle - 3\langle -1, -10 \rangle = \langle 1, 27 \rangle$

(c) $5\mathbf{u} - \mathbf{v} = 5\langle -2, -3 \rangle - \langle -1, -10 \rangle$

$\qquad = \langle -9, -5 \rangle$

12. (a) $2\mathbf{v} + \mathbf{u} = 2(6\mathbf{i} + 9\mathbf{j}) + (\mathbf{i} - \mathbf{j}) = 13\mathbf{i} + 17\mathbf{j}$

(b) $\mathbf{u} - 3\mathbf{v} = (\mathbf{i} - \mathbf{j}) - 3(6\mathbf{i} + 9\mathbf{j}) = -17\mathbf{i} - 28\mathbf{j}$

(c) $5\mathbf{u} - \mathbf{v} = 5(\mathbf{i} - \mathbf{j}) - (6\mathbf{i} + 9\mathbf{j}) = -\mathbf{i} - 14\mathbf{j}$

13. (a) $2\mathbf{v} + \mathbf{u} = 2(-\mathbf{i} - 2\mathbf{j}) + (2\mathbf{i} + 3\mathbf{j}) = -\mathbf{j}$

(b) $\mathbf{u} - 3\mathbf{v} = (2\mathbf{i} + 3\mathbf{j}) - 3(-\mathbf{i} - 2\mathbf{j}) = 5\mathbf{i} + 9\mathbf{j}$

(c) $5\mathbf{u} - \mathbf{v} = 5(2\mathbf{i} + 3\mathbf{j}) - (-\mathbf{i} - 2\mathbf{j}) = 11\mathbf{i} + 17\mathbf{j}$

14. Unit vector $= \dfrac{\mathbf{v}}{\|\mathbf{v}\|} = \dfrac{1}{\sqrt{49 + 16}}\langle 7, 4\rangle$

$= \left\langle \dfrac{7}{\sqrt{65}}, \dfrac{4}{\sqrt{65}}\right\rangle = \dfrac{\sqrt{65}}{65}\langle 7, 4\rangle$

15. $12\dfrac{\langle 3, -5\rangle}{\|\langle 3, -5\rangle\|} = \dfrac{12}{\sqrt{34}}\langle 3, -5\rangle = \left\langle \dfrac{36}{\sqrt{34}}, \dfrac{-60}{\sqrt{34}}\right\rangle$

$= \left\langle \dfrac{18\sqrt{34}}{17}, \dfrac{-30\sqrt{34}}{17}\right\rangle$

16. $250(\cos 45°\mathbf{i} + \sin 45°\mathbf{j})$, first force

$130(\cos(-60°)\mathbf{i} + \sin(-60°)\mathbf{j})$, second force

Resultant: $\left[250\left(\dfrac{\sqrt{2}}{2}\right) + 130\left(\dfrac{1}{2}\right)\right]\mathbf{i} + \left[250\left(\dfrac{\sqrt{2}}{2}\right) + 130\left(-\dfrac{\sqrt{3}}{2}\right)\right]\mathbf{j}$

$= \left(125\sqrt{2} + 65\right)\mathbf{i} + \left(125\sqrt{2} - 65\sqrt{3}\right)\mathbf{j}$

Magnitude: $\sqrt{\left(125\sqrt{2} + 65\right)^2 + \left(125\sqrt{2} - 65\sqrt{3}\right)^2} \approx 250.15$

Direction: $\theta = \arctan\left(\dfrac{125\sqrt{2} - 65\sqrt{3}}{125\sqrt{2} + 65}\right) \Rightarrow \theta \approx 14.9°$

17. $\mathbf{u} \cdot \mathbf{v} = \langle -9, 4\rangle \cdot \langle 1, 3\rangle = -9 + 12 = 3$

18. $\cos\theta = \dfrac{\mathbf{u}\cdot\mathbf{v}}{\|\mathbf{u}\|\,\|\mathbf{v}\|} = \dfrac{-8}{\sqrt{53}(4)} \Rightarrow \theta \approx 105.9°$

19. No, the dot product is 24, not 0.

20. $\text{proj}_{\mathbf{v}}\mathbf{u} = \dfrac{-37}{26}\langle -5, -1\rangle = \left\langle \dfrac{185}{26}, \dfrac{37}{26}\right\rangle = \mathbf{w}_1$

$\mathbf{w}_2 = \mathbf{u} - \mathbf{w}_1 = \langle 6, 7\rangle - \left\langle \dfrac{185}{26}, \dfrac{37}{26}\right\rangle = \left\langle \dfrac{-29}{26}, \dfrac{145}{26}\right\rangle$

$\mathbf{u} = \mathbf{w}_1 + \mathbf{w}_2$

21. $|z| = 2\sqrt{2}$

$z = 2\sqrt{2}\left(\cos\dfrac{3\pi}{4} + i\sin\dfrac{3\pi}{4}\right)$

22. $100(\cos 240° + i\sin 240°) = -50 - 50\sqrt{3}i$

23. $\left[3\left(\cos\dfrac{5\pi}{6} + i\sin\dfrac{5\pi}{6}\right)\right]^8 = 3^8\left(\cos\dfrac{40\pi}{6} + i\sin\dfrac{40\pi}{6}\right)$

$= 3^8\left(-\dfrac{1}{2} + \dfrac{\sqrt{3}}{2}i\right)$

$= -3280.5 + 3280.5\sqrt{3}i = -\dfrac{6561}{2} + \dfrac{6561}{2}\sqrt{3}i$

24. $(3 - 3i)^6 = \left[3\sqrt{2}\left(\cos\dfrac{7\pi}{4} + i\sin\dfrac{7\pi}{4}\right)\right]^6$

$= 5832\left(\cos\dfrac{42\pi}{4} + i\sin\dfrac{42\pi}{4}\right) = 5832i$

25. $128\left(1 + \sqrt{3}i\right) = 256\left(\dfrac{1}{2} + \dfrac{\sqrt{3}}{2}i\right) = 256\left(\cos\dfrac{\pi}{3} + i\sin\dfrac{\pi}{3}\right)$

Fourth roots: $\sqrt[4]{256}\left(\cos\dfrac{(\pi/3) + 2\pi k}{4} + i\sin\dfrac{(\pi/3) + 2k\pi}{4}\right),\ k = 0, 1, 2, 3$

Four roots are: $4\left(\cos\dfrac{\pi}{12} + i\sin\dfrac{\pi}{12}\right) \approx 3.8637 + 1.0353i$

$\qquad\qquad\quad 4\left(\cos\dfrac{7\pi}{12} + i\sin\dfrac{7\pi}{12}\right) \approx -1.0353 + 3.8637i$

$\qquad\qquad\quad 4\left(\cos\dfrac{13\pi}{12} + i\sin\dfrac{13\pi}{12}\right) \approx -3.8637 - 1.0353i$

$\qquad\qquad\quad 4\left(\cos\dfrac{19\pi}{12} + i\sin\dfrac{19\pi}{12}\right) \approx 1.0353 - 3.8637i$

26. $x^4 = 625i$

Fourth roots of $625i = 625\left(\cos\dfrac{\pi}{2} + i\sin\dfrac{\pi}{2}\right)$

$\sqrt[4]{625}\left(\cos\left(\dfrac{(\pi/2) + 2\pi k}{4}\right) + i\sin\left(\dfrac{(\pi/2) + 2\pi k}{4}\right)\right),\ k = 0, 1, 2, 3$

Four roots are: $5\left(\cos\dfrac{\pi}{8} + i\sin\dfrac{\pi}{8}\right)$

$\qquad\qquad\quad 5\left(\cos\dfrac{5\pi}{8} + i\sin\dfrac{5\pi}{8}\right)$

$\qquad\qquad\quad 5\left(\cos\dfrac{9\pi}{8} + i\sin\dfrac{9\pi}{8}\right)$

$\qquad\qquad\quad 5\left(\cos\dfrac{13\pi}{8} + i\sin\dfrac{13\pi}{8}\right)$

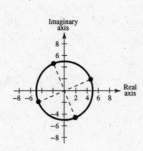

Chapters 4–6   Cumulative Test Solutions

1. $\theta = -150°$

(a)

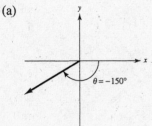

(b) $-150° + 360° = 210°$

(c) $(-150°)\dfrac{\pi}{180°} = -\dfrac{5}{6}\pi$ radians

(d) $\theta' = 30°$

(e) $\sin\theta = -\dfrac{1}{2}$ $\qquad\qquad$ $\csc\theta = -2$

$\quad\ \ \cos\theta = -\dfrac{\sqrt{3}}{2}$ $\qquad\quad$ $\sec\theta = -\dfrac{2}{\sqrt{3}} = -\dfrac{2\sqrt{3}}{3}$

$\quad\ \ \tan\theta = \dfrac{\sqrt{3}}{3}$ $\qquad\qquad$ $\cot\theta = \dfrac{3}{\sqrt{3}} = \sqrt{3}$

2. $2.55 \text{ rad} = 2.55\left(\dfrac{180°}{\pi}\right) \approx 146.1°$

3. $\sec^2 \theta = 1 + \tan^2 \theta = 1 + \left(-\dfrac{12}{5}\right)^2 = \dfrac{169}{25}$

$\sec \theta = -\dfrac{13}{5}$ (since θ in Quadrant II)

$\cos \theta = -\dfrac{5}{13}$

4. $f(x) = 3 - 2 \sin \pi x$

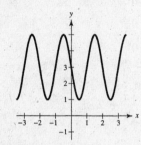

5. $f(x) = \tan(3x)$

Period: $\dfrac{\pi}{3}$

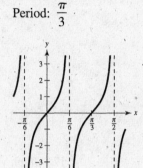

6. $f(x) = \dfrac{1}{2} \sec(x + \pi)$

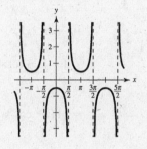

7. Amplitude: 3

Cosine curve reflected about the x-axis

Period: $2 \implies h(x) = -3 \cos(\pi x)$

Answer: $a = -3, b = \pi, c = 0$

8. Let $\theta = \arctan \dfrac{3}{4}$.

$\tan \theta = \dfrac{3}{4}$

$\sin\left(\arctan \dfrac{3}{4}\right) = \dfrac{3}{5}$

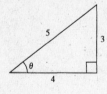

9. Let $\arcsin\left(-\dfrac{1}{2}\right) = \theta$.

$\sin \theta = -\dfrac{1}{2}$

$\theta = -\dfrac{\pi}{6}$

$\tan\left(-\dfrac{\pi}{6}\right) = -\dfrac{\sqrt{3}}{3}$

10. Let $\theta = \arctan 2x$.

$\tan \theta = 2x$

$\sin(\arctan 2x) = \sin \theta = \dfrac{2x}{\sqrt{1 + 4x^2}}$

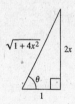

11. $\dfrac{\sin \theta - 1}{\cos \theta} - \dfrac{\cos \theta}{\sin \theta - 1} = \dfrac{\sin^2 \theta - 2 \sin \theta + 1 - \cos^2 \theta}{\cos \theta (\sin \theta - 1)}$

$= \dfrac{\sin^2 \theta - 2 \sin \theta + \sin^2 \theta}{\cos \theta (\sin \theta - 1)}$

$= \dfrac{2 \sin \theta (\sin \theta - 1)}{\cos \theta (\sin \theta - 1)} = 2 \tan \theta$

12. $\cot^2 \alpha (\sec^2 \alpha - 1) = \cot^2 \alpha (\tan^2 \alpha) = 1$

13. $\sin(x + y)\sin(x - y) = [\sin x \cos y + \cos x \sin y][\sin x \cos y - \sin y \cos x]$

$$= \sin^2 x \cos^2 y - \sin^2 y \cos^2 x$$

$$= \sin^2 x(1 - \sin^2 y) - \sin^2 y(1 - \sin^2 x)$$

$$= \sin^2 x - \sin^2 x \sin^2 y - \sin^2 y + \sin^2 y \sin^2 x$$

$$= \sin^2 x - \sin^2 y$$

14. $\sin^2 x \cos^2 x = \dfrac{1}{4}(2 \sin x \cos x)^2$

$$= \frac{1}{4}(\sin 2x)^2$$

$$= \frac{1}{4} \cdot \frac{1 - \cos 4x}{2}$$

$$= \frac{1}{8}(1 - \cos 4x)$$

15. $\sin^2 x + 2 \sin x + 1 = 0$

$$(\sin x + 1)^2 = 0$$

$$\sin x + 1 = 0$$

$$\sin x = -1$$

$$x = \frac{3\pi}{2} + 2n\pi$$

16. $3 \tan \theta - \cot \theta = 0$

$$3 \tan \theta - \frac{1}{\tan \theta} = 0$$

$$3 \tan^2 \theta - 1 = 0$$

$$\tan \theta = \pm \frac{1}{\sqrt{3}} \implies \theta = \frac{\pi}{6} + n\pi, \frac{5\pi}{6} + n\pi$$

17. Graph $y = \cos^2 x - 5 \cos x - 1$ on $[0, 2\pi)$.

Roots are $x \approx 1.7646, 4.5186$.

18.

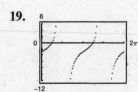

Zeros: $x \approx 1.047, 5.236$

Algebraically:

$$\frac{1 + \sin x}{\cos x} + \frac{\cos x}{1 + \sin x} = 4$$

$$\frac{1 + 2 \sin x + \sin^2 x + \cos^2 x}{\cos x(1 + \sin x)} = 4$$

$$\frac{2 + 2 \sin x}{\cos x(1 + \sin x)} = 4$$

$$\frac{2}{\cos x} = 4$$

$$\cos x = \frac{1}{2}$$

$$x = \frac{\pi}{3}, \frac{5\pi}{3}$$

19.

Zeros: $x \approx 0.785, 3.927$

Algebraically: $\tan^3 x - \tan^2 x + 3 \tan x - 3 = 0$

$$\tan^2 x(\tan x - 1) + 3(\tan x - 1) = 0$$

$$(\tan^2 x + 3)(\tan x - 1) = 0$$

$$\tan x = 1 \implies x = \frac{\pi}{4}, \frac{5\pi}{4}$$

20. $\sin u = \dfrac{12}{13} \Rightarrow \cos u = \dfrac{5}{13} \Rightarrow \tan u = \dfrac{12}{5}$

$\cos v = \dfrac{3}{5} \Rightarrow \sin v = \dfrac{4}{5} \Rightarrow \tan v = \dfrac{4}{3}$

$\tan(u - v) = \dfrac{\tan u - \tan v}{1 + \tan u \tan v} = \dfrac{(12/5) - (4/3)}{1 + (12/5)(4/3)} = \dfrac{16}{63}$

21. $\tan(2\theta) = \dfrac{2\tan\theta}{1 - \tan^2\theta} = \dfrac{2(1/2)}{1 - (1/4)} = \dfrac{4}{3}$

22. $\sec^2\theta = \tan^2\theta + 1 = \dfrac{25}{9} \Rightarrow \sec\theta = -\dfrac{5}{3} \Rightarrow \cos\theta = -\dfrac{3}{5}$ (Quadrant III)

$\sin\dfrac{\theta}{2} = \pm\sqrt{\dfrac{1 - \cos\theta}{2}} = \pm\dfrac{2}{\sqrt{5}} = \pm\dfrac{2\sqrt{5}}{5}$

$\dfrac{\theta}{2}$ in Quadrant II: $\sin\dfrac{\theta}{2} = \dfrac{2\sqrt{5}}{5}$

23. $\cos 8x + \cos 4x = 2\cos\left(\dfrac{8x + 4x}{2}\right)\cos\left(\dfrac{8x - 4x}{2}\right)$

$\qquad\qquad\qquad = 2\cos 6x \cos 2x$

24. $\tan x(1 - \sin^2 x) = \dfrac{\sin x}{\cos x}\cos^2 x = \sin x \cos x$

$\qquad\qquad\qquad = \dfrac{1}{2}(2\sin x \cos x) = \dfrac{1}{2}\sin 2x$

25. $\sin 3\theta \sin\theta = \dfrac{1}{2}[\cos(3\theta - \theta) - \cos(3\theta + \theta)]$

$\qquad\qquad = \dfrac{1}{2}(\cos 2\theta - \cos 4\theta)$

26. $\sin 3x \cos 2x = \dfrac{1}{2}(\sin(3x + 2x) + \sin(3x - 2x))$

$\qquad\qquad = \dfrac{1}{2}(\sin 5x + \sin x)$

27. $\dfrac{2\cos 3x}{\sin 4x - \sin 2x} = \dfrac{2\cos 3x}{2\cos 3x \cdot \sin x} = \dfrac{1}{\sin x} = \csc x$

28. $\sin B = \dfrac{\sin A}{a}b = 0.2569 \Rightarrow B \approx 14.9°$

$C = 180° - 46° - 14.9° = 119.1°$

$c = \dfrac{a}{\sin A}(\sin C) \approx 17.0$

29. $A = 32°, b = 8, c = 10$

$a^2 = b^2 + c^2 - 2bc\cos A$

$\quad = 64 + 100 - 2(8)(10)\cos 32°$

$\quad = 28.3123 \Rightarrow a \approx 5.32$

$\sin B = \dfrac{b\sin A}{a} \approx 0.7967 \Rightarrow B \approx 52.82°$

$C = 180° - B - A = 95.18°$

30. $B = 180° - 24° - 101° = 55°$

$b = \dfrac{a}{\sin A}\sin B \approx 20.14$

$c = \dfrac{a}{\sin A}\sin C \approx 24.13$

31. $\cos A = \dfrac{b^2 + c^2 - a^2}{2bc} = 0.8982 \Rightarrow A \approx 26.1°$

$\cos B = \dfrac{a^2 + c^2 - b^2}{2ac} = 0.8355 \Rightarrow B \approx 33.3°$

$C = 180° - 26.1° - 33.3° = 120.6°$

32. $A = \dfrac{1}{2}bh = \dfrac{1}{2}19 \cdot 14\sin 82°$

$\quad \approx 131.7$ sq. in.

33. $s = \dfrac{12 + 16 + 18}{2} = 23$

Area $= \sqrt{s(s - a)(s - b)(s - c)}$

$= \sqrt{23(11)(7)(5)} \approx 94.10$ sq. in.

34. $\mathbf{u} = \langle 3, 5 \rangle = 3\mathbf{i} + 5\mathbf{j}$

35. $\mathbf{v} = \mathbf{i} - 2\mathbf{j}$

$\|\mathbf{v}\| = \sqrt{1 + 4} = \sqrt{5}$

Unit vector: $\dfrac{1}{\sqrt{5}}\mathbf{i} - \dfrac{2}{\sqrt{5}}\mathbf{j} = \left\langle \dfrac{\sqrt{5}}{5}, \dfrac{-2\sqrt{5}}{5} \right\rangle$

36. $\mathbf{u} \cdot \mathbf{v} = 3(1) + 4(-2) = -5$

37. $\mathbf{u} \cdot \mathbf{v} = 0$

$\langle 1, 2k \rangle \cdot \langle 2, -1 \rangle = 0$

$2 - 2k = 0$

$k = 1$

38. $\text{proj}_{\mathbf{v}}\mathbf{u} = \dfrac{8 - 10}{26}\langle 1, 5 \rangle = \left\langle -\dfrac{1}{13}, -\dfrac{5}{13} \right\rangle = \mathbf{w}_1$

$\mathbf{w}_2 = \mathbf{u} - \mathbf{w}_1 = \langle 8, -2 \rangle - \left\langle -\dfrac{1}{13}, -\dfrac{5}{13} \right\rangle$

$= \left\langle \dfrac{105}{13}, -\dfrac{21}{13} \right\rangle$

$\mathbf{u} = \mathbf{w}_1 + \mathbf{w}_2$

39. $|z| = 3\sqrt{2}$, $\theta = \dfrac{3\pi}{4}$: $3\sqrt{2}\left(\cos\dfrac{3\pi}{4} + i \sin\dfrac{3\pi}{4} \right)$

40. $8\left(-\dfrac{\sqrt{3}}{2} + \dfrac{1}{2}i \right) = -4\sqrt{3} + 4i$

41. $[4(\cos 30° + i \sin 30°)][6(\cos 120° + i \sin 120°)] = 24(\cos(30 + 120°) + i \sin(30° + 120°))$

$= 24(\cos 150° + i \sin 150°)$

$= 24\left(-\dfrac{\sqrt{3}}{2} + i\dfrac{1}{2} \right) = -12\sqrt{3} + 12i$

42. $2 + i = \sqrt{5}(\cos\theta + i \sin\theta)$, where $\tan\theta = \dfrac{1}{2} \Rightarrow \theta \approx 0.4636$

Square roots: $z_1 = 5^{1/4}\left(\cos\dfrac{\theta}{2} + i \sin\dfrac{\theta}{2} \right)$

$z_2 = 5^{1/4}\left(\cos\left(\dfrac{\theta + 2\pi}{2} \right) + i \sin\left(\dfrac{\theta + 2\pi}{2} \right) \right)$

$z_1 \approx 1.4533 + 0.3436i$

$z_2 \approx -1.4553 - 0.3436i$

43. $1 = 1(\cos 0 + i \sin 0)$

$\cos\left(\dfrac{0 + 2\pi k}{3} \right) + i \sin\left(\dfrac{0 + 2\pi k}{3} \right)$, $k = 0, 1, 2$

$k = 0$: $\cos 0 + i \sin 0 = 1$

$k = 1$: $\cos\dfrac{2\pi}{3} + i \sin\dfrac{2\pi}{3} = -\dfrac{1}{2} + \dfrac{\sqrt{3}}{2}i$

$k = 2$: $\cos\dfrac{4\pi}{3} + i \sin\dfrac{4\pi}{3} = -\dfrac{1}{2} - \dfrac{\sqrt{3}}{2}i$

44. $x^5 = -243$

Five fifth roots of $-243 = 243(\cos \pi + i \sin \pi)$ are:

$\sqrt[5]{243}\left(\cos \dfrac{\pi + 2\pi k}{5} + i \sin \dfrac{\pi + 2\pi k}{5}\right)$; $k = 0, 1, 2, 3, 4$

$3\left(\cos \dfrac{\pi}{5} + i \sin \dfrac{\pi}{5}\right)$

$3\left(\cos \dfrac{3\pi}{5} + i \sin \dfrac{3\pi}{5}\right)$

$3\left(\cos \dfrac{5\pi}{5} + i \sin \dfrac{5\pi}{5}\right) = -3$

$3\left(\cos \dfrac{7\pi}{5} + i \sin \dfrac{7\pi}{5}\right)$

$3\left(\cos \dfrac{9\pi}{5} + i \sin \dfrac{9\pi}{5}\right)$

45. $\tan 18° = \dfrac{h}{200}$

$\tan 16° \, 45' = \dfrac{k}{200}$

Hence,

$f = h - k = 200 \tan 18° - 200 \tan 16° \, 45'$

$\approx 4.8 \approx 5$ feet.

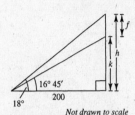

Not drawn to scale

46. $y = 4 \cos\left(\dfrac{\pi}{4}t\right)$ or $y = 4 \sin\left(\dfrac{\pi}{4}t\right)$

Amplitude: 4

Period $\dfrac{2\pi}{(\pi/4)} = 8$

47. Add the two vectors:

$500(\cos 60°\mathbf{i} + \sin 60°\mathbf{j}) + 50(\cos 30°\mathbf{i} + \sin 30°\mathbf{j}) = \left(250 + 25\sqrt{3}\right)\mathbf{i} + \left(250\sqrt{3} + 25\right)\mathbf{j}$

$\tan \theta = \dfrac{250\sqrt{3} + 25}{250 + 25\sqrt{3}} \approx 1.56 \implies \theta \approx 57.4°$

Direction: N 32.6° E or 32.6° in airplane navigation

Speed $= \sqrt{\left(250 + 25\sqrt{3}\right)^2 + \left(250\sqrt{3} + 25\right)^2} \approx 543.9$ km/hr

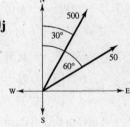

48. $\cos A = \dfrac{60^2 + 125^2 - 100^2}{2(60)(125)} = 0.615 \implies A \approx 52.05°$

$\cos B = \dfrac{100^2 + 125^2 - 60^2}{2(100)(125)} = 0.881 \implies B \approx 28.24$

Angle between vectors $= A + B \approx 80.3°$

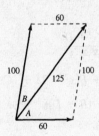

Chapter 7   Chapter Test Solutions

1. $x - y = 6 \implies y = x - 6$. Then

$3x + 5(x - 6) = 2 \implies$

$8x = 32 \implies x = 4, y = 4 - 6 = -2.$

Answer: $(4, -2)$

2. $y = x - 1 = (x - 1)^3 \implies x = 1$ or

$1 = (x - 1)^2 = x^2 - 2x + 1 \implies x^2 - 2x = 0.$

Thus, $x = 1$ or $x(x - 2) = 0 \implies x = 0, 1, 2.$

Answer: $(0, -1), (1, 0), (2, 1)$

3. $x - y = 3 \implies y = x - 3 \implies 4x - (x - 3)^2 = 7$

$$4x - (x^2 - 6x + 9) = 7$$
$$x^2 - 10x + 16 = 0$$
$$(x - 2)(x - 8) = 0$$

$x = 2, 8$

Answer: $(2, -1), (8, 5)$

4. $\begin{cases} 2x + 5y = -11 & \text{Equation 1} \\ 5x - y = 19 & \text{Equation 2} \end{cases}$

$-\frac{5}{2}$ times Eq. 1 added to Eq. 2 produces

$-\frac{27}{2}y = \frac{93}{2} \implies y = -\frac{31}{9}$.

Then $2x + 5\left(-\frac{31}{9}\right) = -11 \implies x = \frac{28}{9}$.

Answer: $\left(\frac{28}{9}, -\frac{31}{9}\right)$

5. $\begin{cases} 3x - 2y + z = 0 \\ 6x + 2y + 3z = -2 \\ 3x - 4y + 5z = 5 \end{cases}$

$\begin{cases} 3x - 2y + z = 0 \\ \quad\quad 6y + z = -2 \\ \quad -2y + 4z = 5 \end{cases}$

$\begin{cases} 3x - 2y + z = 0 \\ \quad\quad y - 2z = -\frac{5}{2} \\ \quad\quad\quad 13z = 13 \end{cases}$

$z = 1$

$y = 2(1) - \frac{5}{2} = -\frac{1}{2}$

$x = \frac{1}{3}\left(-1 + 2\left(-\frac{1}{2}\right)\right) = \frac{1}{3}(-2) = -\frac{2}{3}$

Answer: $\left(-\frac{2}{3}, -\frac{1}{2}, 1\right)$

6. $\begin{cases} x - 4y - z = 3 \\ 2x - 5y + z = 0 \\ 3x - 3y + 2z = -1 \end{cases}$

$\begin{cases} x - 4y - z = 3 \\ \quad\quad 3y + 3z = -6 \\ \quad\quad 9y + 5z = -10 \end{cases}$

$\begin{cases} x - 4y - z = 3 \\ \quad\quad y + z = -2 \\ \quad\quad -4z = 8 \end{cases}$

$z = -2$

$y = -2 - (-2) = 0$

$x = 3 + 4(0) + (-2) = 1$

Answer: $(1, 0, -2)$

7. $6 = a(0)^2 + b(0) + c \implies c = 6$

$2 = a(-2)^2 + b(-2) + c$

$\frac{9}{2} = a(3)^2 + b(3) + c$

Hence, $\begin{cases} 4a - 2b + 6 = 2 \text{ or } 2a - b = -2 \\ 9a + 3b + 6 = \frac{9}{2} \text{ or } 9a + 3b = -\frac{3}{2} \end{cases}$

Solving this system for a and b, you obtain

$a = -\frac{1}{2}, b = 1$. Thus, $y = -\frac{1}{2}x^2 + x + 6$.

8. $\dfrac{5x - 2}{(x - 1)^2} = \dfrac{A}{x - 1} + \dfrac{B}{(x - 1)^2}$

$5x - 2 = A(x - 1) + B = Ax + (-A + B)$

$\begin{cases} A = 5 \\ -A + B = -2 \implies B = 3 \end{cases}$

$\dfrac{5x - 2}{(x - 1)^2} = \dfrac{5}{x - 1} + \dfrac{3}{(x - 1)^2}$

9. $\dfrac{x^3 + x^2 + x + 2}{x^4 + x^2} = \dfrac{A}{x} + \dfrac{B}{x^2} + \dfrac{Cx + D}{x^2 + 1}$

$x^3 + x^2 + x + 2 = Ax(x^2 + 1) + B(x^2 + 1) + (Cx + D)x^2$

$\quad\quad\quad = (A + C)x^3 + (B + D)x^2 + Ax + B$

$\begin{cases} A + C = 1 \\ B + D = 1 \\ A = 1 \\ B = 2 \end{cases}$

$A = 1, C = 0, B = 2, D = -1$

$\dfrac{x^3 + x^2 + x + 2}{x^4 + x^2} = \dfrac{1}{x} + \dfrac{2}{x^2} - \dfrac{1}{x^2 + 1}$

10. $\begin{bmatrix} 2 & 1 & 2 & \vdots & 4 \\ 2 & 2 & 0 & \vdots & 5 \\ 2 & -1 & 6 & \vdots & 2 \end{bmatrix}$ row reduces to $\begin{bmatrix} 1 & 0 & 2 & \vdots & 1.5 \\ 0 & 1 & -2 & \vdots & 1 \\ 0 & 0 & 0 & \vdots & 0 \end{bmatrix}$.

Infinite number of solutions. Let $z = a$, $y = 2a + 1$, $x = 1.5 - 2a$.

Answer: $(1.5 - 2a, 1 + 2a, a)$, where a is any real number

11. $\begin{bmatrix} 2 & 3 & 1 & \vdots & 10 \\ 2 & -3 & -3 & \vdots & 22 \\ 4 & -2 & 3 & \vdots & -2 \end{bmatrix}$ row reduces to $\begin{bmatrix} 1 & 0 & 0 & \vdots & 5 \\ 0 & 1 & 0 & \vdots & 2 \\ 0 & 0 & 1 & \vdots & -6 \end{bmatrix}$.

Answer: $(5, 2, -6)$

12. (a) $A - B = \begin{bmatrix} 1 & 0 & 4 \\ -7 & -6 & -1 \\ 0 & 4 & 0 \end{bmatrix}$

(b) $3A = \begin{bmatrix} 15 & 12 & 12 \\ -12 & -12 & 0 \\ 3 & 6 & 0 \end{bmatrix}$

(c) $3A - 2B = \begin{bmatrix} 7 & 4 & 12 \\ -18 & -16 & -2 \\ 1 & 10 & 0 \end{bmatrix}$

(d) $AB = \begin{bmatrix} 36 & 20 & 4 \\ -28 & -24 & -4 \\ 10 & 8 & 2 \end{bmatrix}$

13. $\begin{bmatrix} -2 & 2 & 3 & \vdots & 1 & 0 & 0 \\ 1 & -1 & 0 & \vdots & 0 & 1 & 0 \\ 0 & 1 & 4 & \vdots & 0 & 0 & 1 \end{bmatrix}$

reduces to

$\begin{bmatrix} 1 & 0 & 0 & \vdots & -\frac{4}{3} & -\frac{5}{3} & 1 \\ 0 & 1 & 0 & \vdots & -\frac{4}{3} & -\frac{8}{3} & 1 \\ 0 & 0 & 1 & \vdots & \frac{1}{3} & \frac{2}{3} & 0 \end{bmatrix}$.

$A^{-1} = \begin{bmatrix} -\frac{4}{3} & -\frac{5}{3} & 1 \\ -\frac{4}{3} & -\frac{8}{3} & 1 \\ \frac{1}{3} & \frac{2}{3} & 0 \end{bmatrix}$

$A^{-1} \begin{bmatrix} 7 \\ -5 \\ -1 \end{bmatrix} = \begin{bmatrix} -2 \\ 3 \\ -1 \end{bmatrix}$

Answer: $(-2, 3, -1)$

14. $\begin{vmatrix} -25 & 18 \\ 6 & -7 \end{vmatrix} = (-25)(-7) - 6(18) = 67$

15. $\det(A) = \begin{vmatrix} 4 & 0 & 3 \\ 1 & -8 & 2 \\ 3 & 2 & 2 \end{vmatrix} = 4(-16 - 4) - 0 + 3(2 + 24)$

$$= -80 + 78 = -2$$

16. $\begin{vmatrix} x_1 & y_1 & 1 \\ x_2 & y_2 & 1 \\ x_3 & y_3 & 1 \end{vmatrix} = \begin{vmatrix} -1 & 1 & 1 \\ 4 & 11 & 1 \\ -1 & -5 & 1 \end{vmatrix}$

$$= -1(16) - 1(4 + 1) + 1(-20 + 11)$$

$$= -16 - 5 - 9 = -30$$

Area $= |-30| = 30$ square units

17. $x = \dfrac{\begin{vmatrix} 3 & -2 \\ -1 & 4 \end{vmatrix}}{\begin{vmatrix} 2 & -2 \\ 1 & 4 \end{vmatrix}} = \dfrac{10}{10} = 1$

$y = \dfrac{\begin{vmatrix} 2 & 3 \\ 1 & -1 \end{vmatrix}}{\begin{vmatrix} 2 & -2 \\ 1 & 4 \end{vmatrix}} = \dfrac{-5}{10} = -\dfrac{1}{2}$

Answer: $\left(1, -\dfrac{1}{2}\right)$

18. Upper left: $400 + x_2 = x_1$

Upper right: $x_1 + x_3 = x_4 + 600$

Lower left: $300 = x_2 + x_3 + x_5$

Lower right: $x_5 + x_4 = 100$

$$\begin{cases} x_1 - x_2 & = 400 \\ x_1 \quad\;\; + x_3 - x_4 & = 600 \\ \quad\; x_2 + x_3 \quad\;\; + x_5 = 300 \\ \quad\qquad\qquad x_4 + x_5 = 100 \end{cases}$$

Solving this system:

$$\begin{bmatrix} 1 & -1 & 0 & 0 & 0 & \vdots & 400 \\ 1 & 0 & 1 & -1 & 0 & \vdots & 600 \\ 0 & 1 & 1 & 0 & 1 & \vdots & 300 \\ 0 & 0 & 0 & 1 & 1 & \vdots & 100 \end{bmatrix} \rightarrow \begin{bmatrix} 1 & 0 & 1 & 0 & 1 & \vdots & 700 \\ 0 & 1 & 1 & 0 & 1 & \vdots & 300 \\ 0 & 0 & 0 & 1 & 1 & \vdots & 100 \\ 0 & 0 & 0 & 0 & 0 & \vdots & 0 \end{bmatrix}$$

Letting $x_3 = a$ and $x_5 = b$ be real numbers, we have:

$x_5 = b$

$x_4 = 100 - b$

$x_3 = a$

$x_2 = 300 - a - b$

$x_1 = 700 - b - a$

Chapter 8 Chapter Test Solutions

1. $a_n = \left(-\frac{2}{3}\right)^{n-1}$

$a_1 = \left(-\frac{2}{3}\right)^{1-1} = \left(-\frac{2}{3}\right)^0 = 1$

$a_2 = -\frac{2}{3}$

$a_3 = \left(-\frac{2}{3}\right)^2 = \frac{4}{9}$

$a_4 = \left(-\frac{2}{3}\right)^3 = -\frac{8}{27}$

$a_5 = \left(-\frac{2}{3}\right)^4 = \frac{16}{81}$

2. $a_1 = 12, \; a_{k+1} = a_k + 4$

$a_2 = 12 + 4 = 16$

$a_3 = 16 + 4 = 20$

$a_4 = 20 + 4 = 24$

$a_5 = 24 + 4 = 28$

3. $b_1 = -x$

$b_2 = \dfrac{x^2}{2}$

$b_3 = -\dfrac{x^3}{3}$

$b_4 = \dfrac{x^4}{4}$

$b_5 = -\dfrac{x^5}{5}$

4. $b_1 = -\dfrac{x^3}{3!} = -\dfrac{x^3}{6}$

$b_2 = -\dfrac{x^5}{5!} = -\dfrac{x^5}{120}$

$b_3 = -\dfrac{x^7}{7!}$

$b_4 = -\dfrac{x^9}{9!}$

$b_5 = -\dfrac{x^{11}}{11!}$

5. $\dfrac{11!\,4!}{4!\,7!} = \dfrac{11!}{7!}$

$= \dfrac{11 \cdot 10 \cdot 9 \cdot 8 \cdot 7!}{7!}$

$= 11 \cdot 10 \cdot 9 \cdot 8$

$= 7920$

6. $\dfrac{n!}{(n+1)!} = \dfrac{n!}{(n+1)n!} = \dfrac{1}{n+1}$

7. $\dfrac{2n!}{(n-1)!} = \dfrac{2n(n-1)!}{(n-1)!} = 2n$

8. $a_n = n^2 + 1$, $n = 1, 2, 3, \ldots$

9. $a_n = dn + c$

$c = a_1 - d = 5000 - (-100) = 5100 \implies$

$a_n = -100n + 5100 = 5000 - 100(n-1)$

10. $a_n = a_1 r^{n-1}$, $a_1 = 4$, $r = \frac{1}{2} \implies a_n = 4\left(\frac{1}{2}\right)^{n-1}$

11. $\displaystyle\sum_{n=1}^{12} \dfrac{2}{3n+1}$

12. $2 + \dfrac{1}{2} + \dfrac{1}{8} + \dfrac{1}{32} + \cdots = \displaystyle\sum_{n=1}^{\infty} 2\left(\dfrac{1}{4}\right)^{n-1}$

$= \displaystyle\sum_{n=0}^{\infty} \dfrac{1}{2^{2n-1}}$

13. $\displaystyle\sum_{n=1}^{7} (8n - 5) = 8\left(\dfrac{7(8)}{2}\right) - 5(7) = 224 - 35 = 189$

14. $\displaystyle\sum_{n=1}^{8} 24\left(\dfrac{1}{6}\right)^{n-1} = 24\left(\dfrac{1 - (1/6)^8}{1 - (1/6)}\right)$

$= 24\left(\dfrac{6}{5}\right)\left(1 - \left(\dfrac{1}{6}\right)^8\right)$

$\approx 28.79998 \approx 28.80$

15. $\displaystyle\sum_{n=0}^{\infty} \dfrac{(-1)^n 2^n}{5^{n-1}} = \displaystyle\sum_{n=0}^{\infty} 5\left(\dfrac{(-1)2}{5}\right)^n$

$= 5\left(\dfrac{1}{1 - (-2/5)}\right)$

$= 5\left(\dfrac{5}{7}\right)$

$= \dfrac{25}{7}$

16. (1) For $n = 1$, $3 = \dfrac{3(1)(1+1)}{2}$.

(2) Assume $S_k = 3 + 6 + \cdots + 3k = \dfrac{3k(k+1)}{2}$.

Then $S_{k+1} = 3 + 6 + \cdots + 3k + 3(k+1)$

$= S_k + 3(k+1)$

$= \dfrac{3k(k+1)}{2} + 3(k+1)$

$= \dfrac{k+1}{2}[3k + 6]$

$= \dfrac{3(k+1)(k+2)}{2}$.

Therefore, the formula is true for all positive integers n.

17. $(2a - 5b)^4 = (2a)^4 - 4(2a)^3(5b) + 6(2a)^2(5b)^2 - 4(2a)(5b)^3 + (5b)^4$

$= 16a^4 - 160a^3b + 600a^2b^2 - 1000ab^3 + 625b^4$

18. $_9C_3 = 84$

19. $_{20}C_3 = 1140$

20. $_9P_2 = \dfrac{9!}{7!} = 9 \cdot 8 = 72$

21. $_{70}P_3 = \dfrac{70!}{67!}$

$= 70 \cdot 69 \cdot 68$

$= 328{,}440$

22. $4 \cdot {}_nP_3 = {}_{n+1}P_4$

$4\dfrac{n!}{(n-3)!} = \dfrac{(n+1)!}{(n-3)!}$

$4n! = (n+1)!$

$n = 3$

23. $26 \cdot 10 \cdot 10 \cdot 10 = 26{,}000$ ways

24. $_{25}C_4 = \dfrac{25!}{21! \, 4!} = \dfrac{25 \cdot 24 \cdot 23 \cdot 22}{24} = 12{,}650$ ways

25. There are six red face cards $\Rightarrow$

probability $= \dfrac{6}{52} = \dfrac{3}{26}$.

26. $\dfrac{1}{_{11}C_6} = \dfrac{1}{462}$

27. (a) $\left(\dfrac{30}{60}\right)\left(\dfrac{30}{60}\right) = \dfrac{1}{2} \cdot \dfrac{1}{2} = \dfrac{1}{4}$

(b) $\dfrac{11}{60} \cdot \dfrac{11}{60} = \dfrac{121}{3600} \approx 0.0336$

(c) $\dfrac{1}{60} \approx 0.0167$

28. $1 - 0.75 = 0.25 = 25\%$

Chapter 9   Chapter Test Solutions

1. $y^2 = 8x = 2(4)x$

Parabola

Vertex: $(0, 0)$

Focus: $(2, 0)$

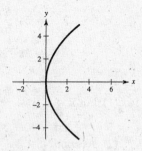

2. $y^2 - 4x + 4 = 0$

$\qquad y^2 = 4x - 4$

$\qquad y^2 = 4(x - 1)$

Parabola

Vertex: $(1, 0)$

Focus: $(2, 0)$

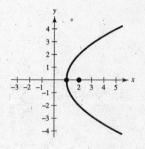

3. $\qquad x^2 - 4y^2 - 4x = 0$

$x^2 - 4x + 4 - 4y^2 = 4$

$\qquad (x - 2)^2 - 4y^2 = 4$

$\qquad \dfrac{(x - 2)^2}{4} - y^2 = 1$

Hyperbola

Center: $(2, 0)$

$a = 2, b = 1, c = \sqrt{5}$

Vertices: $(0, 0), (4, 0)$

Foci: $\left(2 \pm \sqrt{5}, 0\right)$

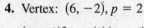

4. Vertex: $(6, -2), p = 2$

$(y + 2)^2 = 4(2)(x - 6)$

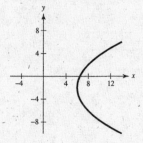

5. Center: $(-6, 3)$

$a = 7, b = 4$

$\dfrac{(x + 6)^2}{16} + \dfrac{(y - 3)^2}{49} = 1$

6. $a = 3, \dfrac{3}{2} = \dfrac{a}{b} \Rightarrow b = 2$

$\dfrac{y^2}{9} - \dfrac{x^2}{4} = 1$

7. $x^2 - \dfrac{y^2}{4} = 1$

$\dfrac{y^2}{4} = x^2 - 1$

$y = \pm 2\sqrt{x^2 - 1}$

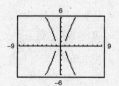

8. (a) $\cot 2\theta = \dfrac{A - C}{B} = \dfrac{1 - 1}{6} = 0 \;\Rightarrow\; \theta = \dfrac{\pi}{4}$ or $45°$

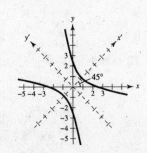

(b) $B^2 - 4AC = 36 - 4 = 32 > 0 \;\Rightarrow\;$ Hyperbola

$y^2 + 6xy + (x^2 - 6) = 0$

$$y = \dfrac{-6x \pm \sqrt{36x^2 - 4(x^2 - 6)}}{2}$$

9. $x^2 + 2y^2 - 4x + 6y - 5 = 0$

$\qquad\qquad x + y + 5 = 0$

$y = -x - 5$:

$\quad x^2 + 2(-x - 5)^2 - 4x + 6(-x - 5) - 5 = 0$

$\quad x^2 + 2x^2 + 20x + 50 - 4x - 6x - 30 - 5 = 0$

$\qquad\qquad\qquad\qquad 3x^2 + 10x + 15 = 0$

This quadratic has no real solutions. Therefore, no solution.

10. $x = t^2 - 6$

$y = \tfrac{1}{2}t - 1 \;\Rightarrow\; t = 2(y + 1)$

$x = [2(y + 1)]^2 - 6 = 4y^2 + 8y - 2$

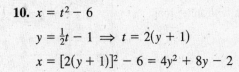

Parabola

$x = 4y^2 + 8y - 2$ or $(y + 1)^2 = \tfrac{1}{4}(x + 6)$

11. $x = \sqrt{t^2 + 2}$

$y = \dfrac{t}{4} \;\Rightarrow\; t = 4y$

$x = \sqrt{16y^2 + 2}$

$x^2 = 16y^2 + 2$

Right-hand portion of hyperbola

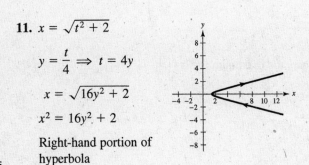

12. $x = 2 + 3\cos\theta$

$y = 2\sin\theta$

$\left(\dfrac{x - 2}{3}\right)^2 + \left(\dfrac{y}{2}\right)^2 = 1$

$\dfrac{(x - 2)^2}{9} + \dfrac{y^2}{4} = 1$

Ellipse

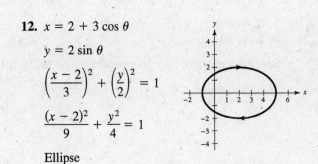

13. $x = t, \qquad y = t^2 + 10$

$\quad x = -t, \quad y = t^2 + 10$

14. Sample answers:

$x = 4 - t^2$

$y = t$

$x = 4 - 4t^2$

$y = 2t$

15. Sample answers:

$x = \pm\sqrt{16 - 4t^2}$

$y = t$

$x = \pm\sqrt{16 - t^2}$

$y = \tfrac{1}{2}t$

16. $(r, \theta) = \left(-14, \dfrac{5\pi}{3}\right) \;\Rightarrow\; (x, y) = \left(-7, 7\sqrt{3}\right)$

17. $(x, y) = (2, -2), r = \sqrt{8} = 2\sqrt{2}, \theta = \dfrac{7\pi}{4}$

$(r, \theta) = \left(2\sqrt{2}, \dfrac{7\pi}{4}\right) = \left(2\sqrt{2}, -\dfrac{\pi}{4}\right) = \left(-2\sqrt{2}, \dfrac{3\pi}{4}\right)$

18. $x^2 + y^2 - 12y = 0$

$r^2 - 12\,r\sin\theta = 0$

$r^2 = 12\,r\sin\theta$

$r = 12\sin\theta$

19. $r = 2\sin\theta$

$r^2 = 2r\sin\theta$

$x^2 + y^2 = 2y$

$x^2 + y^2 - 2y + 1 = 1$

$x^2 + (y-1)^2 = 1$, Circle

20. $r = 2 + 3\sin\theta$

Limaçon with inner loop

21. $r = \dfrac{1}{1 - \cos\theta}$

$e = 1 \Longrightarrow$ Parabola

22. $r = \dfrac{4}{2 + 3\sin\theta} = \dfrac{2}{1 + \frac{3}{2}\sin\theta}$

$e = \dfrac{3}{2} \Longrightarrow$ Hyperbola

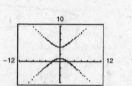

23. $r = \dfrac{ep}{1 + e\sin\theta} = \dfrac{(1/4)(4)}{1 + (1/4)\sin\theta}$

$r = \dfrac{4}{4 + \sin\theta} = \dfrac{1}{1 + (1/4)\sin\theta}$

24. $r = \dfrac{ep}{1 + e\sin\theta} = \dfrac{(5/4)(2)}{1 + (5/4)\sin\theta} = \dfrac{10}{4 + 5\sin\theta}$

25. $r = 8\cos 3\theta$

The maximum value of $|r|$ occurs when $|\cos 3\theta| = 1$. Hence, the maximum is $8 = |r|$.

$r = 0 \Longrightarrow \cos 3\theta = 0$

$\Longrightarrow 3\theta = \dfrac{\pi}{2} + n\pi$

$\Longrightarrow \theta = \dfrac{\pi}{6} + \dfrac{n\pi}{3}$

On the interval $0 \le \theta \le \pi$, $\theta = \dfrac{\pi}{6}, \dfrac{\pi}{2}, \dfrac{5\pi}{6}$.

Chapters 7–9   Cumulative Test Solutions

1. $\begin{bmatrix} -1 & -3 & \vdots & 5 \\ 4 & 2 & \vdots & 10 \end{bmatrix}$ row reduces to $\begin{bmatrix} 1 & 0 & \vdots & 4 \\ 0 & 1 & \vdots & -3 \end{bmatrix}$.

Answer: $(4, -3)$

2. $2x - y^2 = 0$

$\quad x - y = 4 \implies x = y + 4$

$\quad 2(y + 4) - y^2 = 0$

$\quad y^2 - 2y - 8 = 0$

$\quad (y - 4)(y + 2) = 0$

$\qquad\qquad y = 4, \; x = 8$

$\qquad\qquad y = -2, \; x = 2$

Solutions: $(2, -2), (8, 4)$

3. $\begin{bmatrix} 2 & -3 & 1 & \vdots & 13 \\ -4 & 1 & -2 & \vdots & -6 \\ 1 & -3 & 3 & \vdots & 12 \end{bmatrix}$ row reduces to

$\begin{bmatrix} 1 & 0 & 0 & \vdots & \frac{3}{5} \\ 0 & 1 & 0 & \vdots & -4 \\ 0 & 0 & 1 & \vdots & -\frac{1}{5} \end{bmatrix}$.

Answer: $\left(\frac{3}{5}, -4, -\frac{1}{5}\right)$

4. $\begin{bmatrix} 1 & -4 & 3 & \vdots & 5 \\ 5 & 2 & -1 & \vdots & 1 \\ -2 & -8 & 0 & \vdots & 30 \end{bmatrix}$ row reduces to

$\begin{bmatrix} 1 & 0 & 0 & \vdots & 1 \\ 0 & 1 & 0 & \vdots & -4 \\ 0 & 0 & 1 & \vdots & -4 \end{bmatrix}$.

Solution: $(1, -4, -4)$

5. $3A - B = \begin{bmatrix} -7 & -10 & -16 \\ -6 & 18 & 9 \\ -12 & 16 & 7 \end{bmatrix}$

6. $5A + 3B = \begin{bmatrix} -18 & 15 & -14 \\ 28 & 11 & 34 \\ -20 & 52 & -1 \end{bmatrix}$

7. $AB = \begin{bmatrix} 3 & -31 & 2 \\ 22 & 18 & 6 \\ 52 & -40 & 14 \end{bmatrix}$

8. $BA = \begin{bmatrix} 5 & 36 & 31 \\ -36 & 12 & -36 \\ 16 & 0 & 18 \end{bmatrix}$

9. (a) $\begin{bmatrix} 1 & 2 & -1 \\ 3 & 7 & -10 \\ -5 & -7 & -15 \end{bmatrix}^{-1} = \begin{bmatrix} -175 & 37 & -13 \\ 95 & -20 & 7 \\ 14 & -3 & 1 \end{bmatrix}$

(b) $\det(A) = \begin{vmatrix} 1 & 2 & -1 \\ 3 & 7 & -10 \\ -5 & -7 & -15 \end{vmatrix} = 1(-105 - 70) - 2(-45 - 50) - 1(-21 + 35)$

$\qquad\qquad\qquad = -175 + 190 - 14 = 1$

10. $\begin{vmatrix} 0 & 0 & 1 \\ 6 & 2 & 1 \\ 8 & 10 & 1 \end{vmatrix} = 44 \implies$ Area $= \frac{1}{2}(44) = 22$ sq. units

11. (a) $a_1 = \dfrac{(-1)^{1+1}}{2(1) + 3} = \dfrac{1}{5}$

$\qquad a_2 = -\dfrac{1}{7}$

$\qquad a_3 = \dfrac{1}{9}$

$\qquad a_4 = -\dfrac{1}{11}$

$\qquad a_5 = \dfrac{1}{13}$

(b) $a_1 = 3(2)^{1-1} = 3$

$\qquad a_2 = 6$

$\qquad a_3 = 12$

$\qquad a_4 = 24$

$\qquad a_5 = 48$

12. $\displaystyle\sum_{k=1}^{6} (7k - 2) = \dfrac{7(6)(7)}{2} - 2(6) = 135$

13. $\displaystyle\sum_{k=1}^{4} \dfrac{2}{k^2 + 4} = \dfrac{2}{1 + 4} + \dfrac{2}{4 + 4} + \dfrac{2}{9 + 4} + \dfrac{2}{16 + 4}$

$\qquad\qquad\qquad \approx 0.9038$

14. $\displaystyle\sum_{n=0}^{10} 9\left(\frac{3}{4}\right)^n = 9\left(\frac{1-(3/4)^{11}}{1-(3/4)}\right) \approx 34.4795$

15. $\displaystyle\sum_{n=0}^{50} 100\left(-\frac{1}{2}\right)^n = \sum_{n=1}^{51} 100\left(-\frac{1}{2}\right)^{n-1}$

$$= 100\frac{1-(-1/2)^{51}}{1-(-1/2)}$$

$$\approx \frac{2}{3}(100) \approx 66.67$$

16. $\displaystyle\sum_{n=0}^{\infty} 3\left(-\frac{3}{5}\right)^n = \frac{3}{1-(-3/5)} = \frac{3}{8/5} = \frac{15}{8}$

17. $\displaystyle\sum_{n=1}^{\infty} 5(-0.02)^n = 5(-0.02)\frac{1}{1-(-0.02)}$

$$= \frac{-0.1}{1.02} = \frac{-5}{51}$$

18. $\displaystyle 4 - 2 + 1 - \frac{1}{2} + \frac{1}{4} - \cdots = \sum_{n=0}^{\infty} 4\left(-\frac{1}{2}\right)^n = \frac{4}{1-(-1/2)} = \frac{8}{3}$

19. For $n = 1$, $3 = 1(2 + 1)$ and the formula is true. Assume true for k, and consider

$$3 + 7 + \cdots + (4k - 1) + (4(k + 1) - 1) = 3 + 7 + \cdots + (4k - 1) + (4k + 3)$$

$$= k(2k + 1) + (4k + 3)$$

$$= 2k^2 + 5k + 3$$

$$= (k + 1)(2k + 3)$$

$$= (k + 1)(2(k + 1) + 1)$$

which shows that the formula is true for $k + 1$.

20. $(x + 3)^4 = x^4 + 12x^3 + 54x^2 + 108x + 81$

21. $(2x + y^2)^5 = 32x^5 + 80x^4y^2 + 80x^3y^4 + 40x^2y^6 + 10xy^8 + y^{10}$

22. $(x - 2y)^6 = x^6 - 12x^5y + 60x^4y^2 - 160x^3y^3 + 240x^2y^4 - 192xy^5 + 64y^6$

23. $(3a - 4b)^8 = 6561a^8 - 69{,}984a^7b + 326{,}592a^6b^2 - 870{,}912a^5b^3 + 1{,}451{,}520a^4b^4 - 1{,}548{,}288a^3b^5$

$$+ 1{,}032{,}192a^2b^6 - 393{,}216ab^7 + 65{,}536b^8$$

24. $\dfrac{5!}{2!2!1!} = \dfrac{120}{4} = 30$ **25.** $\dfrac{6!}{3!} = 6 \cdot 5 \cdot 4 = 120$ **26.** $\dfrac{10!}{2!2!2!} = 453{,}600$ **27.** $\dfrac{10!}{3!2!2!} = 151{,}200$

28. Hyperbola with center $(5, -3)$ **29.** Ellipse with center $(2, -1)$ **30.** Hyperbola with center $(0, 0)$

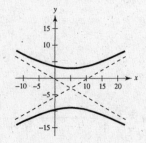

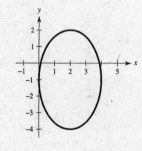

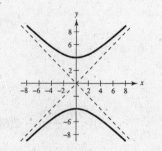

31. $(x^2 - 2x + 1) + (y^2 - 4y + 4) = -1 + 1 + 4$

$(x - 1)^2 + (y - 2)^2 = 4$

Circle

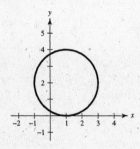

32. $(x - 2)^2 = 4p(y - 3)$

$(0, 0)$: $4 = 4p(-3) \Rightarrow p = -\frac{1}{3}$

$(x - 2)^2 = 4\left(-\frac{1}{3}\right)(y - 3)$

$(x - 2)^2 = -\frac{4}{3}(y - 3)$

33. Center: $(1, 4)$

$a = 5, b = 2$

$\dfrac{(x - 1)^2}{25} + \dfrac{(y - 4)^2}{4} = 1$

34. Center: $(0, -4)$; $a = 2$

$\dfrac{(y + 4)^2}{4} - \dfrac{x^2}{b^2} = 1$

$(4, 0)$: $4 - \dfrac{16}{b^2} = 1 \Rightarrow \dfrac{16}{b^2} = 3 \Rightarrow b^2 = \dfrac{16}{3}$

$\dfrac{(y + 4)^2}{4} - \dfrac{x^2}{16/3} = 1$

35. $B^2 - 4AC = 16 - 8 = 8 \Rightarrow$ Hyperbola

$\cot 2\theta = \dfrac{1 - 2}{-4} = \dfrac{1}{4} \Rightarrow \theta \approx 38°$

Graph as:

$2y^2 - 4xy + (x^2 - 6) = 0$

$\qquad y = \dfrac{4x \pm \sqrt{16x^2 - 8(x^2 - 6)}}{4}$

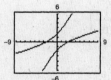

36. $x = 2t + 1, y = t^2$

(a), (b)

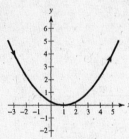

(c) $t = \dfrac{x - 1}{2} \Rightarrow y = \left(\dfrac{x - 1}{2}\right)^2 = \dfrac{1}{4}(x - 1)^2$

37. $x = \cos \theta, y = 2 \sin^2 \theta$

(a), (b)

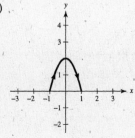

(c) $y = 2 \sin^2 \theta$

$\quad = 2(1 - \cos^2 \theta)$

$\quad = 2(1 - x^2), \ -1 \le x \le 1$

38. $x = 4 \ln t,\ y = \frac{1}{2}t^2$

(a), (b)

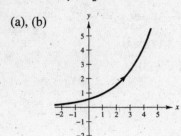

(c) $t = e^{x/4} \Rightarrow y = \frac{1}{2}e^{x/2}$

39. $y = 3x - 2$

Sample answers:

$x = t$

$y = 3t - 2$

$x = -t$

$y = -3t - 2$

40. Sample answers:

$x = \pm\sqrt{t^2 + 16}$

$y = t$

$x = t$

$y = \pm\sqrt{t^2 - 16}$

41. $y = \dfrac{2}{x}$

Sample answers:

$x = t$

$y = \dfrac{2}{t}$

$x = -t$

$y = -\dfrac{2}{t}$

42. Sample answers:

$x = t$

$y = \dfrac{e^{2t}}{e^{2t} + 1}$

$x = \dfrac{1}{2}t$

$y = \dfrac{e^{t}}{e^{t} + 1}$

43. $(r, \theta) = \left(8, \dfrac{5\pi}{6}\right)$

$\left(8, -\dfrac{7\pi}{6}\right), \left(-8, -\dfrac{\pi}{6}\right), \left(-8, \dfrac{11\pi}{6}\right)$

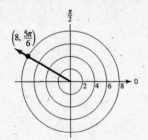

44. $(r, \theta) = \left(5, -\dfrac{3\pi}{4}\right)$

$\left(5, \dfrac{5\pi}{4}\right), \left(-5, \dfrac{\pi}{4}\right), \left(-5, -\dfrac{7\pi}{4}\right)$

45. $(r, \theta) = \left(-2, \dfrac{5\pi}{4}\right)$

$\left(-2, -\dfrac{3\pi}{4}\right), \left(2, \dfrac{\pi}{4}\right), \left(2, -\dfrac{7\pi}{4}\right)$

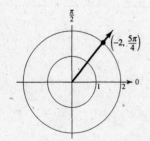

46. $(r, \theta) = \left(-3, -\dfrac{11\pi}{6}\right)$

$\left(-3, \dfrac{\pi}{6}\right), \left(3, \dfrac{7\pi}{6}\right), \left(3, -\dfrac{5\pi}{6}\right)$

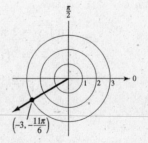

47.
$$4x + 4y + 1 = 0$$
$$4r \cos \theta + 4r \sin \theta + 1 = 0$$
$$r[4 \cos \theta + 4 \sin \theta] = -1$$
$$r = \frac{-1}{4 \cos \theta + 4 \sin \theta}$$

48. $r = 2 \cos \theta$
$$r^2 = 2r \cos \theta$$
$$x^2 + y^2 = 2x$$
$$x^2 - 2x + 1 + y^2 = 1$$
$$(x - 1)^2 + y^2 = 1, \text{ Circle}$$

49.
$$r = \frac{2}{4 - 5 \cos \theta}$$
$$4r - 5r \cos \theta = 2$$
$$4(x^2 + y^2)^{1/2} - 5x = 2$$
$$16(x^2 + y^2) = (5x + 2)^2 = 25x^2 + 20x + 4$$
$$16y^2 - 9x^2 - 20x = 4$$

50. $r = -\dfrac{\pi}{6}$, Circle

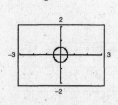

51. $r = 3 - 2 \sin \theta$

Limaçon

52. $r = 2 + 5 \cos \theta$

Limaçon

53. $32{,}500 + 32{,}500(1.05) + \cdots + 32{,}500(1.05)^{14} = \displaystyle\sum_{n=1}^{15} 32{,}500(1.05)^{n-1}$

$$= 32{,}500\left(\frac{1 - 1.05^{15}}{1 - 1.05}\right) \approx \$701{,}303.32$$

54. There are two ways to select the first digit (4 or 5), and two ways for the second digit. Hence, $p = \frac{1}{4}$.

55. Let $y = -ax^2 + 16$.

$(6, 14): \quad 14 = -a(6)^2 + 16$

$\qquad\qquad 36a = 2$

$\qquad\qquad a = \frac{1}{18}$

$\qquad\qquad y = -\frac{1}{18}x^2 + 16$

$y = 0: \quad 16 = \frac{1}{18}x^2$

$\qquad\qquad x^2 = 288$

$\qquad\qquad x = 12\sqrt{2}$

Width: $24\sqrt{2}$ meters

Chapter 10   Chapter Test Solutions

1.

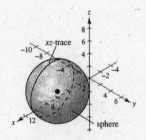

Points: $(-2, -2, 3)$, $(5, -2, 3)$, $(-1, 2, 1)$

2. $AB = \sqrt{(8-6)^2 + (-2-4)^2 + (5+1)^2} = \sqrt{76}$

$AC = \sqrt{(8+4)^2 + (-2-3)^2 + (5-0)^2} = \sqrt{144 + 25 + 25} = \sqrt{194}$

$BC = \sqrt{(6+4)^2 + (4-3)^2 + (-1-0)^2} = \sqrt{100 + 1 + 1} = \sqrt{102}$

No. $\left(\sqrt{76}\right)^2 + \left(\sqrt{102}\right)^2 \neq \left(\sqrt{194}\right)^2$

3. Midpoint $= \left(\dfrac{8+6}{2}, \dfrac{-2+4}{2}, \dfrac{5-1}{2}\right) = (7, 1, 2)$

4. Diameter $= \sqrt{(8-6)^2 + (-2-4)^2 + (5+1)^2}$

$\qquad\qquad = \sqrt{4 + 36 + 36} = \sqrt{76}$

Radius $= \sqrt{19}$

$(x - 7)^2 + (y - 1)^2 + (z - 2)^2 = 19$

5. $\mathbf{v} = \langle 4 - 2, 4 - (-1), -7 - 3 \rangle = \langle 2, 5, -10 \rangle$

$\|\mathbf{v}\| = \sqrt{2^2 + 5^2 + (-10)^2} = \sqrt{129}$

6. $\mathbf{v} = \langle 3 - 6, -3 - 2, 8 - 0 \rangle = \langle -3, -5, 8 \rangle$

$\|\mathbf{v}\| = \sqrt{(-3)^2 + (-5)^2 + 8^2} = \sqrt{98} = 7\sqrt{2}$

7. $\mathbf{u} = \langle 6 - 8, 4 - (-2), -1 - 5 \rangle = \langle -2, 6, -6 \rangle$

$\mathbf{v} = \langle -4 - 8, 3 - (-2), 0 - 5 \rangle = \langle -12, 5, -5 \rangle$

8. (a) $\|\mathbf{v}\| = \sqrt{(-12)^2 + 5^2 + (-5)^2} = \sqrt{194}$

(b) $\mathbf{u} \cdot \mathbf{v} = (-2)(-12) + 6(5) + (-6)(-5) = 84$

(c) $\mathbf{u} \times \mathbf{v} = \begin{vmatrix} \mathbf{i} & \mathbf{j} & \mathbf{k} \\ -2 & 6 & -6 \\ -12 & 5 & -5 \end{vmatrix} = \langle 0, 62, 62 \rangle$

9. $\cos\theta = \dfrac{\mathbf{u} \cdot \mathbf{v}}{\|\mathbf{u}\| \, \|\mathbf{v}\|} = \dfrac{84}{\sqrt{76}\sqrt{194}} \approx 0.6918 \implies \theta \approx 46.23°$ or 0.8068 radians

10. (a) $x = 8 - 2t, y = -2 + 6t, z = 5 - 6t$

(b) $\dfrac{x - 8}{-2} = \dfrac{y + 2}{6} = \dfrac{z - 5}{-6}$

11. $\mathbf{u} \cdot \mathbf{v} = 0 - 2 - 6 \neq 0$ and $\mathbf{u} \neq c\mathbf{v} \implies$ neither

12. $\mathbf{u} \cdot \mathbf{v} = -2 + 3 - 1 = 0 \implies$ orthogonal

13. First two points: $\mathbf{v} = \langle 4, 8, -2 \rangle$

Last two points: $\mathbf{w} = \langle 4, 8, -2 \rangle$

Opposite sides are parallel and equal length.

Adjacent sides: $\mathbf{v}$ and $\mathbf{u} = \langle 1, -3, 3 \rangle$

Area $= \| \mathbf{u} \times \mathbf{v} \|$

$$\mathbf{u} \times \mathbf{v} = \begin{vmatrix} \mathbf{i} & \mathbf{j} & \mathbf{k} \\ 1 & -3 & 3 \\ 4 & 8 & -2 \end{vmatrix} = \langle -18, 14, 20 \rangle$$

$\| \mathbf{u} \times \mathbf{v} \| = \sqrt{18^2 + 14^2 + 20^2} = 2\sqrt{230} \approx 30.33$ square units

14. $\mathbf{u} = \langle 0, 8, -1 \rangle$, $\mathbf{v} = \langle 4, 5, -4 \rangle$

$$\mathbf{n} = \mathbf{u} \times \mathbf{v} = \begin{vmatrix} \mathbf{i} & \mathbf{j} & \mathbf{k} \\ 0 & 8 & -1 \\ 4 & 5 & -4 \end{vmatrix} = \langle -27, -4, -32 \rangle$$

Plane: $-27(x + 3) - 4(y + 4) - 32(z - 2) = 0$

$-27x - 4y - 32z - 33 = 0$

$27x + 4y + 32z + 33 = 0$

15. Let $A(0, 0, 5)$ be the vertex.

$\mathbf{u} = \overrightarrow{AD} = \langle 4, 0, 0 \rangle$, $\mathbf{v} = \overrightarrow{AB} = \langle 0, 10, 0 \rangle$,

$\mathbf{w} = \overrightarrow{AE} = \langle 0, 1, -5 \rangle$

$$\mathbf{u} \cdot (\mathbf{v} \times \mathbf{w}) = \begin{vmatrix} 4 & 0 & 0 \\ 0 & 10 & 0 \\ 0 & 1 & -5 \end{vmatrix} = 4(-50) = -200$$

Volume $= |-200| = 200$ cubic units

16. $2x + 3y + 4z = 12$

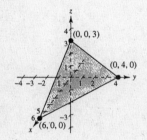

17. $5x - y - 2z = 10$

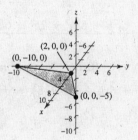

18. $\mathbf{n} = \langle 3, -2, 1 \rangle$, $Q = (2, -1, 6)$, $P = (0, 0, 6)$ in plane, $\overrightarrow{PQ} = \langle 2, -1, 0 \rangle$

$$D = \frac{|\overrightarrow{PQ} \cdot \mathbf{n}|}{\| \mathbf{n} \|} = \frac{|8|}{\sqrt{14}} = \frac{4\sqrt{14}}{7}$$

Chapter 11   Chapter Test Solutions

1. $\displaystyle \lim_{x \to -2} \frac{x^2 - 1}{2x} = \frac{(-2)^2 - 1}{2(-2)} = -\frac{3}{4}$

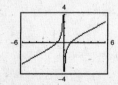

Limit is -0.75.

2.

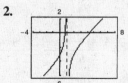

The limit does not exist.

$\displaystyle \lim_{x \to 1} \frac{-x^2 + 5x - 3}{1 - x}$ does not exist.

$x = 1$ is a vertical asymptote.

3.

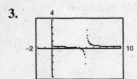

$$\lim_{x \to 5} \frac{\sqrt{x} - 2}{x - 5} \text{ does not exist.}$$

4.

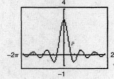

$$\lim_{x \to 0} \frac{\sin 3x}{x} = 3$$

$$f(x) = \frac{\sin 3x}{x}$$

5.

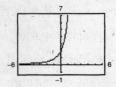

$$\lim_{x \to 0} \frac{e^{2x} - 1}{x} = 2$$

$$f(x) = \frac{e^{2x} - 1}{x}$$

6. (a)
$$\frac{f(x + h) - f(x)}{h} = \frac{3(x + h)^2 - 5(x + h) - 2 - (3x^2 - 5x - 2)}{h}$$

$$= \frac{3x^2 + 6xh + 3h^2 - 5h - 3x^2}{h}$$

$$= 6x + 3h - 5$$

$$f'(x) = \lim_{h \to 0} [6x + 3h - 5] = 6x - 5$$

$$f'(2) = 6(2) - 5 = 7$$

(b)
$$\frac{f(x + h) - f(x)}{h} = \frac{[2(x + h)^3 + 6(x + h)] - [2x^3 + 6x]}{h}$$

$$= \frac{2x^3 + 6x^2h + 6xh^2 + 2h^3 + 6x + 6h - 2x^3 - 6x}{h}$$

$$= \frac{6x^2h + 6xh^2 + 2h^3 + 6h}{h}$$

$$= 6x^2 + 6xh + 2h^2 + 6, \ h \neq 0$$

$$f'(x) = \lim_{h \to 0} [6x^2 + 6xh + 2h^2 + 6] = 6x^2 + 6$$

$$f'(-1) = 6(-1)^2 + 6 = 12$$

7. $f'(x) = \lim_{h \to 0} \dfrac{f(x + h) - f(x)}{h}$

$$= \lim_{h \to 0} \frac{5 - (2/5)(x + h) - (5 - (2/5)x)}{h}$$

$$= \lim_{h \to 0} \frac{-(2/5)h}{h} = -\frac{2}{5}$$

8. $f'(x) = \lim_{h \to 0} \dfrac{f(x + h) - f(x)}{h}$

$$= \lim_{h \to 0} \frac{2(x + h)^2 + 4(x + h) - 1 - [2x^2 + 4x - 1]}{h}$$

$$= \lim_{h \to 0} \frac{2x^2 + 4xh + 2h^2 + 4h - 2x^2}{h}$$

$$= \lim_{h \to 0} (4x + 2h + 4) = 4x + 4$$

9. $f'(x) = \lim_{h \to 0} \dfrac{f(x + h) - f(x)}{h}$

$= \lim_{h \to 0} \dfrac{\dfrac{1}{x + 3 + h} - \dfrac{1}{x + 3}}{h}$

$= \lim_{h \to 0} \dfrac{(x + 3) - (x + 3 + h)}{h(x + 3 + h)(x + 3)}$

$= \lim_{h \to 0} \dfrac{-1}{(x + 3 + h)(x + 3)}$

$= \dfrac{-1}{(x + 3)^2}$

10. $\lim_{x \to \infty} \dfrac{6}{5x - 1} = 0$

11. $\lim_{x \to \infty} \dfrac{1 - 3x^2}{x^2 - 5} = -3$

12. $\lim_{x \to -\infty} \dfrac{x^2}{3x + 2}$ does not exist.

$f(x) = \dfrac{x^2}{3x + 2}$ decreases

without bound as $x \to -\infty$.

13. $0, \frac{3}{4}, \frac{14}{19}, \frac{12}{17}, \frac{36}{53}$

$\lim_{n \to \infty} a_n = \frac{1}{2}$

14. $0, 1, 0, \frac{1}{2}, 0$

$\lim_{n \to \infty} a_n = 0$

15. Width of each rectangle: $\frac{1}{2}$

Heights: $8, \frac{15}{2}, 6, \frac{7}{2}$

Area $\approx \frac{1}{2}\left[8 + \frac{15}{2} + 6 + \frac{7}{2}\right] = \frac{25}{2}$

16. Width: $\dfrac{4}{n}$,   Height: $f\left(-2 + \dfrac{4i}{n}\right) = \left(-2 + \dfrac{4i}{n}\right) + 2 = \dfrac{4i}{n}$

$A \approx \sum_{i=1}^{n} \left(\dfrac{4i}{n}\right)\left(\dfrac{4}{n}\right) = \dfrac{16}{n^2} \sum_{i=1}^{n} i = \dfrac{16}{n^2} \dfrac{n(n + 1)}{2}$

$A = \lim_{n \to \infty} \dfrac{16}{n^2} \cdot \dfrac{n(n + 1)}{2} = 8$

17. $f(x) = 3 - x^2$, $[-1, 1]$

The width of each rectangle is $\dfrac{2}{n}$. The height is

$f\left(-1 + \dfrac{2i}{n}\right) = 3 - \left[-1 + \dfrac{2i}{n}\right]^2$

$= 2 + \dfrac{4}{n}i - \dfrac{4}{n^2}i^2.$

$A \approx \sum_{i=1}^{n} \left[2 + \dfrac{4}{n}i - \dfrac{4}{n^2}i^2\right]\left(\dfrac{2}{n}\right)$

$= \sum_{i=1}^{n} \left[\dfrac{4}{n} + \dfrac{8}{n^2}i - \dfrac{8}{n^3}i^2\right]$

$A = \lim_{n \to \infty} \left[\dfrac{4}{n}(n) + \dfrac{8}{n^2} \dfrac{n(n + 1)}{2} - \dfrac{8}{n^3} \dfrac{n(n + 1)(2n + 1)}{6}\right]$

$= 4 + 4 - \dfrac{8}{3} = \dfrac{16}{3}$

18. (a) $y = 8.79x^2 - 6.2x - 0.4$

(b) Velocity = Derivative = $17.58x - 6.2$

At $x = 5$, velocity ≈ 81.7 ft/sec.

Chapters 10–11 Cumulative Test Solutions

1. $(-6, 1, 3)$

2. $(0, -4, 0)$

3. $d = \sqrt{(4 - (-2))^2 + (-5 - 3)^2 + (1 - (-6))^2}$
  $= \sqrt{36 + 64 + 49}$
  $= \sqrt{149}$

4. $d_1 = 3, d_2 = 4, d_3 = \sqrt{4^2 + 3^2} = 5$
  $d_1{}^2 + d_2{}^2 = d_3{}^2$

5. Midpoint: $\left(\dfrac{3 - 5}{2}, \dfrac{4 + 0}{2}, \dfrac{-1 + 2}{2} \right) = \left(-1, 2, \dfrac{1}{2} \right)$

6. Center $= (2, 2, 4)$
  Radius $= \sqrt{2^2 + 2^2 + 4^2} = \sqrt{24}$
  $(x - 2)^2 + (y - 2)^2 + (z - 4)^2 = 24$

7. xy-trace: $(z = 0)$
  $(x - 2)^2 + (y + 1)^2 = 4$, Circle
  yz-trace: $(x = 0)$
  $4 + (y + 1)^2 + z^2 = 4$ or $(y + 1)^2 + z^2 = 0$, Point
  $(0, -1, 0)$, Point

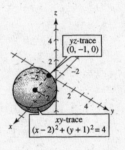

8. $\mathbf{u} \cdot \mathbf{v} = \langle 2, -6, 0 \rangle \cdot \langle -4, 5, 3 \rangle$
  $= -8 - 30 = -38$

$\mathbf{u} \times \mathbf{v} = \begin{vmatrix} \mathbf{i} & \mathbf{j} & \mathbf{k} \\ 2 & -6 & 0 \\ -4 & 5 & 3 \end{vmatrix} = \langle -18, -6, -14 \rangle$

9. $\mathbf{u} \cdot \mathbf{v} \neq 0, \mathbf{u} \neq c\mathbf{v} \implies$ neither

10. $\mathbf{u} \cdot \mathbf{v} = -8 - 12 + 20 = 0 \implies$ orthogonal

11. $3\mathbf{u} = \langle -3, 18, -9 \rangle = -\mathbf{v} \implies$ parallel

12. $\overrightarrow{DA} = \langle 0, 2, 0 \rangle, \overrightarrow{DC} = \langle 2, 1, 0 \rangle, \overrightarrow{DH} = \langle 0, 0, 3 \rangle$

$\begin{vmatrix} 0 & 2 & 0 \\ 2 & 1 & 0 \\ 0 & 0 & 3 \end{vmatrix} = 12$ cubic units

13. (a) Vector is $\langle 5 + 2, 8 - 3, 25 - 0 \rangle = \langle 7, 5, 25 \rangle$.
  $x = -2 + 7t, y = 3 + 5t, z = 25t$

  (b) $\dfrac{x + 2}{7} = \dfrac{y - 3}{5} = \dfrac{z}{25}$

14. $\mathbf{v} = \langle 2, -4, 1 \rangle$ and $P = (-1, 2, 0)$
  $x = -1 + 2t$
  $y = 2 - 4t$
  $z = t$

15. $\mathbf{u} = \langle -2, 3, 0 \rangle, \mathbf{v} = \langle 5, 8, 25 \rangle$

$\mathbf{u} \times \mathbf{v} = \begin{vmatrix} \mathbf{i} & \mathbf{j} & \mathbf{k} \\ -2 & 3 & 0 \\ 5 & 8 & 25 \end{vmatrix} = \langle 75, 50, -31 \rangle$

Normal to plane

Plane: $75x + 50y - 31z = 0$

16.

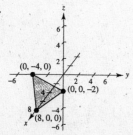

17. $\mathbf{n} = \langle 2, -5, 1 \rangle$, $Q = (0, 0, 25)$, $P = (0, 0, 10)$

in plane, $\overrightarrow{PQ} = \langle 0, 0, 15 \rangle$

$$D = \frac{|\overrightarrow{PQ} \cdot \mathbf{n}|}{\|\mathbf{n}\|} = \frac{15}{\sqrt{30}} = \frac{\sqrt{30}}{2} \approx 2.74$$

18. Normal to plane containing: $(-1, -1, 3)$, $(0, 0, 0)$ and $(2, 0, 0)$ is

$$\langle -1, -1, 3 \rangle \times \langle 2, 0, 0 \rangle = \begin{vmatrix} \mathbf{i} & \mathbf{j} & \mathbf{k} \\ -1 & -1 & 3 \\ 2 & 0 & 0 \end{vmatrix} = \langle 0, 6, 2 \rangle \text{ or } \mathbf{n}_1 = \langle 0, 3, 1 \rangle$$

Normal to front face is: $\langle 1, -1, 3 \rangle \times \langle 0, 2, 0 \rangle = \begin{vmatrix} \mathbf{i} & \mathbf{j} & \mathbf{k} \\ 1 & -1 & 3 \\ 0 & 2 & 0 \end{vmatrix} = \langle -6, 0, 2 \rangle \text{ or } \mathbf{n}_2 = \langle -3, 0, 1 \rangle$

Angle between sides: $\cos \theta = \dfrac{|\mathbf{n}_1 \cdot \mathbf{n}_2|}{\|\mathbf{n}_1\| \|\mathbf{n}_2\|} = \dfrac{1}{\sqrt{10}\sqrt{10}} = \dfrac{1}{10} \implies \theta \approx 84.26°$

19. $\lim\limits_{x \to 4} (5x - x^2) = 5(4) - 4^2 = 4$

20. $\lim\limits_{x \to -2^+} \dfrac{x + 2}{(x + 2)(x - 1)} = \lim\limits_{x \to -2^+} \dfrac{1}{x - 1} = -\dfrac{1}{3}$

21. $\lim\limits_{x \to 7} \dfrac{x - 7}{(x - 7)(x + 7)} = \lim\limits_{x \to 7} \dfrac{1}{x + 7} = \dfrac{1}{14}$

22. $\lim\limits_{x \to 0} \dfrac{\sqrt{x + 4} - 2}{x} \cdot \dfrac{\sqrt{x + 4} + 2}{\sqrt{x + 4} + 2} = \lim\limits_{x \to 0} \dfrac{(x + 4) - 4}{x(\sqrt{x + 4} + 2)} = \lim\limits_{x \to 0} \dfrac{1}{\sqrt{x + 4} + 2} = \dfrac{1}{2 + 2} = \dfrac{1}{4}$

23. $\lim\limits_{x \to 4^-} \dfrac{|x - 4|}{x - 4} = -1$

24. $\lim\limits_{x \to 0} \sin\left(\dfrac{\pi}{x}\right)$ does not exist.

25. $\lim\limits_{x \to 0} \dfrac{\dfrac{1}{x + 2} - \dfrac{1}{2}}{x} = \lim\limits_{x \to 0} \dfrac{2 - (x + 2)}{x(x + 2)2}$

$$= \lim\limits_{x \to 0} \dfrac{-1}{(x + 2)2} = -\dfrac{1}{4}$$

26. $\lim\limits_{x \to 0} \dfrac{\sqrt{x + 1} - 1}{x} \cdot \dfrac{\sqrt{x + 1} + 1}{\sqrt{x + 1} + 1} = \lim\limits_{x \to 0} \dfrac{(x + 1) - 1}{x[\sqrt{x + 1} + 1]}$

$$= \lim\limits_{x \to 0} \dfrac{1}{\sqrt{x + 1} + 1} = \dfrac{1}{2}$$

27. $\lim\limits_{x \to 2^-} \dfrac{x - 2}{x^2 - 4} = \lim\limits_{x \to 2^-} \dfrac{x - 2}{(x - 2)(x + 2)}$

$$= \lim\limits_{x \to 2^-} \dfrac{1}{x + 2} = \dfrac{1}{4}$$

28. $f'(x) = \lim\limits_{h \to 0} \dfrac{f(x + h) - f(x)}{h}$

$ = \lim\limits_{h \to 0} \dfrac{4 - (x + h)^2 - (4 - x^2)}{h}$

$ = \lim\limits_{h \to 0} \dfrac{4 - x^2 - 2xh - h^2 - 4 + x^2}{h}$

$ = \lim\limits_{h \to 0} \dfrac{-2xh - h^2}{h}$

$ = \lim\limits_{h \to 0} (-2x - h) = -2x, \ \text{Slope}$

At $(0, 4), \ m = 0$.

29. $f(x) = \sqrt{x + 3}$

$m = \lim\limits_{h \to 0} \dfrac{f(x + h) - f(x)}{h}$

$ = \lim\limits_{h \to 0} \dfrac{\sqrt{x + h + 3} - \sqrt{x + 3}}{h} \cdot \dfrac{\sqrt{x + h + 3} + \sqrt{x + 3}}{\sqrt{x + h + 3} + \sqrt{x + 3}}$

$ = \lim\limits_{h \to 0} \dfrac{(x + h + 3) - (x + 3)}{h\left[\sqrt{x + h + 3} + \sqrt{x + 3}\right]}$

$ = \lim\limits_{h \to 0} \dfrac{1}{\sqrt{x + h + 3} + \sqrt{x + 3}} = \dfrac{1}{2\sqrt{x + 3}}$

At $(-2, 1), \ m = \dfrac{1}{2}$.

30. $f(x) = \dfrac{1}{x + 3}$

$m = \lim\limits_{h \to 0} \dfrac{f(x + h) - f(x)}{h}$

$ = \lim\limits_{h \to 0} \dfrac{\dfrac{1}{x + h + 3} - \dfrac{1}{x + 3}}{h}$

$ = \lim\limits_{h \to 0} \dfrac{(x + 3) - (x + h + 3)}{h(x + h + 3)(x + 3)}$

$ = \lim\limits_{h \to 0} \dfrac{-1}{(x + h + 3)(x + 3)}$

$ = \dfrac{-1}{(x + 3)^2}$

At $\left(1, \dfrac{1}{4}\right), \ m = -\dfrac{1}{16}$.

31. $f'(x) = \lim\limits_{h \to 0} \dfrac{f(x + h) - f(x)}{h}$

$ = \lim\limits_{h \to 0} \dfrac{(x + h)^2 - (x + h) - (x^2 - x)}{h}$

$ = \lim\limits_{h \to 0} \dfrac{x^2 + 2xh + h^2 - x - h - x^2 + x}{h}$

$ = \lim\limits_{h \to 0} \dfrac{2xh + h^2 - h}{h}$

$ = \lim\limits_{h \to 0} (2x + h - 1) = 2x - 1, \ \text{Slope}$

At $(1, 0), \ m = 1$.

32. $\lim\limits_{x \to \infty} \dfrac{x^3}{x^2 - 9}$ does not exist.

Function increases without bound.

33. $\lim\limits_{x \to \infty} \dfrac{3 - 7x}{x + 4} = -7$

34. $\lim\limits_{x \to \infty} \dfrac{3x^2 + 1}{x^2 + 4} = 3$

35. $\lim\limits_{x\to\infty} \dfrac{2x}{x^2 + 3x - 2} = 0$

36. $\lim\limits_{x\to\infty} \dfrac{3 - x}{x^2 + 1} = 0$

37. $\lim\limits_{x\to\infty} \dfrac{3 + 4x - x^3}{2x^2 + 3}$

does not exist. Function decreases without bound.

38. $\sum\limits_{i=1}^{50} (1 - i^2) = 50 - \dfrac{50(51)(101)}{6} = -42{,}875$

39. $\sum\limits_{k=1}^{20} (3k^2 - 2k) = 3\dfrac{20(21)(41)}{6} - 2\dfrac{20(21)}{2}$

$= 8610 - 420 = 8190$

40. $\sum\limits_{i=1}^{40} (12 + i^3) = 12(40) + \dfrac{40^2(41)^2}{4}$

$= 480 + 672{,}400 = 672{,}880$

41. Area $\approx \dfrac{1}{2}[1 + 2 + 3 + 4 + 5 + 6]$

$= \dfrac{21}{2}$ square units

42. Area $\approx \frac{1}{2}[4.875 + 4.5 + 3.875 + 3]$

$= 8.125$ square units

43. Area $\approx \frac{1}{2}\left[\frac{9}{16} + 1 + \frac{25}{16} + \frac{9}{4}\right] = \frac{43}{16} = 2.6875$

44. Area $\approx \dfrac{1}{4}\left[\dfrac{1}{1 + \left(-\frac{3}{4}\right)^2} + \dfrac{1}{1 + \left(-\frac{1}{2}\right)^2} + \dfrac{1}{1 + \left(-\frac{1}{4}\right)^2} + \dfrac{1}{1 + 0} + \dfrac{1}{1 + \left(\frac{1}{4}\right)^2} + \dfrac{1}{1 + \left(\frac{1}{2}\right)^2} + \dfrac{1}{1 + \left(\frac{3}{4}\right)^2} + \dfrac{1}{1 + 1^2}\right]$

$= \dfrac{1}{4}\left[2(0.64) + 2(0.8) + 2(0.941176) + 1 + \dfrac{1}{2}\right]$

≈ 1.566 square units

45. $f(x) = x + 2$, $[0, 1]$

The width of each rectangle is $1/n$. The height is

$f\left(\dfrac{i}{n}\right) = \dfrac{i}{n} + 2.$

$A \approx \sum\limits_{i=1}^{n}\left[\dfrac{i}{n} + 2\right]\dfrac{1}{n} = \sum\limits_{i=1}^{n}\left(\dfrac{1}{n^2}i + \dfrac{2}{n}\right)$

$A = \lim\limits_{n\to\infty}\left[\dfrac{1}{n^2}\dfrac{n(n + 1)}{2} + \dfrac{2}{n}(n)\right]$

$= \dfrac{1}{2} + 2 = \dfrac{5}{2}$

46. $f(x) = 8 - 2x$, $[-4, 4]$

The width of each rectangle is $8/n$. The height is

$f\left(-4 + \dfrac{8i}{n}\right) = 8 - 2\left(-4 + \dfrac{8i}{n}\right) = 16 - \dfrac{16i}{n}.$

$A \approx \sum\limits_{i=1}^{n}\left(16 - \dfrac{16i}{n}\right)\left(\dfrac{8}{n}\right)$

$= \sum\limits_{i=1}^{n}\left(\dfrac{128}{n} - \dfrac{128i}{n^2}\right)$

$A = \lim\limits_{n\to\infty}\left[\dfrac{128}{n}(n) - \dfrac{128}{n^2}\dfrac{n(n + 1)}{2}\right]$

$= 128 - 64 = 64$

47. $f(x) = 2x + 5$, $[-1, 3]$

The width of each rectangle is $4/n$. The height is

$$f\left(-1 + \frac{4i}{n}\right) = 2\left(-1 + \frac{4i}{n}\right) + 5 = 3 + \frac{8i}{n}.$$

$$A \approx \sum_{i=1}^{n}\left(3 + \frac{8i}{n}\right)\left(\frac{4}{n}\right) = \sum_{i=1}^{n}\left(\frac{12}{n} + \frac{32}{n^2}i\right)$$

$$A = \lim_{n\to\infty}\left[\frac{12}{n}(n) + \frac{32}{n^2}\frac{n(n+1)}{2}\right]$$

$$= 12 + 16 = 28$$

48. $f(x) = x^2 + 1$, $[0, 4]$

The width of each rectangle is $4/n$. The height is

$$f\left(\frac{4i}{n}\right) = \left(\frac{4i}{n}\right)^2 + 1.$$

$$A \approx \sum_{i=1}^{n}\left(\frac{16i^2}{n^2} + 1\right)\left(\frac{4}{n}\right) = \sum_{i=1}^{n}\left(\frac{64}{n^3}i^2 + \frac{4}{n}\right)$$

$$A = \lim_{n\to\infty}\left[\frac{64}{n^3}\frac{n(n+1)(2n+1)}{6} + \frac{4}{n}(n)\right]$$

$$= \frac{64}{3} + 4 = \frac{76}{3}$$

49. $f(x) = 4 - x^2$, $[0, 2]$

The width of each rectangle is $2/n$. The height is

$$f\left(\frac{2i}{n}\right) = 4 - \left(\frac{2i}{n}\right)^2.$$

$$A \approx \sum_{i=1}^{n}\left[4 - \frac{4i^2}{n^2}\right]\left(\frac{2}{n}\right) = \sum_{i=1}^{n}\left(\frac{8}{n} - \frac{8i^2}{n^3}\right)$$

$$A = \lim_{n\to\infty}\left[\frac{8}{n}(n) - \frac{8}{n^3}\frac{n(n+1)(2n+1)}{6}\right]$$

$$= 8 - \frac{8}{3} = \frac{16}{3}$$

50. Width: $\frac{1}{n}$, Height: $f\left(\frac{i}{n}\right) = 1 - \left(\frac{i}{n}\right)^3$

$$A \approx \sum_{i=1}^{n}\left(1 - \left(\frac{i}{n}\right)^3\right)\left(\frac{1}{n}\right) = \frac{1}{n}\sum_{i=1}^{n}1 - \frac{1}{n^4}\sum_{i=1}^{n}i^3$$

$$= \frac{1}{n}(n) - \frac{1}{n^4}\left[\frac{n^2(n+1)^2}{4}\right]$$

$$A = \lim_{n\to\infty}\left[1 - \frac{1}{n^4}\left(\frac{n^2(n+1)^2}{4}\right)\right] = 1 - \frac{1}{4} = \frac{3}{4}$$